DAUNTLESS

THE 1ST & 2ND FILIPINO INFANTRY REGIMENTS
UNITED STATES ARMY
FOUGHT TO FREE THE PHILIPPINES
A SECRET GROUP RETURNED BEFORE THE ALLIES

★★★★★

MARIE SILVA VALLEJO

DAUNTLESS: The 1st and 2nd Filipino Infantry Regiments United States Army fought to free the Philippines. A secret group returned before the allies.

Copyright © 2023 by Marie Silva Vallejo

Published by Eastwind Books of Berkeley: 2023
2022 University Avenue, Box 46
Berkeley, California 94704

www.AsiaBookCenter.com

Email: eastwindbooks@gmail.com

ISBN: 978-1-961562-00-4 (Paperback)
ISBN: 978-1-961562-03-5 (Epub)

1.   Philippines —History— American period, 1898-1946.
2.   United States. Army. Filipino Infantry Regiments, 1st. I. Title.

Editor – Susan Evangelista   Layout – Michael Ambion
Maps/Cover – Norman Tioco   Bibliography – Xenia Romero

(Originally Published by Marie Silva Vallejo: 2023. Bonifacio Global City, Taguig, Philippines)

Second Printing: September 2023
Third Printing: July 2024

Printed in the USA

DEDICATION

My Family
My father, Saturnino Ramos Silva
His children

The 1st and 2nd Filipino Infantry Regiments
The 1st Reconnaissance Battalion
United States Army

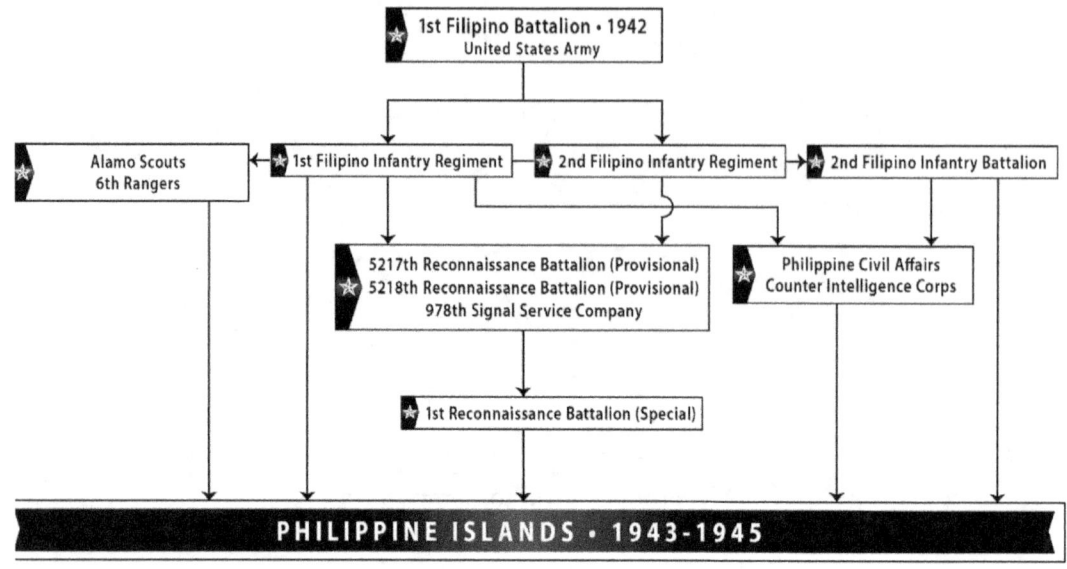

1st Filipino Battalion · 1942
United States Army

Alamo Scouts
6th Rangers

1st Filipino Infantry Regiment

2nd Filipino Infantry Regiment

2nd Filipino Infantry Battalion

5217th Reconnaissance Battalion (Provisional)
5218th Reconnaissance Battalion (Provisional)
978th Signal Service Company

Philippine Civil Affairs
Counter Intelligence Corps

1st Reconnaissance Battalion (Special)

PHILIPPINE ISLANDS · 1943-1945

*1st and 2nd Filipino Infantry
Regiment patch*

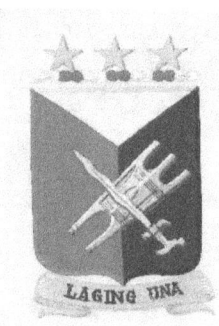

1st Filipino Infantry Regiment crest

2nd Filipino Infantry Regiment crest

*1st Reconnaissance Battalion
Airborne Patch*

*1st Reconnaissance Battalion
(Special) patch*

United States Army Pacific patch

PREFACE

1941. December 7. At almost eight in the morning, the Japanese launched a surprise attack on Pearl Harbor that ended with thousands of American casualties. Less than ten hours after, the Japanese invasion of the Philippines began.

President Franklin D. Roosevelt dubbed the attack "the date that will live in infamy." It was the mainspring for the entry of the United States into World War II and catalyzed the formation of the U.S. Army's First Filipino Battalion.

With the Philippines attacked, Filipinos living in the United States of America up and volunteered for the military—burning with the desire to return to their motherland and repel the enemy. I wondered why they joined when their entire life in California was debased with systemic and rampant anti-Filipino sentiment and legislation. *What complicated patriotism had made my father and thousands of Filipino men in America volunteer into the Army despite decades of horrible inequities inflicted upon them? Was it even worth it?*

The First Filipino Battalion was formed, and around 7,000 eagerly volunteered. From the unit came the First and Second Infantry Regiments that returned with the liberation forces. From the Regiments formed the First Reconnaissance Battalion (Special) and the 978th Signal Service Company, secret military units highly trained and known only to General Douglas MacArthur and a few of his staff. Why were these suicidal missions sent to the Philippines in the two years before his return? And why were they a secret?

Discovering that my father was a member of this clandestine unit, I was compelled to find out. But the secrecy of the covert missions to the islands challenged the creation of their chronological history. Large gaps about their exploits in the Philippines had to be filled in largely from guerrilla records to complete their story. The children of the men who were so proud of their fathers, shared their stories that fleshed out the daunting experiences they underwent in the jungles of the islands.

In the hunt for information, my family expanded with the wonderful discovery of my other siblings, Isabel and Saturnino Jr., that kept the momentum in me. I like to say that my father was a busy man during the war.

To be clear, depending on the events, I sometimes write "Caucasian" or "White" soldiers to differentiate and give proper credit to the Filipino soldiers in their missions and assignments.

The lives of the Filipinos of my father's generation in America, reaching for a better future, was of exceptional hardship, pain, and sacrifice in a time of intolerance and racism. It had to take a war to slowly break the ignorance, to make them "equal" and enable their own triumphs to emerge. What they underwent begat better opportunities for the Filipinos in America.

I wrote this book thinking of my father and the men. I had to tell their story.

Marie Silva Vallejo
Manila, Philippines
2023

ACKNOWLEDGMENTS

A fateful and chance meeting with Alex Fabros, historian of the Regiments, for opening the military world of my father and his unit. His pioneering research work with the Filipino American Experience Research Project provided the framework of the book.

James Zobel, director and archivist of the MacArthur Memorial Library for providing several feet high copies of records before cell phone cameras were handy in the several times I visited. And he answered every question and email I sent.

Peter Parsons, son of Charles Parsons of the SPYRON submarine missions that brought the men including my father to the islands for allowing me to go through the 40 boxes of his father's papers that provided many key information.

Professor Rico Jose for encouraging me to finish the book when the diverse paths of the men and scarce records in various mediums scattered in numerous locations overwhelmed me.

The Filipino American Historical Society (FANHS) of Fred and Dorothy Cordova for preserving the men's military reports.

Pelagio Valdez, who promised his father to keep the memory of the Regiments alive, and sharing the photos and stories given to him by the families of the men.

Philippine Veterans Bank and Michael Villareal for their invaluable contribution to Philippine WWII history.

Most important, the support of my family and the many others who kept my motivation strong and alive. My husband, Gus, wished me well wherever research brought me.

TABLE OF CONTENTS

1930s – Filipinos in California dressed for photos to send to relatives in the Philippines. Saturnino Silva, my father, is second from the right. Source: Marie S. Vallejo Collection

U.S. Nationals

Because of the vast distance across the Pacific Ocean, the United States was initially hesitant to include the Philippines with Cuba and Puerto Rico among her acquisitions from Spain in 1898. However, the riches of natural resources and raw materials for manufactured goods convinced her to keep the islands as a new market for her products.

Meanwhile, the Filipinos had waged a revolution against their Spanish colonial rulers and declared independence in 1898. The Spanish-American War brought the Americans to the Philippines, where they, working with their Filipino allies, defeated the Spanish forces. But it did not take long for the Filipinos to see America's colonial ambitions on the islands. Wanting to preserve their independence, the Filipinos went to war against the Americans in 1899. When the four-year war ended, America had its first colony and a period of building the islands in its own image followed. Although a naval base was secured in the islands, America's outlook for the Philippines was still unclear after 25 years. A U.S. Supreme Court justice commented that "the Filipinos are neither alien, subjects, nor citizens of the United States." To its neighboring countries, the Philippines posed "an awkward problem... identified Filipinos as 'Asians but not Asians, Westerners but not Westerners.'"

The American way of living given prominence in the local newspapers and magazines convinced young Filipinos to go to America. The stories were not completely accurate, yet they were hard to ignore. Countrymen and relatives wrote of a good life and never mentioned the hardships they were going through. Filipino townmates in America sent photos of themselves dressed in suits and hats, looking well-dressed and well-groomed, reassuring families and friends that they were doing well—that their dreams were realized. Little did their families back home know, many of them rented or borrowed the suits.

Narciso Caliva, a Filipino photographer, maintained an extensive wardrobe for rent for his customers who could not afford to buy suits. He made sure that they always looked good and prosperous in their postcard pictures. It was the easiest way to keep in touch back home—much easier than a letter when there was not much to say. A letter meant that money was inserted, Tony Lasitonia recalled.

Travel to America did not require much paperwork or time. Filipinos were allowed entry as "nationals," even though they were not citizens. Virtually all the Filipino immigrants who came to the United States before 1920 came to earn and save money to bring back home. Others were students, nicknamed "fountain-pen boys" because they used fountain pens in school to impress others, while hoping to advance their life situations with an American education. Some came from well-to-do families while others were *pensionados* — young men from the rich and influential upper class in the Philippines qualified for government sponsorship to higher institutions. A 1920 census reported the presence of over 5,000 Filipinos in the United States. Most were those who worked in the fields, *pensionados,* and students who had stayed on. Others were recruited for the Navy as mess boys, and then discharged, having served their enlistment period.

When the Immigration Act of 1924 barred Japanese immigration to Hawaii, a major recruitment program brought Filipinos to Hawaii from the mainland and the Philippines. Young men flocked to recruitment offices in Manila, not really caring if it was to Hawaii or the mainland, if it was in America.

Special lectures with movies, were shown throughout the islands to induce male Filipinos to sign three-year employment contracts. Steamship agents knowingly misrepresented facts as they distributed leaflets giving glowing accounts of idyllic working conditions on Hawaiian plantations. Some even overcharged the emigrants. A particular labor agent reported that he selected uneducated men from the provinces "because he can work with them better." The recruitment pace quickened when another piece of legislation pending in the U.S. Congress threatened to cut off the supply of Mexican labor.

Saturnino 'Tony' Ramos Silva, my father, was one of many young Filipino men who was convinced there was a better life in America. It was a long shot, because he was not a *pensionado* who could study in America at the expense of the U.S. government. Their travel tickets were free and were often for first class accommodations. My father had not attended the Philippine Military Academy (PMA), eliminating his chance to get a scholarship to go to West Point and stay with an American host family. That would have allowed him entrance to places that otherwise might have been closed to him due to discrimination. He had to work and save for all the requirements for passage.

Silva graduated from the Normal School in his home province of Pangasinan on the island of Luzon. American teachers in the school taught courses in health, sanitation, and democratic governance, which were important in the preparation of much-needed teachers and leaders for the country's administration.

Answering the Philippine government's call for teachers to go to Sulu in Mindanao, he taught in Tawi-Tawi, a southern island in the Philippines for two years to save for his passage. His school was under the jurisdiction of Edward Kuder, the School District Superintendent who spent 20 years with the *Moros*, the Muslim inhabitants of the area.

Travel to America

After two years, Silva returned to Manila filled with ambition and the spirit of adventure. He believed that with his diploma, he could easily find work in America. He could also speak Tagalog, a widely-spoken language, and three other languages of the province he came from.

By then, he had saved enough money for a ticket, but it is possible that his parents sold off some land or animals to help pay for his fare, like many others did for their sons.

As a "national" or "ward," he had legal entry to America. He boarded the S.S. *Empress of Asia* in May 1929, clutching a ticket for Steerage class — the lowest deck on the ship. It was packed with people and the conditions were stifling. Boarding the ship, he caught glimpses of his well-to-do countrymen, educated Filipinos, government- sponsored *pensionados,* and cadets chosen for West Point reveling on the luxurious top decks. A month later, the ship arrived at Victoria, British Columbia before finally docking in the United States.

Teodoro Taluban, one of those who travelled to America to work, recalled: "I needed a birth certificate showing I was over 18 years old, a statement from the court that there is no case pending against me, and another that I do not owe (money) to anyone... In Manila, we went to the ship and showed the papers. They showed me the compartment where I should go to. I realized that I never looked back, and that was the last time I saw my father. In America, I saw my uncle who was working in the fields. We worked in the field."

The words, "American Citizen" were stamped on Felipe Dumlao's passport. This was misleading because Filipinos were not given citizenship. It was an uncertain situation for the new migrants to be labeled "ward" or "national." They had no idea what they were in for.

Dumlao, another Filipino migrant, remembers: "Immigration in Manila took photos of my passport and proper papers. They asked me what I wanted to do in America but not when are you coming back to the Philippines. I carried only a *cedula* (identification card) that showed I paid taxes when I turned 21 years of age. If I lost it, I would have to show my birth certificate as identification. I received my passport but only received my papers when I arrived in the United States. I stayed at San Miguel Hospital (Manila) for two weeks to get a physical examination, vaccination, and blood tests. If I was not there when my name was called, I could not board the ship that day. I slept on the floor to stay there. They fed us and, if late for meals, we had to wait until dinner. I could only bring my clothes on board the ship. It was one big storage area, no air, without ventilation. People got sick and just throw up [sic]. There was no janitor. A fireman just ran water on the floor. We were hundreds of Filipinos in steerage. We arrived at Victoria (Canada) at sunset and the next day transferred to a ferry to Seattle, where we were given a list of hotels in Chinatown. Some of us had no money and slept on the streets with the risk that we might get picked up by the police. I was able to find work at 10 cents per hour for one year at Monroe. There were hundreds of Filipinos there, but there was free board and a bunkhouse. I made one dollar a day all by myself with no contract."

Another migrant, Leonard Aliwang, described the conditions on board the ship bringing them to America: "Each bunk was four beds high in one compartment with around 100 people. There was no air conditioning, and many got sick and died. Their bodies were dumped into the sea. Manila to Honolulu took over 20 days."

Mariano Angeles, a medic, added, "I had a passport, but no visa was needed. I could stay as long as I wanted. There were many stops in between; Hong Kong, Shanghai, Nagasaki, Yokohama before Vancouver. Each stop was overnight, and the ship loaded more passengers, mostly Chinese and Japanese. Some Filipinos left in Shanghai and did not want to go to America,"

For Leovigildo Giron, the trip to America entailed a lot of sacrifice. His parents sold the only piece of land they had to buy his ticket. When he arrived in San Francisco, he only had $5.00 in his pocket. He was only sixteen years old.

In America

The men who arrived in the late 1920s were not the first Filipinos to arrive in the U.S. In 1906, twenty years earlier, fifteen Filipinos were sent from the rural lands of northern and central Philippines to work in the sugar plantations of Hawaii. This was after migrant labor sources from China, Japan, and other countries had been stopped and the Hawaiian Sugar Plantation Association began recruiting *sakadas* (workers) from the Philippines. When their three-year contracts expired, around 150,000 workers returned to the Philippines or worked for higher wages in the mainland. On the West Coast, they found work as canners in Alaska and fruit and vegetable workers in Washington and California.

In 1929, the S.S. *President Lincoln* arrived in San Francisco with Philip Ordoño onboard, eager for new opportunities. His first job was picking berries in Mountain View south of San Francisco. It was also his first time to see berries. "He ate more berries than he could pick," a friend laughed.

He hopped on boxcars in freight trains to places wherever crops were ready for harvest. He learned that it was easier to find a job with people of the same Philippine province he was from because they spoke the same language and had the same cultural habits. The Ilocanos, from the north, worked mainly in the fields while the Visayans, from the islands in the center of the archipelago, and those from other islands found work as busboys, hospital helpers, and service workers. *Shortcuts,* an article published in the November 15, 1929, issue of the newspaper *The Three Stars* in Stockton, California recommended that men hide their age in photographs by tinting their hair and lifting their faces by lying on their backs, because jobs for men over 40 were very scarce, if not non-existent.

Silva arrived in the same year as Ordoño. It was the start of the Depression, and unemployment among the white workers was running high. Any hope for opportunities upon arrival quickly vanished. Racism and discrimination relegated the Filipinos to menial jobs or work in the fields. But Filipinos kept arriving in America at the rate of 4,000 a year. A thousand of the recruits were even sent to Idaho, outside of the Western states to work in the farmlands.

Racism was raging and attacks on Filipino field workers in the vegetable fields of California happened almost every week. For years, many of the citizens in California were indifferent to the Filipinos and the racial issues they had to face. When the Depression struck, many white Americans lost their jobs but were unwilling to work in the fields. American farmers turned to foreign laborers to do the back-breaking work. Filipinos were familiar with farm work and had a reputation for being skilled and hard workers while willing to accept lower wages. However, the Americans saw their jobs taken away. Thus, ranchers were continuously threatened to employ only white workers. On many occasions, discrimination in the vegetable fields resulted in deaths when white men attacked the camps.

Agustin Santos, another teacher like Silva, crossed the Pacific Ocean to America on the S.S. *President Jefferson* and arrived in Seattle in 1926. His high school teachers were former *pensionados* who told stories of their studies and life in the U.S. It was easy to come to America, but when the ship docked in Seattle, no one was there to meet him. He and the other Filipinos, who arrived on the same ship, were brought by a taxi driver to Chinatown where they were left to find work on their own. He realized that "No matter how many degrees you had at that time, [you] could not get a job, just work in hotels."

The racial bias against Filipinos reached a legal climax when a San Francisco Municipal Court judge declared Filipinos "scarcely more than savages, come to San Francisco (to) work for practically nothing, and obtain the society of these (white) girls... decent white boys cannot get jobs." Labor groups and racists endorsed laws that excluded and discriminated against Filipinos whom they believed lowered American morality and were "health hazards." Almost every week, there were attacks on Filipinos, but no arrests were made. Sympathetic police officers would quell an attack or arrive early to warn the Filipinos, allowing them to escape before a mob arrived. Many found refuge working in the fishing and canning industry of Alaska and the West Coast. Others who found employment in the cities even, as domestics, fared better when it came to salary and living conditions. Even if they had saved some money, their colonial status prevented them from owning property and there was no path to citizenship.

Silva began his first job picking apples in the central valley of California hoping to eventually work in the cities to go to school. He read the recruitment ads for apple pickers from *The Philippine Independent News,* the first Filipino newspaper in the mainland U.S. published in Salinas, California, where many Filipino families and field workers lived.

In the fields, he met the *Manongs* — an Ilocano title of respect for an older brother — mostly bachelors of over 30 years old. They first arrived in the mainland around 1910 to replace the Japanese and Chinese workers who had dominated the farm labor in California. Unlike the *sakadas* of Hawaii who were brought in by the Hawaiian Sugar Planters' Association, the *Manongs* came as independent workers and paid their way to the U.S. Over half of them were from the provinces. Many were without formal education, spoke neither Spanish nor English, and lacked any training. They signed up for the labor camps in hopes of gaining an education or continuing their education in America. But they found out that they could not earn enough money to support themselves and go to school at the same time. Work on the farms and low-paying menial jobs were the only ways to survive but these would not earn them enough money to save. Two labor camps offered night school three nights a week. These were heavily attended by Filipino laborers. "After working all day in the lettuce fields, they devoted their time in the evenings to obtaining what education they could." Silva may have attended these sessions or taught one himself.

Filipino women were a rare sight then. This made it nearly impossible to settle down and hope for a family. The traditional belief that a woman's place was in the home curtailed their migration. Filipino parents believed it was dangerous for their daughters to go to the U.S. On the West Coast, there were more than 28,000 Filipino males and fewer than 2,000 Filipino women. Exploitation in the farms made work dangerous for the women.

The laborers did not stay long in one place. They had to go where the crops were, moving around according to the harvest seasons. During asparagus season, for instance, the Filipino population swelled into the thousands—as they were sought- after as asparagus workers—but fell to one thousand during winter. Thus, they could not build more permanent Filipino communities.

They frequented the easy entertainment establishments wherever they moved: pool halls, restaurants, cigar stands and the like. "There were no other places to go except billiards [sic], restaurants, and hotels owned by the Japanese, and gambling dens owned by Chinese dealers that provided free doughnuts and coffee whether you are gambling or not. I never went hungry." The presence of Filipino women and family could have reduced the time spent in gambling.

Mingling with white women was viewed as "corrupting their morals." Many of the riots were triggered by Filipinos being seen with white women. That was also used as an excuse to drive them out of the fields. But that did not stop the men from dancing with white women in local dance halls. Filipinos owned the dance halls and white girls, lots of girls wanted to get some extra money, charging 10 cents a dance. "Every night in front of the dance hall, there were fights. When you date a girl, they pick you up and put you in jail. I met a girl at a dance hall who told me that Filipinos were not allowed to go with white women. But there were hardly any Filipinas in those days," complained Felipe Dumlao. Of course, on the other side of the Pacific all-too- commonly white American men danced with Filipino women in the cabarets of Manila, and no one bothered them.

Associating with white women was a crime under the anti-miscegenation laws. These laws declared it illegal for non-White groups to marry Whites. Filipinos were included in the ban. More than 100 marriages performed after 1921 were invalidated. Those who wanted to marry traveled to Seattle, Vancouver, or Gallup, New Mexico, with their white fiancées. Martin Mamuyac, Sr. recalled friends who married two or three times to prove their marriages were legal.

Others like Francisco Taggaroa had no illusions. "My life, like those of my countrymen, could have been better...." He came to the United States because it was the thing for young men to do. His "primary purpose was to work, save, and go home." He continues, "When I arrived, I secured a busboy job, saved, and sent home sixty percent of my wages. Three years later, my wages were so low it was almost impossible to save anything."

Looked upon as not having equal rights granted to citizens, the *Manongs* tenaciously appealed for wages, better living conditions, the right to own property, and job opportunities equal to those of the Whites. Made aware of their plight, the Philippine Resident Commissioner in Washington, appealed to the U.S. President for an end to the persecution and mob violence against Filipinos. But in the Philippines, rabid nationalists stirred up intensely negative feelings against their countrymen by calling Filipinos in America "a menace because majority of the cases are the worst type of Filipinos [sic]. They belong to the riffraff class who decided to migrate to America and work in farms, canneries, and domestic service sectors of urban areas. Their mode of living, behavior, intellectual level is a menace to the Americans and the good name of the Filipinos in the US". This was published as an editorial in the May 25, 1925 issue of *The Manila Tribune*.

By August 1933, Japanese farmers began buying up farms and adopted a code of fair practice, including a schedule of wages and working hours. *The Philippine Mail* published an article entitled, *Wage Increase in Sight Now for Field Laborers*. By 1941, the laborers organized themselves into militant labor unions in Hawaii, California, and Alaska, calling for the reversal of racially-biased laws. In response, anti-Filipino employment laws were passed, further squeezing their job prospects. In Watsonville, California, the farmworkers were called "a menace" as they went on strike to enforce the wage increase while vegetables rotted in the fields. White workers then replaced them.

A demand to deport the Filipinos surfaced. Mob violence against Filipinos greatly increased and swept across farm fields to the cities of San Francisco and San Jose. Dozens of Filipino farmworkers were assaulted. Their homes and barracks were shot at and ransacked. The racial antagonism in California had become so unbearable that thousands of Filipinos were anxious to go home. Exclusionists who wanted Filipinos to go home, championed the Filipino Repatriation Act of 1935 that paid for passage back to the islands. Some people called this humanitarian, but it was a one-way ticket, and those who took advantage of it could not return. Of the 45,000 Filipinos in America, only around 2,000 chose to return home. Those who stayed took on work no matter how menial despite the homesickness, hardship, and discrimination. The majority remained unable to return to families and hometowns because they had no savings and nothing tangible to show for their years of hard work. Silva was determined to stay.

The Western States pushed for Philippine independence from the United States, hoping that the Filipinos would return to the islands. There would no longer be a reason for the so-called "entitled" Filipinos to come to California. The date was set for the colony's independence with the passing of the Hare-Hawes-Cutting Act of 1933. The act meant that Filipinos could no longer migrate to the U.S. Across the Pacific, nationalists in the Philippine government continued to push for independence from the United States. A 10-year transition period towards independence began with the 1934 Tydings-McDuffie Act that changed the Filipinos' status in the U.S. from American Nationals to Aliens ineligible for citizenship. The Act subjected the Philippines to the Immigration Quota Act of 1924 that allowed only fifty Filipinos per year to enter the US. It was practically an exclusion law.

San Francisco, California

To become a US citizen was now an almost impossible dream. Farm owners did not want to hire educated Filipinos as they might incite other workers to unionize. Fearing for his life and facing the general problems of Filipinos on the farms, Silva set out for San Francisco to find work. On a work application, he wrote "secretary" as a past position, and he listed his birth year as two years younger than he was.

Men who fled to San Francisco believed no one would bother them in the city as they were protected by the presence and support groups of the Filipino organizations. The February 5, 1930 issue of *The Three Stars* published the article, *The Desirability of the Filipino*. It claimed that according to studies of the Commonwealth Club in San Francisco, 15,000 Filipinos lived within a radius of 100 miles from the city.

While Silva was looking for work in the city, Japan had withdrawn from the League of Nations. Its relentless expansion to build an "Asia for Asians" brought Imperial Tokyo closer to the Philippines. Fearing that the islands would be next, the United States exercised a provision in the Philippine Commonwealth constitution that permitted the United States to draft organized military forces of the Philippine Commonwealth to defend United States' interests. Thousands of Filipinos enlisted. Ironically, Filipinos in the U.S. could not join and bear arms.

Saturnino 'Tony' Silva is on the left with a friend.

With their shiny hair slicked back, the pants of their vested suits are noticeably long as it touched the ground. Each held a fedora at his side. He must have borrowed the suit for the photo.

The general attitude in San Francisco was not encouraging for those looking for work. Filipino organizations, however, came together to keep their countrymen from standing in bread lines or living on welfare. The work available was usually for busboys, cooks, dishwashers, domestic help, and gardeners. Opportunities in business and professional positions were very limited, even for those qualified.

An American education would prove that he was just as good as the white man in the face of condescending attitudes, thought Silva. He was already a graduate of a school for teachers taught by American teachers in the Philippines and had taught for two years. He could easily graduate from an American school. Sadly, he realized that fulfilling admission requirements did not matter if he could not support himself. He needed to find decent work.

Those who found domestic work in private homes were able to save for tuition. They learned to cook American meals by reading American cookbooks and learned to clean a western house which was so different from their spartan living quarters in the farms properly. Despite low wages and irregular working hours as live-in house

San Francisco delegates to the Institute of Pacific Relations conference, Santa Cruz, California, 1933. It was an NGO to provide a forum for discussion of problems and relations between nations of the Pacific Rim.

servants, they were glad to escape the deadly riots in the fields. They managed to stay out of trouble from the racism in the cities and kept abreast of the times by reading locally-published Filipino newspapers.

Amid the violent racism directed towards Asians, Silva managed to enroll at San Francisco State College. He was a member of the Institute of Pacific Relations for three years. The institute's systemic study laid the foundation for Asian and Pacific studies in the West. The school sent him and eight others as delegates to the Institute of Pacific Relations Conference in 1933. A group photo showed a slim Tony, almost underweight, sitting on a long wooden bench on the left under Redwood trees with two other men and seven women.

He persevered to finish his studies. It took him six years to graduate with a major in Economics and a minor in English and Government. The college still had his 1937 transcript showing grades that were average or passing in some subjects.

Soon after his graduation, Bertha H. Monroe, a social science professor at San Francisco State College, sent a letter of recommendation to the President of the National University in Manila. Monroe was responsible for initiating a series of courses for the school's Department of Immigrant Education. The letter was an application for a fellowship. In it, she mentioned Silva's "dynamic personality coupled with control which inspires others to test their capacities." She listed his positions in the International Club and his presidency in both the Oriental and Filipino Clubs, among others.

Did the prevailing anti-Filipino sentiment make him want to return? The enforcement of the Filipino Repatriation Act was ongoing with full passage back to the Philippines. He may have wanted to take advantage of this and have a position waiting for him in Manila. If he had wanted to start a family, he knew that it was almost impossible to find a Filipina in California. He was one of over 50,000 bachelors in California at that time.

Silva chose to stay in the US, opting to enroll in law courses at the University of San Francisco. Going back without assurance of earning a living and no savings meant failure. He was determined to stay and find work, though he knew that with the still rampant racism he may not find a teaching position. To support himself, he had to choose between staying in the city and applying as a hotel or restaurant help or going back to working in the fields. He may also have gone to work in the canneries of Alaska during the summer, which paid more. If he needed help while looking for work, there were Filipino organizations in the city that provided living support. The Filipino value of *Bayanihan* or mutual assistance helped many until they found work, especially during the Depression.

Working as a waiter in San Francisco, plus doing general office work in the Filipino organizations, Silva wore tailored suits and made many friends. His ability to speak in Ilocano and other languages of Luzon as well as English and some Spanish made it easy to move around the various Filipino groups. Some friends who were cooks gave him free meals that saved money. He was then a member of the Filipino Community of San Francisco and the Philippine Commonwealth Club partnering with privileged Filipinos and those with university educations to help needy Filipinos with food, clothing, and funds for medical and burial expenses, returning the help extended to him when he was looking for work after college. Being single, he was President of the San Francisco Bachelors Club. No longer wanting to return home, he declared his intention to become a U.S. citizen in December 1940.

According to Alex S. Fabros, Jr., Silva developed a close friendship with his father, Alex Sr., another bachelor at the time whose couch he frequented whenever Filipino celebrations were held in the farming community of Salinas. Fabros Sr. arrived the same year as Silva, but he was fortunate to have his own father, a farm labor contractor, welcome and guide him.

As the ominous winds of war were filling the airwaves and newspapers, the U.S. continued to pass restrictive laws against non-U.S. citizens. News from the Philippines described a desperate ramping up of the Philippine Army even as civilians basked in the embrace of "peacetime" with American goods, culture, and entertainment.

Asia was on the brink of war as efforts and negotiations to control Japan's brutal conquests of its southern neighbors failed. She needed raw material from her occupied territories to build a Navy at par with the United States and Great Britain to defend these same territories. By 1940, she was at the doorstep of the Philippines. With the threat of war looming, Filipinos in the U.S. were afraid of being mistaken as Japanese. Many followed the tactic of the Chinese community and wore pins that read "I am a Filipino" to avoid being mistaken for Japanese.

In the cold early morning after an Easter Mass, Silva decided to enlist in the Army. He signed up at the Army recruitment center with the Filipino Community of San Francisco as his employer on the registration card. He was 31 years old, his citizenship status read "Philippine," and his unique physical characteristic was a mole by his left eye. He weighed only 140 lbs. As a Christmas gift to himself, he signed a Declaration of Intent to become a U.S. citizen. His address was Lombard Street in San Francisco, and occupation was "waiter." Four months before the attack on Pearl Harbor, he received his draft card. Records listed him as a Warrant Officer, having similar responsibilities as a commissioned officer. He was assigned to the 27th Division in Hawaii for several months before reporting to the 1st Filipino Battalion for Basic Training in San Luis Obispo in California.

The expectation of an attack would be on the Philippines, a colony of America, since it was in the Japanese' path to the oil reserves in the Dutch East Indies (Indonesia). Rear Admiral William J. Galbraith who was the air defense officer on the battleship USS *Houston* based in the Philippines in late 1941 reported that "they expected an attack on the bases in the Philippines and had been on full-scale alert since Thanksgiving, 1941. The focus was to fend off an attack in the Philippines and not Hawaii."

The Japanese surprise attack on Pearl Harbor in the early morning of December 1941 stunned the Filipino communities. What will happen to our families back home? Across the U.S., every eligible man was told to report for duty. Some did not even know where Pearl Harbor or the Philippine Islands were.

Mindanao Island Guerrillas.
Source: National Archives and Records Administration (NARA)

The Resistance

Across the Pacific Ocean, the Philippines was America's presence in Asia and would be a significant prize for Japan. The archipelago's seas were passages to the oil of Borneo and the Southeast Asian countries that were key to Japan's becoming a major power.

The threat of war arrived at the Philippines' doorstep with the fall of Indochina to the Japanese Imperial Army. America bolstered its military position in the islands by establishing the United States Army Forces in the Far East (USAFFE), a new military command post in Asia. The military forces of the Commonwealth Government were placed under the command of the U.S. Army. Under General MacArthur, around 200,000 enlisted men and 12,000 Philippine Scouts were stationed in the islands.

On December 7, 1941, the Japanese aircraft carrier *Ryujo* was already in the waters off Mindanao when Pearl Harbor was attacked. The *Ryujo's* dive bombers and fighter planes strafed and burned two U.S. seaplanes and the destroyer-seaplane tender, USS *William B. Preston,* in Malalag Bay in the Davao Gulf of Mindanao. It was not the Philippines' war, but after forty years, the beleaguered nation was again in the middle of a conflict between two countries — reminiscent of the Spanish-American War.

The prevailing belief was that Japan would attack in April of the following year. U.S. Army Chief of Staff General George Marshall's telegram to MacArthur that read, "Giving you everything we possibly can," gave the hopeful impression that aid would arrive sooner or later. It would be too late.

The U.S. was committed to the defense of the Philippines, but funds were gravely lacking. The defense posts of Corregidor and Bataan had not been upgraded since they were built before WWI. There were not enough soldiers to defend the archipelago. In 1935, to build an army, the Philippines passed the National Defense Act which required every male over 21 years of age to complete military training and serve some time in active service. With only 30,000 regular American troops, the Philippine Scouts, and the Philippine Constabulary to face the formidable enemy, President Franklin D. Roosevelt issued a presidential order calling into the service of the United States of America all the organized military forces of the Commonwealth Government of the Philippines. Douglas MacArthur was then called into active duty to command the USAFFE and American officers were flown in as instructors. Many arrived only months before the attack. Enlisted Filipinos trained with mostly WWI equipment. Others had to make do with weapons made of bamboo.

The outbreak of war was anticipated at any minute. There were signs almost two weeks before of the possibility of an attack on Pearl Harbor. Army intelligence reported five Japanese army divisions in fifty transports had embarked from Shanghai and sailed southward; this observation was acknowledged by General Marshall after the war. Leadtime in Hawaii could have led to some preparation. The headlines of the November 30, 1941, issue of *The Honolulu Advertiser* and *Hilo Tribune Herald* screamed *Japanese May Strike Over the Weekend.* However, instead of heeding this warning. Private Paul Brown of the 52nd Field Artillery Battalion who was recuperating at the Schofield Barracks Base Hospital witnessed the front pages of these newspapers being ripped off by Army personnel as "ordered by the top brass." All copies found were confiscated. The following week, Pearl Harbor was attacked.

Another warning on the morning of December 7 was the sinking of the unarmed American merchant steamer, *Cynthia Olson,* by the Japanese submarine I-26. None of the crew which included 23 Filipino sailors survived. According to Stephen Harding's article, *Prelude to Pearl,* that would make the incident Japan's first attack of the Pacific War.

A partially submerged Japanese submarine had gone ahead of the enemy fleet and attempted to enter Pearl Harbor. The destroyer, USS *Ward*, sank it ninety minutes before the first bombs fell on the harbor.

The Japanese force relentlessly bombed the battleships, but the Pacific Fleet's aircraft carriers were not in the harbor that day. Belinda Aquino's article *Pearl Harbor Revisited* highlighted Bob Sigall's research revealing the Japanese' grave 'tactical error' that enabled the U.S. Fleet to recover quickly. Leaving the Pearl Harbor dry docks and its ship repair facilities unscathed enabled the repair of the damaged ships quickly and rebuild the U.S. Fleet in less than three years. The Japanese bombers had left undamaged the fuel reserves that would have wiped out the entire base to avoid the black clouds of smoke that would have obscured their view of the ships.

Valentin Villafuerte saw the carnage of hundreds of unsuspecting sailors blown up in the bowels of the battleships. He worked in the Red Hill Navy project, an underground fuel storage for submarines. On that fateful day, he sat outside the cafeteria and saw planes repeatedly dive-bombing the ships in the harbor. He thought a movie was being made. Thick smoke blanketed the area. Some ships were trying to maneuver out of position. As ships overturned, he saw a sea of white overturning with the ships which he dismally realized were the sailors onboard. The sailors floated to the top as they dislodged from the ships. He rushed inside when shrapnel began raining down on the lawn. When night fell, the harbor was smoldering and smoking from the burning hulks of the sunken warships. Occasionally, he saw a bomber take off to go to the mainland to be fully armed. The unforgettable sight prompted him to enlist.

TUESDAY 11/29/16 $1 Oahu, $1.25 neighbor islands — Partly cloudy. High 81, low 71 » B8   staradvertiser.com

DAY OF INFAMY

88-PAGE COMMEMORATIVE EDITION COMING WEDNESDAY

Star ★ Advertiser

Sailors watched Pearl Harbor burn after it was attacked by Japan 75 years ago on Dec. 7, 1941. At left is the submarine USS Narwhal.

Honolulu Star Advertiser. November 29, 2016. Sailors watch Pearl Harbor burn after it was attacked by Japan 75 years ago on Dec. 7, 1941. The USS Narwhal is on the left.

Four submarines, including the USS *Narwhal*, were docked for refitting at the harbor when the Japanese struck. She was across from Ford Island, the direct site of the attacks. The Japanese were focused on eliminating the battleships and ignored the submarine base. That was a grave miscalculation. Within minutes after the first bomb explosions, *Narwhal's* gunners opened fire with two of her four machine guns and destroyed two torpedo planes.

A photo taken during the attack captured in the foreground the submarine *Narwhal* tied up at dock while the skies were dark with towering black smoke shooting up from exploding and sinking ships at Ford Island. The submarine would later be transformed from a torpedo-carrying pigboat to the lead submarine carrying supplies and cargo to the Philippines.

With General MacArthur having no direct authority over the Navy, Admiral Thomas C. Hart of the Pacific Fleet followed orders from Admiral Harold Stark, Chief of Naval Operations to move his fleet, from Manila to Java, out of range of the Japanese bombers that had attacked the Cavite Navy Yard. Only nine motor torpedo boats, Philippine Q-boats, and U.S. P.T. boats were left to defend Manila.

U.S. High Commissioner to the Philippines, Francis Sayre appealed for America's help in a radio broadcast: "Come on, America!" President Manuel L. Quezon issued the order to fly the Filipino flag with the red field up, announcing that the Philippines was in a state of war. MacArthur reiterated President Roosevelt's order four months earlier calling all personnel of the Philippine Army to duty and all active units into the service of the USAFFE.

Ricardo Galang, enroute to a scholarship to Boston University described a stopover in Honolulu. "It was a war zone... grim face... display windows taped or papered... sandbags on almost every street corner... military police on every sidewalk... barbed wire around all restricted areas." Instead of accepting the scholarship in Boston, he wound up in California and was pressed into military service. He believed that he was sent by the Philippine government to study in the U.S. because of an article he had written in *The Evening News* about the presence of Japanese radio and spies in Baguio, a city in northern Luzon. He would later be called on a mission.

Before the end of the year, President Roosevelt broadcasted to the Filipino people giving, "my solemn pledge that their freedom will be redeemed, and their independence established and protected. Entire resources in men and material of the U.S. stand behind that pledge." But the formal policy to support Europe first was already in place. There were not enough resources to fight two wars. There was no set date for the Philippines' redemption.

1941 ended with much Filipino blood spilt. Only two divisions of the Philippine Army went head on against the invaders' army of 40,000 troops on the beaches of Lingayen Gulf. In Manila, there were no Christmas celebrations. With the Japanese forces only several hours away, Manila was declared an "open city" the day after Christmas to protect it from further destruction. The residents of Manila were urged to evacuate the burning city, but many stayed.

MacArthur transferred his headquarters to the island fortress of Corregidor at the mouth of Manila Bay. President Quezon, Vice President Sergio Osmena, U.S. High Commissioner Sayre, and their staff sailed with him on the interisland ship *Mayon*.

SS *Mactan*

On the eve of the fall of Manila, MacArthur cabled the American Red Cross asking for a ship to evacuate about 200 of his wounded men. Amid the chaos at Dry Dock No.1 in the port of Manila, dozens of volunteers hurriedly painted the rusty, decades-old, interisland blockade runner, SS *Mactan* with the colors of a hospital ship–white with a large red cross. She was no longer bringing supplies to the defenders of the powerful coastal artillery guns on Corregidor; she was now a chartered hospital ship set to leave with the wounded of Bataan. She was the only ship left after all the others were deliberately sunk to prevent their use by the enemy.

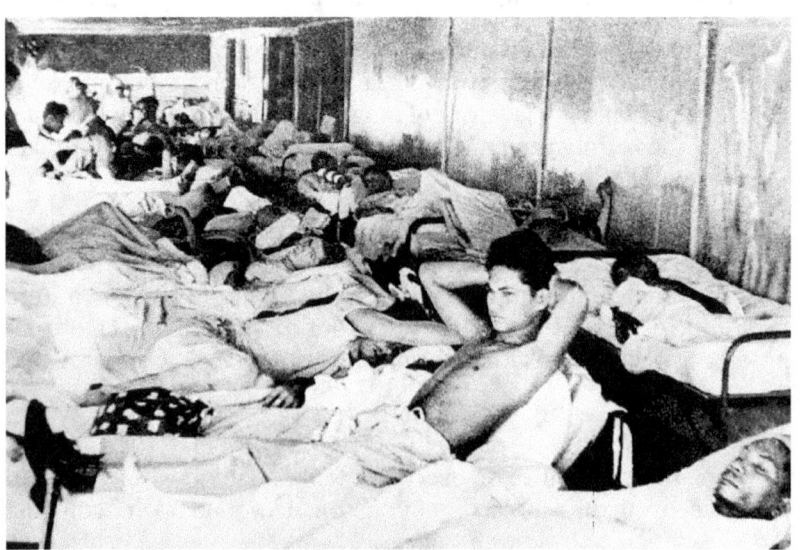

The Baltimore Sun – MANILA-OUTWARD BOUND—Wounded soldiers of the American-Filipino forces of General Douglas MacArthur lie on cots on deck of the SS Mactan, hurriedly outfitted by the Red Cross to evacuate the casualties from Manila. They were taken on a perilous voyage to Australia. (AP Wirephoto) 1942. Photographer: Irving Williams of Patchogue, Long Island and released through the Red Cross.

Over 250 wounded soldiers, mostly Filipinos, medical and Red Cross personnel, and the Filipino crew members boarded her close to midnight of New Year's Eve. A nurse was heard commenting that the wounded seemed to all be Army soldiers. What happened to the wounded Navy soldiers? Six doctors volunteered to accompany the wounded on board. Dr. Gregorio Chua volunteered as a junior physician in response to the American Red Cross's plea for help.

The ship cautiously sailed out of Manila Bay through a minefield while artillery shells from Corregidor fired overhead. A wounded soldier onboard wrote, "Roars of explosions and hot, white-blue flame of fires quickly faded to angry red under umbrellas of black smoke spread out over the harbor." The oil storage and supply dumps in Manila were being destroyed before the enemy arrived, turning the sky orange. It looked like Manila was dying.

Heading south, the *Mactan* was halfway through the archipelago on New Year's Day. She was moving slowly on her own to navigate without proper nautical maps and to ensure that her engines held up. She maintained radio silence to avoid detection from enemy ships and planes that could attack any minute. Crossing the Celebes Sea, a Dutch cruiser guided her through minefields. After six days in the open sea, she ran silent to stop at Macassar, Indonesia for fuel, water, and clean laundry. A fire in the engine room was doused by a wounded soldier with only a gas mask and a fire extinguisher. Under blazing heat for over a week, ants and cockroaches brazenly crawled on the wounds of the injured. Pills for pain were limited. Boils and sores developed on the patients and festered inside casts. A Filipino soldier, depressed over his condition after his left arm was amputated, jumped overboard. Water and food were rationed. Everyone lost weight. The high waves of a powerful storm lifted and dropped the ship, popping rivets in her hull. The passengers and crew gritted their teeth and prayed that she would not break apart.

When they reached Darwin in Australia, the enthusiastic welcome they received boosted their morale. There was no medical support for the wounded so they continued to Sydney, passing Townsville and Brisbane, picking up whatever water, supplies, medicine, and clean linens the towns could spare. Captain Julian Tamayo brought and used seven sacks of cement at Townsville "to stop the leaks in the rivets of (her) shell plating." The Australian Red Cross had given them up for lost days ago. Several patients did not survive the month-long journey.

In Australia, the *Mactan* was declared unseaworthy and the crew stranded. Some left on the Dutch ships. The Filipino doctors and nurses, however, could not stay and practice in Australia. They were socially accepted, "but not on the professional level by the Australian and U.S. Army," wrote the American Red Cross Director onboard the ship. Instead, five of the doctors joined a Filipino battalion being formed in California.

Over half of the wounded returned to active duty after months of recovery in the Australian hospitals. When the wounded soldiers heard of a Filipino unit being formed in California, they wanted to join as soon as they were well. With iron braces for his wounded legs and six months in the American hospital, Sgt. Policronio Sanchez did not wait for his legs to heal to join. Recovered from their wounds, Lt. Genaro Javier, a Philippine Army officer, and seventeen-year-old Sgt. Camilo 'Rammy' Ramirez, a Philippine Scout and a former prisoner of war, signed up. Ramirez was captured on Bataan. Sick with fever and malaria, he escaped with other prisoners through a gap in the barbed wire surrounding the prison camp and followed a ditch in the general direction of Manila. In an empty house, he changed into civilian clothes and hid his dog tags inside his shoe. He met an ambulance on the road to Manila and got on a truck to the pier where he boarded the *Mactan*. Exhausted, he fell asleep without knowing where the ship was headed. Fortunately, the quinine given to him on the ship lowered his fever.

Lt. Patricio Ubarro, chief of radio communications of the 26th Cavalry Regiment, Philippine Scouts, was inducted into the U.S. Army and assigned to join the operation of the first U.S. Army radio station in Brisbane. Lt. Delfin Yuhico, a graduate of the Philippine Military Academy worked with the military police investigating deaths of soldiers. He also vetted Filipino soldiers like Lt. Gregorio Martinez who somehow escaped from the Philippines after the surrender on Bataan, survived the harrowing seas on a raft after the boat he was on sank and was picked up by an Australian ship. Yuhico was trained for infantry and was praised for his excellent performance of duties by his commanding officer. But he was released for a better use of his skills. A hand- written note described him as an "officer of a very high type. His performances of duty must be rated at least Excellent. This (transfer) request was approved despite the shortage of officers in the command due to the fact that it is not deemed advisable to assign an officer of his nationality the command of American Troops. It is believed that his services would be of more benefit to the service in a combat unit of his own nationality." Yuhico would later be called to serve on a mission.

Australia under Attack

A devastating air raid on Darwin in February 42 sunk the blockade runners *Don Isidro, Florence Dy,* and *Don Esteban* that were bringing supplies to Bataan then to Corregidor. The survivors added to the number of Filipinos stranded in Australia.

Guillermo Romillo reported onboard the *Don Isidro* that the ship was torpedoed off the north coast of Australia during the full-scale Japanese air raids on Darwin. Eleven Filipino crewmen were killed on the *Don Isidro* and two more died in the Darwin hospital.

On board were crewmembers Dominador Gobaleza and Petronio Huerto. Patricio Jorge, an apprentice mate recalled, "When we were trying to beach the ship, we looked at the map. On the seashore was quicksand, then the cannibals. They never crossed the swamp because of the quicksand. A U.S. bomber from Darwin saw us and signaled that they were going to send (a ship) for us." The Australian ship *Warmambool* brought them to Darwin, including the executive officer Rodolfo Ignacio and Philippine Army officers Toribio Crespo and Irineo Ames, whose ships had also been destroyed. Ignacio and Crespo were former officers of the Off-Shore Patrol Reserve that defended the shore installations on Bataan. Later, these men would go on secret missions.

Over a hundred survivors from the sunken ships were transferred to the Small Ships Command, United States Army Forces in Australia (USAFIA) in Sydney to wait for orders. A need for radio operators brought many of them to technical schools in Melbourne. Their training included physical drills under an Australian Army officer. Gregorio Martinez from the *Don Isidro* could not keep up with the physical training. His exposure for a long period in an open boat without food or water following the sinking of his ship had greatly weakened him. He was returned to his unit. The men's unfamiliarity with the language, weak physical condition, and low educational attainment resulted in a high failure rate.

Meanwhile, Australia followed the news of short-lived victories in Bataan against an offensive by fresh Japanese reinforcements. "Men were like rats burying themselves in the ground. No replacements in the front lines while the Japanese had. The same troops stood in front from January to April round the clock till lines were finally broken by the constant firing, bombing, and shelling of Japanese forces. The only thing keeping the spirits of the boys alive was the hope of a convoy," wrote Dr. Victor Buencamino, who treated the wounded. "Eighty percent of enlisted men of the Philippine Army (were) using (WWI) Enfield rifles without extractors and (had) to use a bamboo rod to extract the shell after each shot," narrated Gen. Basilio Valdes, Chief of Staff of the Armed Forces of the Philippines.

The Bataan resistance caught the enemy by surprise. Lt. General Masaharu Homma, commander of the Japanese Imperial forces did not expect his plan to conquer the Philippines would be delayed by three months.

Some efforts were made to help the beleaguered men on Bataan. The inter-island *M/V Doña Nati* and the Chinese registered *Anhui,* converted to blockade runners, successfully delivered some 5,000 tons of rations, arms, ammunition, and medical supplies from Brisbane to Cebu in March. The following month, eight ships loaded with rations, medicines, ammunition, gasoline, and oil waiting at the ports of Cebu and Iloilo in the Visayan islands were ready to leave for Bataan. But the planes from Australia that were to escort them had not arrived. The approval was given only three days before Cebu fell to the Japanese.

Fearing MacArthur's possible capture, President Roosevelt ordered him, his family, and staff to leave for Australia. He was "to lead the counter offensive in the Southwest Pacific Area (SWPA)," signaling America's commitment. But MacArthur wasn't intending to leave. His message of February 11-12, 1942, to President Roosevelt: *I am deeply appreciative of the inclusion of my own family (to evacuate) ... but they and I have decided that they will share the fate of the garrison.* It took a month before the general finally agreed to go.

With General MacArthur's departure, Gen. Johnathan M. Wainwright crossed over from Bataan to Corregidor to assume command of the United States Forces in the Philippines (USFIP). He was replaced by Gen. Edward P. King, Jr. during Bataan's final doomed weeks. Over 60,000 men were cornered at the southern tip of Bataan. King was intent on saving the remaining starved and dying men from total annihilation. After the war, a Bataan veteran defiantly declared, "The Japanese did not defeat us. We were defeated by hunger and diseases."

Wainwright, in desperation, proclaimed, "Corregidor can and will be held." Earlier, he had released the Visayas and Mindanao commands. But, threatened with destruction and with no aid in sight, he surrendered the island fortress and directed the Visayas-Mindanao Forces to surrender with authority from President Roosevelt. MacArthur then deployed his contingency measure "that should Luzon fall into enemy hands, Mindanao would be the base of guerrilla operations." The plan for guerrilla warfare saw the use of submarines, the same way that submarines continued to run the Japanese blockades to bring supplies to the Bataan-Corregidor garrison before it fell. On the way out, the pigboats evacuated important personnel, equipment, and the treasury of the Philippine government.

Gen. William Fletcher Sharp, commanding officer of the Visayas-Mindanao Forces, received MacArthur's order to "initiate guerrilla operations," release his subordinate commanders, and expect no immediate aid. He was to lead the resistance and execute a defense plan of the four major coasts of Mindanao to delay Japanese actions until "the aid arrived." Sharp would finally surrender his forces but the delay of his communications with Wainwright and MacArthur by several days gave his men time to escape, hide, or join the guerrillas.

With the surrender of the Filipino American forces, further bloodshed was avoided. But Japan now had a broader path to New Guinea, a short distance away from Australia. Fearful thoughts about the fate of being prisoners of war were magnified. Riled with the devastating news of their country, Filipino men in Australia enlisted in droves into the Army of the United States (AUS).

The Filipino situation descended into chaos after the country's surrender. Local governments were left on their own as old political rivalries in the islands   as well as the centuries-old conflicts between *Moros* and Christians on Mindanao emerged. Independent guerrilla units formed to fight the Japanese adding to the political turmoil.

Allied Intelligence Bureau (AIB)

Darwin was in ruins from a Japanese bombing raid when General MacArthur arrived. He had been appointed Supreme Commander of the SWPA, but there was no army waiting for him to mount a counteroffensive. Over a quarter of a million Japanese garrisoned the SWPA area that reached as far as Port Moresby—Papua New Guinea's largest city and only 500 miles from Cairns, the closest city in Australia.

To cover the wide expanse of Japanese-occupied territories, the Allied Intelligence Bureau (AIB) was created with directives agreed on by the Americans, Australians, and the Dutch. The bureau was to report enemy information and conduct sabotage missions behind Japanese lines. Repeated requests for any reinforcements from the Allied command were not met but the U.S. Navy was given nearly 25,000 more troops in the Central Pacific.

But even without the needed reinforcements, the forces commanded by GHQ-SWPA successfully drove back the Japanese to the north coast of Dutch New Guinea to isolate Australia. Hollandia, the Japanese main supply dump, was cut off. As each stronghold of the Japanese was captured, the Naval Constructions Battalions— the "Seabees" constructed airfields, hastening the march towards the Philippines.

The Resistance

Unknown to GHQ, a budding resistance was forming in the Philippines. Civilians banded together with soldiers who refused to surrender to form guerrilla groups. The only thing to do now was to attempt to contact the American forces in Australia for support. The next few months would be a struggle for the guerrillas to contact the Allied forces with clandestine radios to let them know that the fight against the enemy was still on.

The first clue to GHQ came two months after the surrender of Corregidor – a radio signal that was picked up by two radio stations one in Australia and one in San Francisco. A Royal Australian Air Force (RAAF) station by the Adelaide River in Darwin heard the same message and forwarded it to MacArthur's headquarters in Brisbane. The message was suspicious because it claimed the uprising was happening in Java.

The next radio message came from Lt. Col. Guillermo Nakar, commander of the 14th Infantry, Philippine Army in Northeast Luzon. He reported that there were Filipino and

American forces who had not surrendered and continued to raid Japanese-occupied towns. He added that Japanese radio censorship and propaganda were brainwashing the Filipinos. His transmissions were in various codes, and the call signs were constantly changing. "Intelligence report [sic] reveals that enemy has detected the existence of our radio station, possibly by geometric process, and detailed a large force to look for us." Nakar's radio soon fell silent. He was captured in a cave, tortured, and executed. As many as five thousand of his men surrendered or dispersed.

Once more, the occupied Philippines fell silent. But the SWPA command now knew of a potentially powerful resistance movement behind enemy lines that needed organization and arms. Two months of deafening silence went by before another faint call for MacArthur was again intercepted from a Lt. Col. Ralph Praeger, an officer with the 26th Cavalry. Cut off from Bataan, he was hiding in the mountains of the Cagayan Valley of Luzon. He had formed a resistance group in Luzon and had salvaged a radio transmitter from a mine that allowed him to send messages to Australia. His message was being transmitted in clear text. GHQ's experience with Nakar's communications made it imperative for them now to use secret codes and to avoid transmitting in clear English. Quickly, codes were sent to secure Praeger's messages. Reports on prison camps in Luzon and the confirmation of guerrilla leaders sent earlier by Nakar were transmitted. But soon Praeger reported that the Japanese had caught on with his transmissions, scattering his unit. MacArthur sent Praeger a message of praise but there was no reply.

The contacts with Nakar and Praeger were short-lived, but these roused GHQ quickly to help the growing resistance in the Philippines. MacArthur issued an urgent directive to find a way to continue communications — no matter what it took. The task fell on AIB head, Col. Allison Ind. He was conducting intelligence work in the Philippines when the war broke out and survived Bataan. Conflicting reports listed him with General MacArthur's party leaving Corregidor. He understood that accurate information on the guerrilla resistance was needed and intelligence on the Japanese from the interior part of the islands was important in the planning for MacArthur's successful return to the Philippines. He believed that the AIB's many duties included "delicate ones which Filipinos could best accomplish."

The return to the Philippines was a strategic part of the Pacific campaign. It was clear that to be able to continue sending messages concerning the Japanese disposition and conditions of the population, the resistance needed weapons and ammunition to fight the enemy. The civilians needed supplies and medicine, or they would have to turn to the Japanese to survive. But the bits of intelligence needed should be useful and understandable if deciphered. Messages sent in the clear without codes could be intercepted by the enemy. Even more necessary was proof that the messages did come from the guerrillas, and not from Japanese decoys. Trained men were needed to establish radio networks and report back on the enemy.

In the meantime, physical proof of the resistance now stood in front of GHQ. Capts. William L. Osborne and Damon J. Gause escaped from Corregidor right after the surrender and reached Australia by sailing through the southern islands of Palawan and Tawi-Tawi. Helped by civilians, they managed to avoid Japanese patrols and crossed the narrow Makassar Strait. Barefoot and in tattered clothes, they reported to an amazed MacArthur in October.

GHQ's desire to get more information out of the Philippines was affirmed with the arrival of Lt. Frank H. Young, a Philippine Army soldier who had joined the guerrillas of Col.

Claude Thorpe on Luzon, and Albert Klestadt, a German who worked in Tokyo providing the British with intelligence on the Japanese Navy. With six *Moro* sailors, and using a Dutch atlas, they sailed from Mindanao and reached Croker Island off Northern Australia before the year ended. GHQ was elated to learn that a strong resistance did exist and that the civilians were fiercely loyal to the Allies.

The transmissions from a radio with the call sign "Free Panay Calling" began to reach the RAAF station in Darwin and the message was relayed to GHQ in Brisbane. Col. Macario Peralta, a former officer in the Philippine Army's 61st Division had united the larger guerrilla groups of Panay. While hiding in the hills, he found a "powerful transmitter from an abandoned British freighter" and used the distress frequency for fear of interception by the enemy. His messages described his organization, leaders, and internal affairs. He also added that his group had the cipher device M-94.

GHQ did not immediately respond because of suspicions that it was an enemy decoy transmission. The messages were not in code, and the claim that they had the M-94 cipher device made it even more doubtful as all codes were ordered destroyed before the surrender.

After weeks of persistent radio calls that went unanswered, Peralta asked that his identity be verified by Lt. Col. Jaime Velasquez, his classmate at Division Staff School in Baguio City, who was with Philippine President Quezon in Washington D.C. He requested that the message in reply should use a keyword in the M-94 cipher, followed by a double transposition of the answer to "the name of the place where Philippine President Quezon and Panay Governor Confesor last dined together." With the correct reply, the decoded message read: *Can you rendezvous (with) a submarine? If so give five places in order or preference. Signed War Dept.*

An excited GHQ directed Peralta to continue his "command with the primary mission of maintaining his organization and securing the maximum amount of information. (Combat) activities should be postponed until ordered from here (SWPA). Premature action of this kind will only bring heavy retaliation upon innocent people... victory will come. We cannot predict the date of our return to the Philippines, but we are coming." Station WBA was assigned to Panay. If no other guerrilla units had contacted Australia, Peralta relished the possibility of assuming command of the guerrilla forces across the islands. (Panay Island is discussed in a separate chapter.)

GHQ's confidence was bolstered by the messages signifying the presence of a growing resistance in the Philippines. It was imperative to send arms, radios, and supplies to the guerrillas so they could continue with their intelligence work on the enemy. Any connection with guerrilla units on the islands had to be completely undetected by the enemy. But considering the number of guerrilla leaders on the islands, it was crucial to select those who could be trusted enough to receive the supplies and arms.

Radio operators were now needed to transmit to Australia the Japanese activities on the islands. A long-range plan to send trained United States Army Signal Corps (USASC) personnel and equipment to the Philippines was drawn up and sent to Washington. At the very least, a unit of signalmen was needed to run KING ABLE ZEBRA (KAZ), MacArthur's radio station, and future incoming radio traffic from the Philippines.

Lt. Col Allison Ind with the AIB immediately submitted plans to MacArthur's headquarters to train Filipinos in radio work and to establish a radio network in the islands for the guerrilla intelligence. He turned to Maj. Joseph R. McMicking, a Filipino who had left Corregidor for Australia with MacArthur's staff. Ind relied on him to identify names mentioned in the guerrilla radio messages and describe the Philippine terrain. McMicking recommended ways to best land on and penetrate the islands.

Desperate to find Filipinos for their plans, Capt. Alan Davidson of the Australian Military forces, McMicking, and Maj. Rafael J. Cisneros of the blockade runner *Don Isidro* surveyed the Filipino men in Australia, disqualifying those with bad health and debilitating wounds. Captain Tamayo, former captain of the hospital ship *Mactan* was tasked to investigate his crewmates' qualifications for further training. Around 100 men were discovered scattered in Red Cross Centers around Queensland. They were assigned there while waiting for further orders and it was proving to be an unpleasant experience for them. They were called by their first names and carried out menial duties not usually done by a soldier. They expressed their dislike for Australians and especially women giving them orders. No longer a unit, they had lost their military bearings.

Immediately, McMicking had them transferred out of the Centers to be soldiers again and declared to the Red Cross that any attempts to 'borrow' Filipinos would be reported to the higher authorities. Cisneros was disappointed with the response from his former crew members. "The men had deteriorated and [sic] inclined to have a rather apathetic outlook towards the war and life in general."

Ind developed an evaluation and classification program for advanced training:

A. Leaders and sub-leaders of secret AIB parties.
B. Agent members of secret AIB parties.
C. Leaders and sub-leaders of coast watcher parties.
D. Members of coast watcher parties.
E. Leaders of supply running parties.
F. Members of supply running parties; and administrative personnel in the Philippine Regional Section (PRS) schools and Milton Staging camp.

The Advanced Training Course for all groups lasted 56 days. Some were held at Cairns' staging base for operations, and others at various specialty schools. When the selection for the final mission parties was made, the officers and men ate, slept, and worked with their parties to adjust to each other. They were also assigned new identities, code names, and nicknames.

Once given their assignments, Group A – leaders and sub-leaders underwent further training on added responsibilities and skills: combat intelligence, fifth column identification, cryptography, transmitting radio signals and Morse code, small boat navigation, Japanese and English language lessons, map reading, sketching, ship and aircraft recognition, and first aid. There were optional courses on marching, weapons firing, and cooking. Later, weather forecasting and aircraft warning detection were added to the program.

Around thirty men passed the training and transferred to AIB in Brisbane to be further trained and outfitted for the secret missions. The first qualified men selected were: Juan Rodriguez, Pedro Carriaga, Dominador Gobaleza, Felino D. Buencamino, and Susano Amodia. Rodriguez was a ship's carpenter on the blockade runner *Don Isidro* and Carriaga, the quartermaster. Gobaleza, Buencamino, Amodia, and Carriaga were graduates of the Philippine Nautical School. Their diverse backgrounds reflected the need for additional training in intelligence and radio operations so they could pass the rigorous USASC requirements. An officer at the Melbourne Technical College for radio operators supervised their activities to ensure their full cooperation, as it was known that the men disliked receiving orders from Australians. In the end, only three qualified for the Signal Corps. The others were assigned to the Infantry and Reconnaissance.

The Australian Military Forces readily offered the training at the Milton Staging Camp in Sydney. The camp was equipped with a radio room and complete facilities for coast watchers. Instructions included "intelligence gathering, commando skills, amphibious operations, radio operation, ship and aircraft recognition, weather observation, and other skills essential to combat intelligence work." Another hurdle was passing the training for jungle combat and land warfare at the Canungra Jungle Training School. Under ruthless discipline, they went through intense physical fitness courses for three weeks, twelve hours each day, learning to live with only their weapons and one canteen. One man reported having only a rope. They were fit for jungle warfare if they made it through the fourth week at the deep and rough MacPherson Range. The range closely resembled the jungles of New Guinea. The rigorous training took a toll on many of the men. Poor health, bodies battered by the war, and lingering effects of wounds from fighting in the Philippines caused many to fail.

From the men who made it through the rigorous training, the first six submarine missions were formed. The first mission was to depart in the last week of December 1942.

The severe lack of qualified men to train jeopardized the plans to send more missions sooner to penetrate the islands. An urgent request was sent to the Chief Signal Officer in Australia for eighteen radio operator-technicians and radio equipment. American signalmen were assigned until more qualified Filipinos could be found and trained.

PHILIPPINE MAP

LUZON

CAMP
CAMARILLA
MANILA

MINDORO

BURIAS
ISLAND

TICAO
ISLAND

SAN BERNARDINO STRAIT

BIRI ISLAND

CAPUL
ISLAND

ALLEN   MONDRAGON   CATUBIG

CATARMAN

DALUPIRI
ISLAND

CALBAYOG   SAMAR

CATBALOGAN

MAUD

JINAGNATAN   TACLOBAN
VILLABA
PALOMPON
ORMOC

GUIUAN ISLAND

CEBU

DULAG

CAMOTES
ISLAND   LEYTE

DINAGAT ISLAND

PANAON ISLAND   SURIGAO STRAIT

LUZON

MANILA

VISAYAS

MINDANAO

1st & 2nd FILIPINO INFANTRY REGIMENTS

MISSION: February 1945 - November 1945

A Filipino Battalion
1st Filipino Infantry Regiment

News of bloody fighting on Bataan and the relentless bombing of Corregidor raised the agitation levels of all Filipinos. In the U.S., thousands of angry Filipino men wanted to join the Army and defend their motherland.

Since the country was fighting on two fronts, the U.S. Selective Service Act of 1940 was amended so that it required the enlistment of all citizens and "every other male person in the United States."

But the U.S. Armed Forces was plagued with race problems. The Marines did not accept non-whites. The Navy took non-whites in the late 1930s but only as mess stewards. Of course, Filipinos wondered how the white-dominated military establishment would treat them. Their civilian life was a constant battle against racism. In joining the Armed Forces, they were to risk their lives to defend U.S. interest and remove the invaders from the Philippines but were still classified as aliens or nationals. They were not to be granted the civic benefits and privileges of citizenship, owning property, and equal employment opportunities.

On the United States mainland, Filipino communities clamored for the formation of a Filipino unit, sending petitions to President Roosevelt to allow Filipinos in the United States to join the U.S. Armed Forces. Telegrams were sent to Philippine Commissioner Joaquin Elizalde in Washington D.C. to spearhead the movement.

Without waiting for a reply from Washington, some Filipinos formed their own militia units in conjunction with the California State Militia. News of the people's sufferings and atrocities committed by the enemy fuelled recruitment.

In Washington D.C., Abdon Llorente and Ernesto Ilustre, a former journalist and public relations man for the Philippine Resident Commission in Washington D.C., discussed the idea of creating a military unit that would help in the liberation of the Philippines.

Ilustre's stint as a 1st Lieutenant in the Philippine National Guard during WWI gave him the idea of an all-Filipino unit. The two men circulated petitions within the Philippine communities to have the U.S. Congress change the law to allow Filipinos to join the service.

Heeding the clamor from the persistent Filipino communities, Commissioner Elizalde met with the sympathetic Secretary of War Henry Stimson, who had formerly served as governor-general in the Philippines. His initial reply was that "under our present law a person who is not a citizen of the United States may not be commissioned or enlisted in the Army. A change in the law will be required if Filipinos who are not citizens are to be permitted service in the Armed Forces of the U.S." Ilustre quickly went into action circulating petitions in the Filipino communities to have the U.S. Congress change the law.

In February 1942, the U.S. War Department approved the creation of a Filipino combat battalion to enable Americans of Filipino ancestry and resident Filipinos to serve in the U.S. Army. President Roosevelt made it official with a letter to Elizalde expressing his unconditional approval of the activation. It was not to be a segregated unit by law. The Filipinos had their own unit and only Filipinos who volunteered were sent to join it.

Secretary Stimson announced the 1st Filipino Battalion as a "new unit (formed) in recognition of the intense loyalty and patriotism of those Filipinos who are residing in the U.S. It provides for them a means of serving in the Armed Forces of the U.S. and the eventual opportunity of fighting on the soil of their homeland." Filipinos flocked to enlist. Wearing a uniform showed their loyalty to the U.S. and when sent to the Philippines, they would be able to see their families back home. The majority had not seen their families in decades.

The War Department instructed all Philippine military personnel stranded in the U.S. to report to the Battalion. Philippine government scholars enrolled in the American universities and Filipinos graduated from West Point were instructed to join. Sons of wealthy Manila families on pleasure tours or enrolled in universities and stranded by the war enlisted. The common bond was the eagerness to contribute whatever they could to help win the war.

From Australia came wounded soldiers who luckily escaped on board the hospital ship *Mactan* and joined as soon as they were able to walk. Pvts. Celestino Galavasa and Alfonso Gil were two of the wounded from an airplane attack on central Luzon while they were defending Lingayen Gulf. After several weeks of recuperating from their wounds, they were sent to the U.S. to join the Filipino unit. Sgt. Eustaquio Corpus had both legs in plaster casts while Sgt Policronio Sanchez was still wearing iron leg braces when they arrived at the Battalion. Lt. Genaro P Javier joined after four months of recovering from machine-gun wounds while defending the beaches of Lingayen.

In the European theater, Filipinos were allowed to serve in the American white units but were strongly encouraged to join racially-organized units. Some had duties in California and inside military bases or warships far from combat zones. Others joined the Army Air Corps, the Navy, the Coast Guard, and the Merchant Marine. The Philippine Resident Commissioner investigated to clarify and rectify the Navy's policy of assigning Filipinos only to service jobs.

1st Filipino Battalion
Camp San Luis Obispo, California USA

The 1st Filipino Battalion was activated on April 8, 1942, at Camp San Luis Obispo, California. It was a distinct Filipino unit of the U.S. Army – Filipinos in the service anywhere in the United States were transferred to the unit for training. They were encouraged to enlist, especially those who believed that they could pass the rigorous basic training despite advanced years. The Filipino community rejoiced as the influx of volunteers increased.

A week later, the news of Bataan's surrender doused the men's joy with the cold reality of defeat. Yet, this only increased their anger. The relentless artillery barrage on the Filipino and American troops on Bataan coupled with lack of food and disease sealed their fate. They stood defenseless against Japanese reinforcements of twenty thousand men with new artillery pieces and bombers. Jose Reyes, a civilian, wrote, "... all hopes died. Finally, we knew there wasn't going to be an American convoy coming. We all died a little when Corregidor surrendered."

All mail sent to the Philippines were stopped and the letters returned. Filipinos from both sides of the Pacific feared the worst. When President Roosevelt ordered General MacArthur to leave for Australia, they wondered what would be left of the Philippines.

At U.S. induction stations, men lined up for physical examinations. Those who were accepted received notices that they were to become members of the Army of the United States (AUS). If they refused or failed to take the oath, their recruitment status remained. They had to arrive clean and sober. Postal cards were provided so that they may notify a relative or friend of the station's location, and the assigned unit as the destination and permanent address. A short list of the Articles of War on court martial penalties imposed on desertion, absence without leave, disrespect toward superior officers, drunkenness on duty, and provocative speech or gestures was given. They could bring only one bag, athletic equipment, musical instruments, a small photo, and one pair of shoes. Firearms, weapons, and liquor were disallowed. There were no parking lots for cars or storage for belongings.

Although they were not standard Army equipment, many brought their *bolo,* a large knife used as a tool and weapon with a curved wide blade that narrows towards the hilt, to Basic Training. Those who did not have one ingeniously shaped new ones from automobile leaf springs to use for practice.

Capt. Abdon Llorente arrived from Washington D.C. and reported for duty with the Battalion making him the fourth Filipino officer in the outfit. He was close to 50 years of age but passed the rigors of physical training.

Filipino newspapers encouraged more men, especially those who had "college education or military experience to join with the enticement of wearing shining bars." With only seven commissioned officers in the beginning, hundreds of officers would be needed if the plans materialized for the Battalion to become a division with three regiments. The units organized and underwent training in California.

As the Army rapidly expanded during the first two years of the war, officers were needed to lead platoons, companies, or sections. Volunteers' personnel files were reviewed. The

1st Filipino Battalion. Camp San Luis Obispo. May 1942 Source: Unsung Heroes

A newly inducted Filipino soldier stands in front of his single-story barracks building at Camp San Luis Obispo, California. He was assigned to the U.S. Army's 1st Filipino Battalion for Basic training.

first group of qualified candidates was shipped to Officer Candidate School (OCS) at Fort Benning in September of 1942. It was the first formal racial integration of men replacing the selection of officers from the social and intellectual elite that was done in WWI. All candidates attended the same classes together learning to fire many weapons from pistols to cannons. Practice targets had faces of Japanese soldiers. After 13 weeks (91 days) of intensive advanced leadership and military training, only one of every three candidates graduated as a 2nd Lieutenant, an effective and well-prepared junior combat officer in the Army and a well-grounded military instructor. They were called the '90-day wonders.'

As a private, Silva's active service in the AUS began in February 1942. He had enlisted six months earlier under the Selective Service Act 1940. Silva, Sergio Solidarios, who was the Oriental Club President at San Francisco State College, and other bachelors in the city, were soon wearing Army uniforms with the 1st Filipino Battalion at Camp San Luis Obispo. After passing Basic Training at Fort Ord, Silva was assigned to the 27th Infantry Division in Hawaii which was directed to defend the outer islands from amphibious attacks after Pearl Harbor. He then attended the OCS 90-day crash course and, after three months, graduated as a 2nd Lt. By the time he returned to camp, the Battalion had swelled to thousands of Filipinos clamoring for a chance to fight for their motherland. But since not enough men graduated from OCS, the plan to have five Filipino officers per company was not realized.

Notes taken in an interview written on a stationery of the Mark Hopkins Hotel in San Francisco listed Silva's serial number, graduation date from OCS, birthdate, place of birth, education, relatives, departure date from the islands, and arrival date in the U.S. Another form showed he could swim, navigate a small boat, use a typewriter, operate a car, a machine gun, a pistol, and a rifle. Also added to his list of qualifications were 'houseboy, gardener, *lavandero* (laundry washer).' Was he asked if he had performed such jobs before enlistment, or if he was willing to do such work? The Army liked the fact that he had no political connections in the islands. In Company I, he was assigned as a platoon leader with Infantry as specialty. His performance was rated as Excellent.

Ruben Songco was also interviewed at the same hotel and notes taken on the same stationery. Thirteen years younger than Silva, he came to the U.S. the year before the war broke out and graduated from the U.S. Naval Academy in 1943.

At OCS, Toribio Luna and other Filipinos in the State of Georgia were allowed because of the war, to use "White" Facilities in restaurants, theatres, restrooms, etc. Although he had experienced similar discrimination in California, Luna found it hard to understand because they were all American soldiers now. "In the military base, ... the training classes were conducted with all the races together, but after, all races were separated and had their separate mess facilities and barracks. People thought they were American Indians... had no idea these soldiers were Filipinos." He returned assigned to the Battalion's Service Company.

OCS graduates Gregorio Minodin was assigned as a Personnel Officer, and Lt. Don Mendoza to head Company C. Lt. Alberto C. Elefano was studying animal genetics and had been teaching at the University of Maryland when the war broke out. Due to his ROTC training in the Philippines, he was sent to OCS. He applied for active duty wanting to be able to go back to his wife and child in the islands. At 5'3" in height, he was the shortest officer in the company but kept up with the bigger men and proudly finished the obstacle course.

Rufino Cacabelos did not allow his height to deter him. He returned from OCS to command a unit. The OCS magazine, *The Bayonet,* featured him as having "prompt and willing obedience to orders of his superior officers... morale now low because he could not carry out an order to the satisfaction of the officer instructor from the infantry school. He did immediately stand up when given the order but standing straight up brought his head to only about the level with those of the rest of the class sitting. He is 5 ft one inch in height."

In the California fields, Teodoro Taluban was harvesting vegetables when he received a draft letter telling him to report on a specific date. He was processed and put up in a hotel with other men overnight. The following day, a bus brought them to Camp San Luis Obispo. He was assigned to Company H and issued all the equipment he needed. He underwent many tests, both physical and written, he recalled. Basic training began, and he was picked to attend OCS. He led men in his company when he returned.

As one of the first to enlist, journalist Ernesto D. Ilustre, followed the men's progress and wrote of their experiences in Basic training. He wrote that they looked up to "their superior officers, paid close attention to their military training, liked the California climate, and relished the Filipino dishes served them Army style with rice and all." He quoted Pvt. Jose Dulay, Jr. "We Filipinos are in this war to stay until we attain our dual objective: to help America with the war and to drive the Japs away from our homeland."

The Battalion rapidly grew as volunteers joined from Army posts across the U.S. Ten thousand volunteered to serve. Around seven thousand were accepted, representing ten percent of the known Filipino male population in the U.S. in 1941.

In California alone, a total of 16,000 men, forty percent of the state's Filipino population registered. About 1500 Filipinos in the Bay Area of California were reported as eligible. They pledged to fight "wherever they want to send us." Their strong feelings came from personal

1st Filipino Battalion. Camp San Luis Obispo. July 1942.

A soldier stands at attention in full gear in front of his barracks.

grudges inflamed with thoughts of families starving, violated, or fighting on Bataan. They left their jobs on the labor farms—hotel waiters, barbers, cab drivers, elevator boys, and hospital attendants. Professionals, office clerks, businesspeople and Federal employees flocked to enlistment centers to sign up. Some joined to fulfil requirements of the Second Powers Act 1942 that granted eligibility for naturalization to non- citizen aliens who served three years in active duty in the U.S. Armed Forces.

The Battalion was segregated by choice, with Filipinos wanting to form their own unit, unlike the Army's policy of segregating the Blacks to their own unit. "We never thought of ourselves as being segregated. We were just happy to be with our own, " recalled Clemente Joe San Felipe. Many of the non-commissioned officers had extensive military experience. Sgt. Isidoro Dacquel celebrated his twentieth year in the U.S. Armed Forces, and Sgt. Francisco Morales had thirty years of Army training behind him.

The *Manongs* from the central valley of California that joined made up a large significant group mostly from the Ilocos province of Luzon. The men from the southern islands of the Visayas and Mindanao who usually found jobs in the cities' service sectors did not want to be identified and called a *Manong* who worked in the fields. The *Manongs* were one of the earliest Filipinos to arrive in the 1920s and were now in their mid-30s. They subtracted a couple of years from their age to be accepted. They burned with the desire to return to the Philippines, fight the invaders and see loved ones they had not seen in decades. With low wages earned following the harvest seasons of fields and orchards along the west coast and the Alaskan canneries, they "could not save enough money; it was a way to return." Frank Cali, a *Manong*, worked in the canneries every summer until he joined the Army as a corporal in the Battalion's Company H. He hoped to return to Bolinao, a cape overlooking Lingayen Bay in his home province of Pangasinan.

Leovigildo Giron was working in a labor camp in Los Angeles when he was inducted and sent to San Luis Obispo. In a large room, he passed an aptitude test that many failed. He followed orders to attend courses on Morse code, flag signals, cryptography, and paraphrasing. He did not know what the Army had planned for him. He went through Basic Training excelling in hand-to-hand combat because, as a child in the Philippines, he had learned *Eskrima,* a Filipino martial art that used sticks for fighting.

In San Francisco, Magno Cabreros related: "When we were drafted (sic), the people here in Monterrey said, 'Boys, you are the only outfit who knows where you are going. You are going to a suicide unit.' Our American contemporaries were proud of us because we all volunteered. We wanted to be there."

Telesforo Atienza, who had owned an art shop since 1937, padlocked it in May 1942 to join the Battalion. While recovering in a hospital ward at camp, he created pencil sketches and exhibited his artwork at the Red Cross station hospital.

In Chicago, Patricio Ollada was drafted in June 1942 before arriving at Camp San Luis Obispo, He trained for two months at Monterrey, California in advanced reconnaissance, intelligence, navigation, ships and planes recognition, sketching, and radio. Lt. Taluban recalled that at the Presidio in Monterey, the most extensive subjects were "to learn to speak and understand Japanese, radio communications by using morse code, and other things like how to manipulate a sailboat in the ocean."

But not all were eager to enlist, and some tried to avoid the draft. Philip Ordono escaped to Seattle. He was found and served notice to sign up. He received insults of "'Hey monkey,' while walking down a street. Go back to where you came from." He changed his mind when a co-worker on top of his bunk at the labor camp was shot and killed. He dared not travel with a white woman he met at a dance hall. But "things got better with the draft and army training," he recalled.

Frustrated with the debilitating racism that restricted him to menial jobs as a janitor, dishwasher, and a farm laborer to survive, Pvt. Manuel Buaken joined. He was studying law when the war began after switching from the ministry, a scholarship he received at the young age of 14.

As news of the Battalion spread, the men began to think deeply of their reason for joining. The prospect of the Army's living conditions of food, shelter, and clothing, so much better than the camps in the back-breaking work fields of California, enticed them. Most joined the Army to defend their homeland, and to prove that they were willing to shed their blood for a country they wanted to be a part of despite rampant inequality. Wearing Uncle Sam's olive-drab army uniform, they felt equal to American boys. A soldier's wife noted that the war and military service gave Filipinos a chance to show themselves as 'soldiers of democracy,' as men, not houseboys.

Doroteo V. Vite explained that his reason to fight was for his home, family, and all the things dear to him as a boy now in the path of the Japanese war machine. Before joining the Battalion, he chose to finish the six weeks of Basic Training in a regular Army unit with White Americans who admitted to him that they wanted "to meet and see with their own eyes a representative of those little fellows who are showing them dirty Japs how to fight." He wanted to understand the nature and attitude of the white Americans "which I could never have got(ten), anytime, outside the Army." He had the "opportunity of close association with men of all classes and from any place." He recognized that his attitude was more intense than the average American, for he was fighting to free his invaded country and his family, even though he did not know if they were still alive. The Army's assignment of soldiers to units where their skills would be most useful erased his prejudice against the military.

Brothers Aurelio and Amado Bulosan enlisted while their well-known brother Carlos Bulosan continued to write articles and books. Aurelio joined Company C as the Battalion was forming. Although the anti-Filipino laws radicalized him, he was hopeful about his situation in the military, declaring the "beauty of the human spirit and its realization in Mother America" in a diary.

As a student at Stanford University, Fernando A. Taggaoa joined in the first year of the war. He looked forward to "doing something which up to now I have never been allowed to – serving as an equal with American boys... we have always been subject to many restrictions."

Eager to join, Felipe Dumlao volunteered in Seattle. "We were seven Filipinos: first... Patricio Tarludo, Estepa, Humgoleno, Corpus, Rodrigo, and me." The men were probably *Manongs'* from the labor fields. They were surprised that there were only two small tents and one captain at Camp San Luis Obispo when they arrived. Seven American cadres arrived to train the men, but they were only seven men at first, one cadre per volunteer before more men arrived and a regiment was formed within two months," he recounted.

William P. Gale arrived with a cadre from the 7th and 35th Infantry Divisions to form the Headquarters company and train the men in military discipline. Support units were the 155th Station Hospital at Camp Roberts, the 1st Medical Regiment, and 7th surgical Hospital at Fort Ord as the Battalion expanded. Gale had served in the G-4 Planning section in the Philippines and was familiar with the Filipino culture. He reported to Gen. Charles Willoughby, head of G-2 Intelligence on General MacArthur's staff and who had earlier expressed doubts on the success of using Filipinos for secret missions behind enemy lines. When the cadre departed, he stayed on as Communications Officer for a limited time.

The cadre supervised and instructed basic training for the Filipino recruits. For invasion tactics, they were trained on throwing grenades; handling anti-tank guns for beach defenses; leaping from assault boats through smoke screens; and, rushing for protective cover on the beach. The men learned to operate the trench mortar, water- cooled and air-cooled machine guns, and the Browning Automatic Rifle (BAR). They spent countless hours mastering the Garand rifle, keeping firmly in mind their trainer's instructions – "If the trigger was not released correctly, you would fire the entire clip. Thumbs were smashed by the bolt, if not fast enough as it snapped forward at you. The rifle had to be held firmly against the shoulder and cheek, or the recoil would beat you up."

Felipe Dumlao was the liaison officer and became skilled in bomb demolition. He instructed on how to avoid gas released on the battlefield. Deep inside, he wanted to go back to the Philippines.

In April, a troop commander with the Philippine Scouts assumed command of the fledgling Battalion. Colonel Robert B. Offley, a West Point graduate, understood the Filipino's military history of revolts against their Spanish colonizers and the American presence in the islands. His father was an American Army officer who was appointed the first governor of Mindoro Island. Spending his youth in the island, he spoke Tagalog, a widely spoken language, which made him ideal to manage situations that provoked the men. When he arrived in California, his staff began to work with the initial cadre of white officers and Filipino West Point officers stranded in the U.S. Abdon Llorente was brought into the unit to "assuage grumblings about the lack of a Filipino officer in the command."

As the Battalion grew, Alex Fabros Sr encouraged more men to join. As a columnist for the *Thorns and Roses* of *The Philippine Mail,* he wrote in the May 31, 1942, issue "The best time to join the First Filipino Battalion is NOW. Those who have college educations or those who have had military training have all the chances in the world to become a wearer of shining bars. Presently there are only seven Filipino commissioned officers. There is a movement afoot to make the battalion into a division which will be manned by more than one thousand officers. There is your chance! The Filipino Battalion in San Luis Obispo needs many things for its recreation hall. Have you got anything in mind to offer? The Filipino fourth estate (the Press) is well represented at the First Filipino Battalion. Ernie Illustre, Ygnacio Sarmiento, Frank Osalvo, Saturnino Silva, and Sergio Solidarios are now wearing army uniforms."

Numerous newspapers across the U.S. published press releases of films about the Battalion to show the American public who were hearing and wondering about the Filipino unit who volunteered to fight for America and that such a unit was being formed to help the war effort. Local officials, who had earlier expressed disdain, helped administer repatriating Filipinos back to their country and used police force against them, were now organizing rallies and parades that included a large contingent of the Filipino soldiers. Anti-Filipino publicity decreased significantly.

The article, *They'll Make Trip with General MacArthur"* in the April 10, 1942 issue of *The Palm Beach Post* described "a small band of grim and determined men, who hope to accompany Gen. Douglas MacArthur when he goes back to the Philippines, form the nucleus of a U.S. Army Filipino Infantry Battalion. Only recently activated and still small but expecting to grow. CO is son of a former governor-general of one of the Philippine Islands, Col. Robert H. Offley, son of Col. Robert S. Offley, U.S.A., who served as gov-gen of the island of Mindoro. ExO Capt. Tirso G. Fajardo said Filipino men are transferring rapidly from other army units to be with us in the army when we retake the Philippines. They want to be with Gen MacArthur on his return trip there. Other officers are Filipinos including Lt. Roberto Lim, son of Gen Vicente Lim, Filipino commander now fighting in Bataan."

The volunteers thought of Offley as the "great white father who took care of them and gave Filipino officers opportunities to lead." He did not forget his time growing up in the Philippines. His understanding of the men's cultural differences from the islands they came from developed a common goal in them as new soldiers. Although they spoke different dialects, the forty-year American presence in the Philippines bound the men together. They were proud of their Filipino heritage and revelled in Offley's acceptance of their food and customs. More than just a commander to the men, he was affectionately called *Tatay* to mean 'father' in Tagalog because he spoke their native Tagalog. He joined the men in eating Filipino food and made sure all his white officers did too. He became the protector of their welfare on the base as well as off. When a white officer refused to serve under Filipino 'niggers' he was banished from the unit.

NEW YORK BUREAU. Training of 1st Filipino Infantry Battalion at Camp San Luis Obispo, California. Private Doroteo F. Corpuz and Private Cornelius S. Daur, both of the Philippine Islands, clean their new Garand rifles underneath the sign reading, "REMEMBER BATAAN" which they have placed over their bunk. CREDIT (U.S. Army Signal Corp FROM ACME) 4-23-42

FIGHTIN' FILIPINO

CAMP SAN LUIS OBISPO, CALIF. – Private Gregorio Mante, member of the recently activated 1st Filipino Infantry Battalion, is about to hurl a hand grenade during rigid training here. Under the command of a U.S. Army officer the new Battalion is shaping into form to fight for the restoration of their homeland. CREDIT (U.S. ARMY SIGNAL CORPS PHOTO FROM ACME) 4-23-42

Training of 1st Filipino Infantry Battalion at Camp San Luis Obispo, California.

1st Sgt. Robert Simmons, of Alameda, California, rear, instructor. Pvts Mario Ramos and Felix Hondalero (sic) in machine gun operation. Signal Corps Photo by BPR. (Mr Carl Gaston) Original neg. received April 1942.

The men gave their best in appreciation for being treated as equals. "The *Manongs* out-ran, outdanced, out-hiked, out-gambled, and out-classed us, "said Antonio Campos, who was half their age. The years of backbreaking work in the fields and canneries hardened the *Manongs* but that later became an asset in the Philippines' hostile jungles and enemy- infested islands. Campos remembered learning about "trust" no matter the reason after an incident after a visit to his wife when he went AWOL after missing the bus and hitchhiking back to camp. His battalion commander, Col. Jaime Velasquez – a former aide to President Quezon – transferred him to 3rd Battalion under Tirso Fajardo because "if he could not rely on him on stateside duty, how could he (be relied on) in combat?"

The U.S. Army Signal Corps Press' attention to the men boosted their confidence.

One of the Signal Corps photographs on the cover of *American Rifleman* (June 1942) magazine featured Private Gregorio Mante, who cut asparagus in the farms before joining, posing with a grenade clutched in his right hand and holding up a Springfield rifle in his left hand.

Tactical training was held on the beaches of Morro Bay and Point Sal. The Army Signal Corps photographed the men with steel helmets carrying Garand rifles as they disembarked from rubber boats and pointing a machine gun towards an imaginary aircraft while in a running Army jeep. Leading them was Lt. Roberto Lim, a graduate of the U.S. Naval Academy. On the back of a photo was written: *Filipino Battalion - The Nation's One and Only. Somewhere on the Pacific Coast is a group of very determined young men. They comprise the nation's first and only Filipino Infantry battalion, of which some officers and men were recruited from the United States while others are former members of the Philippine Army. With friends, relatives, and families in the Jap held Philippine Islands, these Filipino warriors attack their tasks of preparatory war training with an added incentive to get at the Japs. In training, the men had to carry a 40-lb pack and rifle over obstructions in 5 minutes.*

Most of the recruits were civilians and unaccustomed to military life. Lim had to teach and discipline them. As their leader, he led the men on "their physical build-up exercises at Morro Bay." Under the eucalyptus groves, he lectured to them. Only 22 years old, he oversaw one hundred men, most of them twice his age, to make them ready for battle. "It was a challenge, and I was enjoying meeting the daily challenges," wrote Lim. Learning about being a good teacher surfaced when he demonstrated the .45 cal pistol. While talking about safety and handling of guns, he showed the trigger and gave a slight pull thinking it was empty. To his surprise and embarrassment, the gun went off. Luckily it did not hit anyone.

He could not understand the men's love of gambling and forbade them with punishment of extra duty. He had to give that up when the entire company was under penalty doing cleaning, kitchen police, or double-timing around the camp.

Officers watched their feet as they proudly marched in crisp khaki uniforms with pants wrapped in leggings, soft overseas caps, and neckties. "In early 1942, boots did not rise high, so khaki canvas leggings or gaiters were wrapped around the person's leg and over the hiking boots to protect from branches, snake bites and mud."

The men underwent intense training under the white cadre officers. They faced each other in practice with fixed bayonets on their rifles. This may have been a propaganda photo since they wore neckties.

Staff Sergeant Roberto Sarmiento gives the soldiers some pointers in close order drill as part of their training.

CREDIT (U.S. ARMY SIGNAL COPS PHOTO FROM ACME) 4-23-42

(Below) SHAPING FILIPINO BATTALION INTO FIGHTING FORM CAMP SAN LUIS OBISPO, CALIF.

Under the command of Lt. Col. Robert H. Offley, U.S. Army, the 1st Filipino Infantry Battalion, which was activated April 1st, here, is now undergoing rigid training. The new Battalion, which will be commanded by American officers and by officers of the Army of the Philippines attached to the Battalion, will give Filipinos a chance to fight for the restoration of their homeland. This picture is one of a set depicting the training of this Battalion, which is shaping into fighting form. W 646220. NW York BUREAU

June arrived with the decisive victory at the Battle of Midway, a turning point in the war. The enemy's plan to surprise the Americans in an area near their homeland backfired partly because US Navy codebreakers intercepted and decoded the plan to attack the naval base on Midway Island. Buoyed by the victory, hundreds more Filipino men from all over the U.S. joined.

Meanwhile, Julius Ruiz, editor, and publisher of *The Philippine Newsletter* in Washington, D.C. visited the ailing Philippine President Quezon in the summer of 1942. He would later join the Filipino unit. For the next two months, he went around the U.S. talking to Filipinos, their leaders, and service members to form a clear picture of their lives and problems. The intense loyalty to Quezon's government stood out. The most severe problems were on the Pacific coast but despite this, the men here were more nationalistic than those in other parts of the U.S. Ruiz reported Filipino workers took the farm jobs in California with meager wages as they were unable to get any other work. They were also blamed for taking white men's jobs and were driven from camps. If they found work, it was in the war factories as janitors or kitchen help. The lack of Filipinas did not promote family ties. Discriminatory laws limited recreation centers and association with the white communities especially with their women. Adding to the isolation was the regional ties with different Philippine dialects that discouraged unity. But all of them were intensely loyal to the Philippine government-in-exile. They were counting on the Quezon government to improve their conditions.

1st Filipino Infantry Regiment:
Salinas Rodeo Grounds, California USA

In six months, the command grew to an equivalent of two regiments with thousands of men from all over the U.S. Orders were given to move the overflowing unit to the Salinas Rodeo Ground as volunteers continued to arrive.

The 1st Filipino Regiment activated on July 13, 1942, with personnel and equipment from the disbanded 1st Battalion. The Regiment did not have a Table of Organization and Equipment (TOE) as it was not under an Army division. If it had reached a regular division, its daily activities would have been recorded and archived. It was an independent unit and its records which were kept separately later proved to be a problem in creating the unit's history.

Salinas Rodeo Grounds. 1942

FIRST FILIPINO INFANTRY BAND
CAPT. ASHCRAFT-COMMANDING

1st Filipino Infantry Band Laging Una (Always First) was formed with 45 members

Al Lim, from Hawaii, tried to enlist in the U.S. Navy but was turned away because he was a 'national' and not a citizen. But he could be drafted into the Army, so he joined the Regiment in Salinas.

Unlike various combat units that shared service units, the Regiment was unique as it was self-supporting with its own supplies, cooks, service company, cannon company, tank company, headquarters, and personnel separate from the regular U.S. Army. Quarters were four-man tents with wooden walls and screened windows lined up around the horses' stables. A cool breeze penetrated through the slats of the thin walls. Rolled tarps hung on the windows and lowered when it rained.

The Filipinos' enjoyment of music and gatherings sent out a call for a band for each of the Regiments. Marching music was the priority. More than the needed number of musicians answered. Former musicians were accepted if they maintained their soldier skills to stay in the band. The Filipino community and friends gladly donated the instruments.

The *Laging Una* band's popularity spread to nearby communities with their repertoire of patriotic songs and popular Big Band songs of the 1940s. The U.S. National Anthem, Filipino songs, and pop tunes were popular. They had a talent for rendering 1940s popular songs in marching cadence. Sgt. Francisco Morales took the task of putting the men into long periods of rehearsals. They played at various units, organizations and regularly at the Station Hospital in the Red Cross building. They accompanied local Filipino talents and soldier pianists in performing at USO Centers to the soldiers, invited friends, and relatives. It was only when Filipino dignitaries visited especially during social events with the Filipino community that *Lupang Hinirang,* the Philippine national anthem, was played. Not a dry eye was in the audience when the anthem ended.

At Offley's request, the band gladly played one more song before putting their instruments away. It was the American Army song, "As the Caissons Go Rolling Along" but with the lyrics substituted with "As the Filipinos go marching along." Offley strongly encouraged participation in local parades, war bond rallies, and social events to publicize the fighting Filipinos' war training in California.

There must have been more musicians available because a third smaller band was assigned at Camp Beale in early 1944 to the 185th Army Ground Forces' parade events and social gatherings.

Offley's main challenge was to turn these middle- aged men with diverse cultures into disciplined, motivated, and focused professional soldiers that GHQ would want to send into battle. Many were in their 30s and a few over 40.

Camp Beale, California. 185th Army Ground Forces Band. 1944. Source: Harold Liban

His concern for the men's inability to finish training due to their advanced age was unfounded when they showed their physical agility and stamina equal to most men ten years younger. Their hard work in the fields and eagerness brought them through. But there were other problems unique to the men that were uncovered during basic combat training. They had difficulties reaching the trigger of Springfield rifles which affected their marksmanship until the release of Garand rifles with shorter stocks that made   it easier to handle. The increase of Filipino commissioned officers who could give instructions in their native language solved their difficulty in understanding military commands. But the commands were given in English. So, in return, they taught their American officers Tagalog, the most common Filipino language, during their off-hours. The old army fashion of 'bawling out' did not motivate the men. Pointing things out worked instead. Class distinctions based on education and social position disappeared when Filipino officers whose parents in the Philippines lived in debt servitude shared quarters with sons of Manila's oldest and richest families.

The communal living in the farms developed their reliance on each other and allowed them to adapt quickly to military life. They became an extended family for each other. Their regional differences from back home disappeared as they trained together. Philippine martial arts and *bolo* knives were integrated into their training. They shone as excellent soldiers and were no longer seen as troublemakers or a threat.

"Their enthusiasm and discipline are far superior to any I have seen in my Army career. The minute you put one of these boys in uniform he wants a rifle. The minute he gets a rifle, he wants to be on a boat. He can't understand why we don't ship him out right away, so he can start shooting Japs," said Offley.

Depressing news arrived a week after their celebration as a Regiment. The Japanese juggernaut seemed unstoppable. The enemy had landed near Gona in northern New Guinea going towards Port Moresby facing northern Australia, only several hundred miles away. Port Moresby was to be held at all costs. The men wondered how this would affect General

MacArthur's promise to return to the Philippines. The enemy came closer when a Japanese floatplane from a submarine dropped incendiary bombs on the forests of Oregon, the next state north of California, and started a forest fire. The U.S. newspapers did not publish the attack to prevent panic in the population.

Nevertheless, basic training in the surrounding open fields of the Salinas rodeo grounds proceeded to make soldiers of the men. They drilled around the horse stables wearing steel helmets and in full gear with rucksacks hanging from their necks. The place was a painful reminder to those with Japanese wives. Japanese civilians arrested as enemy aliens had previously occupied the horse and animal stalls before they were sent to the internment camps.

Two months after the Pearl Harbor attack, Japanese submarines torpedoed coastal targets near Santa Barbara in southern California, triggering fear of a West Coast invasion. That swayed the decision of the Western Defend Command to order the rounding up of Japanese Americans to be placed in internment camps. There were reported cases of double suicides by Filipino Japanese couples who were forced to break up by the Japanese woman's

FILIPINO REGIMENT IN U.S. MANEUVERS.

The United States Army's all-Filipino regiment is engaged in extensive maneuvers in the western section of the country. They are training for the day when they'll take a part in driving the Japs from their homeland. F-NR-X

family. Despite efforts of the American Civil Liberties Union and the Filipino American community of San Francisco, there were no exceptions. Offley could not convince his superiors to exempt the Japanese wives and children of his men from the order. The best he could do was to allow the men to visit their families. He encouraged the men to leave the military and take their families to the East coast.

Andres Lumicao Velasco wrote in his autobiography that he joined in Salinas. He completed basic training at Camp Callan near San Diego and was sent to an Anti-Aircraft Radar Searchlight Company, where he was the only Filipino amongst all-Italian soldiers. He almost left when he saw that his quarters were a filthy barrack, without water and toilets. A cousin, Domingo Lumicao Logan, a recruit from Chicago, convinced him to be patient and stay until he returned to the Regiment.

Allen Dawang, a *Manong* who worked as a farmworker following crops up and down central California, volunteered at the Salinas draft board. He was inducted in San Francisco and trained as a radioman for Company L. He proudly related that "At camp, not one Filipino in the Regiment was ever court-martialed or had jail time." According to Gregorio Chua, "the guardhouse proudly stood empty all the time. And there were telephone booths where hours were spent lying to girlfriends, ever faithful when in fact (they) were frequent customers in Salinas' Chinatown red-light district."

At the Salinas Rodeo Parade, the men sponsored a float that depicted a soldier with a wound on his chin, a bandage around his head, and holding a *bolo* and a Philippine flag. Behind him was a hut with a bamboo roof and sitting beside it was a mother holding a child with an arm being bandaged by a nurse. The *Laging Una* band led the float playing mainly John Philip Sousa marches but, nearing Chinatown, they switched to playing Filipino old melodies.

Soon the men exhibited a new confidence in their attitude and the way they carried themselves with heads up high. They were hungry for combat. In their neatly pressed, tailor-made uniforms to fit their shorter physique and mandatory crew cuts, they marched in the July 4, 1942, parade through downtown Salinas, the same town that harassed, insulted, and targeted them with mob anger for years. Thousands watched as they paraded in perfect cadence to the regimental marching song 'On to Bataan' composed by Francisco Urbano. The Colors led the way as they proudly belted out pop tunes. The hatred now turned towards the barbaric enemy.

As the town dignitaries and thousands of onlookers watched in amazement, Fox Movietone and Pathe News filmed the parade. A mass of radio broadcasters, photographers, and newspaper and magazine reporters jostled with the crowd to document the lively parade. Media coverage changed the attitudes toward Filipinos as they talked of men being trained for combat duty in the Philippines.

An early short film entitled, *America's First All-Filipino Army*, introduced the Regiment with Offley as commanding officer and Tirso Fajardo as executive officer. The *Army and Navy Movie Journal* No. 13 entitled *On to Bataan* showed the men in actual training, getting ready to avenge the USAFFE's loss in Bataan. It was a commercial undertaking but sponsored by the Philippine government in Washington D.C. President Quezon, who had been following the Regiments' progress, agreed with Offley that the film would carry the message of the Philippine government and the Filipino soldiers' determination to return soon to the islands. The men flocked to the theatres to see themselves on the screen.

The men did not forget the few who were able to escape from the fighting in the Philippines. A 'Salinas Nite' was held in honor of the officers and men from Bataan at the Salinas Armory. A grand march led by the Bataan officers and their respective partners opened the event. The full orchestra of the *Laging Una* Band accompanied American and Filipino songs. Speaking on behalf of Col. Offley, Lt. Fred C. Guerrero urged the audience 'to forget for the duration that there is such racial distinction as Americans and Filipinos. In this crisis, we are all Americans.'

Camp San Luis Obispo, California USA

A month after activation, the 1st and 2nd Filipino Infantry Regiment's shoulder sleeve insignia was approved. Capt. Talman Budd, who had first reported to the Filipino Battalion at Camp San Luis Obispo and became head of Regimental Headquarters, designed the patch initially made of felt and later of cloth. It was also reported that MSgt. Rudolph Corpuz was also involved in the creation of the patch according to his son Arthur Corpuz. The patch was unique as shoulder patches were normally issued to a division-sized unit composed of three-line regiments with supporting elements. There was a 3rd Regiment planned but it did not materialize. "The men stormed the main PX for a couple of days to get their patches and rapidly sewed them on to wear proudly."

1st and 2nd Filipino Infantry Regiments Shoulder Sleeve Insignia Approved 6 August 1942.

On a yellow disc 3-1/4 inches in diameter, leaving 1/8-inch edge, a conventionalized black volcano emitted smoke. The latter charged with three yellow mullets in fess. The gold and black colors represent the black of the volcano erupting; the gold background being the golden opportunity of restoring the country to its rightful owners. The volcano is a representation of the Mayon Volcano – normally inactive but now seething with the rage of conquest. The three stars are the Filipino symbols taken from the Philippine flag and symbolize Luzon, Visayas and Mindanao, the three principal islands. Source: 1st Filipino Infantry. Camp Roberts Trainer. Col2. No.3

Camp San Luis Obispo 1st Filipino Infantry Regiment, Regimental Headquarters. The coat of arms Laging Una (Always First) is on the barracks. On the ground, white stones formed the volcano's shape from their shoulder patch, the sun and three stars of the Philippine flag, and the stripes of the American flag.

One of Talman's duties was to work with the Inspector General to ensure that military requirements were followed. He issued findings covering June 30 to July 1, 1942 to company commanders listing violations that needed rectification: ID tags not worn; irregularities in unit funds; improper handling of mail; incomplete duty rosters; correspondence files not kept; incomplete morning reports; incomplete property accountability; unauthorized changes to daily sick report, unauthorized abbreviations, and illegible entries, among others.

Finding qualified officers continued to be a problem as more men joined. Offley's fledgling staff was composed of qualified Filipinos from military schools and Philippine Army officers who were caught by the war in the U.S. Around 50 Philippine Scouts who had been discharged and were in the U.S. before Pearl Harbor and seeking civilian employment joined the Regiment. Many were qualified but being over thirty-five many did not pass the rigors of training. The officers' ranks were filled with professionals from the East Coast who were not limited by the anti-Filipino laws in California from obtaining a higher education.

The Filipino officers successfully managed the different life experiences, dialects, and diverse cultural backgrounds of the men from the various parts of the islands. They trained and eventually went into combat with the men. James Wingo, a Washington correspondent of the *Philippines Free Press,* reported, "Under him (Offley) are a number of Filipino officers. The War Department commissioned from civilian life, two medical men, a trade assistant from the Philippine Resident Commissioner's office in Washington, and a former F.B.I. agent, who organized the Philippines' D.I. (Department of Intelligence). Some Philippine Army officers who happened to be in this country at the Pacific War's outbreak were taken into the United States Army and assigned to the Filipino Regiment."

From West Point arrived Tirso Fajardo, Leon Punsalan, and Atanacio Chavez, Pedro FlorCruz, Eduardo Suatengco, Rafael Ileto, and Jaime Velasquez. From the U.S. Naval Academy came Roberto Lim, who joined and trained the men for several months before transferring to the Air Corps. Fajardo offered his services as the most senior Philippine Army officer in the U.S. at that time. The *Camp News* published the names of the eight men who were the "first arrivals of (the) First Filipino Battalion soldiers to report to officers Tirso Fajardo, Atanacio Chavez, and Roberto Lim." The names of the men were unreadable.

Camp San Luis Obispo. 1st Filipino Battalion officers

From left to right: Roberto Lim, Tirso Fajardo, CO Robert H. Offley, and Atanacio Chavez

Chavez and Lim were 3rd Lieutenants in the Philippine Army and became among the first officers of the 1st Filipino Battalion when they were ordered to Camp San Luis Obispo. With no other ranking officers present, they were assigned as company commanders but without soldiers yet. Their rank posed a problem as the U.S. Army did not have a 3rd Lieutenant grade. To continue as officers in charge of men, they had to be promoted to 2nd Lieutenants. With Fajardo who was also under the Philippine Army, they were informed at the Philippine Embassy in Washington D.C. that the orders could not be issued because the Philippine Army headquarters was in the thick of fighting on Bataan. The problem was solved in July 1942 when a radiogram from the War Department authorized the enlistment of officers and enlisted men from the Philippine Army into the Army of the U.S. (AUS).

In 1914, twenty-seven years before war broke out, America began allowing one Filipino from the educated upper-class families to enter West Point yearly. Other countries including

Japan, soon followed. Later, two candidates were chosen and sent. After graduation, they returned to the Philippines to train and lead units in the Philippine Army. If more than one eligible Filipino was allowed, the Philippine Army might have benefited more from qualified officers in the islands' defense preparation. During the building of the Philippine Army, General Vicente Lim Sr., of the Philippine Scouts and the first Filipino graduate of West Point, wrote a concerned letter to General MacArthur lamenting the lack of officer training.

The elite 'pensionados' in West Point were not immune to the racial antagonisms levelled against them. They belonged to a group of "little brown brothers" and allies in the war against Japan. Although Rafael Ileto and fellow West Pointer Eduardo Suatengco were mistaken for Japanese on several occasions, they commanded respect as American officers.

Those who graduated from the United States Naval Academy but were denied commissions joined the Army. The Academy restricted Filipinos to the rating of office stewards and mess attendants and these restrictions did not exclude even those with higher education. In disgust for being denied a commission, Lt. Ruben Songco, joined the Regiment in Salinas. When the Regiment moved to Camp Beale, he was assigned to direct the S-3 Operations section.

Short in stature but with a booming voice, Fajardo became the executive officer, charged with S-3 Plans, Training and Operations and commander of the 1st Battalion. His adjutant and men in charge of Supply, Personnel, Transportation, and Maintenance were white officers. Lt. Fred C Guerrero was the Intelligence Officer. Other Filipino and American soldiers handled the administrative and personnel work.

A memo signed by Col. Offley listed his first staff as the following: Exec Officer – Maj. Tirso Fajardo; Adjutant – Robert Thomas; Intelligence Officer – Flaviano Guerrero; Plans and Training – Tirso Fajardo; Supply Officer – George Walker, Jr.; Personnel Adjutant – Julius Klein; Communications Officer – William Gale; Transportation Officer – Don Yancey; and Maintenance Officer – Don Yancey.

Camp San Luis Obispo, California. Here are the first arrivals at Camp San Luis Obispo of soldiers for the 1st Filipino Infantry Battalion. They're here to get in fighting trim to help take back the Philippines from the Japanese. List A—4-9-43. Maj. Tirso Fajardo speaks to the 1st Filipino Battalion soldiers. Behind him are, from left: Lts. Atanacio Chavez (left) and Roberto Lim.

Maj. Tirso Fajardo at Camp San Luis Obispo sitting on a jeep with a license plate that boasted of a bolo under three stars. Source: Camp San Luis Obispo Historical Museum.

The medical unit began when Drs. Gregorio Chua and Francis Roman from the hospital ship *Mactan* joined and underwent officer training when the regiment was still a battalion. Chua did not stay long, however, as he was sent to serve other units in the Pacific that had a shortage of doctors. When the war broke out, Guillermo Rustia, a surgeon with the Philippine General Hospital, and Francis Dy, who was taking postgraduate studies at Johns Hopkins University were ordered to join the Regiment. Six doctors were commissioned and stayed in the camp with their wives. Concerned for his men's health, Offley sent men for advanced medical training to the Medical Field Service School at Carlisle Barracks, Pennsylvania. T4 Esteban Simangan and T4 Moddy DeFiesta underwent training in surgery from the Fitzimmons General Hospital. T4 Michael Bermudes and T5 Hilario Unciano returned from training in a medical school in Colorado. But as more Filipino doctors joined or returned from further training, they were ordered to other Pacific operations and this practice worked to the detriment of the Regiment's formation. It took a year for the unit to have a full complement of medical personnel.

Camp San Luis Obispo. Roasting pigs (lechon) on wooden spits over coals.

An article appeared in the *Shot 'N Shell* newspaper entitled It's *Barbecue Time Among Soldiers of the 1st Filipino Infantry* with photos of *lechon* (roasted pigs) turning on spits over coals. Whenever the men transferred camps, they always managed to find a pig to roast. It took them two days to dress and prepare the pig between their duties and training. Civilian friends would arrive early to help in the preparation and contribute food for the feast, adding to the dishes from the camp mess hall. By Sunday, the feasting began. Members from the Filipino community in nearby farms joined with their children in tow bringing fruits and vegetables. White soldiers from nearby camps were welcomed to the feast. The Filipino custom of cooking for more people than expected made sure that there was enough food. And if the weather was just right, there was an entertainment of nostalgic Filipino dances, songs, and homegrown comedy.

The 'I am an American Day' celebration highlighted the sale of war bonds sponsored by the U.S. Treasury Department in many cities. The largest bond event kicked off at the nearby Sta Maria City participated by the various military units stationed at the camp. *Adobo* and *lechon* parties drummed up sales establishing records in Central California's Filipino communities as compared with the rest of the state's population. The Regiment's patriotism placed it in the top ten highest sales.

There was competition in buying war bonds amongst the men with Arcangel Baniares buying forty-one thousand in a bond rally. Concerned about his wife in the Philippines, Lt. Alberto C Elefano of Company L bought over $700 worth of bonds. The regimental sweetheart, actress Carole Landis, spent a day at camp for the War Bond Drive tour she was conducting for the Regiment. The men roasted a whole pig to celebrate her visit and she watched the preparations with a slight smile of wonder.

Unfortunately, the War Department was still unsure of what to do with the regiment. Not wanting his men to do duties as cooks, messengers, or become part of a labor group like the Black units in the Army and Navy, Offley resolved not to command a Service Regiment. When the flag was hand-sewn by the Regiments' wives, he insisted that the field colors be blue, the color of the Infantry and the word 'Infantry' be added to all official correspondence. That word told the Army that the men of the regiment were "soldiers trained, armed, and equipped to fight."

The *Philippines Free Press* reported: "The men of this Filipino regiment are taking the business of sudden death seriously. Their American officers have commended their amazing conscientiousness and ardor and have encouraged them to add a purely Filipino tactic to the orthodox warfare methods. In simulated jungle fighting, these sons and grandsons of guerrilla warriors who held the American Army of Occupation (Philippine-American War) in the Philippines at bay for three and a half years, like to creep close to the enemy with bayonets tightly gripped across their mouths, and then jump at him, wielding their bayonets as if these were their native *bolos*. The officers know Filipino psychology and have handled their men with unusual skill. All appreciate the fact that they are served Filipino dishes, like steamed rice and 'adobo' (pork or chicken stewed in vinegar with a heavy dash of garlic). The California climate agrees with them. The guardhouse is used for storing supplies."

Sunday duty was a penalty imposed on most outfits as was commonly practiced. But for the Filipinos, it was another training day that they wanted. They began showing up with muddy shoes, incomplete uniform or whatever would cause a penalty, so Sunday duty was replaced with more training. After continuous hours of training, a cold beer was most welcomed. With the biggest smiles on their faces in their crisp khaki uniforms and overseas caps, they raised their beers for the camera outside the Post Exchange.

Philippine Resident Commissioner Joaquin Miguel Elizalde and Commonwealth-in-exile National Defense Secretary and Philippine Army Chief of Staff Major General Basilio J. Valdes, visited the expanding regiment. It was a reunion for Elizalde with his brother, Lt. Angel M. Elizalde, who had formerly commanded a unit of the Philippine Air Corps. Thousands of men paraded in front of the two VIPs. It was the first time

The U.S. Army's first all-Filipino Regiment is undergoing intensive training "somewhere in California" at present. Here, after a hard day's drill, the men enjoy refreshments in front of the post exchange and try to look "photogenic" for the cameraman. The islanders look forward grimly to a day of retribution against the Japs. Aug 14, 1942. AP.

Resident Commissioner of the Philippines Joaquin Miguel Elizalde visited the Regiment at Camp San Luis Obispo. October 1942.

Camp San Luis Obispo. The National Broadcasting Company (NBC) broadcasts interviews with the newly formed Filipino Battalion. An NBC microphone was set up as the men practiced shooting machine guns.

they saw the growing strength of the Filipino combat unit, formed of men with Filipino blood and Filipino nationals in the U.S. after a petition by American citizens. A 13-gun salute greeted their entrance. A guard of honor, commanded by Lt. Atanacio Chavez, led the guests to a passing review as the men stood in attention in full dress uniform.

In his speech, Elizalde exclaimed that "Bataan is no longer just a rugged peninsula of mountains and jungle. It is sacred ground where the best of the youth of the United States and the Philippines have given their lives for our two flags." A prayer was offered to honor the dead of Bataan. Afterwards, the Office of the Resident Commissioner published *The Filipino Fighting Spirit* extolling the unit as "much more than just another regiment in the army. To us, it is the symbol of revenge, the continuation of the glorious and heroic spirit of Bataan. It is proof that the spirit of our brothers at home burns also in the hearts of Filipinos here."

As the press gave more exposure to the Regiment, America changed its attitude towards the men. The National Broadcasting Company (NBC) interviewed Offley, Fajardo, and several staff officers regarding the progress of the newly- formed unit. The Army had not anticipated that so many inductees would arrive. The United Press International (UPI) published that members of the 1st Filipino Infantry Regiment were constantly training to absorb and use new techniques learned by the United States Armed Forces in combat. The Filipinos were acknowledged as the "hardest-working troops in the Army." "They never know when to quit," one officer said. "When the five o'clock whistle blows, other troops tend to drop everything for the day. Not these boys, though. They spend their evenings studying and work just as hard during their hourly 10-minute rest periods as they do the rest of the hour. The Filipinos take more pride in their uniform than do most troops, it seems. A visitor to their barracks is impressed by the spotlessness of their clothing and their knife-edge creases. Naturally very self-reliant, they fit right in with the Army's training program for infantry troops, which places a premium on men who can work singly or in small groups. A small percentage of the regiment's officers are Filipinos, and some of them have seen service in the Philippines as recently as the fight put up on Bataan by American and Filipino forces."

Fort Ord, California USA

The Regiment swelled to over five thousand officers and enlisted men activating the 2nd Filipino Infantry Regiment on November 22, 1942. A separate chapter discusses the 2nd Regiment.

There would have been a 3d regiment consisting of Filipinos from the Hawaiian National Guard, but the Hawaiian Sugar Plantation Association argued that the cheap labor was needed to support the war effort and the Tydings-McDuffie Act prohibited Filipinos from entering the continental United States.

The move to Fort Ord in October provided the space needed to continue proper training of both Regiments. The new quarters were low pyramid tents on the ground. During inspection, all equipment, folded uniforms, mess kits, helmets, canteens, gas masks, etc., were laid out in front of the tent in a specific order. Rifles on the side separated each soldier's area.

The later move to the Main Garrison quartered the men in two-story multi-platoon size buildings painted white with pitched green roofs and awnings over the first-floor windows. The buildings were laid out end-to-end with a road running between them. Grass grew around the buildings. There was a chapel specifically for them to pray for their safety and for their family in the Philippines unmolested. To be more efficient in combat,

1st Filipino Infantry Regiment. Ft. Ord, California

they were attached to the 4th Army, headquartered at the Presidio in San Francisco. The 4th Army was not a combat unit and primarily responsible for the defense of the West Coast.

The Filipino community continued their support with stage shows that featured Filipino radio and stage stars. They were proud of the men's patriotism and willingness to go into combat to liberate the islands. In the towns near to the camp, they cooked familiar Filipino dishes and provided entertainment boosting the men's morale.

"Friday and Saturday nights at Fort Ord had buses filled with Filipino soldiers in uniform scrambling for seats to go to nearby Salinas," recalled Gregorio Chua. They came to join in the many picnics, weekend entertainment, and dances with Filipino girls from all over California. Filipinas were a rare sight then. *The Panorama* newspaper was a bit more suggestive in their report "Filipino USO to open in Salinas with a plan to "import girls from neighboring cities to entertain the boys every weekend."

1st Filipino Infantry Regiment. Fort Ord. October 1942. The same men carried the regimental flags throughout the drills and the war.

May the new Year bring Victory and Freedom

1st Filipino Infantry Regiment, Company I. Fort Ord. Christmas 1942

The visit of Philippine Commonwealth Vice-President Sergio Osmena, along with military and civilian dignitaries, boosted the men's morale. The parade of the regiments to honor him was reported as the biggest ever held at the post. Thousands of men stood at attention in the open field facing the platform as they listened to the VIPs' remarks. The *Laging Una* band played the national anthems and Filipino songs. The ceremonial parade was held on the dry field and the Regiment's Color Guard kicked up clouds of dust as they marched. They firmly held the American and Philippine flags with mounted eagles on top briskly waving in the air. With rifles slung on their right shoulder, they marched as 'Swing' music was played.

The Christmas season arrived, and the support of the Filipino community was again evident. The Filipino Community of Stockton treated the men to a two-hour musical variety show staged by the most "attractive *Pinays* (nickname for Filipinas) ever to invade the camp." Filipino folk dances were shown, and piano selections played accompanied by the band.

Company platoons hosted holiday celebrations in their building. Company I sent out invitations showing the *Laging Una* crest against a background of entwined flags of the United States and the Philippines. At the bottom was a wish saying, "May the New Year bring Victory and Freedom.' One of the four company officers listed was Lt. Silva.

Capt. Talman Budd led his company in an Officer's Supper and Dance at the

American Legion Club sponsored by the 1st Regiment. Guests were the officers of the 2nd Infantry. All members of the officers' staff were present except for those on duty. Col. Robert H. Offley of the 1st Infantry and Lt. Col. Charles L. Clifford of the 2nd Infantry and their wives received the guests.

A photo at the Regiments' Christmas party at Fort Ord showed Cpl. Alex de Leon Fabros in uniform dancing a mean boogie with a smiling lady friend. His granddaughter Michelle summed up his recollection of that day.

"Christmas day 1942, 1st Filipino Infantry Regiment and parts of the 2nd Filipino Infantry Regiment were still stationed at Fort Ord, CA. Today was going to be a big affair since many believed that this would be their last Christmas in the U.S. before they shipped out in the summer.... Because all the cooks were Filipinos, recipes weren't needed since the cooks worked from the heart to feed the soldiers. The (Regimental) band entertained the guests with music, and many of the guests got up to perform Filipino folk dances or songs for the white officers and their families. Col. Offley remembered that this was the first time for many white officers and their families to experience Filipino cuisine. For the Colonel (Offley), it brought back fond memories of his time living in the Philippines with his father in the years before the war."

The aroma of Filipino food and the lechon turning at the outside charcoal pit wafted into the hall. Dozens of soldiers of the Headquarters Company in crisp uniforms, dress ties, and big smiles on their faces sat side-by-side on long benches across each other at tables covered with white tablecloths. Platters of oranges, apples, and peaches decorated the table. An American officer stood with a long knife ready to cut into the *lechon* on the table.

Fort Ord 1942: Christmas celebration with a lechon, (roasted pig).

An unsolved dilemma continued to haunt the camp in spite of the festivities and camaraderie: What use was all the combat training if they could not bear arms because they were non-citizens? However, the Philippine Commonwealth Constitution allowed the U.S. to draft organized military forces of the Philippine Commonwealth to fight the invaders and defend American interests. It was rare for Filipinos at that time to have joined and served in the U.S. military for three years, making them eligible for naturalization. Those who became eligible had joined the Navy when President McKinley issued an Executive Order in 1901 that allowed 500 Filipinos to enlist as part of the insular force.

In the meantime, while the men were celebrating the Christmas holidays, the Allied Intelligence Bureau (AIB) in Australia was desperately trying to maintain contact with the guerrilla resistance in the Philippines. The radio contacts with Nakar and Praeger on Luzon had stopped when they were captured and there was the risk of also losing contact with Panay Island. Who else was trying to reach Australia? The first of secret suicidal missions would soon be sent to the Philippines to find out.

By the end of the year, around 7000 hardworking troops were undergoing intense training crowding Ft Ord. The 2nd Infantry Regiment moved to Camp Roberts and Camp Cooke.

Camp Beale, California USA

The news of Japan granting independence to the Philippines in exchange for Filipino cooperation troubled the men. How will the cooperation be obtained? Anything American would be purged. Reacting vigorously, the Philippine government- in-exile in Washington D.C. broadcasted on shortwave from San Francisco, urging the Filipinos to continue their resistance against Japanese propaganda and assured them that the Allies would not lose the war.

The transfer north to the cooler climate of Camp Beale which was surrounded by pine trees provided some distraction. The two-story barracks were laid out side-by- side with a road running around each block. The center was an open space and used for company and battalion formations or the parking lot for vehicles. The outside walls were unpainted wood and did not provide much insulation from the low temperatures. Inside, their footlockers occupied the foot of their wooden cots, ready for surprise inspections. Bed covers were pulled with hardly a crease and blankets neatly folded per Army regulations. Supplies may have been slow in coming because a question was written in *We Wonder Department:* "How come Filipino enlisted men do not have bedsheets yet?"

The range office controlled three-fourths of the area comprising Camp Beale. The site had artillery, mortar, miniature anti-aircraft ranges with pop-up targets as well as moving targets for rifle and anti-tank practice. As a replacement depot, soldiers trained while waiting for assignments. Men were prepared in small units because movement overseas could happen anytime. An essential small-unit training was the rifle platoon attack on a fortified position from a landing on a beach. High priority was given to tests for the Expert Infantryman Badge and most men earned the decoration. Physical conditioning was enforced through weekly nine-mile cross-country marches in two hours, and monthly twenty-five-mile marches in eight hours for assault troops. With the men's years of physical labor in the fields, there were no dropouts.

1st Filipino Infantry Regiment. Source: Maximo V. Maglangit. He is in the front row, second from the left.

The Quartermaster's nightmare was tailoring the standard uniforms for smaller than average men with small size shoes. Even their denim work uniforms had to be fitted. Appearances were essential to the men long before when they were dressed in their best suits and attracted the white women at the dance halls. Their fatigues had to be cleaned and pressed instead of only washed. They had to look just as impressive as the postcard pictures, they had sent back home to their families.

Filipino food fueled their enthusiasm. The mess sergeants spent a good deal   of time trading potatoes and spaghetti for bags of rice which was a mainstay of the Filipino diet. Theirs was the best food in the Army, the men boasted, because their Filipino cooks were formerly chefs of famous men and other well-off businesspeople and celebrities. On weekends, Service Club No. 2 at 5th and D Streets in the camp was packed, serving sumptuous Filipino dinners. The *Ford Ord Panorama* reported the main camp of Post Exchange No. 2 was probably the "only one of its kind in the whole U.S. Army specializing in spicy Philippine dishes for the gastronomic delectation of Filipino soldiers stationed here. There is hardly any standing room in the café during rush hours at noon and in the evening when the boys, tired and hungry, come to partake of their national dishes."

Some Filipino soldiers like Emiliano Francisco had chosen to join the white "American outfit" believing that there might be better accommodations and benefits. "We had no fear of dying during the training period. We were dedicated to go and fight in the Philippines." But after almost a year of Basic Training in a white outfit and finding out that the Filipino Regiment was eating Filipino food, he immediately requested a transfer.

Company C, 1st Regiment held an *adobo* picnic to welcome the arrival of their genial commander, Captain Abdon Llorente, from advanced training at OCS. Ft. Benning's *Bayonet* stated that "Captain Abdon Llorente is mighty proud of his Regiment, and he has good reason to be for it is the only kind in the world organized at the request of the Philippine government and Philippine people in this country (US). Captain Llorente, the first Filipino to be commissioned in the Army of the United States, was instrumental in the formation. The regiment was organized for overseas duty only. Ten of its officers are at Benning now in the Advanced Classes and 95 percent of all their officers are from Benning." Lt. Silva graduated from the same class. His D code sheet listed him as a 2nd Lt in November 1942.

In camp, the men went through various phases of attacks and defense in a three- day tactical exercise in the simulated jungles of 14th Street and Smartsville Road. Three battalions moved against other Filipino troops acting as Japanese troops using blank rounds and firecrackers. They marched to control an area for an attack at dawn to seize an objective and prepared defensive position to withdraw at night. The exercises prepared them for the jungles of New Guinea and the Philippines. Drills on interrogation and search of POWs honed their skills in finding valuable information.

The men showed such "intense hatred for the enemy and desire to return, that they voluntarily increased their training program. They requested seven days of training instead of six. When a man was assigned to a typewriter, he brought his rifle with him and, when not typing, took the rifle apart or studied his Soldier's Handbook. At night, they practiced Morse code knocking on the floor while others listened. The 10-minutes rest period every hour during training usually for smoking or conversing were used instead for rapid-fire questions directed to instructors, giving commands to each other, and correcting each other's mistakes. They held informal seminars in their barracks on all phases of military training. A rule they remembered was 'always zigzag through the grass. The enemy easily spots a straight path.' The Filipinos had so few violations of military rules and orders that only two had ever stayed in the guardhouse, and those two for speeding while off the post. Eventually the Prison Officer's job was abolished, reported James Wingo. The guardhouse was relegated to storing supplies.

The newspaper *Bealiner* of the nearby city of Marysville followed the Regiment's appearances before service clubs, civic groups, and in both national and local radio audiences. The *Laging Una* band played as goodwill representatives of the men. Band member Daniel Palmeto said that the band's rendition of pop tunes and nostalgic Filipino songs contributed immensely to the men's morale.

Major Tirso G. Fajardo was featured by the *Bealiner* as the highest-ranking officer at Camp Beale. He was in command of S-3 (Operations) throughout the Regiment's time in the U.S. Held in high regard by the men, he had their confidence and shared their desire to liberate the islands. In an interview, he expressed his pride for the men. "I agree entirely with the sign over our door – Under these portals pass the best damn soldiers in the world." And the Regiment lived up to it.

2nd Lt. Saturnino R. Silva finished the 90-day course for officers (OCS) at Fort Benning, Georgia. 1942

From OCS, Silvino Tallido and Al Hernandez returned as 2nd Lt officers to Camp Beale. Before the war, Hernandez and his wife were a champion team of Latin American dance. He worked in the Hollywood film industry producing military instruction films. The *Bayonet's* January 23, 1943 issue wrote about him in an article *Famed Dance Master is Infantry School Student* with a description of his costume "now olive drab of the Army and his partner, an M-1 rifle." Whenever his wife joined him at camp, their Rumba numbers entertained the men.

In the April issue of *The Bayonet*, Lt. Paul Mauricio was featured in the article *Philippine Vet Explains How to Distinguish Japanese* instructing the largely White soldiers in his OCS training class.

The vigorous support of the Filipino community for the right to bear arms influenced the passing of a law for naturalization. It was a morale booster for the men. As U.S. nationals, they carried U.S. passports but could not vote, own land, or apply for citizenship. Now that the law had changed, they were urged to apply for citizenship. Lt. Fred Guerrero was granted U.S. citizenship in the Superior Court of San Luis Obispo and assigned as the naturalization officer for the Regiment in addition to his other duties.

Immigration and Naturalization Service (INS) officials arrived at the camp to complete the required petition papers and answer questions. Emiliano Francisco related that the judge came down and said, "Okay, boys. Sign right here." He felt real comradeship with the Army.

It took a while before the date of the swearing-in ceremony was determined. On that historic day of February 20, 1943, a naturalization judge opened a court session at Camp Beale and swore in 1,200 soldiers of the 1st Filipino Infantry Regiment. The *Lupang Hinirang* was played just before the troops marched onto the parade field and stood in a wide V formation. The Star-Spangled Banner was then played to begin the official ceremonies. They removed their caps and proudly stood in attention wearing full dark uniforms with neckties on the wet parade ground of Camp Beale. They had accepted a wider citizenship with the Philippines in their heart and American in their mind. Before film cameras and radio microphones broadcasting the ceremony across the continent, the largest group of Asian immigrants were naturalized. The men sang 'On to Bataan,' their inspirational regimental song. Col. Offley then delivered a brief speech as the Regiment's commanding officer, "They are fine soldiers, and I am proud to be their commander." Before they were released to join their friends, families, and wives, the band played 'God Bless America'. The 2nd Filipino Infantry Regiment at Camp Cooke travelled in groups to Santa Barbara to take their oath of citizenship.

A strong media presence documented the historical event. Photos and films by *Movietone News* showed the naturalization ceremony around the world. *The Des Moines Register* reported that "before the present hostilities, Filipinos born outside the continental U.S. were not qualified for naturalization even though the Philippines were under the flag of this country. They were neither aliens nor citizens but were classified as Nationals. When World War II broke out, a chance to become citizens was offered to those who are in the service."

The men were now naturalized as soon as they joined. Some were granted citizenship upon arrival at the stations without the formality of applying. By the time they left California a year later, over half of the men in the unit were American citizens. They proudly arrived in the Philippines with thoughts of "fighting under the American flag to protect the Philippines from the Japanese."

SWEARING IN OF 1000 MEN AS CITIZENS OF THE UNITED STATES, CAMP BEALE, CALIFORNIA.

Camp Beale. 1200 Filipino Soldiers Become U.S. Citizens. A federal court clerk, hand upraised on a platform, administered the oath of allegiance to 1,200 Filipino soldiers who became American citizens in a mass naturalization ceremony Saturday. The Filipinos are members of the first Filipino Infantry regiment in the United States Army.
WIREPHOTO AP

But becoming a citizen did not allow them to own land in California. Alfredo Mendoza said the citizenship committee told him that he could. Many believed that they could now marry their white girlfriends, provide them with allowances, and make them beneficiaries of their insurance. Marrying a non-white was allowed but the 1934 California law that made marriage to white women illegal was still in place but not strictly enforced. Published in the February 22, 1943 issue of the *Philippines Mail Society Notes* was the notice that "Pfc. Raymundo married in church ceremony to Agnes Lara... Bride is native of Colorado of Mexican extraction."

But not all the of men were eager to join. Julius Ruiz who had lived in the U.S. for many years, had mixed feelings running from "boiling anger to excessive pride." There were doubts and questions in their minds. Some were unsure if they wanted to become Americans and wanted to wait until after the liberation of the Philippines before making their decision. Mariano Angeles wondered why they were asked now when they were described before as barbarous and troublemakers. He resented that citizenship was given only to service members and not all Filipinos. His refusal to sign the petition had him "called to the intelligence officer twice in fifty minutes." He had enlisted when he was unable to find office work and deeply felt the discrimination of not being served in restaurants. A separate area in church had made him stop attending. But he wanted to return to the Philippines. Assigned to the medical unit, his commanding officer wanted him to become a citizen. "Now that when my future is to die, that is the time when they give me the privilege to become an American citizen?"

Unsure and anxious about what to expect from the U.S. if he accepted citizenship, Jose Trinidad expressed his dilemma. There was still "limits to the status due to pervasive racism and existing regulations, particularly anti-miscegenation laws aimed at Filipinos." Citizenship only answered their desire to go into combat against the Japanese. Still, others joined after some officers threatened them with "KP (kitchen duty) for the rest of your life."

Although they were now U.S. citizens wearing the Army uniform and with the status of fighting men, the Filipino soldiers continued to experience the discrimination that was still rampant in the small towns of California. Worse, some of them were mistaken for Japanese men in the larger cities of Sacramento and San Francisco.

The group of Antonio Dixon Campos, then a young, hot-headed teenager, wanted to charge into town and fight with the locals who had posted 'No Filipinos Allowed' signs, refused them service at a restaurant and barred them from a USO dance. Signs like "Japanese and Filipinos – Do not enter after sunset" were posted around the town limits of Marysville. In downtown Salinas, the business establishments posted signs that read "No Mexicans or Filipinos Allowed." Barbershops in San Francisco refused to cut their hair, and hotels barred them and their families. Although they were wearing Army dress unforms, the men were told that it was a city ordinance and they had to follow it.

Reporting the matter to Offley, their commanding officer, he handled it. Offley faced the Marysville's city fathers and gave them a choice. He gave them his first proposal. "My men are American soldiers. You will treat them as such, or I will place Marysville under martial law. Filipino soldiers would be patrolling its streets. There would be a curfew from sunrise to sunset, and anyone found violating it could be shot." As the city fathers fell into a shocked silence, Offley then told them the other choice, "Take down the signs and give service to these men just like anybody else," he ordered. Offley succeeded. The Filipinos could now enter a restaurant or a movie house, where people begrudgingly served them. The men could now eat Chinese food in Marysville during the months they were at Camp Beale.

There were still some incidents, of course. When Clay Pardo returned to Camp Beale after three months at OCS training, one of his sergeants was refused housing in Marysville until he intervened. The Filipinos were called 'Gooks,' he related.

Eager to join, Ricardo Galang wrote he was single when he was not and subtracted a couple of years from his age. He reported at Camp Beale to the 3rd Battalion after his ninety-day OCS training. His tent mate was Lt. Silva, whom he described as an "old resident of San Francisco" and whose snoring kept him awake. Later, when the unit moved to Camp Roberts at Hunter Liggett, he was appointed adjutant to Lt. Jaime Velasquez and concurrently company commander of Headquarters & Headquarters Co. He reversed the chow line by putting the privates first, then the non-commissioned officers and officers. He thought, 'Take care of our men, and they will take care of us, especially in battle.' This was not a popular decision and was rescinded.

Offley appreciated the uniqueness of the unit and its important contribution to Philippine and American history. He asked Cpl. Alex Fabros, to join as the official Regimental historian acknowledging his experience as a journalist and advocating for legislation to protect the Filipinos in California. Fabros wrote for the *Philippine Mail* whose feisty editor fought for the rights of Filipinos in America as "human beings." His son, Alex Jr, related that his father was assigned to the anti-tank company but was not required to attend boot camp. Instead, he was kept busy organizing Filipino community events to support the troops starting at Fort Ord and the next camps of the men. At age 37, he was legally too old to join and changed his age on paper to be four years younger. He turned down a request from Maj. Fajardo to attend OCS, saying that he was too old to go through the intense rigors of the school's crash course. As the battalion's chronicler and Public Information Officer, he was on the Regiments' staff and provided a deeper insight to the men. In return, he shared with the men the commander's perspective of them. All-out support for the Regiment from the nearby Filipino communities at the camps was mainly due to his efforts creating social events for the soldiers.

By April, fifty Filipino commissioned officers, including graduates from West Point, Annapolis, and a medical unit were on duty with the Regiments.

A rare photo taken at Camp Beale showed a unit of the 1st Regiment in their specially tailored 'pinks and greens' uniforms for the winter. Wearing brownish-green jackets with lighter-colored trousers of a slight pinkish hue, they posed in front of peaked-roof barracks.

First Filipino Infantry Headquarters. Camp Beale. Source: Alex B. Lucas

April 9, 1943. On the first anniversary of the Fall of Bataan, Lt. Velasquez, commanding officer of the 1st Battalion read General MacArthur's prayer to the several thousand soldiers as a solemn tribute to their American and Filipino comrades who had given their utmost in the defense of the islands. Before the entire Regiment standing at rigid attention with the flags snapping in the stiff breeze, Velasquez read the prayer: *I was the leader of that lost cause, and from the bottom of a seared and stricken heart I pray that a merciful God may not delay too long their redemption, that the day of salvation be not so far removed that they perish, that it be not again too late.*

Good news soon arrived that lifted the men's spirits. On April 18, Japanese Admiral Isoroku Yamamoto, the mastermind of the Pearl Harbor attack was killed when his bomber plane was intercepted by U.S. Air Force fighters and shot down. U.S. code breakers pinpointed his location near Bougainville in the Solomon Islands and sent eighteen P-38 fighters after his aircraft.

News of the Japanese brutalities against the prisoners of war tempered the good news and shocked the world. Ten escapees from the Davao Penal Colony in Mindanao were evacuated to Australia and there, they exposed the savage treatment of POWs.

In Manila, agents sent disturbing news of the increased presence of the enemy in the city. Intelligence reports place them at around 8,000 consisting of Japanese soldiers from neighboring occupied countries; those who came back from the battlefields of Bataan and Corregidor; and conscripted Koreans and Taiwanese. Numerous buildings and homes were occupied and used as soldiers' barracks. The Nielsen Airfield became the main base for their bombers, fighters, and transport planes. Edgar Krohn, a young man in Manila, learned early that a Japanese-occupied building was easily identifiable if there was a table in front where a sentry checked identification.

Secret Missions

In May, Col. Courtney Whitney, head of the Philippine Regional Section (PRS) arrived at Camp Beale with an urgent authorization from MacArthur to choose men from the Regiment for advanced training in Australia. He had proposed an approach to the Philippines that included a plan to obtain Filipino personnel for missions: "... bring Filipinos to Brisbane from the U.S., put them through a course of appropriate and rigorous training, and as transportation by air, surface or submarine becomes available, dispatch parties composed of highly selected men to assume areas of responsibility at the positions indicated...each of which parties to maintain direct communication with this HQs." The missions were 'Top Secret' known only to MacArthur and a few of his staff.

Three missions had already left Australia and there were only a few qualified Filipinos left who had the right skill, military experience, and physical ability for the next planned missions. There was a great need for radio operators and signalmen. Two Filipino officers reportedly arrived with Whitney – Maj. Mariano A. Eraña, Judge Advocate General of the Philippine Army in Washington D.C. and Col. Alejandro Melchor who was instrumental in creating the pontoon bridges extensively used during the war.

Whitney reviewed the Regiments' service records, interviewed, and tested over eight hundred men. Eighty-seven were officers including Lt. Silva. Two hundred fifty-two soldiers did not have their age listed. Only ninety-five of the men were listed as under thirty years old. Five hundred were over thirty years old, with the oldest at fifty-four.

The men who passed into the three upper brackets of the IQ Tests were given radio aptitude tests. More than 75% of the selected group qualified to become radio operators. They were dispersed to Signal schools at Camp Kohler in California, Camp Crowder in Missouri and Camp Murphy in Florida before they were shipped off to Australia. Sgts. Artemio Ibea and Wilfred Regala were sent to Camp Crowder.

A large group of officers and men were sent to the Army Language School at the Presidio in Monterey, California, to learn elementary Japanese, Japanese ship and aircraft recognition, and sailing at the Navy school.

Rufino Cacabelos was an early recruit joining the Army two months after Corregidor surrendered. He recalled when "American radio operators with sophisticated equipment arrived at Camp one day in four trucks. Immediately Filipinos with high school education and up were told to take the radio operator's test. It took one day, over ten hours, to test four hundred Filipinos."

Message Center: Radio training operations. Source: Camp Roberts Trainer Vol 2. No 3

He had only 15 minutes to decide if he wanted to volunteer for a special job. Lorenzo Pimentel was in a meeting with other men who were interviewed. Informed that General MacArthur wanted very much to see them, they were told to pick from a pile of papers at the door and read it. "If you like it, sign. If you do not like it, do not sign," he was told. He read it, liked it so he signed. Called back inside, he was asked why he signed it. "Because I am a soldier wherever the army wants me to go or to do, I do it." He was told to go back and tell his company commander that he was going on furlough for 15 days before departing to Australia.

Some of the men had already gone through numerous trainings and further drills might cause low morale. To counteract this feeling and enhance the importance of their future use, the men were quickly shipped out for special training in facilities in Australia.

Offley feared the loss of his best officers. The 1st Regiment greatly suffered from the loss of around four hundred newly commissioned officers, some who graduated as 2nd Lts from OCS while the rest were the best of its enlisted men. The strength of the regiment was no longer up to par. Around three hundred men were pulled from the 2nd Regiment to refill vacated ranks and retrain. Reorganizing and rebuilding the unit's strength had to be done before orders arrived to deploy to the Philippines.

Another crushing directive dropped both Regiments' strengths. Men over 38 years old or those who had medical issues were allowed to leave the Regiments. This also gave the Army a way to discharge misfits. But if specific skills were needed by the Army, the application for a discharge was denied. The average age of many of the men was over 30 years old. The dearth of younger men was due to a decade-old law that restricted immigration from the Philippines during the Depression. Men who were sons of *sakadas* born and raised in Hawaii were not allowed to join to replenish the Regiments. The Hawaii Sugar Planters Association convinced the government that they were needed to support the war effort. Many requests were submitted by men enticed by higher wages in jobs essential to the war effort and the hope of better treatment in the California farms. The news that Filipinos could own real property also accelerated requests for discharge from those who refused naturalization. But that law was not fully enforced.

One-fifth of the strength of each Regiment departed. The 1st Regiment's combat strength depleted to 2,600 officers and enlisted men — more than 600 below its authorized strength. Men from the 2nd Regiment were ordered to replenish the 1st to 125% of its strength.

Fajardo wanted to keep the older men as support – cooks, clerks, truck drivers – so the younger men could be assigned to combat. But some of these men were not physically qualified or unable to adapt to military life. Pvt. Carnuto Galindez, one of the earliest to enlist, was discharged Pacifico Juanitas, who was assigned with the medical detachment of the 1st Regiment, was determined to stay but eventually took the discharge due to high blood pressure and his age. There were exceptions, like platoon officer Valentin Abuan, who enlisted at forty-two but went on to Manila to fight.

Camp Roberts, California USA

The 1st Regiment transferred south to an Infantry and Field Artillery Replacement Training Center in Camp Roberts to complete their training. Field and live-fire drills were conducted at the adjacent Hunter Liggett Military Reservation. The camp had an expanse of fourteen football fields with one lone tree in the parade ground. An area called 'The Soldier Bowl,' which can be seen from the main road, had a stage and seating area used for regimental muster. It was a perfect location for the chaplains who said Mass for thousands of men.

The blistering heat rose from the asphalt and windstorms whipped the camp. Thousands of men from various units dotted the cleared grounds quartered in rows of canvas two-man or squad tents. The mechanics of the Regiment's vehicles were assigned to wooden buildings at the south end of the main post. In the motor pool,

Camp Roberts: 1st Filipino Infantry Saturday morning inspection in front of their pup tents. Blankets and gear were meticulously organized behind them according to regulations. Some were lucky to set up tents under large shady trees.

vehicles were inspected regularly especially the all-purpose one-fourth ton jeeps that had to be combat-ready when conducting field training exercises and for deployment overseas. Some men were assigned to the supply warehouse like Pvt. Andres Darang.

All personnel, including the medical detachment that had been with the Regiment since the beginning, were directed to do both field and garrison duties. Skills were always kept in a ready state. They were trained as infantrymen first before doing administration work. The Service Calls provided information and guidance, beginning with Reveille followed by Mess and other calls as needed - Sick, Drill, Church, Recall, Mess, Recall Guard Mount, Retreat, to Quarters, and Taps.

Aside from garrison duties, men were dispersed to various training schools. The Chemical Warfare School graduated men after a two-week session. The Service Company attended a specialization course in auto mechanics.

The new Basic 17-week training cycle at Camp Roberts included weekends which kept them in a ready state. As riflemen, they mastered the Springfield rifle, M1 Garand, 60mm mortars, and the Browning Automatic Rifle (BAR), bayonets, and grenades. The BAR weighed twenty lb so the men had to be physically fit to carry it for long distances. Training in hand-to-hand fighting with bayonets and grenade-tossing were also done. Field training in air-ground tactical maneuvers to infiltrate behind enemy lines were done in full pack uniforms and alongside other army units preparing for deployment to the European theater.

Morale-boosting signs dotted the training grounds. A photographer snapped a group of soldiers with fixed bayonets on a gravelly path going uphill reading a sign through sweat-burned eyes. The sign: 'On the close-combat battlefield, there are only the quick and the dead. I intend to be one of the quick. The bayonet is a close-combat weapon, and I intend to be a bayonet expert.'

They ran obstacle courses that Capt. Galang excelled in. Nighttime walks in the woods were to see if they could find designated points. They learned to dig trenches properly and

Four men of the 1st Filipino Infantry demonstrate their technique in waiting to rush the enemy with the wicked bolos. From left to right are: S-Sgt Laurelio Alcaide, Pvt Nemesio Gemendige, Cpl Isaac Gabatu, and Pvt Quintin Varanon.

Camp Roberts, California. 1st Filipino Infantry Regiment training with their bolos. Source: Camp Roberts Trainer. Col2. No.3

surround them with protective barbed wire in preparation for an enemy attack on the camp. The men caught a snake while they were on patrol, and although tired and hungry as they were, brought it back to camp and figured out a way to cook it. At night, they wrote letters, played cards, washed clothes, polished shoes and brass, and listened to radio broadcasts. Last wills and testaments were written to take care of their wives. Cussing became common and nicknames were given to some favorite personnel.

Squad leader Ponce Cazem used his mesh-covered steel pot helmet as a water bowl to shave outside his tent at Camp Robert before starting field training exercises at Hunter-Liggett. Hot days had him bare-chested and quenching his thirst with a cold bottle of Coke.

In July, the Regiment commemorated the unit's activation in Salinas the year before with a celebration that opened the camp to the public. Every effort was made to make their spartan tents presentable. Many took the time to hang colorful banners on their tent poles. The commanding officer of the II Armored Force, Maj Gen. William Henry Morris, presented the regimental colors to the Regiment in a formal ceremony. A photo is preserved of that day with an inscription on the back written by Morris: *Col. Offley and I started the regiment from scratch.* After a brief speech citing the men of the Philippine Scouts who fought on Bataan, company commanders gave out awards and good conduct medals to deserving soldiers. The men proudly marched with a pass-in review in perfect alignment. Offley proudly watched as his men showed that they were battle-ready.

1st Filipino Infantry Regiment. Ponce Cazem after field exercises wearily reclines under a tree. The volcano patch can be seen on his left sleeve. 1943. Source: Mark P. Cazem

"The celebration of the Colors continued with an American-Filipino night," reported the *Camp Roberts Dispatch.* Several soldiers who saw action in the Philippines and on Bataan stood at stiff attention beside a bust of MacArthur as the band played the national anthems of both countries. For entertainment, Yo-Yo champions Pvt. Albert Vierenes and Pfc. Selmo Saladino showed their skills that earned them the title in the Philippines.

A Trainer was published in August. Photos of the Regiment's 1st Anniversary were included.

The Trainer showed photos of the officers in Col. Offley's staff. Tirso Fajardo oversaw S-3 Operations. Heading companies were Filipino officers: Velasquez as head of Headquarters Staff, Chavez in charge of Company B, Llorente of Headquarters Company, 2nd Battalion, and FlorCruz in command of Company L, 3rd Battalion. Other Filipino officers in the photos were unidentified. A schedule of regular masses held by the chaplain was published as an announcement. A page described the 1st Filipino Regiment's coat of arms approved on Oct 20, 1942, and the shoulder sleeve insignia designed by Capt. Talman Budd, commanding officer of Company D.

Offley wrote a poignant message to his men. "A day never passes that I do not become more and more impressed with the realization of the great honor bestowed upon me as Commanding Officer of the First Filipino Infantry. It is a source of pride and admiration to see the enthusiastic efficiency displayed by the officers and men alike. It is my prayer that I may be allowed to round out my service in close association with the same loyal people whom I learned to love when a young boy – the Filipino People."

A local newspaper published an article entitled *Filipino Regiment Turns Out Paper.* With Ernesto Ilustre as editor, the first edition of the monthly *Bolo News* rolled off the camp's mimeograph machine. Various stories and articles from "personal items

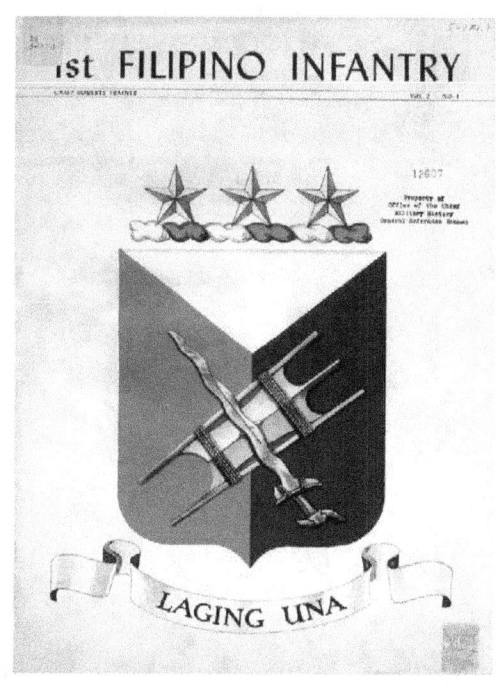

1st Filipino Infantry Regiment. Camp Roberts Trainer. August 1943.

about the boys in the camp to a list of some slang native to the Regiment" were featured. Fabros continued publishing *The Bolo News* while Ignacio Sarmiento distributed a weekly mimeographed sheet featuring some interesting news of activities in and around the camp.

Aware of the biases and consequences of the times, Offley reminded the men of maintaining cooperation with the American officers and cordial relations with the locals around the camp and nearby King City. The Regiment's popularity grew when they were encouraged to represent the U.S. Army in the many community activities and parades along the California coast.

> To the Members of the Regiment....
> I have considered it an honor to command this Regiment. You have made my work easy by your fine spirit and your devotion to duty. Such qualities will bring us victory... God willing, our history will fill many more books (Camp Roberts Trainer No. 2 Vol. 3) of this size and be a great source of pride to all....... It is my hope that this book will help to cherish the memories of days spent together during the formative period of our history. A glance at the pages will show that we have worked hard. May the work not have been in vain, but instead, help us to conduct ourselves with glory on the battlefield.
>> Robert H. Offley
>> Colonel, Infantry
>> Commanding

To the Members of the Regiment....

I have considered it an honor to command this Regiment You have made my work easy by your fine spirit and your devotion to duty. Such qualities will bring us victory... God willing, our history will fill many more books of this size and be a great source of pride to all . . . . .It is my hope that this book will help to cherish the memories of days spent together during the formative period of our history. A glance at the pages will show that we have worked hard. May the work not have been in vain, but instead, help us to conduct ourselves with glory on the battlefield. .

Robert H. Offley
COLONEL, INFANTRY
COMMANDING.

Col. Rob't H.Offley
COMMANDING

Maj. Fajardo
S-3

Lt.Col. Mullinix
EXECUTIVE OFFICER

Capt.Ashcraft
ADJUTANT

Maj. Mueller
S-2

Capt. Spiotto
S-4

Chaplain Noury

Lt.Col.Mann
2nd Bn.

Maj. Iverson
1st Bn.

Lt.Col.Kessner
3rd Bn.

Camp San Luis Obispo, 1st Filipino Infantry officers. Col. Robert H. Offley, Commanding; Lt. Col Mullinix Exec Officer; Maj Tirso Fajardo S-3; Maj Mueller S-2; Capt Spiotto S-4; Maj Iverson 1st Bn; Lt Col Kessner 3rd Bn; Lt Col Mann 2nd Bn; Chaplain Noury. Source: Camp Roberts Trainer Vol 2. No 3

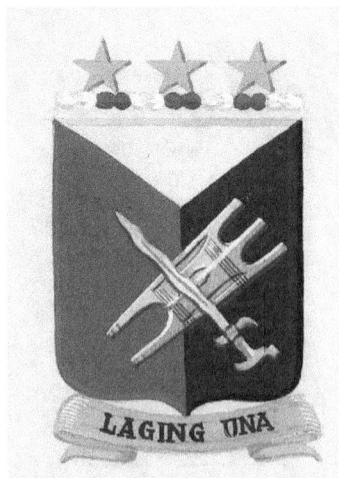

1st Filipino Infantry Regiment Coat of Arms. The colors of the shield represent the national colors of the Philippines and the United States. The crossed kris and Igorot war shield represent the two dominant war-like pagan tribes. The three stars represent Luzon, Visayas, and Mindanao, the three principal islands. Motto: LAGING UNA (Tagalog: "Always First"). Approved on Oct 20, 1942. Drawn in watercolor by Talman Budd.

Americans took notice of the Filipino unit as they continued to parade in the towns. The band marched through Bradley's main street, a town near Camp Roberts. Many onlookers lined the street to see these men proudly marching in uniform, the same men who worked the fields around their town and were targets of hate mobs.

The men worked details wearing dungarees, fatigues, or work uniforms with four-inch brim soft hats. But weekend passes saw them looking well-groomed after morning inspection to get a ride to King City and Jolon. Some hitchhiked all the way to San Francisco but were required to get back before Sunday night's curfew.

The *Chicago Sun* published a unique drive to cast the 1st Regiment's coat of arms – a shield with three stars on top and the motto *Laging Una* – in sterling silver. Filipinos and their friends in Chicago sent in their "discarded trinkets and silverware." "Less than one hundred crests were made, and only for officers that contributed some silver privately," reported the *Shot 'N' Shell*. The men reported that Offley creatively obtained the silver in an unusual and successful boxing wager.

Few newspapers reached the camp, but the radio receivers kept them informed on current developments in the Philippines. The radio program *Vox Pop* brought entertainment highlighting people and the job that the Armed Forces and all the people of America were doing. Radio and news films continued to broadcast stories about the camp directed mainly at the Filipino people in Japanese-occupied Philippines. Under the sponsorship of the Office of the War Department, Station KSFO San Francisco made recordings of about one hundred officers and men, with each man offering his message of patriotism and encouragement to their countrymen to keep the spirits high until the liberation. Their identities were concealed to avoid reprisal against relatives and friends on the islands. Filipino selections played by the *Laging Una* band were broadcasted. These were all sent on shortwave stations that Filipinos in the homeland were able to listen to on carefully-hidden radio sets. Owning a radio was a dangerous activity. The Japanese had confiscated and kept track of all radios. If caught, the owner would be tortured and killed with his family.

When orders arrived to transfer back to Camp Beale, a gigantic 'Summer Fiesta' was staged as a break from a year of rigorous war training. The 2nd Battalion of the 1st Regiment sponsored the event in August with a dance and a 'lechonada' in a community hall at nearby Mission San Antonio to escape the blistering heat. To reciprocate, Ilustre donated twenty-five dollars to the Filipino Press Club of California from the Athletic and Recreation Fund of the Regiment.

Sgt. Sergio Solidarios chaired the event and gave the welcome address, followed by Capt. Llorente, who introduced the special guests. Two thousand soldiers posted at the camp attended. Buses were chartered for the girls in the surrounding areas to come and join the dance, picnic, and *lechon* feast. A mass at the historic San Antonio mission in the nearby town of Jolon began the program. The band played the stirring 'On To Bataan.' A female singer belted out tunes and a soldier showed his guitar skills while another played the violin. A tap-dancing and *sipa* (a ball made of woven bamboo strips kicked back and forth) exhibition closed the entertainment on a high note.

1st Filipino Regiment, 2nd Battalion. Summer Fiesta Aug 1943.

November 1943: 1st Filipino Infantry, Company D participated in a Veterans Day parade on Market Street in downtown San Francisco, California.

Men paraded in San Francisco and surrounding cities. Source: Ray Bisares

Camp Beale, California USA

REGIMENTAL FIELD MASS
In commemoration of the Eighth Anniversary of the Philippine Commonwealth and in Memory of the Filipino and American Martyrs of Bataan ..... November 14, 1943.

Regimental Field Mass. In commemoration of the Eighth Anniversary of the Philippine Commonwealth and in Memory of the Filipino and American Martyrs of Bataan ..... November 14, 2943.

1st Filipino Infantry Regiment. Regimental Headquarters. Camp Beale. Christmas Day 1943.

The Mass was held in the open field commemorating the 8th Anniversary of the Philippine Commonwealth and honoring the memory of the Filipino and American soldiers of Bataan. The *Shot 'N' Shell* newspaper reported infantrymen assembled on the main parade ground before the altar. Guards of honor commanded by Lt. Chaves stood at attention with their rifles flanking the altar. Two buglers, Sgts. Alphonso Conde and Jose Pastrana sounded 'Taps,' one beside the altar and another far from the field answering with his bugle as the echo. Col. Offley spoke about the glory of the American and Filipino soldiers who stood and died together. "Our brotherhood was sealed with blood in this heroic struggle." In conclusion, the 'Star-Spangled Banner' was played.

The men shared in celebration of Christmas and were ready to go into combat.

*California 1943. 1st Filipino Infantry Regiment. Browning Automatic Rifleman.
Trained in the U.S., wearing the standard temperate-issue U.S. Army wool uniform, M
1941 'Parsons' jacket (named after its designer), M 1938 leggings over Type 2 leather
service shoes., M 1937 BAR ammunition belt with infantry suspenders and 1942 M 1917
A1 Kelly helmet. Source: Mandirigma by Albert Labrador. 2021*

1st Filipino Infantry Regiment New Guinea and The Philippine Islands

Oro Bay, New Guinea

Orders arrived to depart for New Guinea. Men continued training to be ready in anticipation of available transportation. Equipment was reviewed and packed. Liberty passes were granted for a job well done in training.

The weekend before the Regiment left California saw the men stationed in San Jose at the State College Spartan Stadium, displaying their equipment, and performing demonstrations. They joined a parade and proudly marched down one of the main streets with raised *bolos* one more time while the band played. Lt. Luna realized that the crowd thought they were watching Filipino troops from the Philippines and not American soldiers.

Departing from Camp Beale, 157 officers and around 3,000 enlisted men travelled to Camp Stoneman, in Pittsburgh, California for deployment to the Pacific theatre. The camp was a major staging area before departure. From there, the SS *Catalina* took them to San Francisco to board the troop ships. One of the men narrated their first landing in Port Chicago in Suisun Bay before they disembarked in San Francisco.

Joseph Agbanawag recalled: "In San Francisco, we marched off to Fort Mason to board the troopship. The Regimental band was the last to board. The other soldiers helped load the gear and equipment of the bandsmen onto the ship. The bandmembers then climbed aboard the ship, reassembled their formation on the ship's fantail, and started to serenade their friends and families who waited on the dock. The next-to-last song played was 'Dahil Sa Iyo,' (Because of You), a song from a late 1930s movie in the Philippines. Many in the crowd below began to sing along, and soon all the men on board joined in. The song was played a second time around, and this time almost everyone was singing and starting to wave farewell. When the ship began to leave the dock, and the ship blew its mighty horns, the band broke out into 'Auld Lang Syne,' and one of the soldiers on board said that at first, they sang it lustfully but by the time the second verse came along the men grew silent. I looked back once more toward the dock and the fast-disappearing faces of my friends. I had to choke back tears."

The trip to New Guinea took twenty-two days onboard the USS *Gen. John Pope*. The Navy gave out pocket guides to New Guinea. The Filipino Regiment made up the bulk of the over five thousand passengers. Lt. FlorCruz wrote that they were fed twice a day and told that it was 'nutritionally sufficient,' although he was hungry between meals. Attendance at Mass onboard increased. Many contemplated the war they were embarking on. The debilitating heat increased as the ship sailed closer to the equator. Men sweated in the closed compartments, and many went bare-chested with the limited laundry facilities. It was a taste of the humid and treacherous jungles they would soon face. The ship docked at Noumea in New Caledonia, debarked passengers, and embarked other units. There was no combat on Noumea, a large American base with facilities, hospitals, supply depots, and airfields. It allowed for training and logistical support that was invaluable in assisting the major campaigns towards the invasion of Japan.

The Regiment did not have the full contingent of Filipino officers upon arrival in Oro Bay, even with the transfer of men from the 2nd Battalion. Twenty-five Army 'Ducks' (amphibious trucks) debarked all 3,053 officers and men in three hours, even with their small capacity and the difficulty in landing the cargo. The white officers were disappointed when they had to train new officers to lead platoons and companies. But they were enthusiastic about being with the Filipino unit believing that they would see some action by being one of the first to land in Leyte in October.

Oro Bay suffered from the low priority of supplies and service troops. It was the war in Europe first. But the Liberty ships (the Navy's cargo ships) regularly docked with rations, ammo, and rows of jeeps. Warehouses and buildings were being constructed to house ordnances, equipment, supplies, and vehicles. The weather was clearer than in other areas making it a frequent target of Japanese planes as hundreds of soldiers unloaded ships.

The Filipinos and other troops units of what was then considered racial minorities were viewed as service personnel. They were immediately assigned to unloading war supplies from incoming merchant marine ships and performing personal services. When the Port Commander overruled a Filipino officer's protest, a message to General MacArthur stopped the practice immediately.

Quarters were larger canvas tents that were taller than the pup tents in California with side vents to pass air. The troops carried out orders in maintaining security around areas and providing personnel in maintenance, repair, storage, and transport services. Julito Balolong later related to his son that the "1st Filipino Regiment wherever they were assigned, operated and maintained a lot of heavy machinery and helped transport equipment and personnel."

The men experienced learning how to survive in an actual jungle in New Guinea and fighting tactics from Australian experts in jungle warfare. Bare-chested and with their *bolos*, they trained in the jungle clearings. Lt. Luna was assigned to security detail while undergoing training. He was issued a Thompson submachine gun to carry while driving his vehicle. Pfc. Alejandro Soria was assigned as an anti-tank gunner. They learned not to leave any trash behind for the enemy to find and follow their trail. They peeled the white cigarette paper from its butt after smoking a cigarette, stuffed it in their pockets and grounded the tobacco into the dirt with their boots.

1st Sgt. Melquisades Ventura of Los Angeles, Calif. and Cpl Bregado Ratanan of Stockton, Calif., (AT Co, 1st Filipino Regt) walking down their company street at (Oro Bay, New Guinea). May 17, 1944. Photographer Ernani D'Emidio.

Sign on the arch read, "Anti-Tank Co., 1st Fil Inf."

T/5 Albert Viernes of Chicago, Ill., a former vaudeville yo-yo artist, amuses men of (Co I, 2nd Bn, 1st Filipino Regt.) with his skillful handling and dexterity with the yo-yo at (Oro Bay, New Guinea). Photographer T/4 Ernani D/Emidio SWPA-SigC-44-12748. May 17, 1944

The men were eager to learn and were sharp at their work, reported the Australian training battalion. They were "average on map reading and compass... outstanding on wonderfully happy dispositions, and extreme loyalty to each other... absolute personal cleanliness, a dirty towel could never be found." More emphasis was given on knowing the rate and method on patrol and moving into fire position after being fired upon.

Men caught tropical diseases. Insects crawled everywhere in the hot and humid weather alternating with rainy days. But many of the enlisted men were better at surviving training in the debilitating weather and jungle terrain than the white officers. They were hardy men used to the hard physical work in the fields of California. "An officer went with us this morning, and he could not keep up with our pace in the homeward trek. The special way of walking in the jungle and the heat must have worn him out," wrote FlorCruz to his wife.

They watched movies when available on crudely set up screens in the open. *The Song of Bernadette* appealed to many of the men who were Catholic.

One of the projects the Regiment did was a bridge in the dense jungle. Over thirty men stood over the bridge they built over a rocky river to show its sturdiness. Several New Guinea natives in loincloths seen on the right helped in the construction. The bridge was made of wood logs strapped together with a single rope strung across as a handrail. Against a backdrop of the jungle, men smiled, and some waved their hands over their heads. A man in the foreground raised his *bolo* as a salute.

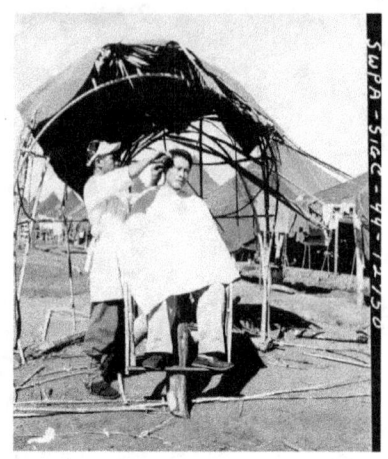

Pfc. Leon Ventura of San Francisco, Cal., gives Pfc. Mauro Laguna of Stockton, Cal., a haircut on an improvised barber stand in the camp area of Co. B, 1st Bn of 1st Filipino Regt., Oro Bay, New Guinea) Photographer T/4 Ernani D'Emidio SWPA-SigC-44-12750 May 17, 1944

A former barber gave haircuts to his comrades. Although in non-stop training, the men wanted to look kempt.

Several Signal Corps photos were taken of the men when they were stationed in Cape Sudest, an island south of New Guinea. The seemingly sturdy quarters of pyramid-like tents they lived in on the open field belied the everyday struggle to keep the tent flaps shut against the rain as water seeped in. The plus side, however, was that mosquitos could not land on them in the gusty wind.

The island-hopping plan of GHQ to the Philippines captured weakly-defended enemy areas, cutting off supplies to the islands with large Japanese garrisons. With the U.S. government's "Europe First" policy and limited resources to the Southwest Pacific, it was the fastest way to reach the Luzon-Formosa area with the least casualties as directed by GHQ.

1st Filipino Infantry Regiment constructed a bridge in New Guinea. Two of the natives are shown in the far right. Source: Talaugon Collection

The 31st Infantry Division of the Eighth Army arrived in Buna by Oro Bay at the end of April at the same time an assault on Hollandia was happening. The Regiment carried Eighth Army patches on their shoulders making them part of the Army unit. This regulation caused their activities to be listed with that of the divisions and made it difficult to follow the Regiment's history. Orders given to the Division to maintain security around the captured Dobodura airfield area deployed the men to conduct patrols while continuing advanced combat training.

Looking down the main street of 1st Filipino Regt area at Camp Sudest, New Guinea) Photographer: T/4 Harold Newman. May 15, 1944, Signal Corp 16328

Camp Sudest, Oro Bay:. Company A, 2nd Battalion, 1st Filipino Infantry Regt. May 18, 1944.
Source: Cedric Jasmin

While in New Guinea, many of the men earned the Combat Infantry Badge (CIB) including Cpl. Alex Fabros, Joe Agbanawag, and Johnny Albano.

From the action reports in the personal papers of Cpl. Alex Fabros, the 124th Regiment Combat Team of the 31st Infantry Division left Oro Bay to take part in the general offensive of Aitape in the bloody battle at Driniumor River to cut off the Japanese 18th Army from their main force. Elements of the Regiment were assigned to the 126th and 128th Infantry Regiments of the 32nd Division in unit actions of hand- to-hand combat within the Division's sector.

Col. Offley gave Capt. Punsalan command of the 1st Battalion of the 1st Filipino Infantry Regiment. It was the first time in the history of the United States Army that an Asian-American commanded white troops in combat. Over one thousand Japanese was reported killed in these actions. The enemy were hiding on cliffs and in caves that overlooked the beaches. They would not readily surrender. Pfc. Benny Ibon, a heavy machine gunner, vividly recalled, "After we stopped firing into the caves, we heard screaming." Several Filipinos were wounded or killed. In the Sarmi-Wakde Island area, the men built bridges, roads, and docks while engaging in skirmishes with the enemy.

Lts. Eduardo Suatengco and Atanacio Chavez fought battles in New Guinea. Lt. Jaime Velasquez cheated death when his battalion was attacked and many of his men killed by the Japanese. He laid motionless as the enemy walked past him. Antonio Campos went through his first combat with mixed memories of fear and excitement. In his letters to his wife, he wrote "I was scared as Hell; my heart kept pounding." There were encounters with the enemy he could not forget. In combat, he said you can't think, "it's shoot or be shot." Waiting for the enemy in bad weather was uncomfortable for the men. Rain seeped into their slit trench, but they could not leave as the enemy could be waiting. Their orders were to shoot anything that moved. They were constantly aware of any noise and movement around their outposts.

With Hollandia captured, the 24th Infantry Division and the 32nd Infantry Division prepared for deployment to the Philippines. The 31st Infantry Division captured Sansapor and Morotai, the closest to Mindanao. Airfields were captured or built from which planes could easily reach the Philippines.

As soon as the enemy heard of the fall of Morotai, the Japanese quickly rushed reinforcements into Manila and the capital cities of the southern islands.

Back at the camp, the traveling Immigration and Natural Service (INS) officials arrived to naturalize those who had changed their minds and wanted to become American citizens. In August 1944, a swearing-in of the applicants from the 1st Regiment was recorded. A court session was opened at camp for the soldiers to swear the Oath of Allegiance. Henry Hazard, the U.S. Immigration and Naturalization Service (INS) representative naturalized 87 applicants, the first group he naturalized in the Southwest Pacific Area. Sgt. Herman Brotonel was naturalized that day in Oro Bay.

Keeping the Regiment intact with combat training ongoing was a continuous challenge for Col. Offley. The men were combat-ready as a unit when they left California for New Guinea. But there was no way to prevent further losses of the Regiment's strength.

An order from General Walter Krueger, commander of the Sixth Army, arrived wanting men who were familiar with geographic regions and could speak the languages to act as guides and interpreters for his Alamo Scouts. They would be further trained in reconnaissance, combat, and penetrating enemy lines as Scouts.

Another several hundred officers and enlisted men who spoke the various dialects of the islands were pulled to form the Philippine Civil Affairs Unit (PCAU) several months before the Philippine landings. The PCAUs were to provide basic needs of the civilians and re-establish local government administration until the Philippine Commonwealth government was able to take over. Col. Offley recounted that "they needed Filipinos, and so they took my California-trained officers and made them PCAU commanders." He lost nearly 300 men to the PCAU, the Counterintelligence Corps (CIC), the Alamo Scouts, and the 6th Ranger units. It took almost two months to bring the 1st Regiment back to strength with men from the 2nd Battalion. (The PCAU and CIC are discussed in a separate chapter.)

By mid-1944, U.S. bombers from captured airfields and the U.S. 3rd Fleet raided Palau and the southern islands of the Philippines. The lack of resistance from the Japanese surprised the Americans. The invasion of the Philippines was then reset to an earlier date for October in Leyte instead of Mindanao in December. The plan to attack Formosa first was scrapped. The sudden decision created a logistical nightmare with hundreds of thousands of men, hundreds of ships, equipment, and support units that had to be mobilized at double time.

In the ten months they were in New Guinea, the men followed the progress of Gen. MacArthur's army along northern Guinea and Admiral Chester Nimitz's fleet as it captured island after island in the Philippine Sea. Without newspapers, there was no news of the outside world except for several radios in camp and radio stations on the island that carried the husky voice of Tokyo Rose's enticing invitations. As the month of the planned Leyte landings neared, everyone was on the alert for the order to mobilize.

The orders to move to Hollandia, a bustling supply base closer to the equator, arrived. The heat and humidity in the area proved to be worse than in Oro Bay. It affected ammunitions, weapons, optical, and other equipment sensitive to the dampness. MacArthur's headquarters was built on a higher and cooler area bulldozed flat with switchback roads leading to it. He did not spend much time at the headquarters as he was leading the Task Forces to Morotai which was already at the doorstep of the Philippines.

The Regiment's new campsite was near Centani Airstrip about 40 miles from the pier. The area was sandy, surrounded by trees, but the heat kept clothes wet with sweat and disrupted sleep at night. The site was reminiscent of Camp Tabragalba except for the more intense heat that did not encourage outdoor activities. Everyone worked and built up the camp in a month. Some tents were set up on wooden floors while others were on ground that collapsed in flash floods of mud. Some of the men made "walls of nipa leaves clipped between whole rattan reminding them of the Philippines." To the Americans, "it is an object of consuming curiosity" Lt. FlorCruz wrote. A kitchen, supply buildings, and a post exchange were erected but there was no refrigerating system, so meals did not have fresh meat. When ham was served in one meal, criticisms that it was "the best food we have had in this island so far" were heard. Tiring of rations, some of the men asked to be allowed to hunt for wild pigs, but hunting was forbidden. Only the natives of New Guinea could hunt for game. To make it legal, kill the animal in self-defense if attacked. Orders were to guard posts, battalion supplies, and motor pools with over a dozen vehicles. A nearby creek was much visited as the red clay soil of New Guinea stuck into hair and clothing like mud. Mail appeased the loneliness. A censor department worked overtime to review the letters before having them reduced to film for transmission, then enlarged and printed in 4 x 6-inch letters.

Luzon Island, Philippines Landing - January 1945

The regiment wanted to be part of the first landings in Leyte in October, but it was understrength even after men from the 2nd Battalion were transferred over. The Navy bombarded the beaches of Luzon prior to landings. Intelligence crowded the airwaves to KAZ on the enemy and conditions on Luzon.

Leyte and Samar Islands, Philippines – February to September 1945

The Regiment would be assigned as a unit in the clearing of Leyte, Samar and surrounding islands. The Regiment of 137 officers and 2525 enlisted men shipped in two groups from Oro Bay and arrived in Tacloban. All troops assembled at Alangalang, ten miles west of Tacloban. They did not wear metal helmets because they made too much noise attracting the enemy when they hit tree branches. The white soldiers wore them, and some muddied their pale faces. Attached to the Sixth Army, they were ordered not to divulge orders issued to them. Maj. Fajardo and Filipino officers landed, each commanding a battalion. It was a safer arrival than the earlier landing in October with slight resistance.

The strict timetable directed the Sixth Army to capture Manila. The Eighth Army was charged to clear out pockets of the estimated 30,000 Japanese soldiers who were still entrenched in the northwest mountains of Leyte and willing to fight to the death. General MacArthur had declared the Leyte campaign as closed except for minor mopping operations but Japanese reinforcements continued to arrive at the port of Ormoc City on the west side of the island. It was a Japanese base and garrison. Many houses were occupied, and the residents ran to the mountains with the warning of beach bombings. The starving Japanese was pushed into a narrow pocket near Palompon. When the civilians returned, food stuff and livestock were eaten by the Japanese. Starving and sick, they ran to the U.S. military camps on the beaches.

Attached to the Eighth Army, elements of the Regiment provided security for GHQ of the Far East Air Force (FEAF) and the 7th Fleet. A company was assigned to secure the Tanuan airfield that was taking a while to be completed in the constant rains and muddy grounds. When fighter planes could not take off from the airfield to provide support to Luzon activities, Mindoro Island at the mouth of Manila Bay was taken and airfields constructed. (Mindoro Island is discussed in a separate chapter).

Samar, Biri, Capul, Ticao, and Burias Islands, Philippines – February 1945

The Eighth Army assumed command with troops from the Americal Division and the 1st Filipino Infantry Regiment. Information received from captured Japanese documents and POWs indicated that as of February, more than one thousand Japanese troops were garrisoned in the central and northern part of Samar. It would be the Regiment's first large-scale battle.

As the 1st Cavalry was entering Manila, signaling the beginning of the liberation of the capital city, the 1st Filipino Infantry Regiment was ordered to capture nearby Samar. The San Juanico Strait separated the island from Leyte at only 2 kms at its narrowest point.

The Eighth Army ordered, "the commanding officer of the 1st Regiment to assume responsibility for military operations on Samar and its contiguous islands with the order to continue mopping up operations of enemy forces and to maintain garrisons in Northern Samar and the surrounding islands of Biri, Capul, Ticao, and Burias." The Regiment was tasked to root out retreating Japanese troops in the mountainous jungle areas filled with diseases. Maj. Fajardo was reported to be a battalion commander on Samar.

The objective on Samar was to secure the supply route to Manila by clearing the islands along the Visayan passages and enemy forces in Northern Samar and the smaller islands. The Naval Forces would then be given unhampered use of the San Bernardino Strait between Southern Luzon and Northern Samar. The Strait was the passage where Japanese forces passed earlier at the destructive Battle of Leyte Gulf. Securing it would prevent a repeat attack from the enemy. Clearing of the Mauo area in the northeastern part of the island was imperative.

Landing ships loaded the men and their equipment at Tolosa, Leyte, and brought them to Catbalogan, Samar. The commands of the Army Task Force and the 1st Regiment were moved from Leyte to Catbalogan. A provisional task force was formed under the Americal Division Artillery Team composed of the 1st Battalion of the 182nd Infantry, X Corps and the 1st Filipino Infantry Regiment (less 1st and 2nd Battalions). Medical and service personnel accompanied the mission to relieve personnel of the XXIV Corps stationed at Catbalogan.

When the American Task Force landed on Samar, the older population still remembered the Philippine-American War, which was called an insurrection by the U.S. government, only forty years earlier. At that time American troops were killed by the locals. In retaliation, all Filipino male civilians over the age of ten were ordered killed. Now the Americans were welcomed as liberators.

The men were nervous with their first combat assignment. On their first night on Samar, they set up headquarters in a school area with a perimeter defense. Clay Pardo recounted that at 1 am, "all hell broke loose. Our machine guns and 81 mm mortars opened up." One of the outposts had heard rustling in the undergrowth. Three men were wounded, and one killed by friendly fire. No Japanese or land crabs were injured.

Men fell in the Samar jungles. The enemy was still strong and retaliated viciously. The Hawaii boys, Filipino soldiers born and raised in Hawaii, were called in to reinforce them. In small teams, they probed an extensive cave system and bunkers, many times using flame throwers. Pedro Visaya was wounded. Pvt. Felix D. Pequit, a former professional bantamweight boxer, was ready. He had fulfilled his desire to return and fight but four months before Japan's surrender, his leg was raked by Japanese machine gun fire that proved fatal.

The Regiment's 2nd Battalion reinforced the Americal Division Combat Team and were referred to as the Northern Task Force and Provisional United Army Force. They established garrisons in Northern Samar. Intelligence reported around 1,000 Japanese in the central and northern part of Samar. The interior of the island was mountainous terrain, dense growth, with many rivers and streams. Road systems were unpassable until repairs were made. Every night the rain poured. The only dry place was "inside the top of one's helmet, a place reserved for matches and toilet paper." Heat and humidity added to the load of a full backpack, a weapon and its ammo, three days' supply of K rations, and two canteens of water. A salt tablet and Atabrine to fight malaria was a daily dose. Their training in California and New Guinea prepared them for situations like this.

The 93rd Division guerrilla forces on Samar under guerrilla commander Juan Causing composed of the 97th, 98th, and 99th Infantry Regiments with headquarters in Calbayog City was attached to the 1st Regiment. The Americal Division reported around 500 Japanese in the Gandara area by the town of Calbayog. The 2nd Battalion with the guerrilla 99th Infantry patrolled and set up roadblocks on the highway north of Calbayog. Guerrilla patrols reconnoitered the enemy in the Acerida-Trojillo area. The 3rd Battalion with the guerrilla 98th Infantry that occupied the northern coast of Samar drove the enemy inland.

Daily combat patrols reported hunting and engaging the enemy with several hundred killed. Civilians were reported capturing stragglers. An additional large number of Japanese died from disease and wounds because of their relentless pursuit and mortar fire. To survive, small foraging groups roamed. Company K advanced unopposed twelve miles north on Highway 1 from Catbalogan to Calbayog where they cleared the area of Japanese stragglers and maintained garrisons. Valentin Villafuerte's unit caught up with Company K to replace them. He saw many dead bodies from the fighting the night before. They were his comrades, recognized from their uniforms and the protective leggings on the feet. They were just left on the trail. Their duty was to hunt, not clean up the dead, a lieutenant barked at him. They were to hunt stragglers, control them, give them a chance but the Japanese fought back.

Company I captured the northeastern town of Mondragon killing several hundred Japanese who crossed the Bugko River. They went inland to clear Catubig pushing the enemy further to the area of Allen. After small sporadic arms and artillery fire from enemy positions in the town, the beachhead of Allen was secured. Patrols began to operate between the villages of Catarman and Allen.

An entry dated March 12 — April 19, 1945 in the captured diary of Japanese Major Kono was translated into English and distributed to the men of Company I. He desperately wrote of "disappearing, always looking for food. Tried to fight out towards North, enemy encircling but heavy rain stopped them. Continued advancing through a secret trail under two mountains. The enemy could not find them. Continued because wanted to find a place to sleep. Stayed in the forest two days without food. Followed trail along river. Farewell to the Emperor. Can do nothing else. I am not fit to be a commander. Can do no more."

The Regiment's Service Company ran supplies and equipment to the combat units under heavy enemy fire. While transporting and guarding Japanese POWS, Lt. Luna learned to speak a few words in Japanese. He could not forget the smell of POWs and thought it was the worst odor he had ever experienced. The extreme suffering, he saw caused him sleepless nights.

Friendly fire was unavoidable and kept men on the alert. Company L was on the east coast hunting Japanese stragglers when they exchanged fire with another company that were wearing Japanese hats they had picked up as souvenirs, recalled Allen Dawang. No one was killed at the encounter after a quick radio contact was done before anyone got hurt. Lt. Luna was careful because he had heard that "the Japanese put out rumors in Leyte and Samar that the so-called U.S. 1st Filipino Infantry were Japanese dressed as Philippine Scout soldiers. So friendly fire erupted between U.S. soldiers, Philippine Scouts, and the people of Samar."

By the end of February, a battalion of the 182nd Infantry reassembled in Allen in preparation to clear the Mauo sector north of Calbayog. Elements of the Regiment were to reinforce them. Plans called for a three-company attack from the north by 182nd 1st Bn, together with support attacks by companies F and K of the Regiment. That would envelope Mauo and Mount Bermodo in the eastern interior. A garrison of three hundred Japanese was in an area of dense forests and undergrowth. A network of trails led Company K with a company of the 182nd to the east of Mount Bermodo but also gave the Japanese a covered route for withdrawal. Camp in the evening was a short distance from the hill. At the appointed time, Company K swung around to dig into the southeastern side of the mountain to prepare to attack the Japanese. They met bursts of enemy machine-gun fire and short bursts of 75 mm gunfire. The unit moved to the foothills at the north banks of the Mauo River for the night to prevent any Japanese attempt to escape. In the meantime, Company F had moved to Calagundian, south of Mauo, with orders to cut off any attempted enemy movement to the south along the coast. They pushed towards the south banks of the Mauo River mouth. Turning east, they moved inland and joined the other attacking forces. They passed around a deep mangrove swamp blocking their path before establishing a perimeter for the night.

The next two days were continuous fighting until the commanding general of the provisional U.S. Army Forces sent a radio message (Feb 23, 1945): "The command of Samar passes to the commanding officer, 1st Filipino Infantry Regiment effective March." According to a communique from General MacArthur, capture of the islands off the coast of Southern Luzon prevented the Japanese envelopment of the area and sea lanes vital to American shipping from Leyte to Manila. The 3rd Battalion with the 93rd Division guerrilla forces attached was given the mission to remain and continue mopping up operations on Samar and its adjacent islands. The Battalion was given the additional task of imparting their skills to the 93rd Division guerrilla when it was redesignated as the 42nd Regiment, Philippine Army.

The Eighth Army withdrew after the turnover of Samar to begin invading Zamboanga in Mindanao. The Regiment was left to continue securing Samar, the smaller surrounding islands, and the waters around from enemy shipping. Often, they were without supplies for weeks. As rations became short, clothes were bartered – one undershirt for one chicken, one can of corned beef for one hundred bananas, and one khaki uniform for a pig. Civilians who barely had any food for themselves gave rice and sweet potatoes. Money had little value. The soldiers eagerly awaited the mail that arrived regularly from Leyte during the four months they were isolated in Samar. It did not matter that some letters were months old.

The 1st Battalion minus Co A sailed south to provide security for the naval airstrip in Guiuan Island at the southernmost tip of Samar. The 2nd Battalion with Company A relieved elements of the XXIV Corps on guard details for vital installations on the eastern coast of Leyte. They eliminated Japanese stragglers from Jignatan Valley and across the mountains of Ligaya and Ormoc.

Turning to nearby Capul Island, the U.S. 182d Infantry made an amphibious landing on the southern tip of the island against slight opposition and garrisoned the village. The Filipino Regiment's Company L, a rifle platoon, and Company A reinforced the 182nd. Company I, 3rd Battalion led by Capt. Vaughan P. Moore landed and captured two Japanese 75 mm guns. Antonio Campos with the 3rd Battalion related that "on patrol, his unit came upon thirty Japanese soldiers. With a strong offensive, all were captured for intelligence questions." Operations at Capul were eventually turned over to the Regiment. The island was scoured until Japanese soldiers were seen fleeing in canoes. They were pursued and killed by machine guns in landing crafts. In the five days of operation, 34 men were killed in action, five were missing in action, and 179 wounded.

Three days later, the 182nd crossed the waters to Biri Island in the north and landed unopposed. A platoon from the Regiment's Company L quickly took over control of the island. Further north, the Regiment sailed to replace the guerrilla 87th Infantry force on Ticao and Burias Islands destroying the remaining enemy garrisons.

The maneuvers in Samar were reported completed in April 1945 but they lasted until September.

After Japan's surrender in August, Company I conducted a recon patrol around the town of Nenita and attacked around five hundred heavily-armed Japanese troops. The Japanese commander and his staff were killed while the rest scattered and were tracked down.

Conflicting reports of the Regiment's casualties and the number of enemies killed soon surfaced. The *Report of the Commanding General of the Eighth Army on the Leyte-Samar Operation including the Visayan passages' clearance (26 Dec 1944 – 8 May 1945)* listed the 1st Filipino Infantry Regiment as having one officer and thirteen enlisted men killed in action. Three officers and thirty-five enlisted men were wounded in action. One report stated the Regiment lost 18 men and more than four hundred Japanese were killed. Another reported five men killed from the Regiment and 1,572 Japanese killed.

Col. Offley was seen with his infantrymen patting them on the back. General Eichelberger, commanding general of the Eighth Army wrote that he "used the 1st Filipino Infantry Regiment in the subjugation of Samar, and its record was excellent. The troops had sound training."

In the meantime, while Europe was celebrating the collapse and surrender of Nazi Germany, General Joseph Stilwell was quoted by the *Free Philippines* newspaper saying that at least half a million men was needed to invade Japan. GHQ was set to begin Operation Olympic on November 1, the first phase of Operation Downfall, the invasion of Japan.

The war with Japan was estimated to last for two more years. A million and a half Purple Heart medals were ordered in anticipation of the enormous casualties. The *Free Philippines* published *Win or Die says Emperor.* The Japanese People's Volunteer Corps members were told not to be taken alive or to surrender in an Allied invasion of the homeland. They were to win, get killed, or commit suicide.

Troops from the European theatre had begun arriving in Manila and reluctantly trained against an enemy who would rather die than surrender. Four thousand five hundred American veterans from Italy ready to go home were diverted to Manila from the European theatre to join the invasion.

Ormoc, Leyte; Camotes; Panaon; and Dinagat Islands, Philippines

Intelligence summaries estimated between 1500 to 3000 enemy troops were still in Leyte after five months of combat. Despite the numerous Japanese transports and naval vessels sunk by the Air Force, Japanese troops and supplies succeeded in landing on the western side of the island in Ormoc to stop the Americans. The Japanese had ordered their best troops to defend Leyte. Their command post in Ormoc had to be destroyed.

The rain and mud slowed down the U.S. Task Forces driving overland from Leyte to Ormoc. Other forces met resistance going north along the west coast. Landing crafts had to be used to transport men around the island.

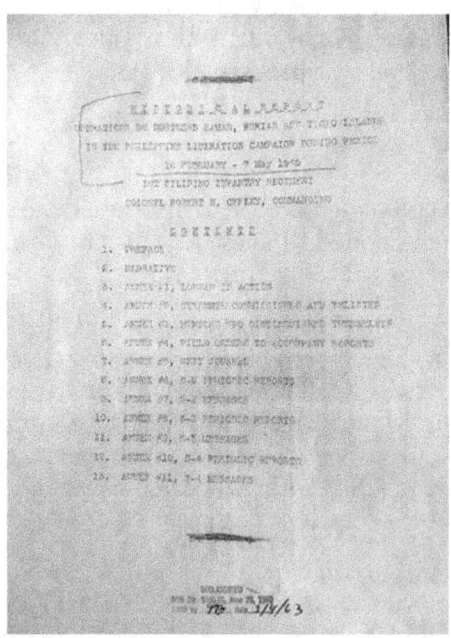
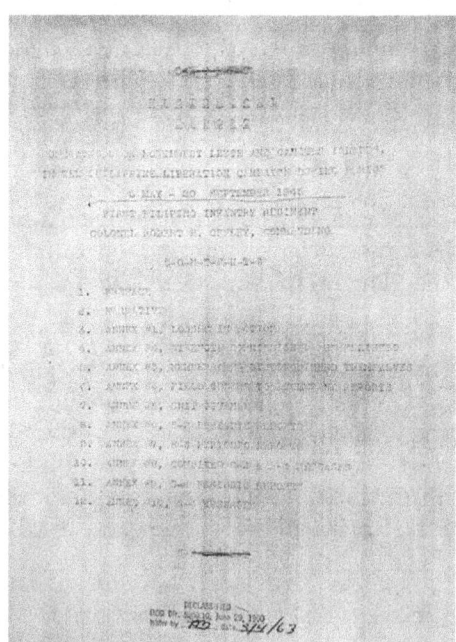

HISTORICAL REPORT

OPERATIONS ON NORTHWEST LEYTE AND CAMOTES ISLANDS, IN THE PHILIPPINE LIBERATION CAMPAIGN DURING PERIOD 8 MAY – 20 SEPTEMBER 1945

FIRST FILIPINO INFANTRY REGIMENT COLONEL ROBERT H. OFFLEY, COMMANDING

By this time, the Regiment had 137 officers and 2,035 enlisted men. Orders arrived in May for the 1st Regiment's 1st and 2nd Battalions to join the Task Force for operations to destroy the enemy forces in Leyte, and the nearby islands of Camotes, Panaon, and Dinagat. The *LSM 18* made two trips to embark the men and their equipment at Catbalogan, Samar and disembarked them at Panalian Point at Ormoc. With the Composite Anti-aircraft Artillery Battery #1, the 25th PCAU, the 43rd Quartermaster War Dog Platoon, and the guerrilla 92nd Division force attached, the Regiment executed a 'search and destroy' mission on Leyte. Pfc. Modesto Lagura was with the 81 mm mortar crew. During the mission, the guerrilla division was inactivated and replaced with the 41st Infantry Regiment, Philippine Army (PA).

The Regimental command post was in Ormoc. The 1st Battalion less Company A battled for two months beginning May in the Villaba area west of Ormoc with two anti-aircraft guns for support. Company B cleared the Palompon area with the support of the 41st Infantry (PA).

Ormoc-Leyte Island–Philipines. Mid -1945. This is a picture of the Regimental Headquarters building. First Filipino Infantry Regiment.

Intelligence data revealed around 2,000 scattered enemy troops of the original Japanese Leyte defense forces remained in the mountainous region of the Palompon-Libungao-Pinamopoan Road.

Elements of the Regiment shipped to the

First Filipino Infantry Regimental Headquarters. Ormoc, Leyte. 1945

south of the Road and took over from the U.S. 77th Division task force to eliminate the enemy. *LSM 315* arrived from Villaba and *LCI 970* from Ormoc with troops and equipment of the 1st Filipino Regiment. The men cleared caves, manned remote checkpoints along the coastal highways, and went on endless jungle patrols eliminating remnants of Japanese resistance. Most of the time, they hunted the enemy one by one. Official reports of the Regiment told of the enemy as disorganized but armed and continuing to resist. Isolated groups were looking for food and clothing, attacking civilians and causing deaths and casualties.

The war dog platoon did not fare well in scouting under the intense heat, so they served as sentinels manning perimeters.

Aggressive combat patrols secured the Limon-Lonoy Highway in the north eliminating the Japanese stubborn resistance from their positions on high grounds. It was one of the bloodiest battles in the islands and reported an average of forty enemies killed and thirty captured daily.

Felipe Dumlao's battalion was armed with an anti-tank rifle, two flame throwers, and three bazookas. His platoon lost seven men in one day and kept losing men until they became a skeletal platoon. "Young Filipino kids from Hawaii were brought in to replace the dead Filipino soldiers," he said, "We engaged as a Filipino battalion with a Japanese battalion and won. An American division was called in to investigate because they did not believe it possible to kill one thousand people in one day. They counted and could not believe it was possible. They called a bulldozer to bury all those people." Around 1500 Japanese soldiers were killed while only five of the Filipinos were lost, stated a report.

The Voice of America (VOA) broadcasted the Allied forces' push of the Japanese into a narrow pocket in Palompon on the western coast of Leyte. The enemy was desperate to escape. Hunting pockets of enemy stragglers continued until the official reports of the Japanese surrender. An additional two thousand enemy were killed. The broadcast reported an average of 40 Japanese soldiers killed and 32 captured per day. An article, *First U.S. Filipino Regiment Proves Battle Mettle on Leyte* in *The Free Philippines* newspaper summarized their achievement.

Ormoc, Leyte Island, Philippines: Communications group, Regimental Headquarters Company, First Filipino Infantry Regiment. 1945

The Americal Division had taken Cebu City and the airfield in April. The Eighth Army directed the Regiment to assemble and stage in Ormoc for a tactical troop movement to the island of Cebu in July. A sizeable Japanese force occupied Cebu with its large harbor and an airfield in the capital city. American intelligence had underestimated the size of the enemy force. The 1st Filipino Regiment with the Americal Division's 132nd and 182nd Infantry Regiments neared the shore under limited Japanese resistance, but the beach was heavy with mines buried in the sand causing casualties. While mine sweepers combed the sand, landing transports crowded the beach. Some did not completely dock on the beach to avoid the mines. The men short in height and weighted down with full combat load were submerged, struggling to keep above water and make it to shore. Inland, scattered resistance slowed down the march to the city. Not much of the city was left after the last three years of Japanese occupation and the Navy's pre-landing bombardment. The Lahug airfield was taken. Many died taking Babag Ridge entrenched with Japanese defenders in their caves and tunnels.

Japan Surrender – August 1945

In August, elements of Company C departed Catbalogan in three landing crafts for Ormoc and Poco Island off northern Bohol. The Company cleared Camotes Island, a part of Cebu. "There were Japanese soldiers and their wives in caves. We got an interpreter to tell them to surrender, and they refused. We piled up dynamite and blew up the cave to close," Philip Ordono recalled.

On the same day that the first atomic bomb destroyed Hiroshima, Teodorico G. Galicia was on his final patrol in the Villaba-Palompon sector when he was shot. He later died of his wounds in a field hospital. He was the last man in the Regiment to die in combat.

The announcement of surrender by Japan's emperor did not stop the hunt for Japanese stragglers because they refused to surrender and continued to harass civilians. The situation was made more difficult with the *Huks* (HUKBALAHAP guerrillas) who were initially anti-Japanese but continued to confiscate food or whatever they desired from civilians.

The Regiment regrouped at Liberty, a village north of Dulag in Leyte after the surrender. A landing craft brought the 3rd Battalion from their command in Samar. While waiting for transportation to Manila, passes some as long as a month were given to visit families. Others found women who later became their wives to bring back to the U.S. The younger soldiers mostly the men who were born in Hawaii, were learning the culture, language, and customs they vaguely knew.

Pfc Teodorico Galicia, 1st Filipino Infantry Regiment. He had his bolo clipped to his belt aside from the rifle he was carrying. Source: Jessie G. Valdez

Months after Japan's surrender, some of the men were still on assignment. In November, T5 Felix Savella and Pfc. Louis Gosmo were onboard separate C-49 planes that tried to take off from Tacloban and the other from Manila to Australia under heavy rain and wind. Running out of fuel, the planes crashed into the sea. Savella was buried at the Manila American Cemetery and Gosmo's body was never found.

The men found local musicians who survived. They brought out instruments they had hidden, and a band quickly formed that continued to play at Sunday gatherings in the company mess hall while waiting for the next assignments. Many brought their newfound loves and families.

The Regiment was attached to the Western Pacific Area Command headquartered in Manila for disposition. They reported with the 2nd Filipino Battalion at Camp Camarilla. Philip Ordono was assigned to the Military Police. He must have known to speak Spanish because GHQ requested Spanish-speaking American troops to serve as military police in the city.

In Manila, those assigned to guard duty by the Pasig River breathed the foul smell of dead bodies all over the city. Fay Hendricks' unit collected "body parts that were floating down the river, and when enough parts for a complete body were amassed (i.e., head, two arms, torso, and two legs), they were buried together no matter the sex, age, or nationality." At Camp O'Donnell where survivors of the 65-mile Bataan Death March where held, San Felipe excavated bodies. He vented his frustration that "many of the guys from the Midwest could not distinguish Filipinos and Japanese." That made it dangerous when several hundred Filipino-Americans were assigned to guard the gas pipeline from Clark to Bataan.

Felipe Dumlao's order was to guard a prisoner-of-war camp in Manila that dealt with the hanging of condemned Japanese officers. They were to keep the Filipinos from killing the prisoners.

Groups of twenty-five men in weekly shifts, were sent to Tagaytay, south of Manila to clear and guard a former signal company buildings until disposition. It was a welcome respite in the cooler altitude overlooking Taal Lake. Off shift, they mingled with the residents of nearby villages and partook of the dishes they missed.

Manila, Philippines: Men of the Regiment gather in front of destroyed buildings. 1945

It was a challenging time to be a Filipino-American in the Philippines. There were separate lines for ice cream, one for the U.S. Army and two for the Philippine Army. Mess lines and theatres had signs 'Exclusively for White' or 'Exclusively for Black Troops'. The segregated U.S. 93rd Division of 'colored' men fought in the Visayas and Mindanao. Filipinos were brown and belonged to neither race, so they were free to join whatever lines. But that did not erase the feelings of discrimination.

A white American looked at San Felipe and his fellow soldiers and asked if they were in the right line. It "drove him up the wall when they were not treated as either white or black GIs. We should have been sent to Europe," San Felipe said. After being told to wait until the GIs were finished eating, he showed them his stripes and got in line for the U.S. Army separate from the line of Philippine soldiers. When told that his group was not in the right line, he began talking fluidly with an American accent. No one could understand why there were Filipinos in the U.S. Army. "All they knew were blacks and whites."

Campos spoke English with an American accent and was assigned to work on supplies. For a frustrating two months, he went around explaining that there were Filipinos in the U.S. Army. "The white and black troops did not understand the presence of Filipino troops in the U.S. Army. All they saw was a Filipino face, so they thought 'Philippine Army' until I talked with an American accent that proved there was a Filipino Regiment."

At times, the Filipino unit was kept separate from the white soldiers. Some locals were suspicious of the "American born that did not speak Tagalog." In the free time waiting for the next assignment, the men clashed with the Philippine Army soldiers over differences in pay, culture, and attention of the local women.

Hardly a building escaped the liberation of Manila. Roofs were gone and rebars exposed from the concrete frames. American colonial buildings crumbled under heavy artillery fires.

The site of so much devastation must have made the men rethink their future. While waiting for further orders, those who did not accept citizenship at Camp Beale in California decided to be naturalized. They feared being unable to return to the U.S. They had fulfilled their desire to join and liberate their homeland and seen the devastation and destruction of it. Life in the U.S. was what they had gotten used to despite the years of discrimination and limited job opportunities.

The newspaper *Free Philippines* reported about Lt. Agustin Santos on Bataan. His draft card had stated 'Filipino alien' and he now returned as a company commander after 10 years in America. He wanted to be a teacher when he returned and was one of the few that took the oath of citizenship in Manila after the war. He felt assured that he would get better treatment in uniform. "People shook our hands and wished us luck."

News arrived that Col. Offley was transferred from the Regiment to command the 45th Infantry, Philippine Scout (PS), under the newly-formed 13th Infantry Division (PS). Soon after, he relinquished his command and took a medical leave to return to the US. He was replaced by Col. William Hamby, a Philippine Scout from the 26th Cavalry, who went around in a jeep called 'Tatay.'

Several hundred of the men re-enlisted in the American or Philippine Army and elected to stay in the Philippines. They were assigned to train and lead units in the Philippine Army. Many from the Regiment joined the new Philippine Scouts (PS) units. Pedro FlorCruz imparted his skills as Executive Officer of the 3rd Battalion of the 57th Infantry (PS) and headed the Regimental S-3 Training in the South Pacific theatre under Admiral Halsey. The Huks turned against the Philippine government after the war and Atanacio Chavez led the 20th Battalion Combat Team to subdue them.

Eugene Balanza with Company A, 1st Battalion was one who did not return to California to be discharged as many did. He remained in the Philippines with the U.S. Army and was discharged from the Regiment in January 1946 at Base K in Leyte. The following day he reenlisted and went on to serve worldwide until his retirement.

Richard T. Arca was assigned to the 12th Division, Philippine Scouts (PS), the Eighth Army Stockade in Japan, duties in U.S. camps, and in Korea.

Men who got married and had not taken the oath of citizenship did so before discharging, afraid that they might not be able to bring their wives to the U.S. Col. Hamby cut the red tape for the men to marry allowing the marriage ceremonies in camp after personal interviews. A tent city was built in Tacloban for the married men and their wives. The men warmly called it the 4th Battalion.

Those who did not have enough service points for discharge or chose to remain in the Philippines were attached to the Filipino section of the 86th Infantry Division. Patrol and guard duties brought them throughout Luzon.

Men were interviewed about their experience. They related that physical endurance was of utmost importance in the encounters with the enemy and hunting stragglers. Jungles were thick and mountain trails rugged and included having to climb cliffs to the caves. Shooting back at a sniper's gun flash might be too late if a soldier had been hit. Being alert to surroundings and conscious of snipers camouflaged in trees and caves saved lives. Small fighting units with flame throwers effectively destroyed and sealed large caves to prevent reoccupation. Additional men were needed to carry food, water, and ammo into enemy territory to free the soldiers for combat. Many of their experiences were incorporated in the training for soldiers in the later wars, Korea and Vietnam, the U.S. was involved in.

Hawaii Boys

The Hawaii Boys were young men born and raised in Hawaii. Their parents migrated to the Hawaii as contract labor for the sugar cane plantations in the early 1900s. They were unfamiliar with Filipino culture and customs except for what they had learned from their parents. They were not prevented from joining as they did not work in the pineapple and sugar fields of Hawaii. In their early twenties, most men of the Regiment were 15 years older than they.

As Japanese bombers attacked Pearl Harbor, they were the first to see Ford Island with the target ships obliterated. In the chaos that followed, many volunteered to drive ambulances and dig foxholes.

It was in late 1943 when they were able to join and underwent 17 weeks of Basic Training at Camp Roberts in California, a Replacement Center to train men to send to all Army units. "There were Latin Americans, Filipinos, Hawaiians, Asians, and Whites but no Blacks," Joe San Felipe noticed. Miguel Taoy was shipped to the Japanese 442nd as the white officers could not tell the difference. But some of the Japanese boys from Hawaii recognized him and told him to go with the California boys.

At camp, they did not recall experiencing blatant discrimination because it was not common in the Hawaii plantations. Unlike in the California labor fields with many competitions from other ethnic groups including white workers that caused much trouble and discrimination, the sugarcane and pineapple fields did not experience such things.

They were young and aggressive, ready for a fight. The boys spoke broken English and thought of the men from California as "more white." Despite the age difference and backgrounds, they got along and called many of the older soldiers "Pops" or "Grandpa." They showed respect for the men from California, whom they considered more mature and "wiser than us anyway. You learn a lot from them. Most of them immigrants. Younger boys are always drinking, too wild, " related Domingo Los Banos.

Occasionally, fights broke out between the 2nd Filipino Battalion men, the white boys, and Hawaii boys in the beer hall. When they received orders to ship out, the White soldiers asked where they were going. "We knew we were going to the Filipino Regiment. The Whites would say 'Oh yeah, you're all buddies of MacArthur.' You got the inside track," related Joe San Felipe.

When the first 50 boys had their training cut to thirteen weeks, they thought they would all be sent to Europe. Domingo Los Baños was training at Camp Roberts when it was cut by four weeks when the Germans launched a major counteroffensive in December 1944 at the Battle of the Bulge. It was the last counteroffensive of the Germans and caught the Allies by surprise. They were issued winter clothing, and many were shipped to Europe and killed. The unit went down to 50 men continuing to Fort Ord for jungle training. Their winter uniform was replaced with clothing for the Pacific tropics.

The boys joined the 2nd Filipino Battalion when the unit shipped to New Guinea. From Hollandia to Leyte, 300 Hawaii boys departed with the Battalion. The first batch of 50 boys were put in a Dutch ship as part of a convoy recalled Angel Ibanez. The ship broke down and was separated from the convoy. For almost a month, they stayed onboard the ship eating sandwiches for meals until they joined a convoy of two hundred ships zigzagging to the Philippines. Two destroyers on either side escorted the USS *Butner* that carried troops and nurses to reinforce the medical companies in the islands. Meals were served twice a day from two different kitchens. To keep in shape, men rotated with other units for an hour of exercise and running on the top deck. The nurses gravitated to them as they played their ukuleles. The officers wondered who they were because they looked different and did not know that Filipinos were on board. The ship travelled slowly in the daytime to avoid smoke coming out of the exhaust that would alert Japanese submarines patrolling the ocean. While they were on their way to the Philippines, the Japanese emperor announced the surrender.

As their convoy approached White Beach in Leyte, they could hear shootings in the mountains. Mendiola heard a call for Catholics and a priest on deck for confession and communion. A landing barge came beside the ship with a cargo net they climbed down on. They saw only stumps of coconut trees left onshore after the naval bombardment. They hauled cargo into the jungles to hide as emaciated people barely covered with torn clothes rushed to them begging for food, medicine, blankets, or anything to cover themselves.

At a replacement depot in Leyte, they requested to join the Filipino Regiment which had arrived earlier. They saw the *Manongs* of the 1st Regiment who were mobilizing between assignments in Samar and Leyte, looking grim and battered from combat. Thomas Otogero who was on his way to Dulag described torn buildings and dead bodies everywhere. He was coughing hard and could not breathe properly from the heavy smoke of firepower. A medic had to clean his nostrils. In the commotion of military activities, morning reports had to be published to keep track of the Regiment and Angel Ibanez was chosen to publish aside from guard duty with the quartermaster supplies and POWs.

Manuel Lagrimas was sent to the Philippines with Los Baños. At first, he thought he was placed with the Philippine Army when he arrived at the Filipino Infantry camp. It was only then that he knew of the outfit and saw the men much older than he. They were barely out of their teens and the *Manongs* treated them that way. "It was a nice relationship," he said.

Filipinos did not forget to celebrate their festivals with whatever they had. It was a semblance of peace time before the war and for a few hours made them forget of their suffering. Lagrimas saw a short procession for Flores de Mayo that made him understand where some of the customs he grew up with came from. Many of the Hawaii boys like him were establishing their roots. When he threw soap at the young girls who smelled it as if it was heavenly, he learned to appreciate basic common items that he took for granted in Hawaii. The soap was also good for bargaining when he had his clothes washed by the women who followed the men's camps. Many became girlfriends of the older men in the Regiment with the approval of the commanding officer who saw it as wholesome. There were no prostitutes.

Rice was not served in the Mess Hall and Lagrimas missed it. Seeing Philippine Army soldiers across the street eating rice, he snuck under the barbed wire fence to join.

More groups of Hawaii boys who finished Basic Training at Camp Roberts joined convoys directly to the Philippines. It was not clear to them where they were going as they arrived at night in Manila. They marched to report to their location past destroyed buildings. They saw flickering of fire amongst the rubble and heard ricocheting bullets whizzing by. The stench of death hung in the air. At first, they were brought to the Philippine Army camp by mistake. Assigned first to KP duty serving thousands of men at chow line, they ate pieces of meat as they served causing double time for the cooks. That quickly ended their KP duty.

The Hawaii boys were attached to the Eighth Army. The largest contingent was sent to Calbayog and Catarman in Northern Samar to reinforce the 1st Regiment who were losing men. The Japanese were in the mountains of Samar and the men pursued to hunt them down. The fighting was vicious as the Japanese refused to surrender. Before the war started, even though they were Filipinos, they did not know anything about the Philippines except what people told them. Landing in Samar, they remembered the bygone stories from their parents but sad to know that they had to fight a war.

Some of the boys were uncomfortable with jungle warfare despite the training they underwent. The battle-scarred *Manongs* would take over when snipers pinned them down. Hitting the ground, they fired at will using up all their ammunition.

The first Japanese Miguel Taoy saw was a prisoner of war he had to guard. It took three guards and four hours to drive a prisoner back for interrogation. He shot quite a few rather than bringing them back alive. "There were 12 men in a squad and if it took three men to guard one prisoner, they would lose firepower," he reasoned. With a Japanese interpreter holding a megaphone, he removed Japanese from caves, brought them to the beach, and told the men to drop them in the middle of the ocean. Easier that way, he reasoned. He reported that he was the number one scout and when number 2 scout was about to get hit, he turned around and shot the Japanese. From that time on, he did more shooting as number one. Simeon Amor was deeply wary of being shot by Japanese in their foxholes as he tried to convince them that the war was over to give them a chance to surrender.

Domingo Los Baños had close friends in California who were Japanese, so when he saw the enemy, he hesitated. But when he pulled the trigger and saw "blood coming out of the mouth with each breath," he puked. His sergeant told him "to blow his brains out, and I disobeyed the order because I could not kill him.... told the officer that there is a Geneva Convention law not to do this to POW... thought they were going to court-martial me... felt good because I let him die in peace." But later, he vividly remembered when he was on patrol and surprised by an enemy soldier, he shot the soldier first. With only six men in his group, they threw grenades while camped overnight on a ridge. He cut off a cloth bag from around the neck of a Japanese he shot and found a lock of the man's wife's hair with her photo. As an 18-year-old, the inhumanity of war hit him. From the ridge, his unit followed the enemy down to the valley where he found a Japanese in a hut fanning himself. Coaxing him to give up, Los Baños shot him when he reached for a grenade.

While on assignment to deliver messages to Capul Island, Los Baños recounted a soldier named Velez who fell in love with a girl and neglected his rifle that he prided as always clean. Upon returning, he showed his ruined rifle blaming the girl for his neglect.

Recalled to Dulag, he saw the *Manongs,* many of them with different patches, Scouts, recon teams, CIC and PCAUs. That was when he realized the depth of what the men from the Regiments did in the Philippines years before him. "The Hawaii boys had no idea that they *(Manongs)* preceded General MacArthur in the Philippines, collecting data, and watching the enemy and ship movements."

Pfc Alex Aguinid related that the 1st Regiment had already departed for New Guinea when he was drafted and training at Camp Roberts with white American units. He proudly earned his combat infantry badge. From San Francisco he boarded the troop ship USNS *Buckner,* a converted liner. There were ten thousand troops and five hundred nurses in the officers' quarters on the upper deck off-limits to enlisted men. He "did not know who they were ... American Indians, whites, lots..." All were assigned to different compartments. All were young men of his age. The ship stopped at Guadalcanal and Hollandia to drop off mail for the GIs before heading to the Philippines. They knew that they were replacement troops.

The Regiment was fighting on Samar when Aguinid arrived on a landing barge at Catarman. With his 3rd Battalion company, they "liberated the northern part and midway towards the south where they joined the 2nd Battalion." They tried to stay dry as it rained almost every night while they were sleeping on the ground. San Felipe was also assigned to the 3rd Battalion then to the Quartermaster because he could speak English with a San Francisco accent that the white soldiers understood. There were problems understanding each other between the Philippine Scouts, Philippine Army, and American Army.

They carried sixty-pound backpacks while fighting in the jungles of Samar. Men took care of each other. Rank did not mean anything. As a group of forty men, they "ate C-rations sometimes while squatting depending on what we did.... Atabrine daily for breakfast to keep from getting infected ... got Dengue fever and walked deliriously with high fever for two weeks ... will yourself to walk, you walk." The *Manongs* taught them what edible plants to eat in the jungles. When Aguinid was bitten by a snake, he became delirious. "They tied a telephone wire from me to the guy in front and the guy in back so I can march with them." He experienced attacks while back at camp and was sent to the hospital in Tacloban. He was given the last rites twice.

Aguinid felt lucky that he got Company M, a Heavy Weapons Co, and not a rifle company. He recounted: "I had to carry the pieces of heavy .50 cal machine guns and 81 mm mortars, assemble and position them. Then fire as well as fight the enemy with our weapons like Carbines and M1. That was my first experience in combat. We used mortars when we knew where the enemy camp was before going to attack. We penetrate with mortar shells and then machine guns with the First and Second platoons. They were positioned to protect us as we went into the camp in a firefight and hand-to-hand combat. We were up against seasoned soldiers. We mopped up Samar from the north to over two-thirds of the island. So many combat scrimmages that sometimes we boarded landing crafts to go to different parts of the island. I was involved in around two dozen combats. Lots of battles."

In one of his last combat missions, Aguinid's unit was in the jungles for two weeks as each man carried only two weeks' worth of ration. Beyond two weeks, they had to live off the land. The carriers knew what to eat in the jungles. "We boiled camote in our helmets." In combat operations, there were forty to sixty men plus forty civilians to carry supplies. All were hungry. A grenade blasted fish to the surface, feeding the hungry men.

A report mentioned that some of the Hawaii boys with the 2nd Filipino Battalion were with the liberation forces in Mindanao, the last island to be liberated. When they returned to Manila, the first atomic bomb had dropped on Hiroshima. They could not believe that the war was over, and their lives spared.

They wanted to look smart and trim in their uniforms when they met relatives and visited women they were attracted to. But most of the time their uniform fatigues sewn for white American soldiers were oversized for them. The shirts and pants were too long and too big. They went to a seamstress to fit their uniforms.

Although Hawaii and America were the Hawaii boys' homeland, they were curious about their Filipino heritage. Many were eager to look for their relatives. When his unit was recalled to Leyte, Aguinid took the offer to go on leave. Los Baños found his father's large family.

Aguinid wanted to experience life in the Philippines and recounted that he travelled with people, pigs, and chickens crossing Leyte Island to Cebu. Told to protect themselves while on leave, he checked out a pistol with two magazines. He found "the peaceful beauty of the islands to drown out the horrible memories of Filipina women having their breast cut off and used for bayonet practice." He went overland to the southern tip of Cebu, where he took a sailboat to the small island of Siquijor to visit his parents and relatives. He was so thrilled with the rousing welcome that he stayed for thirty days. An aunt said a special prayer while killing a pig because "I had come from war and my hands were bloody with the killing. It worked because I have had the urge to kill people, but I keep myself from doing it," he joked. All the relatives came to her house to celebrate. It was a wonderful time in my life to see my cousins ... had to be careful because some of my cousins were good-looking girls. Auntie said to be careful because they are your cousins." He returned to Dulag, happy to have known more of the islands' culture.

In Manila, the fuel tanks depots in Lamao, Bataan that had fuel going to Clark Field in Angeles, Pampanga was guarded for six months by Aguinid's group of ten men. A four-mile perimeter was secured around the pipeline from the *Huks* and civilians tapping the gas. Another important site was the Insular Ice and Cold Storage Plant in Manila that Angel Ibanez's unit guarded.

The unit of 500 Hawaii boys stayed on as occupational troops in Manila. The majority did not have enough points to go home. They were sent to nearby Marikina and formed into the 1st Filipino detachment of the 86th Infantry Division (Blackhawk) which was a White unit. They relished the familiar food served, Vienna sausage and rice for most meals with onions and tomatoes. The White soldiers complained and finally were given a separate mess hall.

When there was no longer the Japanese Imperial Army to fight, the Hawaii boys began to clash with the Philippine Army soldiers. They were young, physically trim, and had more money. The pay for a Philippine Army lieutenant was the same as that of a private in the U.S. Army, and thus they could afford to take out the local girls. The attractive young women gravitated towards them. Los Baños said that although they could not speak Tagalog, they still got all the girls. Jealousy and gun battles ensued. "We were cocky. That is when we knew that the war was over because now, we were fighting Philippine Army soldiers." Some of the clashes were from the boys' lack of knowledge of Filipino customs and traditions. They had a crash course on 'being Filipino.' It was their chance to learn more and experience the traditions and practices they were familiar with in Hawaii. They could not speak the language but knew the latest American songs. Los Baños said, "We would walk down the street, looking into windows, hoping that somebody would invite us in."

Aguinid was beginning to enjoy the Philippines and did not want to return, he confessed. He enjoyed the month-long guard duty for a WAC barrack in Manila while waiting for a troop ship. Once he was onboard the USNS *Aiken,* just outside the islands, a typhoon hit making the ship creak loudly every time enormous waves slapped its sides. He thought he would not make it. After two years in the Army, he was discharged at Camp Beale and returned home to Hawaii.

American soldiers get acquainted with local belles in Malasiqui, Pangasinan. 17 Jan 1945. SWPA- SigC-45-518 Filipinas in their native dress mingling with the Filipino and American soldiers.

On July 4, 1946 while on leave, Los Baños stood in uniform amidst the huge crowd that gathered to witness the official ceremony ushering in the newly-proclaimed Republic of the Philippines. With a camera he carried with him throughout the war, he got close to the grandstand and snapped photos of General MacArthur and the newly-elected Philippine President Manual Roxas. What thoughts went through him the moment he saw the U.S. flag go down and the new Republic's flag go up?

1946

In Tacloban, the *Canao News,* a weekly newspaper of the 1st Filipino Infantry continued to conduct a lively recruiting campaign to get every man to re-enlist in the U.S. Army. "As of February, six hundred have re-enlisted at the rate of one hundred a week. Another two hundred are in limbo because they are not citizens and have no intermediate opportunity of becoming naturalized. The men hoped that the outfit would be incorporated in the regular Army instead of deactivated." A representative from Immigration and Naturalization Service (INS) arrived and accepted around 500 applications for citizenship.

The Regiment was ordered to move across to Dalupiri Island to occupy a former Naval installation with two-story wooden buildings. The tent city for married couples in Leyte was closed and another was built on the island.

Leyte Island. The staff of the Canao News, issued by the First Filipino Infantry regiment.

Left to right standing: Col. William Hamby, Lt I. M. Sarmiento, adviser. Kneeling left to right: T/4 Peter A. Aduja, associate editor; T/4 Benjamin Menor, editor; Cpl Benny Sacramento, news reporter; and T/5 William Pasco, production.

The orders arrived for the deactivation of the 1st Regiment. Most asked for discharge in the U.S. and were shipped back. In the early morning of March 23, a large barge embarked the men and brought them to the USAT *General Callan* anchored three miles away from shore. Friends, comrades, and civilians lined the shore with sobbing wives who would return to their families to wait for their husbands' return. Knowing that music always soothed the Filipino, members of the Regimental band found or borrowed musical instruments and played what they could on the open deck of the ship. At sea, the trip was marred with an explosion of an undersea volcano that created destructive waves and battered the ship losing much of the cargo.

Arriving in the early morning off San Francisco under the Golden Gate Bridge, Captain Clay E. Pardo recalled the welcome of fire craft spouting streams of water and dancing girls. *Powerful tugboats maneuvered the large ship beside Pier 2. Soldiers three to four deep on the side railings peered out to find a familiar face in the waving crowd. Hundreds of friends and relatives were waiting for their ship to dock,* was written on the back of a Signal Corps photo. A building with white stars in the front and large letters 'ARMY' had the American flag flying on the roof's peak welcoming the men. *The New York Times* reported that only 555 men arrived, and a Capt. Gene Frias, one of the officers, "listed the total tally of dead Japanese cut down by the Regiment as one thousand seven hundred sixty-eight. The Regiment's fatalities were relatively low." Army buses brought them to Camp Stoneman near Pittsburg, California to disembark for discharge. Others like Capt. Pardo were discharged at Camp Beale.

It had been eight months since the surrender of Japan and four years after the activation of the Filipino Battalion (Special) in Camp San Luis Obispo. A cloudy day blurred a photo taken of the men at Camp Stoneman in front of the wooden chapel. A sign on top of the front door read 'Separation Ceremonies'. They saluted their Regimental colors before encasing them and starting the deactivation process. The Regiment's flag sewn by the wives of Filipino officers was folded for the last time. They wore their green uniforms, overseas caps, with big smiles, waving high their separation papers. It was a quiet affair for the men knowing that "they have earned a place in America in their will to fight." Some silently anticipated their brides that would follow to start a family, an education from the GI Bill, and hopefully new professions they could attain as America's Filipino soldiers. Some went back to the familiar labor fields.

For many, it was the end of their military careers. Antonio Campos remembered that "when he returned home from the war, it was a foggy day and he felt great. There was a band and a 'Welcome Home' banner, and of course, crowds of loved ones."

Later, a member of the unit secretly took the flag from the mailroom before it was shipped to the Army archives. It was sewn by the men's wives, so it was theirs. The flag continues to be flown at unit reunions and veterans' events throughout the U.S. thanks to Pelagio Valdez's promise to his father, Sgt. Pablo S. Valdez to keep the memory of the Regiments alive.

In August 1952, the U.S. Army unit was officially disbanded per a memo signed by Lt Col. George Kelly, Chief, Heraldic.

PHILIPPINE MAP

LUZON

CAMP CAMARILLA
MANILA

MINDORO

BURIAS ISLAND

TICAO ISLAND

SAN BERNARDINO STRAIT

BIRI ISLAND

CAPUL ISLAND

ALLEN • MONDRAGON • CATUBIG
• CATARMAN

DALUPIRI ISLAND

CALBAYOG •

SAMAR

CATBALOGAN

MAUD •
JINAGNATAN TACLOBAN •
• VILLABA
• PALOMPON
• ORMOC

GUIUAN ISLAND •

CEBU

• DULAG

CAMOTES ISLAND

LEYTE

• DINAGAT ISLAND

PANAON ISLAND •

SURIGAO STRAIT

LUZON

• MANILA

VISAYAS

MINDANAO

1st & 2nd FILIPINO INFANTRY REGIMENTS

MISSION: February 1945 - November 1945

A Filipino Battalion
2nd Filipino Infantry Regiment

Fort Ord, California USA

The overflow of volunteers at Fort Ord was moved from the Main Garrison to the East Garrison. When their number equaled that of a regiment, they were activated as the 2nd Filipino Infantry Regiment as of November 22, 1942. *Sulung* (Charge) was formally adopted as its slogan. Recuperated men from the hospital ship *Mactan* and survivors of the sunken blockade ships joined the Regiment. Maj. Charles L. Clifford was assigned as its commanding officer.

As more men joined, there was a need for Quartermaster personnel that would be responsible for quarters, rations, clothing, and supplies. Men were sent to the Quartermaster Replacement Training Center at Camp Lee, Virginia. An obstacle course at Camp Lee had soldiers going through metal cylinders covered with branches as part of the course located in a wooded area. On one of the trees, a white sign was posted with large letters 'Remember Bataan.'

Lt. Luna joined the Regiment at Fort Ord and underwent Basic Training. Trained as a rifleman, he was assigned to the Motor Pool (Transportation) and was sent to OCS for further training in the storage, servicing, and maintenance of military vehicles.

Ramon Vitorio arrived in a convoy of Army trucks. He was relieved at the site of the new quarters since in Salinas they were crowded in "one small hut built for five

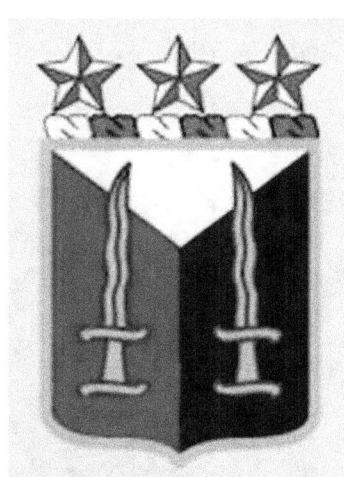

The 2nd Filipino Infantry Regiment insignia as drawn by Ted Visaya.

The Shield was the same as that of the 1st Filipino Infantry Regiment except that there is a kris in each of the red and blue panels. A charging carabao is reported to be on the white field but it is not shown here.

but then occupied by six men. Ft. Ord had bigger barracks, and I liked it because we are nearer to San Francisco. Our barracks here are better equipped, air-conditioned, showers, and a comfortable bed all in one roof. The streets around the camp are all paved. We have Service Clubs, theatres, Army stores, a post office, and recreation halls where you can enjoy and relax after a long day's march. Every other day we have fog around the camp thick as a pea soup in the early morning hours. I am planning to have a 36-hr pass starting at six o'clock Saturday afternoon. If I can get it, I'll probably be in San Francisco before midnight Saturday. It is awfully nice to see San Francisco after two months' absence."

Carlos Arguelles, who had graduated with an architecture degree in the U.S. and had been caught by the war recalled "that either you were "born American or an old-timer..." when he joined in the winter of 1943. The camp was filled with almost all Filipino soldiers. Most of the enlisted men were *Manongs* who left the fields to be in battle. Philippine President Quezon wanted a regiment, and eventually, two were formed from their enthusiasm, but there were not enough qualified Filipino officers. Arguelles became commanding officer of the 1st Battalion, Headquarters Co. then Company C.

There were times when the Filipino psychology confused the White officers. When the men were told to dig foxholes on the firing line, they wondered if the order was a mistake. They loudly informed the officers that digging a foxhole to hide in is a sign of cowardice.

A Cannon Company specific to certain types of targets was added to the Regiment for additional fire power. They trained as a unit morning and night in full gear, wearing warm jackets and gloves. They had their photo taken with a white officer in front. Andres Soria, an anti-tank crewman, staffed an anti-tank company with soldiers and conducted night and live-fire field training. The Howitzer – a cannon mounted on two wheels -- had a powerful recoil that required several men to sit on top of it. Sgt. Mamerto Ilumin was listed with the Cannon Company and may have taken a turn sitting on the cannon. The diameter of the gun barrel was almost the same width as the men's helmet. The shells were over half their height. On the side of the cannon, they stenciled the inscription *Inang Bayan* (Motherland).

Cannon Company, 2nd Filipino Infantry Regiment. 1943. The cannon has 'Inang Bayan' stenciled on it.

Simulated enemy fire was part of the training. In complicated river-crossing operations, their rubber rafts and assault boats were rocked by actual dynamite explosions in the water on all sides, winning high praise from observing officers for their efficiency. The *Bealiner* reported, "Advancing in three waves from the muddy bog on their side of the river, lit by enemy flares in the early dawn, the gallant *Pinoy* (slang for Filipino) ... made the 300-yard crossing in the space of a barrage of gunfire from the opposite bank. Only one raft capsized in the crossing, and all men were saved when amphibious jeeps dashed to the rescue. Working as a tactical unit with perfect coordination, the men with engineer troops made a landing and went on to overcome the enemy."

Camp Roberts and Camp Cooke, California USA

A belated Christmas present arrived in the form of a high commendation by the Major General of the II Armored Corps "for outstanding devotion to duty and fine conduct both in garrison and in the surrounding communities." The commendation stressed that "they made a most enviable record while at Fort Ord."

After the 1st Regiment left, the 2nd moved to Camp Roberts and underwent the same air-ground tactical maneuvers to complete their field training. Quarters were also in one or two-person pup tents in long rows. They laid down half inside at night because of the heat. But in the morning, it was full uniform with rifles and helmets for Saturday morning inspections. During patrols and marches, they had to be careful not to step on snakes that abounded in the dry terrain.

2nd Filipino Infantry Regiment Band marched through nearby Bradley, California

Camp Roberts, California. Attending Mass outdoors. 1943.

Since most of them were Catholics, the men looked forward to attending Mass on Sundays. When they were at San Luis Obispo, they "could not completely worship in the century-old mission although they volunteered to defend America." At Camp Roberts, they regularly attended Mass outside Mission San Miguel close to the camp. If there were seats left in the church after the local families entered, the officers and their families followed and finally the enlisted men. With not enough seats, they knelt in front of a makeshift altar under the shade of the trees in the summer heat.

2nd Filipino Infantry Regiment Headquarters, Camp Cooke. Large maguey plants can be seen in the foreground

Volunteers continued to arrive at Camp Cooke. Sgt. Pablo Valdez, an early draftee, was assigned to supervise the new arrivals. The camp was a dry and hot place with palm and cactus plants as tall as the men. Barracks were temporary single- story buildings laid out in blocks to accommodate between one or two battalions. Exterior walls were made of plywood and the ends had vents to cool the attics. The buildings were gradually painted white during the period the Regiment was there. Donated flowering plants from the nearby flower fields of Lompoc gave color and diffused the two sharp-tipped Maguey plants in front of the company room. Sgt. Richard Tuzon, an avid gardener, and a veteran of World War I, made sure that the flowering plants were watered daily. Men of the Headquarters Co. of the 2nd Battalion took turns caring for the garden planted around their tents.

Camp Cooke: 2nd Filipino Infantry Regiment, 3rd Battalion Regimental Bolo Parade. 1943

When 19-year-old Antonio Dixon Campos answered the question of racial descent that his mother was Scottish and his father Filipino, the draft board did not know which unit to send him. Only pure Filipinos went to the Filipino Infantry. Mestizos could choose to go to the Filipino unit or a regular white unit, not a Black or a Japanese unit. So, off to Camp Cooke he went for Basic Training with the 2nd Filipino Infantry Regiment.

News of the Filipino unit undergoing combat training to return to the islands reached the upper echelon of the Army. A special visit by the Army Chief of Staff, Gen. George C Marshall was greeted with an impressive regimental parade and banquet.

The regiment's publicity continued with the public showing of films about them. *On to Bataan* featured men of the 2nd Regiment training and practicing with their *'bolos.'* The Anti-Tank Company was featured. Alex Fabros, the Public Information Officer, coordinated the documentary and enjoyed being in a few of the scenes.

A film crew from Warner Brothers Studios arrived to film a movie short, *Filipino Sports Parade,* depicting Filipino cultural activities, including games, dances, and favorite dishes. Company B willingly obliged to provide a *lechon* party as the centerpiece.

Under the direction of Howard Hill, a world champion archer and sports buff, shots were taken of the unit's activities like wrestling and fencing with *bolos* while those who were more athletic were filmed playing volleyball, *sipa,* pingpong, and badminton. More than thirty soldiers led by Fabros and six Filipino women wearing Filipino dresses led by Pacita Todtod, participated in the movie. Todtod was an activist who tenaciously campaigned to allow Filipino men to serve in the U.S. Armed Forces. She was popular with the men as she regularly coordinated with the camp's entertainment committee to bring shows and talents to the soldiers. The mess halls held the dances and parties. The film was released just in time for the men to watch before their departure for New Guinea.

One of the Filipinas, Helen Ragsac Sanchez reminisced: "At Camp Cook we had to go through a lot of rehearsals of the *Karinyosa* (a Filipino folk dance) with Filipino soldiers as

1942
"BOLO"
KNIFE
FIGHTING
DEMO IN
California

Cpl. Fabros made sure that the film showed a scene of two men dressed in khaki uniforms showing their adeptness in using their bolos in front of nervously smiling Filipinas.

our dance partners. The soldiers selected to be in the movie were among the first wave of Filipino immigrants *(Manongs)* who settled in California in the 1920s and '30s. It took several days with many takes. After Camp Cooke, (we) went to Camp Roberts for more filming. I

Camp Cooke, June 1943

remember Howard Hill asked the soldiers to play the Filipino game called *sipa*. As I recall, the soldiers were in two lines facing each other, so the ball made of bamboo (woven bamboo strips) was kicked back and forth. The roasting of the pig was also included in the movie."

Some men met their future wives and went on three-day passes to marry Filipinas in the Regimental chapel officiated by Chaplain Torralba. If the bride was non-Filipino, they travelled to a state that did not have the prohibitive law. Others went on leave to see a new baby. To visit his family, Sgt. Mariano Navarro from Headquarters Co. hiked the two- hour roundtrip to nearby Lompoc. He justified that it was good for his health.

The Army Recreation Center at nearby Pismo Beach provided a convoy of trucks loaded with men and their cooking utensils for weekends by the sea. Stripping down to white undershirts, they ate fresh clams and abalone.

Fabros reported the "Satisfactory" rating the men received during the three-day Air-Ground Tests conducted by the Training Test Section of the 4th Army. Noted were the high morale, excellent blackout discipline, and physical condition of the boys. Celebrating their score, the men invited fellow soldiers at Camp Beale and from the Fourth Army to a dinner serenaded by the *Sulung* Band.

The *Sulung* band's popularity got them invited to play at a newly opened Filipino USO Club in downtown San Luis Obispo where the U.S. and Philippine flags were hoisted to the top of the courthouse across the street. Jose Borromeo, a pianist who had conducted orchestras over radio stations in Chicago and New York, gave frequent concerts at the various Service Clubs.

The largest formation of Filipino troops carrying out day and night maneuvers with combat firings were conducted at Camp Cooke. The parade ground was dry, and the wind was strong.

2nd Filipino Infantry Band Sulung (Charge)

Soldiers in full gear with rifles travelled in jeeps with a .50 cal machine gun. Opposing sides conducted reconnaissance and war games, erecting fortifications that included minefields, barbed wire fences, and gun emplacements. Exercises were in attacks on fortified areas combined with frontal assaults. Sweating in helmets, field jackets, and heavy ammunition belts, they kicked up dust drilling on the parched terrain.

In the classroom, the 3rd Battalion was the first group to learn the fundamentals of military teaching and proper instructions. They learned to speak, demonstrate, and explain various topics to their fellow soldiers. The course was helpful to the many men from the labor camps and non-professional fields who had difficulty with English. Instructions were on using weapons, personal survival, personal confidence, and cohesiveness as a fighting unit.

"The day rooms for relaxing, listening to music, writing letters to loved ones were infiltrated with gambling of all kinds," wrote Gregorio Chua. After duties were done, the Filipinos taught their American officers the languages of their hometown. The cafeteria served ten-cent cheeseburgers. "We are in the Army now, and you are alive this minute, but you may be dead soon," the men would say.

Around the Sta Maria mountains, company units went on tactical field marches wearing protective gas masks. Trails passed deep ravines surrounded by thick trees. One man with a sense of humor wrote "15 days furlough. 1-2-3 go" knowing that they had to live off the land for fifteen days. But throughout the training exercises, foremost on their minds was their hometown and whether their families were still alive. They painted a name of a hometown province on each of the ten trucks in the motor pool. A photo captioned: *Ilocos Norte (a northern province in the Philippines) went to the city of Santa Maria carrying a crack rifle squad from the 2nd Regiment's 1st Battalion.*

Company 'I'

Second Filipino Infantry

Camp Cooke, California

Saint Patrick's Day

Dinner and Dance

1 9 4 3

For Saint Patrick's Day, Company I, 2nd Filipino Infantry celebrated with a Dinner and Dance at Camp Cooke

The *Camp Cooke Clarion* newspaper published news on social and religious activities, war news, sports, and special events that maintained the men's high morale. Visiting guests and enlisted men were featured. An article on the awarding of thirteen soldiers for outstanding scores named Sgt. Maximo Cacas and Pvt. Benjamin Marcelo proudly receiving their awards. With the regimental band playing and the colors flying, the 2nd Regiment's commanding officer and staff conducted a military review. An article also featured Laureto Lagpacan as the secretary at the Unit Personnel Office. To prepare for his future after the war, Cpl. Conrado Fores of Company D took correspondence courses from the U.S. Army Armed Forces Institute. Winners of weekly quiz tournaments participated in by enlisted men from three departments on general information about the Philippines and international topics were published. The recruitment of Julius Ruiz, former editor and publisher of the *Philippine Newletter* in Washington, D.C. into the 2nd Regiment was announced.

Sports were also encouraged. Cpl. Fabros wrote about Sgt. Regala of E Company on the winning softball league team and Lt. Daniel D. Dizon as the youngest officer to enter the Golden Gloves Tournament held at the Hollywood Legion Stadium.

Pvt. Pagsilang Rey Isip served as an artist in the Office of the Special Services creating patriotic and instructional posters. He designed the masthead of the *Camp Cooke Clarion* depicting a grinning Allied soldier charging with a bayonet fixed on his rifle at Adolf Hitler, Benito Mussolini, and Hideki Tojo. When a piano with eighty rolls of modern swing and classical music was donated, his hidden talent surfaced banging the keys almost every night to the delight of his comrades.

A stirring poster, *The Fighting Filipinos,* portraying the fighting spirit of the Filipino people in the cause of freedom brought fame to his brother, Manuel Rey Isip. He painted the original in February 1944 with Seaman Aurelio Palafox, U.S.M., a relative of his wife that drew much attention. The Office of Special Services of the Commonwealth of the Philippines distributed fifteen thousand copies reported the *Camp Cooke Clarion.* President Quezon expressed that it exemplified the "determination of the Filipino soldiers in the field and the civilians behind the lines to fight the invader until death and expel him from the Philippines."

A Camel Cigarette Caravan arrived to promote their product to the men. At the end of the show, shapely young 'Camelettes' dressed like drum majorettes with short skirts and white boots walked through the audience of hundreds of men distributing free cigarettes that were always welcomed.

Actress Carol Landis returned to Camp Cooke as she was their sponsor for the War Bond drive. She cut a large cake that can feed all the men crowded around the table.

The icing had 'Welcome all visitors. The Second Filipino Regiment Camp Cooke.' The men mingled around her in their crisp dress uniforms with broad smiles as the *Sulung* Band with their music sheet holders was seen playing in the background.

Daily training was physically exhausting. To liven up the men after a gruelling day, Col. Clifford, held a contest for the best-shined shoes with a prize. The number of 'dawgs' – the name given to the men who neglected to shine their shoes, hang their clothes properly, keep their bedding well-fixed, clean their guns and keep their barracks intact, became scarce as competition heated up.

The USO Service Clubs kept morale high by providing a friendly and surrogate home to the soldiers. They could relax, read, and write letters. The *Sulung* band regularly played when camp shows were held. It was open to all races, but if the local white communities preferred segregation,

"The Fighting Filipinos – We will always fight for freedom."

War Bond Drive. Carol Landis, unit sponsor, at Camp Cooke

they followed to avoid conflicts even though it prevented the men from entering some of the clubs.

"For most Filipinos, especially those from California, being denied service outside ethnic communities was not a new experience," recounted Pvt. Buaken. Before the war, he and his wife lived on his "houseboy meager wages because all skilled trades, professions, and industries have been closed to Filipinos." His wife wrote of having "lived in slums, among prostitute and drunks and assorted derelicts, for only in slums have Filipinos been allowed to rent living quarters... members of my husband's race have not been allowed to own real property, so there was no escape from the slums for us." But Pvt. Buaken proudly pointed out that "we were in the Army now and the Army will go to bat for us."

Filipino newspapers continued to follow the men's' progress. *The Philippine Mail* announced the arrival of Lts. Richard Arca and Jose Mendoza from OCS to lead a company. Arca, a Philippine Scout before the war, joined the 2nd Filipino Regiment at Fort Ord and was a platoon leader. Four medics arrived from medical school. Sgt. Lorenzo U. Pimentel finished training in radio operations. He was with an American tank destroyer outfit in Texas to study mechanized outfits when he read a circular that wanted all Filipinos in a regiment. Assigned to the 2nd Filipino Infantry Regiment, he said that he trained old guys.

Washington Sycip tested as overqualified for the Infantry. He was a student caught in the U.S. when Pearl Harbor was attacked. With grave concern about the fate of his family in Manila, he joined and finished Basic Training at Camp Cooke. The Army decided that he would be of much better use in intelligence work and thus he was sent to a special language-training program in Denver for six months "to live and breathe Japanese" which was then followed by a course at the Signal Corps Cryptographic School at Vint Hill Farms Station in Virginia. Throughout the war, he was at a cryptography base near Calcutta, India on the banks of the Hooghly River, where the British 14th Army had set up a codebreaking operation. He was the only Filipino in the group of thirty cryptographers under a (White) commander. Narrating his story during the war years later, he said that "as cryptographers, they were never allowed to be flown over enemy territory because if captured and forced to talk, the Japanese could switch codes, complicating and prolonging the war effort."

Gen. Valdes, Chief of Staff of the Philippine Army, arrived at Camp Cooke. His last visit was at Camp San Luis Obispo when the unit was still a fledgling battalion. Clifford walked beside him as he inspected the men proudly standing at attention in full uniform and rifles at their right side. The volcano patch was on their left shoulders. Valdes read a special message from the Philippine President after inspection.

When the 2nd Regiment invited President Quezon to attend a ceremony at camp, his precarious health did not allow him to travel. In his stead, Manuel Nieto, senior aide-de-camp to the President's War Cabinet in Exile, delivered his message at Camp Cooke: "I see in your abundant hopefulness which augurs well for you and for our brothers — a race of brave men — who are still fighting in our mountains." Nieto told of the men dodging Japanese shells and burning steel in Bataan and Corregidor.

The Regimental Colors with the Regimental Shield was brought forward and blessed by Chaplain Torralba. On behalf of the President, Nieto presented the Colors. The crack platoon went through intricate maneuvers that ended in forming a V for victory with the letter in Morse code. In front of onlookers on the arid field, with the *Sulung* Band playing, the men proudly passed in review, snapping eyes right with their vehicles displayed behind

Camp Cooke, California. Gen. Basilio Valdes, Chief of Staff of the Philippine Army arrived to inspect the men.

them. A strong wind went through the open field against the flags held tight. At the Officers' Club, guests and officers with their wives were entertained. A dance and entertainment program at the recreational hall were held for the men. A team played a softball tournament with the 76th Medical Battalion.

The Midwick Country Club in the city of Pasadena requested a parade on their expansive grounds. The *Sulung* Band led the parade with the Filipino National Reserves of California. Mrs. Pilar Hidalgo Lim was in the crowd. Even with her husband's dangerous plight troubling her mind, she gladly took charge of entertaining the men who would return to fight the Japanese invaders. Her husband, General Vicente Lim commanded the 41st Infantry Division, Philippine Scout of the U.S., the division that was valiantly fighting on Bataan.

Training continued, broken up by a mid-summer party and dance at the end of August aptly called 'Fun in Wartime' held by the Service Co. More than three hundred guests and visitors attended and savored the American and Filipino dishes served in the company's mess hall, reported Fabros. Soldiers showed their talents in solo presentations on the saxophone and piano. Comedy with a Filipino flair elicited laughter. A jitterbug contest was jammed with participants for the chance not only to display their dancing skills but also to win a cash prize. It was not difficult to hold large events when the Filipino

Filipino cooks, Camp Cooke 1943.

cooks enjoyed preparing the requested food and delicacies for the men which were different from the daily diet of the White units.

At Camp Cooke, the Regiment had their own cooks preparing their own Filipino dishes separate from the White dishes. Men called them 'wag-wags' because of the wagging of their heads in agreement when the men requested larger portions of food.

To get more support from the American public, Offley dreamed up a publicity stunt with the Army's approval. It was the presentation of a *bolo* to each soldier. He remembered from his father's collection when he was military governor of Mindoro that Filipinos historically used their bolos for everyday activities such as clearing vegetation and protection. The *bolos* were not standard Army equipment, but the Regiments were authorized by the War Department to carry them. They were shorter than the machete, with slight modifications to make them unique for the regiments. For publicity purposes, the men practiced with the *bolos,* which was "much better on film than shots of short Filipinos in bayonet practice against the taller White soldiers," Offley said.

The Los Angeles City's Chamber of Commerce raised the money to have the *bolos* made as a gift from the people of Los Angeles for their service. *The Philippine Mail* requested the public to donate $5.00 or whatever amount to provide Filipino soldiers with *bolos.* Over four thousand were made. The media filmed *Filipino Regiment Receive Bolos* at an awarding ceremony at Hunter-Liggett with Gen. Valdes presenting the *bolos* to the men.

The 2nd Regiment travelled to Los Angeles to celebrate the eighth anniversary of the Philippine Commonwealth and to show the Americans a fighting Filipino infantry unit training for war. On the steps of City Hall, they held a rally in front of onlookers. The *Sulung* Band played patriotic marches as the American and 2nd Filipino Infantry Regiment flags flew. With rifles slung over their shoulders, they paraded along Broadway Street in downtown Los Angeles in full uniform with fixed bayonets and helmets. The Press filmed the event as they waved *bolos* passing the reviewing stand and pledging to drive the Japanese from their motherland. Anxious concern ran through the crowd that the Filipino unit was

LOS ANGELES, Nov 15, 1943 –FILIPINO SOLDIERS PLEDGE REDEMPTION FOR MOTHERLAND—Holding long-bladed bolo knives aloft, these United States Filipino troops pledge themselves to drive the Japanese invaders from their motherland, during the celebration here today of their own Independence Day–the Eighth Anniversary of the Philippine Commonwealth.

Mayor Fletcher Bowron who addressed them, said that Filipino Regiments are the only ones authorized by the War Department to carry bolo knives. (AP Wire Photo) 1943

Filipino Infantry Ready for Japanese

Camp Cooke: Filipino Infantry Ready for Japanese. Members of Co. B, 2nd Filipino Infantry brandish bolo knives – all-purpose jungle swords– in anticipation of the day they will meet the Japanese and avenge the overrunning of their island homeland. AP July 24, 1943

a "suicide battalion armed with only *bolos*." Nevertheless, the men wore big smiles as they proudly showed their volcano patches and *bolos* to an interested crowd. Mayor Fletcher Bowron spoke: "Through the presence here of our brave Filipino soldiers, we give concrete evidence they are sharing the responsibilities and sacrifices of their American brothers and are willing to give their lives, if need be, to establish the principles of democracy... We know with those knives the last Jap will be driven out of the Philippines."

Back at camp, they continued to drill, brandishing their bolos. The attitude from the decades before the war when the Filipinos were called "the most worthless, unscrupulous, shiftless, diseased, semi-barbarian(s)" to come to the U.S. was changing.

On the second anniversary of the Japanese attack on the Philippines, a solemn Mass was held by Chaplains Monleon, Noury, and Verceles to honor those who died defending their country. As the men formed a guard of honor around the altar, the *Sulung* band played. At Camp Beale, a Mass was also held for the 1st Infantry.

Alarming news arrived in October. The Second Philippine Republic under Japan was inaugurated. What would change in the Philippines? What would happen to their families? Would the men be forced to join the Japanese army? How much more suffering would there be?

It had been over a year of learning military skills at Camp Cooke and the 2nd Regiment was anxious and raring to go into combat. Morale had to be kept up in between intense training sessions. A dance and entertainment program were prepared by the Anti-Tank Co. and Company E in honor of Filipinos serving in the Armed Forces, reported Fabros. Highlights of the affair was the selection of the men's 'sweetheart,' a dance contest, and door prizes. Prominent Army and Navy guests were introduced. Fabros mentioned Pvt. Eduardo Almalel with Company C who was the official engraver and had worked long hours engraving the many cups, medals, war weapons, and watches as prizes and awards aside from the *bolos* owned by the officers. He went on a much-needed furlough to rest.

The morale of the American people was just as important to keep up for their continued support of the war factories building arms. A movie, *This is the Army,* was filmed on location at Camp Cooke. In return, an impromptu show for the Filipino soldiers was packed with two thousand men. It was a memorable night for the performers and a morale booster for the men in a gymnasium "without a stage, footlights, or any of the magic of the theatre." It was a pleasant surprise to the men who had been at remote boot camps for months. On constant alert for orders to ship out, they seldom received passes into town. But requests from the Filipino community were granted to speak about the Regiments' status. The *Fresno Merry Go-round* published Lt. Bringas of the 2nd Filipino Regiment as a guest speaker.

Four representatives from the Office of the War Department arrived again at Camp Cooke to record a smaller group of soldiers' words of encouragement to the Filipino people. By now, the Philippines had been under Japanese rule for two years and starvation and sickness had taken over hope for a coming liberation.

The 2nd Regiment had already been depleted of men chosen to go to Australia and the directive that allowed men over the age of 38 years to discharge. Now men were transferred to the 1st Regiment to bring it up to 125% strength before departing for Oro Bay, New Guinea. Capt. Leon Punsalan, was placed in charge of reorganizing, disbanding the unit, and activating the 2nd Filipino Battalion (Separate) in its place in March 1944. A memo by order of Punsalan listed the total strength of 929 men. A memo marked 'Secret' from the War Department to ComGen, III Corps formalized the reorganization of the 1st Battalion, medical detachment, and attached a Chaplain to the 2nd Filipino Battalion (Separate). The memo listed 41 officers, one warrant officer, and 899 enlisted men as of July 21, 1943. Two temporary companies, the 1st and 2nd Provincial Companies were formed from the overstrength of 400 men when it downsized to a battalion. There was no reduction of grade for the enlisted men. Maj. Edwin L Sallman accepted the command of the new Battalion with 36 officers and 1064 enlisted men. Members of the *Sulung* Band merged with the *Laging Una* band.

Men who were transferred to the 1st Infantry were glad for the chance to be deployed to the islands. Many believed that if left with the Battalion, they would never get the opportunity to fight and instead be doled out to various units as supply and service troops. Only Filipinos from the 2nd were to be sent as replacement in the 1st. The "110%" strength of the 1st Regiment was required in anticipation that at least 30% of the men would be killed in action or seriously wounded. At 80%, a unit could still function in most cases."

The departures would not have affected the Regiments' strength if the original plan for a 3rd Regiment was approved, creating a Filipino Division. The 3rd Regiment was to be composed of Filipinos from elements of the Hawaiian National Guard and the excess soldiers from the two Regiments. The fear of a Japanese invasion in Hawaii, however, kept the men there. But when the threat dissipated, the Hawaiian Sugar Plantation Association (HSPA) successfully lobbied the War Department to keep the two thousand Filipinos who worked in the plantations to continue harvesting sugar cane and pineapples at substandard wages frozen for the duration of the war. Exempted were those who did not work in the fields. Called the 'Hawaii boys,' they were born in Hawaii, over 10 years younger than the *Manongs,* and did not work in the fields like their Filipino parents. Several hundred men enlisted.

A civilian clerk from the Department of Naturalization at Santa Barbara, California was called to help with the many pending requests for naturalization. Thirty-six soldiers at Camp Cooke were granted naturalization papers after pledging their allegiance and loyalty to the flag before a Superior Court judge. The new citizens were Sgts. Robert Sarmiento, Severo Berces, Jesus Mangrobang, and Marciano Ramos; T4s Felipe Oliveros, Bernardo Donaldo, and Arsenio Soriano; Cpls. Carlos Calacal, Salustiano Segui, Patricio Ramirez, Carlos Lapitan, Celso Moreno, George Gorospe, and Jose Ouano; T5s Licerio Duco, Benny Guieb, Nicasio Asuncion, Leoncio Tactancan, and Antonio Gabriel; Pfcs. Faustino Adona, Alejo Galeon, Francisco Reclosado, Genaro Celaya, Conrado Robles, Anastacio Herrara, Jacinto Duro, Remegio Santos, Restituto Tigson, Potenciano Halabaso, and Marcelo Tamsi; and Pvts. Modesto Osoteo, Ireno Furtogaliza, Aguilino Santos, and Simplicio Lilisan.

Promotions surprised deserving men as they beamed at the compliments heaped on them. Pfc. Nicholas Dalompinis proudly received the Soldier's Medal for heroism in his attempt to save the life of Sgt. Santos Bisquerra. While he was standing on a rock out with a group of soldiers, a powerful high wave swept Bisquerra into the ocean, hitting his head and knocking him unconscious. Dalompinis managed to swim through the rocky shore and rough sea to hold on to him. The powerful surf overpowered his grasp as he endured head and body bruises.

Oro Bay, New Guinea – May 1944

The 2nd Battalion followed the 1st Regiment two months after the latter unit departed for New Guinea. They travelled north from Camp Cooke along the west coast of California onboard the SS *Catalina* to Camp Stoneman. After two weeks, the unit boarded the USAT *Noordam* without a convoy or escort and sailed to Oro Bay passing Port Moresby to pick up supplies. The entire Battalion was onboard, related Al Lamata. Eager to see old comrades, Capt. Serapion Ledesma, a former doctor with the unit who had been stationed at Port Moresby, rowed alone to the anchored ship to greet the men. Under Col. Edwin Sallman, the unit arrived at Oro Bay and camped at Dobudora located 10 miles west of the 1st Regiment's camp. The men plowed through training under the debilitating heat. As a respite from the humidity, the 2nd visited the camp of the 1st by the ocean to go fishing. It took twenty minutes by jeep to get to the other camp. recalled Lamata. Tragedy struck Company E of the 1st Regiment when one platoon was almost wiped out by a rogue wave during a fatal fishing expedition.

Activities in the Battalion were moving fast. Men left in the middle of the night. The 2nd continued to be stripped of men to bring up the 1st Regiment's strength as men were assigned to PCAU and CIC. Most men assigned to PCAU and CIC came from the 2nd Battalion, according to Lamata.

A new Table of Organization and Equipment of the Battalion authorized a total strength of 929 men. Ranks of the men were not affected.

Local diseases abound. The two major diseases that New Guinea mosquitos spread were malaria and filariasis – an incurable disease that brought about and 'enlargement of the testicles.' John Maraggay contracted jungle rot on both feet while Paulino Bersamin was reported ill with malaria.

Six months later, the Battalion proceeded to Hollandia, only some 1200 miles to Mindanao, southern Philippines, onboard the USNT *Lurline*. Jungle training and assignments occupied the men until their deployment to the Philippines. Going AWOL was not an option as the natives of New Guinea practiced headhunting and cannibalism. Treated well, the natives provided labor as guides for building roads, carrying supplies, cutting timber, and clearing land for airfields.

Lt. Richard Arca, a platoon leader with the 2nd Filipino Regiment went into combat with his unit. In his letters to his wife, he narrated how his group departed San Francisco in March 1944 onboard the S.S. (sic) *Lurline*, a former luxury liner for Milne Bay, New Guinea and transferred to another ship where "he was traveling alone under secret orders." He did not write details of the battles. "The company had experienced some action losing one man and another in the hospital while in New Guinea. Fighting at Madang and Wewak (Feb-April 1944) north of the Regiment's base was heavy. You will not read this in the papers as Filipino activities are kept secret. Two large columns of Australian and New Zealand units supported my troops' advance. We were leading the combat column and had four machine guns and eight mortars. I received a very nice compliment from my commanders. Other units (Whites) lost a lot of men. We ran into some bombings, strafing, artillery fire and machine guns."

Leyte Island, Philippines Landing – October 1944

After several days of naval bombardment to clear the beaches, the Sixth Army landed on Leyte. Officer Jaime C. Velasquez was onboard as Asst G-2 with the Sixth Army. He worked with the guerrillas and civilians to obtain the intelligence needed for the surprise liberation of the Santo Tomas Internment Camp (STIC) prisoners in Manila and to track civil-military operations.

The men under PCAU and CIC, landed with the U.S. Task Forces under swarms of Kamikaze suicide planes pouncing on the invasion convoy. Hundreds were killed and ships sunk. As the landing forces pressed inland, they knew of the enemy's location and strength from the radio messages the signalmen sent in by the guerrillas.

Leyte Island, Philippines – May 1945

The 2nd Filipino Battalion (Special) arrived in Leyte three months after the 1st Regiment. At eighty percent unit strength, the Battalion never went into combat but instead provided personnel to higher echelons and to areas where Filipinos familiar with a certain language were needed. The Battalion was ordered to Dulag before moving on to their headquarters in Quezon City adjoining Manila. The unit's orders were to support the PCAU teams as they landed with the Task Forces in the various islands. They carried out support services after Manila was cleared of the enemy. Their Standing Guard orders were for the Officers' Club and tents, Battalion mess, supply building, and motor pool. The unit's daily schedule began at 0600 hours Reveille, Mess call, Drill, Retreat, Call to Quarters, and ended with Taps at 2300 hours. Suspicious persons were challenged by the sentries. Pvt. John F. Marias with Company C, 2nd Filipino Battalion was ordered to duty with G-1 Personnel as clerical assistant to Gen. Basilio Valdes. An early casualty was Sgt. Rubin Gregory assigned to G-4 Logistics, who died of food poisoning at the 73rd Field Hospital on Leyte.

In May, elements of the Battalion provided support to the Eighth Army in their liberation of Mindanao until August at Sarangani Bay

1946

July 4th, 1946, was a memorable occasion when the men in Manila witnessed the Philippine independence ceremonies at the Luneta. The Battalion disbanded at the end of the year. The men's lives dramatically changed as they chose to return to the U.S. to be discharged, join the new Philippine Scouts, transfer to the 86th Infantry Division (PA) or be discharged in the islands to become civilians.

Camp X (Tabragalba), Australia: 1st Reconnaissance Battalion
(5217th Reconnaissance Battalion and 978th Signal Service Company). Source: Don Ruiz Collection

VI

Top Secret

The American Forces marched through northern New Guinea towards the Philippines with hardly any information on the resistance and enemy situation in that country. GHQ in Australia needed the invaluable information in the planning for a successful return. The Filipinos were being pressured for food and support while Japanese propaganda was brainwashing them. The guerrillas were struggling to consolidate and organize against the Japanese. GHQ was looking for strong leaders who could unite factions in the resistance so supplies could be sent in return for intelligence on the enemy. GHQ believed that the offer of food, medicine, and arms would divert the guerrillas from their internal squabbling for men, territory, and civilian support and focus on gathering intelligence on the enemy. The constant question in the messages from Panay and Mindanao was when are the Americans coming to destroy the enemy?

Philippine Regional Section (PRS)

The Philippine Regional Section (PRS) was separated from the Allied Intelligence Bureau (AIB), a joint special agency of the United States, Australian, Dutch and British Intelligence for operations behind Japanese lines, to focus on coordinating guerrilla activities in the Philippines. It bypassed and avoided Army regulations to provide the intelligence needed by General MacArthur in his march towards the Philippines. Its plans and schedules were in parallel to General MacArthur's.

Heading PRS, Col. Courtney Whitney focused his attention on penetrating Luzon in the north to obtain the most intelligence on the largest Japanese concentration in the islands. There were many underground networks operating in the Manila area and the liberation of the capital city would be the highlight of a successful Philippine campaign. He envisioned the missions to be composed of Filipinos but mostly led by White Americans. Having lived in Manila, he believed that "Americans could provide a stronger and neutral leadership than (the) politically and socially divided Filipinos."

Earlier, the AIB had sent six submarine missions to the recognized guerrilla leaders in Negros, Panay, and Mindanao. The request for eight combat submarines patrolling the SWPA waters to be available for the next penetration parties was refused by the Navy. The smaller-sized patrol submarines would be put at risk having to take many trips to carry the large cargo and men. The patrol subs' objectives were to hunt Japanese shipping and not deliver cargo.

Finding submarines was not the only problem. AIB had earlier identified around 130 Filipinos only in Australia composed of the crew and wounded of the hospital ship *Mactan*, survivors of sunk blockade ships, and others who were living in the country at the time. From this number, some were still recovering from their wounds while around 30 were expected to pass the training and further selection. After the last of the six submarine missions left in August 1943, there were no qualified Filipinos left in Australia for future missions. The urgent need to fill up the shortage meant looking at the Filipino Regiments in California where they were training in the conventional Army.

An essential part of Whitney's plan was to bring highly selected Filipinos to Brisbane from the U.S. and put them through a course of appropriate and rigorous drills. Unlike in the Army, training was to be in small units to execute special advance missions for which they could easily blend in as Filipinos and successfully carry out their duties.

Whitney sent his initial recommendation to GHQ on the organization and training of Filipino manpower for the secret missions: 1. Infiltrate into enemy-occupied areas to organize, encourage, and lead native resistance, 2. Counter Japanese propaganda, 3. Sabotage enemy installations, 4. Engage and provide support to guerrillas, 5. Prepare landing and beach facilities, 6. Act as guides and interpreters organizing locals in strategic areas to facilitate landing operations and cross-country advances once American troops landed in force, and 7. Develop information on enemy installations and movements and maintain such contacts as may be desired with people of several areas.

Absolute secrecy was ordered of PRS activities. The entire operation was labeled 'Top Secret' and all related memos were stamped as such. The missions, men, the training camps in Australia, and everything related to it was to be wrapped in strict security. The missions were devoted exclusively to coordinating the scores of scattered bands of guerrillas supplying them with field radios and a systematic basis for collecting intelligence. Only a few on General MacArthur's staff were privy to PRS' activities. Guerrilla messages went directly to PRS and General MacArthur's desk bypassing Major General Charles Willoughby's G-2 Intelligence numerous times.

They were the "advance echelon in his campaign of liberation." They were called the "eyes and ears of MacArthur." The men were not to go into combat but set up radio networks and gather intelligence on enemy dispositions to be used in the planning of the return. The personal arms and ammunition of the men were for defending the radio stations.

It was a dangerous assignment with no promises of a safe return. Each mission party faced danger from unexpected turns-of-events, guerrilla responses, or loss of supplies. Once on the island, they could not quickly turn back. They had no advanced knowledge of the reception they would receive from the guerrillas and the local population. The guerrillas were expected to provide security, messengers, intelligence agents to obtain information and even food but that was not guaranteed. They risked their lives in disease-ridden jungles, crossing open rough seas, and penetrating enemy lines to set up radio networks to monitor

Japanese activities. As coast watchers, they established and maintained radios in the most remote places overlooking the shipping lanes used by the enemy, constantly on the alert to quickly move to avoid discovery. They trained members of the resistance on military tactics to fight the enemy, operate the weapons they brought, and gather useful information on the enemy. To protect the radio, they lead combat teams. If captured, they would have no protection under the Geneva Convention as Japan had signed but not ratified it.

California, USA

The 1st and 2nd Filipino Infantry Regiments were training in California when Whitney arrived to execute an order from the War Department. Over half of the Regiments had become American citizens three months earlier in a mass naturalization ceremony at Camp Beale to show their loyalty and desire to fight. The continuous reports of hardships inflicted on the civilians fuelled the men's desire to be naturalized and fight to throw out the invaders from their homeland.

The large number of volunteers eager to avenge the attack on their homeland overwhelmed Whitney even after they were informed that there was almost no chance of ultimately accomplishing this. They were to use their own judgement and devices to accomplish the objectives. They were not going to do combat missions. The majority would be radio operators in a signal service company to enable the guerrilla resistance to provide intelligence to Australia and communicate within their network. The rest would be infantrymen accompanying the radio operators to reconnoiter the areas as protection of the radio networks and signalmen. With the help of Jaime Velasquez, commander of the 1st Battalion of the Regiment, Whitney chose to interview around 800 men undergoing training in infantry tactics at Camp Beale, Camp Cooke, and Camp Roberts. The men's composure, training records, education level, background in the Philippines, and spoken dialects were scrutinized. Sending a man to an area where he spoke the same dialect was helpful since different dialects were spoken and understood in various provinces. After an intensive interview, the men had to take a severe physical test which was probably not that difficult for some of the men who had been hardened by the back-breaking physical labor in the vegetable fields they came from. Lt. Silva probably passed the test out of sheer determination since he came from the city where he had been working as a domestic to support himself to pay for his college fees.

Four hundred men and 30 Filipino officers were selected. Whitney's memo to G-1 Personnel stated he was sending 491 men from the ranks of sergeants to privates and that the total would increase with the officers. Reasonably sure of the chosen men following the very severe screening method, Whitney still "granted human error of 25% for training to disclose resulting in around 300 men to further train and send to the Philippines. It takes 6 to 8 weeks to train men once in Australia properly... get the U.S. inculcated softness out of their system." Less than a quarter of the selectees were White Americans. Movement of the men in one group was to avoid leakage of the missions to Tokyo. Those who failed the final selection for missions could be assigned to the 298th Infantry Regiment of the 24th Division which presently had a strength of 606 Filipinos.

The selected men underwent a separate radio training schedule before departure to lessen the load of men to be trained in Australia. Over two hundred enlisted men were sent to Camp Kohler in California for radio training. Cpl. Julius Ruiz, a journalist, asked Sgt. Fred Villarta to pose for a photo for the camp's newsletter. He recalled Ruiz predicting "that the group will be a nucleus of operations that will be instrumental in Allied victory in the Pacific theatre."

They began departing for Australia as soon as transportation was available. Some went by way of New Guinea while others landed directly in Sydney. There were also others who were flown to Brisbane. They were to be trained as one compact unit. Interest at GHQ ran high on the arrival of the Filipinos who were going on secret missions. They were called 'The Avenger' but there are no records to suggest that the name was official.

Whitney earnestly recommended that no Filipinos be assigned to positions in PRS that would give them access to reviews, comments, or recommendations from it. "One of our primary jobs will be to develop the strengths and control the weaknesses through frank character appraisal unrestricted by considerations of delicacy to any individual and without pulling punches," he said. Initially, he envisioned White officers to lead most of the missions.

Whitney believed that "Filipino guerrillas were likely to be impulsive and cause crushing Japanese reprisal before more arms and ammunition could be sent." He thought American officers elicited confidence in their leadership as shown from the Philippine Commonwealth Army's divisions that were commanded by White American officers. But he also realized that it could create mistrust. Depending on the need at the time of each mission, American and Filipino officers both led parties with the majority composed of Filipinos trained in the infantry, radio, intelligence, and later, weather, demolition, and aircraft warning.

Brisbane, Australia

Before the first large group of men departed California, four Filipino signal men arrived in Brisbane in response to General MacArthur's priority request to Washington. They reported to the Milton Tennis Club that was converted to a Staging Camp for further training. The AIB took over the 14 tennis courts as supplies warehouse and training grounds. Sgt. Catalino Laanan, Pfc. Ray Punzalan, Cpl. Peter Oandasan, and Pfc. Fermin Consolacion were quartered in tents in the parking lot.

Irregular availability of transportation delayed the arrival of men to Australia. Training facilities at Milton were ready but there were hardly any men. Whitney appealed to the

Brisbane. The Milton tent area to house transient AIB personnel

Navy to include these vital personnel in any shipping going to Australia, or in cargo ships from other ports. As soon as they arrived, the AIB set to work conducting radio training while Australian soldiers began their commando training. Soon, Milton Camp was crowded with tents and men.

It was later learned that the growing organization needed a support unit. A thousand men was expected to undergo the training, urgently needing an administrative staff. Lt. Patricio Ubarro, a wounded former Philippine Scout on the *Mactan* was brought in as adjutant to Capt. Alan Davidson, commandant of the Australian Army's training program. Ubarro was chief of radio communication in the 26th Cavalry, Philippine Scouts and when he had fully recovered from his injuries, operated the first U.S. Army Radio station, USASOS. At first, he had some difficulty managing the camp using the Australian system of administration, but he soon overcame the problems.

The staff expanded with the arrival of Sgt. Juan P. Dahilig who was with the 5th Air Force in New Guinea and Sgt. Jose J. Ramos from the Ship and Gun Crew Command #1 in Australia. Both immediately reported to AIB.

Submarines, Supplies, and Parsons

Enter a key person in the plan to return to the Philippines - Commander Charles 'Chick' Parsons. The Philippines had been his home for two decades and he had been raising a family when the war began. He managed to avoid internment and safely moved his family to the U.S. using documents showing him as a Panamanian consul. His penchant for adventure brought him to General MacArthur in January 1943 bringing notes of what he observed in Manila – a budding underground resistance and the treatment and location of prisoners. Well-travelled in the island and with   his stocky built and white skin burnished by the sun, he blended well with the local population. It also helped that he could speak the local languages and could run fast. He bravely risked moving around unarmed.

Parsons worked with AIB supplying the early missions to the Visayas and Mindanao. When PRS was formed, he focused on the needs of the guerrillas and GHQ's requirements from them to maintain the critical connection. But Parsons action went beyond those of a supply officer as he related in his report when he was in Mindanao that in early 1942, he had organized a Manila Intelligence network to gather useful information.

After his first trip to Mindanao in March 1943, he returned nine months later and went on a month-long assignment informing guerrilla leaders of GHQ's emphasis on gathering intelligence on the Japanese and civilians as most valuable in helping the Allies in the plans for their return. He had the authority to find the best-suited guerrilla leader to unify an island. It was difficult to convince the guerrillas not to openly attack the Japanese or succumb to enemy propaganda when there was no definite date on the Americans' return. The Japanese atrocities drove them to strike back no matter how inadequate their weapons were.

By the time Parsons departed Mindanao from his second trip in December 1943, there were four coastwatcher posts on Davao with one overlooking the Gulf to watch Japanese ships' movements. The Japanese had received information that he was in the country and had put a $50,000 gold bounty on his head. The radios were established close to Davao City where the Japanese were entrenched comfortably with the 20,000 Japanese residents in the city, the most significant colony outside of Japan. With a Navy signalman, a station on

Dinagat island had a view of the Surigao Strait between Mindanao and Leyte, a vital passage into the archipelago. Stations on Camiguin Island, off the northern coast of Mindanao and western Zamboanga monitored the Mindanao Sea. Stations in Zamboanga monitored the Basilan Straits and in Cotabato overlooking the Moro Gulf. Outside Mindanao's jurisdiction, the Verde Island Passage, and the Sibutu Passage were also watched since they were heavily used by Japanese transports.

Parsons reported what he noted during his observation mission: The Japanese bombarded areas on Mindanao not under their control with propaganda in radio, schools, press, and billboards. The guerrillas countered by publishing sheets of newsprint that reported daily the Americans' progress from radio broadcasts in news bulletins to the public. The increase in morale of the population also meant support for the guerrillas. The people requested that the Philippines be mentioned now and then, showing that they were not forgotten, and not to mention guerrilla activities   to avoid bringing the enemy to their area. Parsons also observed Mindanao's mobile combat units commanded by real fighters influencing GHQ to send only personnel selected for combat leadership or training troops in future missions. On Panay, combat troops were formed. He was hopeful that the infighting in Negros amongst the guerrilla leaders would be resolved with the appointment of Villamor, the first mission party's leader as district commander. Guerrillas had difficulty identifying pro- Japanese Filipinos and executed civilians and influential people for allegedly supporting enemy activities. He did not get involved in the executions since his main mission was intelligence gathering. From Luzon agents or his clandestine meetings with the underground Manila network, he reported the northern Luzon guerrillas as organized and disciplined, while those in central and southern Luzon were disconnected. He suggested holding off recognition except in northern Luzon. The policy of GHQ to supply Mindanao with arms and medicine, in exchange for limited aggressive actions against the Japanese was to continue. They were to concentrate on developing coast watcher stations and gathering intelligence.

The earlier AIB mission men with their healthy looks, good clothing, shoes, modern equipment, and stories of luxurious treatment in Australia had affected the morale of the guerrilla troops. Sending a White American would be less suspicious and not regarded as an agent of the enemy. Friendly civilians would be expected to protect him. But a white American would not be able to blend in and pass through Japanese lines to gather intelligence. He would constantly be in hiding because capture meant instant torture and death. It was decided that qualified Filipino men trained to lessen the bravado were best suited to blend in with the population and work with the resistance to obtain intelligence on the enemy.

On the development of coast watcher networks in the islands, Parsons convinced GHQ to steadily send larger amounts of supplies to keep the resistance strong and for the survival of the missions' remote outposts. Without their security, message couriers, and equipment carriers, it was almost impossible to carry out their tasks. Parsons convinced GHQ that Mindanao was the strategically important distribution point   of supplies to other islands before submarine missions could arrive in those islands. Mindanao could also relay radio messages from the northern parts of the islands that could not reach Australia.

From Parsons's report, Willoughby concluded that the next step was "to capitalize on the guerrilla workforce, loyalty, and available resources in the Philippines to the end that our conquest of the islands will cost us the least in manpower and time. Must take advantage of the assistance which organized guerrilla forces in the Philippines can give us."

The Navy's 7th Fleet did not want to divert their combat patrol submarines from the waters around the islands for the two-week roundtrip missions from Australia to designated landing areas. But since they were provided with cargo submarines, the operational submarines were not diverted from their hunting missions. Parsons obtained naval support for two cargo subs that could carry ten times more than a patrol sub. On the way back, evacuees could be taken out of the islands.

When a Davao coast watching station reported a ship sunk offshore, the Navy was convinced to provide submarines that could carry signalmen, radios, and cargo. The Navy sent little books of silhouettes of Japanese ships for the watchers. Submarines had limited observation range and could not see enemy ships in their area. With more watchers and their radios, the subs were able to sink enemy ships when their locations were relayed to them from the coastwatchers' flashes to KAZ, General MacArthur's radio at GHQ. They could then also receive confirmation of the damages within minutes. The Navy then delightfully received the credit for the destruction.

The Navy turned over the USS *Narwhal* and USS *Nautilus* with their huge 100-ton capacity. The two subs were not suited for combat with their slow speed and limited torpedo capacity. Because of their size, maneuverability was also limited, and they were easier to detect at a depth of 65 feet when submerged. But they were excellent for carrying cargo to the guerrillas and evacuating people.

With the SPYRON operation approved, Parsons worked closely with Whitney on future missions and gave recommendations to General MacArthur that influenced policy decisions concerning the guerrilla forces in the Philippines. Private codes for areas north of the 12th parallel (like Samar Island) were authorized. Parsons had his secret code for radio messages whenever he was in the Philippines. Unknown to G-2 Intelligence, it leaked out when two signal officers handling Philippine traffic received incoming messages that they did not know how to decode. Willoughby found out and confronted General MacArthur, but nothing changed according to Chief Supply Officer, Bobb Glenn. He remembered hearing Parsons' familiar low-pitched voice that was very persuasive when they personally met for supply requests.

The *Narwhal's* early missions departed from Brisbane but when departures were moved to Darwin, the *Nautilus* joined her. From Darwin, their round trips were shortened by two weeks as it was closer to Mindanao. The number of supplies that could be carried doubled. Tons of supplies were stored on open platforms and several more rooms were commandeered in Camp Milton's recreational halls when the warehouses did not have enough space. As missions were ordered, gear and supplies were flown to Darwin. The heavy demand to supply the cargo submarines in Darwin "made up 90% of the entire AIB's supply efforts" creating a separate supply section out of Darwin explicitly to fill up the two cargo subs for the Philippine missions. Lt. Arsenio Arellano was assigned as Battalion Supply Officer holding office at Heindorff House in Brisbane while Sgt. Estanislao Quitoriano headed a group of twenty enlisted men as warehousemen.

The missions were exempted from the standard allowances for supplies and budgets. Approved by General MacArthur, there was no limitation for the missions. "Col. Ind had obtained and delegated to me the authority to draw without question anything except War Department-controlled items (like tanks and airplanes)," Glenn said. He shipped machine guns, mortars and bazookas aside from the cannon the Navy shipped earlier. Whitney wanted heavier weapons such as an anti-tank gun that could be broken down for shipment as it had to be wedged into the submarine structure because it could not get into the pressure hull of the sub.

The shipments were never enough. Wherever the submarines landed in the islands, the boost to the sinking morale of the guerrillas and civilians made up for the limitations. But Whitney may not have thought of the problem of transporting and hiding the heavy weapons in the islands and the training required to operate them. It was the guerrillas' responsibility to transport and use that equipment with the help of the mission men.

As the shipments rapidly increased, Parsons closely monitored and prioritized supplies. It was difficult to procure, pack, and make ready as soon as a mission was ordered. Radio equipment for the expanding communication networks was a primary procurement item. The sets and parts had to be packed in marked boxes that could fit through a submarine hatch for quick set-up. A separate and dedicated section was created to prevent mishaps and ensure badly-needed supplies reached the island. A standard mode of crating and shipping was followed for convenience and security. Each piece of cargo had to be waterproofed and easily handled. From reports of earlier penetration parties, preparation and packing of equipment was modified in the technical training of the men. "The character of dispatching supplies to advance parties had become as complex as the enterprise was suicidal," wrote Whitney.

With the missions under the Army and the submarines under the Navy, mishaps were unavoidable. A 20 mm gun was unloaded by the guerrillas but it had no shells. The guns were Navy requisitions and the shells under the Army were not yet available. Whitney wrote "Occasional slip-ups happened as in a large, complicated organization, but Filipino cleverness quickly resolved them... Several cases whose markings indicated that they contained highly prized submachine guns instead turned out to contain antiquated cavalry sabers... many a dead Japanese proved that the guerrillas had put even those outmoded weapons to good use. Difficulties in transshipments from central Australia to Darwin to the Philippines of either whiskey (medicinal), pesos, or other essential items (were resolved when) these necessary items were wrapped to look like military rations." The plan worked. But some shipments landed without rations. Worse were rations that arrived after months but did not reach the remote outposts as some of the boxes were looted or the carriers, who were always hungry because of their work, ate the contents.

The quartermaster scrounged around to procure basic cargoes of radio, weapons, ammunitions, medicine, and various guerrilla requests per island. Extra efforts were made to fulfil guerrilla leaders' special requests for soap and shoe polish. Kangleon even put in a request for denture paste. Hundreds of standard tins called 'CW tins' after Courtney Whitney, weighing some 55 lbs. and made to look like Red Cross packages contained medicine, toilet articles, and food with some spices. Miscellaneous items like flashlight, bulbs, pocketknife, cigarettes, clothing, and some books and magazines were included to be given out to guerrilla outposts as part of Whitney's plan to supply the Philippine resistance.

It would take months to make customized or find non-standard equipment for the missions. Various types of radios, ranging from bicycle-powered ones to those weighing one ton, were found and prepared for shipment. Messages were transmitted using a bicycle-powered radio.

A British-made transmitter that could be carried on a man's back and a Dutch- made radio set that could withstand the debilitating humidity of the jungle were procured. Important parts such as oscillator coils for the radio sets, frequency crystals, and magazines for the guns, were crated and shipped.

In each mission, the submarine skippers sailed under 'sealed orders' to be opened once at sea. The ships' logs entered the mission parties as 'secret cargo.' As cargo submarines, they were not to attack enemy ships, but this could not be prevented – captains' logs sometimes mentioned sinking enemy ships spotted while on mission runs. Spotted Japanese vehicles were at times tempting targets.

Mindanao Island. These radio operators of the 978th Signal Service Co., U.S. Army, operated radios for a year before the invasion of Leyte by pos- ing as Filipino citizens. They are American citizens of Filipino ancestry. Aug 10, 1945. Signal Corp 216372

The increasing volume of messages coming into KAZ and PIMC, the Philippine Message Center in Brisbane from the resistance forces in Negros, Mindanao, and Panay pressured the operators who were working in shifts to keep up with the information load. American radio operators were sent to reinforce until trained Filipino radiomen arrived.

Meanwhile in California, the selected men prepared to depart for Brisbane. Whitney had sent his early plans for the Luzon penetration (subject to change) that pinpointed certain locations for radio setup, and the list of officers and men most qualified to undertake missions with the least amount of preparation needed.

At Camp Roberts, California, a list of officers and men was drawn up. The mission was given high priority and the orders for the earliest possible shipment by the quickest available means of transportation was issued.

Special orders were cut on July 10, 1943 for Capt. Ricardo Galang and Lts. Carlos Arguelles, Ruben P. Songco, and Saturnino R Silva to depart from Hamilton Field. Others on the list were Lt. Pedro FlorCruz, Sgts. Arcangel Baniares, Restituto Besid, James Gallemore, and Gerardo Sanchez. T5 Ramon Vitorio and T4 Leo Marquina joined. All went on missions except for Lts. Arguelles and FlorCruz who were given other assignments. On the list from the 2nd Regiment were Lts. Richard T Arca, Roberto Lim, Edmundo Marfori, Ceferino R Rola, Silvino B. Tallido; Sgts. Eustaquio Cabais, Roberto N. Sarmiento, Benjamin M. Harder, Buenaventura Dolendo; and Pvt. Juan Palac.

Galang's group departed for Australia. Before the flight, he was given an envelope to open once in the air. Galang remembered the temperature lowering and his ears hurting from

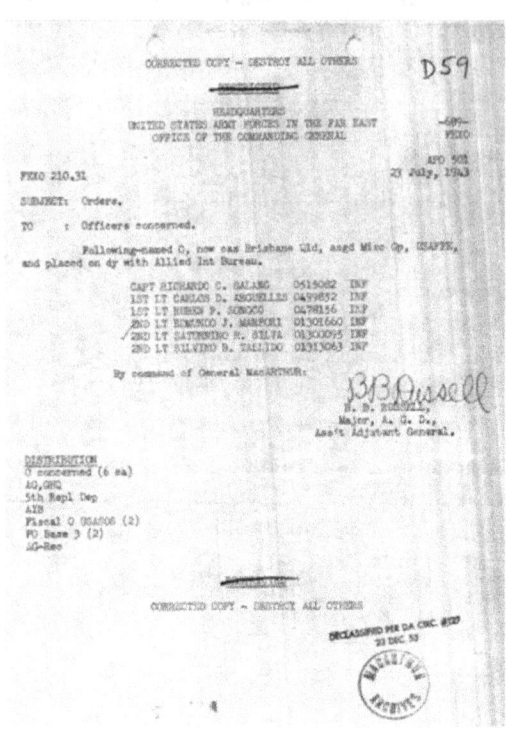

The orders dated 23 July 1943 for officers Ricardo Galang, Carlos Arguelles, Ruben Songco, Edmundo Marfori, Saturnino Silva, and Silvino Tallido to depart California for Australia by command of General MacArthur. Source: MacArthur Archives

the air pressure as the planes quickly soared after take-off to a high altitude. He opened the envelope. On it were written the words– "Destination: Amber Field." No one knew where that was. The engine's drone made everyone sleep and he slept leaning on Lt. Silva. They had several hours in Honolulu, which had become a soldier- sailor town with uniforms of Army and Navy seen on the streets. They stopped at Kanton Island west of New Guinea, a secret airplane base with individual hangars hewn into the rocks. They arrived at Nandi in the Fiji Islands, where they ate a lot of papayas and bananas. On the same day, they landed in New Caledonia and left again. A lieutenant met them in Brisbane, took their temperatures, and brought them to Ascot, an old stable as their temporary headquarters, before moving to Camp Milton. The Australian locals were suspicious of them. Who were these brown-skinned people? Were they Indonesians, Chinese, or Japanese? Those doubts were assuaged when they learned they were Filipino soldiers of the U.S. Army.

The next group that left in July from the 1st Regiment listed Sgts. Isidoro D. Dacquel, Amado S. Corpus, Aniceto C. Manzano, Faustino B. Bacus, Baldomero De Leon, Vidal Garcia, and Vicente A. Pinuela; T4s Faustino H. Habon, Crispulo G. Robles, and Daniel B. Sabado; T5s George R. Herreria, Juan E. Hidalgo, Aniceto S. Kintanar, and Ruperto Ruiz; Cpls. Felix R. Lopez, Clemente C. Udasco, David D. Cardenas, and Paciano S. Racho; Pfc. Ladislao De Dumo; and Pvts. Aayas B. Inocencio, Carlos S. Placido, and Guillermo S. Ventura.

From the 2nd Regiment, were Capt. Abner K Pickering; Lts. Claudio M. Tamayo, Carlos F. Ancheta, Jose P. Dulay, and Jose T. Mendoza; Sgts. Eulogio Reyes, Florentino L. Lucero, Cipriano L. Miguel, Val R. Tabios, Gregorio V. Querubin, Marcelo B. Abalos, and Jose D. Mamaril; Cpls. William L. Hiquiana and Frank B. Bueno; T5 Serafin Buenavista; Pvt. Apollo P. Tullo; and Pfcs. Washington Z. Sycip and Robert C. Minder.

Not all the men listed departed after orders were cut. Some arrived later or were transferred to other military units like Richard Arca, Roberto Lim, and Washington Sycip.

Meanwhile, GHQ eagerly anticipated the arrival of the selected men. No other able and trained Filipinos could be found in Australia. The last of the six AIB parties had left in August and several months had passed with no mission sent out.

The men were placed on duty with the AIB. Whitney met the men and delivered the message that they were chosen on their merits and willingness to go on this mission. "General MacArthur does not consider you simply as a group of men but as individual officers and men. He will follow your progress and activities with personal interest."

Wasting no time upon his return to Brisbane, Whitney appointed Major Lewis Brown III as executive officer of PRS and commanding officer of the men arriving from the U.S. He had submitted a training plan entitled, *Organization and Training of Filipino Manpower,* that was approved by the War Department. Patterned after the Australian program, Brown was to deploy it with the arriving men. The unit's objectives were like the 'M Special Unit' of the Australian Army that gathered information on Japanese shipping and troops movements in New Guinea and the Solomon Islands by landing behind enemy lines. The Australian Army instructors "presumed that personnel selected for this course will be trained soldiers with a high mental and physical standard." Ind justified that there would be a better chance for success with Filipinos due to the knowledge of the language, country, and being with his people.

978th Signal Service Company, Brisbane, Australia

A new organization, the 978th Signal Service Company, was activated in Brisbane on July 1, 1943, with Capt. Harry Croell commanding, to produce radio operators for the missions. Men from the 832nd Signal Service Company in Brisbane added to the new company until Filipino radio operators were trained.

Groups of fifty men began arriving from all walks of life, a mixture of musicians, engineering graduates, journalists, waiters, farmers, field workers, former Philippine government employees, and even WWI veterans. The slogan was to have a 'Whiskey and Soda at the bar of the Manila Hotel.' But in the meantime, they had to get used to the Australian diet of beef stew at almost every meal or bully beef (corned beef) with hard biscuits.

Capt. Enoch G. Jones of the Signal Corps oversaw the training of the signalmen. As men graduated from Milton Training Camp, they were placed on the staff of KAZ and the Philippine Intelligence Message Center (PIMC) that received messages from the islands. Only one-fifth were non-Filipinos.

A cadre of thirty-two White enlisted men arrived on the SS *Mariposa* after they were outfitted on Angel Island in California to join the Filipino soldiers.

Camp X (Tabragalba), Australia

Camp Milton quickly became crowded with the continuous arrivals from the U.S. Soon, there was no space for more tents. The men were moved to an area in Beaudesert outside Brisbane that was still being cleared and structures being built. It was called Camp X stamped 'Top Secret' in all related documents indicating it was under heavy security. Only General MacArthur, his staff, and Whitney knew about the men and the camp's location. Whitney had earlier scouted the place and saw that the former cattle station and Australian camp had enough space to accommodate almost a thousand men that would be activated around a signal company. AIB had negotiated to acquire it from John B. Chandler, mayor of Brisbane. The area was strictly highly secured so that even Chandler, who had access to almost anywhere in the camp, had to produce a pass to see his former property.

The place had to be transformed into a camp before the specialized part of the training program like drills, marches, and practice use of weapons could be properly conducted. Physical training required an obstacle course. The men suffered through the debilitating Australian summer seven days a week to clear the area. The commanding officer down to the last enlisted man participated in removing layers of dried cow manure before tents could be set up and wooden bunks erected. Old, abandoned buildings were updated into headquarters where personnel would train and live for at least a year.

A supply unit was formed to administer to the men's needs, including rations, cigarettes, and even "homemade American toilet paper." The warehouse for supplies, including Post Exchange items, was in Brisbane, requiring daily round trips of trucks. A problem with requisitions for Army cots necessitated the men to sleep in tents on the ground for a while. Rows of pyramid tents for the estimated twelve hundred men were set up within a barbed wire fence perimeter.

The men supplied the labor and Engineering, the materials, tools, and key men. Electrical poles and wires with fittings and light globes replaced old hurricane lamps that the Australians once used. Using work lights, construction continued day and night, and the camp was completed sooner than expected. The Australian civilians watching the fast progress from the other side of the creek, connected their power to the poles, and enjoyed the electricity their new neighbor brought.

Temporary huts were built for the field kitchen, latrines, showers, and laundry area while old buildings were repaired, cleaned, and painted into recreation halls and mess halls. A dispensary was established to accommodate around 1,000 men. The Chief Engineer constructed "a warehouse to store the men's belongings, two hot water systems for new showers, a covered shed for maintenance of motor vehicles, a metal building for storage of motor fuel, and a building for an ammunition dump." The supply of drinking water gathered from roof buildings and from the lagoon by a pumping system into a storage tank was supplemented by a drilled well. Water pipes were buried in the ground, and a telephone system strung out. Roads and graveled sidewalks were laid out and defined areas, giving the camp a semblance of a small village.

Chicken and equipment to raise, feed, and consume them in the camp's mess were requested. Sgt. Ciriaco Castro, a former Philippine Scout, was a welcome addition when he took charge of the Mess and prepared the familiar Filipino dishes. The Army food made the men miss the Filipino dishes they prepared for themselves in California. Sgt. Artemio Ibea, an avid farmer from California, led a detail of farmers operating an improvised system of purifying liquid refuse that flowed into a pit near the camp to be used as fertilizer. The vegetables, eggplant, bitter melon and okra, common to several Filipino dishes, grew abundantly partly shielded by the flaps of their tent, to the men's delight. Well-fed, the men felt they were ready for any challenges sent their way.

A creek separated the men's camp from the Australian camp. Vegetables planted where the water passed caused quizzical looks among passers-by who wondered what those vegetables were. The Australians received American cigarettes and rice while the men received cheaper Aussie cigarettes and many potatoes. Galang regularly swapped the potatoes for American cigarettes and rice for the men.

Classes resumed between construction "to make up for the lost time in digging latrines and painting fences." Men had gotten used to walking around barefooted and developed a deeper tan under the sun. Class rings and watches were removed for they left white rings around fingers that would be noticeable to the enemy in a mission. Many of the men's height required mastery in handling their rifles as the tip of a fixed bayonet on the rifle came to right above their chins.

Introduction of new weapons required additional training, and the men welcomed it. The Quartermaster announced the availability of self-loading Carbines that were gas-operated, magazine-fed from a box, air-cooled and weighed only 5 lbs. It was an individual weapon that could replace a rifle, pistol, or revolver. For their size, the light weight of the weapon appealed to the men.

By August, the strength of infantry and other non-radio related personnel was at 125 officers and enlisted men at camp. The 978th Signal Company reached 8 officers and 200 enlisted men.

It took a grueling five months to the end of 1943 for the thirty-eight buildings to be built. An officers' club and an enlisted men's service club were morale boosters since passes to the city were prohibited due to the secrecy of their unit. A motor pool with forty-two vehicles, including trucks and an ambulance, provided the transportation needs. More trucks were requisitioned for training and maintaining the camp as the number of men doubled. Street signs pointing to Manila and Tokyo were installed. When the camp was considered done, a basketball court, a softball diamond, a tennis court, and a picnic ground were among the finishing touches.

Whitney sent a memo to the Chief of Staff on the completion of the camp. The "American flag was raised on August 22, yesterday, at our new Filipino camp near Beaudesert. Hopeful out of this camp will emerge useful Philippine service and that   the best there is in these men will be developed and made ready." He sent an invitation to the Commander-in-Chief for the formal opening of the new club to "unveil a plaque   at the entrance which contained... *Bahala Na* (Tagalog for 'Come what may') chosen by them to govern their service." GHQ proudly acknowledged the significance of the unit's purpose to the progress of the Pacific War. An all-day inaugural celebration was filled with athletics, dancing, singing, and fellowship. Doubtless, there were eats and plenty of beer which flowed from the bar from 4 to 11 P.M. The New Guinea natives training as scouts and Australian soldiers on the other side of the creek eagerly accepted the invitation to come over.

The camp was still called Camp X. Lt. Arguelles suggested 'The Bastard Outfit.' Trained as a tank commander with the 32nd Armored Regiment in the Nevada desert to be sent to Europe, he was assigned instead to the Filipino Regiment. He may have wanted to be sent to Europe which what he may have meant when he replied "they should have kept me with the Americans" in an interview after the war.

5217th Reconnaissance Battalion (Provisional)
5218th Reconnaissance Company (Provisional)
Camp X, Tabragalba, Australia

By mid-1943, messages were flying between PRS, Washington, and GHQ to send more missions to penetrate the islands. The only Filipino men available for future missions were in Camp X undergoing special training. With so many details involved, a formal structure of training and forming of missions were developed for efficiency and planning. As fighting intensified in New Guinea, the War Department ordered the 5217th Reconnaissance Battalion (Provisional) activated in October 1943 with the 978th Signal Service Company as its core. Maj. Lewis Brown was commanding officer. Lt. Patricio Ubarro, a Philippine Scout, was

Caloundra, Australia. Carlos D. Arguelles, 5217th Exec Officer–Scout Co. Allied Intelligence Bureau. June 1944

Arguelles headed the 5218th Reconnaissance Battalion Company in charge of the men's training.

battalion adjutant and oversaw the Headquarters and Headquarters Service Company. A medical unit with Capt. Maurice N. O'Connor as surgeon was attached to the unit.

To have men available as a ready reserve for the 5217th' to form missions to the islands, the 5218th Reconnaissance Company (Provisional) was activated to train the arriving men. Commanding the company was Lt. Carlos D. Arguelles, a Philippine Army Reserve officer who had graduated with an architecture degree in the U.S. before the war broke out. He administered the challenging task of executing the training program. The Scout Company, composed mostly of sergeants to drill the men, reported to him. "After I trained the men, they were turned over to PRS for operations," recalled Arguelles. The 5217th was authorized for 520 men but was regularly understrength by 100 men. Silva was the Executive officer, his second in command.

The Battalion was assigned to the United States Army Forces in the Far East (USAFFE) and reported directly to PRS. Procuring and training selected Filipinos who volunteered for hazardous Philippine service was the unit's objective. The battalion was set up as a 'provisional unit' so as not to require approval from the War Department. No Table of Organization and Equipment (TOE) was provided because it would restrict the unusual conditions under which personnel were organized, trained, and committed to missions. Fulfilment of the missions' objectives were to be in line with GHQ's timetable of the Philippine campaign. But without a TOE, promotions of the men were negatively affected. The Battalion's secret purpose sanctioned by General MacArthur, greatly shortened the time to obtain equipment and supplies for the missions.

Men chosen from the 1st and 2nd Filipino Infantry Regiments training in the jungles of Australia for the secret missions to the Philippines that they would be called for.

Seventy-five percent of the men at camp came from the Regiments. Once orders arrived, the men ate and slept together to adjust to each other and allow any temperament unsuitable for the missions to surface. The close quarters were not a problem to the men who were used to living in barracks with hundreds of other men in the crop fields. They filled out questionnaires about their hometowns and spoken languages to be sent on missions to islands with the same language to ease the communication with the locals. A soldier was asked what work he thought best fitted him – radio operator, clerk, runner with military forces in the islands, interpreter, guide in the islands, handling explosives, mechanic, and others.

The secrecy enforced on the camp complicated the administration of hundreds of men. They were a unit created with urgency for a specific need. Camp X was 40 miles away from Brisbane at Beaudesert. And orders for administrative tasks emanated from the GHQ and HQ, USAFFE in Brisbane. The Army procedure of documenting all orders and changes complicated administration of the Personnel section especially when stamped 'Secret' in bold letters. Short emergency orders from PIMC at all hours kept the Personnel Office open 24 hours. Daily Reports, however, were all submitted every 10 days and did not include all radio activities due to the secrecy.

Knowing that music was essential to the Filipino, the 45-member *Bahala Na* orchestra was formed, beginning with only string instruments. Familiar tunes from their homeland and the latest American band sounds were played. One of the officers of the Special Services office, Lt. Edmundo Marfori, sang with the band. The orchestra gained popularity when they were allowed to leave the camp and play at children's hospital and wherever they were invited for special occasions. The Australians enjoyed the Filipino songs and the American type of music they had not heard before.

Marfori, one of the first officers to arrive, was perfect for the job. The secrecy of the missions kept the men in camp, and he recognized the need for recreation to release the tensions of physical exhaustion from the field activities and crammed schedules. Under his guidance, the Special Services office created a full schedule of recreational activities that lifted the men's spirits. Tournaments in softball, volleyball, ping-pong, tennis, and horseshoes were held regularly. Major Brown, the camp's commander, played volleyball with the men and always played 3rd base with a whistle in the softball games. Indoor games of ping pong, darts, cards, checkers, and chess were available. Officers teamed up with the men. A bar was set up to sell Australian beer brought over from the Post Exchange. The men eagerly looked forward to the dances that were held. It was rare to meet a Filipino woman in Australia. "We held parties for our neighboring outfits as well as civilian people around the town of Tabragalba. We had symposiums and directed our energies into sports," recalled Sgt. Cacabelos.

In their zeal to support the war effort, the Battalion made a record purchase of War Bonds. The Post Exchange published record sales for a unit their size. Lt. Ceferino R. Rola, the battalion adjutant, was the War Bond officer and Assistant Commanding Officer of the Hqs & Hqs Service Company. He also served as Personnel officer attending to various interests and personal problems affecting morale. His staff tracked meals per officer at the Officers Mess, created duty rosters, worksheets, and strength reports for the personnel office. The beer, Coca-Cola, and cigarettes he requisitioned with other goods were great morale boosters. There was even a mobile PX that followed the men while training.

A weekly camp newspaper, *Bahala Na,* was launched with Julius Ruiz as editor. From the 2nd Filipino Infantry Regiment, he was assigned to the 978th Signal Service Company to publish the newsletter. As a former editor and publisher of the *Philippine Newsletter,* he was fit for the job. He requisitioned a mimeograph machine to print the newspaper covering the activities of over twelve hundred Filipino and American soldiers at that time in the camp. The men eagerly looked forward to news of the Battalion and Company. Rola was advisor while two 978th men, Sgt. Jose Candelario and Pfc. Marcelo Ovalles, who were commercial artists before the war, drew the cartoons and designs. The men took turns making portrait sketches of their favorite Filipina pin-up sweethearts, born in the Philippines or in America.

The men and officers contributed articles and interviews. An article, *Pvt Paez Takes a Bride,* described the ceremony at Corpus Christi Church in Brisbane. The execution of close-order drill of the graduated troops from the Canungra Jungle Warfare School was announced. An invitation to visit the camp library stocked with newspapers, magazines, and books from the mainland was published. Lt. Al Hernandez ordered five hundred books for the new library. The Red Cross provided reading materials too. The easy access encouraged a love of reading in the men especially those who came from the labor camps and did not have access to books. Hidden talents surfaced and were shared. Painting and acting were popular activities.

An average of five mailbags a day of letters written by men in the camp had to be read by the Censor Section. The piles of letters mounted as more personnel were added. The most demanding Censor Officer was Marfori, who required the greatest number of letters to be rewritten. Letters carelessly mentioning about war plans, military installations, troop movements, and even morale were returned for rewriting in case any of the letters fell into the hands of the enemy. "We wrote so many letters which the censors hated so much because they were deluged with outgoing mail," Sgt. Cacabelos recounted.

At the Enlisted Men's Club, discussions on various topics were held in the Friday night symposium hours. The newspapers and magazines in the library kept the men current on topics such as Filipino-American relations. Problems and prospects in post- war Philippines were a favorite topic. Sgt. Leandro Tanato and Sgt. Juan Dahilig led the discussions. Men who had expertise from their pre-war professions spoke on the subject. Veterans who fought on Bataan shared their experiences on guerrilla tactics: secretly robbing enemy garrisons, using floating mines or unexploded bombs to fill empty shells, and 'ambushcades' (hit and run) as the guerrillas called it. Another topic of interest was comparing the educational system in the islands during the Spanish and American periods and the need to continue the more practical vocational training that was interrupted when the signs of war became imminent.

Paramount in the men's minds was the retaking of the Philippines. The discussions aroused much interest from the men and officers. Lt. Bartolome Cabangbang, a Bataan veteran who was evacuated in one of the submarine missions, spoke on the history of the guerrillas and their situation in the areas they occupied. He spoke about the civilians' cooperation with the enemy to survive, the food situation, and their lives in the Japanese-occupied regions of the islands. Sharing his knowledge of Japanese customs in the islands, propaganda, treatment of the prisoners of war, and spies helped prepare the men's' expectations. He recommended in a document marked 'Secret' supplies and equipment for the infiltration parties. Aside from the personal equipment and supplies of clothing, shoes, arms, ammunitions, and medicine, he listed binoculars and food enough for a month. After that, they would have to fend for themselves. An interesting item with their personal equipment was a housewife outfit as disguise, a *bolo* for the jungle, and a wristwatch which was not allowed in the early missions. Other supplies included a tent for a make-shift office, a rubber boat, a machete, a shovel, a lister bag to purify water, canvas for a boat sail, and fishing equipment. Radio equipment included tools, spare parts, batteries, and battery equipment. Office supplies for transmitting messages such as a portable typewriter with extra ribbons, typewriting paper, envelopes, coding papers, maps, charts, aircraft and ship identification devices, old Philippine money, and even materials for secret ink were recommended to keep records.

The spiritual needs of the men were not forgotten. Most of the men were Catholics and were grateful for the Sunday Masses regularly served by Capt. Herbert Leger and later Capt. Alphonse Paulekas. That increased their resolve to accomplish their mission.

Training at Camp X – 1943 to 1944

As more men arrived, more supplies and equipment were needed. Lt. Ubarro confronted the disruptive problem of paper tracking personnel movement with men shipping out before proper orders were issued. Once issued, correcting the changes in triplicate forms and distribution took up much more time.

In the planning of Philippine intelligence operations, Whitney acknowledged that the men would be under constant fear of enemy capture and influenced by their family and people. To counteract this kind of thinking, he looked to develop further the strength of the "do or die" ideal exhibited by those who had gone on earlier missions. To maintain the high standard of services, the men chosen from the Filipino regiments were further culled after each training course in Australia.

The camp's strength increased as men continued to arrive. Arguelles and his staff, Tallido, and Marfori juggled the training program and progress of hundreds of men. Lectures, instructors, multiple locations of various specialized schools were monitored to ensure that men were not idle. As each course was completed, another was quickly assigned. Men were moved to another required course when limited facilities accommodated fewer men. It was imperative that qualified men were available and ready for a mission as soon as possible and at any time.

The Australian Army taught the intelligence and commando courses with complete facilities. The head instructor was a British commando, Capt. Alan Davidson of the Royal Australian Air Force and had graduated many commandoes of the British and Australian Army. He was assisted by staff officers from AIB. Instructors were veterans of jungle warfare in New Guinea, Malaya, Java, and some European theatres, who gave instructions on infiltration, sniper fire, and hand to hand combat. "We lost many men in the beginning of the war due to our lack of knowledge of the methods and tactics used by the Japanese in the front lines. Now all tactics that the Japanese had used in jungle warfare are taught in this school. It is up to you to learn the best you can now that you are in training." They learned to organize, train, and lead Filipino guerrillas. They learned tracking Japanese boots with a separate enclosure for the big toe and that they ought to watch out for the Japanese returning to carry out their dead on a pole or burn them at the combat site. The men learned to operate Australian infantry weapons and were familiarized with Japanese military equipment and weapons. Lt. Don Mendoza demonstrated various weapons used by his infantry unit in the U.S. Those who tried to avoid classes were removed. Once, Davidson sent a memo to Whitney asking him "to dismiss Lt. Virgilio Felix due to evasion of training at the school and breaking orders."

No information was given to the reasons for the training or possible ultimate destination. Their physical fitness, mental alertness, initiative, and genuine desire for reconnaissance work were evaluated as they went through the rigorous training. Their performance was tracked to determine which individuals qualified as leaders or sub-leaders in a penetration party. The question, "Are you willing to volunteer for service as an individual?" was asked. Once individuals of leadership caliber were identified, they were separately interviewed and familiarized with the training objectives and the nature of their missions.

The men's bodies hardened in the four weeks at Canungra Jungle Warfare School, a remote training camp. They underwent tough, realistic battle scenarios in the jungle and guerrilla fighting to survive. It was a different mode of combat in the thick and dark jungles where rare streaks of light hit the ground. Combat was with small hidden targets and close hand-to-hand while avoiding snipers everywhere. Only a canteen per day was allowed. There was no leave. Discipline was strict, and a physically-fit body was needed to survive the trekking of miles. "The first week was twelve hours of hiking daily, 2nd week was 24-hours out in the field and the 3rd-week general training with the daily twelve-hour hikes. The culmination was the 4th week of river crossing and climbing the McPherson Ranges that resembled New Guinea jungles carrying their own food."

The first group of 16 officers with lieutenant ranks attended Canungra before the end of 1943: Bartolome Cabangbang, Enrique Torres, Carlos Ancheta, Pascual Capiz, Don Mendoza, Jose Mendoza, Luis Padilla, Isidro Samforna, Ignacio Sarmiento, Saturnino Silva, Claudio Tamayo, Joe Tudtud, Paul Mauricio, Manuel Tobias, Frank Ventura, and David Vivit. Back at camp, they were ordered to bring 100 enlisted men to Canungra for training.

Almost all of the men who successfully went through Canungra Jungle Warfare School were commended for their eagerness and determination in preparing for the job. Officers commented that the Filipinos were the best-disciplined soldiers they had encountered in the School as they trekked throughout the terrain. The men were proud of setting "new hiking endurance records through mosquito and leech-infested mountains and rivers." The 10-hour day back-breaking work in the crop fields of California certainly helped their endurance in surviving the course. This was not entirely new to them, since the majority were born in the Philippines' provinces, farms, and mountains. To those who were in the professional field or worked in the city, it was their first time to experience jungle and rugged terrain. But they persisted, realizing their lives depended on their stamina.

Sgt. Esteban Diputado declared that the training at Canungra Jungle School was not "new from their basic training with hiking day and night in full gear. It was a rugged school and too strenuous for our age." The best part was the 3-day passes afterwards to Brisbane even if the Australians mistook them for Javanese, Chinese or even Japanese, he said. They were identified with the U.S. Army by their American uniform and called 'Yankees.' Brisbane had become "a city the Yanks have taken over," described by the Australians.

Eight weeks at Fraser Commando School for amphibious training at Fraser Island followed. Instructions were given on intelligence, Japanese habits, beach landing operations, small boat navigation, raiding, machine guns, attacking small war ships, signal, and cryptography. Around 150 men underwent the two-month course. Officers Arguelles, Cabangbang, Torres, and Tamayo completed the course. Silva began training but did not finish as he was called on a mission in early February 1944. On a list of graduates were Sgts. Isidoro Dacquel, Val Tabios, Isay Real, Custodio Bolivar, and Amado Corpus. Lt. Maynard Hawley headed a team with T5 Larry Soliven, Cpls. Hilario Lobendino, Victor Marquez, Baltazar C. Agbunag, and Felix Bermudes. Other early trainees were Sgts. Florencio A. Baldos and Alfred Despy; and Cpls. Mariano Pilar and Servando Abel.

After the physical hardening of the body, the men were divided into two specialized training programs although they would work together in a mission. Radio operators in the 978th Signal Company focused on radio operations. Infantrymen trained in reconnaissance, sabotage, infiltration, and heavy weapons. They also served to protect the 978th Signalmen and radio to keep it on the air. They were not to be a combat mission but if needed, they knew how to handle weapons and attack. They were to provide support and training to the guerrillas.

The progress of the best radio operators from California was carefully monitored since they were to set up radio networks that would cover as much of the Philippines as possible. They learned to assemble and reassemble radio equipment with special emphasis on the type of radio they would bring until they could do it blindfolded, encoding, decoding messages, and sending them quickly. Most important, they learned how to connect the radios into a network. They learned to devise cryptographic systems if those they carried became compromised or captured by the enemy. Only about a fifth of the group, mostly cryptographic specialists, were non-Filipino. The rest were the selected Filipinos from California "who had volunteered for this peculiarly hazardous duty." Only those who passed all the strict requirements of the training program went on missions. Others were sent to manage KAZ and PIMC radios.

Ind personally instructed three men, Lt. Oates, Sgt. Fred Rodriguez, and Sgt. Manuel Faragan, on ciphers and codes for two intensive weeks to become instructors in the signal school. He later told them, "They will give you as much and as fast as you can assimilate, but whatever information and skill you shall have gained here will be the result of your effort. Work hard so we can drive the enemy hard. The Philippines is waiting for you. Good luck."

A radio station was set up to communicate with the Milton Camp to practice sending coded messages and decoding received

Message center installations at the Philippino (sic) Signal Training Center, Camp X, Tabragalba, Queensland, Australia, 20 August 1944. Photographer: Walter Drexler

message. "All weekday classes were from 4 AM to 4 PM but most ended at midnight. The training...was strenuous...interesting to all of us. The methods which were taught to us were the same as those used in actual operation. (Soon), we were using only two syllables in our private conversations: dit and dah," Galang said. Men were further divided into smaller groups and assigned to various small islands scattered along the Australian coastline to simulate radio field operations.

Galang recalled some of the courses to pass - physical hardening, individual combat with bare hands and with weapons, survival in the forest by recognizing edible roots and poisonous plants, treatment of insect and snake bites, river crossing, Morse code, coding and decoding, identifying enemy aircraft and watercraft, servicing and maintaining radio transmitter/receivers, traveling at night, and signals including the use of lighted cigarettes in the dark. Map and compass reading, judging distance, smoke, and semaphore were added to the instructions. There were lectures on security. Any breaches meant dismissal from the program.

Filipino Signal Training Center

The PX building in camp housed the Filipino Signal Training Center with radios and a practice code room. Photos of Japanese weapons, flag, maps, and military gear covered the walls. Beside a table of training publications stood a life-size dummy of a Japanese soldier in full uniform. The Allied Geographic Section, GHQ, frequently visited the Battalion headquarters and interviewed men familiar with certain territories in the Philippines to create the maps for the Operations Room and the aircraft bombers.

Capt. Alvin Bone headed radio signal training. Men under him were chosen to be instructors: Sgts. Marcos Lopez, Fred N. Rodriguez, Manuel Faragan, Vincent Visaya, Jose B. Parcasio, Leo V. Pomares, and Marcelino C. Alcantara along with Cpls. Simeon R Guzman and John R Alvarado. Shifts of men heavily used the center to learn operating radio sets using Army Signal Corps procedures. These procedures would later become an issue as the men would find out at their remote outposts in the islands' jungles. The guerrillas used a streamlined commercial fixed radio method different from that of the Army. Relearning on the job caused some delays in providing full radio support to the guerrillas. That may have also caused many security lapses in the first six AIB missions.

Men were selected for specialized training and to be instructors. G-3 Operations sent out a memo dated February 1944 listing officers to attend the Australian Military Intelligence School in Maryborough, Queensland. Silva may have been on the list but had already left on a mission. Richard Arca, with his experience as a Philippine Scout, was assigned to the Battalion's Training Group. Capt. Thomas Dionolo and Lt. Silvino V. Tallido trained for four weeks at Sydney's Central School of Ship and Recognition and returned as instructors. Two groups comprised of Capt. Gene Frias and Sgts. Modesto de los Santos, Jimmy R. Esguerra, and Celestino Valdez (first group) and Capt. Rizalito Abanto with Sgts.

Louis Palmejar, Valentine Jugance, Salvador Yotoko, Miguel Paraiso, and Jose Carpio (second group) attended a five-week course at Mount Martha Research School to specialize in espionage and intelligence work. Four officers with Lts. Carlos Ancheta and Jose Mendoza attended camouflage for concealment at the South Port's Camouflage Wing School.

Health and body had to be kept in top form as the men could be called at any time. The intense toll of training for long hours interspersed with the physical work in building the camp necessitated bi-weekly visits by doctors from the 155th Station Hospital and the 42nd General Hospital to monitor their health.

As each man passed the intensive training program in radio, jungle, commando, and specialized courses, instructors wrote down their progress that included emotional stability, initiative, resourcefulness, endurance, perception, and the ability to instruct before transfer to the 5217th Battalion in reserve for mission assignments. By the time they were graded ready for a mission, each man had three specialized schools to his credit combined with battalion training. When Lt. Arguelles asked when he could go on a mission, the reply was he was needed to continue command of the 5218th Battalion to prepare the men for missions.

When an order arrived for a special mission, the 978th chief instructor, the chief of radio training, the commander of the 5217th, and the commander of the 978th Company met to review the roster of qualified men. Orders were sent out emblazoned 'Top Secret' on top. The mission members were called and reminded of the importance of the radio networks they would set up to gather information on Japanese strengths, activities, shipping, and military installations to transmit back to Australia. The information was emphasized as critical in the planning of the return. They were reminded that there was no guarantee that all the guerrillas would welcome them. They were sworn to tell no one and report back with packed gears. An M1 carbine or a Thompson SMG, a .45 cal pistol, and a trench knife or *bolo* was each man's weapon depending on rank. Not all the men went on missions. Some were assigned to KAZ radio station and PIMC for guerrilla radio traffic.

They began to be called 'mission men' those who were ready to be ordered anytime to a mission. Five parties of roughly twenty-five men each were kept in continuous radio and combat training to maintain morale while waiting for a mission order and an available submarine.

Men began to wonder when they would be called on a mission. Unable to leave camp due to maintaining the secrecy of the unit and missions, thoughts of their families' sufferings in the islands intensified. They could not participate in the planned activities of the Red Cross in Brisbane, or the parties, dances, and excursions held by the Australian volunteers for the American forces. If there was room in trucks to Brisbane, small groups of enlisted men were allowed a rare pass privilege.

The officers were given passes to go into Brisbane, making security challenging to control. Various offices of GHQ in Brisbane were open 24-hours all week as Army vehicles crowded the streets with free rides for the men in uniform. Trucks were dispatched to the city twice a week after reminders were given of not talking about their camp and unit. But after a few drinks at bars, dances with local ladies, and the like, some lips loosened up. A Filipino sergeant was reported to G-2 Intelligence for spreading "the rumor of troops trained in sabotage and 5th Column activities landing on rubber boats from submarines with the mission to work from the inside (Philippines)." Lt. Silva was in Brisbane enough times to court an attractive Filipino Australian woman with the Australian Women's Army

Service (AWAS). The *Bahala Na* newsletter published a wedding announcement: "*Lt. Silva Takes Bride.* A flower- bedecked altar in the camp's lawn was the scene of a beautiful wedding January 8 of Lt. Saturnino R. Silva and Miss Priscilla Conanan, Australian-born Filipino and a member of the AWAS. The bride in white organdie looked gorgeous and lovely. She was given away in marriage by Major Brown, Post commander. Lt. Carlos D. Arguelles commanding officer of the Recon company was best man. Gifts of all kinds were given by officers and men of the camp. – Sgt. Vidal F. Alvarez."

Keeping the camp secret was a challenging task with the hundreds of men. An officer of the Signal Corps reported to PRS of two officers of the 978th Signal Co. who openly talked about the submarine landings in the Philippines to a person outside of the company who was not supposed to know this information. A Counterintelligence Corps (CIC) report concluded that the high level of boredom with the continuous training and anticipation of a mission amongst the men caused loose talk outside the unit. A Filipino CIC agent was ordered to investigate. Brown was ordered "to take action to impress upon all the urgent need for security in their public conduct." A series of security lectures were inserted in the training schedule. But that did not prevent leaks to other units.

A large group led by Lt. George T Baker from the 2nd Filipino Infantry Regiment arrived in November. All immediately began intensive training in anticipation of a mission: Sgts. Aguilino M. Corpus, Frederico R. Villarta, Robino Africa, Alfonso Ancheta, Jose E. Candelario, Abamile L. Mauricio, and Carlos A. Ortiz; T4 George T. Pajara; Cpls. Eugene M. Ergas, Materno P. Peralta, and Leon V. Pomarez; Pfcs. Juan D. Alcaras, George U. Arce, Romain S. Carrasca, Numeriano D. Davin, Sergio B. Estavillo, Alfred T. Flores, Antonio Maravel, Marcelo C. Ovalles, Bartolome M. Padilla, Joe E. Peredo, and Gregorio Santos; and Pvts. Arsenio B. Bumanglag, Frank A. Hufana, Teodoro M. Jose, Fred M. Madariaga, Peter C. Roduta, and Leoncio S. Sapon.

Those who later went on missions were Sgts. Emiliano A. Almero, Conrado A. Gustilo, Salvador B. Fortun, Gerardo B. Nery, Angel T. Sarmiento, Daniel B. Begonia, Ramon F. Cortez, Vincent C. Goloyugo, Gaudencio V. Guyot, Donato E. Herreria, Jack Montero, Paterno A. Ortiz, Carlos S. Placido, and Jamie R. Reynoso Jr.; T4 Jose T. Gonzales; Cpls. Urbano B. Barcenas, Enrico M. Bugarin, and Teodoro J. Rallojay; T5s Richie D. Dacquel, Fortunato C. Dagandan, Hermenegildo L. Villalon, and Julio C. Advincula; Pfcs. Eddie C. Holgado, Patricio Ollada, Angelo E. Pascual, and Alipio C. Toribio; Pvts. Roberto O. Angcos, George M. Bulatao, Edward M. Burgos, Leodegario O. Nuevo, Gaudioso L. Patubo, and Rudolph Santos; and Pvt. Patricio Ollada.

As hundreds of men underwent training, Whitney requested authority to land parties close to Luzon. In November, PRS ordered the first mission under its control to Mindoro, the island closest to Luzon to watch the entrance to Manila Bay. A second mission to Samar soon followed in the next month. (The missions to the islands are discussed in its own chapters).

The continuous influx of personnel as the camp was being built, the men undergoing continuous training, and the fact that no passes were issued caused problems requiring an inspection of the Battalion. The Inspector General's report declared that the unit was on the station list as USAFFE troops but was directly responsible to AIB, causing some confusion. An American planning and training officer in direct charge of training the 5218th unit was relieved. His use of humiliation, threats of consequence, and abusive language, made the men lose confidence in him. "No man will be placed under the leadership of another who is not willing to subordinate himself to that leadership." None of the men wanted to be on the same mission with him.

A significant finding was the Battalion's lack of a Table of Organization and Equipment (TOE) which precluded promotions. The TOE provided organizational staffing, and equipment of units. The men had left the 1st and 2nd Filipino Infantry Regiment where promotions were possible, to do a job of a hazardous nature for which they expected recognition. In their present situation, future advancement was lacking.

Another finding was the case of fifty men who were recent arrivals waiting for their orders to start a training program. In the haste to bring them to Brisbane, complete uniform and equipment were not issued causing embarrassments of segregation while working on details around camp.

Men expressed inconsistencies with what they were told about the mission and how training and administration were handled upon arrival at camp. Major Brown was cited as a capable officer but "handicapped through interference, lack of complete authority, and the temporary situation the Battalion was organized and attempting to function." He had to obtain approval from Whitney in all his decisions. Whitney's response to the low morale of the men was due to their intense desire to be sent on a mission which he considered a healthy sign. Other deficiencies cited by the Inspector General were in the record-keeping of maintenance and storage equipment, receipts, Morning Reports, Daily Sick Book, Duty Roster, and sales transactions which were missed in the chaos of camp administration.

Building the camp non-stop seven days a week brought complaints to the Inspector General of discrimination in giving passes. Some individuals who had returned from earlier missions due to sickness or per orders and were permitted to address the members negatively affected the men's morale. They proudly stressed their own individual accomplishments to the men who had been training hard and waiting to be sent on missions. Within two weeks, Brown responded to the irregularities. The response to not allowing Filipino enlisted men into Beaudesert, the closest town, was for security reasons. To defuse the lack of passes a movie projector was brought in for movie nights. "For reasons of security, the men of this unit are seldom permitted to leave the post on other than training missions, resulting in a morale problem that the frequent showing of motion pictures will help to solve. Signed by Courtney Whitney."

Before the end of the year, Camp X reported 350 men at camp. Two missions under PRS had departed for Mindoro in November and Samar via Mindanao in December. Three months later, a memo from Lt. Rola, the Assistant Adjutant, showed the Battalion's total present and absent men at 616. At camp, 424 were present while the rest were on missions and assignments.

Christmas holidays were the men's first time to celebrate it overseas, and it was unusual. They were issued heavy winter uniforms in preparation for cold months and some snow, but Australia was hot and humid due to its geographical location.

1944

Men continued to arrive in large groups throughout the first part of 1944. In January, more qualified men arrived at Camp: Sgts. Leandro P. Tanato, Fred N. Rodriquez, Emil J. Sibonga, Marcos Lopez, Manuel F. Faragan, Raymundo E. Agcaoili, Mariano T. Arce, Fred

Balcena, Leovigildo M. Giron, Balbino R. Padilla, Semeniano D. Acenas, Julio V. Balleras, Jacinto Cutaran, and Fred C. Ignacio; T5s Julius B. Ruiz, Emilio A. Alegre, Don J. Veray, Daniel S. Palmeto and Al C. Aglubat; Pvts. Isidoro C Marin, Manuel V. Andres, Lope G. Batara, Victor I. Vicente, Anaceleto Biado and Alex A. Nagtalon; Cpls. Atanasio L. Alcala, Alejandro A. Buccat, Telesforo Asuncion, Isidro P. Magale, Felix T. Reyes, Porfirio Balino, Agrifino J. Duran, Ambrosio Navarro, Domiciano O. Boncavil, Angelo Calmorin, and Pete L. Luz; T4s Domingo Logan and Mariano A. Medina; and Pfcs. Alejandro C. Aquino, Juan F. Rimando, Selmo C. Saladino, Eustaquio L Juan, Florendo R. Pedro, and Geronimo M. Dela Pena.

In the same month, enlisted men who had arrived earlier and had been on duty with AIB transferred to 5217th and 5218th Recon Battalions for further training. All orders were stamped 'Secret' and signed by General MacArthur. Those who arrived at Camp X were Sgt. Alex B. Legaspi; Cpls. Cirilo G. Lopez, Paul P. Villote, Pete S. Guion, and Ceferino P. Sison; T5s Toby R. Marinas and Amadeo C. Cendania; Pfcs. Ray H. Abadilla, Lorenzo M. Guieb, Filomeno A. Laurina, Remy B. Cabrera, and Dionisio H. Ventura; and Pvts. Simeon R. Guzman, Ricarto S. Marzam, and Fulgencio L. Mejia. Those who later went on a mission were Sgts. Ronald A. De Torres, Henry E. Runtal, Doroteo V. Vite, Marcelo B. Umipeg, and Carl Wright; Cpls. Cipriano C. Fernandez, Lucas D. Runes, and Astor C. Parong; T5s Rosendo M. Palafox; Pfcs. Cleodoaldo J. Isaac and Aquilino B. Ramos; and Pvts. Edward E. Goroza, Juan C. Marcelino, Carl M. Mateo Roberts S. Corpus, and Anselmo D. Moscoso.

By March, the last group of Filipino volunteers arrived: Pfcs. Estanislao Peralta, Ben V. Silverio, and Feliciano Rodriguez; Pvts. Don Liberato, Ambrosio Pabros, Calixto Cortez, Estanislao Martin, Bernardo P. Aquino, Aproniano E. Calip, Armand Manalo, Segundo Q. Albalos, and Frank Salvador. A truck transported them to Camp where they met men burnished by the sun. Showing their calloused hands, they were "informed of the grueling five months of hard labor." But they proudly pointed out the neat rows of tents, gravel walks, recreation building, supply building, basketball courts, and baseball park, they had erected.

A memo announced the arrival of the following men: Sgts. Rufino F. Cacabelos, Ray F. Castro, and Miguel E. Ignacio; Cpls. John A. Ilusorio, Roberto A. Randall, and Isaias T. Torio; T5s Namaso Nebre and Doroteo M. Pena; Pfcs. John R. Alvarado, Victor P. Enderiz, Pete G. Felizar, Juan L. Gonzalez, Catalino U. Morales, Victor L. Piana, Doroteo A. Rosales, Andrew S. Avila, and Pedro P. Tabofunda; Pvts. Edward J. Padua, Martin Amaral, Philip G. Milana, Dick V. Obrero, Bernardino B. Abang, Marcelino C. Alcantara, Adolfo C. Cayabyab, Telesforo B. Astronomo, Mauro G. De Peralta, Chris G. Domantay, Norberto C. Campos, Artemio C. Espiritu, Hipolito M. Gatchalian, Marcelito H. Garcia, Arthur H. Garcia, Arthur V. Gonzalo, Presciliano B. Gurrea, Manuel E. Riotoc, Juan A. Saria, Jr., Pascual P. Seares, Lorenzo F. Taban, Pedro R. Penalver, Alberto U. Ulivas, Frank G. Valmonte, Jose F. Viduya, and George R. Dulay. With them were Sgt. George B. Bueno, Pvts. Carlos D. Udani, Dale Y. Ogoy, Regino T. Patacsil, Benjamin B. Marcelo, and Vincent Balinton who later went on missions.

Pfc. Edward Goroza, a signalman recalled: "We traveled in a big transport ship together with members of different services from the Army, Air Force, and WACS. We sailed along the coasts of North and South America to avoid enemy detection, especially from submarines on the prowl. Halfway along South America, we turned west, passing by New Zealand and landing in Port Moresby, New Guinea in the early part of 1944. We disembarked and camped in the area for several days waiting to be transferred to Australia." After a few days, his group arrived at Camp X. "We had our briefing and further underwent polishing of our skills learned back in the States." His group of 70 communications operators were soon ready for assignments.

Officers on General MacArthur's staff who knew of the special unit and skills, requested a visit to the camp. Whitney was quick to warn that it was a training camp for missions and not a replacement center to provide men. The men had already been selected and assigned to parties whose departure on approved mission were on schedule. They can look the camp over but "any plans to utilize the personnel should submit a plan to determine feasibility for discussion."

In the first six months of 1944, missions were sent out to Panay and Mindanao for supplies to Tawi-Tawi and Cebu, Samar, and Palawan. Two of the missions carried only supplies. One reason was men who arrived from the U.S. had not yet passed the advance rigorous training in radio and reconnaissance in time for the missions. In a group of 99 men slated for Luzon penetration parties, only 27 passed. Of those who arrived in Australia, "the remainder (would not fulfill) our purposes. The regimental commanders probably substituted to preserve the better men for their employment," said Whitney justifying the low number who arrived and passed. In another list of 60 men for 978th Signal training, only 17 passed the intensive training.

The Allied Geographic Section, GHQ needed maps, information on the terrain of the islands and surrounding waterways to decide on where best to send the next missions. Men familiar with various islands were interviewed at camp for location of possible landing sites. The planners were cautious and wary as the first mission to Mindoro was a tragedy. (The Mindoro mission is discussed in a separate chapter).

Capt. Croell, commander of the 978th Signal Company, sent a roster of radio operators to PRS listing 11 officers, 4 warrant officers, and 159 enlisted men. They were identified as absent from the area on special assignment or in training. The enlisted men were divided into 37 White Americans and 122 Filipinos.

Whitney sent a memo in March to the Chief-of-Staff saying "they now have 200 qualified men who have undergone jungle warfare training and now ready to go when transportation is available. Spirit of the men proven since the camp established Aug 43, no court-martial, and no cases of V.D... half of the men in the earlier mission were infected and caused the last-minute delay." The only company punishment was meted out to T5 Julius Advincula for drunkenness.

At this time, sabotage training was added for the troops to destroy bridges, roads, and ammunition dumps. The policy was against sabotage, but if GHQ required it, they were ready. A cadre of six officers arrived from the 82nd Airborne Division in Italy to train a group of the men on sabotage and communication disruptions.

A course on weather forecasting essential to accurate bombing of enemy targets was added. Men chosen to be weather observers or instructors had some meteorology experience or Army aptitude grades above average as well as good educational qualifications. Six groups of fifteen each attended a month-long training at the Weather Observer's School at Amberley Airfield in Brisbane with fieldwork at Red Lands Bay. Forecasting typhoons, cloudy skies, and heavy rainfalls were crucial in planning transports, equipment, and supplies for a successful operation. A combination of good weather observation and radio skills provided hourly weather forecast information to GHQ for the bombers to accurately bomb their targets.

The Battalion received American officers of the Coastal Artillery Corps with coastal, harbor, and aircraft defense skills. They quartered with the men at Camp   to acquaint themselves with Filipino ways and habits before going on missions with them. Lts. Edward T. Pompea, Robert L. Ungvary, Sidney S. Rexford, and Leon C. Tinnell, from the 583rd Signal Aircraft Warning (SAW), arrived to instruct and form groups of SAW ground observers to report on enemy air traffic. Two American platoons from the 597th SAW Battalions joined the men at camp for future missions.

Changes in plans and orders multiplied coming from GHQ. In May 1944, orders were issued to release men from the 5217th to the 1st Filipino Regiment for various reasons: Capt. Luis P. Morgan, Lts. Rosalino A. Abaya, Pascual Capiz, Enrique J. Aquino, Vicente D. Carag, Marcial A. Edillon, Pedro A. Ginete, Raymond B. Licudine, Don Mendoza, Isidro B. Samporna, and Vincent F. Zerda; T5 Julius D. Adricula and Pvt. Mercelito H. Garcia. No reason was given in the order. But Capiz did not transfer out as he was sent on a mission in September. The officers may have been transferred to Philippine Civil Affairs (PCAU) and Counterintelligence (CIC) departments that were part of the Philippine landing plans.

Unable to leave camp easily, the men created their recreation. Tabragalba was their home in a foreign country, and many friendships developed with the Australians and Americans who came to enjoy their social gatherings. Mothers from neighboring farmlands brought the men cakes, sandwiches and tasted the Filipino dishes they cooked.

From camp, the men hiked across the mountains to buy pigs and chicken from the McVey family. A roasting pit for the lechon (roasted pig) was set up between Biddaddaba creek and a lagoon to share with the family. Children watched, mesmerized at the crackling skin of the pig as it rotated over hot charcoal. They brought candy and Coca-Cola for the children and opened 'billy-cans' (large metal pails) of beer for themselves, which by the time they arrived at the farm had grass seeds floating on top.

The Cawley family that lived near the camp wrote that the men enjoyed riding their horses, learning to milk a cow, and "being taught our dances – especially the Gypsy Tap. They treated us to Coca-Cola, candy, Camel cigarettes, California walnuts, and other goodies from their PX stores – these treats were unavailable to us because of rationing and shortages." A photo showed the men playing their guitar and ukulele beside a female member of the family strumming a banjo.

The *Bahala Na* newsletter published the Mothers' Day picnic in camp sponsored by the 978th Signal Service Co. A large crowd of American and Australian civilians, officers, and enlisted men attended. Mothers and their families were invited

MOTHERS DAY *Program*

Speed the Victory
BUY WAR BONDS
COURTESY OF
978 Signal Service Co.
MAJ. H. T. CROELL, *Commanding*

SUNDAY. MAY 7th 1944

from neighboring farms. Under towering eucalyptus trees near the creek, speakers paid tribute to mothers, the foundations of real family homes. The menu featured corn-fed pork lechon, a Signal Corps special and delicious hot rolls prepared by the Bataan Camp Bakers. Visitors were told not to bring their ration books as a bounty of dishes were served.

In an interview, Dulcie Cunningham, a neighbor whose cows would wander into the camp and eat everything in the food patch next to the men's tents until it was bare ground, related "The men from Camp Tabragalba were not allowed to go to town so they invited many townspeople. They constructed a big platform as a dance floor... pigs in a spit... 'lollies' and cigarettes. On Mother's Day, they pinned a flower on every mother that came out there. Food was rationed, and you had to have coupons, but we never had any shortage of anything in the country because we had all these camps and all these people bringing in food to the various houses. Several of the Beaudesert girls married the men. Officers did not go into dances in Beaudesert because they had their officers club in Brisbane, one hour away by car."

Lt. Arguelles had a dancing pavilion erected and decorated with fronds, colored paper, and lanterns to look like a small-town fiesta. "Jitterbug contests were held on the portable dance. I was small, sort of their height, and had this tap dance. I conducted kissing games and undignified musical chairs," related Cunningham. Filipino music was played. The officers gave morale boosting speeches as the dance floor became crowded with Australian girls, nurses, and AWACs.

In the week after the Mother's Day celebration, 25 Filipino radio operators, officers and reconnaissance men left on two missions, with White American weathermen accompanying them.

Availability of the men continued to be tracked as more missions departed. In June, G-3 Training published that there were 30 officers and 238 enlisted men in camp and requested 100 men to attend military intelligence school. At Canungra Jungle Training School were five officers and 107 men. Six officers and 30 men were at Fraser Island. Eight men were training as weather observers at Amberley Field and Archerfield weather stations. Nineteen officers and 81 men were left at camp of which 60 were already formed into missions within the next 30 days. Therefore, the request to immediately assign 100 men to attend the Intelligence School was impossible and could only start with officers and men as they became available.

GHQ's initial decision to land in Sarangani Bay in Mindanao followed by a strong offensive on Leyte and Samar in December changed. Air attacks by the 3rd Fleet over the Visayas and Mindanao produced hardly a counterattack from the enemy. Guerrillas in Leyte and surrounding islands confirmed last-minute intelligence on the weak enemy presence. Leyte was also closer to Manila, the prize on Luzon. When the landing date was changed to Leyte two months earlier in October, the pressure to produce qualified men at a faster rate for missions greatly increased. Men were on high alert as they continued to drill and maintain readiness.

The new landing date readjusted the missions' timetable. More radios and men were needed to be sent to trusted guerrilla leaders in Panay, Negros, Mindoro, Samar, and Mindanao to expand coverage of enemy dispositions. Luzon had still to be penetrated. Luzon guerrilla leaders had been trying all this time to get a radio from Panay and Mindanao, which had contact with KAZ. From May to October, multiple missions with radio, demolition, aircraft warning, and weathermen departed to various islands. The *Narwhal* and *Nautilus* were landing men and supplies non-stop without undergoing the vessels' scheduled overhaul. Skipping their scheduled yard overhaul placed the submarines at great risk. The urgency to set up more radios and gather more intelligence in a shorter time convinced the Navy to divert four operational submarines to augment the cargo subs.

Whitney became concerned with possibly losing a submarine. He wrote a memo to G-2 Intelligence asking them to stop sending cigarettes, chocolate bars, *Newsmaps, Life* and *Victory* magazines he had ordered, or other American-made items with the missions "It would be a direct invite to increased Japanese patrol activity," he reasoned. Later, he rescinded when Japanese propaganda was overpowering the morale of the Filipino population. The Japanese controlled all media and purged anything American or in English. A concerned War department cabled the missions to maintain security and take every possible precaution.

Propaganda materials: Cases of matchboxes, cigarettes, and chewing gums printed with "I shall return."

A request was sent to the Office of War Information for two million books of matches treated with wax to resist dampness and tropical conditions. The matchbooks were printed in color with one side a crossed American and Philippine flag and on the other side, the phrase 'I shall return MacArthur.' This was overwhelmingly approved by General MacArthur. Hundreds of thousands of *Free Philippines* were printed in Australia and shipped to the islands in submarines. No one could miss 'I shall return' in bold letters across the front page. Photos and news of the war's progress filled the pages. The same phrase appeared on cases of cigarettes, chewing gum, candy bars, and sewing kits. Almost every village had a sewing machine and generator, but needles were needed as they easily wore out from sewing the heavy *abaca* (leaf fiber) material for the people and guerrillas' clothes.

The 5218th Reconnaissance Co. organized a *Salo-Salo* (a get-together). An early morning Mass in the Recreational Hall began the celebration, followed by an Assembly of Command in the tent area, the reception, and a luncheon. The national anthems were played before the program began. A note from Arguelles, who designed the program cover, welcomed the attendees, and expressed thanks on behalf of his command. A short story of the *Salo-Salo* in

the program explained the Filipino custom that symbolized "the Filipinos love for fun, comradeship, feasting, and everything beautiful."

As a community affair, everyone brought whatever they wished. Music and games were played while cooks prepared dishes in abundance. A barbecued pig, roasting on a spit patiently turned by the cooks, was the main course with 'Arroz con Pollo,' a Spanish rice dish. According to the men, there had to be rice on the menu or it would have been an incomplete meal. The highlight of the fiesta was the selection of a *Lakambini* queen, the prettiest among those present. The celebration was not only "to perpetuate Filipino customs but to forget for the moment, the stark and brutal realities which have brought many of us here to this friendly and hospitable country." A festive dance ended the celebration.

Behind the happy occasion was the tense anticipation of orders for a mission at any time. Seventy-one 978th signal, officers and reconnaissance men missed the *Salo-Salo* celebration as troops had been departing every two weeks on the eight missions until September, the month before the Leyte landing.

The good news of the victory at the Battle of the Philippine Sea in June boosted morale. The capture of the Marianas Islands and Guam sliced the Japanese supply lines. In the spirit of comradeship and patriotism the camp celebrated the Fourth of July and Jose Rizal Day with Americans and Australians.

Gen. Basilio Valdes, Chief of Staff of the Philippine Army, visited the Camp. He was impressed with the "cleanliness of the camp, clean and orderly appearance of tents, and particularly the enthusiasm of the Filipino soldiers undergoing training. He was assured that the men would render the same efficient service when the time comes." Sgt. Sotero Libatique was sent to him fulfilling his request for a soldier with photographic experience.

Anniversary issue August 1944 celebrating a year at Camp Tabragalba.

Paratrooper training was added to the men's packed schedule for air drop assignment in the islands. Capt. C. E. Walters from the 503rd Parachute Infantry led the many eager volunteers, including Arguelles, head of the 5218th Recon Company. (A separate chapter discusses the paratroopers).

Dozens of men continued to disappear from camp at all hours, sworn to not inform anyone of their orders. Five more submarines departed loaded with men and equipment as the landing date drew near. Everyone was anxious to be called.

It had been a full year of many accomplishments in creating Camp X (Tabragalba), training the men, and seeing the mission parties sent out. To raise morale, Brown ordered an August anniversary issue of the paper *Bahala Na* commemorating the Battalion's first year at Camp. It was a historical edition, depicting the progress made in a year of the Battalion and remembering their comrades in the remote islands risking their lives behind enemy lines. The editor's closing words noted the men's willingness to die for their homeland and America. It was a special issue dedicated to the lasting relationship between the Filipinos and the Americans. Julius B. Ruiz was the editor-in-chief, Juan F. Dahilig managing editor, Jose E. Candelario as art editor, Sidney M. Malitovsky as production editor, and Marcello Ovalles as associate editor. Artemio I. Ibea was the society editor. Poems written in their various languages extolled devotion to America and the Philippines.

Officers and men contributed the articles on the camp's activities and subjects of interest. Although published in a limited facility, the issue covered many of the headquarters' activities and the men's training program. There were articles highlighting the daytime classes held by the Enlisted Men's Club on various topics, games at night, and Masses on Sundays. The USO Club had sponsored holiday parties, bought Cokes, books, and even razor blades for the men. The *Bahala Na* band was featured playing Filipino, Western, Hawaiian songs, and touring with men with good voices. When they could get a three-day pass, they played at dances, special gatherings, and hospitals in Brisbane. There is a memo from Gen. Willoughby, head of G-2 Intelligence, thanking for the use of the band at celebrations resulting in the enjoyment of guests and success of the affairs. Marfori was lauded for "splendid work that assisted materially in cementing close and friendly relations."

The anniversary issue also described volleyball, badminton, and cards among the many recreations for the men. How to play *Sipa* was described. The *sipa* ball was made of thin bamboo strips woven into a ball about the size of a grapefruit, and teams kicked the ball in pairs, passing from partner to partner and not letting it hit the ground. A softball team exhibited their skill against teams from various units. Benjamin Marcelo, Marcelo Umipeg, and Segundo Bucol displayed winning streaks making their team win all the games played in Australia. Not to be outshone, Carmelo Sison, Alfonso A Peralta, and Patricio Ubarro starred in ping pong matches. Emiliano A Almero and Leodegario Nuevo also won ping pong matches. The horseshoe tournament was the specialty of Alvin Bone, Melecio Montesclaro, Albert Halla, and Arthur Gonzalo. Fortunato Santos and Eustaquio L. Juan who represented the Philippines in the Far East Olympics held in Tokyo before the war, headed the chess tournaments.

Native Filipino game of Sipa in progress at recreation field of Co. C., 2nd Bn. (1st Filipino Regt at Oro Bay, New Guinea). Object of the game is to keep wicker ball in air as long as possible while kicking it to other players. Photographer T/4 Ernani D'Emidio. SWPA-SigC-44-12721. May 18, 1944

An article described the pioneers of Camp X. Maj. Lewis Brown III, commanding officer of the Battalion arrived first followed by Galang, Arguelles, Songco, Marfori, Tallido, and Silva. A bond developed between the men as they were together from the beginning of the Battalion and underwent the back-breaking construction of Camp interspersed with intensive training. They split up when Capt. Galang, Lt. Songco, and Lt. Silva were called on separate secret missions.

The men appreciated the treatment they received from the officers. They wrote about the company's accomplishments that reflected the capable administration of Arguelles. He was highlighted with his "dashing personality (winning) the admiration of all personnel in the company." In assuming his duties, he balanced the men's health, recreation, and

1943 – Camp X (Tabragalba), Australia. From left: Ricardo Galang, Ruben Songco, Edmundo Marfori, and Saturnino Silva. Except for Marfori, the men went on secret submarine missions to the Philippines.

diversions to promote training efficiency. Deserving men who complied with existing army regulations received passes and furloughs. He laid stress on the absolute cleanliness and orderliness in the troops, their tents, and tent areas. Sunday morning inspections were expected of every man and tent of the company. He endeavored to make the best possible use of each man's occupational skills, abilities, and aptitudes in the company. Each man was assigned to the type of training he could most readily absorb, which allowed each man to render the greatest value to the service. His leadership fostered goodwill and harmony among the men, and the company with the neighboring population around the camp.

The Unit's first anniversary was commemorated against a backdrop of U.S. Task Forces deep in the taking of Morotai, the closest island to Mindanao. Immediately two airstrips were prepared for the assault on the Philippines.

The warehouse on Logan Road in Brisbane bustled day and night with the loading of equipment and cargo into transport planes and military trains to Darwin for the submarines. Missions departed every month, sometimes two, ahead of the October landing.

Messages from the Philippines poured in, concerning mounting Japanese attacks. Emilio Quinto, who had returned from Negros (January 1943 mission) to recover his health from malaria, set up a station in Darwin with a group of Filipino operators in the advance supply depot for future missions. More missions were to depart and travel time had to be shortened. The move to depart missions from Darwin cut the route and time in half as compared to departures from Brisbane.

Back at Camp, efficiency ratings had to be filled out for the Battalion officers as it had been a year since activation. But the ratings had to wait for ten officers who were in the Philippines on a mission. Silva was one of them.

The devastating news of the Kill-All policy from the Tokyo War Ministry unnerved GHQ. All remaining POWs in the Philippines would be killed. The landing date in Leyte became more imperative and final.

Announcing their return after three years, the U.S. forces began bombings over the Philippines with the first raid over Mindanao followed by Manila and its harbors. Looking up and seeing dozens of bombers without the menacing red dot must have brought mixed feelings of relief to the people as they sought shelter. History books state raids started in September, but some residents of Davao insisted it was in August. KAZ in Australia received the 'flash' reports from the coast watchers on the ground within five minutes after sighting the planes. The net control station in Mindanao was on 24-hour standby and used a separate circuit direct to U.S. Air Force headquarters. Messages were in code but in short and complete clear text for faster distribution.

September brought devastating news. The submarine *Seawolf* failed to return on a mission run and was believed to have been sunk with the loss of everyone onboard. The news shook the men with the realization of the great danger they faced. They might not return from an almost suicidal mission. Sgt. Ibea, the avid farmer who created a vegetable garden around the camp and the society editor of the *Bahala Na* Anniversary edition, was onboard.

Marfori tried to ease the tension and fears by hosting a dance two weeks later for the men. He was acting commanding officer of the Battalion while Arguelles was hospitalized for a broken leg during paratrooper training.

Brown sent out a memo to PRS with the list of available radio operators and volunteer reconnaissance men trained and ready to go on a mission. To the men, he emphasized reminders to concentrate on intelligence work. At their outposts, they were to limit the messages on the plight of the civilians as they took up precious time on the air and diverted the focus away from the enemy activities which KAZ needed for planning. The Office of War Information requested 18 recon men and 8 radio operators but only eight operators were qualified. All others had departed on missions.

From the comments of earlier missions, the Chief of Staff received the latest changes of supply shipments to the islands: 60% Ordnance (arms and ammunition), 10% Signal and engineering (radio and parts), 10% Sundry (specific requests), 8% Medical, 7% Quartermaster special requests and propaganda items, and 5% Currency tools (plates, paper, ink).

As the return to the Philippines neared, the scope and scale of PRS activities had expanded to the point that it needed to coordinate with other SWPA operations in the Philippine campaign. PRS was split up. Intelligence matters came under G-2 Intelligence's direction of guerrilla activities, Personnel passed to G-3 Operations and Supply to G-4 Logistics. Whitney continued as overall active head but was now assigned to planning the Philippine Civil Affairs Unit (PCAU) services for the Philippines. A memo to the Chief of Staff listed the PRS objectives that were met as well as its accomplishments and duties. It had successfully coordinated with the scattered bands of guerrillas in supplying them with supplies, radios, and gathering intelligence. To accomplish this, it had formed and trained a secret Battalion of Filipino radio operators and reconnaissance men and sent them behind enemy lines in the Philippines. The duties of PRS were described to ensure the smooth continuation of support to the missions in the Philippines. It was of utmost importance to the men at their remote outposts in the islands that the consistent and responsive support they received from PRS continued. It was their lifeline to be able to continue their mission.

But with the split of PRS, the mission teams on the islands highly suffered. They were not fully informed of the changes. Some did not receive their requests   for supplies to maintain their outposts for months before the Leyte landing when radio traffic was at its height. Starving and sick, they reused radio and cipher supplies. Low in ammunition they ran and hid to protect their stations. Numerous times, they desperately asked the guerrillas for protection and food.

After PRS's inactivation, the Filipino soldiers who departed on missions during the October landing in Leyte were identified in reports as radio operators although they may have been 978th Signal, demolition, weather, or aircraft warning men with radio training. The next missions carried Filipino and White American men assigned to the 597th Signal Aircraft Warning (SAW) Battalion and Weather Squadron.

Preparations were made for the problems the Army would face after landing. The Secretary of the Interior, Harold Ickes, in charge of U.S. colonies and Commonwealth wanted to take over the Philippine government and execute those who had collaborated with the Japanese no matter the reason or situation. MacArthur rejected the order by creating the Philippine Civil Affairs Unit (PCAU) and Counter-Intelligence Corps (CIC) teams made up mostly of soldiers from the 2nd Filipino Battalion. These groups handled the basic needs of the civilians and the control of collaborators until they were turned over to the Philippine government once it was restored. (The PCAU and CIC are discussed in a separate chapter.)

Leyte Island, Philippines Landing – October 1944

While General Yamashita was arriving in Manila to take command of the Japanese Army, the convoys of Allied Forces were zig-zagging their way to Leyte. Much of the location and activities of the Japanese forces were already known and destroyed by the guerrillas before landing. "By the time our forces hit the beaches at Leyte, we had 134 radio stations, 46 on Mindanao, 23 on Panay, 21 on Luzon, 13 in Negros, 11 in Leyte, 6 on Mindoro, 5 on Palawan, 3 each on Cebu and Samar, and 1 each in Bohol, Masbate, and Tawi-Tawi. Twenty-three weather observation stations operated from southern Mindanao to northern Luzon, from western Palawan to eastern Samar. All were in regular communication with our headquarters," proudly wrote Whitney.

In parallel, four missions landed in the surrounding islands carrying radio operators as well as recon, aircraft warning, weather, and demolition men plus tons of supplies. Left at Camp X were 22 officers, 84 enlisted men, 12 paratroopers and 63 signalmen.

In a memo dated November 1, 1944, highlighting the submarine missions, Whitney wrote: "In the first twelve months of its operations, PRS oversaw thirteen submarine missions and twenty-seven insertions into the Philippines to extend SWPA control in the islands."

Hollandia, New Guinea – November 1944
The 1st Reconnaissance Battalion (Special)

The rest of the 5217th Battalion staged in Hollandia in preparation to go to Leyte. An Administrative and Troop Movement Directive dated November 9, 1944, ordered the 5217th Reconnaissance Battalion (Provisional), including the attached 978th Signal Service Co. to pull out of Camp X. The Training Section with all their equipment, moved with the Battalion. Brown asked Whitney for the TOE as the unit was still a provisional battalion. That would allow the entire unit with their vehicles and equipment to ship intact in one vessel. The unit completely shipped.

While on the high seas onboard the USS *Booth* to Hollandia, Dutch New Guinea, the 5217th Battalion was inactivated. An order released the men and assigned them to the 1st

Reconnaissance Battalion on November 20, 1944. The total strength was 67 officers and 388 men, below the authorized strength of 530 men due to the difficulty of securing trained volunteers for the dangerous missions. Another report listed a total approved strength of 85 officers and 445 enlisted men. The new Battalion composed of a Scout Company and Column that replaced the 5218th Battalion; paratrooper platoon; Headquarters, Headquarters and Service Company; Medical Detachment; and the cadre Scout Company that was the training unit of the 978th Signal Co.

The men celebrated their arrival in Hollandia with a Mass held by Father Dietzel. They were now in the area of combat operations. Building their headquarter camp in Hollandia was not a totally new experience after Camp X. The new camp near Sentani Airstrip (Jayapura) was close to the one used by the 1st Filipino Infantry Regiment who had arrived earlier. It was under big, tall jungle trees with a nearby creek. The trees afforded protection against the blistering heat of the New Guinea sun and camouflage for the camp. The creek was inviting for swimming and washing away the red clay soil that stuck to everything. In a month, tents were erected for the enlisted personnel and huts for the officers. Simple structures were erected to house a kitchen, orderly room, supplies, dispensary, motor pool, and a post exchange.

It was a smooth transition to the new Battalion with the men having the same assignments. The main purpose was to continue training the men for missions. Ubarro was assigned in charge of the Headquarters & Headquarters & Service Co and was later replaced by Melecio J. Montesclaro as acting adjutant. Arguelles led the new Scout Company & Column. Original members of the 5218th Reconnaissance Company were made initial members of the Scout Company & Column. The Medical Detachment had Serafin G. Diana, Cling M. Baker, Conrad A. Barba, and Peter A. Calustro on special duty. A Special Order appointed Ceferino R. Rola as acting S-2 Intelligence and S-3 Operations officer until one was appointed. Arsenio Y. Arellano became Battalion Adjutant. Juan Dahilig was identified with his technical experience in personnel work and given the job to head the Personnel office, but he had to qualify as a Warrant Officer first. Cecil E. Walter, Jr., who led the paratrooper training, took over the job as intelligence and planning officer reporting to Brown.

When word reached Whitney that the 1st Reconnaissance Battalion was now in Hollandia, he immediately looked for transportation to the Philippines. There was a great need for Filipino operators throughout all areas. The Sixth Army at this time had liberated and taken over the radio network of Leyte and Samar Islands and was "holding on to the radios like grim death."

The long-awaited TOE was approved allowing promotion of the men. But personnel could only be promoted if there were vacancies. Brown wanted to gather as much information on the men as there were few vacancies. He wanted to know the status of Lt. Ruben Songco, Sgt. Amado Corpus, other casualties, the circumstances of their deaths, and if their bodies were recovered. Waiting for replies delayed the promotions. The pending requests of mission leaders at their outposts recommending promotions of their men who did exceptional work had to wait until after the war.

It was the "only time I could promote. Some of the men had been there for two years and not promoted. Some of the men were embarrassed by no promotion," recalled Arguelles. Whitney concurred to immediately work on promotions now with the TOE approved. All party leaders wanted their men promoted two or three grades due to their hazardous assignments but there were only a small number of vacancies in the higher grades and promotion to officer required a classification test score of 110 to attend OCS. Signalman Guyot was promoted to 2nd Lt. with his score of 110 making him eligible for OCS. A temporary promotion of officers was awarded. There were many privates who had gone forward (missions) from the Signal Co. and "many undoubtedly deserved a boost. The 978th has sent 98 enlisted men north," reasoned Brown.

By a series of Battalion Special Orders beginning 13 December 1944, mass promotions of enlisted personnel brought the organization close to the standard of the TOE. "Everyone in the Battalion received at least one promotion; those deserving jumped two grades; others were given second promotions later." Records showed three staff officers, the company commanders, and two others, Marfori and Tobias, promoted to the next higher grade. Fifty-one other officers were promoted to the next higher grade by January 1945.

The men surviving in the islands were unaware of the promotions. It was probably the farthest thing from their mind, with the enemy's constant pursuit reminding them that every day they woke up, it might be their last. Lt. Silva was in the jungles of northern Mindanao advising and teaching heavy weapons to guerrillas in preparation for the Americans' liberation of the island.

There were disappointments. Sgt. Dominador Malic, who was on the first AIB mission expressed that while he was on an assignment in Panay, the 7th MD headquarters promoted him to Staff Sergeant, 3rd Lt. in April 1944, and to 2nd Lt. in September. Without authorization from GHQ, the promotions were not valid.

With the rapid pace of the Allied Forces' movements, plans were quickly changing. Several hundred men had already gone on missions or were assigned to KAZ and PIMC stations. The possibility that the signalmen might be taken over by another unit would be bad for morale since they had been part of the Battalion from the beginning. In a letter to Whitney in December, Brown expressed that the men wanted to be in the operations. Fifteen Filipino officers in training could be used for civil affairs. Nine were parachutists.

Morale jumped when their shoulder patches, drawn by Marcelo Ovalles arrived from Australia. Immediately, the men sewed the patch shaped like a shield with a gold carabao head and five stars, signifying the southern cross. It was a complement to their motto, "*Bahala Na.*" Quick messages describing the patch were sent to their comrades at their outposts. The mission men never got to wear the 1st Reconnaissance Battalion patch. Silva was one of them. It did not matter as rank and insignias were not worn in the jungles to avoid enemy and pro-Japanese attention. Survival was more important.

1st Reconnaissance Battalion (5217th Bn, 5218th Bn, and 978th Signal Service Co) - Shoulder Sleeve Insignia

The dark blue originally purple indicated intelligence, the stars theatre of service, and the canton with the 'carabao' head (insignia of the Philippine Division) signifies the battalion's intelligence gathering mission. Coat of arms: Approved by the U.S. Army of SWPA but was never submitted to the Quartermaster General.

Motto: BAHALA NA (Tagalog: "Come What May.")

Luzon Island, Philippines Landing – January 1945

The SPYRON project was disbanded soon after the landing in Luzon. The USS *Nautilus* and the USS *Narwhal* performed fifty percent of the special missions bringing supplies and mission men. The *Narwhal* accomplished the most missions with nine landings making her the leading submarine supporting the guerrilla movement. The last submarine mission with personnel left for the islands in January 1945. The *Nautilus* continued and brought cargo twice to Linao Bay and Baculin Bay in Mindanao, the last island declared liberated in June 1945. Other supplies were airdropped over airfields and designated areas or in small amphibious ships to ports and beaches held by guerrillas.

Total numbers varied. The Navy claimed 1325 tons delivered while the Army claimed 1,600 tons between 1943 and 1945. One record listed that the submarines landed around 1400 tons of supplies with 331 specialists including over 200 Filipinos on special missions. From the islands, 472 refugees were evacuated.

The development of communications in the northern area of Luzon was proceeding too slowly in support of the landing. Five missions with large contingent of men and cargo were sent to Luzon several months before the liberation forces arrived. (The Luzon missions are discussed in a separate chapter.)

As the date of the landings neared, communication to KAZ increased from over 130 stations in the islands. Radio traffic increased with intelligence on enemy activities, guerrilla attacks on enemy installations, disruption of enemy movements, and downed pilots.

Men at camp and at their outposts followed the Luzon landing. When a memo from G-3 Operations to the Chief of Staff reported a "call on the enemy to remove forces and installation in the city to save Manila...make it an 'Open City'" faltered, they pushed for Brown to send them to Luzon.

By this time, some of the men from the earlier missions began returning to camp in Hollandia when they could no longer continue. Sick and malnourished, they were happy to be alive. The 1st Filipino Infantry in the nearby camp, the same outfit that the men were picked from in California, received orders to proceed to the Philippines in February. The 2nd Filipino Battalion in the next camp soon moved out.

In April, Brown received the order to report to the USAFFE in Leyte before proceeding to Manila. His presence was needed to plan the utilization of the 1st Reconnaissance Battalion personnel left in Hollandia. Would there be new assignments now that Manila had been taken? With his departure, Arguelles took over as Executive Officer. Montesclaro took over the Scout Company & Column. In addition to his other duties, Arellano handled the Headquarters & Headquarters & Service Company relieving Ubarro, who was transferred to USAFFE headquarters. Alberto C. Elefano was appointed as his assistant.

By mid-year, the Army wanted an accounting of the funds disbursed by PRS for the missions. GHQ analyzed and kept track of dollars, Philippine pesos, and printed Japanese currency dispatched to the guerrilla commanders to pay for their men's salaries and the maintenance of their camps. Civilians wanted to be paid in Japanese currency for their support as carriers, couriers, security men, or other services to keep the radio outpost running. They needed to buy food. They dare not be caught with American money. Samples of Japanese currency sent to the War Department in Washington were approved for printing nineteen million worth, which were aged to look used. Philippine currency was sometimes used. Operational money was given to each mission's leader to distribute to his men to buy food. But some men did not receive and asked help from friendly locals or scoured the jungle for edible plants to eat at their remote outposts.

Rola was assigned to gather the records and Arguelles to bring the records and all remaining funds to GHQ in Manila. The request for an accounting proved complicated due to lack of receipts and scarcity of paper. Paper receipts in the islands risked the lives of men whose names were on it. Lists of names of individuals who had been paid for their services had to be kept hidden from the Japanese. Other guerrilla leaders, like Robert Lapham, refused to receive the money knowing that they would have to account for it.

Records showed as of September 1945, various currencies were sent to the military districts for maintenance and where they were accepted. $150,000 was sent to Peralta (6th MD), Rowe (4th MD), and Fertig (10th MD) for maintaining their organization. Php1,000,000 were sent to Fertig, Smith (9th MD), and Cushing (8th MD). Japanese currency worth 2,300,000 was sent to Cushing, Fertig, Abcede (7th MD), and Peralta.

As the liberation unfolded and the enemy was destroyed in each island, radio messages began to slow down. Orders were sent to inform the mission men at their outposts to report to Manila. Tracking their location was difficult, especially when they were constantly on the run from the enemy. Some of the men received orders from their commanding officer to report to headquarters in Manila. Others who did not receive the orders turned over the radio to the guerrillas they trained when they felt their service was no longer needed and their mission done. Some reported to the American Task Force that landed on their island. They had been in the remote jungles and impoverished areas for many months and some for years.

They began to drift into Tacloban and Manila not knowing who to report to, looking for Whitney, KAZ, or PIMC stations that were their lifeline. Their tattered uniforms hung on their emaciated and sick bodies from many months in the jungles. Some, causing stares of disbelief, were barefoot because their boots had rotted. Due to the secrecy of their unit, no one knew or believed that such a Battalion with Filipino soldiers existed. USAFFE did not know about them, or what to do with them. That made it difficult to go home. It took some time to find Lee Telesco in charge of weather observations and assistant to Whitney who vouched for them. Some found Whitney or Willoughby, head of G-2 Intelligence who assigned an officer to provide them with funds and shelter. Only then did they have a decent meal, new uniforms, and boots.

Their new patch was the U.S. Army Pacific (USARPAC) with 12 stars and a red arrow pointing left, some say towards Japan. The insignia of the U.S. Army Pacific was originally designed and approved for U.S. Army Forces Pacific Ocean Areas on October 18, 1944.

U.S. Army Pacific (USARPAC)

The patch contains a red arrow and white stars on a blue field. The red arrow of war denotes the valor and self-sufficiency of the forces of the command. The blue field represents the vast expanse of the command area. The white stars portray the North Star, Big Dipper and the Southern Cross, which locate the command headquarters.

The USARPAC Insignia

The 978th men were directed to San Miguel, Tarlac on Luzon where General MacArthur had set up an advance headquarters on the way to Manila. A large tent in San Miguel was constructed to serve as headquarters for the arrivals. A separate mess area fed the several hundred civilian construction workers. After having been gone for decades, they ate with the construction workers to know more about their hometowns. Living conditions in Luzon were much improvement over the jungles of Hollandia and muddy grounds of Leyte. There was an orderly and supply room. Some of the men formed a temporary pool of drivers for the eight vehicles. They took charge of camp guard duty at around fifteen posts for twelve hours at night.

Whitney was present when the men began arriving. Galang who narrowly escaped the ambush of his mission party in Mindoro was assigned as commanding officer of the men. Embraces were exchanged, with the men thankful to see comrades they had not seen since Camp Tabragalba still alive as they recounted to each other their experiences. They understood the full significance of their hazardous work. They described to each other how they transmitted enemy information always on the alert, on the run, starving, and sick. The men in KAZ station thought back on how they took over as inexperienced operators and closed with confidence of their contribution.

With enemy air and ground activity not as active as in Leyte, the men relaxed more, and morale improved. Thankful to be alive, they celebrated with Mass. With the Filipino penchant for celebrations, a safe return and completion of their invaluable task for the liberation of their homeland meant a dance with food, and drinks. It was a chance to meet with young Filipina women. Truckloads of Filipino women just as eager to meet them were brought in from neighboring villages to the dance.

Departing San Miguel, the reconnaissance men reported to Camp Murphy (Camp Aguinaldo) in Manila and were assigned to ordnance support of the task forces liberating the southern islands. Segundo Bucol with his comrades maintained the trucks and jeeps in the company motor pool.

Julius Ruiz was assigned to compile the history of the Battalion. He spent weeks at San Miguel and Camp Murphy interviewing the men to obtain an after-action report of their missions. If not for his work, Rola, Adjutant of the 1st Reconnaissance Battalion, would not have been able to write the U.S. Army's official history formally documenting the men's invaluable contribution to the liberation of the Philippines. But the document did not provide many important details on the men's work with the guerrillas.

Some who had extensive knowledge of the Philippines, were assigned as aides to Philippine government officials in the rebuilding of the country. Galang (Mindoro mission, November 43) became an aide to President Osmeña in the post-war reconstruction of the Philippines. Since the Philippines was one of the original signatories of the UN Charter in 1945, a Philippine delegate was sent to the United Nations Organization Conference with him as the aide. In his book, *Hammer and Anvil*, he mentioned that in a conference, he met and talked with Ambassador MacArthur, uncle of Gen. Douglas MacArthur who told stories about the General's father, Arthur MacArthur. As military governor of the Philippines, he spent a few thousand dollars to build elementary schools – and then was nearly court-martialed for spending money for the education of Filipino children.

The men welcomed a proposal to make the 1st Reconnaissance Battalion and 978th signalmen into a cadre in the USAFFE Training Group at Camp Camarilla to instruct and guide units of the reorganizing Philippine Army. They felt they were particularly qualified as instructors and advisors with their military training and experiences. But the plan was abandoned. Later, some of the officers joined the New Philippine Scout to train and lead companies.

The rest of the 978th Signal men from Hollandia arrived with Croell, commanding officer of the 978th Signal Service Co. in July. Some of the men were assigned to the Signal Intelligence Service (SIS), the code-breaking section for several weeks.

Goroza who had returned to Hollandia and re-enlisted wrote "Then we were ordered to proceed to the Philippines for a new assignment. We camped in (San Miguel) Tarlac near the Luisita sugar factory. At that time, there were still sporadic engagements between American troops and remaining Japanese soldiers. From Tarlac, we were dispersed to several outfits. I was assigned to the 86th Infantry (Black Hawk) Division as a communications operator in Marikina. Several months later, I was transferred to Camp Dau with some of my buddies to provide cadre training for the reformed Philippine Scouts. When their training was completed, I was assigned as an MP at Base M in Poro Point, San Fernando, La Union which was a supply depot during that time."

The Battalion was to be disbanded in August, only six months after activation. The Commander in Chief received the formal memo of disbandment from Whitney, head of PRS.

"1st Reconnaissance Battalion (Special) formerly 5217th Reconnaissance Battalion disbanded composed mainly of selected Filipinos who volunteered for hazardous service behind enemy lines in the Philippines in early part of 1943. Elements of this Battalion acting in cooperation with recognized guerrilla units or separately drove communication nets throughout the Philippines to provide channels through which military intelligence might be provided this headquarter for planning purposes. In addition, they established a wide-spread weather system to provide us with a weather map of the Philippine area daily during a period when such information was most necessary to get the weather picture of the entire pacific area. In addition, they performed numerous other services such as supply, sabotage, special intel, etc. A total of 36 officers and 221 enlisted men were dispatched to the Philippines by submarine for these operations. Casualty rate was approximately 10%. So dedicated were the men to the purposes for which it was formed that the two years since activation, show 1 court-martial and 2 cases of VD, a record probably without parallel in any unit in the entire army."

Brown, the commanding officer was reassigned and Rola, his adjutant and members of his staff were left to attend to all matters of disbandment. But many unavoidable irregularities in paperwork, due to the secrecy of the missions and being headquartered in Hollandia throughout the Philippine landings, had to be resolved before the disbandment date. Applications for leaves to see if families they had not seen in decades were still alive, were disapproved, except for a few emergency cases. Promotions were affected. The enlisted men were transferred to the 1st Filipino Infantry Regiment or to the Replacement Command, U.S. Army, Western Pacific in Manila for disposition. Those who

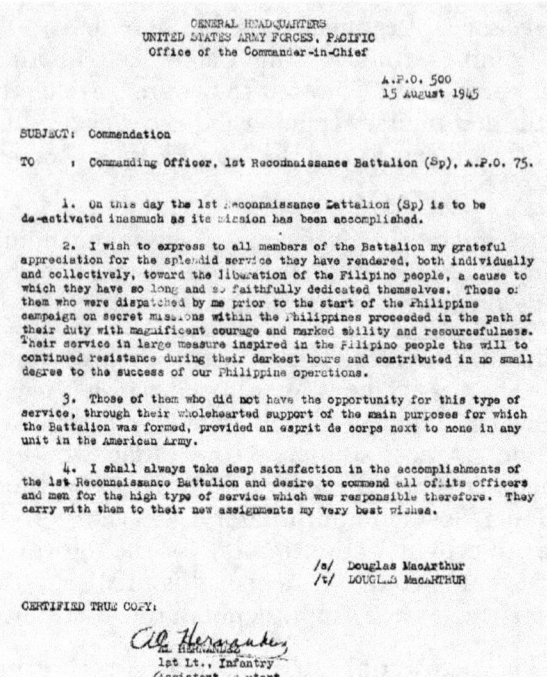

had enough points were eligible for discharge. Men wanted their pay after one to almost three years on mission and sometimes their services were unaccounted for.

The KAZ and PIMC signal men remained active. Their skills and experience did not go unnoticed. Around fifteen men were assigned for message and crypt work with the Signal Intelligence Section (SIS). The Chief Signal Officer decided that the less than two dozen of the original signal men left who were not eligible for discharge were to continue to Japan. Radio operators from other units replenished the signal unit. Henry Campos and Leodegario Nuevo who were reported discharged enlisted in the regular Army. At the end of November, the 978th Company's strength stood at 4 officers and 73 enlisted men under the Eighth Army.

Commendation and deactivation of the 1st Reconnaissance Battalion. 15 August 1945

A new training program began in preparation for further duties in Japan. The men' training were as field radio operators (portable), not as fixed station operators requiring them to master Morse Code. Physical training and close order drills kept them in shape.

Before the year ended, the 1st Reconnaissance men held a second *Salo-Salo* to bid farewell to their comrades, remember 17 who had fallen, and to offer best wishes to those

who would follow the occupation forces to Japan. The program listed all the men, the unit commendation of General MacArthur and citation by his chief-of-staff, General Sutherland.

Orders arrived in December to follow the occupation forces to Japan. Lt. McGiveney who replaced Croell crated their equipment and supplies to travel with the men. They left camp San Miguel in a convoy of trucks and arrived at North Harbor in Manila. Their ship docked at Yokohama, and they were convoyed to their assigned area in Tokyo. After a few days in an old warehouse, they moved to the San Shin building in the heart of the city. Men operated water and land-based transmitters, repaired radios, and undertook other administrative duties at the Daichi building.

Goroza recalled "I was sent to Tokyo to be a communications operator for MacArthur's GHQ. My buddies and I had a great time seeing the cities of Yokohama, Osaka, and Nagasaki. One cannot be lonely in Japan during the occupation. A pack of cigarettes will go a long way if one is looking for a nice time."

After six months, the company inactivated in Tokyo ending with 4 officers and 80 enlisted men. "We were again recalled after several months to the Philippines and assigned to the 738th Military Police Battalion responsible for securing Manila and Quezon City, apprehending criminals involved in theft of U.S. supplies, and as guards to Japanese POW" Goroza said.

The successful liberation of the Philippine Islands was significantly affected by the secret submarine missions of the 1st Reconnaissance Battalion (Special). For over 2 years until the Leyte and Luzon landings, they worked with the guerrillas to be the "the eyes and ears" of General Headquarters, SWPA. The men's powerful contribution in working with the Resistance, landing personnel and over 1500 tons of radios and supplies behind enemy lines in the major islands, providing knowledge of the enemy's activities and location in the planning of the return saved countless lives of the soldiers, guerrillas, and the Filipino people.

General MacArthur paid tribute to the Filipino guerrillas. *We are aided by the militant loyalty of a whole people – a people who have rallied as one behind the standards of those stalwart patriots who, reduced to wretched material conditions yet sustained by an unconquerable spirit, have formed an invincible center to a resolute overall resistance.*

978th Signal Service Company, 1st Reconnaissance Battalion. Men training in radio operations at Camp X (Tabragalba), Australia. Source: National Archives and Records Administration

The Radio

Messages on the enemy situation streamed from the Philippines as GHQ wanted desperately to maintain the communication for future planning. KAZ, the radio unit of General MacArthur's headquarters and the Philippine Intelligence Message Center (PIMC) were formed to ensure its continuation. MacArthur expected all incoming messages from the islands to be analyzed with comments and outgoing replies on his desk first thing every morning. Planners of the Philippine campaign pored over the reports for a triumphant return.

A priority message was sent to Washington for "Filipinos to operate 15 words per minute, install and maintain complete stations, battery charger or power units, and receivers under remote field conditions. Essential these men possess courage, reliability, and intelligence." Sgts. Catalino Laanan and Ray Punzalan along with Cpls. Peter Oandasan and Fermin Consolacion arrived in Brisbane in April 1943. They had passed radio training in the United States and immediately reported to Milton Staging Camp for more intense radio training. They would help to organize the training school for the arriving Filipinos.

Messages coming into KAZ from the Philippines were handled by eight cryptanalysts and two radiomen from GHQ working in shifts led by Lt. Charles Ferguson. They manually coded and decoded messages sent to and from the Philippines using the Double Transposition (DT) system. One mistake and the entire message could not be decoded. The average message was about 30 words, and some ran as long as 200 words. Limited personnel caused the same men to code and retype the messages in clear text for distribution. Urgent messages were immediately dispatched.

The infrastructure was in place to handle the incoming messages but there was difficulty in understanding the guerrillas' responses and needs. As a solution, the Philippine Regional Section (PRS) was established independent of AIB to focus on communication only with the guerrillas.

Col. Courtney Whitney, a lawyer in Manila before the war, took charge of the PRS, reporting directly to General MacArthur's headquarters. The organization's purpose was to penetrate the islands and coordinate with the guerrillas in gathering intelligence on the enemy and supplying them with arms and radios to relay the information back to Australia for the planning of the future return to the Philippines. With the guerrillas' help, coastwatchers set up radios to monitor and flash information on Japanese shipping in the major waters around the islands for the Navy to earmark for destruction. Whitney believed in the potential value of the guerrillas and disagreed with G-2 Security who initially did not want to send arms to them to fight the Japanese.

By this time six AIB missions had departed and there were not enough eligible Filipinos left in Australia to train in radio for future missions. Filipinos were preferred because they could blend in and easily escape detection whereas a White American could not. Plus, the Filipinos could speak the language if sent to the island of their hometown. After rigid interviews, selected men were shipped to Brisbane after basic radio training at Camps Kohler, Crowder, and Monmouth radio schools to undergo advanced training in radio, reconnaissance, and commando.

The first group to arrive were Sgts. Ambrosio Rolluda, Pio B. Borillo, and Artemio J. Ibea; T4s Vincent A. Agbayani, Alfred A. Alberto, and Daniel B. Sabado; Cpl. Vincent L. Visaya, T5s George R. Herreria, Leo Marquina, and Ramon D. Vitorio; Pfcs Andres S. Savellano, and Querubin B. Bargo; and Pvt. Nicomedes Pacificar

The training program at Milton Staging Camp in Brisbane quickly became overcrowded. Camp X, a former Australian camp in Beaudesert outside of Brisbane became the new base. Strict security was enforced – the camp was only identified on records as Camp X and the paperwork was marked 'Top Secret.'

From July to August 1943, ten American officers and warrant officers from Milton Training Camp were transferred to Camp X to help build the camp and give instructions in radio operations. A cadre of White American crypt men joined them. The setup of the camp buildings and facilities interspersed with the men's training. Buildings left by Australian troops were still being renovated when they arrived. Supply rooms, mess halls, dispensary, latrines, and electricity had to be installed. They lived in tents on the ground and ate their meals in dusty open field kitchens. White officers took charge of mess, transportation, property, utility, and served as platoon officers.

Men of the 978th Signal Service Company are training in radio and crypto operations to provide support to the guerrillas' gathering intelligence on the enemy in the Philippines Islands. Source: National Archives and Records Administration

With the increasing traffic from the islands, the 978th Signal Service Company

(978th SSC) was directed on Sept 6, 1943, to handle all the communication networks of the Philippine resistance movement. The unit was conceived in the War Department six months earlier and activated as the core of the 5217th Reconnaissance Battalion in October 1943. It had the strength of eleven officers, five warrant officers, 280 enlisted men in one HQ platoon, five radio platoons, and one Signal Intelligence Section from the 832nd Signal Service Company. The White crypt men reported to the 978th.

The vital importance of communication with the guerrillas to the planning of the return prompted the Chief of Staff to issue a memo entitled 'Operation of Philippine Intelligence Message Center.' It laid out the maintenance and operation of the 978th Signal Message Center under the Chief Signal Officer, GHQ. Its purposes are the following: 1) reception, coding/decoding, and delivery of messages received from or to be transmitted to such agencies as established in the Philippines, and 2) handling of incoming/outgoing messages to and from the Philippines.

Men in the 978th Signal Service Company were given advanced training in radio operations, jungle warfare, and amphibious landings. Once completed, memos announced their availability for a mission as they were transferred to the 5217th Reconnaissance Battalion (Provisional) for review of their grades and qualifications as radio operators.

As men passed the advanced training in radio, they were assigned to missions, PIMC or MacArthur's KAZ station at his headquarters at the AMP building in Brisbane.

A member of the 978th Signal Service Co. training in radio and cryptography. Source: National Archives and Records Administration

The PIMC was stationed in one large room at the Heindorff house in Brisbane as a section of AIB. A net central radio station and code room operated to handle intelligence, operational, and administrative functions between the guerrilla stations and KAZ. The unit decrypted incoming messages, encrypted outgoing messages, and distributed copies to appropriate departments.

The very existence of the PIMC was kept so secret that the staff members were usually accompanied by armed guards and were kept separate from the AIB. They worked in shifts of one officer and two or three enlisted men. Men could not talk about what they were doing. They were sworn to tell no one of their organization except that they were with the Signal Service Company. Pfc. Allen S. Dawang, one of the radio operators, remembered always having a bodyguard around him.

The men lived at Yeronga Park and took buses and the train to work. Some traveled between the office and Camp X. They were attached to HQS, USAFFE for rations and quarters but managed by the Personnel office at Camp X. Due to the shortage of personnel, men handled 24-hour shifts as the radio traffic increased. Money was allotted for meals as they would miss the mess hall hours due to their heavy workload, but this did not matter to them as Australian dishes were not appealing. One Filipino operator commented, "beef stew for breakfast, lunch, and supper or bully beef with hard biscuits, strictly Australian diet."

They preferred the American food served at the American Red Cross and the American Center for Service Men which was closer to the message center. They missed outings planned by the Red Cross due to the exhausting long shift work. It was rare to be able to attend social events or visit bars to drink Australian beer.

In a large room, PIMC received the messages from the islands through KAZ using Teletypes operated by Agbayani, Alcala, and Alegre, with Salgado as reliever. According to technician Bob Stahl, a teletype connected the desks in the code room to "some mysterious location where a radio station transmitted and received our messages. The guerrilla movement was just beginning, and radio traffic was sporadic." Guerrilla messages were manually decoded with paper and pencil, typed in clear text before they were sent to PRS for analysis. The replies to the guerrilla messages were carefully encoded for transmission. One mistake in the code word such as a misspelling hindered the decoding of a message. The typed reports in clear text with Whitney and GHQ staff's comments were distributed to select departments. G-2 Intelligence was bypassed many times for comments. The AIB Chief Supply Officer, Bobb Glenn, noticed that MacArthur was sending handwritten notes to Whitney through his Chief of Staff outside the regular channels.

Summaries and 'Daily Philippine Message Sheets' were distributed. A highly classified Special Intelligence Bulletin containing evaluated radio intelligence from Ground, Air, and Navy sources both inside and outside SWPA to be used for future planning was distributed to a very limited number of recipients.

Lt. Alvin V. Bone was appointed acting CO of the 978th to begin acquiring the best equipment for the radio training room until the 978th Signal Company commander, Capt. Harry T. Croell arrived. Three White Americans were assigned to Croell's staff and another three as signal instructors with the 978th. Capt. Abner Pickering who oversaw S-3 Operations began training Filipino soldiers in reconnaissance to accompany and protect the signalmen with the radios on the missions. Sgt. Emiliano A. Allermo oversaw supply, motor pool dispatches, motor mechanics, aides to the sergeant of the administrative room, and company affairs in the headquarters.

In November 1943, the Royal Australian Air Force (RAAF) operators by the Adelaide River in Darwin suddenly stopped operating their station. This was the station that was intercepting the guerrilla messages from the Philippines. The disruption of the lifeline with the resistance alarmed GHQ in Brisbane. Immediately, two White Signal men and four operators from the 126th Radio Detachment were temporarily tasked to carry on the work until Filipino radio operators could complete the 978th Signal Company training program and be assigned to the station.

As messages for arms and requests for assistance steadily poured into GHQ, assurances were needed that these came from loyal guerrillas and not from Japanese transmissions. The missions were carefully assembled to send arms, radio, and operators to any guerrilla leader who could organize and control the factions on his island. In return, he was to provide information on the Japanese that would directly impact the islands' liberation.

KAZ, Brisbane operators, complained of guerrilla radio operators, saying they were "unreasonable, extremely impatient and very nasty" with them. An official investigation was launched, carried out by listening to both sides of transmissions. Only the stations of the 10th Military District of Mindanao were determined to be uncooperative. A review of the most qualified operators at KAZ station showed that messages tapped out at 12- 16 words per minute were not fast enough to handle messages coming in from several control stations in the islands. Schedules were not met, and radio traffic not cleared. The KAZ signal men were trained in field station procedures and reception speed was at an average of 15 words per minute while traffic from the Philippines using fixed station procedures was at almost double the rate. The guerrillas' streamlined procedure significantly reduced traffic between stations and shortened their time on the air limiting the chances of enemy interception. Under constant pressure from enemy discovery, messages were pushed faster to avoid a backlog. Patience grew short waiting for a reply from KAZ. KAZ operators were found "equal to Army standards and that the 10th MD operators often sacrifice accuracy for speed from findings of the results of actual tests." The guerrilla commander of Mindanao was directed to instruct his radio operators to be more "considerate and cooperative in their dealings with KAZ operators. His recent radiograms on this subject (were) considered to be very unmilitary." Facilities and equipment in Brisbane were expanded. Oscillators and headsets were added aside from keys to the code practice equipment. Four adept Filipino operators were added from Camp X to learn the new procedure quickly and tap out 18 words per minute. Sgt. Alberto C. Franciso and Cpl. Santiago Q. Jaramilla, were temporarily sent to KAZ to learn the new procedure and instruct men for future missions once they returned to Camp.

On the islands, signal men reported their frustration over their inability to operate the guerrilla radios immediately. They had to quickly learn the procedures the guerrillas used to man the radio.

Proud of the men, PRS informed GHQ and all Philippine control stations to "advise all Filipino operators in your command, that this station (KAZ) is now staffed by Filipino operators and that their signals are received here by their own countrymen. General MacArthur's confidence in thus entrusting to Filipinos so large a share of the responsibility in the maintenance of his vital channels of communications with the Philippines, is a challenge to maximum Filipino effort, both there and here, which must be fully met."

The KAZ transmitter in Darwin was in a small, prefabricated house cramped with radio equipment with barely space to move. The operators worked with one transmitter that was difficult to tune and had only one workable frequency. The other transmitter required hourly tuning to a different frequency causing further transmission delays. The receivers were in a mobile truck with barely room for them to fit in until they were moved into a house. Two more receivers were built from salvaged radio parts showing the men's ingenuity. The 24-hour work shifts allowed immediate contact with the Philippines despite the numerous handicaps confronting the signalmen.

A rehabilitation inspection by Ind provided more powerful radios and equipment to meet increasing traffic with less problems. The men welcomed the construction of a new, larger Darwin station with their own living quarters.

As men arrived to report to the 978th SSC, the administration staff was kept busy with the increase in the company's strength. The training program was further developed and enhanced from communication with the Philippines and the experiences of earlier mission parties.

Increasing reports from the signalmen in the islands kept the men busy as they decoded news of increasing Japanese brutalities. They knew the dangers that their radio counterparts at the other end were facing just to be able to send their messages they could not stop once a message was transmitted except for air raids until the message was sent or received in full. They knew that the reconnaissance men with them posed as fishermen, farmers, or merchants to blend in with the population to procure information on Japanese movements and situations. They were aware that a radio message might be sent while desperately running from the enemy or the message being transmitted may be their last one.

By the end of 1943, GHQ was delighted to receive direct messages from the islands of Mindanao, Negros, Panay, Mindoro, and Tawi-Tawi in the Sulu Archipelago. Delays were avoided as the messages were no longer passing through network stations. Couriers from Luzon were delivering intelligence reports to Mindanao, Panay, and Negros to relay to KAZ. Eight submarine missions had brought radios, supplies and signal men to the islands. Training at PIMC was updated as problems were encountered and dealt with by each mission. Emphasis was given to properly encoding messages to avoid wasting time breaking them. PIMC proved they were ready when they handled 32,000 code groups in December. A code group was equivalent to a letter or a group of letters or numbers.

1944

Submarine missions departed for the Philippines every month. Three missions arrived in each of the two months leading to the landing on Leyte Island. As the radios were distributed and communication networks established, the guerrilla organizations expanded. More operatives provided the guerrillas with intelligence increasing the traffic to Australia. KAZ and PIMC received, decoded, analyzed, and distributed the reports every 24 hours.

With more radio men scheduled to arrive, Ind ordered Sgt. Vincent Visaya to be instructed on cryptography while T4 Leo Pomarez learned to instruct the streamlined commercial radio procedures used by the guerrillas. For practical experience in coding and encoding, a radio station was set up to transmit messages to the Milton Staging Area. Groups of men were assigned to various small islands scattered along the Australian coast transmitting practice messages. While men were at field operations, others graduated from Canungra Jungle School and Fraser Island for amphibious training. Long marches over hundreds of miles physically hardened them.

Larger groups of 50 radio men from the 1st Filipino Infantry Regiment arrived from California. Trained in the fixed station procedures used by the guerrillas, they readily reinforced the radio stations in Brisbane and the Adelaide River in Darwin. Men took over KAZ operations after only 30 days of training. The majority were sent with mission parties.

Pvt. Padilla took a photo of his comrades at the Adelaide River in Darwin. The caption listed *Lt. G. T. Baker, T4 Agbayani, T5 Andres, Sgt. Nagtalon, T4 Santos, T5 Sapon, T4 Carrasca, Sgt. Ancheta, Cpl. Alcala, T5 Alegre, T4 Jose, Sgt. Ergas, T4 G. Arce, T5 Davin, Sgt. Corpus, Sgt. R. Africa, T4 J. Peredo, Sgt. Young, Sgt. Mauricio, Pfc. Marapao.*

Camp X strength now stood at eight signal officers and 211 EM. Around 60 men had been sent on missions to the Visayan Islands and Mindanao bringing much-needed radio and supplies.

A new cipher machine (M-209) was added at PIMC to resolve the lack of code and radio security that plagued the earlier missions and lead to discovery by the enemy. White signal men with cryptographic experience were assigned. From their remote locations in the islands, messages were manually coded until missions brought the new M-209 cipher machine. It was the size of a shoebox, weighed only six pounds, and could be carried in a backpack while running from the enemy. Stricter security procedures were now followed than in earlier missions.

An interesting note here were the private codes unknown to G-2 Intelligence used by the mission party leaders, transmitting messages only for PRS and General MacArthur's eyes, and thus upsetting Willoughby of G-2.

A message about the release of Filipino prisoners-of-war to their families arrived at KAZ. These men had to undergo a Japanese indoctrination course to become members of the Japanese constabulary force. However, many of the former POWs quickly joined the guerrillas as operatives or fighters, adding to the radio traffic of intelligence information.

The direct contacts resulted in 46,000 code groups handled in June. Intelligence coverage in the islands was rapidly expanding with the pending Leyte landings in October. PIMC's request for additional personnel brought five White operators to Brisbane in August and September. Activities in the message center were bustling and the secrecy of their support of the underground movement was paramount, thus exempting them from regular publication of their 'Daily Sheets of Activities and Duties.'

By mid-year, PRS carefully tracked the available men for missions as the landing date neared. Eleven officers, four warrant officers, and 159 enlisted men were on the PRS roster. Those on missions were identified as absent on special assignments and all others were in training. There were 37 Whites and 122 Filipinos. Although most of them were naturalized Americans they were listed separately as Filipinos to identify that a mission had the proper number of men with Filipino heritage.

Meanwhile, General MacArthur had convinced President Roosevelt that it was easier to free the Philippines before Formosa because of the loyalty and support of the Filipino people who were fighting and dying while waiting for America's promise to return. He also added there was a powerful armed resistance persisting in fighting and wearing down small groups of the enemy. A convincing reason was the nine secret missions that had already penetrated the Japanese-occupied islands and collaborated with the resistance movement to send intelligence reports on the enemy.

When the landing date was changed to October instead of December, more signalmen had to be trained and additional missions sent. Luzon was still to be penetrated. As soon as men passed radio training, they quietly disappeared, carrying a new cipher system for their new station as their orders arrived for a mission. More radio men as coastwatchers to monitor the waters around the islands were sent. Weather observers, saboteurs, and anti-aircraft warning watchers were added to the parties. Weather observations were needed to support the accuracy of bombing raids.

From June to October, twelve submarine missions with reconnaissance and radio men heavily laden with cargo left for the islands. Landings were made on multiple islands within days of each other. KAZ and PIMC braced for the increasing traffic as the number of networks increased.

The Brisbane message center bravely handled the dramatic increase to 103,000 code groups in July. By the end of August, it leaped to 123,000. Even with improved organization and schedules, White radio operators were temporarily added to PIMC in August and September to ensure messages were regularly cleared.

The increasing traffic taxed the message center, but a commendation from the Naval Intelligence Office boosted the men's morale. The Navy could not make publicly known their contribution due to the secrecy of the 978th Signal organization. The important messages forwarded by the operators resulted in their victory at the Battle of the Philippine Sea that crippled the Japanese Imperial Navy in June. Admiral Halsey gave 75% credit of the naval operations to intelligence passed to them from coastwatcher stations in the Visayan islands.

The need for more Filipino radio operators became acute with the news of General MacArthur moving his headquarters from Brisbane to Hollandia in September. His forces were now halfway across northern New Guinea cutting off supply lines of the Japanese on their path to the Philippines. The communication network moved with his headquarters as they were the link to the guerrillas' networks.

An advance team from KAZ in Darwin and PIMC in Brisbane went forward to Hollandia. "All signal equipment ... to Detachment 978th ... Must arrive by September 17, 1944." Filipino operators led by Sgt. Young were Pfcs. Roman S. Carrasca, Theodore M. Jose, Sergio B. Estavillo, Fred M. Madariaga, and Toby L. Marinas. The KAZ station in Hollandia was in a temporary location while an office with equipment and quarters with kitchen for 30 operators was being built. Five receivers were on 24-hour operation schedule.

A memo to the 978th commanding officer ordered 13 radio men to follow: Sgt. Miguel E. Ignacio, T4 George T. Pajara, Cpl. Alejandro A. Buccat, T5 Nicomedes Pacificar, Pfcs. Victor L. Piana and Pedro P. Tabofunda, and Pvts. Bernardino P. Aquino, Aproniano E. Cayayab, Lorenzo F. Taban, Albert U. Ulivas, Frank G. Valmonte, and Jose F. Viduya. The advance group was relieved when they arrived as they were working double time and seldom given a day off. An additional group of White operators had temporarily reinforced the advance group. The radio training program was reviewed and updated. All KAZ operators underwent intensive training in high-speed point to point and were expected to handle any KAZ circuit within 30 days. Lower rank men were to pull KP duty, but Whitney overrode the order keeping them on the radio schedule.

Pfc. Doroteo A. Rosales, and Pvt. Telesforo R. Astronomo were transferred to the naval radio base in Woendi Island, Biak.

Men from Camp X were sent to Darwin to replace the advance team who had moved to Hollandia. Lt. George T. Baker oversaw the radio men. Reconnaissance men with their infantry skills were assigned for indefinite service. The Darwin station became crowded with men – Sgts. Aquilino Corpus, Robino Africa, Alfonso Ancheta, Abamile Mauricio, Florentino Lucero, Clemente Gascon, Simeon Marapao, Alberto Santos; T4 Vincent Agbayani; Cpls. Eugene Ergas and Materno Peralta; Pfcs. George Arce, Numeriano Davin, Bartolome Padilla and Joe Peredo; and Pvt. Teodoro Jose. Arriving later were Sgts. Leandro P. Tanato, Alex A. Nagtalon, and Anacleto Biado; Cpls. Atanasio L. Alcala and Roberto A. Randal; T5 Emilio A. Alegre; Pfcs. Florendo R. Pedro, and Gregorio Santos; and Pvts. Manuel V. Andres, and Leoncio Sapon.

The Darwin station was ordered closed when the rest of the equipment and personnel were shipped to Hollandia. KAZ station's new location was at Lemot Hill, a few miles from

Queen's Hill, General MacArthur's headquarters. PIMC personnel and their equipment, codes, and ciphers were located at Queen's Hill, which had a cooler climate at a higher altitude. The men's tents were over concrete floors avoiding the rains that flowed down the hillside. The Army engineers constructed a long-framed building for the office. Additional equipment was added for the anticipated increase in traffic. A Teletype machine was installed to connect KAZ and the message center. With KAZ and PIMC just a few miles from each other, fewer messages were garbled, and relays between them speeded up. An advantage being beside GHQ was the quicker distribution of vitally important messages of enemy disposition in the planning of invasion plans. Messages poured in daily as the target landing date in Leyte was less than a month away. The eight missions that arrived in the islands from March to August increased traffic to around 174,000 code groups in September.

The major concern was the slow development of communication in Luzon, which was the next landing site. Communication was particularly substandard in the northern area. There was "much work to be done and little time available," expressed a PRS communique. Three missions were swiftly formed and shipped to Luzon before the Leyte landings.

According to the *Army's Calendar of Shipments to the Philippines,* 120 978th Signal men were sent on missions to the Philippines prior to the October landings in Leyte. This did not count the men on the earlier six AIB missions and those labeled as radio operators in the later missions. Those labeled radio operators in the later missions when PRS turned over duties to AIB were most probably 978th men.

PRS sent assurances to GHQ that the robust guerrilla networks were more than ready to support the October landings. One hundred thirty-four radio stations throughout the Philippines were in communication with GHQ, SWPA. Mindanao had forty-six stations, Panay twenty-three, and Luzon twenty-one. In the western outer reach of the islands, Tawi-Tawi and Palawan had stations. Forty-three weather stations and stations without weather instruments providing only visual observation, reported four times daily to GHQ.

As General MacArthur moved, so did KAZ and PIMC. Excitement swept through the men to know that they were included in the convoy to the Philippines. Their hard work in operating with the main control stations of guerrillas had paid off. They knew the contacts were mostly their comrades in the remote jungles. Their messages would now allow the invasion forces to strike with almost complete knowledge of Japanese activities. But that did not prevent the ruthless resistance of the Japanese Imperial Army.

On October 13, the advance group of KAZ and PIMC Filipino operators left Hollandia onboard the radio ship *PCE(R)-849* and joined the convoy of the landing force to Leyte. Sgt. Leandro P. Tanato led KAZ operators T5 Sergio Estavillo, Cpl. Roberto Randal, Pfc. Fred Madariaga, and Sgt. Anacleto Biado. They brought all their equipment and cipher systems. They had the ability to receive 18 words per minute handling KAZ messages, but the old Navy receiver sets did not work well with the high frequencies used in the guerrilla circuits. They had to fiddle with the set to find two low frequencies used by island sub-stations for contacting their net control stations to take   a complete message. Another ship carried five White Signal Corps men under Capt. Charles Ferguson and a third ship carried Signal Corps men serving the Sixth Army.

After the advance group left Hollandia, ten White operators arrived as replacement. They kept communication with the islands when the advance group could not receive messages while on the high seas and setting up after landing.

As the Allied forces were on their way, Japanese patrol ships swarmed around the islands to defend against the approaching convoy. Sightings of Japanese air patrols searching for U.S. ships increased, adding to the radio traffic. The coastwatchers were now told to report on these flights to protect the convoy. It was of high priority and sent in a "brevity code which needed no ciphering and had to be passed to KAZ within 10 minutes of the sighting, or they were useless."

Leyte Island, Philippines Landing – October 1944

The battleships pounded the beaches of Leyte for several days. About 1100 Signal Corps operators onboard saw stumps of coconut trees through a thick curtain of smoke covering the island. They switched on their equipment to establish radio contacts. News reporters broke the news to the world that the invasion had begun from long-range transmitters on the radio ships *FP47* and *Apache*. The Press added to the ever-increasing radio traffic as they broadcasted their scoops to the world.

Tension filled the air as the radio men entered Leyte Gulf with the Sixth Army. A mortar hit a LST (landing ship tank) beside theirs filled with troops killing signal men. For several days, they were kept onboard until it was deemed safe. Four Japanese planes were shot down by their LST as they nosedived towards their ship. They were in the thick of the Japanese forces preventing the landing – The Battle of Leyte Gulf.

Leo Gangl, a cryptanalyst with the 978th Signal was on board and lived to tell his story. As their ship entered Leyte Gulf, their destroyer escort began flashing a signal. All guns pointed upwards to three Japanese torpedo bombers diving low towards the *British Columbia Express*, a converted fruit vessel, filling the air with explosions. Pvt. Donald Higgins recounted that the men were on the ship's deck. Men rushed down below deck as the air raid bells blared and machine guns began firing. Japanese torpedo dive bombers dropped two bombs exploding behind the ship. The ship heaved violently as another bomb exploded overhead at the exact moment. A shell fired from their anti-aircraft cannon by the gunner on deck struck the belly of one of the Japanese planes. It burst into flames as it flew by the ship. The others flew into the clouds as anti-aircraft bursts chased them.

Under mortar fire, a landing craft dropped its ramp and a mobile radio unit on a truck bravely sped ashore to establish communication with the radio ships. Under bombardment, General MacArthur sent out a message to all radio stations operated by coastwatchers and guerrilla leaders to rise up in arms with him against the enemy. All naval and air movement of the enemy was to be immediately reported to his headquarters. The men excitedly punched out the long-awaited message.

The Japanese attacked nonstop while wooden boats transported the men to shore. Their location with GHQ Signal Section was in the open amid piles of equipment, arms, and heavy weapons on the beach. They were vulnerable to mortar fires and snipers. But they were on the air the following day. Moving inland, they sank in the muddy ground kept flooded by the rainy season. "Living conditions were not good. In addition to inclement weather, food was bad, and men suffered from stomach illness making them lose weight," said Sgt. Tanato.

A move to the Price house in Tacloban, the capital city of Leyte, was good news as the radios were getting soaked in the open. It was the chosen radio camp and headquarters of General MacArthur. Armed guards secured the perimeter of the house. The PIMC held office in the big room on the main floor relaying messages to KAZ operating nearby by Teletype. The 978th signalmen's room was next to the weather section headed by Lee Telesco, executive officer of PRS. He oversaw the urgent weather traffic flashed from the weather observers in the mountains and quickly transmitted the information to the 5th Air Force.

The radio networks of Leyte with the code and cipher system were handed over to the Sixth Army as planned to contact the guerrilla forces directly on the island. Names and contacts of coastwatchers, mission men, and guerrillas operating on the island were passed on. This procedure was followed in each island's liberation. KAZ and PIMC now acted as backup if communications between the guerrillas' networks and the Sixth Army signal men went down. While on Leyte, Simeon Marapao requested a 20-day furlough in December to visit his family whom he had not seen for 16 years in neighboring Bohol Island. His father was the Bohol provincial governor. He was terribly disappointment when it was denied as Bohol was not an authorized area for furlough.

At the Price House, they saw General MacArthur dine with his officers in the large dining room under nightly air raids. A bomb dropped on the roof creating a large hole in his bedroom wall above the bed's headboard. General MacArthur slept through the nightly air raids, according to reports.

A photo taken in the Price house preserved the men's presence. Atanacio Alcala was a teletype operator of the 978th Signal Company and stationed in Tacloban until April 1945.

The rest of KAZ and PIMC with Cpl. Eugene Ergas, Sgt. Robino Africa, and Pfc. Joe Peredo, who were left behind in Hollandia, boarded whatever transportation was available to move north. The battle to capture Ormoc to the west of Tacloban was at its height when they arrived. Ormoc was overland at the back of the landing sites and Japanese reinforcements were pouring in to attack the landing beaches. The Japanese were determined to stop the invasion of the Philippines.

Leyte, 1945. Price House. "American and Filipino signalers in radio transmitting and receiving room"

When all the KAZ and PIMC personnel were together in Tacloban, Signal Operations officer Lt. Emilio Quinto took over KAZ station. Quinto was the chief operator of GHQ's floating wireless station, Arcturus, off New Guinea that carried the bulk of the radio traffic between GHQ and Washington D.C. before ordered on the first mission to the islands in January 1943. He evacuated to Australia due to severe malaria and had now recuperated.

General MacArthur prematurely declared the Leyte campaign over when there were still pockets of the enemy to be destroyed. The Eighth Army was called in to take over the clearing up of Leyte while the Sixth Army focused on liberating Mindoro adjacent to Leyte to establish airfields and a base in preparation for the capture of Luzon.

An advance group prepared to accompany the Luzon landing force. Six names were submitted by Sgt. Tanato: Sgt. Fred Randal, Cpl. Eugene Ergas, Pfc. Joe Perido, Pfc. George Arce, Pvt. Leoncio Sapon, and Pfc. Jose Viduya for radio maintenance. White operators for the message center under Capt. Ferguson joined.

Meanwhile, the 978th headquarters and Training Section at Camp Tabragalba crated and loaded their equipment into ships for Hollandia. With the rest of the 5217th, they arrived in December. While on the high seas, the 1st Reconnaissance Battalion (Special) was activated and the 5217th disbanded. The 978th remained as the core of the new Battalion with Croell commanding. The training of signal men for missions and message center support continued in Hollandia until the major units were called to Leyte.

Traffic reached another peak as the year ended and the planning of the Luzon invasion was gearing up. Five more radio contacts were added from guerrilla radio networks on Luzon and the Bicol Province. Their reports covered enemy activities, dispositions, and defense preparations. Reports of ship sightings were especially wanted by the Navy.

Luzon Island, Philippines Landing – January 1945

The KAZ and PIMC groups on Leyte prepared to work three shifts daily to handle the added traffic of the new contacts on Luzon. The messages had to go through despite rainy weather and the limited personnel handling the increased traffic.

A large flotilla of radio ships, dubbed the 'Signal Corps Grand Fleet' joined the convoy to Luzon. Capt. Ferguson, Sgt. Carter, and Pfc. Bent of the Philippine Island Message Center group were on General Headquarters communication ship with Lt. Quinto, Sgt. Ergas, and Sgt. Peredo, intercepting messages throughout the voyage as best as they could shifting frequencies. Three men attached for temporary duty from the 125th Radio Intelligence Company joined. On board an LST, Lt. Card oversaw Lt. Tanato with the six Filipino operators and ten White operators from the old message center.

It was not a smooth landing at Lingayen beach. Landing crafts damaged by enemy bombings collided with each other. The shoals of the gulf hampered maneuverability. Throughout the days of bombardment to clear the beaches, Kamikaze planes flying out of Central Luzon attacked the landing force hitting battleships and destroyers. Anti-aircraft fires dotted the sky over Lingayen Gulf while billowing black smoke enveloped ships that suffered direct hits. The suicidal raids lasted for a tense month.

Their LST missed the landing twice on the Lingayen beach, recalled Pvt. Higgins. With no stern anchor, the LST was pushed to Green Beach, but it continued to drift in the shallows until it rammed into an LST that received a direct hit from a Japanese mortar. The following evening, Japanese coastal guns opened fire nearly hitting them, half a mile from shore. But once onshore, enemy opposition from ground troops was lighter than in Leyte.

The advance group set up camp in the coastal town of Dagupan, close to the area of the troop landings. For two weeks, they handled traffic on the invasion and worked as a relay station between the remaining KAZ group in Leyte and GHQ on Luzon. Codes were turned over to the Army to contact guerrilla stations as radio traffic increased. Guerrilla leaders attached their radio to the landing force providing intelligence on Japanese activities, defense preparations, ship sightings, and minefields in the Lingayen Gulf, like the Leyte landing.

Meanwhile Leyte's radio traffic began dropping as traffic increased in the retaking of Luzon and planning for the Visayan and Mindanao invasions.

Luzon was not unified under one guerrilla leader as was the case in the southern islands. GHQ accepted supporting several guerrilla commanders on Luzon who maintained their men and territory. KAZ had to verify duplicate intelligence reports from the guerrilla commanders' overlapping boundaries.

Two weeks after the landing, a request for ten radio operators, preferably 978th Signal Service Company men, came in. They were to be attached to GHQ for duty with the 978th Detachment Message Center at Dagupan. Status of available Filipino 978th radio operators who could tap out 14 words per minute wherever they may be was requested. White operators from the 3170th Signal Battalion supplemented the support.

General MacArthur was again on the move, eager to arrive in Manila, and very concerned about the 'Kill all' orders from Tokyo. He pushed ahead of the Sixth Army to establish headquarters at Hacienda Luisita, a sugar plantation in Tarlac halfway to Manila. His radio KAZ and PIMC personnel followed. So did PRS headquarters.

The road was hazardous with Japanese patrols and snipers. The men were not to attack the enemy but should be on the alert and ready to protect the radios in keeping up with GHQ. A Mobile Communication Section was established. Around 200 vehicles of a mobile land message center were spread out to set up a GHQ signal center. It was reported that more communication wire was laid out on Luzon than any other Pacific operation. Following a wire always led to a radio.

It was much better living in private homes, eating at GHQ's mess hall, doing recreational activities in good weather. The constant bombings and sniping that pressured them in Leyte was hardly felt in Luzon, allowing them to recuperate physically and mentally.

Problems with missing code groups surfaced. Sgt. Leandro Tanato reported, "Our main difficulty at this time was the inefficient service from the transmitter site. Leyte was heard calling but was unable to receive the message. The undependable transmitter and the changing frequency kept the station out of service for hours. It was puzzling because the equipment used was the same as that used in Hollandia and Darwin. The Signal Corps mobile trucks were close to the message center. Another site was chosen near San Miguel across the road from a large sugar central processing plant. Once that was working, the station in Dagupan was closed and the rest of the operators moved to San Miguel."

Back in Leyte, the remaining KAZ and PIMC moved south from Tacloban to Tolosa. Work continued in 24-hour shifts. The stations had new equipment in separate locations, and messages were relayed via Teletype wires. There was a post exchange, and men were allowed to attend local dances with the local girls, which they eagerly welcomed. When traffic began slowing down with the focus on Luzon and the Visayas, men were given day passes between shifts.

Manila – February 1945

The progress of retaking Manila tested the patience of General MacArthur. He assigned the Eighth Army to take over the liberation campaign of the Visayas and Mindanao while the Sixth Army concentrated on clearing Manila and Luzon. The Eighth Army had set up their own net control station while still in Leyte, between Tolosa and Dulag separate from the KAZ networks. Guerrilla communications were turned over to them. KAZ operators continued to manage smaller stations outside the guerrilla networks not picked up by the Eighth Army and relayed them to proper Army stations.

On the march to Manila, Japanese encounters were sporadic as the enemy troops were ordered by General Yamashita to fortify the roads and trails moving north. The advance group of Filipino and White signal men entered Manila right behind the 1st Cavalry's combat teams. It was a cause for celebration when the internees at the Santo Tomas Internment Camp were rescued. But the news was tempered with the beginning of indiscriminate killings of the residents and the rampant burning of their homes in southern Manila by a fanatical enemy.

When the entry to Manila Bay was opened, the 'Radio Grand Fleet' were the first vessels to enter and handle the heavy load of traffic for the Press in the next two months. The Press averaged nearly twenty-three thousand words per day to inform the world of the momentous unfolding of events in the Philippines Eventually, a commercial station was set up ashore to take over the load. The station kept moving to maintain the radio on the air while avoiding Japanese snipers. KAZ personnel docked and proceeded to set up a mobile communication site at the Trade and Commerce Building. By the end of February, 142,000 code groups were processed.

Through the destruction, they saw the burning rubbles of a once-cosmopolitan city with art deco buildings and impressive churches reminiscent of Europe. That was their last memory decades ago before crossing the Pacific to America for work and better opportunities. Seeing the incomprehensible destruction of the city motivated them to finish their job. They set up their tents at the GHQ detachment area near the clock tower of the City Hall that was still miraculously standing, although pocked with artillery shell holes.

Military bulldozers hummed, scraping the streets of rubble, and disposing of corpses. Dazed residents asked the men for help in Tagalog or English as they sifted through what was once in their homes. The military civil affairs received reports of people in tattered clothes lining up to get rice and other sustenance. Other lines just as long were for medical treatment of wounds, sickness, and diseases. But even with such a pitiful sight, the men were glad to mingle with civilians who survived the destruction. Soon enterprising civilians began to set up tables and chairs in front of their demolished homes selling whatever would appeal to the servicemen so they could buy food.

While the clearing of Manila was ongoing, the Eighth Army looked south to liberate the Visayan Islands and Mindanao. The Task Forces began landing in Palawan and, five months later, ended their final battles against the enemy in Mindanao. As the Eighth Army task forces liberated each major island, codes and radio instructions were turned over to contact the guerrilla leaders directly. Net control stations providing intelligence to Australia were turned over. The stations set up by the original six AIB mission on Negros, Panay, and Mindanao, had been operating the longest time, for three years providing intelligence on the enemy.

The code groups drastically dropped to 46,000 by April. Traffic volume continued to decrease, dropping by another 10,000 code groups as southern islands were liberated. White radio operators from the 3170th Signal Battalion were released from duty with the 978th Signal Service Company. Encounters with the enemy lessened to hunting down scattered units by the guerrillas in coordination with elements of the Army.

Total radio messages sent per island to GHQ peaked during the October Leyte landing and the January Luzon landing.

MONTHLY TOTALS OF RADIO MESSAGES SENT TO AUSTRALIA FROM GUERRILLA COMMANDERS AND AGENTS (1942 – 1945) (a)													
Guerrilla Leader/Radio	Dec42 Jan43	Feb Mar(a)	Apr May	Jun Jul	Aug Sep	Oct Nov	Dec43 Jan44	Feb Mar	Apr May	Jun Jul	Aug Sep	Oct Nov(m)(k)	Dec44 Jan45 (n)
PRAEGER (Northern Luzon)	21	15		24	10 (c)								
PERALTA (Panay)	38	58	155	337	247	15 (d)	68	110	185	176	290	305 (g)	347
AUSEJO (South Negros)		8	(b)										
VILLAMOR/ANDREWS/ BENEDICTO (S. Negros)		28	32	35	47	69	35	27	20	22	29	50	52
FERTIG (Mindanao)		67	66	61	235	172 (f)	288	270	577	317	395	758	1,193
HAMNER/YOUNG /SUAREZ (Tawi Tawi)				7	19	58	12	49	14	10	15	31	29
ABCEDE (Negros)				19	15		45	20	33	53	71	174	280
PHILLIPS (Mindoro)						6	75	17 (h)					
INGENIERO (Bohol)							22	8	11	7 (i)	(j) 12	20	37
SMITH (Samar)							5	42	50	41	63	210	(l)
KANGLEON (Leyte)								6	8	25	30	68	(l)
CUSHING (Cebu)									50	40	84	124	101
ANDERSON (Central Luzon)										57	52	113	198
HALL (Central Luzon)										17	50	67	55
ROWE (Mindoro)											130	131	122
VOLCKMANN (Northern Luzon)											45	255	610
CABAIS (North Palawan)											14	37	46
Cabangbang (Central Luzon)											32	127	1,061
RAMSEY (Central Luzon)													60
STAHL (Bondoc Peninsula)										10	17	17	63
CORPUS – PLACIDO (Southern Palawan)											19	12	20

Total radio messages do not include the hundreds of messages from sub-networks on each island.

Aircraft sighting messages not included in totals after Jan45, Luzon landing
a) Only a few ship sighting messages were included in totals after Dec44-Jan45
b) AUSEJO – radio taken into VILLAMOR net
c) PRAEGER – off air due to capture of radio and personnel
d) PERALTA – off air due to enemy in mountains near headquarters
e) Not used
f) FERTIG – main radio bombed and destroyed
g) PERALTA – back on air

h) PHILLIPS – killed and party dispersed
i) INGENIERO – off air due to capture of radio
j) INGENIERO – back on air with new radio from Leyte
k) Leyte landing
l) SMITH and KANGLEON – turned radio net over to Sixth Army
m) Aircraft sighting messages not included in totals after this date
n) Luzon landing

Source: MacArthur Archives

Meanwhile, feverish preparations for Operation Olympic – the invasion of Japan – were at their height. GHQ's communication unit in Manila had a huge mobile communication section to serve the Army's advance echelon in the invasion of Japan. A new training program for additional duties in Japan was planned to train the men. They were to be part of the occupation force. Only 19 men were eligible to go to Japan as many had met the 60-points required for discharge. Men from other units were brought in to replenish the radio unit.

All preparations for a destructive invasion of Japan halted after the sudden surrender of the Emperor of Japan in August. But KAZ and the message center in Manila continued to operate until large-scale fighting in the Philippines ceased, and the radio traffic was almost negligible. Six men continued to operate the message center in Malacañan Palace for the Philippine government while the rest were called to their headquarters in San Miguel, Tarlac. Malacañan was the official residence and office of the Philippine President. The radio office was one of the last to close.

At the end of operations in Luzon, KAZ had two officers, Lts. Emilio F. Quinto and George T. Baker, both of whom were overseeing Filipino radio men: Sgts. Clarence R. Young, Aquilino M. Corpus, Eugene N. Ergas, Leandro P. Tanato, Alfonso Ancheta, Abamile Mauricio, and Alex A. Nagtalon; T4s Vincent A. Agbayani, Nicomedes Pacificar, George Pajara, and Joe E. Peredo; Cpls. Atanasio L. Alcala, Alejandro A. Buccat, and Roberto A. Randal; T5s Emilio A. Alegre, Sergio Estabillo, Teodoro M. Jose, Toby Marinias, Bartolome Padilla, and George U. Arce; Pfcs. Manuel V. Andres, Bernardino P. Aquino, Romain S. Carrasca, Pete G. Felizar, Fred Madariaga, Simeon P. Marabas, Dick Obrero, Florendo Pedro, Victor Peana, Leoncio Sapon and Jose Viduya; and Pvt. Numeriano D. Davin.

Commendations from the Sixth and Eighth Armies recognized and appreciated the men's work. Looking back, Lt. Ferguson, head of PIMC, had high praises for the men as did Signal officer Lt. McGiveney who said, "During the period from the invasion of Leyte up to the invasion of Mindanao and the Visayas, it is doubtful whether any other comparable message center and radio station ever handled such a volume of this peculiar type with the limited personnel available, and with less services, and which contributed so much by furnishing intelligence information to the success of the Leyte, Luzon, Mindanao, and the Visayan operations."

Over the radio, in the thousands of messages that passed through the radio nets, high praises were given to the achievements of the guerrillas. Manning their networks were the Filipino volunteer signalmen enabling KAZ to function as the net control station for the entire guerrilla network of secret communications. The message center handled the traffic of nineteen network control stations from all over the islands clearing thousands of messages of military intelligence. From July 1944 to April 1945, the busiest times were the peaks before the Leyte and Luzon landings.

An order from the Eighth Army to follow General MacArthur's forces to Tokyo arrived before the year ended. Crated equipment and supplies were loaded on trucks in a convoy and onto railroad cars to the embarkation point at Manila harbor. Equipment was loaded on an LST, and personnel boarded a camp boat to Subic Bay and Japan. Onboard they carried out kitchen duties and guard detail until Yokohama. The new camp was at the Sanshin building in Tokyo with other signal units of the Eighth Army while General MacArthur's headquarters was in the Daichi building. Classified signal equipment was heavily guarded.

The unit was further depleted by two officers, one warrant officer, and 30 enlisted men operating in the communications facilities for GHQ. The following month, the signal units were strengthened with the arrival of the 71st Signal Service Battalion. Fire drills were held to ensure evacuation of radio equipment and the vehicles in the motor pool. For recreation, a snack bar was opened in the basement selling beer, Cokes, and hamburgers.

As more military units arrived to occupy Japan, additional medical teams were sent to control the increase of reported venereal disease and bring it down to the authorized rate of 50 cases per 1,000 men per year. Men were directed to attend classes that were "to instruct men unfortunate enough to contract VD."

Captain George Baker, head of the Filipino radio operators, remarked almost two years later: "Their devotion to duty was high, and despite great handicaps, they did their level best to carry out their duties."

1st Reconnaissance Battalion (Special) Paratroopers at Camp X (Tabragalba), Australia.

1st Reconnaissance Battalion Paratroopers

The addition of Airborne training at Camp Tabragalba in June of 1944 to participate in the Philippine operations brought many volunteers.

Lt. Cecil E. Walters was assigned to organize the paratrooper school for the 5217th Reconnaissance Battalion. He had men from the 503rd Parachute Infantry Regiment (PIR): TSgt Matthew T. Rini, T5 Lawrence McGreer, Pvts. Wilbur T. Burns and Horace W. Gladney, veterans of the New Guinea campaign, who were camped at Cairns to come to Tabragalba and help train the volunteers. He submitted a training plan to the Chief of Staff for approval from the War Department. The unit's purpose was "to establish a reserve from which may be drawn experienced personnel to drop in any area of the Philippines with equipment where immediate spot formation is later desired, but which is outside the coverage of existing nets." The three-week program included all aspects of Ft Benning's Parachute School Program except for the formal tower training that required 250-ft jump towers.

They were all volunteers to start with, and they volunteered again for parachute training eager to be chosen for a mission. A total of 90 men were very willing to do parachute jumps. The majority were 978th Signalmen. Called the 1st Parachute Platoon, ninety percent were *Manongs*, men from the Central Valley farms in California. Informed that they were trained for missions with almost no chance of returning, they had taken "the fatalistic unit motto *Bahala Na* – 'Come what may.' An incentive was the extra pay for paratroopers.

Two classes of sixty-six officers and enlisted men completed the three-stage course of instructions in September and October 1944. Although the average age was 36 years old, the men's excellent stamina carried them through broken legs and fractures to graduation filling a ward of the 85th Station Hospital. Lt. Walters proudly described the men as "excellent men, unbelievable men. I would take nine out of ten of them into hell."

Queensland, Australia: 1st Reconnaissance Battalion (Special): Paratroopers move out the door during training jump near Camp X, Queensland, Australia Sept 29, 1944. NARA

The first group boarded RAAF planes from Brisbane and came near the camp to practice jumping in a nearby open field. They lined up for Walters to check their chutes before the jump. Mouths dried up, and bodies sweated in the jumps. There were not enough chutes, and they were given two jumps to qualify. Some had four daylight jumps and two night jumps from an aircraft to qualify assisting airborne operations in the Philippines.

The only problem they encountered was the paratrooper's body weight. "Two- thirds of them didn't weigh enough, and they would float around up there... something that added about 20-30 pounds to their weight got them down," related Walters. To drop down quicker than 15 minutes, the Filipino men were given heavy packs. Camilo Ramirez said he used only one grip cord because he was lighter and thus avoided being dragged into the tail of the plane. Two grip cords were for the bigger men.

In his autobiography, Andres Lumicao Velasco wrote that he and seven soldiers from the 2nd Filipino Battalion signed up for paratroop training. Six of them performed three jumps from an airplane and qualified to earn their wings.

The officers who proudly completed the course were Capts. Gene Frias and Isaac R. Unciano along with Lts. Robert C. Allen Jr., Kenneth H. Oates, David Vivit, Felix Hondolero, Joe Tudtud, Pacifico Agpaoa, Ignacio Sarmiento, Edmundo Marfori, and Frank B. Smith. Arguelles may have also completed the course as a photo showed him wearing parachute gear.

First Qualified Paratroop Group		Second Qualified Paratroop Group	
Sgt Arsenio Turqueza	Cpl Moses A. Anderson	Sgt Jimmy Esguerra	Pfc Custodio P. Alerta
Sgt Carmel A. Ilapitan	Cpl Santiago Q. Jaramilla	Sgt John K. Constantino	Pfc Angelo M. Malvar
Sgt Cipriano B. Pablo	Pfc Anton T. Garlitos	Sgt Narciso T. Arias	Pfc Francisco H. Formoso
Sgt Don S. Ruiz	Pfc Antonio L. Maravel	Sgt Paulino A. Rosales	Pfc Narciso N. Arcia
Sgt Emil J. Sibonga, Jr.	Pfc Floro O. Javier	Sgt Sergio V. Solidarios	Pvt Alfredo H. Despy
Sgt Hermenegildo R. Caoili	Pfc Francisco I. Ouano	Cpl Antonio M. Quesada	Pvt Eduardo Balderama
Sgt Luis P. Padilla	Pfc Frank C. Cachapero	Cpl Aurelio Dupitas	Pvt Servando A. Abel
Sgt John A. Bolante	Pfc Restituto I. Nalangan	Cpl Basilio A. Quinitio	T4 Benjamin B. Deleon
Sgt Juan O. Javonillo	Pfc Santiago Kinanahan	Cpl Primo T. Mendoza	T4 Benjamin C. Bulatao
Sgt Melvin D. Fortes	Pfc Camilo Ramirez	Pfc Agapito B. Serrano	T5 Alfredo S. Baylon
Sgt Pastor A. Rillera	Pvt Crisanto Modellas	Pfc Celestino D. Alayu	T5 Bartolome Javier
Sgt Roque Velasco	Pvt Hermenegildo A. Aguilar		
Sgt Santiago D. Abrenica	Pvt Joseph Alabanza		
Sgt Teodorico B. Sabinianao	Pvt Serge M. Mendoza		
Cpl Cipriano C. Peneyra	Pvt Victor I. Vicente		
Cpl Fermin B. Costales	T3 Carl G. Hilario		

Lt. Felix Hondolero beamed with pride when he received his wings Unfortunately, the new paratroopers received sobering news of a tragedy when two of their companions, Cpls. Moses A. Anderson and Magdaleno S. Loma were killed in a motor accident on the New England Highway north of Camp X.

Camp X, Australia: Men of the parachute platoon receiving Airborne wings from Capt C.E. Walters. Oct 1, 1944. NARA

The unit was shipped to Hollandia to wait for further orders. Proud of the Filipino units, Whitney sent photos of the first group of qualified Filipino Parachutists at Camp Tabragalba and the 1st Reconnaissance Battalion to GHQ.

It was a disappointment when their request to wear their patch, '1st Reconnaissance Battalion Paratroopers,' was disapproved. The reason given was that only a small segment of the Battalion was airborne.

Walters followed up with PRS on the deployment of the *Bahala Na* paratroopers. The original plan was for the paratroopers to be attached to airborne units that were going to be used to invade the islands. They were to be guides or would be dropped behind enemy lines as reconnaissance details preparatory to airborne landings. To keep in shape, the paratroopers continued training in the jungles of Hollandia and practiced rubber boat operations at Sentani Lake. Weekly night infiltration operations behind enemy lines were held. Aside from athletics and the use of firearms, educational subjects such as Accounting, History, Social and Political Science, and Cooking, among others, were taught. This was a boon to the men from the crop fields of California who had not had the chance to improve their educational level.

1st Reconnaissance Battalion Paratroopers patch

Luzon Island, Philippines Landing – January 1945

There was a "plan to be dropped in Luzon to reinforce the guerrillas at a mountain pass where the Japanese would be forced to go through, but the Japanese retreated earlier so it was cancelled," recalled Walters.

By March 1945, after Manila was retaken, it was a disappointment when a few of their comrades began returning from missions in the Philippines.

Seven paratroopers assigned to the Counterintelligence Corps (CIC) were attached to the 11th Airborne Division. They did not jump into the Philippines but became part of amphibious landings.

Shangri-La, New Guinea

But it wasn't long before the men's courage was exhibited but not in combat. Three months before the surrender of Japan, a U.S. military plane with twenty-four people on a sight-seeing flight flew too low and crashed into a mountainside of a valley, one hundred twenty miles south of Hollandia. It was in the middle of New Guinea marked 'Unknown' in the maps. Twenty-one passengers perished. The Marines had earlier captured the area but there were still Japanese scattered ten miles around the crash site.

A volunteer rescue team was called for. All the paratroopers stepped forward. Ten enlisted men were chosen to take part in the rescue of three crash survivors. Walters proudly bragged that the men who volunteered to do the rescue "exemplified courage, perseverance, and ability of a very high order."

A week after the crash, a rescue team parachuted into the uncharted 'Hidden Valley' area. Sgts. Benjamin C. Bulatao and Camilo Ramirez, both trained medics made the first jump near the wreck. The following day, eight men jumped with Walters - Sgts. Santiago D. Abrenica, Hermenegildo R. Caoili, Fernando L Dongallo, Juan O. Javonillo, Don S. Ruiz, Roque M. Velasco, Alfredo S. Baylon and Cpl. Custodio P. Alerta. They established a base camp and cleared out scrub bushes for a glider strip to pick them up later. Gliders were the only means of evacuating from the inaccessible terrain. No plane could land because of quicksand.

Ramirez, unfortunately, hit a rock when he landed but could still walk. The parachutists were also spotted by natives, who hurried to the site. They surrounded the Filipino soldiers armed with spears and bows and arrows. The men, at first, thought they were hostile and cocked their carbines. One of the natives who looked like the village chief began talking to them but none of the soldiers could understand him. Through words and gestures, Ramirez

Shangrila, New Guinea: 1st Reconnaissance Paratroopers who volunteered to rescue survivors of a plane that crashed in May 1945 in uncharted territory in New Guinea. Camilo Ramirez is seated.

explained that a plane had crashed, and they were there to help. The natives were primitive but turned out to be friendly. There was a "horrible smell emanating from the natives due to pig grease and charcoal smearing."

For two hours, Ramirez and his group trekked through the jungle guided by two young natives who nimbly moved through the undergrowth like rabbits as men tried to keep up. At a clearing, they saw the three survivors, two men and a woman. They examined and treated their wounds. They were starving and as rations were opened, Ramirez roasted sweet potatoes growing wild in the area. In boiled water, he added scrapes from his block of chocolate to make hot chocolate. Ramirez thought that the survivors were from the city and only saw fruits but did not know how they grew.

From what they saw at the crash site, the plane's tail had broken off and, in the fire that ensued, the survivors went down the plane slide to escape the flames. Bulatao and Ramirez spent a week slicing off the thick gangrene on the survivors' burns and washing them with peroxide. One survivor had a cut on his head and a dislocated shoulder that was immobilized with a splint from the branch of a tree and wrapped with bandages. From the looks of it, the survivors were in no condition to walk through the jungle. The men contacted their base, set up a camp, and waited for the survivors to recover. Walters arrived with five of the men leaving Abrenica, Velasco and Baylon at base camp. The dead were buried, and tents were set up. For sanitation, a hole was dug and covered up.

It took almost two months for the survivors to be strong enough to walk back down to the base. During that time, the camp was supplied with food and water through sporadic air drops when the sky cleared of fog. The men were on constant alert because there were natives in the area who were known to be headhunters and were zealously guarding their sweet potato farms from outsiders.

Once, a message was sent for clothes when the survivors' uniforms began to shred. But "all the clothes that came in were for the WAC Corporal. I think it was three or four dozen bras, and we all howled."

The hike back to base camp was going to be a shorter route but still a forty-mile walk. For two weeks, supplies were dropped daily at base camp until the glider came. A war correspondent, Alex Cann, representing the Dutch East Indies Government parachuted into the valley with a movie camera for a scoop of the unusual rescue. Celebrating their last night, a pig was roasted Filipino-style and the natives joined blowing on their bamboo harps.

The glider circled several times until it was low enough to slide onto the bumpy runway. When a plane arrived, the survivors and medical team quickly boarded the glider that was then towed out to the sky. The rest followed when the glider and plane returned.

Lt. Walter commended the men for the rescue job and for volunteering to go into a country where no white men had ever gone, hiking through virgin jungles and almost impassable mountain trails. "How little fear the men had, all gung-ho and should I say 'Bahala Na.'"

The rescue team received commendation letters from the Major General Commanding, U.S. Army, for "rendering invaluable services in the Hidden Valley rescue." Among men on the rescue team, Ramirez had fought as a Philippine Scout, was wounded, and escaped onboard the *Mactan* to Australia. After recovering from his wounds in the U.S., he reported to the Regiment. Before the war, Ruiz had worked on a farm in exchange for room and board, graduating from high school before enlisting.

Months after the successful rescue, publicity of the men continued. Ramirez signed autographs at war bond rallies. The film clip taken was released and it was the first film record of the Dani natives of New Guinea. Scenes showed the Filipino men preparing a lechon, a pig on a spit stuffed with sweet potatoes. The men's interaction with the natives and placing white sheets on the ground to mark the glider landing were shown.

Ramirez thought he would be sent back to camp to rest. But the following day he received orders with two others to go on a secret mission without being informed of the purpose or location. Once they had been picked up in a small plane and then transferred to a bigger plane, an officer told them to put on parachutes and gave them a list of names. It was a list of ten Filipinos who were reported to be Japanese collaborators. Ramirez recognized the names of a neighbor friend and a man he went to school with. They were to jump and guide a party of officers into his hometown. However, he was not to contact or enter his hometown to avoid being recognized or be tempted to go home. They split up on the outskirts of town. A captain told him it was the end of his mission, so he turned over the map that pinpointed the houses of the men on the list.

From Hollandia, he boarded a plane to Luzon and jumped with his weapons into the outskirts of Tokyo after Japan's surrender. Returning to Manila, his request to see his family that he had not seen for three years since he left the Philippines as a wounded soldier was approved. But he felt so uneasy going to his parents' house that he asked to be assigned elsewhere. The list of collaborators he was given earlier bothered him. He had pinpointed their houses on the map except for the two he knew. He did not want to go home and be recognized. For a year, he was stationed at the 22nd Hospital in Guam before returning to the U.S. He really wanted to see his parents but was still afraid that someone was waiting for him to avenge the guidance he had given. A re- enlistment sent him to Germany, and he was promoted to sergeant. Feeling homesick, he went back to his hometown but hid for several hours. It was then he learned that his house had been carried out to sea by typhoons. The U.S. was now his home. He wrote that he was proud to be a disciplined soldier who followed orders without question.

U.S. SUBMARINE MISSIONS TO THE PHILIPPINES 1942 – 1945
978th SIGNAL SERVICE CO., 1ST RECONNAISSANCE BATTALION
RADIO, WEAPONS AND SUPPLIES FOR THE GUERRILLAS

Submarine	Location	Arrival Date	Men onboard
Allied Intelligence Bureau Missions 1942 – 1943			
Gudgeon-211	Catmon Pt. Negros	January 1943	6
Tambor-198	Tukuran, Pagadian Bay, Zamboanga	March 1943	5
Gudgeon-211	Pucio Pt. by Pandan Bay, Panay	April 1943	4
Trout-202	Pagadian Bay, Tawi-Tawi, Mindanao	June 1943	5
Thresher-200	Cansilan Point, Negros.	July 1943	4
Grayling-209	Libertad, Panay	Aug 1943	4
Philippine Regional Sections Missions 1943 - 1945			
Narwhal-167	Paluan Bay, Mindoro, Butuan Bay, Mindanao	November 1943	9
Narwhal -167	Cabadbaran, Butuan Bay, Mindanao	December 1943	12
Narwhal -167	Pandan Bay, Panay, Balatong Point, Negros	February 1944	1
Narwhal-167	Tawi-Tawi, Mindanao	March 1944	6
Nautilus-168	Tukuran, Pagadian Bay, Mindanao	June 1944	1
Narwhal-167	Gamay Bay, Samar Tukuran, Pagadian Bay, Mindanao	June 1944	39
Redfin-272	Ramos Island, Palawan	June 1944	6
Narwhal-167	Lipata Point, Antique, Panay	June 1944	4
Nautilus-168	Balatong Point, Negros	June 1944	4
Nautilus-168	Pandan Island, San Roque, Leyte	July 1944	25
Seawolf-197	Tawi-Tawi, Mindanao Pirata Head, Cambari Island, Palawan	August 1944	12
Stingray-186	Bangui Bay, Luzon	August 1944	15
Narwhal-167	Dibut Bay, Baler, Luzon	September 1944	42
Nautilus-168	Eastern Mindanao		
Narwhal-167	Macajalar Bay, Mindanao, Illana Bay, Mindanao	September 1944	40
Seawolf-197	Samar, Batan Island, Leyte	September 1944	16
Nautilus-168	Mindoro	September 1944	2
Stingray-186	Suluan Island, Samar	October 1944	3
Narwhal-167	Tongehatan, Tawi-Tawi, Negros and Cebu	October 1944	52
Ray-271	Mamburao, Mindoro	October 44	3
Nautilus-168	Dibut Bay, Luzon, Masanga Pt., Tayabas, Luzon	October 1944	36
Cero-225	Infanta, Luzon	November 1944	16
Gar-206	Darigayos Cove, Luzon	November 1944	16
Unidentified Landing craft	Dulag, Leyte Mindoro	December 1944	26
Cargo sub	Taw-Tawi, Mindanao	January 1945	4
Ship or plane	Samar for Bicol	December-January 1945	5

Source: MacArthur Memorial Library; Calendar of Submarine Shipments to Philippines Guerrillas and Agents, Dec 1942-Dec 1944; First Reconnaissance Battalion, Special, United States Army Forces in the Far East, 12-Dec-44, Personnel on TD Outside Australia

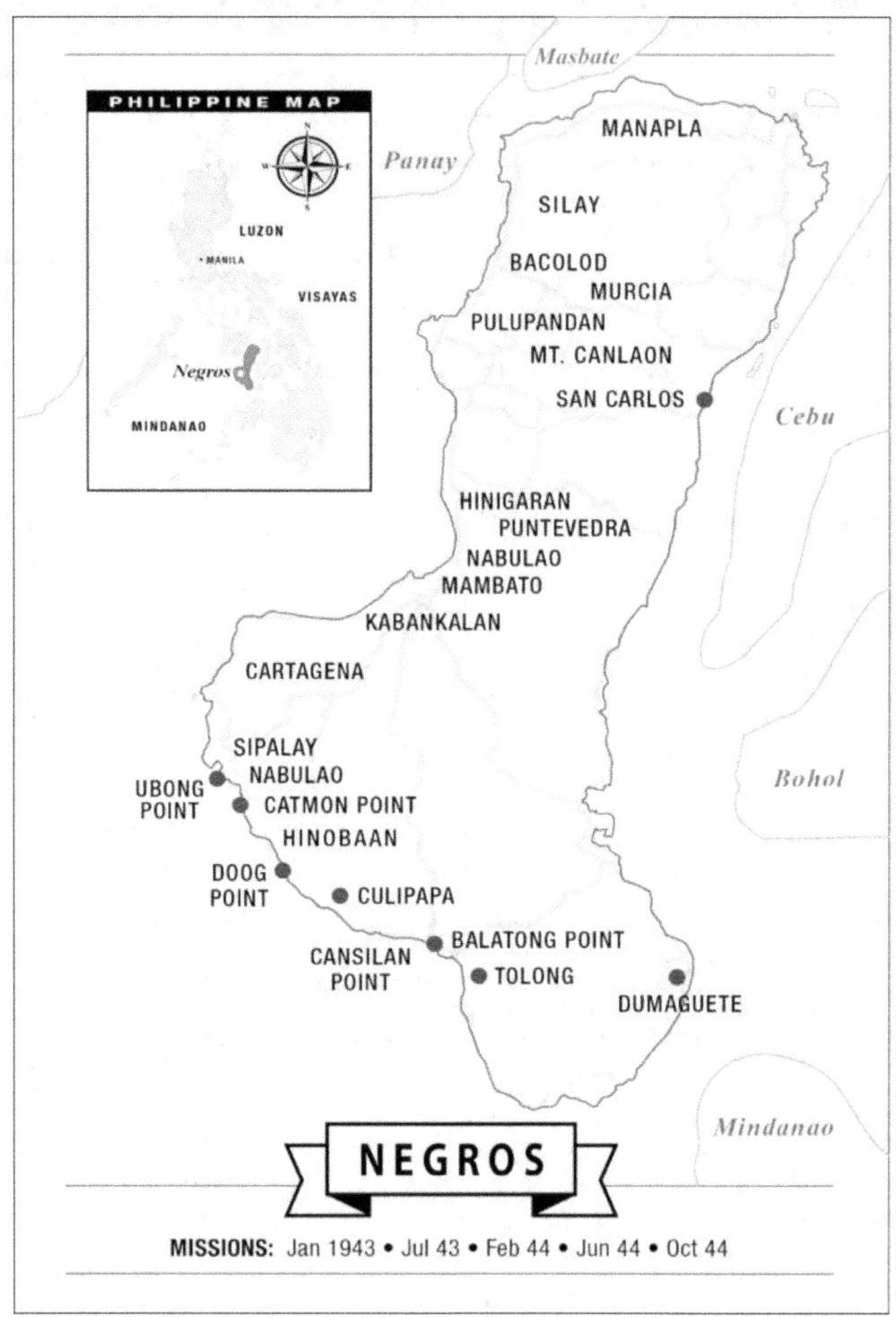

PHILIPPINE MAP

LUZON

• MANILA

VISAYAS

Negros

MINDANAO

Masbate

Panay

MANAPLA

SILAY

BACOLOD

MURCIA

PULUPANDAN

MT. CANLAON

SAN CARLOS

Cebu

HINIGARAN

PUNTEVEDRA

NABULAO

MAMBATO

KABANKALAN

CARTAGENA

SIPALAY

NABULAO

UBONG
POINT

CATMON POINT

HINOBAAN

Bohol

DOOG
POINT

CULIPAPA

BALATONG POINT

CANSILAN
POINT

TOLONG

DUMAGUETE

Mindanao

NEGROS

MISSIONS: Jan 1943 • Jul 43 • Feb 44 • Jun 44 • Oct 44

Negros Island

Soon after the fall of the Philippines to the Japanese, scattered radio messages began coming out of the islands to anyone who could pick them up. Hints of a resistance force in Luzon elated GHQ but communications soon broke off after the leaders of these groups were captured. The radio contact from Panay six months after the surrender of Bataan and Corregidor was the only one left. The transmission was unsteady and in clear text that could easily be intercepted. But to GHQ, it was proof that a hidden resistance force existed in the Philippines ready to fight the enemy. General MacArthur and his staff at GHQ, SWPA recognized the significance of harnessing them into organized and united groups to plan a successful return.

Eager to find out more about these groups, AIB was ordered to send the first of many secret missions to survey the guerrilla and enemy situation immediately. Information transmitted to Australia needed to be vetted to be useful. Allison Ind, AIB controller, recalled, "G2 Intelligence left little doubt in our minds (of the situation): the obvious precariousness of the equipment in the Islands, the lack of secure ciphers and, above all, the contradictory nature of the information that was beginning to come out underscored the need for an observer placed there and controlled by GHQ." Mindanao was the closest island to Australia, but who would be the best person to lead a party?

The person who would be sent to lead the party would be given the mission of deploying AIB's plan to establish a "net for military intelligence and secret services throughout the islands: organize a chain of radio communications, both within the Philippines and to Darwin in northern Australia; locate and contact influential persons known to be loyal; and develop an organization for covert submarine actions and sabotage." If not wholly accomplished, at least begun. What the first party would find out and report back would be used to modify plans in the future.

Australia was scoured for Filipinos, who were then enlisted into the Army of the United States (AUS) for the mission. Candidates were screened, and their crucial role as team members was explained to convince them to volunteer. But they were made aware that they were going into unknown territory without a welcome party to guide them or provide security, lodging, and food. They would not even know who was pro-American and pro-Japanese.

Maj. Joseph McMicking and Ind, the AIB controller, decided on the man they felt could successfully lead PLANET PARTY, the first team to penetrate the Philippines. McMicking knew that Maj. Jesus Villamor had fled to Australia and wanted to go back to his country. He was a decorated pilot who had shot down many advanced Japanese planes from an obsolete P-26 fighter plane he flew after the Japanese attacked Nichols Field. He had evacuated to Corregidor, boarded the blockade runner SS *Legaspi* to Mindanao, then left on a plane to Australia. With his knowledge of the Philippines, Ind handed him the directive initialed with MacArthur's approval.

Villamor submitted a report to Ind he marked 'Secret' detailing a rough plan for the mission's objectives. He recommended a small party of men with a radio to be brought to the Philippines by submarine. All the men would have military backgrounds. He preferred to enlist men he knew in the U.S., but GHQ denied the request.

AIB chose survivors of the ill-fated blockade runner *Don Isidro* sunk off Bathurst Island: Capt. Rodolfo Ignacio, radio officer Lt. Emilio Quinto, and Sgts. Virgilio Felix, Susano Amodia, and Patricio Jorge. Sgt. Dominador Malic, a boatswain on the hospital ship *Mactan* and Lt. Delfin Yuhico, one of the wounded passengers onboard were chosen. Pedro Carriaga and Virgilio Felix were backups.

Villamor gave code names to the men in his party to be memorized and used in an emergency. His own code name was W-10. They had aliases: Villamor was Ramon Fernandez, Ignacio was Carlos Noble, Quinto was Juanito del Rosario, Yuhico was Juan de Jesus, Jorge was Vicente Reyes, and Malic was Dalmacio Macilag. They would live off the land after their twenty-one-day rations were consumed. After six weeks, Villamor stated that he would return to Australia to decide with AIB where to establish the main transmitter and plan for future missions. He was confident that his contacts all over the islands would help in gathering information and organizing groups for future contact in subsequent missions. He asked to be given authority to commission people in the U.S. Army, who would help in the mission's success.

Sgt. Jorge recalled "We were eight men who volunteered to go. We were all trained together in Brisbane three rooms away from Gen. MacArthur's office in the Courier building. Col. Allison Ind trained us in codes, ciphers, aircraft, ship recognition, semaphore, blinker signals and Morse code. We were quartered in Ascot airfield separate from other groups... in a certain room...no one else with us... All the rest of the area did not see us. We were not allowed to go out for recreation. A staff car with darkened windows transported us to Milton Court for physical training in unarmed combat under an Australian officer for a month. Ran five miles and swam for three hours. We were trained on how to kill. Firearms training was in Sydney. Boat handling and landing was at Morton Island for two weeks. Two Australian and New Zealand soldiers, experts in commando and jungle warfare, trained us in Canungra. We went barefoot to form calluses on our feet (to blend in with the Filipino farmers and avoid capture). We had nothing to carry because the Japanese usually inspected the shoulders to check if you carry a rifle. At Eagle Heights in a jungle close to Brisbane, we trained for two weeks in jungle survival, camouflage, living off the land, eating snakes, wallaroo, and even caught a wild turkey before Thanksgiving Day. Many times, we trained

Jungle training in Australia. Patricio Jorge is at front left. He was on the first submarine mission to the Philippines that landed in Negros in January 1943. Source: Patricio Jorge.

naked in the jungle with only our hunting knives. That is the way we were trained into a unit. The last day of training was just before we departed.... We were very young. There was no need to be afraid. If we get killed in New Guinea or anywhere else, they cannot (sic) find our bones, so might as well go to the Philippines."

Lt. Emilio Quinto's code name was 'Sparks.' As the communications officer on the ill-fated *Don Isidro,* he was the only qualified Filipino radio operator found in Australia and was the chief operator of the ship *Argosy,* a floating wireless station off Port Moresby supporting the American forces fighting in New Guinea. The ship carried the bulk of radio traffic between GHQ in Australia and Washington. It took a memo from Maj. General Richard Sutherland, MacArthur's Chief of Staff to release him from his position, emphasizing that the party for special penetration was built entirely around him. Since he was the "only expert native radio operator in this area, the success of use of the entire party depends on including Quinto."

Aside from their personal M-1 Garand rifles, machine guns and grenades, they brought radios used by the Australian coast watchers and spare parts to operate for at least a year. Six crystal channels enabled precise frequency usage. Four metal boxes each weighing less than 200 pounds were packed with a 3BZ transmitter, receiver, battery, a collapsible 40-ft antenna tower, and accessories that could fit through the 24-inch-wide submarine hatch. The ATR4 transceiver and radio weighed only twenty- three pounds and fitted into a backpack. For the messages to cross the roughly 3600 miles between the islands and GHQ in Australia, they brought an 80-watt transmitter specially built by the U.S. Army Signal Corps. They brought maps and the latest pamphlets on intelligence and Japanese ships. An interesting cargo was *bolos* for each of them. Precious stones were brought as recommended by Villamor. He believed that the sale of the stones had greater exchange value and the money would go to carrying out the mission.

The day after Christmas, they were ready. But before their departure, the mission's scope and objectives were expanded. They included unifying under one strong leader the warring independent guerrilla groups in Negros and the outlying islands of Panay, Cebu, Bohol, and Mindanao; working out an "eventual escape route to accommodate evacuation of selected individuals in submarine subsequent missions; locating radio equipment and listening stations; and locating the Bureau of Posts and old Philippine Army transmitters and their operating frequencies. Also obtain general information on "Japanese political, military, civil intentions, military strength, dispositions, equipment, and Japanese operations of future significance." A plan was discussed to create a secret spy organization, particularly around Manila, that included prominent Filipino men. Villamor was to return to Australia with a full report and recommendation for future parties.

What may have been kept from the men by the AIB was that the torpedoes of the submarine they were boarding had mechanical defects that added to the risk of the mission. Months after the Pearl Harbor attack, it was first noticed that the torpedoes either exploded before reaching the target or, if they did hit the target, did not explode. A more severe defect was the torpedoes running deeper and ended up passing below the target. Some merely sank or took wide turns to return and hit the submarine that fired them. It took over a year to fix the defects, but in the meantime, the submarine missions to the Philippines were at risk whenever they fired at enemy ships.

In the last month of 1942, Villamor described his cipher system using Philippine city names. Combinations using his birthday and a city were to be transmitted in a specific sequence in emergencies. At Ind's office, the team signed their last wills and testaments. If the beneficiaries were in Australia or the U.S., Ind sent a letter just like he did for Quinto stating that the latter would be "absent on a mission of considerable importance and which may require his further absence for a prolonged period. He was unable to advise you of his departure as he was forbidden from speaking to anyone about it."

Guns were still bombarding Guadalcanal southeast of New Guinea when the first party written as 'secret cargo' in the submarine log was dispatched on a suicidal mission into a Japanese occupied territory. They had no knowledge of pro-American guerrilla activities for support.

USS *Gudgeon* – January 1943

On Christmas Day, a mission order marked 'Secret' listed the six men – Maj. Jesus A. Villamor was team leader with Lts. Delfin C. Yuhico (codename 'Dagwood and Z-40'), Rodolfo C. Ignacio, and Emilio F. Quinto and Sgts. Patricio Jorge and Dominador Malic – and informed that they were to be sent to an undisclosed destination. It was signed 'By command of General MacArthur.' They were sworn to secrecy before embarking on the submarine.

Ind had hinted to Villamor that they were flying to the port of Fremantle in Perth before boarding the submarine. Sgt. Jorge recalled that before they left, General MacArthur and his staff shook their hands and asked if anyone wanted to withdraw, no questions asked. No one backed out. MacArthur then brought out a brown envelope with orders to Commander Stovall of the submarine *Gudgeon* "that at the landing point if gunfire is heard, they will shoot you first. Yes. Shoot us. Supposing somebody fires?" he thought. He also "carried microfilm with a high-grade cipher system for Fertig, guerrilla commander of Mindanao and Peralta of Panay in a dental alteration and under a patch in his shoes."

For her unique assignment, some of the *Gudgeon's* sixteen torpedoes were offloaded to make room for the party and their equipment. She was a combat patrol submarine with a small cargo space. A dinghy, lashed to the submarine's side for the party to use in the beach landing affected her speed and diving depth.

The men changed into old dungarees, sailors' working clothes, and ordinary shoes to blend in. They packed their gear in water-tight containers and boarded a truck to the dock. Looking like mess boys, they helped unload the submarine of extra weapons before loading their equipment and never came out. Inside, their gear and provisions for seventy-five days were stored with the crew. Crates of food and other supplies were stashed in showers or engine rooms or wherever they would fit. Bunks were fitted throughout the many compartments, including the torpedo rooms.

Departing Fremantle, the *Gudgeon* set its course towards Pagadian Bay in Mindanao, the closest island to Australia. She ran on the surface at night to recharge batteries and diving to avoid Japanese patrols. A heavy bomber escorted the submarine going through narrow slots by the Solomon Islands. Once, the ship's alarm blared at the sight of an enemy ship and dove. It was a first harrowing experience for the men who were just getting used to the constricting life onboard.

The men's bunks were in the forward torpedo room, and their equipment stowed under the floor plates. Yuhico said they signed the torpedo before it was loaded into the tube. Their gasoline supply in five-gallon cans was stowed in the escape trunk, safely sealed off from the rest of the ship. They took pills to supplement their brief exposures to sunlight and walked barefoot to thicken the soles of their feet like farmers. Repeated dry runs of landing in the dark within the restricted confines of the torpedo compartment were executed. They did calisthenics regularly to keep physically fit. Jorge (codename 'Octopus') recalled it was hot inside the submarine as they were underneath the charging batteries. They put the torpedoes between their legs to cool off. The submarine had the best food in the entire Armed Forces, he thought.

Yuhico was seasick throughout the voyage and was the only one with Villamor allowed to go to the conning tower for relief. Breathing the fresh air, he recalled that just over a year ago, he was escorting a convoy of trucks transporting ammunition to Mauban, Quezon to defend the beach when the truck he was riding swerved to avoid some civilians. The truck flipped and fell into a ravine. He was the only survivor, suffering serious bruises and a broken hip. He was brought to the hospital ship *Mactan* for treatment then transported to Australia. Now he was inside a submarine on a suicidal mission to go back to the islands.

While onboard the *Gudgeon*, Villamor wrote a letter for the skipper to deliver to Ind. He gave tips to be incorporated into training plans for future parties. Health and stamina were essential. One of his men was sick the entire trip, and his weakened condition would make him of no use at the landing and the mission itself. Security was to be emphasized as one of his men gave out names and addresses of 'swell people' in Brisbane to some of the submarine crew. Aside from loyalty and devotion to duty, physical courage was needed. They must have mature minds for a unique undertaking like this. Two of his men, who were barely in their twenties had a "jitterbug frame of mind and (were) immature." He opined that the current Filipino men in Australia had no military background and needed the training to cope with the situation they were to face. They would be useful in actual military operations but doubtful in intelligence work. He recommended Filipinos who graduated from West Point and Annapolis to lead future parties.

The plan was to land on the southern coast of Mindanao to establish a radio station, then proceed to Panay where the rest of the men, medicines, and ammunitions would land. Jorge recalled that when the submarine surfaced on the eastern side of Davao, a Japanese submarine was spotted. The horn blared for battle stations. The submarine dove to close in and fired a torpedo but missed. Rounding the southern tip of Mindanao, the submarine passed Pagadian Bay, the rendezvous site. The skipper gave the order to land once the site was clear.

Villamor informed the submarine commander that the new destination was the island of Negros. High winds on the first night kept the *Gudgeon* submerged as it sailed quietly along the coast. The next morning was spent scanning for the new location. Cansilan Point looked clear but, in the afternoon, small boats were sighted along the beach covering all prospective landing sites. At dusk, mysterious lights suddenly appeared on a beach which one of the men identified as night fishermen. The periscope continued to scan the coast to ensure that a landing could be made safely.

A deserted beach at Catmon Point was spotted. The submarine surfaced and headed toward the beach as soon as night fell to avoid late night fishermen. As the men set up their rubber boats, the submarine's deck guns were manned as an added precaution against water and shore attacks. The crew brought the cargo from below to topside. One of the three rubber landing boats did not inflate properly and exploded. Seats were removed from the rubber boats and replaced with plywood to serve as the floor. Villamor refused the offer to use the dinghy lashed to the submarine's side reasoning that they were not trained to use it. It was also too big to bury, and it would take more time inflating it than loading another boat. It was more important to get to ashore fast and undetected. Everything had to go into two rafts to avoid a second trip. Forty numbered water-sealed tin cans measuring 2 x 2 feet and wrapped in burlap were piled into the rubber rafts. They had only their side arms and one radio set for communication with Australia. With one rubber boat less, supplies of medicine, some ammunition, vitamins, candy, cigarettes, and American magazines wrapped around fragile bottles that were earmarked for the guerrillas were left behind. Villamor said that they could not bring the medicines because they were not waterproofed although Ind had explained to him "prior to departure that the medicines were sealed in glass containers and would not be detrimentally affected by moisture or emersion for reasonable periods." He also turned down a British automatic rifle with a 50-round plate magazine offered by the submarine skipper. Jorge cringed at the refusal. In his patrol report, the submarine commander wrote that Villamor changed the rendezvous site from Mindanao to Negros, had trouble with one boat, and made only one trip to shore.

They began paddling with the two rubber boats tied together with a long rope bringing mostly their personal weapons, radio equipment, codes, gems, and money. Jorge recalled that they had their uniform and a few pesos "because no pesos in Australia." They had "a few medicines... our 45s except for my Garand...and only one box with Tasmanian big apples and sandwiches." Maj. Villamor held the diamonds and rubies in a pouch. Later

Jorge recalled that many civilians in Kabankalan where he was assigned were able to get some of those stones.

Paddling through the darkness, a *banca* (small wooden boat) circled them with a "gun pointed at us." Sgt. Malic, code name 'Lapu Lapu,' who spoke the Visayan dialect of the area, shouted out for help for a shark-bitten friend to get a hint of how many were in the banca. "They replied and pointed us in the direction to go to, but instead we turned the other way towards Catmon Point." It was high tide, and they did not see the corals up to the beach. Slowly navigating around the coral reefs, they furiously paddled to reach land. It was a landing into the unknown and they were relieved that they reached shore unscathed. Three flashes to the submarine signified their successful landing and she slipped out to sea.

One rubber boat was buried under a large boulder and the gear stashed in a cave fifty meters from the beach. Malic and Quinto hid with Villamor in the cave with Jorge's gun to cover the others who were outside armed only with pistols. "We had one case of apples in case the guerrillas or anyone" came. Soon, moving lights from many small boats arrived and they were relieved to see guerrilla men in tattered clothes and barefooted. Around ten guerrillas armed with spears, surrounded the three, and confiscated their hunting knives. They gave apples that distracted them from seeing the others in the cave. Two men stayed on the beach while the others brought the three to their leader, Capt. Jorge Madamba in Nabulao. Madamba, an agriculture graduate in the U.S. had armed his men with ammunition and supplies he had salvaged from the SS *Panay* that sunk off the southern end of Negros while bound for Corregidor in January 1942. In the cave, the men stayed crouched all night and hardly slept. A snake had crawled close to the group, but no one was bitten.

Madamba's daughter recalled that the captives spoke in Tagalog, a language they could not understand, nor could the strangers understand their Visayan dialect of southern Negros. They identified themselves as coming with a group of coconut pickers. Their story was not believable from their unstained white teeth, and although barefooted, their toes were not splayed from years of not wearing shoes or climbing coconut tree trunks. They were tied to coconut trees. The following morning, the men were alarmed when they were told to dig three graves. If Villamor had not arrived looking for them with the other guerrillas, they would have been killed. Madamba immediately recognized Villamor from a photo on the front page of one of the last issues of the *Manila Daily Bulletin* in December 1941. In a soft voice, he welcomed the men.

A crowd soon formed, jubilant with the thought that the Americans had arrived. News of the submarine's arrival spread like wildfire fueled by Villamor's popularity. Help had come. Higinio de Uriarte, a Basque Negros guerrilla, recalled, "The general rejoicing and enthusiasm caused by the party's arrival were indescribable and impossible to control."

Villamor, Madamba, and Malic were brought to the headquarters of Col. Salvador Abcede who led the guerrillas in the central and northern part of the island. Abcede was an officer of the Philippine Army at the time of the surrender and took his forces to begin guerrilla resistance. Another Army officer turned guerrilla, Maj. Placido Ausejo harassed the Japanese in the south. Between them were seven thousand organized guerrilla troops, 600 Enfield rifles, and assorted firearms. Abcede's troops were organized in groups of 30 men with 10 rifles conducting 'ambushcades' – quick hit and run ambushes that avoided capture to survive and fight another day. Snipers were ordered to kill at least one Japanese soldier a day. With two small radios probably from Panay, Abcede had contact with Panay who had earlier contacted Australia. The Japanese reportedly wanted the fertile plains of Negros, so he pushed his men to hide in the mountains.

Informed that there were Japanese in Abcede's area, the men set up the radio in nearby Cartagena. It took three days to set up the 80-watt transmitter facing the Sulu Sea. Professor Roy Bell from Dumaguete Island helped Quinto establish radio station 4E7 with the 80-watt transmitter. The first messages were tapped to the station in Darwin operated by the Royal Australian Airforce (RAAF) over 1600 miles away to pass to station KAZ, MacArthur's headquarters. It was on the second night that contact was made. But it was not that simple to decipher. The message arrived in clear part and with the information that the word to break the coded part was "the name of the author of the book that I (Villamor) gave Glenn for Christmas." Bobb Glenn was the Chief Supply Officer of AIB in Brisbane. Elated with the proof that men can be sent into enemy-occupied Philippines and establish radio contact, Ind reported to Sutherland who replied, "The first to go in, but not the last, We're on our way!"

In his message, Quinto explained that their silence was due to enemy reconnaissance planes circling over Negros. He added the reason for landing in Negros instead of Mindanao. But, aside from their ongoing situation, they were, "doing well." Messages at this time reported an infestation of Japanese spies and pro-Japanese police in Manila. Throughout Luzon, morale was low. In the ten days before moving again, Quinto operated the radio while Malic procured food and assisted Villamor.

Abcede provided a dozen men to help build a communication network of three- man teams. Radio skills the men learned at Victoria Barracks were passed on to guerrillas in a training program. Men were carefully selected and personally trained at camp on intelligence and sabotage duties by the mission men before they were sent out to gather data on the enemy.

The Japanese began attacks on the guerrilla camps upon hearing of the submarine arrival. Sea patrols intensified between the islands. Villamor became dependent on his agents to do his field work. He could not travel to Panay and Mindanao because, as he reported, "Travel is very dangerous and at times almost impossible due to severe restrictions imposed by our forces and increased patrol activity of Japanese. Besides am having personal difficulty due to my picture being in many homes, however, will still try to make [Manila]." Villamor always had to go back for fear of being recognized whenever we went out, Jorge said.

For his safety, a hideout was set up in the hills of Culipapa in the Hinobaan area within the safety perimeter of Abcede's headquarters. Deeper into the jungles of Culipapa, Quinto set up the radio outpost to oversee the radio traffic coming from various agents into his net control station.

Call signs between Negros and Darwin regularly changed. Messages from Negros told of enemy concentrations, internment camps, strengths of various guerrilla organizations, their equipment, and arms. The intelligence reports were evaluated, coded, and decoded before transmission.

A two-week period of receiving no messages from Negros concerned GHQ enough for them to issue a directive to the hideout to immediately send a short report and a very short report every four or five days to assure that all is well. Quinto replied that he was directed to send only the leader's messages, but he was contacting Darwin daily. His request for

Darwin to change their daily listening period of Negros' messages was denied unless there was a security breach.

When word arrived that the Cebu command post was captured, Roy Bell was sent to survey the guerrilla situation.

A Manila contact 'MARS' began relaying messages from the Manila area. Intelligence from the city came from groups of formerly USAFFE men and well- known politicians. With so much radio traffic being transmitted, KAZ directed to send less messages on administrative issues including efficiency reports and details of outstanding services that deserved decorations for later to avoid endangering cipher security.

AIB spent a great deal of time choosing the right guerrilla unit to recognize. MacArthur would only recognize a leader who united the guerrillas, was prepared to gather intelligence on the enemy, and would assist in the return of the American forces. Only then would supplies, weapons and ammunition be sent. To determine a leader in each major island to set up an intelligence net required contacting guerrilla groups and surveying their situation. But the order was not to interfere with guerrilla affairs for control of terrain and men. Reports were to be limited to only on their personalities and organizations. It was a balancing act between following orders to stay out of guerrilla affairs and showing enough interest so as not to lose their support.

Three aggressive guerrilla leaders, Salvador Abcede, Placido Ausejo, and Gabriel Gador, a Philippine Army officer who led the guerrillas in Southern Negros, fought for men, prestige and civilian support for food, further delaying recognition. Complicating the situation were the guerrilla leaders of Panay (Peralta) and Mindanao (Fertig) who were attempting to take control of the Negros guerrillas. Peralta supported Abcede while Fertig backed Ausejo. Peralta was "drunk with power impressing all with being appointed head... impressing you with what he had done and can do... making preparations for armed demonstrations against occupied towns in Panay and Negros," Villamor reported. The disagreements prompted GHQ to define the areas of their responsibilities. The Mindanao agents were to work in northern Leyte, and the islands of Masbate and Samar. The Panay agents were to operate in Cebu and Negros. Both were to work in cooperation for their own and the civilians' sake. But Peralta wanted to know more about Fertig's setup because he complained to Villamor that Fertig never sent him information even after GHQ had issued their areas of control. (Mindanao Island is discussed in a separate chapter.)

The distance separating Australia and the Philippines limited the amount of tangible support that could be brought to the guerrillas. Japanese presence close to northern Australia and U.S. support of Europe cut into any help that could be given. A complicated task throughout the missions was convincing guerrilla leaders to lie low and concentrate on gathering proper intelligence to send back to Australia for planning a successful return. It was a dilemma for the guerrillas to lie low to avoid Japanese retaliation on the civilians since their main source of food came from the civilians who expected protection from Japanese brutality. The areas of Negros that were free of the Japanese sent about twenty percent of all food gathered in their area to the guerrillas.

A conference at Obong Point was held to discuss the leadership of the Negros forces. Villamor recommended solutions to local problems and requested KAZ to be given the authority to deal with some of the problems. Unless policies were established, support of the civilians was lost. Afraid that he would be drawn into a power struggle and be distracted from his main objective of intelligence gathering, SWPA refused to grant him the authority to mediate and settle guerrilla and local problems.

At camp, Malic worked on moving the radio and equipment to Daug close to Hinobaan. He went north to Mambato with an extra radio and parts to hide in case the main radio was discovered. Japanese patrols combed the area close by and it wasn't long before the 40,000 rounds of ammunition he hid at Mambato were discovered. He hid in Cartagena before returning to camp where he hid more extra guns, ammunition, and radio equipment in the nearby jungles. In May, Japanese patrols blocked his way as he was going to rendezvous with a submarine at Lipata Point in Panay to send intelligence documents and obtain supplies.

Yuhico was sent to Panay to deliver call sign SOI and confirm intelligence reports sent to GHQ. Codes were sewn into his boots. "My cipher code was in a cellophane embedded in the heel of my shoes," he related. It took him a week to cross by banca paddled by civilians when the weather was good and Japanese patrols were not in the way. In Panay, he had a guide and depended on the people for food. He slept between "big hair roots of trees and plugged his nose with cotton against leeches that stretch very long and might enter."

Suspicions between guerrilla units delayed a meeting with the overall commander. For three months, Yuhico was passed among several units until he met the executive officer of Col. Peralta. He explained his mission to deliver radio codes and signal operating instructions to contact GHQ in Australia and to gather information on the guerrillas and the enemy situation on the island. Using his code, he sent messages to Australia via a radio established on the northwestern corner of Panay that had contacted Australia earlier. MacArthur's message to Z-40 congratulated him "for advising CO (Peralta) of constructing safe codes and procedures with his own net of radios because Washington broke his messages and warned (of) Japanese endangering our master traffic to Australia. Imperative channels between us be protected by tightest local system based on our procedure and frequent key changes. Use Peralta code and use yours for rare occasions when needed."

Yuhico sent a positive impression of Peralta and tried to mediate the ongoing feud between him and Governor Tomas Confesor that divided the people's loyalty. "He has ten thousand men, half of them armed, with machine guns and assorted firearms. Under martial law, he rules with an iron hand, threatening to execute spies by drumhead courts-martial. His agents are everywhere. He fills the airwaves with messages to MacArthur."

Peralta gave Yuhico access to all reports and codes allowing him to send messages to GHQ. Available food had high prices, he reported. One message sent in February was that of a transport with three thousand Japanese soldiers that sunk off Tablas Island, east of Mindoro. Two transports with several thousand enemy soldiers each landed in Iloilo and departed for the south. Yuhico requested for a submarine to bring weapons and ammunitions and named Pandan Bay as the rendezvous site.

He observed Peralta's dislike of GHQ's orders that limited his authority to Panay and some of his agents in Cebu and Negros. His second-in-command, Maj. Leopoldo Relunia had organized the 67th Infantry as a guerrilla group in the Bicol province without his knowledge.

In March, the Japanese were very active in both Panay and Negros. Yuhico reported the Japanese plan of granting Philippine independence and having the government declare war on America. He sent Japanese strength reports from Panay, Romblon, Mindoro,

Masbate and Marinduque. Later, Peralta asked for Yuhico to be transferred to Panay, but Villamor denied it.

"He (Peralta) does not have much faith in you. Your identity and mission are all over Negros and other islands," Yuhico informed Villamor. He was concerned by the Panay commander's plan to control other units by giving them transmitters that would transmit only to him and making the messages come only from him when sent to Australia.

As more agents penetrated more important areas, a network of radio stations was established, with the radio station operated by Quinto at the Headquarters in Culipapa, Hinobaan as the central net station. A trained group that successfully penetrated Manila reported the city swarming "with spies and puppet police. Throughout all of Luzon, morale was low, and elsewhere it was not a great deal better."

President Manuel Quezon in Washington pressed MacArthur on news of the allegiances of the men he appointed to run the government while he was in exile. Manila had to be accessed for answers. He wondered if others continued to support him and the return of the Allies. "How do the common people feel? What do they think of my having come here (Washington) and what I am doing?" Have they succumbed to Japanese propaganda?

After barely two months in Negros, Villamor reported that he could "get things done and wanted to return" to give a complete report and recommendations only as an observer, not as a policy enforcer. He cited poor health as his reason, including a cryptic note that he "suffered from all diseases known including love… to 'MARS' (Manila) chances of being recognized (are) great" in his message.

GHQ replied that Villamor was to stay in Negros and "concentrate on supplying information and advice on areas of responsibility and trustworthiness of main guerrilla leaders in the Visayas and Mindanao." But he could not avoid the Japanese activities affecting the people. They were now forced to convert their sugar cane to alcohol and high-octane fuel while the fields were to be planted to cotton. Where will they plant for food?

People felt helpless and did not care who would win the war. They just wanted it to end. Officials and civilians were reported as loyal but were forced by the Japanese to work for them. Guerrillas, on the other hand, would now target them for assassination for being collaborators. Villamor strongly recommended that President Quezon appoint provincial governors independent of guerrilla leaders to limit abuse and protect the people. The guerrillas needed the support of the civilians for food and the civilians in turn needed protection against the Japanese. Although the loyalty of the people ran deep, it had to be maintained by mitigating the chaos, starvation, and wanton killings. If not, the mission's objectives would not be met.

By mid-year, Japanese patrols pushed into the south and made the camp's location dangerous. Quinto reported to GHQ that "Japanese land, sea and air activities started very suddenly. Location menaced. Expecting word from leader (Villamor). Fighting lasted four days and dispersed the PLANET team." Three days later, Quinto moved the radio to Villamor's hideout in the mountains. "Conditions getting serious daily."

A talkative courier had been freely distributing packs of Camel cigarettes on his way to Panay from Mindanao and talked about the submarine landings of troops and supplies. He asked about Villamor, who was supposed to go to Manila and not hide in the Visayas.

Whitney who had ordered the propaganda materials to be sent expressed concern when he heard of the talkative courier. He had not intended to send anything made in America unless there was any objection from the team. A package of American cigarettes was a dead giveaway of American contact and invited an increase of enemy patrol activity that may compromise the mission's security. Worse, it could lead to the loss of a submarine.

GHQ directed Villamor to stay and proceed to Panay or Mindanao or another place of refuge if needed to accomplish his mission. Villamor insisted that a commanding officer be appointed for Negros to maintain unity and he wanted it clearly stated to the appointee what line of action to take regarding those who threatened peace and order in areas of command. GHQ replied with a message addressed "to all allied military groups leaders" — and there were more than several of them — "that friction existing between military groups in Negros is profoundly disturbing, particularly in a present dangerous situation. All leaders were to cooperate against the common enemy faithfully. The commander of the 7th MD (Negros) directly responsible to me will be designated when the situation is sufficiently clarified. Pending such designation, I hold each leader individually responsible for cooperation."

Casualties from the guerrillas' infighting spread to nearby Siquijor island. The guerrilla leaders of Panay, Negros, and Mindanao wanted control of Bohol Island. They sent their agents to obtain intelligence to report back to GHQ. The reply was to transfer the intelligence reporting of Bohol and Cebu to Negros.

Under pressure to end the divisiveness and expand the radio network, Villamor requested to make him temporary commanding officer of Negros. In May, he was appointed head of the 7th MD with orders to "require all leaders of allied groups in your district to cooperate against the common enemy under your direction." He was to maintain the civilians' morale so they would continue to cooperate with the guerrillas. Immediately, the Negros guerrilla and political leaders, including on Dumaguete Island, were informed of his temporary appointment as district commander until GHQ recognized one in Negros and Cebu. The operatives gathering intelligence on Bohol and Cebu soon began reporting to him.

As military district commander of Negros, Villamor's base support was boosted when he was able "to appoint a civil administrator, establish a civil government with a governor, and regular pay for the guerrillas." But, over his reports of helping in the plight of the civilians, he was reminded to focus on gathering intelligence. Once a strong guerrilla leader was found, he could return to Australia.

Villamor's request for $10,000 for Peralta was approved, but the P1,000,000 monthly budget for Negros, which was twice the amount he asked for earlier, was disapproved. Whitney wanted more clarifications. Funds for salary and allowances must be managed to avoid favoritism. A currency board on Negros was recommended whose members were to be approved by President Quezon. GHQ wanted a complete set of various denominations of Japanese money for the possibility of printing them for distribution. Peralta, who was printing currency, was ordered to stop and account for the money already printed.

Insurance protection was offered to entice cooperation from guerrillas who were not formally enlisted in the United States Army. But GHQ emphasized that these policies had no guarantee yet as their effectivity "was to be decided by post-war relief laws determined by the Philippine government."

In June, Jorge was ordered to bring a radio to Bohol to contact guerrilla leader Maj. Ismael Ingeniero and all the Philippine pilots stranded there. Ingeniero had formed the largest guerrilla unit named 'Boforce' and was eventually recognized in December as the Bohol Area Commander except for Cebu which he tried to claim earlier. The guerrilla commanders of Negros, Panay, and Mindanao also laid claim to Bohol and established agents there to maintain their influence.

Jorge used his own AIB code, 'Kangaroo Fifty' to send and relay trusted messages wherever he went. He was almost shot by the pilots if not for Ingeniero vouching for him. He brought back Lts. Bartolome Cabangbang, Enrique Torres and all the pilots to Tolong, near Culipapa in Negros to meet Villamor. A request was sent to Australia to bring the pilots with him on his return.

Cabangbang was wounded on Corregidor and was among the prisoners-of-war sent to Camp O'Donnell where he was placed on the burial detail digging graves, and burying as many as twenty-five bodies in a grave without names and insignias. Released with the promise to undergo training for a month for the Philippine's part in the Great East Asia Co-Prosperity Sphere, he accepted the appointment to be a 5th class inspector of the Bureau of Constabulary in his home province, Bohol, where he spoke the language. Torres was a Bataan veteran and led a unit under Abcede near the area of Binalbagan in Negros. Maj. Henry Meider was a former airline pilot who later reported that when brought to Negros, he was appointed advisor and headquarter commandant of the Negros network control station. He expressed his frustration that G-2 intelligence reports on the enemy were not regularly obtained. "Our first thought was always the radio and then Villamor. At first, we would keep Major Villamor under cover."

Jorge went south to Col. Ruperto Kangleon's headquarters in Maasim, Leyte to obtain documents listing locations of Cebu and Bohol islands' outposts, alternate headquarters, names of collaborators, and guerrilla supporters to send to Australia. He saw a banner that said, 'Welcome Chick Parsons'." But Parsons had landed with the USS *Tambor* at Pagadian Bay in Mindanao.

On his return to Culipapa, his *banca* was approached by a Japanese patrol boat pulling a 500-ton barge going to Bohol to load mangoes. The *banca* owner wanted to make a run for it but there was no wind. Jorge told the crew to lower the sails. The Japanese boarded, began inspecting the cargo, and signaled by hand if they had a radio. They were shown the boat's contents: rice, pots, and five-gallon cans of sugar and dried fish. A Japanese officer went to the pile of firewood where Jorge's pistol and documents were hidden. Thinking quickly, Jorge offered him a handful of dried fish, but the Japanese shook his head, intending to inspect the pile. Jorge then quickly grabbed a can of *binuro* (pickled fish) and offered it to the officer. The Japanese must have been hungry as he took the can, opened it and began to eat. He then signaled his men to let them go while putting food in his mouth. The boat left with a slight breeze.

Back at camp, when Jorge realized that he was not carrying identification papers, he shook uncontrollably with the thought of how close he was to death. While waiting for the next assignment, he scrutinized guerrillas who wanted to join in case they were Japanese collaborators. He trained agents in gathering intelligence and defensive moves. Once, when talking to a guerrilla he suspected as a collaborator, he noted that the former's statements did not make sense. Not wanting to risk the discovery of the camp, Jorge ordered the man to be shot.

When word arrived that a second submarine was to be sent but this time to Mindanao, Quinto sent a message not to use Pagadian Bay due to the presence of unfriendly *Moros*.

Messages brought by couriers and by radio from Manila continued to arrive. An underground secret group called 'Free Philippines' and a network under pre-war Senator Manuel Roxas was reported operating in the city. A press relations officer in Malacañan Palace, the Philippine president's official residence, broadcasted on Radio PIAM (the former radio station KZRH) coded information for the guerrillas which was passed on to Negros.

At the end of May, Yuhico reported that a man from Manila representing four secret societies was meeting with Peralta. The man said he was authorized by GHQ and President Quezon to set up a phantom civil government in Luzon. KAZ ordered Villamor to meet the same man to compare what was said to Peralta. Yuhico confirmed that the man was a spy, and he was ordered executed.

Whitney expressed his concern that only five messages were received from Negros in the last month as compared to the many Panay had sent. Messages from either side was the only way GHQ could compare and identify true information.

Seven thousand enemy troops from Cebu attacked Negros, landing in the port of San Carlos. They shelled Refugio Island off the port burning the *Princess of Negros* ferry that had carried President Quezon and his party who had escaped from Corregidor to Mindanao. The burning hulk was towed away.

Quinto's fingers flew as he transmitted the arrival of 63 Japanese transports in Manila with 150,000 troops. Many were recruits from countries occupied by Japan and the majority continued south to Mindanao. A month later, 50,000 Japanese soldiers were reported garrisoned in Manila. Clark Airfield had over 4000 enemy planes and Nielsen Airfield 5000 planes with 2000 POWs maintaining them. Guerrilla groups were reported outside of Manila up to the northern Mountain Province and were not to enter the city unless necessary. All the cement produced and the copper, manganese, iron, coal, and chromium from the mines were taken. The sabotage of a lumber mill was planned, but GHQ ordered a halt to it fearful of retaliation by the Japanese against civilians.

By mid-1943, agents had spread out and fed a continuous stream of information from Luzon and the Visayas. There were shortages of food and basic needs on the islands. Civilians unafraid of the Japanese shared whatever food they could with the guerrillas. Many people were malnourished and succumbed to diseases. Villamor got sick and a woman from a well-to-do family nursed him back to health. "People moved about in clothing they had worn for months, men and women walked barefoot or   in shoes with soles made from bits of old automobile tires. A bar of soap was like a precious jewel," he recalled. And everywhere there was the nagging, persistent question when the 'aid' would come.

Villamor again asked to return to Australia before designating a permanent commander. He was willing to return to the Philippines and continue the work after a brief rest. After the war, Meider confirmed in his report of Villamor's political ambitions that made him want to go back to the Philippines.

USS *Thresher* – July 1943

Six months after the PLANET PARTY's arrival in Negros, another party of four men landed to support the radio net and deliver the first shipment of supplies. Led by Lt. Petronio Huerto, the group consisted of Cpl. Juan A. Rodriguez and Pvts. Pedro F. Carriaga and Dominador Gobaleza. Huerto and Gobaleza were also from the hospital ship *Mactan*. The men were trained in radio and commando fighting by the Australian Army to provide security to the party. A concern was that only one of the four men in the party spoke and understood the Visayan language.

The cargo had some of the supplies the Planet PARTY requested. Small and portable Dutch radio sets were for agents going to Luzon and to set up stations in Panay, Leyte, and Cebu. The latest pamphlets on intelligence and Japanese ships, office supplies like typewriters, pencils, bond paper, and chemicals for secret ink and developers were included. For emergencies, a heavy-duty outboard motor with a muffler and cans of gas were included along with more rations, shoes, clothes, socks, field jackets, toothbrushes, and toothpaste. The request for a Japanese interpreter was crossed out. In his memo to Sutherland, Whitney commented that "If he (Villamor) had full support of the guerrillas, which of these requests could have been obtained locally even if clandestinely?"

Maj. Emigdio Cruz, Quezon's personal physician, was also onboard. He had volunteered to find out the conditions of the officials left behind in Manila, the political situation, and the people's reaction on the flight of the Commonwealth Government to the US. He carried the codename 'Gatbiala (sic)'. Before he left, MacArthur warned him that it was "a very tough job" and he had no chance of getting through since his connection with the President was well-known. But Cruz strongly felt that he would succeed. The submarine crew welcomed the party and knew the dangerous mission they were embarking on. They voluntarily contributed their rations of cigarettes, soap, candy, and other personal items that could be of use.

The *Thresher* spotted a four-ship convoy in the Makassar Strait and unleashed two torpedoes scoring direct hits on the passenger-freighter *Yoneyama Maru*. On the fourth day of the trip, she chased a Japanese convoy of three tankers and a destroyer. After releasing torpedoes, she immediately crash-dived to a depth of 95 feet as the ships dropped a dozen depth charges after her. Explosions were heard and concussions felt. The ships propellers became louder as it circled, searching for the submarine. Drenched in sweat from fear and spending 18 hours in the unbearable heat of the submarine, Rodriguez prayed and called on all the saints. Finally, the enemy ships gave up and left. The periscope showed the damaged destroyer *Hokaze* had run aground. In celebration, four bottles of whiskey were consumed with cakes the cooks whipped up.

Four days later, the submarine arrived at Cansilan Point in Negros, not far south from Catmon Point, where the first mission party landed in January. Visibility was low in the early morning as she approached the west coast of Negros. Sighting a Japanese destroyer from the east, she remained submerged all day until the destroyer departed by late afternoon. Through the periscope, Huerto sighted two approaching sailboats and concluded they looked like a welcome party as there was hardly any fishing in this spot. When asked if he would go through with the original plan, he replied, "Oh yes – I must."

Three sailboats came alongside the submarine to unload. Abcede delivered a suitcase of important papers and intelligence reports to the submarine commander. Forty thousand rounds of ammo and five hundred pounds of supplies of medicine, weapons, cigarettes, matches, soap, toothpaste, razor blades, and magazines were offloaded onto the sailboats. Some of the rations were surprisingly more luxurious than the usual dehydrated potatoes, salt and pepper, luncheon meat, sardines, ham, salmon, and the trio of coffee, evaporated milk, and sugar. With the usual rations came dozens of pounds of cheese, lobster, and spaghetti noodles. In less than an hour, Abcede's guerrillas unloaded 500 lbs of supplies to shore where Villamor and Malic were waiting. The submarine slipped out to sea for a major overhaul at Mare Island Navy Yard in California.

Unfortunately, the Japanese had learned of the landing despite the absolute secrecy that was maintained in the preparations made only less than 24 hours earlier. A shootout ensued as the Japanese raided the landing-place, capturing a considerable number of arms and ammunition. But the medicine and the radios were saved. Abcede was almost captured while a number of civilians were killed, and many houses in the area burned. Abcede's forces were only at 85% in strength, seldom armed with enough weapons and had difficulty recruiting men in his area. Despite the raid, the supplies buoyed the guerrilla's morale as there was now proof that the Allies were planning to return. They eagerly welcomed the training from the mission men.

Huerto's party hiked the following day through rugged terrain to Villamor's hideout, hiding parts of the cargo along the way. A Japanese patrol landed in the same spot several days later and took some of the cargo as the guerrillas were burying them. They did not have enough arms to fight back.

Rodriguez assisted Quinto in operating the radios in Kabankalan at Abcede's camp and at Villamor's hideout. He joined Ignacio and Jorge in the mountains of Kabankalan when they went to the camp of guerrilla Major Roberto Benedicto, a former USAFFE officer. With a group of twelve men including civilian carriers for their cargo, they reached the camp of guerrilla Capt. Ramos under coconut groves that had been raided several times by the Japanese. The main route was blocked. For the next two weeks, the guerrillas ambushed Japanese patrols to clear the path, but they were still unable to get through. They had to find another way.

Earlier, Villamor had given up on going to Manila for fear of being recognized. Instead, he sent agents to provide him information which was transmitted to Australia. This may have influenced President Quezon to send Cruz, his personal doctor, who volunteered for the mission. Cruz was eager to go as he had received news that his entire family had been killed and he was desperate to find out more information.

Concerned that Cruz might talk too much, Villamor warned GHQ that "Cruz lacks secrecy, discipline for the mission" and recommended that he not be sent forward as he had already "revealed his mission to many." He could get the information for Cruz, but he would also abide by President Quezon's decision if he wanted Cruz to proceed. The order came back that Cruz was to continue his mission.

A fountain pen popular before the war and a cigarette lighter purchased in the U.S. was in Cruz's possession, which Villamor feared would compromise the mission. If the Kempeitai caught him, he would undoubtedly be tortured to reveal the radio network and agents. He was garbed in a typical farmer outfit with a straw hat, shorts, and a thin shirt. Ready to undertake his mission, Cruz was not given information that might endanger the intelligence nets and guerrilla activities. Villamor did not provide support, so he was left on his own to devise his plan from the information of couriers going in and out of the camp. He observed the intelligence work of the camp but was not allowed access to its workings. Yuhico was ordered to trail him when he left for Panay in case of capture. Cruz travelled for two weeks with a guide overland and across seas to Bicol in southern Luzon. Posing as a trader, he escaped Japanese patrols while carrying personal letters from President Quezon. He hid with friends he knew before the war and, upon reaching Manila, was reunited with his wife. He was assured of loyal people and learned of others who sought positions in the Japanese-sponsored government. The Filipinos were more interested in the return of the Americans than independence from Japan. He became a wanted man as news of an aide of President Quezon in Manila quickly reached the Kempeitai.

To safely return to Negros, Cruz asked for help from Abcede, who gave him a guide and an escort of a dozen men. He dressed as a peddler and "concealed all notes and papers in a bamboo trunk under merchandise" as he evaded Japanese patrols. Crossing the sea, his small sailboat was hit by a typhoon that stranded him on Higatangan island north of Leyte. Through the guerrillas in Leyte, he sent several identical messages of his Manila trip by courier to the Panay and Negros radio nets to be sent to Australia in case he was unable to return. Only three men arrived with him at Governor Montelibano's camp on the northern shore of Negros.

In July, Abcede was appointed acting 7th district commander to give Villamor more time to concentrate on intelligence work. Duplicate reports were sent to KAZ through the Panay and Mindanao radio nets to ensure that they reached Australia adding to the Negros radio traffic.

As the intelligence network expanded, so did the reports covering the strengths of both guerrilla and Japanese forces, airfields, propaganda, ship sightings, and atrocities committed against civilians. Non-involvement in civilian affairs continued to be a balancing act. Without the civilians' support, the guerrillas and the party would not get food, transportation help, security, and guides to accomplish their objectives.

Sightings of downed planes saved many U.S. pilots. Men of the 'Free Negros' command brought, and many times also tended to the pilots' wounds at their command post until they were evacuated by submarine. The *Free Philippines* published a story of an airman who was forced to land his burning plane in a safe area and was embarrassed when women ran to him and began kissing him. After feeding him, a guide brought him to the mayor of Cadiz where a rich dinner was prepared for him with the mayor's attractive daughters serenading him. It was a treat since his airbase in Morotai was in the dense jungles surrounded by unfriendly natives, eating only rations, and quarters exposed to the elements.

The camp was always on the alert for Japanese raids and infiltration by spies. Spies who infiltrated Abcede's camps and were caught underwent a trial that almost always ended in the execution of the traitor. As the Japanese increased their attacks in the south and central parts of the island, food production collapsed. To ease the starvation, the Negros civil government printed Japanese currencies to buy all the rice harvests for distribution to the people. Abcede retaliated with sporadic attacks enough to keep the morale of troops and loyal civilians.

A lull in the resistance convinced the Japanese that they had succeeded in persuading the people to come to their side. Thousands of enemy troops arrived in Panay, Cebu, and Negros. The enemy garrison in central Negros increased to around four thousand soldiers who patrolled the cities and the mountains. But the guerrillas reluctantly followed KAZ's orders to lay low to protect the civilians against Japanese retaliation.

The news of the enemy launching a simultaneous campaign against all the guerrillas immediately after the granting of Philippine independence quickly circulated. A better life was promised with the granting of independence under the Japanese, but the new government would then brand the guerrillas as traitors or bandits.

Messages of the Japanese construction of ammunition storage areas in the major cities of Manila, Iloilo, and Davao arrived. On Luzon, storage sites were constructed on Corregidor, Bataan, Cagayan Valley, and Zambales. Agents reported two thousand more enemy troops landing in La Union heading towards the Cagayan Valley.

A submarine was scheduled to arrive in Mindanao. Villamor requested MacArthur to call a conference of guerrilla commanders to formally recognize a leader for Negros before he left for Australia. The firm reply was for him to hold the meeting and submit a recommendation before reporting to Mindanao to meet the submarine. Whitney, whose expectations were for Villamor to determine a strong guerrilla leader not only in Negros but also in Cebu, insisted that he remain, conduct the conference, and leave on a later submarine. But Villamor departed for Mindanao to meet the submarine without waiting for GHQ's reply.

GHQ ordered a meeting of senior guerrilla officers to develop and submit an organization plan for approval. In a letter to each leader, MacArthur asked for cooperation with all working together towards a decisive victory. He wanted to know what forces were resisting their command, the enemy situation in their areas, and recommendations for future supply and assistance to Negros that were important   to his return. He asked the leaders to convey to their officers and men "his grateful recognition for services of inestimable value to your country and the cause." The letters were addressed to the leaders with their full military titles and signed only with his name without a title. Abcede, as acting Negros commander, called the meeting and was accepted as the Negros leader except by Gador who did not attend, insisting that he was following SWPA directives. Stating that he was the most senior officer, he refused to send reports to Abcede. Whitney expressed disappointment that Gador was not served with an order to submit or be reported for disciplinary action before Villamor left. He feared disunity in Negros, but Abcede successfully dealt with the problems through the rest of the war.

Villamor tried unsuccessfully to reach Mindanao. A Japanese patrol chased his boat while crossing the Mindanao Sea. Sailing back to Negros, the boat had to be repaired for the next attempt. In the meantime, without waiting for approval from GHQ, Villamor assigned Edwin Andrews, a Philippine Army Air Corps pilot, to take charge of the radio station YAF for contact with Australia and WPI for messages from submarine stations. The cipher systems were turned over to him. Andrews managed the radio traffic but had no authority over the agents in the field. Quinto and Malic reported to him, transmitting intelligence while Jorge and Ignacio were with guerrilla Maj. Benedicto in the mountains of Kanlaon. Malic, Quinto and Rodriguez contracted malaria but took turns operating the radio to keep it on the air.

Whitney was greatly concerned by reports that Andrews, who held the codes   in his hand, was anti-American. Andrews was the second in command to Salipada Pendatun, a *Moro* guerrilla chief in Mindanao before he reached Negros. Although the 200 rifles he obtained from a government official, helped defeat the Japanese garrisons in Bukidnon, Mindanao, he was the cause of the delay in Pendatun's joining the 10th MD. In contrast, Frederick Worcester, USN who spent five months in Negros from Mindanao, reported that Andrews "did good work in a responsible position as head of AIB (radio) under unpleasant physical handicaps without complaints…. at times incapacitated with asthma and legs showing signs of beriberi."

When Malic was sent to check on Japanese activities in the Hinobaan area, he met a large Japanese patrol heading towards the camp. The party of 14 men engaged 300 Japanese soldiers in a gunfight, killing 16 and saving the radio. Two days later another encounter with a larger group of several hundred enemy soldiers forced a retreat to an inner camp. But it wasn't long before they were forced to destroy the camp as the Japanese entered their area. Two miles away, in the village of Canturay deeper into Sipalay, the party set up the radio. Malic continued observing enemy activities and coding messages.

Only three pilots, Cabangbang, Torres, and Meider, were allowed to depart with Villamor. GHQ concluded that Cabangbang and Torres' military background would be useful for future missions. The group crossed the seas to reach Mindanao. A radio was in the sailboat to keep contact with Quinto. Sighting a motor launch heading towards a sailboat and another heading towards their boat, they turned back with the wind strong enough to outrun the motor launch.

Whitney replied that the submarine could not advance its arrival as it was on distant patrol and on radio silence. The USS *Cabrilla,* on her first war patrol received the dispatch for a special mission to evacuate four military personnel from Doog Point in the Hinobaan area of Negros. The rendezvous point in the southwest coast of Negros was pinpointed and the evacuees were in a sailboat flying a large white sheet throughout the day as a security signal for the submarine. But the enemy landed three miles from the pickup point and the signal sheet had to be hurriedly taken down. The submarine submerged and laid in wait after failing to see the proper security signal from the shore. Meider recalled "All signals were as ordered by radio. The failure to make contact was due to our pulling down the sail of the boat. The second day we made contact." The enemy was no longer in the area. The submarine sighted the sailboat flying the correct signals and surfaced five hundred yards from it. The American evacuees came onboard. After assurances that the boat crew belonged to an organized guerrilla band, all readily available ammo and a submarine machine gun the submarine could spare were loaded. The crew donated food, cigarettes, matches, toilet articles, clothing, medicine, and other personal items thought to be useful for the guerrillas. The submarine quickly departed.

Yuhico sent a confirmation that the pickup of Villamor and the three pilots on September 12 was successful. He later reported that he could not break a message intended for Villamor because he was not given the latter's cipher.

By this time, strong guerrilla leaders were now identified in the islands of Negros, Panay, Cebu, Tawi-Tawi and Mindanao. (Each island is discussed in separate chapters). There were agents in Manila but strengthening the southern islands meant a greater chance of knowing the situation of enemy and resistance forces on Luzon.

After three weeks in Australia, Villamor gave his report covering his time in the Philippines to GHQ. He included the remark that an American officer would not be successful with the guerrillas as he would be easily recognized. An American might be "unable to obtain all the facts in the field with the difference in his and   the Filipino psychology. His net is operating.... contacts are individuals of prominence and integrity whose usefulness may be considerable," he reported. The materials he brought back showed the net "just starting to pay dividend." He did not divulge to Andrews or anyone else the extent of the net. He reasoned that having the contacts report to someone else other than himself might endanger them or be misunderstood. He strongly expressed his desire to return and continue directing and expanding the net beyond Negros. Morale appeared high on the islands but crumbled with each punitive Japanese expedition in the area. Villamor ended his report with a plea to send desperately needed medicine in the following missions. His recommendations were not entirely implemented whether future missions were to be led by an American or a Filipino. But GHQ believed Villamor's political involvements would make it difficult for him to use the net less for military and more for political purposes. He was not to return. Most troubling was, in his haste to return, he turned over the secret net to Andrews without authority from GHQ.

GHQ concluded that although the "trip had limited usefulness" in completely fulfilling the mission's objectives, it did show that penetration parties could be inserted into the islands by submarine. The mission was difficult with a broad scope, but it began a structured communication with the Philippines. GHQ was confident that plans could be developed and be prepared to send future parties to support a confirmed resistance. But factions in GHQ disagreed on whether an American or a Filipino should lead the missions.

Villamor then departed for the U.S. to meet with President Quezon, who was eager to hear his report. He went into details of agents in Manila and spy rings composed of influential persons. If returned to the Philippines, he would continue contact with Manila. President Quezon, unaware of the full extent of GHQ's plans and communications with the resistance in the islands, had "expressed concerns that GHQ was not doing all it could to support the guerrillas and people of the Philippines."

Back in Negros, Abcede worked with Andrews in sending out ship sighting reports as well as general reports of Japanese atrocities, strength, movements and guerrilla activities in Negros and Bohol. Operatives from Manila including Maj. Edwin Ramsay, a leader of the East Central Luzon area sent couriers with information on Japanese strength, movements in Luzon, and prisoners of war. Seven transports of Chinese soldiers, one large aircraft with thirty carrier planes, two destroyers, and ten gunboats arrived in Manila from Amoy, China. The planes were camouflaged with coconut and bamboo foliage. A consoling piece of news was the attacks of the U.S. and Australian forces on the large Japanese base on Rabaul.

An intelligence report Andrews sent to KAZ on Nov 19, 1943, came from agents in Bohol who reported on a certain Marcos claiming to be in command of the former units of Nakar and Enriquez in northern Luzon after their capture and surrender. Marcos met with Fertig, 10th MD to arrange recognition of his units by GHQ and gave intelligence information on Luzon. Andrew was to receive a copy. Nothing further was reported on the results of the meeting.

By the end of the year, food for the troops had been depleted and prices for commodities had gone up. Abcede met with Fertig, who had closed all banca traffic that carried food between Mindanao and Negros to control the entry of enemy agents into Mindanao. Guerrillas were reported to be losing weight down to an average of about 40% of their body weight. In desperation, carabaos needed for ploughing the fields were killed for meat. Even the blood was congealed over cooking fires and eaten.

It had been two months without any other news since Villamor left, reported Andrews to GHQ. He requested recognition, promotion, and authorization to promote people. The camp was crowded with agents waiting for their next orders since he was told to hold off on their assignments as funds to pay them were exhausted. Air Corps pilots were also waiting to go to Mindanao to be evacuated. Concerned for the radio net, Whitney ordered Andrews to report to Abcede with all equipment and personnel assigned to the secret network.

1944

Maj. Edwin Ramsey, formerly with the 26th Cavalry led the East Central Luzon Guerrilla Area (ECLGA), headquartered in Montalban, Rizal in the Manila area. He was desperate to get a radio and contact Australia. His operatives and patriotic civilians had gathered much intelligence on enemy activities. Guerrilla Modesto Castañeda travelled over 1,000 miles from southern Negros with a radio for Ramsey with the agreement that all his reports passed through Negros' net control station for relay to KAZ.

Rodriguez, a signalman had set up a temporary station at Cadiz in the northern coast of Negros to contact Andrews. For five days, he could not hear anything, so he went to Benedicto's camp which was closer to adjust the radio to work. He sailed to Masbate Island and found locals who were pro-American and used non-Japanese money. He charged his batteries while civilians erected two bamboo poles to set up the antenna. Activities around Masbate area provided by locals were transmitted until the transformer burned out.

Two months later, Rodriguez reported from Tayabas on Luzon setting up camp in the mountain overlooking the open sea. Rodriguez was the first mission man to reach Luzon and set up a radio, but he could not contact the main radio in Negros. After two weeks of trying, he had emptied the charger of fuel. Huerto arrived with radio sets. Broadcast news from San Francisco was translated into Tagalog and distributed to the rejoicing people. Village officials brought down American pilots to the beach outpost until picked up.

USS *Narwhal* – February 1944

The Japanese had known of the earlier submarine landings and were on the alert. The *Narwhal* returned to bring supplies to Panay and Negros. Its master gyro compass broke but the submarine continued and risked a landing in Negros. The gyro compass was the heart of the fire control system and its erratic performance would have been useless if a fire erupted onboard. Navy Cmdr. F. Kent Loomis was onboard as an observer to report back for planning future submarine missions. He stayed onboard throughout the landings.

The submarine dove when it contacted several planes and small boats on the way to Negros. A small ship was detected nearby but the crew decided not to attack it as it was too small a target to risk discovery of the submarine's position. But sometimes, the radar did not pick up small vessels and they were surprised when they encountered boats while surfacing. Approaching Balatong Point, security signals were seen on the beach, and the submarine surfaced at dusk. Lights came from civilians making salt along the beach. Six or more sailboats were seen between the submarine and the beach. The first boat came alongside with Abcede and began unloading forty-five tons of cargo. Over two dozen evacuees boarded, including five service men and one British subject sick with malaria and tropical ulcers. A chaplain was onboard and conducted services in the forward torpedo room at sea. Approaching Darwin Harbor, the submarine joined an escort and disembarked the thirty-four passengers to an Army tugboat before she docked.

Cruz, who had returned from Manila, was one of the evacuees. While waiting for the submarine, he continued to treat people with whatever medical supplies he was given. His report to PRS described airfields, fortifications, and concentration of enemy troops on Luzon, indicating the Japanese decision to defend Luzon. His favorable report buoyed President Quezon's confidence in the officials he left behind.

The USS *Crevalle* was patrolling the area when orders arrived to evacuate forty people. This was a cover to pick up the documents that MacArthur was most anxious to receive. Abcede was ordered to "exercise great caution in executing the mission and to preserve utmost secrecy concerning the documents." Although there were security signals, it was perilous for a submarine to surface a mile off enemy territory. She emerged at Balatong Point in southwest Negros. Men operated the deck guns while on the lookout for enemy crafts as the evacuees arrived. A POW escapee brought the box of important documents from Cebu. Other sources state that the documents were in an empty mortar shell box. These were papers that Admiral Koga, commander of Japan's Imperial Navy, had when his plane crashed in Cebu. (The Koga incident is discussed in the chapter on Cebu).

Quinto, the chief radio operator, boarded to return to Australia. He was struggling with deteriorating health due to malaria attacks, and this caused delays in transmitting messages. He had trained others to take over if he could no longer keep the radio on the air.

On the way back to Darwin, the submarine fired on a boat, forcing its Filipino crew to jump overboard. Quinto questioned them, found out they were *Moro* fishermen, and were released. The *Narwhal* tried to attack a Japanese convoy and dove deep to avoid the depth charges. The periscope was damaged, and the submarine stayed underwater so the crew could repair it. More damage was also found on her deck, so she was forced to remain on the surface for as long as she could. She limped on, diving and surviving depth charges twice from Japanese aircraft and ships. The evacuees were first brought to the Lugger Maintenance Station near Darwin which served as a temporary home for all evacuees, where they underwent physical exams, then were clothed, and fed. Since November 1943 and throughout the war, it was the receiving home for evacuees from the Philippines.

In Bohol, a punitive campaign by the Japanese forced guerrilla commander Ingeniero to flee the island after the transmitter from Negros was captured and his men scattered. A month later, he returned after obtaining another radio and continued sending messages on enemy activities, U.S. plane losses, airfields, harassments of the Japanese, and a general report on neighboring Cabilao Island.

The Japanese propaganda movie, *Liwayway Ng Kalayaan* (Dawn of Freedom), began showing in Manila theatres in May. It was one of a film trilogy produced by the Japanese. MacArthur wanted to see the film, which depicted Americans as the real enemy and boasted of the benefits of the Japanese invasion. American POWs were used as actors. One scene in the movie showed deserting Filipino soldiers fleeing the battlefield in Bataan. The reels were taken by members of the Hunter's Guerrillas and had been hidden in the Manila residence of Yulo, a member of the Manila network.

Jorge was tasked to bring back the films. He was probably the first U.S. Army soldier to enter Manila. He packed the film reels in two cans for the trip to Batangas and across the sea to Negros. The Kempeitai were looking for him. While waiting for a safe time to leave, he met with Calixto Duque, a guerrilla leader of the United States Army Forces in the Philippines-Northern Luzon (USFIP-NL) to inform him and other leaders not to engage their guerrillas until an order arrived from GHQ. At the Nichols Airbase main gate checkpoint, a sentry kicked the sacks and confiscated the two sacks of rice around the cans. Jorge thought it would be his last moments. From Benedicto's headquarters in Murcia, the reels were sent to Hinobaan and picked up by a submarine. The film is presently stored in the U.S. National Archives and Records Administration (NARA) in Maryland, U.S.

USS *Nautilus* – June 1944

The *Nautilus* left Australia on her tenth war patrol with reconnaissance men Sgts. Val R. Tabios and Custodio P. Bolivar and weathermen Sgt. James R. Gallemore and Pfc. Apollo P. Tullo. They brought complete weather equipment and 100 tons of supplies: weapons, ammunition, medicine, propaganda, Aussie rations, candy, cameras, radio equipment, explosives, gasolines, lubricating oil to run the radios, and 20 CW tins. All cargo other than weapons was identified with 'AE' for Abcede or 'CG' for Cushing. Twenty tons were to be sent to Cebu. The weather equipment was for additional stations in Samar, Mindanao, Palawan, Panay, and Negros, providing comprehensive coverage of the southern part of the archipelago.

On her way to Negros, the submarine closed in on a one-hundred-ton schooner and fired her deck guns. They left behind five survivors on two lifeboats.

After sighting the proper security signals at Balatong Point, the *Nautilus* surfaced at dusk. Two boats approached and Abcede embarked with seventeen evacuees and one German prisoner of war. The party was transported to shore.

The submarine's arrival brought much-needed relief. Running and protecting the radio headquarters consumed much oil, ammunition, and dry stores. Most welcome were the new rubber-soled, cloth-top boots with coverlets around the ankles 12 inches high to prevent leeches and thorns. They replaced the boots that had quickly rotted after a few weeks in the humid jungle and mud, leaving the men barefooted like the locals.

The weathermen measured the atmosphere that would disrupt radio transmission and aerial bombing of targets. To obtain good measurements, they set up their weather radios in high and remote locations that were, unfortunately, disease-ridden. Reports were relayed to the weather main station at Dimorok Canyon in Mindanao, that had been set up by an earlier mission.

Food had not been readily available for all the mission men and the guerrillas who provided their security, so the new food supplies were a great relief. The constant movement to evade the enemy and keep the radio on the air had exhausted them. The food supply that included 18 lbs of corned beef and 28 lbs of cured bacon were distributed and the men feasted together. But the twelve pounds of baking powder in the cargo were useless in the remote jungle outposts.

On her return to Fremantle, the submarine fired her guns at a small sailing vessel, sinking it. The eighteen Malayan survivors who were pulled out of the water claimed that the Japanese shaved their heads and took their money. The boat was bound for Ambon under a Japanese charter with twelve tons of pistol and rifle ammunition.   A cargo ship HMAS *Malanda* came alongside to receive the evacuees and survivors before she moored in Darwin.

The pending Allied landing on Leyte increased radio traffic, and couriers arriving from Manila brought dire news of burning buildings stripped of all metal and shipped to Japan. The enemy was forcibly demanding one million sacks of rice from the government. Chinese warehouses were raided and whatever was confiscated were hoarded in stadiums or shipped to Japan. Civilians were left cutting up asphalt blocks to use as fuel because it burned like wood.

USS *Narwhal* – October 1944

Arriving in bad weather at Tolong, in the southwest coast of Negros, the navigators found landmarks hard to identify at first. Security signals were sighted but the sub still had to wait until dark to surface. Abcede met the submarine and half of the ninety tons of cargo were unloaded into sailboats under heavy rain.

The submarine carried the first party of Filipino and White soldiers with expertise in demolition, aircraft warning, and weather observation men with their gear. Medics were also with them. Onboard were Lts. Leon C. Tinnell, and Sidney S. Rexford; Sgts. James E. Ellis, G. Madden, and Raymond E. Caley; Cpls. E. F. Dunat, E. T. Hunter, L. W. MacDonald, R. F. Pollini, D. A. Phillips, L. V. Reddick, and W. P. Rudolph; Pfcs. A. E. Kunkel, G. A. Malenson, M. D. Routson, F. Bill, N. G. Morris, J. A. Nelson, and H. J. Welch; Pvts. J. M. Sooter, D. F. Palmer, B. B. Owens, J. Rubenstein, and M. Rulle; T3 Cesar E. Espinoza; T4s John E. Forrette and C. B. Igmundson; and T5s E. J. Eccleston, G. M. Cook, T. Harris, G. M. Williams, F. E. Yost, H. E. Powers, R. W. Nickay, F. Graw, Q. S. Knight, L. E. Smith, and H. I. Weber.

Boarding from Mios Woendi were Pfc. Martiniano S. Atad and T4 Marcelo B. Abalos handling reconnaissance while Sgts. Pedro L. Garces, Eusebio J. Manuta, Cpl. Antonio M. Quesada, and Pvt. Carlos P. Antikoll handled demolition with Lts. Albert Schmaltz and Maynard C. Hawley. PRS had sent 3,800 pesos for use by the weathermen.

A PBY Catalina almost bombed the submarine when she mistook her for an enemy vessel. At the last minute, the aircraft turned away and signalled, "Good luck, *Narwhal*."

Pfc. Atad was an operator assigned to the radio stations of KAZ and PIMC before he was sent to Biak Island. He wrote in his after-action report. "After five days, I boarded the *Narwhal* from Mios Woendi. Sailing the Celebes waters near Morotai, the submarine suddenly submerged to avoid a bombing by our plane. Nearing Tawi-Tawi, a Japanese patrol boat was sighted, and for the next several hours, we stayed submerged. Two American submarines were sighted, heading in opposite directions on our left flank and the other on the right. There was a delay with a misunderstanding of recognition and security signals, but we were able to unload equipment for the guerrillas." Twenty-six evacuees boarded the submarine. The rest of the men and cargo continued to Cebu.

Atad recalled that many sailboats unloaded the supplies when they arrived in Negros and dumped them in different places along the river. Much of their personal gear was missing. They were served coffee and carabao meat soup before hiking to the command post of the 7th MD eight km away. They stayed and were fed for two weeks while waiting for their supplies and radios and then were left on their own. Atad and T4 Abalos had to find food and swap it with a few clothes. It took the guerrilla carriers two weeks to arrive with the radio equipment, fuel, and other things. Once the radio station was set up, it was always on the air. Upon contacting GHQ and the 5th Fighter Command about the intense air activities in their sector, they received congratulations. They witnessed dog fights – four Zeros and a bomber shot down by a P-38 and P-47. Atad said that an enemy patrol boat headed towards them, and soldiers jumped ashore, so they quickly evacuated with their equipment. When it was all clear, they returned to resume operations, but many of their items were missing. "The guerrillas did not help much (in getting) food. Once we were out of clothing to give them, they became unfriendly."

When he refused to give his paratrooper boots to Abcede, an uneasy relationship developed. Atad said, "We were not armed, and guerrillas had pistols and could shoot us. I could not do anything if picked on. The guerrilla officers enjoyed the supplies from the U.S. Army and hoarded them for their own use and their families, relatives, and friends. The enlisted men greatly in need of food and clothing did not get their share."

They began to train men on demolition missions. There were hardly any arms to practice with. Men were starving and many joined to get more arms and food. Three days was not enough time to train the guerrillas properly, but orders arrived to split the men to coordinate attacks. Manuta and Antikoll went south to the island of Dumaguete. A group headed by Quesada went to Central Negros. Hawley and Schmaltz remained with the camp's intelligence and security section.

The civilians reported that there were two Japanese garrisons at Hinigaran and Kabangkalan. Demolition man Garces and Schmaltz went north with his men, taking a sailboat through stormy weather and landed at Miranda, close to Hinigaran. Garces observed that the guerrillas did much looting instead of uniting to fight the enemy. Cooperation was difficult between units. Weather and aircraft warning men who were assigned to remote outposts for long months far from camp headquarters did not receive their fair share of the cargo. By the time it reached them, the cargo had usually been looted. However, they felt that their presence helped the civilians in their area as the guerrillas decreased their harassment.

At Miranda, the equipment was loaded on a carabao cart and sent to guerrilla commander Maj. Ernesto Mata's Camp Quezon on the side of Mt. Canlaon. When the 81 mm mortar arrived, Garces was thankful for his heavy weapons training to instruct the guerrillas to fire it. He joined the guerrillas at an advanced command post near Murcia occupied by a garrison of two hundred Japanese soldiers. The Japanese crossed the Bago River and burned the rice warehouse in the village of Tan-Ag. Garces and Schmaltz fired the mortar in support of the machine guns. Four shells hit the garrison. The Japanese replied with mortar shells. The machine guns were more effective, and the Japanese withdrew. "I was handicapped because I had no binoculars to check the bursts. We fought for a day without casualty, but over fifty enemy soldiers were killed." They withdrew when ammunition ran low. Before he left for the next assignment, Garces gave the field manual of the 81 mm mortar to the guerrilla commander for more accurate mortar fire in their next operation.

Operatives on Negros filled the network with locations of Japanese depots, communication networks, and fighting positions in preparation for the upcoming landings. Training continued for the guerrillas and specialized units like the machine gun companies.

Leyte Island, Philippines Landing – October 1944

At this time, Jorge requested combat duty although his mission was intelligence. He was assigned to the Flying Column, a small and armed ad hoc unit that was ordered to take a bridge at Pontevedra before the Americans arrived. He fired the bazooka because no one knew how to fire it. He saw four soldiers on top of the bridge go down. After the war, he heard that one of the soldiers he thought he killed had returned 30 years later looking for him but was unable to find him.

A Japanese aircraft carrier arrived at the coast of Pontevedra. Their landing barges docked during low tide and unloaded their attacking forces. Jorge retaliated with the Flying Column and retreated to the mountains knowing the Japanese would not follow for fear of cerebral malaria. Over a hundred Japanese were killed in the intense fighting. His unit moved south and crossed the damaged bridge at Tolong where a sizeable Japanese garrison of four thousand men was located. He proudly reported to have been able to obtain arms, ammunition, and blankets for all his twenty-seven men before attacking the garrison.

On Masbate Island, Rodriguez continued to radio KAZ to pick up downed pilots who were brought to his outpost. Before the year ended, he was directed to go to San Juan in Batangas on Luzon. He kept moving, setting up the radio in the nearby hills of Angono, Rizal and reporting on heavy fighting in the Pasig area. Civilians who worked for the Japanese told him that tunnels were being constructed in the hills. When the Eighth Army liberated the town of Taytay, he sent information on all the locations of the Japanese caves and tunnels to the intelligence officer for the planes to bomb. After the liberation of Manila, he hitchhiked overland to San Miguel in Tarlac to the headquarters of the 978th Signal men. In April 1945, still sick with malaria, he departed Nichols Airfield in Manila for New Guinea, headquarters of the 978th Signal Company.

After the news of the Leyte landing, Andrews moved the radio further north into the mountains. Enemy threats were delaying the operatives' transmission of enemy activities in the waters around southern Luzon. It was imperative that reports on enemy shipping in the San Bernardino Strait between Bicol, Samar, and Leyte were transmitted as it was the shortest supply route for U.S. forces from Leyte to land in Luzon.

A secret order from GHQ sent Malic to Panay to bring Governor Confesor to Negros. The trip took over a month and was lengthened by a week when Malic had to wait for Confesor to finish his affairs before leaving. Was it to meet with Abcede and Andrews to broker a truce between him and Peralta? Their intense confrontation could have cause Confesor to be killed if he stayed in Panay. Malic stayed as an aide to Andrews until he was called to Leyte.

By December, the guerrillas under Abcede's command numbered more than twelve thousand troops.

Luzon Island, Philippines Landing – January 1945

With the anticipated news of the Americans landing in Luzon, Andrews withdrew his men from the Visayas and sent them ahead to Luzon. Benedicto was designated as his replacement to head the network. He continued to send information on the POW situation, guerrilla activities, Japanese atrocities, the rescue of American fliers shot down in combat, and general information on supplies and funds needed to carry on intelligence work until the arrival of the liberation forces in Negros.

To better prepare the men, Hawley and Schmaltz conferred with Abcede to set up a demolition training school and create a special unit called the 'Rangers.' Men were recalled from their outposts to give instructions on demolition and equipment use. Operatives of various towns like Bacolod, Manapla, Silay and those inside Japanese areas in the northern part of Negros were contacted. Garces and Quesada recruited volunteers to train. The school "devised a system of elimination. Volunteers must have a high school education, be physically fit, and have complete knowledge of the Japanese occupied areas." For two weeks, the volunteers, 46 eligible men, quickly learned about demolition weapons, surveillance, commando tactics, unarmed combat, and security.

Ignacio, codename 'Brisbane'could not cross the Sulu Sea after sighting a Japanese patrol. He was to report on an airstrip in Dipolog for the forthcoming Mindanao landing. After a few more days of delay, Jorge went instead. Malaria struck Jorge as he rode overland in Negros to a boat that brought him to the airfield in Zamboanga. He saw the guerrillas place 55-gallon barrels filled with cement and rocks all over the strip. When rolled to the side, the airstrip was serviceable.

Turning north to Mindoro, Ignacio set up a transceiver as a forwarding station that received reports from Manila and relayed to the 10th MD in Mindanao and KAZ.

Meanwhile, Panay civilian radio operator, Modesto Castañeda, memorized how to assemble a radio, disassembled it, then hid the hundreds of radio parts in fruits and vegetables that were loaded into a small boat to Luzon. He and Ignacio entered Manila after convincing the many Japanese checkpoints not to delay the produce for the meals of Japanese officers. Once the radio was assembled, the cipher system was set up. The radio could only reach a maximum of two hundred miles from Manila, so the transceiver on Mindoro Island received the messages and relayed them to Mindanao and KAZ. With the radio set up, Ignacio turned it over to an operative to return to Negros. He was almost exposed when a cousin recognized him. Gritting his teeth, wanting to respond, he brushed him off.

Ignacio was one of the first U.S. Army soldiers to enter the city and the first to set up a radio. Before leaving the city, he contacted Frank Jones, an operative in Intramuros who was forwarding information to his sister Helen Jones' station in Bulacan, north of Manila. She had worked in the office of U.S. High Commissioner Frank Sayre before the war and pretended to be pro-Japanese while receiving and transmitting information from her brother.

Ignacio and Malic were picked up by plane and taken to Leyte. As a former merchant marine officer, Ignacio was proud to add the Philippines to his military service aside from Europe and China. He was able to visit his family in Tacloban for four days.

Negros Island Landing – March 1945

When news of the coming Negros landing arrived, Abcede advised GHQ that of the 14,000 guerrillas under his command, only a little more than half were armed.

The U.S. forces landed at Pulupandan, Negros Occidental with five battalions in anticipation of the estimated 15,000 Japanese soldiers on the island. There was no preliminary naval bombardment and no heavy opposition at landing. Maj. Gen. Rapp Brush formed the nucleus of the assault task force and headquartered in the Aguinaldo-Gamboa house in Silay which was earlier the headquarters of the Japanese Command. The radio under Abcede with its codes and ciphers were turned over to the Task Forces for the guerrillas' direct contact with the Army on the enemy's disposition.

The Rangers, demolition team, joined in the push northward towards the Patag area below Mt. Canlaon where the Japanese garrison fiercely fought back. That prevented the Japanese from going further south. The Japanese retreated to the mountains after heavy fighting. Trapped in the northwestern part of the island, they began launching banzai attacks until they were annihilated.

When operatives could not enter the towns, civilians were paid operational money to carry prepared charges inside. The men's intensive sabotage work was crippling the enemy to the point that the Japanese hired highly trained Filipino spies to shoot them on sight. Five thousand pesos a day was offered to shoot any man new to the area.

From Negros, Yuhico evaded capture by Japanese patrols and braved typhoons while sailing around Masbate and Ticao Islands before reaching Legaspi in the Bicol province. He became a marked man as he tried to contact guerrillas in the area. In the market of Legaspi City, he narrowly escaped recognition, he related. With Jose Crisol, a fellow graduate of the Philippine Military Academy (PMA), they formed the Philippine Air Corps Intelligence and Sabotage Battalion comprised of men from the six Bicol regions. They conducted sabotage missions on enemy facilities. The Japanese Mitsubishi storage facility was burned, cutting off ammunition and food to the Japanese garrison in Bicol. Having made the stationmasters his operatives, a sabotage group damaged the train tracks derailing the movement of Japanese soldiers, food, and supplies. The weakened enemy was forced to retreat to the mountains.

The Alamo Scouts led by Lt. William Nellist of the Sixth Army reconnoitered the Bicol Peninsula area in preparation for the American landing on the island. "One of his operatives, Maj. Russell Barros, snatched a map of the Japanese fortification plans for Legaspi City and the rest of Bicol from the portfolio of a Japanese Marine commander as a woman in a refectory entertained him." Yuhico related that he received the maps through a famous writer, Vicente Rivera, in a bamboo tube. The documents pinpointed target locations of enemy installations in Legaspi. He showed it to Nellist and turned it over to the PBY forces which landed in Donsol Bay. The Legaspi area was saturated with bombings on enemy transportation facilities, communication lines, camps, and storage facilities. The Japanese garrison in the town of Salvacion with five hundred troops was destroyed by the continuous American air raids on Legaspi city. The enemy was so demoralized that there were reports of suicides and desertions. With all the gun and shore emplacements destroyed, there was no shooting at the American ships coming into Albay Gulf and landing.

Yuhico reported back to Tacloban and went on leave. With Ignacio and Malic, he flew to Morotai and was attached to the 2nd Filipino Battalion. They continued to train because the war was still going on. They were about to begin training in parachute jumping to be dropped in China as weather observers but were flown back to Leyte the following month to join the 1st Filipino Infantry Regiment who were mopping up the enemy in Samar and Leyte. After the Japan surrender, Yuhico was battalion adjutant of the 7th Military Police in Manila granting U.S. citizenship to Philippine Scouts. He was the last man, and the American judge told him to raise his hand and he did.

The Planet Party's mission to establish an intelligence network was one of the most active and best connected of any intelligence group in the Philippines, Whitney said. "The potential was enormous but largely unrealized," concluded G-2 Intelligence. Management of the net for only military purposes without political influences was difficult. Despite that, the men in the party went on dangerous assignments to relay intelligence reports to the net control station.

The Party's network was credited with sending some 469 messages from the Philippines to GHQ. "(The) agents provided crucial intelligence on Japanese naval movements before and during the decisive 1944 Battle of Leyte Gulf." In an article, *Six Came Back,* written by Ind, he stated that "Yuhico was the man who got most of the information that we used at headquarters."

When Villamor departed, the extensive contacts in Manila were uninformed and "SWPA had to start from scratch with the succeeding missions to the islands of Mindoro and Samar to find and recreate Manila contacts." It seemed GHQ did not completely know who those Manila contacts were either.

The first mission paved the way for the other missions in the next two years of the war. GHQ would send letters of instructions to the guerrilla leaders informing them of the men's mission and asking the leaders to support them. But that was not always followed. In many instances, the guerrillas did not give the basic support of food to continue their work. Instead, they gave the mission men additional work. Needing security, messengers, couriers, and carriers to transport equipment and supplies they turned to civilians who had to be paid. At times, not enough money was brought by the men. It was a problem if they were sent to an island that did not speak their language. There was not enough medicine when they got sick, and many had tropical diseases like beriberi, ulcers, dysentery, and malaria. The intense physical and skills training they underwent in Australia was the only constant they could rely on to keep going. Their training was on conventional warfare and jungle survival. They had to quickly learn the irregular guerrilla procedures while setting up their radios and evading the enemy.

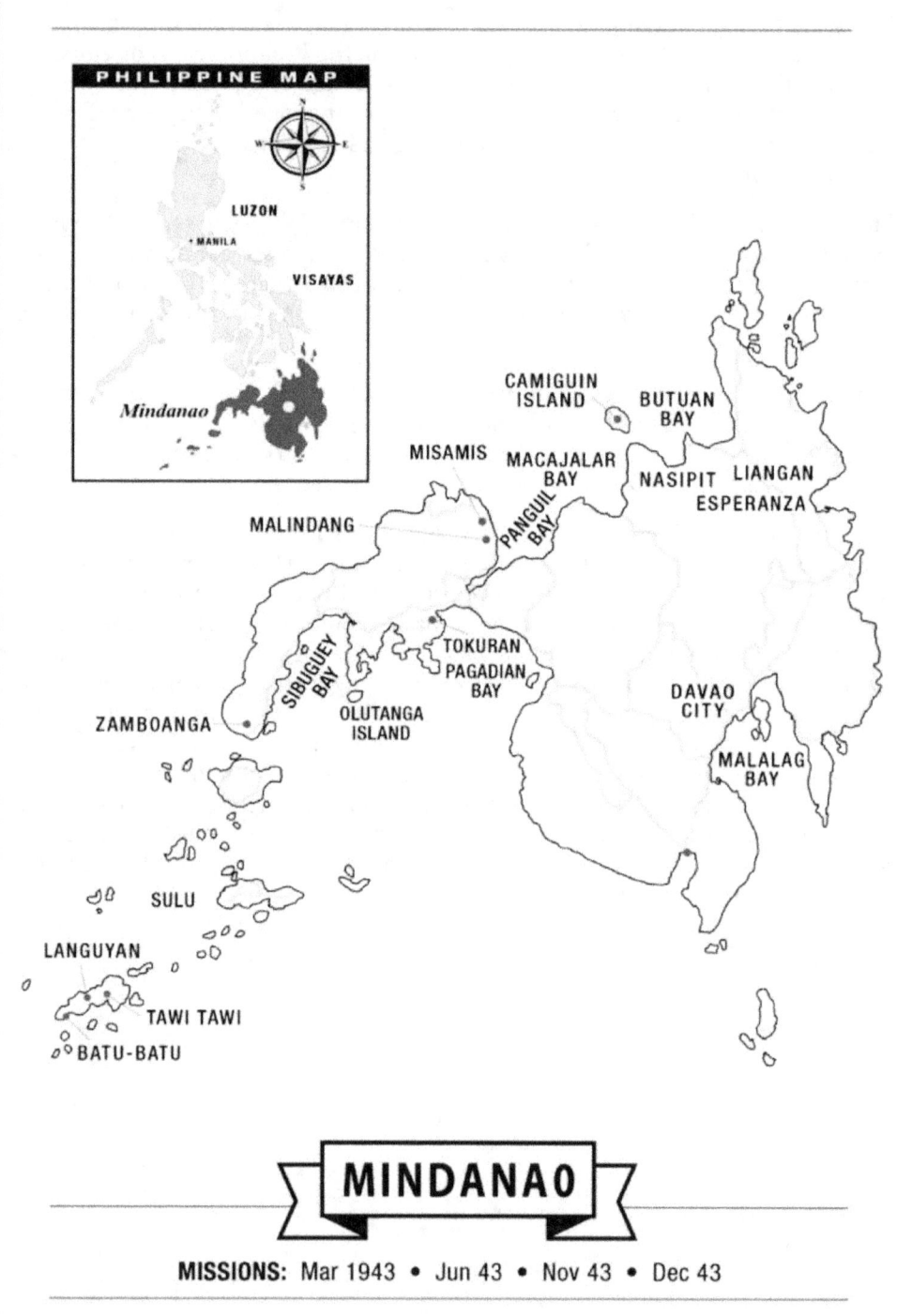

PHILIPPINE MAP

LUZON

• MANILA

VISAYAS

Mindanao

CAMIGUIN ISLAND

BUTUAN BAY

MISAMIS

MACAJALAR BAY

NASIPIT

LIANGAN

ESPERANZA

MALINDANG

PANGUIL BAY

SIBUGUEY BAY

TOKURAN

PAGADIAN BAY

ZAMBOANGA

OLUTANGA ISLAND

DAVAO CITY

MALALAG BAY

SULU

LANGUYAN

TAWI TAWI

BATU-BATU

MINDANAO

MISSIONS: Mar 1943 • Jun 43 • Nov 43 • Dec 43

Mindanao Island February 1943 – March 1944

Two days before Pearl Harbor was attacked, the 19th Bombardment Group and 14th Bombardment Squadron were ordered transferred from Luzon to Del Monte Airfield in Bukidnon Province, Mindanao to be out of range of Japanese bombers in Formosa. Wendell Fertig, who was then a civilian engineer building airstrips for the U.S. Army was stranded in Mindanao.

Several hours after the Pearl Harbor attack, Japanese planes bombed the seaplane anchorage in Malalag Bay, Davao. Two weeks later, the Japanese High Command sent a small battle group from Palau to secure Davao City after becoming worried about the security of the large Japanese population in the City once the invasion landings began. Occupation of Davao would cut off the southern border of the Philippines preparatory to invading the Netherland East Indies.

The American soldiers fled to the hills. Many would later become part of the Force Radio Station (FRS), the main radio network of the 10th Military District.

Fertig surfaced in Kolambugan, Western Lanao and moved to Misamis Occidental where he began recruiting civilians who refused to submit to the Japanese. When the Japanese attacked his headquarters, he fled back to Lanao. He was able to "maintain support among the opportunistic *Moro* (Muslim) tribes in the area through distribution of a LIFE magazine article in which King Ibn Saud of Saudi Arabia allied Islam with the United States." The active support from the Maguindanao and the Maranao *Moros* significantly contributed to the success of the guerrilla resistance on the island.

Around 20 Filipino civilians and the stranded men from the Bomb groups manned FRS together with 16 code men. They would need help as Mindanao became the cargo distribution point to the other islands, the net control station for coastwatcher stations, and the main radio network to Australia.

Small guerrilla groups were consolidated in northern Mindanao by Luis Morgan, a former officer in the Philippine Constabulary, who convinced Fertig to take over command and provide the leadership to organize the guerrillas with community support. An effective resistance movement needed the support or neutrality of the *Moros*, who made up a third of the island's population. Morgan's ruthless attempts to recruit men into the guerrilla ranks and his dislike for the *Moro* people discouraged many from fighting the Japanese. But, to stop the killings, the *Moro* fighters agreed to join the guerrillas. Disagreements in command between Morgan and Fertig, with Morgan issuing conflicting unauthorized orders, combined with a threat to mutiny, and complaints of unfair treatment from the Filipino soldiers broke the relationship between Morgan and Fertig. Convinced to leave for Australia, Morgan reported to Camp Tabragalba and underwent the same training program as all the other men.

It was not good news for Fertig when he heard that Colonel Macario Peralta, the guerrilla leader on Panay, had successfully contacted Australia. Mindanao was only 1400 miles away from Australia, and Peralta, who was further north in Panay Island, was able to reach radio station KAZ in General MacArthur's GHQ. Fertig had been trying for months to contact Australia and inform them of his large organization on the Philippine's second largest island. Like Peralta, he desired to command all the guerrilla forces in the country.

Impatient, Fertig supplied a sailboat and provisions to three of his fellow engineers, Jordan Hamner, Charles M. Smith, Athol Y. Smith, and two other civilians chosen to accompany them, Lakibul Nastail, a *Moro*, and Eugenio S. Catalina, and ordered them to set sail for Australia. Their journey lasted for a month, sailing through the Celebes Sea and Morotai Strait, surviving dehydration and starvation while avoiding enemy patrols and attacks by hostile natives in New Guinea. They arrived at Cape Don lighthouse north of Darwin on New Year's Eve. An Australian patrol boat picked them up and delivered them to Darwin. Their arrival on January 1943 with news of the resistance further gladdened GHQ. Hamner would return to the Philippines to lead a mission in June 1943 and Smith would do so in December the same year.

At Fertig's headquarters, an old Spanish fort in Misamis (now Ozamis) on the coast of Panguil Bay, a transmitter was constructed from pieces of old radio receivers and sound equipment by Filipino boys who had read books from a radio correspondence school. A Filipino radio operator with the former Bureau of Aeronautics Radio brought a radio transmitter buried near his home. Prof. Roy Bell from Negros arrived with a transmitter and with Air Corps radio operator Robert Ball, fixed the crystal and made a makeshift antenna using a wire coiled randomly around a bamboo pole. After much testing, messages were sent out across the broadband with the hope that there would be a response. The transmission used an old code and call sign by the former Air Corps station at Del Monte in Mindanao. The Navy station KFS in California was suspicious and thought the Japanese were jamming the air. Smith and Hamner having arrived in Australia confirmed the message from station WYZB and contact was formally established in February 1943.

Although Panay was able to contact earlier, Mindanao was much closer and the transmission was quickly vetted. KAZ wondered about the extent of the Japanese presence as the island was sparsely populated.

Code words in a new cipher system were set up with station KAZ. Fertig reported that his organization was composed of over ten thousand men who desperately needed arms and ammunition to continue harassing the enemy. He was appointed commander of the 10th Military District to whom supplies would be sent.

KAZ formally declared to Peralta and Fertig in a memo issued in February 1943 its intention to develop guerrilla territorial command based on pre-war military districts led by recognized strong guerrilla leaders. Bases were needed to send supplies, men, and radios and, in return, these leaders were to create intelligence nets in their areas and establish liaisons with organized forces in adjacent districts to transmit intelligence on the Japanese disposition to be used in planning the return of the Americans. The commanding officers were to function under the direct control of GHQ in Australia.

FRS was designated as the central collecting and control center for intelligence reports. Messages were to pass through FRS for relay to Australia. Couriers carried messages to FRS about enemy ships as far as the islands south of Luzon until more radios were set up.

The headquarters in Misamis was within fifteen miles of a Japanese garrison. When Japanese aircraft strafed the area and a destroyer landed close to the camp, FRS was forced to go off the air and hide for a month. A Japanese colonel's Filipina mistress had warned of the attack after a collaborator betrayed their presence and guided the Japanese to the camp. The traitor was later executed after he was arrested by the civilian authorities. After 36 hours, the first message received was from a very angry General MacArthur: "Under no circumstances are you to ever, repeat, ever be off the air again." And they never were off the air again operating for 24 hrs every day for the next two years.

As backup and security, A Corps was established as well as a sub-net control station in the mountains of Malindang under Robert V. Bowler. This control station would handle all radio stations on the western half of the island if FRS was captured. The station was manned with military and civilian Americans and Filipinos in an efficient operation. Radio traffic passed 24 hours daily.

USS *Tambor* – March 1943

AIB quickly dispatched a second submarine mission with code name FIFTY PARTY headed by Navy commander Charles "Chick" Parsons. It was his first mission to the Philippines after the Japanese invasion and this was followed by several missions to Mindanao. The first of these was originally destined for Mindanao but was changed to Negros when the landing site was compromised.

As soon as the mission departed, AIB sent an urgent request to Washington for 18 trained radio operator-technicians of Filipino nationality who were qualified for secret intelligence missions. The bureau was running out of qualified Filipino men in Australia. They soon received a reply: six men were to be available in Australia by May 1943, six in July, and six by September.

Parsons knew the terrain, politics, and the people, having lived in the Philippines for many years. His mission was to observe and assess guerrilla organizations. Instructions from GHQ were to survey developing radio nets, coastwatchers, and construction of airfields. He was to stress the importance of coast watchers at or near strategic sea passages to observe Japanese shipping and report to Australia.

Lt. Col. Charles Smith joined in the setting up of coastwatchers in the Davao Gulf and the waters of the southern islands to monitor Japanese shipping. GHQ wanted no action, only intelligence on the Japanese presence and movement throughout the islands. Fertig realized that aside from mediating fights within his organizations to maintain control and harassing and defending his group from Japanese raids, he now had to contend with intelligence-gathering.

Parsons prioritized radio equipment over medical supplies, but he did make sure to order "especially a quantity of Atabrine for use against malaria.... pills to combat dysentery, pneumonia, and staph infections." He brought five radio sets to gather intelligence and transmit to Australia. The *Tambor* was an operational sub and could not carry more. Seven tons of supplies, including "a set of cipher, weapons, 70,000 rounds of ammunition, medicine and morale boosters of cigarettes, chocolate bars, and oranges," landed with him. Sixty thousand pesos were for the 10th MD for operating expenses. Currency was just as crucial as arms and medicine to control the guerrillas. Fertig calculated that four tons could be carried by 10 men or 25 carabao carts back to camp. Pictorial magazines printed in Australia were distributed.

The sub departed Fremantle and headed towards Pagadian Bay, the aborted landing site of the earlier PLANET PARTY. Two *Moro* brothers Pvts. Hadjula Bairulla and Sabtal Bairulla who knew the Mindanao terrain and spoke the language, and Pvt. Eugenio Catalina were chosen to join. Southwest of Apo Island, the sub fired three torpedoes at a Japanese convoy which chased the sub to drop depth charges. The *Tambor* was dangerously close to shore when she surfaced to get away from the charges. They would learn, three days later, that one of her three torpedoes hit a freighter.

Paddling frantically to the shore of Tukuran from the sub, the three men had to lie low in their boat as tracer bullets, fired from onshore, zinged overhead from the shadows. Their rubber boat hit the beach and they soon found themselves facing the guerrillas of the 108th Division of the 10th Military District. Shouts of joy filled the air without appropriate caution of the Japanese.

GHQ learned that Mindanao had a large guerrilla organization that divided the island into military divisions: 105th, 106, 108th, 109th, and 110th Divisions, each headed by an American commander except for the 105th Division headed by Col. Hipolito Garma. The 107th Division was formed a year later to protect the Davao area, where the capital city was located.

But there was a lack of officers to lead and train personnel in military combat. The one hundred Americans who evaded capture or were stranded on the island were mainly from the Air Corps and Navy with limited infantry training.

The resistance in Mindanao was unique. The *Moros* who had been fighting the American presence before the war broke out, took up arms and formed their guerrilla groups against the Japanese led by their *datu* (Muslim chief). In the past, they had resisted any attempts by the Spanish, Christian Filipinos, and Americans to rule over them. Now they united to fight a common enemy under the 10th MD. *Datu* Gumbay Piang sent a message to GHQ reiterating his pledge of loyalty to the U.S. His 20,000 *Moros* were enlisted as bolomen of the U.S. Army. Even before the surrender, Busran Kalaw led the Fighting Bolo Battalion of Maranao *Moro* swordsmen to resist the Japanese. *Datu* Mindalano, and other Maranao *Moros* later banded together to form the 108th Division, 10th MD.

When Guy Fort, the 81st Division Commanding General they respected and reported to had surrendered together with all those who remained with him (about ten percent of his original force), they were bewildered. To the *Moros*, a fight went on until one or the other of two opponents was killed, and they thought of mass fighting in the same terms. When told to limit their activities to intelligence gathering and not provoke Japanese retaliation, they continued to harass the enemy anyway.

Fertig and other guerrilla leaders agreed that if they wanted supplies from the Americans, they would not confront the enemy. Even if they were provided arms, they were not enough for prolonged gun battles. The guerrillas knew that they needed the support of the civilians who expected to be protected in return. So periodically, they conducted short ambuscades. A few casualties on the enemy here and there eventually totaled into the thousands. The prisoners at the Davao Penal Colony (DAPECOL) were left alone to avoid their being killed. Their important mission was for the security of coastwatchers stations established around the island.

Fertig's influence expanded with Mindanao as the distribution point of supplies for the guerrillas in the other islands. As a result, Mindanao received the most supplies sent by the Americans. Although never enough, the limited supplies gave hope and kept them maintaining their activities as they planned for the return of the Americans to the islands.

Although it was the central drop-off point of supplies, what Mindanao faced was the complicated logistics of shipping men, food, and security in a convoy over land and water to the other islands. It usually took weeks or months for the supplies to reach their destinations. Submarine rendezvous points had to be secured from the enemy. Thousands of cargo bearers, fleets of bancas, with guerrilla security travelling on routes across rivers and mountains had to be planned. And all the men involved had to be fed and paid for their work.

The Sulu area was separated from the 10th MD when Filipino guerrilla commander, Col. Alejandro Suarez was appointed to command the area. He was a Constabulary officer who fought with the USAFFE until the surrender of Bataan and Corregidor. He later escaped and returned to Sulu in January 1943. He obtained supplies from Mindanao and set up stations covering the Celebes and Sulu seas that Japanese troop ships travelled through to send troops to New Guinea. With his organization of guerrillas who were mostly civilians, they harassed the Japanese in the Sulu archipelago. They obtained information on Australian prisoners who had escaped from Japanese custody from Northern Borneo to Tawi-Tawi.

Whitney thought that the messages from Mindanao showed the guerrillas as more resourceful before the submarines arrived with large amounts of supplies. He analyzed that much of their strength came from the ability to "improvise through necessity from the little they have to ingeniously progress despite deprivation such as clothes from hemp, light from coconut oil, … emphasis on crop planting in free areas to ensure adequate food supplies." He tried to control the requests for basic needs. Fulfilling some of Fertig's requests was difficult and Parsons had to remind him that "anything he (the submarines) brings comes from personal interest of General MacArthur, not from the War Department. MacArthur has to beg for anything he gets."

A month after arrival, Smith sailed around the south coast of Mindanao with two American officers to successfully establish a coast watching station near Davao where there was a large Japanese naval base. When informed of their submarine's first sinking of a

Japanese ship coming out of Davao, the Navy realized the benefits of coast watching stations around the islands and became eager to help in providing the men, radios, and supplies. The reports were of great importance because the Japanese were sending reinforcements to New Guinea through the waters of southern Philippines.

Parsons and Smith went north to investigate guerrilla groups on the other islands. In Leyte, Col. Ruperto Kangleon, who had retired from active duty after escaping imprisonment by the Japanese, was convinced to unite and lead the guerrillas. Warring guerrilla leaders were appeased with promotions under his command. (Leyte Island is discussed in a separate chapter.)

FRS station moved to Liangan but after six months moved across the island to Esperanza, a town by the Agusan River just in time as the Japanese landed on Liangan after intense artillery shelling.

In the meantime, ten American prisoners at the Davao Penal Colony escaped. It was a maximum-security prison for hardened Filipino criminals, and the strongest Filipino and American prisoners were sent to tend livestock and cultivate food crops for the Japanese garrisons in the Philippines. The escapees were found by guerrillas and were brought to 10th MD headquarters. The men waited for three months providing support as assigned before being evacuated by submarine. Their somber report on the treatment of prisoners to General MacArthur in Australia strengthened his resolve not to bypass the Philippines. They commented on working for Fertig, who "did not invite them to dine with him and did not share the coconut pie being baked." Most of the time, Fertig ate alone, separate from his men. He had his own cook and separate quarters from his officers. This may have been self-preservation on his part.

As more missions were sent out, the roster of qualified Filipinos in Australia dwindled. Three months had passed since the urgent request for trained Filipino radio operators were sent to Washington. In desperation, Whitney looked to California, where thousands of the 1st and 2nd Filipino Infantry Regiment men were training to be chosen from and sent to Australia.

USS *Trout* – June 1943

On her way to the Philippines, the submarine *Trout* missed hitting a transport with three of her torpedoes. She landed the TENWEST PARTY of five led by Capt. Jordan Hamner at Labangan in Mindanao's Pagadian Bay after security signals were recognized. Three tons of supplies, radio, ammunition and $10,000 for the 10th MD operations were dropped off. The mission was to establish a secret intelligence and coastwatcher unit to report all enemy ship and air movements in the Sulu Archipelago and Zamboanga areas to FRS and survey the area for suitable future operations. Hamner, whose alias was Merle Hart, had sailed six months earlier with Charles Smith to Australia. He returned to Mindanao leading Lt. Frank Young of the Philippine Army (PA) whose alias was Francis Jumlia, and three Filipino *Moros:* Pvts. Sahibad Aliacbar, Jalil Ladiahasan, and Lakibul Nastaio.

The *Trout* departed and went on patrol, damaging the Japanese tanker *Sanraku Maru* with three of her torpedoes, the *Isuzu Maru* cargo ship with four, and sinking two of three coastal steamers with her deck guns. A week later, she received an urgent message to proceed to Olutanga island off Sibuguey Bay in southern Mindanao to pick up five

Americans – Smith, Parsons, and three prisoner escapees from the Davao Penal Colony, Stephen Mellnik, Melvyn McCoy, and William Dyess to return to Fremantle. In front of an amazed General MacArthur, they revealed the first information on the horrors of the Death March on Bataan.

The three *Moros* remained in Mindanao while Hamner and Young continued to Zamboanga and provided communication support to Suarez, the guerrilla leader of the Sulu Area Command. Coastwatcher stations were established with guerrillas trained to handle radio in the north at Languyan and Batu-batu in the south to monitor the Sibutu Passage, British North Borneo, Celebes and Sulu Seas. Extensive data on ship sightings and minefields throughout the Sulu archipelago was transmitted. The critical food situation of the guerrillas and civilians throughout Tawi-Tawi was also reported. When ten messages to KAZ were not received and may have been intercepted by the enemy, Hamner was advised to send the messages through FRS.

The Japanese had established a temporary base in Tawi-Tawi for their fleet, which departed a year later for the Battle of the Philippine Sea. When Japanese planes departed from the base to bomb Tawi-Tawi, the guerrillas fought back "equipped with new firearms brought by the submarine *Narwhal*. The BAR (Browning automatic rifle) remained our most powerful weapon."

Across the islands, disturbing messages of Japanese brutality were streamed to FRS for relay to KAZ. One chilling message described an American soldier, Hayden Laurence, who was captured near Arayat, Pampanga on Luzon on Sept 21, 1943. He was kept in Angeles City without food and water for two days. "Then a detachment of Capt. Nakaseco Company took him to the cemetery where he was tied to a tree with barbed wire and used for bayonet practice until dead. 200 Filipinos witnessed."

Hamner left Young in Tawi-Tawi to set up a coastwatcher station to monitor the Balabac Strait south of Palawan Island. He had to return due to deteriorating health and problems with the *Moros*. He was evacuated on the *Narwhal* when it returned a fourth time. Whitney was disappointed but commended Hamner for making the journey to Tawi-Tawi until the enemy drove him out. In the meantime, Young became ill with malaria and recovered to continue operating the radio station and providing military training to the guerrillas. He pacified the *Moro* Battalions, quelling a mutiny, and this earned him a promotion. He was resourceful growing food with seeds he had been given. The civilian volunteers in his unit were more than willing to carry out ambuscades against the enemy with weapons and ammo from the subs.

USS *Narwhal* – November 1943

From Mindoro, the *Narwhal* arrived at Nasipit, Butuan Bay, landing 50 tons of supplies. It was the submarine's first assignment as a cargo submarine and her first trip to Mindanao. She had bypassed Mindoro when security signals were not sighted. Surfacing at dusk in Butuan Bay, she sighted a launch flying an American flag that had been patrolling up and down in front of Nasipit Bay. After flashing signals, the launch followed the periscope until the submarine surfaced.

To keep the arrival secret, Edward McClish, commander of the 110th Division ordered 500 of his troops to attack the town of Balingasag and passing Nasipit for a supposed rest period. Unknown to the troops, they were to secure the landing site and help unload the cargo. The men were startled when the behemoth *Narwhal* surfaced.

The crew did not expect the bright light of the lighthouse at the harbor's entrance. Canoes lighted up and emerged from a channel by the entrance to the Agusan River and surrounded the sub as it slowly moved towards the dock. Fertig, Lt. Reyes, his executive officer, and Robert Ball boarded and shook hands with Parsons, appreciative of more supplies.

The harbor was deep enough but dangerously crooked. A launch guided the sub to the pier. Suddenly, the sub shook as it hit something in the water. A sailor hollered "NO WATER!" The sub had grounded on a hard sandbank in the harbor channel. The sight of the *Narwhal's* six-inch deck guns had probably unnerved the launch pilot who brought the submarine too close to shore and out of the channels. The skipper quickly lowered the aft of the sub, reversed his propellers at maximum power while the crew ran back and forth on the top deck to rock the ship off the sandbar. In minutes, the submarine moved over ten feet into the channel and was able to maneuver into the Nasipit pier.

Tension rose when a Japanese patrol plane circled the sub a half a dozen times. One of the sailors later told a waiting evacuee that they were close to scuttling the ship and joining the guerrillas in the jungle if the plane attacked. Luckily, the plane left without doing anything.

"Free Philippines" Guerrilla Stamps. Printed by the 10th Military District, Mindanao. They were morale boosters. No fee was collected and used on envelopes sent outside the Philippines. Source: Bobb Glenn

Parsons was onboard bringing more radio sets and 500 *Free Philippines* guerrilla stamps printed in Australia. Couriers used the stamps to impress the people and the enemy of the presence of an organized guerrilla force. Propaganda material of small packs of cigarettes and chocolate bars with the 'I Shall Return' were brought. Part of the cargo was the unusual request of 1500 ping pong balls that the Quartermaster reluctantly fulfilled as space was better taken up by medicine and ammunition. Parsons reasoned that the men played a lot of table tennis with homemade tables and paddles. The games were generally played between shift hours to relieve the stress of constant alertness

The crew began unloading the cargo of ammunition, guns, and medicines with the help of the guerrillas and civilians. It was a festive scene with bright lights run by the submarine's generator. An "endless line of sailors in blue denim carried off the ship cases of sending and receiving sets for the coastwatchers. Shoes and clothing served as dunnage to keep the angular boxes from shifting in the sub's curved interior," wrote an evacuee. "The Filipino boys scurried down the hatches, practically naked, and come up out of the hatches carrying 50 to 65 pounds and wearing 4 to 5 sets of clothes and a pair of jungle boots three sizes too large."

The *Narwhal* patrol log reported that the crew, realizing the genuine need for any supplies in guerrilla-occupied areas, transferred more than were consigned as cargo. Some of the sailors gave their shirts to the guerrillas. It took only four hours to unload the entire cargo compared to the three days it took to load them on the submarine.

While the cargo was being unloaded, Kangleon, Bohol commander Ingeniero, and Fertig had coffee and sandwiches onboard. Fertig spoke of the cargo for Leyte and Bohol to cement loyalty. The evacuees were herded to cots in the forward torpedo room. An escapee from the Davao Penal Colony could not wait to tell horror stories of the beatings he and other men received from the brutal Japanese guards. The submarine quickly moved out to Butuan Bay to return to a bombed-out Darwin. To preserve secrecy, the evacuees were quartered in a remote hotel on the coast of Caloundra where they were interviewed and informed to maintain silence of their time in the islands.

USS *Narwhal* – December 1943

The *Narwhal* had only three days of refitting before going out to sea again. A mission party was ordered to land in Mindanao with 90 tons of cargo to set up radio communication in Samar. Orders arrived to leave the camp in November 1943. They were interviewed and informed of the risks of diseases and the slim chance of returning. Asked if they were volunteering and not being coerced, all of them agreed that they were volunteers to the mission. Each man carried his personal gear in his barrack sack. For two days they helped the submarine crew load the cargo and equipment including four Australian portable radio sets at the Brisbane dock.

Charles Smith returned leading the party comprised of Capt. James Evans, a medic and 978th Signalmen: Sgts. Restituto Besid, Aniceto C. Manzano, and Gerardo A. Sanchez; T5 George R Herreria; T4s David D. Cardenas, Crispulo G. Robles, and Daniel B. Sabado; T3 Robert H. Stahl, and Pvts. Querubin B. Bargo and Andres S. Savellano.

Evans selected the supply of medicine: "sulfa powder, morphine for the wounded, Atabrine to prevent malaria and quinine for treating it, adhesive tape but no bandages because cloth in the islands could be used, and other drugs, supplies, and medical instruments." They found out later that cloth was not readily available as many people were in tattered rags or rough *abaca*, a leaf fiber.

PRS hoped that the Australian-made radio equipment would withstand the humidity and work better than the American-made sets in the Philippines. The old sets would short out when turned on in the morning. The signalmen learned that by leaving the sets on 24 hours a day, it would not fail. But the station would now use twice as much fuel. Condensers failed, and "at this time, three coastwatcher stations (were) not operating due to battery failures." They learned to submerge the batteries in *tuba* (fermented coconut sap), to get several more months of use.

Onboard the sub, the men saw the large mass of supplies they were bringing into unfriendly and unknown territory. Among the cargo were spare parts for generators and radios, propaganda material, heavy weapons, jungle boots, D-rations, thousands of pesos, counterfeit Japanese currency, and printing plates and paper for manufacturing legal currency. Money was packed in water and sand in sealed five- gallon metal containers to age it. Caught with crisp new bills meant death. Each pack had to fit through the submarine hatch and narrow interior hallways for easy transport. Newspapers about Allied victories that boosted people's morale were wrapped around breakable items. The Navy gave heavy weapons and four large transmitters to connect with FRS and flash enemy ship sightings to their submarine command in Fremantle and KAZ without delay. A mountain-type anti-tank gun was broken down for shipment, packed in cosmoline to prevent corrosion and wedged into the sub's structure.

More guerrilla postal stamps printed in bright blue ink and over twelve thousand Carbines were sent, according to AIB supply officer, Bobb Glenn. It took the backs of four hundred carriers to carry ninety tons on trails through jungles, swamps, and mountains. Small lots of a ton each, that took months to be sent from Mindanao by sailboats or overland by carriers, were lost along the unsafe coast of Panay. The last shipments of ammunition to Negros were lost to the enemy. There was always the risk of pilferage, weather, or accidents.

The submarine's large batteries required regular surfacing for recharging to run underwater, so that it took seven days to reach Butuan Bay. Travelling on the surface kept the crew on the alert for enemy aircraft and ships. They often had to perform crash-dives and complete silence to avoid enemy ships passing overhead. Any sound of cracking or knocking noises inside the sub unnerved both crew and passengers. That was a frightening experience for the men who had never been in a submarine. To Sabado, it was an adventure.

The *Narwhal* dove and entered Butuan Bay on the north coast of Mindanao. At dusk, she surfaced off the area of Cabadbaran. The crew could see the lighthouse's intense rays shining at the entrance to Nasipit harbor 15 miles away. Soon a dozen canoes surrounded the sub as it moved slowly through the crooked channel.

At Nasipit harbor, preparations for area security and the sub arrival were in place. Guerrilla troops were prepared to attack if the Japanese from a garrison west of the harbor were spotted coming their way.

When the submarine docked, McClish's 15-member Filipino military band dressed in their crisp white uniform greeted the vessel at the pier, recalled Bob Stahl, a signal man onboard who was bound for his Samar mission. As soon as the hatches opened, the band played "The Stars and Stripes Forever" and "Anchors Aweigh" with much gusto and enthusiasm. They didn't seem to mind the danger that the Japanese were only 10 miles away in the garrisoned towns of Nasipit and Cabadbaran. It had been two years and American aid was finally coming. They were proud to show their appreciation.

A large barge with Fertig and Parsons arrived with copies of *The Tribune* newspaper less than a week old for GHQ. PRS had been relying on Parsons's observations from his trips to the islands to make policy decisions concerning the guerrilla forces and future missions. The guerrillas helped unload the cargo onto the launch. Ten tons were for Smith's men in Samar and ninety tons of arms, ammunition, and medicine for Mindanao. She also delivered three hundred gallons of lube oil and a small number of hand tools. The filled barge slowly travelled 15 kms into the mouth of the Agusan River for storage in the town of Amparo. The heavy weapons of mortars, machine guns, and cannon that the Navy gave were stored until the American Task Forces' return.

Sabado wrote in his report "that members of his party kneeled and kissed the earth" when they set foot on land. They expected firing from the Japanese with the bright lights, people shouting, and the band playing. It was an unexpected but pleasant surprise to be "entertained by Filipina young ladies who came down to see the new arrivals from the other side of the world."

With the unloading completed, seven evacuees with 10 tons of bananas embarked the sub with Parsons. "God Bless America and Philippines, My Philippines" was heard playing as they boarded. The skipper received a message to proceed to Alubijid at Macalajar Bay, an enemy-infested area, and Negros Island. She sank the *Hinteno Maru* (the former Philippine ship, *Dos Hermanos*), with its powerful deck guns off Camiguin Island and quickly left the area. In Darwin, the evacuees and Parsons were transferred to an Army launch. The *Narwhal* docked at Fremantle until her next assignment.

Evans sailed with the men and cargo up the Agusan River to Amparo where part of the supplies was stored. The rest went deeper inland to the swamps and jungles of Esperanza, the new location of the 10th MD headquarters to keep FRS on the air.

But not all the cargo arrived complete. Smith was already aware while onboard the *Narwhal* (which he warned Whitney about in a letter) of the looting of their supplies while they were in Darwin. He sent a request for the personal supplies that did not arrive with them. Robert Bowler, guerrilla leader of the 105th Division, 10th MD, lamented that an estimated 10 tons was lost. When the shipment was opened, Fertig informed PRS that the radio sets had missing pieces such as frequency coils. Radio tubes were broken due to not being placed in sponge rubber tube protectors.

The party spent the next two weeks preparing their equipment. After two weeks of orientation, groups of men went overland to Samar north of Mindanao and crossed inter-island seas to surrounding islands with radios and supplies. They operated independently with guerrillas and carriers for their equipment. Evans stayed in Mindanao.

There were not enough supplies for all at Mindanao headquarters prompting some grumbling that "Americans were given complete new uniforms, shoes, and pistols while they were given none."

The 10th MD was surviving on two meals a day in their jungle headquarters. Some were sick and glad for Evans medical services. On Christmas day, thankful for the contact and supplies from Australia, Mindanao radio transmitted to General MacArthur: "The key pounders and the cryptographers of the Tenth Military District extent their heartiest Christmas greetings to their compatriots in the South. We will sight them, report them and you sink them. Bahala Na (Repeat) Bahala Na!"

About thirty stations covered the coastline of the entire island of Mindanao and the island of Camiguin. Radio traffic increased when watcher stations increased to 80 around Mindanao and surrounding areas gathering and passing information to each other, according to Howard Watson one of the American soldiers stranded on Mindanao and assigned to the radios of FRS.

The seas between Davao Gulf and Zamboanga were watched and sightings flashed to submarines on patrol. Within an hour of receiving the messages, the enemy ships were attacked and sunk. The route between Davao and Zamboanga City became known as 'Torpedo Alley' named by one of the cryptographers on Mindanao. Eight more stations in the Visayas relayed intelligence to FRS for relay to the Navy Submarine Command Center at Fremantle.

From the coast watcher reports to KAZ, three hundred enemy ships were reported sunk, particularly between Davao Gulf and Zamboanga City. With such accuracy, the Navy permanently assigned several combat patrol subs around Mindanao.

Before the year ended, Tokyo published a proclamation that "After January 25, 1944, any American found in the islands whether an unsurrendered soldier or civilian will be executed." With this sobering news, KAZ increased efforts to evacuate people who were not with guerrilla combat units.

Not so good news was coming out of Manila. "After more than eighteen months of occupation, some of the men and women residents of Manila had begun to 'play ball' with the enemy," according to Adalia Marquez, a member of the underground and later with the Counter Intelligence Corps. She and her husband were later arrested and would undergo months of unspeakable torture in the dungeons of Fort Santiago.

Mindanao Island. A large formation of guerrillas march under the U.S. flag. Many were Moros (Muslim). 1945
Source: University of Wisconsin

PHILIPPINE MAP

LUZON

• MANILA

VISAYAS

Mindanao

DUMARAN ISLAND

SURIGAO STRAIT

SURIGAO

BUTUAN BAY

MACAJALAR BAY

DAMPALAN

BUGO

NASIPIT

LIANGA

DIPOLOG

BUKIDNON VALENCIA LANTAPAN

TALACOGON

SIARE

DIMOROK

MANTICAO TALAKAG

LA PAZ

DUMINGAG

DANSALAN

ESPERANZA WALOE

PAGADIAN BAY

DAVAO PENAL COLONY

KAPALONG

TAGUM

MALABANG PARANG

TUKURAN

ISING

CARAGA

DALICAN KABAKAN TALITAY

DAVAO

ZAMBOANGA

DIGOS

DAVAO GULF

MALALAG

BUAYAN

SULU

BATU-BATU

MT. TUKAY

SARANGANI BAY

BOHI GANGSA LANGUYAN

SIASI ISLAND

TALAUD ISLAND

TARAWAKAN

TAWI TAWI

BALIUNGAN

SANGA-SANGA ISLAND

MINDANAO

MISSIONS: Mar 44 • Jun 44 • Aug 44 • Sep 44 • Oct 44 • Dec 44

XI

Mindanao Island
March 1944 – June 1945

As more missions were sent to the islands, Fertig's network grew. He readily accepted funds, supplies and personnel sent to him for distribution to the other islands. He expressed his dislike for the Filipinos from Australia. They looked healthy and well-clothed. He wanted more supplies of clothing, food, and training so that his men would look like the arriving Filipino men.

He complained about the men not knowing the radio procedures that the guerrillas used. In defense, Whitney replied that the "men sent to him were carefully selected, fully trained in field communication and met prescribed U. S. Army qualifications as field station operators." On one occasion, he ordered the recently arrived party to be "divested of their belongings and confined to a stockade to reduce them to the spiritual, moral and material level of guerrillas with whom they would serve," Whitney commented in his copy to the Chief of Staff. He further commented that "Fertig and Peralta resented the arrivals of Filipinos. Resented their good shoes, clothes, their equipment, and their well-fed appearance – have been known to take everything from them and confine them for a period to reduce them to the guerrilla level. The explanation may lie in the guerrillas' desire to discredit any aid given them by the U.S. so that history would record no such aid. Fertig's evaluation should receive zero evaluation." If the complaints were brought up again, he recommended sending the men north to other main guerrilla stations in the Visayas where their support would be welcomed.

Robert Bowler, commander of the A Corps sub-net in Western Mindanao that served as backup if FRS radio net was destroyed also disliked the Filipino men who looked well-fed in their clothes and shoes. He thought they would undermine the morale of his barefoot troops and be able to withstand combat better. PRS reported that "Arriving men were treated harshly."

The headquarters of station FRS was hidden in the jungles of Esperanza in Agusan. Mosquitoes swarmed the camp, spreading malaria. Japanese patrolled the waters and islands around Mindanao. The camp had moved with every danger of discovery. Between mountains, radio antennas had to be carefully located and oriented for messages to reach KAZ. FRS could not be off the air. Radio call changed to WAT and regularly changed to confuse enemy monitoring. Reports on enemy shipping, plane sightings, and land movements had to be maintained.

The thirty coastwatcher stations covering Mindanao, with four covering the Surigao Strait and two around Davao city, added to the hundreds of messages received by station WAT requiring 24 hours on the air. WAT had "two stations assigned to gather all the incoming messages from the guerrilla nets." Another station was used to send out all the intelligence reports to GHQ in Australia.

1944

The year opened with a request from Fertig to send "At least four Filipino Non- Coms radio operators (at this) time. Our net expansion to 24-hour day required this. Operators are not available to fill our needs. Sources of local trained personnel are exhausted. High priority on this." Whitney informed GHQ that "a detachment of four (4) operators under a very competent Filipino officer ... will leave on the next trip of the *Narwhal*. Officer is being sent for purposes of administrative control over Fertig." The officer would prevent the "possibility of these men being pushed around by the guerrillas.... It will ensure greater efficiency from the men sent in."

A mission was hurriedly formed to strengthen the radio network in Mindanao. Secret letter orders were issued for an officer and enlisted men of the 5217th Reconnaissance Battalion - 978th men Sgts. Conrado A. Gustilo, Sgt. Angel T. Sarmiento, and Sgt. Paterno Ortiz with recon man Sgt. Louis Cortez to "points outside Australia to carry out instructions and return at an indefinite date." The men would not be attached to any recognized military unit while on the station. In charge of the men was Lt. Saturnino R. Silva, Infantry. Lt. Montgomery Wheeler, USNR, and Commander Charles Parsons were onboard, but the submarine's log did not list Parsons' name. Instead, a Lt. W. H. Kimbrough. Was this an alias for Parsons? The men were armed with Carbines, 45 cal pistols and trench knives.

Silva had finished training in jungle warfare at Canungra Jungle School and was bringing qualified men from the 5218th Reconnaissance Company to Canungra when he received the order. On an Officer's Qualification Code Sheet dated February 1944, his unit was listed as the 5217th Reconnaissance Battalion (Provisional) with principal duty as Second in Command, 5218th Reconnaissance Company (Provisional).

He left Camp Tabragalba for Darwin without finishing the next course at Fraser Island. His familiarity with Mindanao as a teacher before the war may have influenced his selection. Sharing a tent with Luis Morgan, the former executive officer of Fertig for five months, at Whitney's request, gave him an insight to the island's situation, the military district's operations, and the guerrillas on the island. Learning more about Fertig's character, style of management, treatment of his men, and how to get along with the man was to prove useful in Silva's dealings with Fertig.

USS *Narwhal* – March 1944

Three days before the February 16 departure of the mission, PRS sent a message to G-2 Intelligence, SWPA. "Action now begun to start flow of radio operators into 10th MD with detachment of the 4 operators under a very competent Filipino officer...Future groups will be sent to Fertig to prevent men being pushed around by guerrillas. Greater efficiency from these men sent in."

Fertig received a message from General MacArthur: "Onboard is Lt. Wheeler, a well-qualified naval communicator with excellent records sent to handle naval communications and act as your advisor on cryptography and signal security. Also on board are four Filipino enlisted radio operators in charge of Lt. Silva who is sent to assist you in administering this Filipino signal detachment as augmented from time to time pursuant to your requisitions." Heavy weapons of mortars, anti-tank guns, and machine guns were part of the cargo.

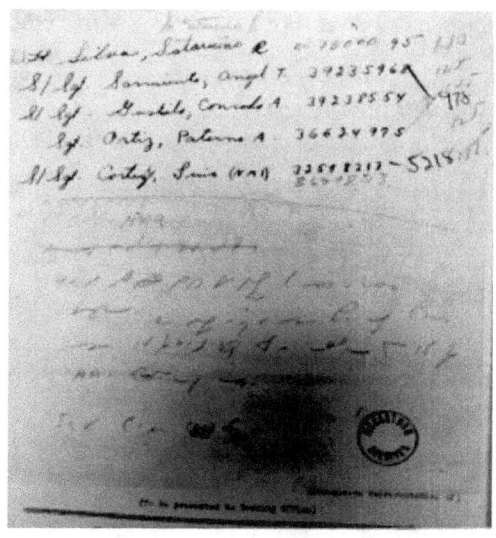

Chick Parsons wrote in Gregg shorthand the departure date of February 1944 for Mindanao under the names of the 5 men (Silva, Sarmiento, Gustilo, Ortiz, Cortez.) The translation was difficult because he had created his own names to replace the real ones to keep the secrecy of the mission.

The *Narwhal* had one day to resupply and refit before departing from Port Darwin. Seventy tons of reinforcement supplies were to be brought to Tawi-Tawi, Mindoro and Mindanao. Twenty tons were to go overland to Cebu from Mindanao. Personnel were to go to Mindanao to strengthen signal communications. Intelligence reports from a mission that landed in Mindanao three months earlier were to be picked up.

Passing by Babar Island, a sailboat was sighted as a light bomber appeared out of a cloud. The sub dove to 500 feet. Scanning through the periscope, the men saw that the plane was gone and may have been a friendly Douglas bomber. An escort ship and several sailboats were sighted on the following day and avoided.

First stop was Tawi-Tawi in the Sulu Archipelago to land supplies. The Japanese had assembled a naval force in the island that could counter attacks from the U.S. in New Guinea and the Marianas. Capt. Suarez received the message to gather boats to unload twenty-five tons of cargo at Mantabuan Island. They waited for four days, as the submarine did not appear. Ortiz onboard reported that for three successive nights, contact failed. No security signals or sailboats were seen onshore. Suarez believed that a Japanese patrol boat might have appeared along the coast, or the sub did not see the white flag they displayed. The guerrillas dispersed when a young officer in their group unintentionally fired, killing a suspected *Moro* spy. Nearby inhabitants heard the shot and hid.

The sub turned north to Mindoro and surfaced at Paluan Bay. No signals were sighted even after five attempts. At this time, Mindoro was reputed to be the most dangerous of all the AIB coastwatcher stations. An ambush on an earlier mission party sent 4 months earlier had shut it down and dispersed the men as they tried to bring intelligence material to a submarine's rendezvous point. There was no movement on the shore. The sub quickly slipped away and proceeded to Mindanao.

Meanwhile, Fertig received another message from General MacArthur: "Latta (Narwhal captain) unable to contact at the first and second spot. Take all supplies (except) courier envelopes and mail for missed spots."

The *Narwhal* entered Butuan Bay under a moonless night and the men were relieved to see security signals and around a hundred people on the beach. When she surfaced near Cabadbaran beach, the locals became alarmed, fearing the engine exhaust noise would bring the Japanese to the area. Filipinos as far as five miles away were awakened. Some rushed to the shore to find out.

No boats came out to the sub, causing suspicion, until a sailboat came alongside with Sam Wilson, Capt. Ernest McClish, and Maj. Clyde Childress, officers of the 10th MD. The Agusan River mouth was too rough to tow the barge into the bay alongside the submarine. They requested permission to bring the sub to the barge at the mouth of the river. The skipper refused as there was no bright moonlight to spot identifiable objects around and the sub had run aground in an earlier delivery. Two of the sub's launches were released to tow the barge beside the sub. Seventy tons of cargo and two 26-ft whale boats were unloaded.

Unloading the cargo took more time because the load for Mindanao was under the shipment for the earlier aborted destinations. The guerrillas went wild at the sight of so much cargo. Everyone was on the alert for the Japanese planes that patrolled the area.

At daybreak, the mission men, Lt. Montgomery Wheeler, Lt. Saturnino R. Silva, Sgt. Louis Cortez and 978th Signalmen Sgt. Conrado J. Gustilo, Sgt. Angel T. Sarmiento and Sgt. Paterno Ortiz disembarked. The 28 evacuees who had been waiting all night on the beach boarded together with Commander Parsons. That meant 28 less to feed at camp as the food supply was becoming a problem. "The evacuees were accommodated in the forward torpedo room that was emptied of torpedoes. The women were in a room with a door for privacy. The sub was immediately infested with cockroaches and lice from the bodies of evacuees who had been living in the jungle for years. Most had tropical ulcers and a unique, intense odor." Meals were served twice a day as there was not enough food for everyone.

At close to midnight, the sub surfaced and headed out on the Mindanao Sea to Tawi-Tawi. Several small sail boats passed. One lowered his sails at the sub's approach. "Five Filipinos stood lined up for inspection and with expressionless faces began solemnly bowing. When they saw white men on the bridge, all of them broke into smiles and cheers. Cigarettes were tossed to them." Mary Maynard, an evacuee, remembered the 'double horn' blaring many times to submerge as enemy ships and aircrafts were sighted swarming the waters of Halmahera. Enemy torpedoes missed the sub.

One time the sub went out of control and plunged to more than three hundred fifty feet during the dive. She creaked from the pressure causing a column in the middle of the ship to bend and paint to chip off before she stopped plunging to the bottom.

Sighting the river gunboat *Karatsu* formerly the captured USS *Luzon,* traveling without an escort, the *Narwhal* fired and hit the bow. Heavy smoke from the explosion shrouded the ship. Breaking noises were heard as the boat sank. An escort appeared, and the sub dove deep, rigging for depth charges. "Lights out, hatch wheels spun, and fuses flew" from the explosions. She shook from the depth charges as the escort ship crossed overhead. It seemed like hours as she waited for the pebble noises precursor to an explosion stopped. In the dark, temperatures rose, and men stripped off their shirts. The conning tower for communication was damaged, said William W. Williams, an evacuee whose plane was shot down. "Passengers said they preferred foxholes."

Meanwhile, the food supplies of Suarez and the radio station on Tawi-Tawi were nearly depleted. The inhabitants of the surrounding islands were anxious to help with food but were afraid of the tight Japanese blockade. The enemy had intensified patrols in the waters regularly passing the islands south of Tawi-Tawi. Pro-Japanese *Datus* (Moro leaders) also patrolled the inlands. A Japanese garrison on Sanga-Sanga Island landed and shelled Batu-Batu, Tawi-Tawi, forcing Suarez, Young with the radio, and civilians to hide in the jungles. Some of the guerrillas with Carbine rifles and cigarettes were captured, thus revealing a sub landing to the Japanese.

The *Narwhal* reached the coast of Tawi-Tawi. Proper security signals ashore were sighted, and she surfaced at Bohi Gangsa. The sub kept running to avoid being slammed into the beach and reefs by the strong current. Hamner, who had led the first mission to Tawi-Tawi a year earlier, came onboard sick with malaria and malnourished, ready to return to Australia. The radio was left to Young. Only two small boats arrived to unload the remaining 20 tons of supply. The constant strafing of Suarez's camp had scared the friendly civilian villages who usually provided sailboats. The Carbine cases were broken open for easy handling. Cargo was loaded into the sailboats and the sub's inflated rubber boats. Three trips were made to shore battling the strong winds and picked up eight evacuees before two enemy destroyers appeared at close to midnight. One turned broadside and fired. People cleared off the sub's top deck, and desperately needed supplies were dumped into the water. The spot was now compromised by the strewn cargo as a future landing location. The engines roared ready as loading hatches spun closed. Six tons were undelivered. Some of the unloaded cargo in the engine room was destroyed when water entered the engine room hatch before it could be closed at crash-dive. She dove deep and ran silent. The destroyer could be heard over the sub, but no depth charges were dropped. Seeing the destroyer moving away at periscope depth, she turned north at full speed away from Tawi-Tawi.

Several men unloading cargo were caught onboard. One was Lt. Jose Valera, who was the liaison officer between the guerrillas and a Python commando team of the Australian Army stationed in Borneo across Tawi-Tawi. Valera would later return to lead a mission to the Philippines. The other was Lt. Konglam Teo, a Philippine Army officer who fought in the Philippines and would also return later to lead a mission. When the hatches opened in Darwin, all gasped to take in the rush of fresh air. The passengers transferred to the Australian Navy tugboat *Chinampa* and were brought to the Lugger Maintenance Station away from the city. The evacuees were told to sign an oath of secrecy so as not to endanger the people left on the islands. But they were reluctant because they had answered the crew's questions and given documents to the skipper. G-2 Intelligence then issued a stern warning to the sub's crews to leave the evacuees alone. The entire sub was disinfected of cockroaches and lice at Fremantle. Mattresses were brought topside to the sunlight for hours.

Back in Butuan Bay, men and supplies in four boats slowly entered the swollen Agusan River. It was the rainy season in this part of Mindanao, and visibility was low. Reaching the city of Butuan, the provincial capital of Agusan, they stayed overnight at the camp of the 110th Division. Part of the cargo was stored at Amparo, a village further up the river. The rest continued to Esperanza, where Fertig was waiting for them. The men probably saw him with a briefcase where he placed maps, codes, and other sensitive intelligence records to grab if he needed to leave quickly. Parsons had brought the briefcase with a hidden switch that would explode into a magnesium fire when opened if not properly activated. Houses were on stilts. They saw men's skin swollen from mosquito bites in the humid jungle. They received their first Mindanao currency of fifty pesos per month as U.S. Army personnel. The guerrillas received forty-five centavos subsistence allowance per day and fifteen pesos cash allowance per month. Twenty tons of cargo for Cebu, Bohol, and Leyte islands were sorted and carried by carriers overland and across waterways. Many of the carriers were *Manobo,* an indigenous group known to carry seventy-five-pound loads all day on their backs without complaint.

After four months at Esperanza, Fertig moved his camp to Talacogon close to the Agusan River and marshland filled with crocodiles. The heavy weapons were pulled out from Amparo and brought to Talacogon. It would have been a daunting task transporting the mortars and machine guns over 200 miles away with no made roads, travelling through swamps, rivers, mountains, and passing the enemy strongholds in Davao city to Sarangani, the first planned landing site when the Americans returned. Furthermore, the guerrillas needed training to use the heavy weapons.

Talacogon was a devastated town from the bombing raids. Most of the people were evacuees from Japanese-occupied areas and had been suffering from lack of food and clothing for more than two years. Seeing the tattered clothes of the guerrillas, they gave some extra pieces of uniform to them. Their quarters for the night was a well-built house formerly belonging to a Chinese trader. The commanding officer of FRS station WAT was a familiar person. Evans was also the chief medical officer of the district. "Everyone of us remembered Doc Evans as the officer who interviewed us and gave us code tests the day after we arrived at Camp Tabragalba."

The men took turns operating the radio and building a camp. Those not on radio duty hacked a clearing with their *bolos* and built huts. Hostile natives who disliked any strangers coming into their river valley watched them from afar. Mosquitoes attacked the men as they diligently reported Japanese troop strength and shipping. The entire base was deep in mud and water when the river overflowed as it was the rainy season.

Radios continued to break down due to the humidity and continuous operation. Buried spare radios were brought out to replace burned sets. An old transmitter sent the messages until the damaged one was repaired so that they could maintain communication with KAZ. Two diesel generators on 24-hour schedule intermittently went down. A replacement was found in an abandoned motor launch at a nearby town.

The new arrivals were immediately given assignments. Naval Intelligence officer Wheeler from the U.S. 7th Fleet was placed in charge of FRS operations and as Chief of Staff. He introduced a new code system to be used by all stations of the 10th MD. An aircraft warning system special code was enforced to speed up transmittal of enemy plane traffic from the watcher station to relay to the Fighter Command Station almost before planes were out of sight of the coastwatcher. Reporting of all Japanese planes and ship movements improved to within ten minutes of sighting, according to Tom Mitsos. Col

Bowler wrote, "on an average of five and a half minutes after sighting." By May 1944, 54 working radio stations on Mindanao were feeding information to FRS.

Silva was assigned to security. In charge of plans and training of the district, he ensured that subordinate commanders and troops knew of their duties and weapons. As the Ordnance officer, he prepared the camp for defense against possible enemy attacks by land or air. There was much reluctance when he required everyone to dig foxholes necessary against aerial attacks.

FRS station was primarily operated by airmen turned radio operators who were caught in Mindanao when war broke out. They were relieved to see the 4 signal men as radio traffic was increasing from the additional radio outposts. Negros and Panay nets were sending flashes to relay to KAZ. At times schedules were not met with receivers running on two frequencies all the time. The Manila networks were sending reports but were concerned when they did not receive replies from KAZ for some time.

Japanese reinforcements poured into Mindanao as news leaked of an Allied return. FRS bristled with the hundreds of intelligence reports that had to be analyzed and coded before sending to KAZ. Men coped with the deluge as messages continued to accumulate without receipt from KAZ. Fertig was told to slow his men down in sending messages. The radiomen began to wonder "why bother risking my life when KAZ Darwin does not care." The promised emergency frequency open 24 hours daily with FRS was not diligently watched by KAZ.

There was difficulty at first learning the guerrilla radio operations, but the men quickly mastered it. "The hours of duty were relatively short at first, but as soon as we could handle the networks, the operators were sent out on assignments outside headquarters. Only two operators in each network were left to handle the heavy traffic from 0530 to 2400 each day," Ortiz recalled.

Gustilo who had finished a radio course in the U.S. before the war, became the chief radio technician of the 10th Military District of Mindanao and Sulu headquarters. Despite few radio parts, he was able to fix the radios. Doc Evans, himself an able radio enthusiast, remarked more than once that Gustilo was the best radio serviceman he had ever seen. Ortiz transmitted Japanese airplane warnings and operated the Navy station ABLE to directly contact the Navy's submarine base in Perth, Australia. Both men operated their frequency stopping only to eat and sleep for barely 8 hours.

Sarmiento and Cortez were assigned to the 'OBOE' station in FRS, the net control station for Mindanao and Visayas stations operating on the frequency for direct contact with KAZ. Although an infantry man, Cortez had radio training.

After several months, Sarmiento was sent north to watch the Surigao Strait. He had trained guerrilla guards at camp on how to handle and fire the Carbines.

A message arrived marked 'Utmost secret for Fertig only.' It was of utmost priority that Claro Laureta and his guerrillas establish radio communication to FRS covering DAPECOL (Davao Penal Colony), Davao City, and establish coastwatcher stations on surrounding enemy installations." Laureta was a former Constabulary Officer who commanded a guerrilla combat regiment operating close to the Penal Colony. After Corregidor surrendered, his men with a large group of civilians moved into the interior and formed the beginnings of a guerrilla organization, the 130th Infantry Regiment. Their activities concentrated on keeping northern Mindanao free of the Japanese. Later the Regiment reported to the 107th Division formed to protect the Davao area. The 'Kill All' policy of Imperial Tokyo had raised the concern for the safety of the two thousand prisoners at the Colony.

Surveillance of Davao Gulf and Sarangani in the south brought in voluminous reports of Japanese activities, strengths, and starving civilians due to enemy confiscation of their crop. The Japanese were observed accumulating and dispersing a considerable quantity of supplies in hillside caves behind Davao city.

The Filipino soldiers who arrived in the last two missions impressed Fertig. In his letter to Whitney, he was convinced that the Filipinos "have excellent training in code work. They were away from the islands long enough that family ties were not as strong as local Filipinos. Can make use of Filipino non-commissioned officers trained by our army as both code men and radio operators. Could use Filipino non-commissioned who have undergone subversive training not to be afraid of explosives. Americans to do work are easily betrayed, so consider this for invasion plans." Whitney replied that the recruitment of the Filipinos instead of sending white Americans was MacArthur's policy. White men would have difficulty adjusting to living conditions and no chance of blending with the local population.

Two weeks after the mission party arrived, a Japanese contingent of one thousand men with air support mounted a surprise attack at Nasipit and Butuan city ports, capturing the town and destroying villages by the riverbanks. The guerrilla company of only 120 men stationed in Butuan slowed down the attack while Japanese aircrafts swooped down on Talacagon dropping bombs and strafing. No casualties, but the first bomb wrecked quarters. "Fortunately, most of our clothing and bedding were in our bags, and only the clothing that was hanging was rendered useless." A second raid by two enemy planes on the same day kept them sweating in their foxholes. Bombs fell dangerously close to the stations and powerhouse. Scared for their lives, most of the people and guerrillas evacuated to the nearby hills. "When more enemy planes returned the following day, the town was almost deserted except for Fertig, the radio operators, cryptographers, and Lt. Silva. Sarmiento begged to fire at the enemy, but orders were not to. Retaliation would bring more planes and possibly land troops to finish up the headquarters."

Headquarters at Talacogon were on highest alert as thousands of enemy troops combed the Agusan area for the radio. They destroyed the 110th Division Headquarters and closed access to Butuan Bay, the only outlet to the sea. They had heard of the stockpile of supplies and heavy weapons in Amparo. Radios operated by the guerrillas were to be destroyed. Using triangulation of several radios from different locations, the Japanese looked for V-beam antennas that the guerrillas may have set up to communicate with KAZ.

"American boys were cracking under strain. Some did not want to do anything except go out (in a submarine). Filipino boys were the answer. They were soldiers, followed orders, and were not given to hysteria," wrote Fertig in a letter to Whitney. Some Americans had been on the island for over two years feeling trapped not knowing what would happen next. He continued "Silva is a good officer and has done wonders with the local politicos. His legal training stands him in good stead. I like him."

Araceli was a young woman whose family lived in the camp because her mother cooked the meals, mostly American canned food and powdered eggs. Sometimes chicken or meat was cooked if found. If nothing else, she boiled *camotes* (sweet potato), or corn kernels. Sometimes there were no meals for the day. Each man received a pinch of salt. Men who arrived or were on their way were fed. Many priests from different orders came and went through the camp and Araceli's mother fed them. She gave injections to a diabetic Jesuit priest until he was evacuated by submarine. A Dutch priest stayed because Talacogon was his area of assignment.

She felt her family was safe at the camp. Araceli recalled that Silva was "not big, just an ordinary person, and had casual conversation with him." At night after their shift, the Filipino soldiers would go to her house and eat the Filipino dishes she cooked. Many stayed overnight, played games, and sang a lot to relax. The radio men wore gray or jungle suit uniforms, boots, and always carried their Carbine rifles with them. She sewed pajamas for Fertig and another officer, Gustilo, who had caught her attention with his extra attentiveness. As the enemy bombings continued, the women feared the Japanese raiding the camp and of being raped, so they left to hide in their farm. She would not see Gustilo again when the camp moved deeper into the Agusan swamps.

An outpost was set on a hill behind Talacogon to warn of impending air attack on FRS while the camp was preparing for a move deeper into the jungles of Agusan. Two men were assigned for plane approaches and to warn by firing five successive pistol shots to give time for cover. The heightened tension produced false warning shots from birds or imagined sounds.

Half of the operators and code men moved deeper to Waloe in the night. As soon as a radio was set up, the rest followed with their radios. It was just in time as six Japanese planes returned to finish off the headquarter area, the air warning post, and the house used as a supply depot. The old water well that Sarmiento used as a foxhole was directly hit. "Only spies who came into the area selling salt, fish, or rice could have known this," thought Ortiz.

With station FRS' deeper inland and closer to Davao, a separate division of men was formed to focus on the station's security. The 107th Division was organized with Maj. Clyde Childress as commanding officer. His command consisted of the 130th Infantry Regiment and two provisional battalions covering the area of Agusan extending to all Davao provinces in the south. Silva was in charge of the outpost and Ortiz operated the radio. They were to warn headquarters of enemy approach and delay their advances to give enough time to move or mount a defense. They took all possible actions to ensure that the flow of food supplies to headquarters continued since all the trails to food and planting areas were blocked by swarming Japanese. The men rotated the night shift but sleeping in the daytime was difficult with the many noises from the jungle. Silva was exhausted in a hammock when an American soldier disrupted his sleep not fully knowing of his night duties. He shot up with eyes blazing and a finger on the trigger of his Carbine. He had to be immediately appeased.

Coastwatchers continued to be sent out to establish radio stations. But there were still forty men at camp to feed and maintain the FRS station. When the Agusan River Valley flooded most of the year, no crop survived. Waloe was a clearing in the jungle surrounded by a swamp. The *Mangahats,* a tribe in the mountains having barely enough for themselves, were unfriendly. With food scarce, a supply officer and a supply route were set up to obtain foodstuff from the Surigao Province to the Agusan area. When couriers arrived with news of 6,000 Japanese landing in Surigao with light tanks, police dogs, pigeons, some artillery, and plenty of ammunition, safer but more strenuous and longer routes were taken to transport the food supplies. Oil had to be found to keep operating the radios.

Although Waloe was in an elevated clearing that was able to transmit the heavy volume of radio traffic, its exposed antennas were easily pinpointed by Japanese triangulation. The 2 powerful 100-watt transmitters established in Loreto and Sagunto near the camp were discovered. The Japanese commander General Jiro Harada in Davao sent six bombers to carry out a bombing raid against the radio stations. KAZ received an urgent message to send planes to raid Davao and divert the enemy planes trying to knock out the radio stations. Throughout the bombing raids, brave couriers continued to evade Japanese patrols through jungles and rivers carrying intelligence on the enemy several days old to transmit to Australia. Sometimes they were on their own unescorted by the guerrillas. They were the unsung heroes as they doubled as scavengers, sentries, and spies. They entered Davao City to bring messages to friendly officials, borrow money, buy clothing, medicine, and salt. Each time they smuggled the items out of the city, they bravely risked their lives passing Japanese inspections.

Ortiz was elated when he was chosen to be sent to the Dutch island of Talaud off southern Mindanao with intelligence money, medical supplies, a radio set, and a USAFFE soldier as cryptographer. He blended in posing as a fisherman. Local traders from Talaud who wanted to return acted as guides. An outpost set up on Talaud would see the shipping lanes approaching the Philippines between Morotai and Mindanao. At only five hundred miles from Mindanao, several airstrips could be constructed over the Talaud's rough coral to launch a counter-invasion to recapture the Philippines. But Ortiz was unable to cross the sea to Talaud as he could not leave from Port Lamon because it was heavily occupied by the Japanese. Radios of guerrillas were off the air. The sailboat prepared for his trip was destroyed and three of the local traders who were to be his guides were killed. Fertig requested the Navy to have a submarine pick him up and drop off at Talaud, but the request was denied. Ortiz returned overland to headquarters.

FRS crackled with reports of Japanese troops increasing to five divisions on Luzon implying the importance of the island in their defense plans. Young Japanese and Formosans were seen training intensely.

Davao Gulf was raided. The radiomen dreaded the reports from the operatives in Davao city that they had come to know. The POWs at DAPECOL could be massacred any time. The enemy closed in on the station monitoring Davao Gulf. Scrambling to escape, two men carried the heavy engine with a generator to operate the radio on a pole as the station hut burned down. One of the American operators had a mental breakdown and was sent back to headquarters. Later, he shot himself.

News from Australia announced the beginning of the liberation of Okinawa from the enemy. The men hoped that the Philippines would be next. Mindanao had still to be liberated.

The radio traffic between coastwatchers, Australia, and the U.S. Navy doubled. FRS alone flashed a total of 214 ship sightings to reach Navy Intelligence at Perth in Australia in the month of June.

Guam was to be invaded in July, but the Japanese naval ships were nowhere to be found. Messages were decoded from coastwatchers in the islands that saw the Japanese task force passing through towards the Pacific Ocean. Pleased with their handling of assigned radiofrequency throughout the Battle of the Philippine Sea, Gustilo, Cortez, Ortiz, and Sarmiento were commissioned to command others.

As Tokyo continued to strengthen its forces in the islands, more operatives infiltrated enemy positions and radio operators ceaselessly coded and transmitted their activities to KAZ. Ten stations expanded to forty-three, and the messages increased sevenfold with the same number of radio operators operating FRS. Messages piled up at FRS. The radio operators in Australia handling the Mindanao traffic were replaced by new personnel who at first had difficulty receiving the guerrilla sending rate of 30 coded groups per minute. The average coded message was about 30 words and some were as long as 200 words. One mistake in the code word and the message could not be decoded. Mindanao complained that KAZ operators did not meet schedules and delayed in completing passing traffic. The Signal office in Brisbane checked KAZ operators and found them equal to army standards. FRS operators often "sacrifice accuracy for speed so that actual morse characters are in many cases not distinguishable one from the other. Extreme impatience by the operators shown to GHQ is destructive." Corrective action was to be taken. But with the new operators able to receive only the rate of ten coded groups per minute as per the Army course in field station operation, radio operators of FRS had to stay on the air three times longer, wasting precious fuel and giving the Japanese more time to discover their position.

GHQ's comments in their internal check sheet recognized that radio traffic in the 10th MD had increased seven-fold the past year and was operating with the same number of operators who "were probably impatient and tired, may even need relief, though none have apparently asked for it." More radios sent would add to the number of messages being sent out. As the transmission problem festered with FRS, the chief signal officer in Australia reviewed the training program of the radiomen at the KAZ installation in Darwin. GHQ immediately added equipment and had KAZ operators undergo intensive training to the competent level of handling any KAZ circuit within 30 days. Radio operator, Lt. Palon brought KAZ operations to the expected high standards. At this time, Lt. Emilio Quinto, the lone signalman with the first submarine mission (January 1943) to Negros had recovered his health from malaria. The operation of KAZ in Darwin was turned over to him. He was the best person to understand what the Filipino operators and guerrillas were undergoing as he was on the first mission to the islands. FRS radio operators were relieved when new and efficient operators took over KAZ. Eighteen messages were passed in less than two hours.

Aside from the regular list of weapons, basic supplies, and medicine, Australia continued to send liberal food supplies that included special requests such as butter, cheese, and spice for Mindanao. The finest equipment procurable was provided to Mindanao. A Signal Corps inspector ensured that all supplies were tested and checked prior to departure. Supplies were specially packed for the damage that would be caused by handling from Brisbane to Darwin, loading into the submarine, and carrying over jungle trails.

Whitney began to be concerned that the "too liberal a supply policy" would "destroy the power of improvisation in which necessity has created the real strength of the guerrilla movement." His concern was not in the condition of the equipment but in the loss of local effectiveness and resourcefulness of the guerrillas. He noted that equipment, weapons, ammo, and medicine, delivered to Negros, Panay, Leyte, and Samar were for immediate military operation upon receipt with no apparent difficulty. It was puzzling that Mindanao had no request for small arms when only less than half of the forces on duty were equipped.

The "high priority" requisitions of Mindanao now included khaki or denim cloth, spices, baking powder, yeast, flour, butter, cheese, prunes, Nescafe, and prepared rations. American evacuees disclosed that a "satisfactory butter had been developed from coconut oil, yeast had been sourced from coconut powder and flour from corn and coconut meal. In Bukidnon some of the finest coffee was grown rendering entirely unnecessary the shipment of Nescafe." Cloth could not be provided for all personnel's uniform. For a few, that would invite the "charge of racial discrimination from the Filipinos and that the 10th MD was favored because it was commanded by an American." Even President Quezon was reported to have demanded that a Filipino replace Fertig, but GHQ found no suitable candidate in Mindanao. Fertig's request to break down supplies into 7 ton lots and delivered by submarine along the Mindanao coast directly to the individual Division headquarters under his command was refused. It was an inefficient use of the submarines' time and diverted it from supplying other islands. PRS' decided to control the items to only practical, badly needed supplies and propaganda items to maintain morale.

USS *Narwhal* – June 1944

The main Philippine weather station was established in Waloe. Men trained in weather forecasting were needed to establish weather stations that would cover all the way north to central Luzon. Reports on the ascension and position of weather balloons and wind readings enhanced the accuracy of aerial bombings. Forecasting weather changes was essential to bombers taking off and landing safely. Without heavy cloud coverings and typhoons, target areas were pinpointed, and planes avoided heavily defended areas.

Weathermen Lucien Campeau, George Finnegan, Charles McGrath and Ray Lozano from the 5th Air Force received orders for a mission. Party members were reconnaissance men Sgt. Jose T. Sitjar, T4 Marcelo G. Gonzales, and T5 Serafin B. Buenavista. The 978th Signal men were Sgt. Salvador B. Fortun, Sgt. Donato E. Herrera, Cpl. Ambrosio Navarro, Sgt. Seminiano D. Acenas, T5 Edward M. Burgos, Pvt. Isidro P. Magale, and T5 Daniel S. Palmeto. Capt. Harold Rosenquist was assigned on a special mission. Before departure, Fortun was ordered to tell of the Mindanao coastlines, terrain, and military value to G-2 Intelligence since he was from the island.

GHQ sent a message. "Onboard next vessel will be ten Filipino radio operators to reinforce Silva detachment (March 1944) and (the radio network in Samar). An officer and six enlisted men, all Americans trained in weather observation with complete equipment (will) establish a weather observatory near your headquarters. Weather observations and forecasts will be made, and all other weather information will be correlated and evaluated for dispatch."

As Silva was the highest-ranking Filipino officer from Australia at the time, Fertig looked to him regarding the Filipinos who arrived. They came from the same Camp Tabragalba. He and Gustilo gave their opinion about the new men's radio abilities. Bowler was wary of the Filipino radio operators and thought the "new American- Filipinos (as) an added burden.' But in the coming months, the men's accomplishments changed his outlook.

From Samar, the *Narwhal* turned south to Mindanao. She picked up some debris and lifebelts from the wreckage of the ship she torpedoed earlier. She did not return to nearby Butuan Bay where she had earlier arrived several times, as the town and every village for 30 km up the Agusan River had been destroyed by the Japanese. Sanco Point on the east coast was the chosen site as it was closest to the headquarters at Waloe. The entire east coast was occupied by the Japanese except for Sanco Point on a peninsula that jutted out. At midnight of the designated date, Ortiz who was on the shore while waiting for orders on the Talaud mission, saw the boats of American evacuees go out to sea to meet the sub. Blinkers were focused on the coordinates and ablaze by midnight. For two hours, the sub did not surface. A repeat the following night also failed. When the sub did arrive, there were neither party nor boats to meet them. The *Narwhal's* earlier attack on a Japanese merchant ship delayed her arrival at the rendezvous point. The sub cruised south to Bislig Bay, but no signals were seen. After two attempts to land, she turned south around the island to Pagadian Bay.

The sub arrived early in Pagadian bay and surfaced at night at the coastal village of Tukuran. Twenty dugout boats with volunteers came along side to unload the 15 tons of cargo. They looked sickly to the men but still they were happy to see them. The unloading was completed by midnight and the party followed wondering if there would be a friendly welcome at the unplanned landing site. Col Bowler was onshore with other American guerrillas to meet them. Hundreds of civilians with carabaos and carts unloaded the cargo onto the beach. To avoid leaks about the landing, the civilians who helped were told that they were just bringing rice to the interior. Their only payment was a few meals of rice and old carabao meat. Palmeto had not been back for a decade. He and the other Filipinos fell to their knees and kissed the ground. The *Narwhal* turned south to return to Darwin.

As the sun rose, the scattered cargo had to be removed from the beach before Japanese patrols arrived. A large portion of the cargo were propaganda items that had General MacArthur's picture and the word 'I shall return,' calendars, 4-piece gum packs, and small Hershey chocolates, waxed to prevent melting.

With a guide, Rosenquist and the weathermen travelled ahead to Bowler's headquarters. Guerrilla Lt. Jose Todtod guided the rest with the cargo on a train of carabao sleds. It was a four-day mountainous hike under heavy rain, mud, and leeches. Herrera got sick and stayed behind for a couple days. Guerrillas holding WWI Enfield rifles guided the Filipino men to a village. Tired and starving, they feasted on rice and carabao meat prepared by the villagers. The people's morale especially the guerrillas were boosted at the sight of their presence. They read *The Free Philippines* and heard of the '*bolo* battalion' of the 1st Filipino Infantry. People from nearby barrios came smiling to see them despite their sick conditions. Fortun related that they tried to share what they could. They gave cigarettes and chewing gum to the officers and guerrillas. Palmeto could not help noticing the flicker of hope in the faces of the people as "we reassured them that we came to help liberate the land of our birth." Passing a secret airfield in Zamboanga, they saw people that were "walking dead, undernourished and in rags." Palmeto wondered what power kept them on their feet as they were all suffering from malaria. Food became scarce and there was still several days of hiking to reach headquarters.

Bowler's camp was a little community in a clearing by a creek with a dentist, chapel, and small huts for offices. Herrera, Rosenquist and the weathermen were at the camp when the convoy of cargo arrived. Among the cargo were two powerful Navy radio sets, station Z4K for Bowler's camp and the other station Z4V for Fertig's headquarters in Agusan. Flashes of enemy shipping would no longer have to be sent to KAZ and the Navy Command Center but directly to the submarines closest to the target plying the waters around the archipelago. A power plant fueled by a diesel engine was carefully maintained to keep the radio on the air.

Ranks were important to issuing orders so patches showing rank were stripped off and the men given higher ranks to lead guerrillas in an assignment.

Planning with Bowler assigned some of the signal men as cryptographers under his command. The men discovered that the local handling of radio traffic, a commercial radio communication system with Q-signals, was different from their Army procedure training. A Q-signal message meant "I am on the move due to enemy raid." After some criticism, they quickly learned the new system to operate the radios.

Campeau, Hoke and Lozano travelled to Dimorok canyon to set up a weather outpost for the Western Mindanao sub-net. Campeau was in charge. He commandeered a horse from a farmer when Lozano's horse weakened on the way. Hoke fell and sprained his wrist. A Chinese merchant in the town of Tambulig who supported the guerrillas fed them well. The last 30 kms trailed up the mountains to a deep canyon with a river at the bottom. Almost waist deep in mud, the short Filipinos maneuvered the carabao sleds. Civilians built shacks for them on stilts and a house with a covered porch for the weather radio and instruments. A pot of boiled monkey was their first meal. Weather and aircraft warnings from Mindanao and 3 Panay stations were tracked and relayed to Australia.

Herrera was placed in charge of the weather and air warning station for western Mindanao. The inter-island network station 3UB received and consolidated weather, ship, and air warnings, into one message, and recoded on a manual code machine before sending them to KAZ for relay to Perth Amboy (submarine base in Australia), Morotai, Halmahera (Air Force), and Sansapor, New Guinea (Naval base). As two-man weather teams arrived with the submarines, they set up weather stations all over the islands and sent weather observations every six hours to Dimorok. When codes ran out in the inter-island network, the cryptographers made up their own from random number selection surprising KAZ when some of the messages were intercepted. Campeau was reported holding on to a briefcase that contained a "gridded map with all eighty-five spotter locations, radio frequencies, list of personnel and a magnesium cartridge which would ignite by pushing hard on a button destroying the contents and the map case."

Signalmen Sgt. Acenas and T5 Palmeto. were assigned as cryptographers to station SL9 in the Western Mindanao sub-net. Station WAN was set up to receive from guerrilla stations in Luzon and the Bicol Peninsula. The area was thick with malaria and snakes as Acenas kept on with his duties encouraging himself that it was for his country's liberation. Acenas and Magale spoke the language and acted as interpreters for Campeau. Magale operated station KAR for three months before moving to Dimorok.

At Dimorok, food had to be procured from miles away. The people had ceased planting since the Japanese would come every harvest time and take away their crops. There were days without meals. Sometimes there was boiled corn and salt, if any. Japanese planes passed overhead morning and afternoon, increasing anxiety. A rear command post in the jungles of Dampalan, Zamboanga, was established. The day came when the guerrillas intercepted a Japanese patrol a few kilometers away. While the radio and weather equipment were being evacuated, Signalmen Palmeto, Herrera, and others continued to encode and decode messages. The peaceful and non-Christian *Subanons,* local inhabitants in the area's mountains, helped to move the heavier equipment like the diesel engine parts and the Navy radio sets through steep and slippery canyons. Both hands and feet had to be used to cross the canyons. Herrera remembered to "pay my reverence to the tough, rugged and weather-beaten instructors we had at Canungra Jungle School in Australia." He was using all he learned trekking through the jungle trails. At the new camp, Palmeto began setting up shacks to house the men and equipment so weather messages could continue to be sent to Australia. An aircraft warning station was set up three kilometers away. The diesel engine worked intermittently and finally died, so couriers ran to bring reports to A Corps headquarters.

After several months, Dimorok was clear of the enemy and the men moved back. Palmeto came down with malaria but continued coding, weak as he was until he was transported to a hospital in Pagadian where he rested for two weeks. Eventually, five stations were on the air, including air warning and weather for Western Mindanao and the inter-island network.

Navarro and Gonzales were at Dimorok for two months before travelling for 5 days to survey the landing airstrip in Dumingag and set up station 8QV to report on the enemy and weather. The landing strip saved many airmen who were forced to land later. After a week, they returned to Dimorok.

Burgos carried his radio to the landing strip at Dipolog. The guerrillas had seized the airstrip and held it while surrounded by Japanese. Food rations and supplies were landed to everyone's joy. In the bombings of Zamboanga, Marine pilots flew out of the airstrip. Weatherman Lozano joined a spotter station on the mountains overlooking Zamboanga City.

Operating the radios was labor-intensive with the coding and decoding at each station but well worth it when the intelligence report reached KAZ. Again, Fertig requested more signalmen. This time he did not insist on Americans. Whitney, who had defended the Filipino radio men, gladly fulfilled the request.

USS *Nautilus* – June 1944

The sister ship of the *Narwhal*, the *Nautilus* was on its first mission to the Philippines. It was a security tactic. PRS believed that the Japanese knew only of the *Narwhal* and would not expect another submarine entry at Tukuran. Fourteen torpedoes were removed to load a hundred tons of supplies. Over a thousand gallons of high-octane gasoline and aviation lubricating oil were loaded in the deck torpedo storage tubes.

Onboard was Lt. J. D. Simmons, a naval officer. He replaced Campeau at Dimorok as head of the weather observation team. Eventually, forty-one weather stations in eighty-five coastwatcher stations from the northern tip of Luzon to Borneo reported weather every six hours.

On the way to Mindanao, the sub dove deep while an enemy patrol craft continuously fired on her. At sea, she sighted a small Japanese schooner and fired direct hits with her deck guns. Arriving off the beach of Tukuran, the crew sighted security signals onshore. Although the scheduled arrival of a submarine was known to give time for preparation, there were times when the number of civilians needed to carry the boxes of supplies to designated areas was not enough. Bowler made sure that around one thousand men were on the beach to prepare unloading the cargo. But they were given instructions on the landing day itself to avoid discovery by the nearby Japanese so as not to endanger the submarine and landing operations. The sub was contacted from shore as she neared the coast. Rafts arrived, and the cargo unloaded before daylight. The captain's log reported disorganization in the unloading of the cargo and stated that one hundred tons was too much to unload in one night. Faster unloading would lessen the danger to the sub.

Meanwhile, Rosenquist refused to reveal his mission to Bowler. At GHQ in Brisbane, he had been working on a plan with Stephen Mellnik, one of ten escapees from DAPECOL, to free the prisoners. He was to contact guerrilla leader Claro Laureta who had helped the escape a year earlier. Disagreements in GHQ had delayed approval of the trip. A concern was that the plan would subordinate Fertig to Rosenquist's orders thus affecting the guerrilla organization. The missed rendezvous and the alternate landing site seriously affected his time-sensitive mission. The party was to be dropped off with radio and weather equipment on the east coast of Mindanao closest to Fertig's 10th MD headquarters. Instead, they were now across the island and would take months to reach Waloe.

Signalman Fortun was requested by Rosenquist to join his mission to free the American prisoners. He received the order to proceed at once on the mercy mission. "He believed he was chosen for his understanding of the language and conduct throughout the trip. Earlier, his strong personality had brought him to Fertig for clarification of his duties. He had tried to tell Rosenquist how to get along with the people but was instead reprimanded. Rosenquist had strongly impressed upon everyone the urgency of his mission from GHQ, G-2 Intelligence, so his transportation had to be prioritized. He made it clear that he was above all ranks, and when carriers were not available, guerrillas were to shoulder the supplies, even at the risk of their lives. He asked for the best food and bed. Men assigned to his group began to grumble, even threatened to desert, and one of them almost committed suicide.

A detachment of the 108th Division located in the Misamis area escorted the group as they sailed north along the west coast to Manticao dodging Japanese patrols. Fortun sensed that the locals and guerrillas providing carriers for their equipment into Japanese occupied areas were uneasy. Continuing to Libertad and inland to Lourdes, he found the town full of spies, which meant it was not a good idea to stay long. The Japanese were only four kilometers away and knew of their presence. They got lost in the mountains dodging Japanese patrols until they reached the headquarters of James Grinstead, commander of the 109th Division in the Bukidnon area. It was the fourth of July, and a little celebration was held amid all the exhaustion.

The Japanese blocked the roads forcing them to head for the mountains. From there, they saw the attack on Talakag in Bukidnon, headquarters of the 109th Division. It was the food granary of the guerrillas. Japanese planes bombed and strafed while troops destroyed and burned what they could not carry. At times, escapees from the Bataan and Corregidor who did not surrender joined them in the mountains, telling the same stories of maltreatment. Civilians fled to the hills and refused to return, including a Bukidnon *datu* with four hundred fifty people refusing to harvest their own corn crop. Living on roots and leaves for a week in the mountains made them lose weight. Snakes and malaria infested the area. The men met the *datu* hoping that his people would harvest their crops if they told him that they had arrived by submarine from MacArthur's headquarters. They showed 1944 issues of American magazines. Having not seen Americans for almost three years, the people were overjoyed to see the group. "It will not be long before the Americans come." They returned to harvest their crop.

A guerrilla officer with a platoon armed with six rifles and five bullets each served as a guide. Fortun recalled, "They were not afraid, so we didn't worry." Hiking at night, they took a secret trail and crossed the Sayre Highway, the main road to Davao and a well-guarded Japanese line of communication. If they crossed the highway, he felt they would be safe. Exhausted, they slept in the first house they came to only to find out in the morning that it was an outpost for the Filipino Constabulary working for the Japanese. They quickly moved on to a mining camp belonging to a guide and rested for two days. "The rest of the trip across the island was full of mishaps, rugged mountains, leeches, hunger, wild animals, and the *Mangahats*," local people of the mountains who had hundreds of rifles taken from retreating USAFFE soldiers during the summer of 1942. The advance camp of the guerrillas was attacked by the fierce people killing civilians and wounding several before they could fire back. Climbing down from a high mountain, they followed the Agusan River and met Ortiz returning to Waloe. Fortun lost all his belongings except his pants crossing the strong river flow that was known to have swallowed many.

After seven weeks of around three hundred kilometers through uncharted territory, the group arrived at the camp in Waloe. The headquarters had moved three times to avoid FRS from being located and bombed. A request for Japanese currency and U.S. dollars to expand the central Luzon and Manila intelligence brought fifty thousand dollars. But anyone caught with U.S. dollars would be tortured and killed, and that would compromise the stations. One million five hundred in Japanese currency was for intensified operations in Davao.

A high priority directive from GHQ to the 10th MD was to summarize periodic reports and transmit daily: airplane and ship sightings, enemy disposition and activities in Davao, roads and bridges connecting the city, beaches and towns, locations of airfields controlled by the enemy or not, availability of food in various districts and cached food supplies, recommended landing sites, possible bombing targets, and information on Sarangani, the first landing site at the time. The details were to be sent by mail with the submarine landings.

Fortun met "my old friends Gustilo, Sarmiento, Cortez, and Lt. Silva, the latter, commanding officer of Headquarters Co." They were operating FRS' radios with stranded Americans and escaped prisoners.

Rosenquist informed Fertig, who was not in favor of the rescue operations, fearing Japanese retaliation on the guerrillas and civilians, of his mission. The plan was to rescue and safely farm out the prisoners to various guerrilla and civilian families in the mountains while waiting for submarines to evacuate them. It was not practical since the cargo subs could each evacuate only 100 at a time taking years and preventing them from bringing supplies to other islands.

On the group's way to Laureta's headquarters, news of a destroyed advance outpost with the death of the American leader and capture of his men, steered them to take another trail. It had been three months of continuous traveling and Fortun was suffering from exhaustion. He treated McGarth one of the weathermen who accidentally shot himself in the knee and sent him back to camp.

Laureta's new camp was in the village of Florida driven by Japanese attacks on his old headquarters. He ruled with a firm hand, commanding 1100 troops of the 130th Infantry Regiment armed with 600 weapons consisting of WWI rifles, spears, and *bolos*. Rifles were badly needed as the front-line companies averaged one rifle, usually the WWI Enfield, to every two combat soldiers with fifteen rounds of ammo per gun. Those not on the active front averaged three weapons per company and five rounds of ammo per gun. Laureta had the heavy weapons of machine guns, mortars, and a cannon available to him but his men needed instructions to operate them. He had one radio.

Laureta was confident with his familiarity of the colony layout. He was ready to attack, having four times guerrillas as the number of Japanese guards inside, and had control of the thick jungles between his camp and the penal colony. All he needed were rifles and ammunition. But it was too late. A mid-June message to KAZ confirmed no signs of prisoners. They were evacuated ten days earlier, relayed Rosenquist with much frustration and mixed emotions.

Agents reported twelve hundred prisoners from the Colony were seen roped and masked, and were boarded onto the *Shinyo Maru,* a Japanese cargo steamer. Filipino military prisoners, the invalid, and the sick were left to die. When the liberation forces with men of the 130th Infantry Regiment reached the colony, the skeletons of those left behind were found with rats scurrying around them. Unmarked, the ship was hit by the USS *Paddle.* Some of the prisoners onboard survived, were rescued, and nursed back to health. Some refused to evacuate to Australia and stayed to join the guerrillas. At Siare on the northern coast, Burgos operated a mobile station with Sgt. Joseph P. Coe, a pilot shot down and one of 84 survivors of the 750 prisoners.

Intelligence reported six hundred POWS remained at Licanan airfield and one hundred fifty at Matina airfield in the south. A rescue was imperative with the orders from Tokyo to kill all prisoners. Letters were immediately dispatched to the operatives at Licanan and Matina to deliver to the prisoners but never reached them. They were also moved out.

With his primary mission aborted, Rosenquist went out to the field to be the liaison between the 10th MD headquarters and the 130th Infantry Regiment which was monitoring the Japanese movement towards the border between northern Davao and Agusan. If the enemy breached the border, FRS headquarters would need to move again. Coordination of intelligence with the operatives was needed.

An intelligence net was established in Davao City with headquarters in Kapalong, an outpost of Laureta's men north of the City. A printed form to guide agents to get more accurate information was distributed. Operatives worked inside the Japanese- occupied city area. Watchers monitored the important Sayre highway and National Highway going north. The net gathered all intelligence from sources in the city and relayed them to FRS. The agents' difficulty moving in and out of the heavily garrisoned city delayed obtaining information. Ten bicycles were purchased with printed Japanese money to speed up delivery. Intelligence operatives doubled as they moved closer to targets to get more information. Chilling news spread throughout the network of two dozen operatives caught and killed.

Fortun operated a network radio station in Magugpo (Tagum), receiving intelligence and weather reports in the Davao area for the aircraft bombers. His work with "the radio and knowledge of the Visayan (language)" impressed Rosenquist. He thought of him as the "best American Filipino he has come in contact." Fortun proudly recalled that "in my spare time, I managed to build a secret headquarter for the 10th MD in case of emergency, and for this, I was later commissioned as an officer." It may have been a location for a backup headquarter to retreat to. When the American Task Forces landed in Mindanao, he took over the net control station of the Davao city area and continued to report on the fighting in the city.

In his letter to Fertig, Rosenquist emphasised the good intelligence work the men were doing. He requested a teleradio, two new batteries, acid, and a hydrometer. The reply was to send carriers with food to headquarters in Waloe to get the equipment. But the cargo never reached Rosenquist's camp. The *Mangahats* had waylaid the carriers.

Silva's opinions continued to be sought. Rosenquist was convinced by Fertig to keep Watson, an American officer, who did not get along with him. "Reyes (Fertig's aide) and Silva seem to have a soft spot for this officer, so maybe there is a lot of good left in him which we can bring out," Fertig wrote.

The headquarters at Waloe, set up to protect FRS, housed radio and support personnel so close that there were bound to be disagreements. Although the Filipino soldiers were born in the Philippines, their outlook had changed with over a decade of living in the U.S. Filemon Lagman, the adjutant of the 10th MD, asked permission to shoot one of the Filipino Americans. They dared to call him a native when his skin was lighter than the Filipino Americans. A stern message to KAZ suggested an indoctrination course for the Filipino men before sending them to the islands. "They are discourteous, even insolent to Filipinos, both civilians, and USFIP (United States Forces in the Philippines) guerrillas. Due to this attitude, they cannot obtain the cooperation necessary to function efficiently. Local Filipinos will not tolerate the superior attitude shown by them toward 'natives' (their expression)." PRS' internal comment was "surprise that Fertig had not indoctrinated them. At one time they took everything away from our men on arrival and put them in a stockade in effort to reduce them to the mental and physical level of the guerrilla. No action indicated."

By mid-1944, the 10th MD reported the present strength of 1850 officers and 24,500 enlisted men to combat the Japanese. In addition, three Regiments of the Maranao Militia Force (MMF) staffed by *Moros* operated in the area of the 108th Division in Lanao. In the Cotabato area of the 106th Division, two *Moro* battalions ambushed Japanese patrols using *bolo* knives.

In Tawi-Tawi, the Japanese continued their heavy air raids and sea patrols scaring the civilians away from cooperating with the guerrillas. Civilians evacuated, leaving their crops unharvested. Young reported a continuous siege of bombing and strafing of the guerrilla camps by Japanese seaplanes along the shore. Japanese warships and barges shot shells. The Japanese tried to land at Batu-Batu and Tarawakan close to Suarez' command post but were repulsed by the guerrillas led by Lt. Alejandro Trespeces, armed with Carbines from the submarine shipments. They dubbed the guns "Anak machine gun" (son of a machine gun) or little machine gun. But 2,000 Japanese were able to land and garrisoned Tarawakan. Suarez reported to Young that the Japanese commanding officer was Col. Takume, who was in Bataan.

The intense pressure from the marauding Japanese made holding their position difficult. Thousands of Japanese had occupied the Sulu Archipelago, and many of Suarez's men were captured, scattered, or surrendered. Young continued to move the radio as the Japanese neared his camp. The guerrillas provided by Suarez as guards stayed with him. KAZ was having difficulty decoding Young's messages and wanted to send another mission with signalmen as reinforcement. He was ordered to silence his radio and to report only large enemy sea or air movements. The Japanese dropped leaflets captioned 'ISLAM,' calling all faithful followers of Mohammed to cooperate with the Japanese as this war was being fought on their behalf.

Civilians were killed, and houses burned for supplying food to the guerrillas. Due to lack of food, Suarez broke down his men into smaller units to continue harassing the enemy. However, they had to look for their own food. Many were in rags, and some died of starvation or malaria. They had no money to buy food, so they ate what was edible or fished for food. All they could hope for was aid from Australia or the district headquarters in Mindanao. Suarez had to keep his men's morale high as the Japanese captured the officers and the soldiers' wives to force their husbands to surrender. He brought his wife with him wherever possible. She could shoot a gun and cook for the men as well.

Suarez wanted recognition of his activities. Some reports relayed to KAZ were not marked as coming from him. To ensure recognition, he requested a separate radio station from the one operated by Young. He requested some of the funds held by Young for his unit's use in gathering intelligence information, obtaining food, transportation of supplies, guarding the radio station, and others. Some civilians and undercover guerrillas provided food and worked for free, but others wanted to be paid. If funds could not be released by Young, then he asked that funds be sent directly to him for that purpose.

At Waloe, the FRS radio operators received news of the plan for the upcoming Leyte invasion instead of the original Sarangani when they received messages with 'Cyclone,' the code name for Leyte. Eight radio men manned FRS in the months leading to the invasion handling 35 messages per day per man with the average message about 30 words and some as long as 200 words. For two years, in the daily ten-hour day, there had been no room for errors according to Tom Mitsos, a 19th Bombardment Group member turned radio operator for FRS. As the Japanese offensive intensified, headquarters were regularly bombed and strafed. One of the large radio transmitters was hit and took a week to repair. A Japanese contingent clashed with the guerrilla guards and broke through the camp's southern perimeter in the Davao area.

USS *Seawolf* – August 1944

Pfc. Edwin Goroza had not left Camp Tabragalba for several months when a rare pass allowed him to tour Brisbane and get to know the locals. He had had a "hard time understanding their variant of the English language."

Orders arrived for 12 Filipino men. They boarded a plane to Darwin close to midnight, camped overnight and packed their equipment: communication gear for radios, batteries, semaphore flags, blinkers, flashlights, clothing, shoes, watches, binoculars, medicine, food, weapons, and ammo while waiting for the submarine.

The *Seawolf* was an obsolete operational sub that had been converted into a cargo sub. She landed six Filipinos led by Lt. Konglam Teo: Sgt. Marciano R. Daelto, Sgt. Balbino R. Padilla, Pfc. Alejandro C. Aquino, Pfc. Feliciano S. Rodriguez, and Pfc. Edward E. Goroza with ten tons of supplies at Tongehatan Point in the northern tip of Tawi-Tawi. Teo, a reserve Officer of the Philippine Army, was an escapee who organized guerrillas in Siasi Island, Sulu before he was brought to Australia. The signal men reinforced Young's net control station at Languyan that supported Suarez's unit. Once the cargo was unloaded, the submarine continued to Dumaran Island to drop off the other Filipino radio operators under Lt. Eustaquio B. Cabais.

Goroza later recounted to his son. "We surfaced in Languyan in Northern Tawi-Tawi at night. Security signals were spotted and exchanged. Several canoes arrived with Capt. Suarez and our local leader and AIB operative Capt. Young, an American mestizo. It took seven boat trips to offload nine tons of cargo safely. We were happy to meet our comrades-in-arms. The guerrillas shed tears of joy. They appeared haggard, malnourished, and almost naked, only wearing pants made from sacks. After eating sandwiches prepared by the submarine's cook, we exchanged stories with them until we retired. Everyone went to seek shelter among the trees and shrubs found nearby. By dawn, we started our journey to a secret camp and further rested. Sgts. Padilla and Rodrigues (were) both radio operators. Our weather forecaster, Sgt. Daelto, (was) to man our secret headquarters under Young and communicate with GHQ in Australia and other branches of command."

Teo was in charge of the sub-station Aquino established at Baliungan in Tawi-Tawi and Goroza at Bas-Bas. Twenty-five guerrillas accompanied them as security and to monitor activities of enemy ships, aircraft and troop movements, especially those from Borneo. KAZ commended the newly arrived radio men for making their messages easily understood. To procure food, they had to determine who was pro-Japanese and could help. Chief Panglima, of the island, helped in finding food.

When Aquino's station was in danger, the group moved to Goroza's station and continued sending messages to the net control station of Young. In September, spies discovered the net control station and 5 enemy aircrafts bombers bombed and strafed for an hour. In the lull, Young ordered the men to move out. Men and equipment reached Mt Sibankat (Dromedario), the highest peak in the island. A station was set up overlooking Sibutu and Alice channels fronting Borneo. For the next three months until the end of 1944, sightings of enemy ships and air warnings were flashed to KAZ. When the outpost observer saw a Japanese patrol approaching, they evacuated the radio equipment to Buan, a small island off the west coast of Tawi-Tawi.

Suarez reported to KAZ that the funds he requested months ago had not arrived with the submarine. The reply was that fifty thousand pesos with a large load of supplies was sent earlier with the *Narwhal* in March, but some of the supplies had been left on deck when the submarine had to crash-dive to avoid two destroyer escorts.

At this time, GHQ abruptly changed the amphibious assault to Leyte and moved the date earlier by two months, setting it in October and bypassing Mindanao. Radio networks jammed with the jump in radio traffic concerning the change. The new landing with the remobilization of hundreds of thousands of men, ships, equipment, and related departments became an almost impossible logistical nightmare to meet the compressed schedule.

Meanwhile in Davao, operatives reported mines, shore-based torpedoes, and anti-aircraft batteries installed by the Japanese. The enemy continued to intercept messages knowing of the brisk communication from the islands with KAZ. They knew of the many weather stations that allowed accurate bombings of their installations.

As the enemy penetrated their area, the 110th Division guerrillas that were a security force to FRS scattered. Many of the guerrillas and their families were looking for food or sick in the malaria epidemic that swept the area.

Forty men at the camp had to be fed to maintain the radio. A total of 100 men rotated scouring the jungles for anything edible. There were no crops or animals. The only means of transportation to get food from the town of Butuan was by small dugout canoes which could only carry a small load. All trails were closed by pursuing Japanese. Men with starving stomachs hung on to keep the radio on the air. Radio traffic was coming in faster and heavier with the pending landing at Leyte, and there was fuel enough for only eighty days and food for ten days. It was the rainy season, and the whole camp was deep in mud and water.

Hopelessness had swept through the camp when over 30 American bombers from New Guinea began bombings over Davao and airfields north of the city. This was in August, insisted one resident in contrast to the September date found in books. The planes flew directly over the camp. As they recognized the planes, the men went wild with joy. Carrier planes hit Davao and the areas of Surigao, Cagayan, and Bukidnon. It was the first bombing in years by American planes and a signal of their return. In Manila, the Japanese Imperial command demanded the Philippines declare war on the U.S.

Over 30 coast watchers were so excited reporting the aircraft over their stations that they could hardly get an exact count. The bombings eased the enemy pressure on the camp. Guerrilla ambushes of the Japanese increased, retaliating with delight. The Japanese retreated, and the Agusan River was open again.

Messages of bombings on the other islands of Cebu, Negros, Leyte, Panay, and Luzon flooded FRS. Japanese response from the Visayan central islands was weak and photographic reconnaissance provided proof of their destroyed defenses.

In retaliation, 50 Japanese bombers flew over Waloe. Within two hours, more than 350 enemy planes were sighted over Agusan, Davao and as far as Lanao.

Fortun, in charge of the Davao network, sent detailed descriptions of the B-24 bombers attacks on the Davao areas. It was the first bombing in the Philippines and happened daily. Residents rejoiced at the sight after two and a half years of waiting. Operatives in the city identified only by number, had been filling the air with reports of the Japanese strengthening their airfields, military installation, and coastal defenses with artillery facing the sea. The Japanese high command believed the Americans' return would be in the Davao Gulf. The operatives watched as bombers dropped over 100 tons of explosives over the city pinpointing the military targets they sent. Some buildings were untouched. FRS crowded with reports of hundreds of enemy planes and ships destroyed or damaged. Nine enemy ships were sunk in the Davao harbor. The bewildered Japanese stopped their hunt for station FRS in Agusan and retreated to the coasts to prepare their defenses for an American assault from the Gulf.

Relentless bombings continued through September. More than fifty *Liberators* returned with clear weather reports to bomb targets but encountered heavy weather over the city. With the low visibility, they dropped their load on the Japanese garrison in Talaud Islands. Reports of two *Liberators* shot down flashed to guerrilla stations, and Filipinos rescued them.

Rescue parties were sent out to planes shot down. Civilians helped rescue downed pilots. They watched during air battles for white parachutes gliding down. If the body went limp before it landed, a sniper might have shot it down. They were "Americans because ... Japanese...(would) rather go down with his plane." It was reported that any Japanese that survived a plane crash or a destroyed ship did not live long. Civilians with *bolos* and weapons hacked them as they landed or waded to shore.

Davao operatives reported that the large garrison of over ten thousand Japanese in Davao City broke up into smaller units and began retreating to the surrounding interior near Mt Apo and the coastal towns. They began storing food and equipment in the caves behind the city to live off the land and establish positions in the hills around the city. The civilians suffered as the food was confiscated from them.

At FRS, Gustilo, Sarmiento, and Ortiz were at the front-line receiving the messages streaming in from the Davao network and radio outposts on the island. They felt the courage and risks the operatives took to bring in information. Many of the operatives who were "poor civilians with no more than grammar school education yet proved to be the best following instructions conscientiously." They risked their lives working in various positions and locations that allowed them to observe and report valuable information. Some worked for the Japanese and kept sending information about enemy positions and defenses. One Filipino surveyor employed by the Japanese passed complete maps of enemy airfields around the city of Davao. Bombed targets of buildings with Japanese aviation fuel and the number of ack-ack guns (anti-aircraft guns) were reported. The men at camp received pencil-and-paper rubbings of identification plates serial numbers as confirmation made by brave operatives who crept close to the guns. An agent who was a truck driver for the Japanese reported three times in a week of large gun emplacements on Samal Island and the exact times crater holes at Licanan airfield were filled just two hours after the bombing. Two days later, it was bombed again.

A Japanese convoy of fifty-two ships was reported destroyed by a Navy task force off Hinatuan Bay releasing tons of food supplies, fuel, and clothing to drift ashore. After more than two years, the "guerrilla boys had something to wear." Weapons and ammunition were enough to arm a guerrilla regiment. Drums of salvaged gas and oil enabled the operation of the river launches. Coastwatchers saw the survivors treading water for a short time until civilians waded to them, waving their glistening *bolos*.

Guerrilla officers were ordered to remove equipment, papers, insignias, and any material from dead enemy soldiers immediately and turn them over to headquarters for evaluation. Laureta ordered his men, who were in rags and barefooted, to take the clothing of enemy soldiers. His 130th Infantry Regiment was now at 63 officers and 1,012 enlisted men. Their weapons included mortars, an ack-ack gun and a bazooka that lacked parts but had not been used as no one knew how to operate it.

Information on aircraft warning and weather changes was imperative to support the Leyte landing. An aircraft warning system that would send enemy aircraft movements to KAZ using the radio circuits already carrying weather and intelligence traffic would overload the channels. Signal Aircraft Warning (SAW) weathermen with separate radios were needed.

USS *Narwhal* – September 1944

The *Narwhal* was due for an overhaul, but after minor refitting, turned around to bring another party to Mindanao. Men at Camp Tabragalba were chosen and trained to be part of 3 Signal Aircraft Warning (SAW) ground observer units called the 597th SAW Battalion. Another group was formed from the 15th Weather Squadron. A separate relay station on Mindanao was to be set up to receive all aircraft warning and weather information and directly flash to Perth (submarine), Morotai (air force), and Sansapor (naval).

The men and equipment were brought to Brisbane early in the morning of September 6. Before boarding the plane to Darwin, Philippine Secretary of National Defense Basilio Valdes, former vice-governor of the Philippines Joseph R. Hayden now advisor to General MacArthur, Whitney and a few officers connected with the operation met them at the PRS office. They were given pep talks and complimented on "what a fine bunch of men (they) were." To keep the harmony in the group, they were "told that we were the first organized American unit to leave Australia on a mission that consisted of Filipinos and Americans (radio operators and infantrymen). That upon completion of our mission, we would receive due recognition." It was a 12- hour flight to Darwin and a three-day wait to board the *Narwhal*. They did not know where they were going in the islands. "That was a secret and not on our papers." They found out days later while onboard the sub. Living quarters were cramped and there were no facilities for bathing. There was not much to do. "Every corner was filled with equipment and supplies...had to sleep on our bags packed like sardines. Some of the signalmen were assigned to operate in the radio room."

The trip was without incident except off the shores of Halmahera where a Japanese convoy was sighted. Pvt. Ogoy remembered "orders barked to clear the bridge with the klaxons blaring." If it was an enemy aircraft the submarine would dive if it got within 10 miles. In this case, the danger was a convoy. Lookouts leaped down the hatch with the captain last. The enemy ship passed directly over while the everyone onboard listened intently to the sound detectors and radar. It was a relief when no depth charges followed. Reaching Illana Bay, the party could not land right away because the submarine commander had not gotten the exact signal from Frank McGee, guerrilla commander of the 106th Division. The sub stayed in the bay for five days submerged in the daytime and coming to the surface at night for fresh air until the "clear signal of a white flag hanging between two trees during the day and two torchlights during the night" was sighted.

The sub moved south to the mouth of the Pangi River close to the town of Kiamba. At night, Frank McGee, came onboard to discuss unloading with the skipper and Lt. Edward T. Pompea, leader of the party. Guerrillas on shore were on the alert for spies and Japanese patrols. With the help of guerrillas and civilians, 55 tons of ration, supplies, radio, and weather observation equipment were unloaded under four hours into large sailboats.

The large party of 25 Filipino and 15 White weather and aircraft warning personnel debarked. Each man brought his rations for only 30 days. Beyond that, they had to live off the land and people. Lt. Pompea who was from the V Fighter Command and had helped train the men in signal aircraft warning, oversaw the Filipinos. The 978th Signal men were Sgt. Doroteo V. Vite, Sgt. Fred C. Ignacio, Pfc. Al C. Aglubat, Pfc. Geronimo De La Pena, Pvt. Vicente Balinton, Pvt. Calixto T. Cortez, Pvt. Martin L. Mapalo, Pvt. Estanislao C. Martin, Pvt. Regino T. Patacsil, Pvt. Gaudioso L. Patubo, Pvt. Del Y. Ogoy, Pvt. Benjamin B. Marcelo, T4 Mariano A. Medina, and T4 Domiciano O. Boncavil. Recon men and technicians were Sgt. Benigno Dumalaog, Sgt. Pascual M. Capis, Sgt. Luis B. Paler, Sgt. Feline R. Bautista, Cpl. Arthur G. Lopez, Cpl. Celestino Valdez, Pfc. Leodegario Lascano, Pfc. Federico Niebres, Pvt. Leon C. Fernandez, and Pvt. Manuel M. Nerez.

The sub continued to Balingasag, Macajalar Bay and delivered three men and twenty tons of cargo. Guerrillas and forces of the Maranao Militia guarded the unloading. Before returning to Mios Woendi, her home port, she was diverted to Siare Bay to evacuate 81 survivors of the transport *Shinyo Maru* torpedoed by the USS *Paddle*. The guerrillas brought the prisoners to meet the sub. In the open sea, she dove to avoid a Japanese anti-submarine patrol plane. She locked at one hundred seventy feet and reversed action to surface. Then, with not enough time to open her deck guns, she dove again to a safe level of ninety feet. She arrived safely at Mios Woendi, her home port.

Mapalo was overwhelmed when he saw the welcome crowd of guerrilla soldiers and civilians on the beach. He felt "deeply appreciated by the people of that area...very proud of us because...the Americans were coming back to re-take the Philippines." He met the townspeople who gathered to welcome them and "appeared practically in rags but looked in fairly good health." The medicine brought in by the earlier submarine landings must have helped them. They learned that the civilians hid from the Japanese and the *Moros*. Before their arrival, *Moros* under *Datu* Bulawan had massacred several hundred Christians on the coast of Milbuk to Mindupok in Cotabato for their clothing, money, and rice. Another *Datu* Maladya had planned the same but was averted by the party's arrival. After attending a talk by Maladya, Boncavil warned him not to start any trouble or he "was going to bring the U.S. Army to wipe them out...bluff worked... became friendly to everybody, especially to us." The men answered questions on "what was taking place in the outside world while they ate the fruits and drank the *tuba* the villagers offered them. The Japanese propaganda had disillusioned them (civilians) with the claim of having taken Australia and the state of California."

A typhoon kept the group inland for three weeks and unable to set up the radio. While waiting for more carriers, they broke up the cargo into smaller loads that a man could carry on trails through the jungles and mountains.

Pompea selected another location for the radio in place of the one assigned by Australia and sent a courier to the 10th MD headquarters stating that he would operate by the end of October. Fertig's message to KAZ ordered Pompea to follow Fertig's orders. The party was split into groups with assignments to set up radio outposts in the 106th and 109th Divisions.

In early December, Pompea sent his first transmission, saying that he had been fully operational since November and had intelligence information to report. Soon after the first contact, enemy patrols were observed heading for several of the signalmen's outposts. "We kept moving and the Japanese kept following," Pompea related.

Recon man Martin and signal man Mapalo went to Malita on the western shore of Davao Gulf to set up station DRA. They reported to guerrilla Maj. Henry Page, commander of 116th Infantry Regiment. Mapalo recalled, "We took a Vinta (sailboat) towards Glan. On our way, we encountered Japanese planes but fortunately reached headquarters safely." Two guerrilla officers, Lts. Nicolas Villamor and Veloso provided protection for five months. "Our job was to report Japanese movement at Davao city, Sarangani island, the Laoyon area, and Balut Island. At these places, the Japanese had their outposts to blockade Malita and capture the large gasoline deposit in the area. They were driven back by the strong forces of guerrilla soldiers that were my bodyguards on our way to Glan." Returning to the 116th's camp, "it took us seven days... passed Latian mountain, third largest mountain in Mindanao...followed the course of the Padido River and crossed it eighty-seven times." For the next six months, his radio station in the Buayan-Batutitik area reported information on the enemy's situation and strength until American task forces captured Sarangani, one of the last areas to be liberated in Mindanao. They saw many dogfights downing planes, some crewmen drifting helplessly down, and targets bombed.

Ogoy and Aglubat hiked for two weeks through the jungle dense with dark swamps and giant trees that covered the sky to reach a Japanese garrison in Western Cotabato and set up a radio close to Parang. Parang was the planned landing site of the U.S. Task forces to recapture Mindanao. The garrison was a Japanese training camp of 5,000 men. They passed pro-Japanese areas, with two guerrilla officers, Lt. Odilon Boens and Lt. Ploris with fourteen of their men as security guards. Forty-four barefooted carriers half-clothed mercilessly exposed to mosquitos and leeches carried their equipment and supplies. Some had given their clothing to the Japanese and *Moros* with the false promise that they would not be molested. If they were caught in their homes, they would be killed, so they left to hide in the forests. Even an old man offered to carry cargo wherever they would go to be paid with a meal.

The trek became more difficult as the days passed. Ogoy was thankful for his two years of infantry training, but he found the hike more strenuous. His shoes were always half-filled with water. Mosquitoes swarmed as they could not build a fire to drive them away for fear of revealing their presence. Two weeks into the journey, he saw a boatload of men approaching the riverbank. He set up his machine gun, fired at the boat, and took prisoners to the 106th Division headquarters. Interrogation revealed they were Filipino soldiers forced to serve with the Japanese Bureau of Constabulary and confirmed about five thousand Japanese troops at Parang. On the same day, while waiting for the change of carriers between villages, they ambushed two Japanese launches netting them baskets of rice. They could not use the launches because that would identify them as Japanese. A hand grenade in each sunk the boats. Waiting for volunteers to carry the supplies would increase discovery by the enemy after the loss of their men, so carriers were immediately recruited. As precaution, the convoy had a man carrying a machine gun in front with four men each carrying old 1903 rifles. Two men with Carbines travelled in the middle with the carriers.

The group reached the camp of an infantry regiment of the 106th Division commanded by a *datu* at the highway heavily travelled by Japanese vehicles between Cotabato and Davao. Ogoy did not reveal too much to the *Moros* as they gathered information on the Japanese garrison. The *datu* and Corelio de Pidro, a civilian ex-Navy man secured information on enemy installations, strength, and troop movements. The group moved closer to a spot on the beach at Sugod about 5 kilometeres inland from the road to Parang. It was under another *datu* so he had to appease both *datus*. He explained that he was only after information on the Japanese and to put aside their grievances until the Americans returned. A civilian runner was sent to the *datu* of Sugod to send men to carry the cargo and provide security as they could not go any further. It was a day's trek to Sugod and a week to contact a local net control station. Ogoy reasoned that civilian operators did not know the Army procedures they used, and he did not know the mixed commercial methods used by the guerrillas. Both sides had to learn from each other over the air. "All kinds of tribes gathered around me. Some of the people crying, so thrilled with thanks...urged the village people to cooperate with me and not to tell anyone of my presence in that area. They must keep it a secret," Ogoy recalled.

He was a signalman, but Ogoy decided to do his own spying. Once the radio was set up at Sugod, he dressed as a *Moro* and pretended he knew nothing. He found out that civilians near the Japanese garrison were being tortured. Several weeks before the U.S. Task Forces landed at Parang, he watched with a big smile the bombers with escort planes dropping their load on the garrison's barracks, ammunition dump, and food storage. The targets were from the reports they had sent. He reported seeing two PT boats, each armed with a bazooka, cannon, and machine guns, with three Americans and twelve guerrillas leading the attack early on the same day as the bombings. "The machine guns did good work, and the attack lasted half an hour." The Japanese were caught unaware as this was usually the time of their morning exercises. The enemy, confused with the direct hits from above and from the water, began to evacuate. The bombings continued twice a week until the landing.

Pompea, Boncavil and Patacsil rowed through a typhoon for four days to Glan in the south. Japanese planes dove on them but did not open fire after seeing them dressed as fishermen casting their hooks into the water. Guerrillas waited for them. On the beach, everyone dug foxholes as Japanese soldiers were seen approaching. Rations and equipment were hidden a few yards inland. Boncavil recalled that he laid on his stomach in battle position with the code and signal operation instructions. When the Japanese arrived, "the guerrillas gave it to them, killing most of them. All the Japanese were survivors of torpedoed ships."

From Glan, Boncavil and Patacsil trekked to the Buayan landing field near the Cotobato-Davao border with station B43. Their camp was one and a half-mile across the Buayan River, where they "could see the Japanese landing field with our naked eyes" and reported on them. They observed planes hidden under palm leaves. Japanese soldiers worked day and night, hiding supplies and ammunition dumps along the river. Trenches were dug around the airfield. When no serviceable planes were left after the bombings, his group moved out leaving two hundred guerrillas to guard the airfield.

Boncavil reported four cargo ships loading and unloading war materials in the Buayan River. At night they could hear very clearly the chatter of the Japanese soldiers. They felt safe knowing there were guerrillas six hundred strong led by Lt. Besana close by. A mile away were 37 men under Datu Magalana. They quickly escaped ambush with their guerrilla and *Bilaan* (indigenous people) security when a Japanese patrol discovered their outpost.

They moved the station to a hill and saw through field glasses the Japanese burning their dead. According to Boncavil, the enemy trailed them with bloodhounds. They kept moving as enemy patrols looked for them. He talked of "a bird spotter called Calao." He may have meant the Kalaw, a Philippine hornbill found in Mindanao.

When the *Liberators* arrived to bomb the islands, Boncavil's group reported, "the whole area, including ours, was machine-gunned and bombed for one and a half hours. Hidden ammo dumps were hit. Supplies, trucks and planes were blown to bits. In Davao city, 2 ships were burned, 1 sunk, and the other slightly damaged. The Japanese tried to repair it, but when the bombers returned several days later, she was sunk."

Itching for some action, Boncavil with a trusted guerrilla, Sgt. E. Romada decided to harass the Japanese at their post. As a signalman, he was not supposed to go into combat except to protect the radio. Seeing six Japanese carrying food, they opened fire at 400 yards. The Japanese ran the other way, yelling and dropping rifles, a sabre, a dagger, a hat, and a sack of rifle ammunition.

Two Japanese regiments were garrisoned ten miles farther north of Buayan at Tinagacan, Conel, and Batutitik. "When we reported this garrison to our net control station, we made an accurate map location and handed it to a Catalina (PBY) crew who landed in Glan," Boncavil said. All the places, buildings, or houses reported were bombed and strafed. "It took them a week to burn their dead. The Japanese commanders called the planes 'the whistling death' or 'the screaming death.'"

Camp X (Tabragalba), Australia: Benigno Duma-laog was with the mission party to Mindanao in September 1944.

When the Japanese retreated to the foot of Mt. Matutum, Boncavil, with a few men, trekked to Kabacan in the middle of Mindanao. It took a month to reach the location and rations had already been gone for a week. They ate green leaves and wild roots of plants. Along the way, an encounter with a Japanese garrison at Maganay was avoided. They crossed the 'Rio Grande,' the Mindanao River, careful not to be seen by two Japanese garrisons close to either side. "We had to watch every move we made... stopped, observed, looked, and listened after every ten steps." Their boat capsized and they scrambled to get back on board. They were more scared of the crocodiles that lorded it over the river than the Japanese. "We saw a display of black flags in all parts of the garrisons."

They reached the camp of the 109th Division where Dumalaog's group of Paler, Valdez, Lascano, and Nerez operated the radio. De La Pena was to be the cryptographer as 109th's messages to KAZ were unsecured. Vite was also reported as cryptographer of the division. After resting for a few days, they started again for Kabacan.

Boncavil recalled that it was "a tough and too hot a place to operate." The Kabacan road junction was the center of all traffic going to all parts of the island. They discovered a sizeable Japanese division camped under the coconut and rubber plantations by the river. Guerrilla forces under Lt. Mata fought the Japanese as they operated the radio. Three days later, from their report on the enemy's bivouac areas, strength, ammunition dumps, motor pools and supplies, the place was destroyed when the bombers arrived, "dropping their 'eggs' on their targets. When their group left, the guerrilla forces continued to drive the disorganized Japanese into the jungles. Traveling along the Pulangi River, they reached the town of Kibawe towards Cotabato City.

A group of signal and aircraft warning men, Capis, Marcelo, Patubo, Balinton and Ignacio hiked to the 106th division headquarters. Marcelo and Patubo, tested the radios they were to bring. The sets had corroded from the saltwater when they were unloaded from the submarine. Either the radio sets were not placed in waterproof containers, or the containers leaked. That had been the constant problem of the missions before them. A guerrilla radio technician, Lt. Olivia, was able to repair the sets. With a party of fifteen men, they sailed to the coastal town of Lebak. Lt. Brooks, communications officer of the 106th Division met them. They set up the transmitter and operated the net control station 3AP6 AND 3PP6 at separate outposts. Balinton and Ignacio operated one of the outposts.

Marcelo and his group reported to the guerrilla 118th Infantry Regiment of Salipada Pendatun, a *Moro* guerrilla. With his brother-in-law Datu Matalam Udtog, he had organized a band of *Moros* into a bolo battalion in the area of Cotabato. His daring attack on a Japanese garrison at Pikit and Kabacan convinced Datus Aliman, Mantil Dilanglan, and Gumbay Piang to join him. His camp was well-hidden in the Liguasan marshland in the middle of Cotabato. Marcelo went inland to the village of Talitay and set up station MIL close to a Japanese garrison guarded by guerrillas. The guerrillas turned a generator's crank to provide the power. When the Japanese heard of a radio station, they sent a patrol of 38 men that was spotted a mile from the station. The radio and equipment were quickly packed, and the men hid for two hours. They returned and sent a message to net control station. The garrison was bombed two days later.

Signalmen Medina and Cortez, and aircraft warning man Schram with recon men Fernandez and Lopez reached their outpost after a week of laboring through overgrown trails. The civilians aided them in setting up a hut for the radio reporting to the 106th Division's net control station. When the radio malfunctioned, they moved to outpost FG2 of Calixto and aircraft warning man McAdams with recon men Bautista and Niebres. When the White soldiers were ordered back to the base camp, the Filipino men stayed to operate the radio and continue gathering intelligence.

An order arrived to move closer to the town of Marbel by the national highway to watch for enemy transports. "As ordered, we continuously moved closer and closer to the enemy's area of operation, which we reported back." Cortez and Medina were the only radio operators, so they took turns exploring the area as the recon men were often sick with bouts of malaria fever. The recon men were ordered back to FRS, leaving Medina and Cortez without security to operate the radio and explore the area.

While Medina was in the town resting and drinking coffee, a civilian tapped him and pointed to 3 dozen Japanese soldiers. He ran and hid for several hours in the bushes. He could not leave as a sentry stood facing him. While hiding, he discovered that he had only one bullet and the clip was missing. He trembled at the thought that he could be easily killed without a fight. Crawling back to camp, he met a boy who was stunned and froze mistaking him for a Japanese in complete uniform until he spoke in his native dialect. Everyone was in rags except for the Japanese. He stayed at the boy's hut until nightfall when the father brought him to his farm and he slept overnight exhausted.

Cortez and Medina were relieved when twenty-five guerrillas led by a Capt. Veloza, escapees from a Japanese concentration area, willingly joined the station. One day, the guards spied the Japanese encircling the station. Oblivious of the situation, Cortez kept pounding the keys until someone shouted that the Japanese had already closed in. Only then did he jump out of the window with the radio leaving the generators. The rest of the equipment were thrown into the bushes. A guerrilla shot the Japanese officer with the raiding party who was shouting and aiming his gun at a civilian. Cortez shot him two more times. Still the officer groggily continued to aim at the civilians. A submachine gun burst finished him. "The guerrilla who fired the first shot got his German Luger, wristwatch, and identification cards while the rest divided his clothing among themselves," Medina said. At the camp in Marbel, Medina continued to operate station WT6 while Cortez returned to join Ortiz at the X Corps headquarters and provide support to the American Task Force when they landed.

Recovered from their malaria attacks, recon men Lopez and Fernandez posing as civilians bravely worked as spies in the Marbel area. They blended in and got close to gather information. Lopez was paid one peso and fifty centavos per day digging foxholes for the Japanese. Fernandez was hired as a carrier. A guerrilla spy, "a cowboy, they called 'Fred' for short secured all information inside the Japanese garrisons, bivouac areas about their activities" and gave them the information.

Capis was concerned that Pompea at his base camp was not sending out weather reports. His order was not to join Pompea as his position in the 106th headquarters flashing weather observations was urgently needed by air operations for the pending Leyte landing. He reported to KAZ that the signal equipment was badly corroded from saltwater, and the food supply was lost during the landing. The radio had a broken generator. But after two weeks, he was able to get the radio on the air and began sending weather observations from two balloon runs morning and afternoon to compute the speed and direction of the wind. The cash of one hundred pesos per man in his group was spent. He sent a message to Lee Telesco, managing the administration of PRS for Whitney, for authorization to draw cash for subsistence from the 106th Division headquarters. All the men were sick and unable to leave again for assignments until food was obtained. The civilians contributed food, and some of the soldiers shot wild deer and pigs in the jungles if they found them. Weak from lack of food and malaria, it was difficult for the men to divide their time between operating the radios and scouring for food. Some civilians who were desperate for their own and families' well-being would not provide food or manual labor unless paid. Pompea, who had the funds heard his pleas and sent money with the clear message that "rations were only for thirty days." But their mission lasted five months.

Before the end of the year, Capis requested KAZ to standby for 24 hours on a specific frequency. He was doing everything possible to get the weather messages out with top priority. He was given the frequency Oboe to connect with FRS directly. He added that there were "lots of planes overhead "and warned KAZ and Telesco that the radio would go off the air without lubricating oil. He desperately requested Telesco to have the 106th Division commanding officer drop oil on the coast of Tibpuan so that he could access it. But the 106th Division's radio was down and unable to send out reports for two weeks.

For the outposts, he sent a list including "8 transmitters and receivers, radio crystals, batteries, field glasses, radio spare parts, spark plugs, two tins sealed Camels (because others did not get), antenna wire, fifty pads of cryptography papers, bond paper, typewriters with ribbons, and a camera if possible." He emphasized that the outposts "did a lot in securing enemy information. With this equipment am positive can do a complete job in efforts to report enemy activity." The reply was "no promise of an airdrop of supplies but will send lubricating oil and other supplies in the early part of January (1945), so make every effort to remain on the air."

Capis radioed Telesco of another weather observation and aircraft warning traffic station and frequency in case his radio went down. He was out of supplies to report weather and code intelligence. He sent another message pleading for lubricating oil and fuel oil. "Doing all in my power to get weather messages out with top priority. Are we to expect anything at all?" he sent in frustration. The reply from KAZ through Telesco was that his appeal for supplies came after the submarine had departed. "Hold tight. It won't be long as we are now flying into several places into Mindanao delivering supplies. Signed Telesco." Two weeks later, KAZ sent a troubling message that "supplies will be sent to you at the first opportunity." On New Year's Day, his radio went down as he was ordered to move to another location. Two months later a small cargo submarine arrived with supplies that took another month to reach his outpost.

As the Leyte landing neared, the Japanese troops in Davao city were observed moving north on the national highway. FRS blasted an urgent order to all stations for the guerrillas to do all possible to prevent the Japanese from moving troops towards Surigao because Leyte was just across the Surigao Strait. The jungles of Agusan were along the way to Surigao and could hide them.

The 130th Infantry Regiment guarded the area from Davao to Surigao. They had to hold the perimeter lines against the Japanese at the three rivers crossing the national highway to stop them. Although the unit was reported to be fully organized and staffed, companies conducted in a military manner, the 107th Division's commanding officer, Clyde Childress, saw the lack of military skills as a cohesive fighting unit to effectively defend the Davao area. The unit had full-time and part-time soldiers who worked with the civilians on farm projects. No one knew how to operate the heavy weapons.

In October, Childress transferred Silva to the 107th Division headquarters in Kapalong as Infantry Advisor to provide military training to the Regiment. A handwritten note on a lined torn paper documented the decision "Am considering transfer Silva to Laureta for heavy weapons." Laureta asked that Silva be assigned there permanently. Silva set up a General Services School patterned after his training at Camp Tabragalba with officers of the regiment to instruct the guerrillas in military procedures while he personally instructed on heavy weapons. All men were required to attend. Civilians, elated with the news of the American forces' return volunteered to fight and went through the training at the established General Service School. "In three weeks of training, the men learned what trainees acquired for six weeks in the pre-war Philippine Army Training camps," related Pfc. Torcuato Cirpo who volunteered at sixteen years old. The handicap was the lack of manuals, stationeries, shoes and clothing, poor diet, and limited medicine. The training was non-stop as men joined to fight until the Americans arrived to retake Davao. Cirpo kept the certificate awarded to him after completing the General Services School course. Yellowed, brittle, and with disintegrating corners, he proudly showed it in an interview after the war and pointed to Silva's signature as Commandant of the school.

USS *Narwhal* – October 1944

While General MacArthur's convoy of over 700 ships zigzagged towards Leyte, the submarine *Narwhal* was charging from Mios Woendi off the northeast coast of New Guinea towards the Philippines. She had a "weak battery, unreliable stern planes, leaky manifolds, and completely rusted engine mufflers." She had just returned from a mission to Mindanao and there was no time to fix her problems as the Leyte landing was happening. She immediately turned around to drop off supplies at Tawi-Tawi and another party to the islands of Negros and Cebu.

Through the periscope, the coastline of Tongehatan Point was visible but did not have a welcome party. The skipper was afraid that the clanking noise would alert any Japanese garrison. Three Japanese patrol vessels suddenly appeared running closer to the beach. She dove deep and went on silent running. She watched the patrol vessels continue toward one of the outer islands. After two hours, she surfaced and received a proper signal from a small boat approaching her parallel to the beach. Suarez and Young came on board. The island had few Japanese with one seaplane and a merchant ship. Most were ordered to Leyte for the pending American landing. There were more than enough boats to quickly unload ten tons of cargo in less than an hour. Bananas and coconuts were loaded into the sub as thank you. The crew reciprocated with generous amounts of flour, baking powder, rice, canned goods, and cigarettes. Exchanging notes, Young said that he left the signal up all day because he wanted to be sure the sub saw them. Suarez was afraid to come out from shore because he might be mistaken for a Japanese and shot. The skipper expressed his admiration for their loyalty and patriotism. The sub's next stop was Negros Island.

Young sent a message of Bolo Battalion Mine Sweeper Units formed by Suarez to sabotage the marine mines planted by the Japanese in the waters around Tawi-Tawi. Within five months, 327 live mines were recovered.

With the pending landing, FRS was to be protected and maintained at all costs. She continued to receive communications from station outposts across the Philippines as far away as Luzon to be relayed to KAZ.

By this time over fifty-one radio stations were operating on Mindanao. A radio was installed in each of the seven airfields, scattered in strategic places on the islands for instant coverage of planes bringing in supplies, needing refueling, repair or when the weather was terrible. Although radio traffic more than doubled, the same number of personnel operated the radios at FRS. Traffic tremendously increased with sightings of an estimated 60,000 Japanese troops building defenses with the highest concentration in the Davao and Cotabato areas.

A week before the American Task Forces landed on Leyte, GHQ estimated the total effective Japanese ground combat forces in the entire archipelago was at 224,000. Later estimates from the Japanese showed a higher number by 64,000. On Mindanao, from FRS reports, the estimate was 58,000 a month after the landing.

Leyte Island, Philippines Landing – October 1944

Mindanao had the most significant number of operating radios, more than Luzon and Panay combined. Gustilo, Sarmiento and Ortiz at their stations, received the message: "Tacloban, Leyte to be invaded 20 October 1944. Evacuate all Filipinos (a) week before."

The U.S. Navy and General MacArthur knew much of the positions, directions, and ships of the Japanese forces in the surrounding seas of Leyte Gulf and Surigao Strait from the continuous reports to KAZ. While bombardment of the shores was ongoing, coastwatchers in the Visayan and southern islands were streaming messages to KAZ.

Pompea related that the biggest thrill was when a signalman tuning the dials thought he heard an announcement from "Radio Leyte… that American authorities had set ceiling prices on food to prevent inflation on Leyte. We knew then the Philippines had been invaded." A jubilant celebration with the civilians followed "with some Japanese whiskey a guerrilla gave us."

Ortiz was sent with a party to Bukidnon to inspect the area around the Umayam River if the radio had to move again and report back. He recalled that "The trail which we traversed was the most difficult I had ever gone through. It was impossible to use the same trail for heavy cargo because a large portion follows a riverbank, some places of which are so steep that hikers must cling to rocks and roots of trees to keep themselves from falling into the rocky and treacherous river." His group arrived in the deserted village of Cabanglasan. The people were hostile to outsiders and hid whatever food they had from the Japanese and the fierce *Mangahats*. They ate wild roots, bamboo shoots, and sweet potato leaves for three weeks. Ortiz wrote, "Our viande consisted of birds and monkeys."

A large formation of planes dropped bombs so close to the camp, signalling that the enemy had discovered the net control station. A Japanese patrol was seen only 8 km away. FRS was hurriedly moved to La Paz, a town by the Adgaoan River, a tributary of the Agusan River. The first group moved to carry two diesel motors, radio equipment, and a small supply of food, leaving the rest to maintain communication with KAZ. The town of La Paz was wide open to enemy planes flying over the area panicking the people already paranoid from earlier bombings at Talacogon. But food was more available, unlike six months earlier. Once a radio was set up in the new site, Waloe went off the air, and the rest of the camp followed.

At La Paz, Cortez held the job as chief operator of the OBOE circuit and Gustilo continued to be the head technician, valued for his service. He had the rare ability to fix radio equipment despite the lack of parts in the shipments they received. All highly appreciated him, and he was eventually promoted. Cortez operated the radio with Ortiz until he was assigned to the guerrilla forces preparing for the U.S. 40th Infantry Division landing in Bugo, Macajalar Bay. Ortiz's job of sending air warning and weather reports to Air Force stations in Morotai and New Guinea was boosted by a new circuit established to handle the traffic of all the Eastern Mindanao division headquarters' networks. Sarmiento returned to camp to help operate the ABLE radio frequency. He trained a guerrilla outfit to handle and fire the heavy weapons of machine guns and mortars. An all-night stand-by station was set up on a 24-hour basis. Every officer in the headquarters below the rank of captain was required to be officer of the day at least once every two weeks. Going on 24 hours per day, two diesel engines running the radios began to falter. Men desperately scoured around and found an engine from a broken launch in the Umayam River. By this time, small quantities of supplies began to be brought in from Leyte by PT boats.

Gustilo, Sarmiento, and Ortiz did not take breaks with the continuous crackling of radio traffic pouring in. Messages to KAZ, averaged between 8,000 to 10,000 per month. It increased by 164% over messages sent in the last six months with no additional radio operators. The Philippines' geographical location was strategic, especially Mindanao, in relation to the Dutch East Indies, New Guinea, and other islands in the Pacific. The networks had become the eyes of the Navy and GHQ in Australia.

A team returned from a briefing with the Sixth Army in Leyte. They brought a trench mortar, ammo in a cargo with "canned turkey, cranberry sauce, plum pudding, some fresh apples and oranges, candy, and nuts," in time for an American Thanksgiving. Only "2 quarts of whiskey (were brought), but that got twenty-six White Americans drunk." The Filipino radio men would have enjoyed the thanksgiving meal if they had participated since they had lived in the U.S. for over a decade.

At the 110th Division area, three separate radio outposts were established to watch the three bays of Gingoog, Nasipit, and Butuan. Providing targets to the Sixth Army at Leyte, Nasipit harbor, and the town was destroyed. But Japanese troops were observed hiding during daytime in the swamp area of the inlet. They were in tunnels on the rocky promontory point of the beach west of the Nasipit lighthouse. Guerrillas were directed to destroy them. A Japanese Navy plane crashed on the Davao Road in the Butuan area, and the guerrillas reported killing the two pilots. The offer of their two heads was refused.

USS *Stingray* – December 1944

A small cargo submarine reported to be the *Stingray* was to land in Tawi-Tawi with four Filipino soldiers and twenty-one tons of arms, ammunition, and clothing for Suarez's men. Thirty-five tons were to be delivered, but nine tons were unloaded at Biak in New Guinea. The names of the Filipino men and the reason for the diverted cargo were not found. The *Calendar of U.S. Submarine Missions to the Philippines* did not list the mission.

In early December, Sarmiento with reconnaissance man Nerez picked up a radio and supplies to head out to the Sayre National Highway in the Bukidnon sector. From their hiding place on a small hill, they reported for six months until their position was discovered and hopelessly surrounded by the enemy. They escaped when the U.S. 31st Division landed and bombed the Japanese positions in the hills clearing northern Mindanao.

As enemy activities eased up, court-martial proceedings of officers and enlisted men who violated the Articles of War during the preceding months were held at headquarters. Knowing the language, Ortiz was appointed as defense counsel for the trial of a guerrilla private charged as a deserter. Later he was appointed as a trial judge advocate in a general court-martial for the trial of a Filipino Signal officer (USFIP) who showed disrespect towards a superior officer. Found guilty, the accused was dishonorably discharged and forfeited of all his pay. Wheeler thought the ten-month hard labor imposed by the court was not enough.

Gustilo went on an inspection trip of the radio stations in Eastern Mindanao. He avoided the retreating Japanese patrolling the Agusan coast in sailboats and reached Magugpo (Tagum) in Davao, headquarters of the 107th Division. He saw the familiar face of Silva, whom he had not seen since the camp in Waloe. He continued further to a guerrilla battalion's camp in Caraga to check on some radio equipment dropped by a submarine. At Lianga in Surigao, he operated the Eastern Mindanao sub-net, as the chief technician and radio officer.

Relieved with the news of the Leyte landing, White American soldiers in the FRS net control station began to transfer to GHQ in Leyte. They had refused to surrender or were stranded when war broke out and they had been on Mindanao for three years. The Filipino operators stayed on to operate the radios.

The Eighth Army's focus was on the next landing on Luzon. Fertig expressed his concern that unless help or air support was given, the Agusan coast, which was not occupied by the Japanese would be infiltrated. The Japanese were constantly on the move, making it difficult for coast watchers to monitor their activities. More mobile radio sets were sent in a man's backpack out to the fields to follow their movements. In the occupied town of Dalican, Japanese soldiers were seen raising the American flag when U.S. aircrafts passed to prevent being bombed.

Luzon Island, Philippines Landing – January 1945

The Luzon landing heralded the end of the submarine missions. Commander Parsons landed on the Labo airstrip in Misamis Occidental with three transport planes bringing fifteen thousand pounds of supplies. He declared that the submarines would no longer be bringing supplies and personnel. In their place would be surface ships and aircrafts.

As the liberation of Manila was ongoing, General MacArthur sent his next plans to Washington. The reply was the Filipino guerrillas, and the army of the Philippine Commonwealth Government were to take care of the rest of the Japanese in the country. Washington was now focused on their ultimate objective, the invasion of Japan. With Japan's objective to delay the Allies and fight to the death, it would be a prolonged and bloody task for the guerrillas. Confident with the valuable intelligence from the many strategic radio stations flooding the circuits to his station, General MacArthur ordered the Eighth Army to capture the rest of the country's major islands. He knew that the Task forces would have advance information on the enemy wherever they landed. The radios set up by the Filipino mission men in the Visayan islands and Mindanao had been providing enemy information, sabotage, demolition, weather, and aircraft warning support for the last two years.

In Sulu, Goroza and Teo proceeded to Siasi Island to set up a new station only a mile away from an enemy garrison. Aquino stayed behind at the net control station with Young. For some time, the guerrillas had been trying to destroy the Japanese camp. "We stayed in a mountain overlooking the garrison. The guerrillas were hopelessly outnumbered and reluctant to follow up with an attack. We sent a message to KAZ for the garrison to be bombed. Soon two reconnaissance planes took off from the Allied Air Force HQ located in the Halmahera Island in the East Indies. The following morning, nine medium bombers came over to strafe and bomb the garrison. One of the planes was shot down, killing a crew member and injuring others. The guerrillas rescued the survivors, and a seaplane arrived to pick them up." They continued to report on U.S. naval vessels sunk in their area and subsisted on Navy rations salvaged from sunken ships brought to them by fishermen.

The following month, with guerrilla guards, Teo and Goroza with Young, left for Parang in Jolo and infiltrated the enemy-infested area of Mt. Tukay to establish a radio station and watch a nearby Japanese garrison. Guerrillas went to great lengths to provide proof of their findings. Sketches of enemy locations and hand-drawn maps provided details that aerial photos could not. So accurate was the information that the Eighth Army did not send their own intelligence agents ahead of the landing on Zamboanga. Young returned to Tawi-Tawi while Teo and his men remained until the Americans landed in Zamboanga.

Young was ecstatic when he finally received the long-awaited message to revolt against the enemy. He had reported the garrison of eight thousand Japanese in the town of Zamboanga controlling the Zamboanga-Sulu region and the waters between Borneo and the archipelago as bombing targets.

The American Task Force landed on Sanga Sanga Island in Southern Tawi-Tawi without firing a shot. Suarez and his men were on the beach to welcome them. The capital city and the airfields at the end of the Zamboanga peninsula were taken against light resistance as the enemy fled to the hills. *Free Philippines* published that over fifteen hundred Japanese troops were dead and a few hundred scattered into bands with little rations and ammo. The *Moros* hunted the Japanese to get their weapons and supplies. A U.S. infantry commander said, "If it were not for the *Moros* we would be fighting Jap detachments all over the islands.

They are eager and fearless. The Japs are afraid of them." Reports of their brute abuses and plundering of non-Moro civilians might seem a contradiction but they certainly helped the liberation forces.

When the news arrived of Manila's liberation, the 10th MD headquarters moved across to Camp Keithley at Dansalan (now Marawi), to prepare for the liberation of the Visayas and Mindanao islands. The former U.S. Army camp was located along the north shore of Lake Lanao. Ortiz was in charge of all radio supplies during the trip from the east coast and arrived with the first group of operators and cryptographers. Silva sent twenty infantrymen from the 107th Division to reinforce FRS. The move was difficult as all equipment had to be carried on the backs of men. The generators needed several men to carry if a carabao was not found.

The western Mindanao sub-net under Bowler took over until the FRS networks were completely set up. Herrera at Dimorok Canyon said that while FRS was on the move, his radio net took charge of all traffic in Mindanao for two weeks. Palmeto was with him working the radios. Eventually, stations V86 and 3UB which he was handling were consolidated into station LW2.

Ortiz was designated chief signal officer and operated on the CHINA frequency that handled the traffic of all the Western Mindanao division headquarters networks. Five circuits were installed at Camp Keithley for all traffic from Eastern Mindanao, Western Mindanao, Cotabato, the Eighth Army, USAFFE headquarters, and the Navy. Air warning and weather reports were sent to the air force stations in Morotai and New Guinea until elements of the U.S. X Corps landed in April in Cotabato. Living quarters were in "houses with dug-in cement foundations... running water and all. Then too, for excitement, the famed *Moros* were there. One can get excited just by looking at them. But they never were to be trusted," Herrera wrote.

The airfield at Manticao in Misamis Oriental was in the area of the 108th Division. Ortiz and Herrera set up a radio and brought trained guerrilla signalmen as operators. Herrera stayed in Manticao operating the radio until ordered to Misamis to operate radio communications for the guerrillas who were bombing a Japanese fortification. Enemy troops were trapped with their backs to the sea.

Navarro hurriedly tapped the keys to the U.S. Task Force on the bombing and strafing of U.S. aircrafts he witnessed. His group moved to the Sayre Highway where the Japanese were concentrated at a main trail leading to the highway. They caught up with the 111th Infantry Regiment, a guerrilla outfit reorganized at Valencia, Bukidnon under Maj. Manuel Jaldon, Capt Victor Hidalgo as Executive Officer, Lt. Stanley Dalman, signal office, and Lt. Latoja, Regimental Intelligence. Hidalgo who was a Scout officer was in a fierce battle with the enemy. Navarro radioed activities on the airfield in Valencia built by the Japanese until his radio died under the daily heavy rain. Food was *camote* and edible weeds. No other supplies were given. All the while, the guerrillas were ferociously fending off the Japanese until they retreated. The muddy trails were littered with the enemy's dead. The Army's 31st Division highly commended the guerrilla regiment.

Herrera moved to the guerrilla camp at Lantapan close to the Sayre highway with a new radio set from the 31st Division to maintain radio traffic. He trained able men on operating the radio so he could concentrate on the gathering and validating of messages. The food situation improved, and clothes were provided by the Army. Navarro reported with his radio in a small house and established lookouts near the beaches for enemy shipping. But his transmitter faltered so couriers brought the messages to Herrera's station for relay to KAZ.

Another mopping operation would have taken the men across the mountains to Agusan, where remnants of the Japanese in Davao and Cotabato were entrenched to prolong the war as much as possible. Herrera, eager to continue, expressed his disappointment when orders to report to the Eighth Army in Leyte arrived.

Mindanao Island Landing – April 1945

Mindanao was the last island to be liberated. The Eighth Army estimated nearly sixty thousand Japanese troops on Mindanao not including the eight thousand in Zamboanga by the time they landed, and the five thousand Japanese civilians armed and converted to military training. They garrisoned mostly in Davao City, coastal towns and major roads afraid of the guerrillas who ruled the countryside. Their combat strength had been greatly reduced by almost half with the transfer of troops and field guns to Leyte. Around 20% were suffering from malaria said a captured Japanese nurse.

The American troops were warned that they had to defend themselves from the *Moros* whose allegiance was questionable and who disliked Christians, aside from marauding guerrillas and Japanese soldiers. The fierce head-hunting *Mangahat* people were to be avoided. In addition, the terrain was inhospitable with swamps, thick jungles, and the roads were mostly trails.

If the Eighth Army was able to get enough ships to carry a corps into Davao Gulf, the harbor would have been the landing site according to General Eichelberger. But the Navy and GHQ were concerned by the heavy build-up of enemy defenses. Coastwatchers monitored the Gulf's entrance. Operatives reported artillery and anti- aircraft batteries circling the coastal shore of the Gulf with the main defense positions inland behind the city to prolong the retaking of the islands.

FRS circuits at Camp Keithley to the Eighth Army were flooded with Japanese fortifications in Bukidnon believing that the American forces would also land in Macajalar Bay.

The landing site was Illana Bay and an advance command post of the 10th MD was set up at the X Corps headquarters. Ortiz was relieved from duty at Camp Keithley and sent as the communication officer to X Corps.

The American Task Forces arrived at Illana Bay ready to land in Malabang. Meanwhile, the guerrillas without enough weapons and ammunitions, in tattered clothes, some sick, malnourished, proved to be a formidable force. With information on the enemy, they launched an offensive of demolition and destruction that cleared Malabang. A message was sent to divert the Task Force to land at Polloc harbor in Parang. Advanced aerial reconnaissance was not needed as radio messages had pinpointed bombing targets. For insurance, the convoy's guns bombarded the beaches before landing.

Small portable radio sets arrived with the Americans and were sent to the four corners of Mindanao - Davao-Agusan, Bukidnon, Cotabato, and Zamboanga - to radio back the latest intelligence on Japanese movements.

As the American forces marched towards Davao City, Ortiz moved alongside them translating intelligence reports that flooded the airwaves. The Corps' intelligence and operational reports were relayed to Camp Keithley through the inter-island radio Ortiz operated with the call sign 'seven queen zebra.' Signalmen pounded out messages of the 24th Division Task Force travelling a third of the way on landing crafts on the Mindanao River to save time. On land, the force double-marched on Highway 1 towards Davao Gulf in 10 days. That was a cause for celebration as there were no maps between Kabacan and Digos, and bridges had been burned out. At Kabacan in the center of the island, the 31st Division drove north against heavy opposition on the Sayre Highway towards the plains of Bukidnon, dividing the 30th and 100th Japanese Army.

At Kabacan, Ortiz was surprised to see Boncavil brought in by two military police who rescued him from a guerrilla compound where he was confined for socking a guerrilla officer. His group had left for Kibawe without him.

The civilians in Kabacan revealed the abuses of the *Moro* guerrillas of the 106th Division on the Christian civilians. The Division was widely known for lack of discipline and continued to loot and plunder civilians even after the landing of the American forces. The Counterintelligence Corps (CIC) men who arrived with the Task Force replied to Ortiz that cases between guerrillas and civilians were outside the scope of their duties. However, Ortiz believed that something was done after he informed Fertig because a week before he left Kabacan, the civilians told him that the *Moros* had stopped bothering them.

The Japanese garrison at Kabacan was bombed ahead of the Task Force double-marching towards Davao. Marcelo related that "after the third bombing in the day, my men and I went to verify and get the casualty counts on the Japanese. We were dressed as civilians. (and were) ...about 600 yards from the garrison when (U.S.) Thunderbolts, ten in all, came again and started strafing. We took cover. Then they started bombing. Somehow, we got out of it alive." He reported eleven hundred Japanese killed, eight hundred casualties, and all emplacements knocked out.

Meanwhile, Marcelo at his outpost in Talitay continued to operate his radio. It was quiet for several months with no bombings of pinpointed targets until February, when heavy bombings and strafing seemed unending. Enemy garrisons in the central and western part of the island were bombed. When the Americans landed in Parang, Marcelo's radio was out of order. The guerrilla scouts and civilians who heard the shelling from the U.S. warships at Parang and Malabang rejoiced. The Japanese scattered.

Orders arrived to go to Digos and Sarangani, south of Davao city to help in the mopping-up operations. The clearing of the beaches and port at Digos opened up to the supply ships of the liberating Task forces. Marcelo's group met the U.S. 10th Infantry Regiment on the National Highway on their way to liberate Davao.

Marcelo's group was diverted to support the U.S. 487th on a combat patrol from Malalag to Buayan. It was a ten-day patrol carrying their radio and equipment. The guerrilla 118th Infantry Regiment under Lt. Quinones covered the patrols from Davao to the Davao-Cotabato boundary. The radio crackled with sightings of the Japanese coming down from the mountains in droves looking for food. When a large Japanese group was seen near a mountain village where supplies were to be dropped, the guerrilla commander radioed to cancel a supply drop to the guerrillas. They were outnumbered and would not be able to easily get to the drops in the mountainous terrain. Marcelo acknowledged that the guerrillas led the tasks and deserved the most credit for his group's success.

Sarmiento and Nerez were in Bukidnon to monitor and stop penetration of the Japanese along the Sayre Highway. American forces rescued them when the Japanese surrounded the hills of Malaybalay. They had been monitoring enemy activities in the last 5 months along the Highway that crossed the island north to south. G-2 Intelligence of the 31st Infantry Division bombed and strafed Japanese positions they pinpointed in the hills.

The orders to attack the twenty-five thousand Japanese soldiers in Davao City kept the Japanese on the defensive and added to the radio traffic. In charge of the radio network, Fortun reported enemy traps in their path gathered from operatives to FRS and the U.S. 24th Division as they approached the city. Detailed maps were relayed of Japanese troops with artillery under coconut trees along the road, and thousands of gas drums under trees. A high-powered transmitter in a high school in the city was mapped. But the communication between guerrilla units was not as reliable, so when "an outfit got attacked, the commanding officer would send runners to the other companies nearby so they could be on the lookout for any attacks."

Airfields in the city area were rapidly cleared. Heavy artillery thought to be too heavy that it might slow down the Task Force were hauled earlier by the guerrillas across the island and fired over the City to the Japanese hideouts. Civilians volunteered to patrol the beaches capturing coastal guns and ammunition dumps. The Japanese began fleeing north. Operatives observed a convoy of hundreds of troops and civilian laborers with truckloads of food, fuel, barbed wire, and galvanized iron moving north to Agusan.

With the accurate bombings of their installations and shipping, the Japanese suspected all Filipinos as spies. Civilians were forced to confess the guerrillas' locations and indiscriminately killed.

News arrived of the Japanese civilians' evacuation to Japan. A brutal march of over 160 kilometers was forced on about five thousand Japanese including women and children from Surigao to Cagayan. Few women and children survived from the disease, exhaustion, and guerrilla attacks. Operatives observed that Japanese children were exposed as shields whenever American planes passed over.

Free Philippines published a statement by a civilian employee of the Japanese Army that told of thousands of sick and starving women and children with retreating Japanese troops. Artillery shells dropped on them as they searched for food and water between the American and Japanese lines. With such intense fear of being captured, "they killed their children then themselves. Others murdered those who would not commit suicide... gave renewed indication today that thousands of the emperor's subjects in the Davao area may resort to mass murder and suicide."

At Mintal, the enclave of the large Japanese community in the city, *Free Philippines* reported raging house-to-house fighting. Airstrikes demolished buildings and tore up streets Strong counterattacks from the Japanese delayed the Task Forces' drive north. Japanese civilians and inhabitants of the city mobbed evacuation routes from the city. Retreating Japanese soldiers fled north where the guerrillas were waiting by the rivers firmly holding their lines.

At the 107th Division's headquarters in Magugpo (Tagum), preparations were underway for the American forces' arrival in Davao City. The Division's 130th Infantry Regiment were ambushing enemy patrols along the Davao-Agusan border, cutting the enemy supply line. They had been ambushing for the last year scattered enemy units from Magugpo in the north to Malalag in the south while waiting for the U.S. to return. They had kept the over thirty thousand Japanese troops and armed civilians inside Davao city.

The radio bristled with requests for food for the Regiment's starving troops to keep up their fight. Send more ammunition. "Silva desired more clips for BARs… mortar ammo and automatic weapons." No need for more Carbines.

A message to KAZ informed of a Japanese garrison on the National Highway at the Ising junction by the river, thirty-six kilometers north of Davao City. As an escape route from Davao, it had caught the attention of the enemy. Japanese troops fortified a kilometer around the junction covering their tents daily with fresh abaca leaves to prevent air attacks. All surrounding buildings were occupied. Cement pillboxes lined the road south of the river. North of the river was the Regiment at Kilometer 41 ready to attack.

The tiny bridge at Ising was a choke point in the movement of the Japanese troops. The Japanese burned the bridge, indicating that they wanted to keep the guerrillas out of the Davao area. From the junction, one road led north to the Agusan River Valley and the other to the northwest past the Davao Penal Colony, where the Japanese could hide and survive for an extended period living among the indigenous people who could give them support. With control of the road, they could continue north and overrun the guerrilla (107th Division) headquarters and the airfield (Farm Seven) from which war supplies were being landed.

The Regiment was ordered to remove the Japanese garrison at Ising. Alfredo Pulmano, a guerrilla trained as a signalman fearfully delivered the flare gun to the front line to signal the troops location. A rag-tag Company H had earlier repelled a Japanese unit moving north along the National Highway at their perimeter line by the Ising River. As the primary defenders of the northern area unoccupied by the enemy called "Free Davao," their perimeters were the three rivers, Bincungan, Tuganay, and Ising.

Silva was relieved of his duties as commandant of the 107th Division General Service School and promoted to Captain assigned as Infantry Advisor and Chief of Staff. The 130th Infantry Regiment was placed under his command with the strength of 67 officers and 1,200 enlisted men. He was to focus on strengthening the Regiment. Memos of his orders transferred, promoted, and demoted men, to strengthen the units. He assigned seventy USAFFE enlisted men and Reserve officers to various battalions and companies. All assignment orders and movements were messaged to the Eighth Army.

Companies in the field moved to strategic locations ready for combat. Maps of principal and alternate defensive positions and withdrawal routes were created and reviewed. Even with the ongoing strict preparation schedules, Silva ordered summary court-martials to promptly resolve minor offenses of the men in the field. The medical company was expanded in preparation for casualties.

Silva had managed intelligence gathering, security to radio units, taught military skills and advised guerrilla units on strategy and planning attacks. Now he was about to lead a regiment to eliminate a Japanese garrison. The heavy weapons of machine guns and mortar were brought to Ising from Agusan. Of all the men who arrived on the submarine missions, he would be the only one to lead a regiment to battle.

Lt. Peregrino Andres, his adjutant, insisted that Silva planned the attack to remove the Japanese garrison and open the National Highway for the coming American forces. Once the garrison was destroyed, the road south to Davao City and north to the 'Free Davao' areas would stop the wanton killings of civilians, women and children at Gatungan and nearby villages.

Other officers no longer wanted to risk their lives now that the Americans had landed and were approaching Davao, Andres related. He understood Silva's ambition to reach the city to meet the Americans and maybe receive a 'star' if he succeeded.

As the Americans entered Davao City on May 1, Silva was promoted to Major as he continued to sign special orders increasing the Regiment's overall strength to 1,500 men. Still understrength, their orders were clear. "The 130th Regiment with a strength of 89 officers and 1292 E/M, aided by the 111th Provisional Battalion, launched the Ising, Davao offensive on 1 May 1945 following orders from the CO, 10th MD, as instructed by the Commanding Gen., 10th Corps. The mission was to remove the enemy from Ising proper and clear the National Highway of this isolated garrison, the only Japanese unit between KM 30 and the guerrilla troops." Units of the Japanese 100th Infantry Division waited for them.

Laureta had returned from his conference with the 24th Division which did not know of the regiment's defense at Ising. An airstrike was dispatched between the Ising River and Tagum River in preparation for the attack on the Japanese garrison. U.S. planes bombed buildings near Ising's triangular junction as the Japanese were having their noon meals. An ammunition depot and encampments towards the Penal Colony were direct hits.

The American forces entered Davao City on May 2 while the Regiment's three combat battalions began marching south for four km to meet the enemy garrison pressing to break their lines. The men bravely went to the frontline of the battle without complete equipment. One guerrilla Private said that he was issued a Carbine rifle, hand grenades, and four magazines with sixteen bullets. Uniforms were not given. They dug in their bunker with their hands as they borrowed shovels from each other.

Couriers carried messages from the front line and the signalmen at camp relayed to the Eighth Army. The 1st Battalion positioned along the road that led to the penal colony. Combat companies G and M lined both sides of the national highway north of the Tuganay River to stop enemy reinforcements. But over a thousand Japanese troops arrived to reinforce the garrison at Ising. Men fiercely fought in hand-to-hand combat repelling the enemy. Fighting intensified throughout the night and casualties mounted on both sides. As dawn broke, the enemy still held Ising but they were surrounded. Exchange of fire continued mostly at night while tracer bullets lighted up the sky.

Silva took charge to aim the 81 mm mortar as mortar shells and ammo were running low. The radio crackled for an airdrop of food rations, mortar ammo, Carbine ammo, and all possible hand grenades at Farm Seven airfield for transport to Ising. They had been fighting all night. "Without supplies will be impossible his men make any further advance," messaged Laureta.

A premature radio message from Fertig to KAZ declared that the 107th Division (130th Infantry Regiment) had taken Ising in the afternoon of the sixth of May. But fighting continued. A large enemy unit remained concentrated on the west side of the National Highway junction and along the south bank of the Ising River. The troops faced them from the north side of the river.

Radio operators hammered away messages for immediate airstrikes on towns south of Ising junction. But the troops had to be moved further north and on the Japanese flanks before the airstrikes. Air support of ten U.S. fighter planes swooped down, bombed, and strafed the Japanese in their foxholes around their headquarters at the Ising Central Elementary School and the towns south of the river.

Over thirty bombs dropped over the friendly area, causing casualties. Some guerrillas tried to hide behind large Lawaan (Red Lauan) trees but were caught in the airstrikes. Operatives reported over fifty Japanese dead soldiers were taken away in trucks.

From across the river, the Japanese watched from the inside of their steel- covered pillboxes. Silva picked up the 37 mm gun and "fired through the pillbox window" of the enemy, recalled Pvt. Jose Cabrera at the battle. Some of the men were ordered to swim across the river. Bodies hit by snipers floated while others, trapped, tried to swim back. A pontoon bridge of wild banana trunks was made with the help of civilians to cross the river. A tunnel dug by the Japanese as their withdrawal route to the forest was discovered.

Silva walked by the burned bridge to check on the men he had ordered to swim across the narrow but deep river to the enemy's side. A Japanese machine gunner sighted his pistol in a thick leather holster attached to his belt and fired. A pistol denoted an officer. He buckled and went down on his shattered left shin bone. His adjutant, Andres, was hit in the toe. Immediately he was taken to the rear. An urgent request was sent to pick up five extremely serious casualty cases at Farm Seven airfield all requiring immediate surgery. A message to KAZ on casualties listed a seriously wounded Silva at the 10th Corps Base hospital at Parang for immediate treatment while waiting for an aerial pickup to Leyte. On his way to the 10th MD advance post, Sgt. Ortiz recalled meeting him badly wounded at the hospital.

By this time, the enemy had scattered to the hills. The elimination of the garrison across the Ising River took over a week, but the fighting continued to pursue scattered Japanese troops. "The entire (Ising) operation lasted for 29 days and proved costly to the Japanese..." The guerrillas were most proud that they succeeded in eliminating the Japanese garrison without American troops on the ground.

After the city was cleared, the U.S. 19th Regimental Combat team marched north and met the exhausted but elated Regiment clearing the highway at Bunawan. Together they pursued the scattered Japanese troops into their fortifications and tunnels in the rugged hills of Kiotoy and Gatungan. It took another month to eradicate them from their tunnels.

In the city, guerrillas and the people who survived the carnage, celebrated except for some of the Japanese inhabitants. The 130th Infantry Regiment assisting in keeping the city and locals secure, reinforced the American troops, and supported the PCAU (civil affairs) in providing relief to the residents. Andres recalled that General MacArthur "was with us in Davao. Took lunch with us." It was confirmed that General MacArthur was in the southern Philippines in June 1945.

Elimination of Japanese resistance on Mindanao was declared accomplished in June but elements of the Eighth Army continued until they landed in Sarangani Bay, the last enemy bastion, in July.

When the Eight Army left, the Regiment continued to hunt for Japanese stragglers. An English-speaking Kempeitai prisoner was ordered to inform General Harada, Japanese commander of the Davao area, that Japan had surrendered.

FRS radio networks had accomplished its primary purpose of relaying intelligence data to Army and Navy higher headquarters. The 10th MD managed the operations of over thirty thousand Mindanao guerrillas.

The signalmen operating FRS' radio networks, the radios in each Division headquarters, in the field supporting guerrilla operations, and conducting survey operations on the enemy themselves ensured that the messages arrived at the Eighth Army Headquarters all at great risk to their lives. General Eichelberger, Eighth Army commander, heaped praise on the guerrilla intelligence he received. He wrote that "we did have considerable information about dispositions of enemy troops since guerrilla forces on Mindanao were the most effective and best organized in the Philippines."

The Task Forces 40th and 31st Divisions left Mindanao closing the campaign. The signalmen left their outposts with orders to go to Leyte then to Manila to report to the 978th headquarters in San Miguel, Tarlac on Luzon. Flying to Leyte was by way of Morotai as aircrafts had to be scheduled in a pattern to land in the heavy traffic of Tacloban.

Navarro, Magale, Burgos, Palmeto, and Gonzales were recalled. Navarro said it was early February when he received his order. Palmeto was recalled in June while Marcelo and Acenas were ordered to Manila in August and the 978th headquarters. The following month, Herrera, Fortun, and Pompea departed after the surrender of Japan. Sarmiento and Goroza were reported to have returned to the 978th camp in Hollandia. Goroza wrote that "among my town mates from Urdaneta, (province in Luzon) Sergeants Aniceto Manzano, Telesforo Asuncion and I were the lucky ones to be alive while Sgt. Nicanor Ambrosio was KIA somewhere in Tayabas." He was in Hollandia for a month and decided to stay in the Army. He signed up again.

Mapalo wrote that "Maj. Harry Croell, commander of the 978th Signal Service Co. congratulated him on his return." Fortun was surprised to receive the memo dated over six months earlier of his promotion to 2nd Lt.

The three radiomen operating the primary radios of FRS were among the last to leave Mindanao. Ortiz continued operating his old circuit at X Corps headquarters. Cortez was in Parang, where the American task forces had landed. He and Ortiz took a plane to Del Monte then Manila. Gustilo was at Lianga on the east coast where he was picked up. The difficulty of securing transportation allowed the last men to leave in September.

The signalmen with their radios connected and united the guerrilla divisions of the 10th MD enabling them to support the civilians and the American task forces in destroying the enemy. There would have been more casualties without the guerrillas' support in the inhospitable terrain of Mindanao.

Soldiers of the 24th Division regarded the post-Davao operations as "the hardest, bitterest, most exhausting battle of their ten island campaigns of the war." Frank Savant, a medic with the 124th Infantry Regiment, commented, "We lost many more people on Mindanao than in New Guinea."

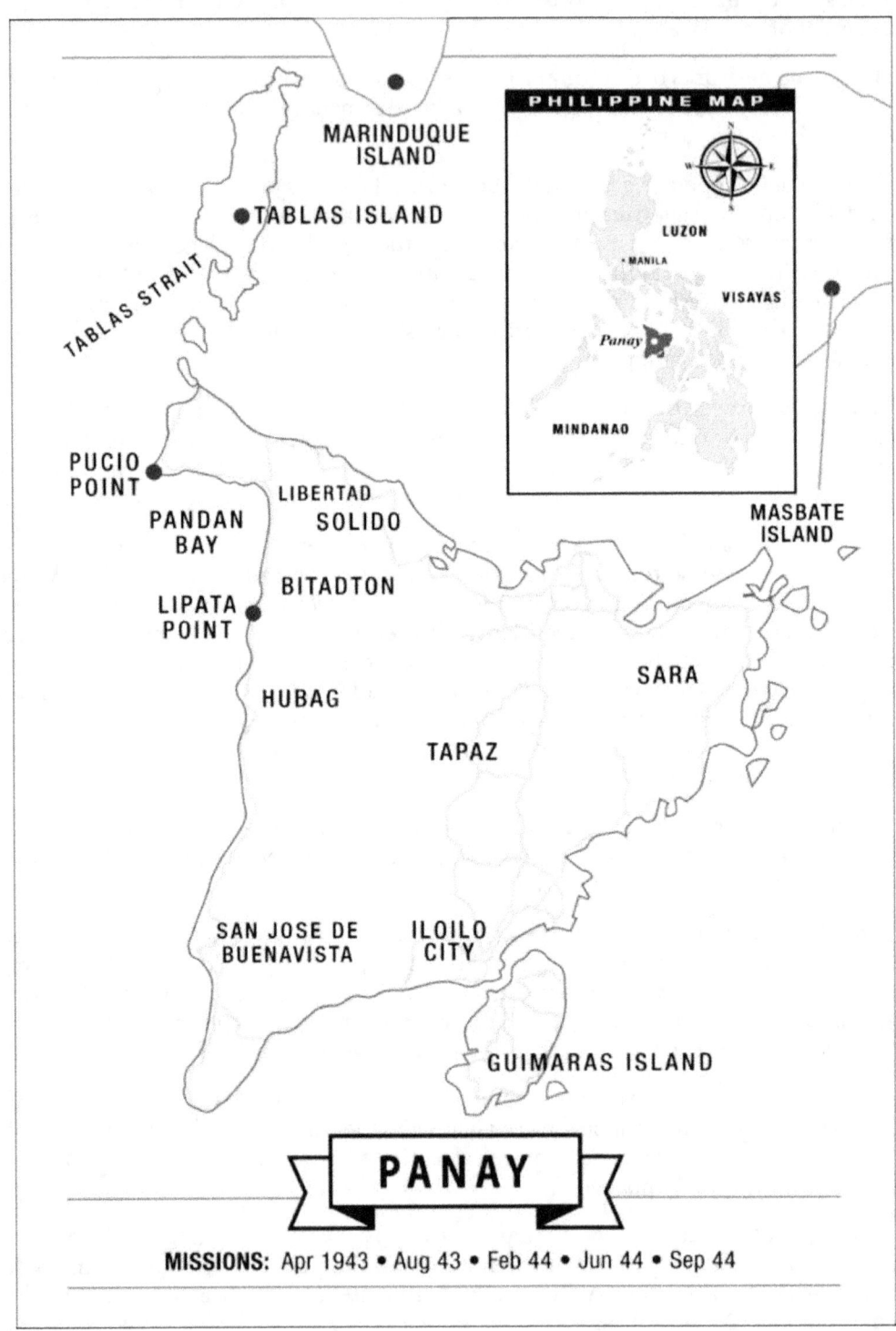

MARINDUQUE
ISLAND

TABLAS ISLAND

TABLAS STRAIT

PHILIPPINE MAP

LUZON

MANILA

VISAYAS

Panay

MINDANAO

PUCIO
POINT

LIBERTAD
SOLIDO

PANDAN
BAY

MASBATE
ISLAND

BITADTON

LIPATA
POINT

HUBAG

SARA

TAPAZ

SAN JOSE DE
BUENAVISTA

ILOILO
CITY

GUIMARAS ISLAND

PANAY

MISSIONS: Apr 1943 • Aug 43 • Feb 44 • Jun 44 • Sep 44

Panay Island

Panay's persistent calls elated Australia after months of silence followed the presumed capture of Nakar and Praeger on Luzon. It was the first evidence that there was a resistance outside of Luzon.

But GHQ was suspicious and did not immediately respond. They feared it was a Japanese ploy because the senders identified themselves as "Free Panay Calling" and used the distress frequency. After many exchanges of radio calls and the sending of a coded message that was deciphered, Macario Peralta was appointed on February 1943 as the guerrilla commander to unify the guerrillas of Panay, Romblon, and Guimaras islands into the Sixth Military District. Peralta's directive from the GHQ was to "continue to exercise command. The primary mission is to maintain your organization and secure the maximum amount of information. (Combat) activities should be postponed until ordered from here (SWPA). Premature action of this kind will only bring heavy retaliation upon innocent people. Victory will come. We cannot predict the date of our return to the Philippines, but we are coming."

Peralta and Wendell Fertig, leader of the Mindanao guerrillas, vied for control of the Visayan islands. When Fertig heard of the Panay appointment, he fired off a message that he was in command of both Mindanao and the Visayas due to his rank. Fertig's guerrilla rank of 'General' was self-appointed and was not cleared with GHQ. It was only later that MacArthur granted him the rank of 'Colonel' over the protests of Col. Whitney. General MacArthur broadcasted that all "command areas will be established based on existing military districts." Peralta was named commander of the 6th MD of Panay while Fertig was commander of the 10th MD that included Mindanao and Sulu. Both leaders would lead their operatives.

Panay was closer to Luzon where Manila was located. It was a strategic area to establish a radio network that would be as robust and expansive as Mindanao. Negros to its east already had a radio network set up by the first submarine mission six months earlier. Peralta sent agents to other islands and as far as Manila to gather intelligence to relay back to

Australia. Although the different island commanders were to cover only their designated areas, GHQ recognized that reports coming in from agents beyond their island commands could be compared and used to validate each other.

Peralta envisioned his organization to be led and accomplished by Filipino officers. He divided his district into combat teams headed by Filipino officers. The few Americans he trusted reported to them. A thorn on his side was Tomas Confesor, the governor of Panay who questioned his actions and his printing of money without verifying a budget. Many times, GHQ had to resolve their disagreements and clarify its position. The ongoing conflict and accusations affected the stability of the guerrilla force and had to be resolved before more supplies were to be sent.

As the war wore on, it became clear in the reports to KAZ that Peralta was feared more than liked except for his officers whom he favored. GHQ tolerated the demands in his messages since Whitney saw his aggressive and ruthless ways as assets.

Peralta's long and flowery messages, like those of many guerrilla leaders vying for attention, prompted the establishment of a priority of information to be followed. Only important information was to be transmitted to limit the time on the air and lower the chance of discovery.

Because of the location of the island and the large amount of traffic of Japanese shipping surrounding it, it was a difficult place to supply. It was a dangerous place for submarines to surface, but supplies, arms, and men had to be brought in for the resistance to continue. Depending on the efficiency of the unloading process, large amounts of cargo over forty tons could take all night, leaving the submarine exposed. If discovered, there would be deaths, loss of shipment, and—the most serious result— the loss of the submarine.

A party was formed after four months of radio contact. One request by Peralta that Ind adamantly denied was "supplying gasoline to guerrilla units was out of the question." But substitutions were worked out. A specially designed and constructed charger like a bicycle were shipped in place of gas-fed battery chargers. But it was later reported that the bicycles were not producing enough power. The men had to pedal for 24 hours to charge the radio.

Toribio Crespo was assigned to the AIB from Milne Bay, New Guinea, when Ind saw him as good material to lead a mission. He was willing to undergo physical training to harden his body, drilling for unarmed combat, hiking for endurance, and remedial training in military basics. In addition, the party was to learn to pass equipment quietly at night and bury it without a trace. The two *Moros* chosen for the party who had no military experience underwent the extensive School of the Soldier class and a concentrated course in the common use of English since they spoke very little English. Crespo, as party leader, was given further instructions in cryptography, Morse code, basic radio maintenance, report procedures, coastwatcher organization, elements of combat intelligence, and propaganda duty. He had his own code to be used in emergency cases. The cipher system was a secure two-key double transposition that required correct encoding or decryption was not possible. The urgency to send another mission cut the training to only three weeks which was not enough time to fully train the men.

The PELEVEN PARTY's mission was to deliver the supplies and directives to the guerrilla commander of Panay and establish two or more coastwatcher stations to cover the traffic in the Cuyo east passage in the Sulu Sea and the Tablas Strait between Panay and Mindoro. Whether or not contact with the 6th MD commander was made, the coastwatcher stations were to be established. GHQ did not want to waste time delaying the setup of radio stations on the first mission to Panay.

USS *Gudgeon* – April 1943

The sub left Fremantle with the PELEVEN party, two tons of radio equipment and 1500 lb of ammo. The men braced themselves as she spotted and sank the *Kamakura Maru*, a former ocean liner with Japanese troops. She sank the Japanese fleet tanker *Toho Maru* with five torpedoes in a night attack and damaged the tanker *Kyoei Maru* in the Makassar Strait. At Pucio Point in Pandan Bay, she landed the first group of men and supplies for the Panay guerrillas. With Lt. Crespo were infantrymen Sgt. Orlando Alfabeto, Pvt. Ali Ladjahasan, and Pvt. Mangona Ladjahasan. The Ladjahasans had sailed to Australia with Frank Young and Klestadt earlier. They were familiar with the Visayan islands and spoke the language. A crew member of the submarine, Ralph L. Wyers, gladly gave his pearl-handled revolver to 'the sergeant' as a souvenir when he asked for it. He thought that these men would not live long due to the danger of their mission. His crewmates donated whatever reading material they had before the sub departed.

The village of Libertad, close to Pucio Point was chosen as the landing beach. The 1st Combat Team headed by guerrilla Cirilo Garcia, who were guarding the rendezvous area, were not prepared for the submarine's arrival. The sub spent all day submerged at periscope depth about a mile off Pucio Point. Crespo and Alfabeto took the risk of paddling to shore in a rubber boat, hoping that the enemy was not waiting for them. When they did not return that night, the submarine submerged and anchored at two hundred yards from shore to wait. The next day the men sailed with Garcia and two of his men in wooden boats alongside the sub. All hands worked to put all the cargo on deck and load into two seven-man rubber boats and the wooden boats for a one-time trip ashore. Two men from shore were evacuated. After the torpedoes were transferred from the aft to the forward tubes, the *Gudgeon* disappeared in the night at full speed. On the way back, she sank the Japanese trawler *Naku Maru* in a surface gun action. The three Filipinos rescued from the sunken trawler said they were forced to work onboard the ship from Manila and were heading south.

Aside from their personal packs, ten of the 40 cases of cargo contained medical items, the 'essential' cigarettes and matches, six radio sets, radio parts, ten machine guns and 120 hand grenades. In addition to the U.S. ammunition, several hundred rounds of Japanese bullets recovered in New Guinea battles were sent per Peralta's request for their captured Japanese guns. Several cases contained ten pairs of boots, one hundred twenty-five pairs of socks, twenty pairs of drawers in "olive drab," twenty- five shirts, five pairs of trousers, and razor blades. They also carried one bolt each of khaki and denim to give or barter. The men in their fitted uniform appealed to the guerrillas who were of the same stature but in tattered clothes, short pants, and barefooted. The American-sized clothes in the cargo were too large and needed to be re-tailored. The large boots were refused as they slowed down escapes.

The aid was not enough, but morale soared. The troops' combat effectiveness was boosted. Crespo was entrusted by Willoughby with P90,000 divided into smaller amounts in five marked sealed tins to deliver to Peralta. Fifty thousand was to be used for military purposes only and the rest to be held in a secure place pending orders for final use. At the meeting with Peralta, Crespo opened his shirt and removed an oilskin pouch tied to his waistline containing the directive. Peralta gave a receipt for the tins of money.

No problems were encountered in the distribution of the cargo, but larger shipments that arrived later were reportedly divided in favor of the officers. GHQ was concerned that the local kinship system that affects proper distribution would damage discipline and loyalty. Favoritism and the chance that these goods would land in the black market proliferated. PRS was aware of the problem but had no choice but to work and send arms and supplies to the strongest guerrilla commander who could unify the scattered groups.

Crespo understood that he was to provide support and secure the radio network of Panay. Instead, he was assigned to guard and secure the submarine rendezvous area. He later reported that he had no duties in connection with the radio or the coastwatcher system but was just overseeing the unloading and handling of subsequent submarine shipments' arrival.

The guerrilla leaders tended to send extensive and flowery reports to get GHQ's attention. Peralta had operatives reporting on whatever they observed in the surrounding islands up to Manila. The messages jammed the air to Australia. Many of the short messages from Panay described accomplishments and situations that were unrelated to the enemy. KAZ had to establish a standard format and priority list for reporting intelligence. The Signal men were greatly relieved since they were responsible for picking out and analysing pertinent data for transmissions so as not to contribute to heavy radio traffic. The priority was established as follows: 1) enemy strength and dispositions; 2) occupied seaports, cities, towns, villages, enemy unit identifications, captured documents; 3) operational airfields, type of aircraft, facilities; 4) location of anti-aircraft defenses; 5) enemy troop movements; 6) treatment of civilians; 7) naval dispositions; and 7) civil administration.

The Japanese were active in Panay. The people's loyalty was divided between Peralta and the provincial governor. To maintain the people's support, Peralta radioed for money in his messages to "combat Japanese attraction policy...paying partial salaries of soldiers and buying food or soldier families starve. The money you sent is reserved for intelligence only. I do not have material to print money. Confident of people's loyalty, but money talks." But Governor Confesor complained that Peralta was printing money without a budget. Reports to KAZ showed Panay printing emergency currency and Japanese military currency to pay for a wider intelligence network, including Luzon. The printing and use of Japanese currency brought up the question of "how far the U.S. was morally bound in the redemption (after the war) of such notes in exchange for services rendered?"

The constant request for more funds concerned PRS as there was little effort to keep records of its use by the guerrillas. Whitney commented that, upon checking of their records, President Quezon had requested Panay Governor Confesor to "release P25,000 per month in the last six months to support forces."

By mid-1943, Panay agents in Luzon reported that combat-ready Japanese troops had increased to 150,000 in Manila. Intelligence on harbor defenses, airfields, and enemy dispositions from the north to the south of the city was transmitted. KAZ had to verify reports from other sources as "reports from Central Luzon and Manila were exaggerated." Helpful civilians were untrained in procuring the correct intelligence.

Guerrilla leaders Capt. Manuel Donato of Masbate Island and Capt. Sofronio Untalan of Marinduque Island were organized in Panay to gather intelligence, increasing the volume of intelligence to Panay's main control station for relay to KAZ. The transmission of guerrilla reports soon became a competition with Mindanao transmissions. The focus of the intelligence reporting changed from uniting to fight the enemy to personal military ambition and vying for KAZ's attention.

From July to August 1943, the Japanese staged a brutal anti-guerrilla campaign in Iloilo, the capital city, and Guimaras Island. Over a thousand civilians were killed in two weeks and the Japanese threatened to kill more if Peralta and Confesor did not surrender. Guerrilla officers escaped to neighboring islands while Peralta holed up in the mountains. The net control station and radio went off the air. However, the alternate station at Hubag, the rendezvous area of the submarines, continued operation. Crespo oversaw the flow of reports coming in and transmissions to GHQ.

The Japanese rampage continued throughout the north-eastern part of Panay. A barrio in the municipality of Sara was massacred. Women were abused in public. The destruction extended to the Aklan area and crossed the Sibuyan Sea to the nearby islands of Tablas and Romblon.

USS *Grayling* – August 1943

The *Grayling* was diverted from its patrol to land the last of the AIB intelligence missions. Lt. Irineo Ames was chosen to lead the party. He was a nautical school graduate and was a member of the Off-Shore Patrol Reserve defending the shore installations of Bataan. He was glad that Sgt. Felipe R. Ginobiagon volunteered for the mission. As a fellow officer, he already had two months of training at Camp Milton. He knew Ginobiagon as hard-working in their trips together in the 35th Transportation Corps Service Group during the New Guinea campaign. Jose J. Ramos volunteered for the mission from his desk job in the AIB Personnel Department. The men trained intensively for only three weeks which might not have allowed enough time for developing their physical endurance for the mission.

Before flying from Brisbane to board the sub at Fremantle, Ames received his directives, a copy of which he had to give to Peralta. His team was to set up one or more coastwatcher posts and report to KAZ once established. GHQ wanted information on enemy strength, ship movements, airfields, routes, and harbor defenses.

The men boarded the sub with two tons of arms, radio sets, medicine, equipment, and U.S. currency. Some propaganda material of 'I Shall Return' cigarette packets, chocolate bars, and others were included. While the sub was running on the surface to recharge batteries, the men opened the hatches to let out fetid air. An officer gave them instructions on their destination. Ramos served as quartermaster and look-out from his previous experience as a civilian. While the sub refueled at Exmouth Bay, the team practiced rubber boat landings.

The sub surfaced at Libertad in the night and waited for the proper security signals. Once the signals were received, the shore party had to locate the sub, which would be impossible to do in the dark of night. So, the cargo was unloaded into rubber boats the following night. They deflated their rubber boats immediately after landing and hid them among the bushes. With the help of volunteer guards, it did not take long to unload the small cargo they brought. A jar of muriatic acid that would have been useful in removing rust from corroded equipment was left behind.

The submarine slipped out cruising the Tablas Strait and "recorded her last kill, the passenger-cargo *Meizan Maru*." That was the last time KAZ heard from the submarine. She was officially reported lost at sea with all hands. The landing party was alarmed when they heard how close they were to die in their mission. The news tempered the eagerness of the men back at Camp Tabragalba. But they were dauntless.

The mission party was brought to the camp in the village of Hubag. Crespo, who had arrived four months earlier, was glad to see them.

Peralta reported to KAZ that his men were envious of the Filipinos that arrived looking well-fed and paid better. Ames' party was placed under his command, and they immediately took charge of the net control station of the 61st Infantry Regiment, 6th MD to relay intelligence reports of enemy shipping and aircraft sightings to KAZ. Ames instructed qualified men at the Regiment's Aerial and Naval Intelligence School for officers on duties and tactics to lead units against the enemy. He inspected a radio outpost on the west coast of the island once a week and successfully penetrated a Japanese garrison on Tablas Island to set up a coastwatcher and air warning station. He commended the operation of his net control station and accurate reports.

Ramos was Ames personal assistant in charge of administrative matters including their security detachment of guerrillas. He kept all secret documents intact. He voluntarily took risks with Ames, trekking mountains, jungles, and crossing seas between islands to accomplish intelligence assignments. From December 1943 to March 1944, Ramos was assigned to Crespo's coastwatcher station. Here he "instructed officers on the use of effective weapons such as the Thompson sub-machine gun, Carbine, and the Garand rifle in preparation for when the guerrillas will receive the order to attack." The guerrillas were also taught "techniques of unarmed defense." Ramos was given command of a platoon of guerrillas to harass the enemy. He displayed a calm disposition against the enemy when the time came for heavy mopping-up operations.

Ginobiagon operated Ames' coastwatcher post and air warning station diligently. In June, he sighted a drifting Japanese motor launch 25 miles away from the observation post. The launch had been hit by a submarine in an attack near the outpost. He took six guerrilla volunteer guards in a sailboat with four Carbines and his pistol to investigate the vessel. Orders were to shoot survivors who would jump overboard to escape. They captured war materials and brought Filipino collaborators and Japanese soldiers back to camp for interrogation.

By the end of the year, reports arrived of 17 Americans captured, beheaded, and burned. And the killings continued into the following year. The Panay District's executive officer analyzed that the "lie low" policy given to the guerrillas did not prevent Japanese atrocities from happening. It, instead, resulted in the disorganization of the men. Since they could not fight back, the situation raised the feelings of insecurity, helplessness, and fear. The civilians in turn no longer helped the guerrillas who were supposed to protect them. Some guerrillas surrendered and were judged as collaborators although some continued their resistance by reporting back to their fellow fighters. To buoy up the sinking morale, the executive officer declared, "... we are not running away from the enemy. Our technique is to run around the enemy... should not fight, neither should you surrender."

One interesting message received by KAZ was "some German mechanics landed in San Jose to assemble airplanes."

USS *Narwhal* – February 1944

The new year found the cargo submarine *Narwhal* returning to Darwin to pick up more supplies for Panay and Negros. Along the way, the master gyro compass failed, and a Mark IX hand compass was used to steer the sub. On board was Commander F. Kent Loomis, U.S. Navy, whose mission was to observe supply runs to the guerrillas and the pickup of evacuees. His observation reports to PRS were for the planning of succeeding missions using Filipinos. Parsons may have been onboard.

Both torpedo and engine rooms were loaded with a cargo of Carbines, rifles, grenades, ammunition, radio equipment, and 20 cases of cigarettes. Three thousand pistol belts, 3,000 Carbine magazine web pouches, and sundry quartermaster equipment were used to wrap and protect the cargo items. Forty-five tons of boxes was marked with white corners for Panay and the same number with black corners for Negros to ease unloading. Interesting to note were certain cargo items— seven volley balls, five footballs, 90 softballs, 100 Ping-Pong paddles, 200 pinochle card sets, and 25,000 phonograph needles.

On the high seas, the *Narwhal* sighted several heavy bombers and quickly dove to avoid being seen. Approaching Panay, she counted twenty-three lights along the beach of Pandan Bay. She surfaced by the coastal town of Libertad and exchanged security signals. A guerrilla officer came on board to begin the unloading. Hatches were opened, and 45 tons of cargo was unloaded into sailboats in four hours. Six passengers embarked. The submarine cleared Pandan Bay for Negros before midnight still steering with the hand compass.

The following month, the USS *Angler* arrived on the west coast of the island to bring supplies and to evacuate 19 U.S. citizens. A large crowd of people crowded the coconut tree-lined beach. The prescribed signal was hoisted between two trees. With the submarine still a thousand yards from shore, guerrilla officer Garcia, who was in charge of the rendezvous area, and American officer, Lt. Maynard Hawley, who had joined the guerrillas in late 1942 came on board. (Hawley would later return to the islands with a mission party.)

By mid-1944, Panay's network was comprised of ten operating radio stations. The chatter in the radio traffic had significantly increased. Whitney, ever vigilant with whatever useful information he could extract from sometimes lengthy guerrilla messages, sent a reminder to "use the radio for essential traffic and periodic contact. All others should maintain silence." The purpose of the radio was to keep KAZ in close contact and exercise control with the military forces and the people. Area commanders were to use it as a means of inter-island communication for local protection and coordination aside from control over their widespread forces. But it was also being used as a morale booster "for working off pent-up feelings and unloading overloaded chests."

Hunger continued to take its toll on the guerrillas and the people. Orders were sent out to limit all food rations inside Panay. The mission men also suffered as they were dependent on the locals when their rations, which was usually good for a month, ran out. All were to plant food in the mountains. Many of the carabaos the civilians used for tilling their farm was shot to feed hungry guerrillas. The only explanation given was that the Japs would have done the same, so the farmer was not losing anything. Everyone tried to keep alive until the Americans returned.

Governor Confesor complained that his messages to KAZ sent through the Panay station were not getting through, so he sent them through Negros's control station. His messages to KAZ told of the guerrillas not having food reserves from the last harvest despite good crops. Many units were subsisting on roots. The guerrillas were ordered to commandeer half of the palay (rice) in the private warehouses. He reported on the guerrillas' lack of discipline, banditry, and the murders they committed.

USS *Narwhal* - June 1944

The *Narwhal* departed Darwin with Sgts. Isidoro Dacquel and Fermin Francisco; Pvt. Alfred Salvacion, and T5 Aniceto Kintanar. These were men trained in weather observation.

Torpedo tubes were emptied to accommodate the 100 tons of Carbines, grenades, machine guns, over a million rounds of ammunition, gallons of gasoline, lubricating oil, 8 tons of signal equipment, medical supplies, and currency materials. A large amount of Japanese currency was to be turned over to Peralta. The *Narwhal's* middle had become bent from carrying large cargoes from earlier missions. Her walls creaked as she departed with an oil slick and a smoke trail. On the way, she fired on the Japanese petroleum facilities on Seram Island.

Two thousand men were stationed to guard the routes to the rendezvous point and an additional reserve of men were placed some two kilometers away. A month earlier at the landing site, an enemy force had advanced north towards the rendezvous area at Lipata Point, interfering with the arrival. Another Japanese column of around 200 troops had advanced west to execute a pincer attack on the site. Another group of enemy soldiers coming from the south was repelled, but around sixty were able to reach Lipata Point.

Security signals were set up for the sub and quickly taken down as a Japanese garrison only three miles in Culasi might see it. At dusk, the signal was put up again, and two hours later, the sub surfaced at Lipata Point. Garcia came on board and informed the captain of the Japanese presence. He left an hour later and did not return. In the dark, there was much confusion, and the unloading of cargo was disorganized, stretching to over nine hours. There were many boats to load one hundred tons of cargo, but the boatmen would only take only a few boxes at a time. Men ran all over the decks trying to get cigarettes, and trade knives and *bolos* for more cigarettes. Some equipment was thrown into the water.

Two evacuees waiting to board related that "they were busy getting cigarettes and personal articles for themselves rather than supervising the job." The *Narwhal's* log reported that "the Filipino officers sent out to direct unloading had no control over the native boatmen" so their boats returned to shore less than half loaded. The captain had to assign a crew member to each boat to make sure it was fully loaded. Some of the supplies were unboxed and dumped into the boats to save space. The unloading had to be finished by daylight so as not to endanger the submarine. Evacuee Pvt. Nathan Talbot did not hold back in his interrogation report in Brisbane saying that the "most of their time is spent pestering Peralta for orders so that they can be sent back to Brisbane."

At dawn, some 30% (twenty-five tons) of the cargo and gasoline drums were still on deck. To get as much of the cargo onshore as possible, the captain "assigned two Filipino men to swim to shore each with a drum of gasoline." When two Japanese troop ships were sighted, the evacuees were dispersed along the beach of the rendezvous area. Fourteen evacuees boarded the submarine, including WOJG Wise (1943 Mindoro mission). Another account said that it was forty evacuees. The concerned captain ordered the cargo thrown overboard if it was not cleared in thirty minutes. He also ordered that those who refused to get into the boats to be "thrown overboard." Other chose to jump. The remaining twelve gasoline drums and thirty-five Carbine boxes hit some of the Filipinos in the water as they were shoved off deck. Fifty tons were thrown overboard according to Sgt. Dacquel. Some of the boxes that reached the shore were opened and not moved into the safety of the forest. Later, Dacquel wrote to express his appreciation of the fighting abilities of Capt. Cadiao's guerrilla unit in the repeated ambushes by the enemy during the unloading of cargo but was "never compensated or even commended."

Departing at daybreak, the sub picked up a vessel on the radar. She sunk the Japanese motor launch *Shinsu Maru* with her deck guns. In the Sulu Sea, she avoided two small escorts and damaged the Japanese tanker *Itsukushima Maru* with two torpedoes.

The party reached ashore in the early morning. They were able to bring two complete radio sets. Boxes were scattered and took several hours to haul away except for the gasoline drums that were buried in the sand. It did not take long for a Japanese fighter plane to spot the strewn boxes of supplies. Hands flew as the men and guerrillas clawed into the sand for a foxhole and bullets ricochet all around them. A bomb dropped, missing their supplies by a hundred feet. For three days, they camped in the hills waiting for their assignments from Peralta's headquarters.

At GHQ, the submarine captain's log and interviews of the evacuees about the disorganized landing caused PRS to rethink about "sending American officers (instead,) though their presence was resented (not totally welcomed), to accomplish in developing a dependable organization on Panay." There was some unity under Peralta, but the discipline was missing amongst his men. Sending Filipino officers was "a waste of manpower because they would soon be submerged in lethargy which prevails throughout the island."

Earlier, Crespo who was assigned to the sub rendezvous area with Ames had sent out a message to KAZ to check on Ames expressing his concern of a change in him.

Ind explained that changes in the character of the men were due to their situation. They were forced to be resourceful, make the best decision, but still had to follow and support the guerrilla leader's decisions. Those whose training was in the infantry and not in radio operations after some time on the islands were to become lax and not as diligent in carrying out the main job and purpose of the mission. Spending more time socializing with the locals resulted in inadvertently revealing secret information. The directive to follow orders from the guerrilla leader would take them away from their mission's purpose causing disillusionment. Ames was the party leader but was "frustrated with (his) assigned job as assistant to Garcia." He revealed that he knew nothing about radio as he was a reconnaissance man trained in infantry.

Ind defended the men. Before the departure of Ames' party, he had recommended Ames and Ramos for a promotion. Ramos possessed qualifications that would entitle him to Officer Candidate School if he remained in Australia. He justified that Ames completed his training with commendable efficiency, delivered supplies and personnel to Panay, and operated as coastwatcher district supervisor. The men's intelligence reports showed diligence to details. Ind acknowledged that men who have taken dangerous assignments were not being brought up for promotion as readily as those still in headquarters because of their isolation.

Dacquel whose patience was exhausted from the unjust treatment of his team, requested in his September 27, 1944, message to KAZ, "to investigate the mission, duties and activities of Lt. Irineo Ames . . .conspicuous in his uncooperativeness causing us more trouble." He added the attention to a wife and the "playing of poker with the enlisted men to the neglect of their duties." Whitney sent an order to Ames to cooperate with Dacquel.

Jose Ramos with the party who arrived earlier, sent word that they were to move to Bitadton, a small village up the coast where a Philippine Army officer, Lt. Mendoza, had "a sort of radio station." They stayed at the station for a week waiting for instructions and spent the time looking for their instruments that were scattered in the scramble for safety during the landing. They were relieved to move to Solido, still further north with a complement of guerrilla security, and set up their stations. Dacquel said that there was "too much time wasted bickering how our reports were to be handled. Garcia and Ames insisted on running it to suit themselves." Dacquel confidently insisted on running it "according to his operational instructions" and to transmit via the weather main station at Dimorok Canyon in Mindanao.

Coordinating radio schedules, frequencies, and time changes had to be sorted out for intelligence reports to be useful. KAZ ordered to send messages in Greenwich time with instructions on calculating for it. Weather reports were to be sent hourly but only at certain hours. His men had to check the cryptographic instructions and review the messages before transmitting. They were not to send another report until it was fully understood. Security violations meant the possible enemy interception of messages.

Supplies began to run low, and Dacquel requested station supplies – forms for codes, ruled pad, twelve cartons of safety matches, and binoculars.

August came, and news of American planes bombing enemy installations over Davao in Mindanao raised everyone's hopes. Bombs fell on Panay the following month. Peralta was waiting for this moment to move his command post from the Antique mountains to the lowlands and openly fight the enemy. Wanting to keep the morale of his men who were with him from the beginning, he stopped recruitment into the 6th MD except for where services were desperately needed. Now that the American forces were near, recruits were looked upon as" opportunistic patriots."

Gathering information on the enemy went smoothly for a while until guerrilla intelligence revealed that the Japanese were hunting for their station. Dacquel sent a message that enemy penetration had caused his group to change locations. In the darkness, they escaped to a village on the northwestern tip of the island and set up a station to report their whereabouts, but they would stay there only for a few days to avoid capture. Finally, he split up his team. Francisco and Kintanar took the overland route through winding trails across the high mountains towards Pucio Point while Salvacion and Dacquel doubled back on the provincial road on the west coast then along the shore northward to the village of Libertad.

In the next three months Dacquel moved his station twice until the end of the year when everyone met up at Pucio Point. He radioed KAZ that he had not received the supplies he requested four months ago. He needed a new motor for charging batteries. KAZ' reply to ask Peralta for the supplies depressed Dacquel. He believed that he would never get those supplies meant for his party after the way they were treated by the guerrillas.

Dacquel and Salvacion stayed at Pucio Point operating "Bahala Na 004" watching the air, weather, and the coast for water transport movements. "Bahala Na 005" was set up in Tapaz, the center of Panay by Francisco and Kintanar for weather observations. Their stations were integrated with the local guerrilla net for intelligence information and the weather control station in New Guinea. More urgent messages for weather station supplies were sent to KAZ.

KAZ believed in the importance of Panay's central location. Operatives reported a hospital and rehabilitation base for the Japanese on the island. The seas around Panay were possible routes for the enemy to reach the Leyte landing site. In spite of this, a small force was left on the island when Japanese troops were sent to Leyte and Negros as reinforcements. KAZ sent a warning to maintain radio silence except in the transmission of essential traffic and periodic contacts.

USS *Nautilus* – September 1944

The submarine surfaced near Libertad at night. She had just escaped a harrowing experience of running aground in Cebu. She opened her hatches and brought on deck 40 tons of supplies, 70 drums of gasoline, and 4 drums of oil. The cargo included flour, sugar, powdered milk, rice, and salt. Three large sailboats arrived with forty-seven evacuees. The sailboats made two trips to bring the cargo and Sgts. Robert Caswell and Robert George who were bound for Mindoro Island to shore. The *Nautilus* cancelled the Mindoro landing as she had only enough fuel left to return to Australia after jettisoning extra load and fuel to break free of the reef in Cebu.

With all the risks taken by the submarine to deliver supplies, Peralta sent a message to GHQ that the subs made "too much noise while surfacing." Supplies continued to be not properly distributed as cargo was picked before the enlisted men were given a share. But KAZ considered this as an internal guerrilla problem. The only concern was the intelligence reports Panay sent.

The submarine USS *Hake* followed the next month, bringing more supplies and evacuating 19 people, including American pilots who had been shot down.

From the first submarine shipment in mid-1943 to October 1944, GHQ delivered 350 tons of supplies, including weapons and ammo, to Panay.

Luzon Island, Philippines Landing – January 1945

Peralta came down from his mountain hideout to Iloilo with the news of the Luzon landings. The Japanese had retreated to an area outside the city although a 3,000 strong Japanese garrison remained. Another 2000 enemy soldiers were scattered around the island. He radioed KAZ that of the 22,000 guerrillas under his command, only half of them were adequately armed. He had repeatedly asked for arms, ammunition and other supplies that would give his men a better fighting chance against the enemy. The arms that arrived with the submarine were not enough. Without uniform, bare headed, they dug foxholes with their trusty *bolos* and anything sharp enough to break through the soil. Unknown to him the American forces on Leyte were also in short supply of war material. The landing forces needed larger quantities of supplies against an enemy that would not surrender.

Panay Island Landing – March 1945

A month before the Americans arrived, Iloilo was under guerrilla control and messages were sent to the naval fleet to spare the city from bombardment. Two-thirds of the city was destroyed partly by the Japanese as they retreated. American bombers destroyed the Japanese garrison on Fort San Pedro and surrounding areas. But the piers on the Iloilo River and nearby houses were spared.

The signalmen operated the guerrilla radios of the 61st Signal Group and sent guerrilla attack plans to KAZ. The guerrillas generated valuable intelligence on Japanese dispositions, served as local guides and, in large numbers fought with the American troops in offensive operations. One extensive operation saw two reinforced battalions of the 61st Infantry Regiment seize the town of Tiring with mortars and bazookas. The guerrillas, trained on combat techniques by the mission men, went on to secure vital bridges, airfields, and road junctions in the U.S. 40th Division's rear areas. KAZ received information on downed Allied aircraft from civilians who often went directly to crash sites to search for equipment and survivors.

After the liberation of Iloilo, Dacquel's party at Pucio Point and Tapaz received orders from KAZ to report to Mindoro. But, almost at the same time, Peralta ordered the radiomen to report to Iloilo. Garcia told their guerrilla guards to leave them so they hired a sailboat to cross Tablas Strait for Mindoro. A plane flew them to Tacloban, Leyte, for the night at a camp for troop movement, where Dacquel's "shirt with all my prized possessions were stolen." They arrived back in Hollandia in May.

Dacquel had strong words about his mission. He had no cooperation on the mission objectives from Garcia and his officers, "who worked against us in all efforts," he related. Peralta's directive was for Garcia to provide protection, food, and quarters to his team. Instead, the rations of Dacquel's team were taken and they were left with nothing to eat the day they landed. From the supplies brought by the *Nautilus* in September, only one jar of Ovaltine, one packet of razor blades, and one carton of matches were all they received at their outposts. The reason given was that supplies were thrown overboard in Cebu waters when the submarine ran aground. Dacquel believed that the boxes were picked through. He had asked that in the "next shipment, please stencil my name on all items meant for me. Food and medicine acute in this isolated station."

PHILIPPINE MAP

LUZON

MANILA

Mindoro

VISAYAS

MINDANAO

MANILA

LUZON

NASUGBU

BALAYAN

TILIK PORT

LUBANG ISLAND

GOLO PASS

GOLO ISLAND

MARICABAN ISLAND

BATANGAS/CALATAGAN/ CAPE SANTIAGO

MINDORO STRAIT

VERDE PASSAGE

BOGIO POINT

VERDE

PUERTO GALERA

DEL MONTE POINT

MARINDUQUE ISLAND

MANANAO

ARO POINT

ABRA DE ILOG

MT. CALAVITE

CALAPAN

NAUJAN

PALUAN TOWN

MAMBURAO

BAGALAYAG POINT

PINAMALAYAN

BARAHAN

Mindoro

APO ISLAND

PANDAN ISLAND

MT. BACO

BANSUD

BONGABONG

MASBATE ISLAND

TABLAS ISLAND

SAMAR

BUSUANGA ISLAND

SABLAYAN

TABLAS STRAIT

ROMBLON ISLAND

SIBUYAN ISLAND

SAN JOSE

LEYTE

PANAY ISLAND

MINDORO

MISSIONS: Dec 1943 • Mar 1944 • Jul 1944 • Oct 1944 • Dec 1944

XIII

Mindoro Island

The radio contacts from the islands of Panay, Negros, and Mindanao cemented the presence of a developing resistance movement in the islands. AIB had sent six secret submarine missions to the appointed guerrilla leaders to set up communication networks in their islands. When the Philippine Regional Section (PRS) took over managing communication with the Philippines, the focus turned to Luzon, where the greatest number of Japanese were garrisoned and Manila, the seat of government, was located.

The PRS missions were primarily composed of Filipino radio operators from the 978th Signal Service Co with reconnaissance men for intelligence gathering and protection. At their remote outposts, they operated to gather intelligence on the enemy to send back to Australia, engaged in defensive combat to protect the radio, and instructed guerrillas in military tactics and intelligence gathering.

Mindoro was the closest island to Luzon and by the mouth of Manila Bay. But the island was one of the most dangerous spots in the country with its surrounding waters swarming with Japanese patrols from the north to their occupied territories in the south. KAZ worried that the tight enemy blockades around Panay, Negros, and Mindanao would spread to Mindoro. A radio station on Mindoro, would be closer for couriers to bring current information on Manila. Japanese shipping in the Strait carrying war material into Manila Bay and further out to Lingayen would be sighted, flashed to patrol submarines, and destroyed. That was the plan.

The submarine *Narwhal* was refitted into a cargo ship and was ready. With a submarine that could carry more cargo at their disposal, the first mission under PRS with Filipino soldiers from the 5217th Reconnaissance Battalion and the 978th Signal Service Company stationed at Camp Tabragalba was formed. Eventually, over two hundred 978th Signal men went on mission behind enemy lines.

USS *Narwhal* – December 1943

The men were recommended for the mission by their instructors. They were told of no guarantee of returning alive. "You have very little chances of living, especially in Japanese hands. If anyone wanted to stay and go on a later mission, speak up and be sent back to Tabragalba. Your backing out will have no official records and will not in the least reflect on you." No one did. "I am very proud of all of you... You will lead us when we enter Manila in a victory march. Are there any questions?" asked Whitney.

The orders were cut. The recon men with infantry skills were Capt. Ricardo C. Galang, Lt. Ruben F. Songco, Sgt. Benjamin M. Harder and Sgt. Vicente A. Pinuela. Radio man Sgt. Arcangel Baniares and WOJB Braynard L. Wise joined. Men from the 978th Signal Co. were T4 Alfredo A. Alberto, and T4 Ramon D. Vitorio. Galang described Wise as a radio expert and Baniares a Philippine Scout licensed as a commercial radio operator before the war. Songco was a graduate of the U.S. Naval Academy, Annapolis and had joined the Army because the Navy would not commission Filipinos. Harder was on a merchant ship when Pearl Harbor was attacked, landed in Capetown and New York before Camp Tabragalba. Pinuela farmed in California, and thus he looked the toughest of them all. Alberto had a radio shop. Vitorio refused to talk about himself but could send thirty words per minute over the radio. Lawrence H. Phillips, a former farm manager at the Del Monte plantation on Mindanao, led the party.

The party had barely a month of "intensive training that included theory and field maneuvers in infantry tactics, and signal communications" with hardly any of the 4 months of physical hardening and amphibious exposure the later arrivals underwent before going on a mission. Galang recalled that he had arrived in Australia only a month earlier from Camp Roberts in California when he was woken up at Camp Tabragalba to leave in 15 minutes. The submarine *Narwhal* was waiting.

Phillips was briefed on the objectives: Establish coast watchers on Mindoro Island to monitor Paluan Bay, Verde Island Passage, and Apo West Pass (across Mindoro Strait) in the western and northern side of the island. From these locations, his men were to monitor enemy shipping entering and leaving Manila Bay. He was to "coordinate the development of information from Manila and central Luzon." A more difficult objective was to unite the warring guerrillas under one leader on Mindoro as the island was the closest to Lingayen Gulf, the back entrance to Manila. It was not an easy goal with guerrilla leaders fighting for men and territory.

Not knowing what reception they would get with no recognized guerrilla leader to prepare the safety of their landing, the men felt it might be suicidal just like the mission to Negros a year earlier. Peralta on Panay Island, across the Tablas Strait, was informed of their arrival, but there was no assurance of any preparations he may have made to secure the landing.

Commander Parsons joined in escorting the group to Mindoro. He would continue to Mindanao with supplies and to Manila although he was directed not to risk entering Manila. But various reports and sightings of him in the city proved otherwise. He may have wanted to find whether members of the Manila intelligence group that Emigdio Cruz contacted several months earlier were still alive and continuing to operate. Throughout his time in Mindanao, he worked with the islands' guerrilla leaders on GHQ's policies.

Parsons requested to have prepared for the trip, the first half of twenty thousand 2 oz chocolate bars with high melting point, Vitamin B, and other vitamins. The American and Philippine flags and the text 'I shall return – MacArthur' were printed on the wrappers. He brought many counterfeit pesos to enable the guerrillas to buy supplies when available and destabilize the Philippine economy. The crisp new bills were placed in drums with chemicals to age them.

General MacArthur was present before their departure from Brisbane across to Cape Moreton Island to board the *Narwhal*. He entered the room, returned their salutes, and shook their hands, calling everyone by name. "Good luck. You will serve the cause of freedom. I am proud of you." Willoughby shook their hands too. Ind walked them out. Galang wondered how General MacArthur knew their names.

The submarine's manifest listed them as mess boys to maintain the secrecy of the mission. Succeeding missions did the same. Inside the sub, they walked the white-painted metal hallway down to the torpedo room, left with only two torpedoes to make room for them. Although the sub was lightly armed to carry more supplies, it was still a warship whose mission was to attack encountered enemy ships. In between boxes of supplies, they squeezed their sleeping cots. They did not see daylight submerged all day passing through enemy patrolled waters. Their training continued under Parsons until the rendezvous site.

Dates and hours were forgotten as there was no sunrise and sunset, Galang recalled. When facing death, titles began to have no meaning. They started calling each other without titles. Wise, a radio expert, made a set of codes for the Navy control station FTAW and WAT to be set up on Mindanao for Parson to deliver. The men studied the little books of silhouettes of Japanese ships from the Supply Depot.

In the Sulu Sea, the sub encountered a convoy of two Japanese oil tankers accompanied by three escort vessels. She released six torpedoes and two of the escort vessels fired back. She zigzagged under shell fire with her four diesel engines christened Mark, Luke, Matthew, and John, at top speed. She went down to 300 feet according to Alberto and stopped all motors until the escorts were believed to have passed over. But when she surfaced, the Japanese vessels were waiting and gave chase firing. In the dark starless night contact was lost with the enemy but it made difficult to identify the surrounding islands. Determining her position, she passed Negros Island and dove to avoid fishing boats. All kept quiet and the men stayed out of the crew's way. They were sweating from the tension as it was their first time to be in a submarine. The knocking and creaking of her old hull magnified as she slipped away.

At daylight, she sailed through the Verde Island Passage between Mindoro and Luzon as Parsons wanted to observe the sea lane. The sub approached the beach off Paluan Bay at dusk after 22 days. The decision to disembark without any security signals from the pitch-black shore heightened the anxiety. Galang wrote in his report that he, Harder, and Pinuela paddled ashore in a rubber boat without any advance information as to whether the enemy was in the area or not. Harder had insisted that he go because "it was part of the game." When the area seemed clear of the enemy, the rest of the party followed. A sailboat, the *Dona Juana* owned by an enterprising civilian, helped unload the ten tons of cargo and submitted a bill for his service. In the cargo list were receivers/transmitters, batteries with a bicycle to charge them, wires for antennas, powerful field glasses, and carpentry tools. A separate box contained medical supplies especially Atabrine tablets, and food, G-1 rations, instant coffee, milk for babies, vitamins, and candy bars with 'I shall return MacArthur' wrappers. The men carried anti-leech powder, clothes, mosquito bars, ponchos, and weapons. Phillips brought Japanese currency and Philippine Treasury notes. There was no welcome party. Without a known guerrilla leader to prepare the security and enforce secrecy, news of their arrival almost certainly reached the Japanese.

With the unloading done, the sub submerged and slowly moved out of Paluan Bay towards the open sea to Mindanao with Parsons and thirty-two evacuees. She surfaced and ran at full speed to leave her oil leaks and avoid being spotted. The old submarine had rivets instead of welded joints, producing the leaks.

It was raining hard as Galang and Phillips made a reconnaissance trip to the island's interior for a possible primary and secondary position for the net control station. Galang was the executive officer and interpreted for the party since he spoke the language. He made contacts with the town officials and secured help with the cargo from the civilians. The town mayor had mixed feelings of joy and fear since the Japanese patrol camp was just in the next town. Everyone wanted a candy bar with MacArthur's photo.

The cargo was stored in the municipal building while they slept in a wooden house guarded by guerrillas in a coconut plantation. They saw civilians with some clothes and a few had shoes. Soon lines of malnourished and sick civilians formed in front of the mayor's office. Phillips gave out 12 cartons of cigarettes to yelling men grabbing at the packs. Coffee was stolen. "So impoverished so as not to fear stealing," thought Galang.

Songco was the recon and security officer for the party. He repeatedly walked through forests and areas, looking for alternate positions and hideouts for the station. At each campsite, he trained civilians and guerrillas in handling weapons and assigned each one to a battle station. Men were instructed including the cook on what weapon to grab and battle station to occupy in case of an enemy raid. Most slept with loaded pistols.

Guerrilla officer Lt. Mosquera guided the party to a new hideout near the top of Mt Calavite. Over fifty *Mangyans,* indigenous people of Mindoro, built the camp houses. Leeches swarmed the camp and slithered into the lace-holes of the men's jungle boots. When the boots rotted, and that did not take long, bare feet were a feast. Phillips proudly flew 'Old Glory' and the flag of the Philippines that he had secretly carried with him. Aghast, KAZ told him to take it down immediately as it might compromise his position.

The net control station RRR named I.S.R.M for "I shall return MacArthur" was set up to receive reports from sub-stations. The men received ten pesos each for food and were instructed to request for more supplies when needed. Wise and Alberto oversaw the radio. The first unpacked radio was shipped with acid that had leaked out. Another radio was set up with the antenna pointed towards Darwin. Wise did all the code work except for the secret messages that Phillips did himself. Alberto kept the main station on the air throughout problems with operations and repairs. He and Wise worked day and night to improvise methods to increase the electrical output when batteries ran low. They found ways to subdue the noise of motor charges. The bicycle- powered generator engine needed three to five men to crank it up with a pole on each pedal aside from the man pedalling. The Dutch radio set required a steady speed of the bicycle generator when operated to send messages, or the frequency would fade. It was waterproof, unlike the Australian and American sets. Battery packs worked better, but the spare batteries had gotten wet and leaked out. A system of screening towards Manila to absorb most electrical waves was devised to avoid detection. Since the headquarters were pleased with his technical skills, Alberto was recommended for a promotion.

"The first radio contact was with Panay," Galang said. A remote station somewhere on Panay Island intercepted their message in the clear and stopped responding due to no security code.

Contact with Australia was made on November 18 but there was no reply. Shorter messages were sent in case they would get through. Six days later, three messages were sent as the men continued to collect information on enemy aircraft. They were lying most of the time on the ground "looking up for Japanese aircrafts painted white-silver with the sun in the back (that) destroyed our eyes," Galang remembered. The first coded message to GHQ was about the stormy weather. Wise reported that "after days of trying, KAZ replied that there was a mistake in code work and to be more careful with paraphrasing our messages." There was a possibility that the Japanese broke the three messages. The radio frequency had to be changed.

Songco was tasked to find a better location for radio transmission. Station RRR moved to higher grounds with a 360-view of Paluan Bay, Verde Island Passage, Cape Santiago in southern Luzon, Calavite Passage, and Corregidor in Manila Bay. The move was also to keep the people of Paluan out of the camp. Too many people were coming to ask for medicine and seeing the radio equipment. The local health officer took over the medical supplies to avoid civilians coming into the camp and talking when they evacuated to other islands. Around 100 *Mangyans* supervised by Lt. Mosquera and Songco built huts of Cogon grass and hideout places for the supplies at the top of the mountain range, south of Mt Calavite.

The new camp of ISRM was on a higher ridge, colder, and in an area with almost daily rain. Wise found the best location for the radio shack and rearranged the antenna for good reception. Alberto took two days to set up the antenna. The clothes and shoes in the boxes they opened did not fit but they wore them anyway as it was cold. During that period of operation, Galang related "that they were exposed to very inclement weather." "It was cold high up in the mountains... wear a lot of clothing. It rained almost daily," Wise said. Food was scarce, so a garden was built to plant food to feed the men and the *Mangyans*. Wet and hungry most of the time with lack of sleep, their health deteriorated. Phillips became thin, Wise looking old, and yellowish. Songco, Harder, Pinuela, and Galang contracted malaria with severe shaking despite using layers of blankets.

The people of Paluan town were given three hundred pesos per month to feed the men, guerrillas and *Mangyans* in the station and sub-stations. Phillips did not want to use the emergency rations. Without food from the people, they would be unable to keep the flow of intelligence and the radios on the air. The guerrilla boys did not like the long hike up the hill to the camp, so the *Mangyans* carried supplies to them.

Phillips initiated policies to ease the civilians' burden of taxation, high prices of food, and guerrilla demands which the guerrilla leader, Capt. Umali, interpreted as interference. Leading citizens of the town who were in jail were released gaining wide support from the locals. To keep the guerrillas' cooperation and protection to allow the station to operate without intrusion, supplies for them were requested of KAZ. Phillips needed to control them.

Galang obtained civilian help whenever the main camp moved. He supervised the construction of hideouts, cut out secret trails, and organized a system of shore patrol among the civilians. A *Mangyan* who was a guide for hunters in the mountains before the war advised them in picking camp sites and talking to the *Mangyans*. On several occasions, Galang volunteered to go into the town of Paluan, a three-hour hike, in disguise to observe Japanese visitors.

The men found out that the Army standard radio training they received did not include the "mixed-commercial procedures" of a field station usually followed in the guerrilla communication systems. Unencoded messages piled up as the operators at KAZ and the men at their outposts learned the new system.

Five guerrilla guards and coast observers accompanied Pinuela and Vitorio to operate the sub-station RGV on the west side of Mt Calavite's summit near the lighthouse. They were to watch the Verde Island Passage and Mindoro Strait of Japanese shipping going in and out of Manila. The remote outpost was barren of food crops and exposed to diseases. The town of Paluan and the Bay was visible. Ship-sightings and general intelligence on the enemy were relayed to station RRR. The men chose fifty strong *Mangyans* who were skilled climbers to carry the equipment on their backs up the mountain. The locals who showed up, mostly old and sickly, were not selected and this angered them. They wanted to earn and be paid with clothes and cigarettes. One of them betrayed the station by revealing its location to the Japanese, as shown in a captured Japanese intelligence report (February 17 – March 7, 1944). The Japanese knew of the party's landing and their radio sets up on the mountain reporting on ship movements. The remote outpost was barren of food crops and exposed to diseases. When Vitorio left for another assignment, Pinuela continued the operation with a civilian and built his camp in a remote part of the forest. For some time, he lived on whatever he could find for food.

Both stations at Mt Calavite sent their messages to ISRM, but had difficulties contacting KAZ, and FRS in Mindanao. Desperate to contact, the men kept the radios on the air too long, sending over a dozen contacts daily and consuming more power from the batteries. Gas to run the generator that charged the batteries began to dwindle and more could not be found.

Another radio ADB operated in the northeast side of Mt. Calavite, at Del Monte Point. It was an area easily accessible to the enemy. Baniares and Harder operated the radio while Songco and Lt Mosquero provided security. A view of Golo Pass by Lubang Island in the north, Cape Santiago in southern Luzon, and Maricaban Island in the east expanded the areas of observation. Another station was set up at Bagalayag Point to the east of Del Monte Point. After the radios were set up, Songco returned to ISRM camp.

Radios in Mindoro set up by operators from other islands relayed their information to their own network station. A ship watcher station in the northern coast of the island relaying to Panay's main radio was set up by guerrilla Capt. Esteban Beloncio with radio operator Lt. Festin and code clerks. When their radio sets died, they brought them to ISRM to be fixed. Later, KAZ directed Panay to withdraw their intelligence net as there was not much Japanese activity on the island. Mindoro agents were to gather their own intelligence and confirm their data from various sources before forwarding to KAZ.

News of atrocities streamed into KAZ. Over six thousand civilians were killed in Tablas Island, southwest of Mindoro.

According to G-2 Security, each mission party leader was briefed on reporting enemy land, sea and air movements, activities, and dispositions. But Phillips requested KAZ for instructions that he would send to the Manila agents on how to obtain military and naval information, documents, and the like to carry out their intelligence work.

Phillips sent messages to KAZ requesting pay for the guerrillas, guidance on guerrilla functions, and duties towards civilians. He wanted an American to handle all the cash to control the finances and pay the over one thousand guerrillas. He could not ignore the guerrillas because they took care of his safety, security, and the means to carry on his intelligence work. He heard that Panay headquarters was collecting money from the civilians for support, so he recommended that one of his men check the organization and supervise the cash.

Reading the messages, KAZ surmised that the party's activities were geared more towards obtaining guerrilla support than on intelligence gathering. But guerrilla support was needed to be able to safely receive and send the intelligence from agents.

Operatives reported over seventy planes and ships on Mindoro. Names of various locations were substituted to hide their identity. Cape Santiago in Batangas on Luzon was Bill, and the Verde Passage was Rose. KAZ repeatedly warned all radio contacts to strengthen the security of codes and vital records. The lost radio contacts upon the capture of Guillermo Nakar and Ralph Praeger on Luzon a year earlier were still fresh in their mind.

A messenger warned of Japanese naval officers in Paluan. At night, Galang and Phillips laid on their bellies by the shore alongside the embankment of a rice paddy the entire evening to observe the movements of motor launches. Before sunrise, a message was flashed to KAZ.

In the last month of the year, heavy enemy bombers over the islands of Romblon and Tablas brought a radio station down, compromising Peralta's position on Panay. To continue intelligence gathering, guerrilla Lt. Mariano Lim with Panay's network in Mindoro was directed to send all intelligence to station ISRM and stay in the town of Abra de Ilog.

When Parsons returned from his observation trips, he cautioned Phillips not to encourage guerrilla actions. But Phillips radioed MacArthur that Parsons gave him full authority of Mindoro activities. He repeated that he wished to work as an advisor to responsible leaders with full power and control of the cash. KAZ reminded him that he was to identify and report the most capable guerrilla leader with a military background to unite the island and receive arms and supplies. He asked for recognition of Maj. Jose Ruffy, commander of the 1st Mindoro PC company, as guerrilla leader with his own intelligence network separate from Panay's network. Maj. Enrique Jurado, formerly an officer of the Offshore Patrol, oversaw intelligence in Mindoro but was sending reports to Panay's network.

Phillips asked for funds to pay the guerrillas to control them. The reply was a warning not to get involved in the disagreements between guerrilla leaders for leadership. He was reminded of his mission to gather intelligence and avoid any moves that might compromise the station's secrecy and mission.

On Christmas Day, Phillips called a meeting with the strongest guerrilla leaders. Without authority from GHQ, he encouraged his choice, Ruffy, to accept the command of guerrillas on Mindoro independent of Panay's influence. The people of Mindoro disapproved because they would have to feed an additional 1,000 men. Phillips tried to keep the unity while waiting for GHQ to reply. The reason for KAZ's negative reply was not found. A report stated that no message came back to recognize the guerrilla leader. He lost his influence with the guerrilla leaders. KAZ, concerned with intelligence gathering, repeated instructions not to get involved in the guerrillas' internal affairs, as they might endanger the mission. He was to continue establishing coastwatcher stations to monitor the Verde Island Passage.

Panay intelligence agents on Mindoro, received supplies and was to share them with Mindoro but Panay with the discipline problem of its officers did not properly distribute cargo shipments, and thus Mindoro did not receive its proper share of supplies. Phillips even paid unpaid agents of Panay in Manila to continue their work.

To help the men celebrate Christmas away from home, a box of American rations was opened in camp and sent to each watcher station.

1944

The year began with reports from Manila on the Japanese movement of supplies to Aparri in northern Luzon. The enemy pitted guerrilla forces against each other by promising privileged treatment if they killed a guerrilla leader.

In January, realizing the necessity and importance of another station across the Verde Passage, Harder volunteered to carry a radio. With Vitorio and guerrilla Lt. Mosquera as security, they crossed the enemy patrol infested passage under cover of darkness and set up a radio at Calatagan near Cape Santiago in southern Luzon. KAZ was thrilled to receive news of their first anchor station on Luzon. Enemy ships spotted entering and leaving Manila, passing through the Verde Island Passage between Luzon and Mindoro were reported to the Navy Station FTAW in Mindanao. The vessels were sunk right in front of their eyes. Wise messaged that some ships were not properly identified since they had been trained on merchant marine and not naval crafts.

The station at Calatagan significantly reduced the couriers' time to bring intelligence from northern Luzon and Manila to KAZ. Since they were now closer to Manila, KAZ wanted more information. Couriers brought back Red Cross distributions and copies of the latest Manila telephone and commercial directories to keep current on the city's residents.

Radio traffic increased with the Calatagan station. KAZ continued to be concerned with ISRM's length of time on the air and the messages transmitted in violation of security procedures. The new directive was to relay all Luzon messages to Panay's main station for relay to KAZ until security problems were fixed. Songco delivered a new set of code with complete instructions to a Panay officer at San Jose. He also brought cash for Peralta who had asked for aid to maintain his net.

In early February, Vitorio's fingers flew at thirty words per minute sending sightings of cargo ships and motor launches from Cape Santiago to Verde Passage. Bombers, fighters, and sailboats were sighted. KAZ immediately advised the Navy which sank or bombed the boats. Survivors of the sunken ships never reached land. The inhabitants met them with their *bolos* and whatever weapon they could use. Phillips recommended a promotion for Vitorio for his handling of the Calatagan station.

By now the sub-stations were sending reports on enemy shipping in the Verde Island Passage, Golo Pass by Lubang Island, Calavite Passage and aircrafts to the net control station RRR, for relay to FRS in Mindanao and KAZ. Panay sent through Mindoro if unable to reach Australia. Parsons in his survey trips around the island sent messages through the sub-stations using his private code. With the heavy traffic, fuel powering the radios was dwindling. The stations sent men out to look for fuel.

A Panay intelligence man arrived telling of a small radio at the edge of Manila and agents in the City's port area. Wise made codes and sent crystals with Lt. Festin to bring to Panay and start communication with the Manila radio.

Problems with code security continued. KAZ received an emergency message from a Dutch network that heard a signal in the clear to KAZ. If the Dutch heard it, the Japanese could too. KAZ sent a message to "obey rules." Improper use of radio codes, messages sent in the open, wrong frequencies, and incorrect keys threatened the integrity of the ciphers. No messages could be over one hundred fifty letters. The station was reminded not to give geographical locations in code names. A lack of punctuation caused confusion. The station went beyond the rule of 150 letters increasing the time on the air and chances for enemy interception. KAZ's stern message: "Imperative (to) exercise absolute control of this matter."

KAZ ordered that intelligence pertaining to Panay and Mindanao be sent directly to their own network to lessen the chances of ISRM's location being found. The enemy could use triangulation to determine the exact location and surround the radio's place. KAZ ordered Station RRR to "Maintain radio silence. Send only vital information and immediate military importance." Later, G-2 Security concluded that "frequent comments and recommendations in his (Phillips) messages (about radio security procedures) revealed that he was not aware of the policy given to him via radio or that he feels our policy is based on insufficient data and should be changed."

Two American escapees from Bataan, Lt. Robert Kramer and Lt. Eugene Jorgenson, and Pvt. Patrick Milady from Corregidor were hiding in Paluan and brought to camp. They were given clothing and fed adding to the food problem. An "Australian ration box was opened every few days, but the main food was still rice and *camotes* (sweet potatoes) and what we could find." KAZ directed not to use the radio camp as a temporary refuge for escapees from Luzon, concerned that more people affected the secrecy of the mission. KAZ wanted a separate camp set up for them.

In January, a U.S. submarine sunk a destroyer escort and a freighter transport in Paluan Bay. A lifeboat with ten Japanese came ashore. The guerrillas shot three before they reached shore. One escaped but was found by *Mangyans* at Itbu Point (Binuangan) and shot by Pinuela's guerrillas. The six captured were tied together behind a carabao ridden by Pinuela to the main station for questioning. They were turned over to guerrilla Capt. Umali for execution. G-2 Security reported that all prisoners were executed by stabbing with a knife to prevent them from bringing in reinforcement. "Their ears were cut off before burial," Wise related. Upon returning to his camp, Pinuela hid his equipment and left the area through the mountains to the edge of Mamburao travelling at night until he reached the camp of guerrilla Lt. Nietura.

At ISRM camp, couriers from the Manila net and people from Luzon continued to arrive. From northern Luzon, a representative of Governor Roque Ablan brought information and planned on sending two radiomen to be trained and return with a radio. Wise reported a Manila contact, Velona, who arrived with a list of publications, maps, sketches of airfields, and other vital data that KAZ requested. He also brought a filled mailbag, some radio parts, and "a jug of rum." Velona warned that the Japanese knew of their location and to be very careful. He had a message for WARES (might be Parsons) which they kept until his return from Mindanao. Parsons "rode the submarines like buses to various islands throughout the war" sending messages using his secret code, Galang said.

When Lt. Festin arrived from Manila, he confirmed that a letter from Manila signed by CIO-12 through Panay's network asking all guerrillas to send their troop rosters was from a spy. KAZ and Fertig in Mindanao who did not know anything about CIO-12 were immediately informed.

As more couriers from Luzon arrived with reports on Manila, the radio was on the air at an average of six hours per day. Before an agent left for Manila, he was given radio crystals and frequency for the Manila net transmitter to send messages to Mindoro avoiding the long and dangerous trips of couriers.

Manila agents reported strong anti-Japanese feelings throughout Luzon. Small enemy garrisons were seen in the northern provinces. Red Cross supplies of American cigarettes were stolen and sold by the Japanese in the city. Sightings of small ships bringing drums of fuel and a German tanker carrying fuel supplies into Manila from Borneo were transmitted. A Japanese hospital ship discharged patients and loaded cargo. A Japanese radio station and landing field on Batan Island, north of Luzon but without a large concentration of troops, was reported. The landing field was of rolled stone and clay but without 'ack ack' (anti-aircraft) defense.

Phillips revealed to Galang that an important mission objective was to rescue General Vicente Lim and Manuel Roxas from Luzon. Lim was commanding general of the 41st Infantry Division, Philippine Army (USAFFE), who fought on Bataan. Roxas was a member of the Philippine government before the war. Both were hunted by the Japanese. They were all to meet at a designated point and rendezvous with the submarine *Narwhal*. Phillips encoded several secret messages after meeting the men's representatives in Paluan. Wise said one message was made in single transposition and KAZ replied asking who sent the unsecure message. As General Lim tried to go to Batangas to cross the channel to Mindoro, he was intercepted. Firearms were found in the sailboat, and he was taken to Ft. Santiago in Manila. Soon after the secret net in Manila was almost entirely arrested.

Not getting enough food from Paluan, the camp moved to a higher ridge between Paluan and the area of Mananao. Wise and a radio man moved first to the new camp. Galang and the rest followed after the radio was set up. A courier arrived that the Japanese knew there was a radio in the mountains and were continuing to look for the main station. They heard the key clicks through triangulation as they patrolled around Cape Calavite.

ISRM moved again to another hideout towards Mananao. The *Mangyans* built a house for the radio in the open field and carried the food and equipment. The guerrillas "did not help carrying anything even food." With 60 men, food did not last 2 days. The next days were spent hiding supplies around the area. Two *Mangyan* boys retrieved the two batteries left behind at the old camp. Guerrillas of Lt. Mosquera left leaving 15 for security. They clutched at the clothes, shoes, rations, machine guns, and the last of the cigarettes before they left. Phillips, Galang, Songco, Alberto, and the three Americans were left. With two meals per day, all were getting nervous and irritable. *Mangyans* brought *camotes* to eat. Phillips was losing weight and was sick but would not give up. He trusted the people and would not move south. "Whatever happens we cannot blame these people. They have helped every possible way." The little left of the rations was moved out of the camp area and buried under a big tree. Beddings were buried further out in the woods. If the guerrillas who left were captured, they would be unable to lead the Japanese back to them at night.

The following morning, the buried radio was brought out to contact KAZ. It had been over a week since the last contact. A *Mangyan* climbed a tree and fixed the pulley for the antenna. Although weak, the men took turns pedalling to charge the batteries to stay on the air and fix broken radios. At night, in the open field on the side of a hill, Wise sent a message to Darwin that a tanker and two transports sank off Golo Island.

The Japanese were so amazed at the precision of the sinking of many of their ships that they sent out a sizeable punitive force to Mindoro to find out who was reporting on their Navy. A force of 300 Japanese troops landed in Paluan and herded the people to the school building in the town's center. ISRM maintained radio silence.

At Baniares' station, a launch with 40 Japanese landed. Two escaped American soldiers were with him at the camp. Harder who had left Baniares earlier testified that Baniares was (brought to Manila) and court-martialed in August at Fort Santiago. Reports showed that the Japanese tortured him to reveal his companions' whereabouts but he gave only his name, serial number, and rank. "We had pledged to one another that in case one of us got captured, he must keep his mouth shut even unto death," Galang said. Civilians reported that a Philippine Scout detained at Fort Santiago, Manila from January to June 1944 was a "Filipino Staff Sergeant captured on Mindoro on February 1943 (and) was a member of Major Phillips' party." Baniares was a Philippine Scout.

A civilian who had helped transport the party's supplies at the submarine landing guided the Japanese detachment at Paluan to the mountain location of ISRM. Songco with Lt. Festin guarding the main trail saw the Japanese heading towards Mananao. When firing was heard, everyone at camp followed Songco's directives to hide around the radio hut and keep quiet. All trash in the radio hut, and copies of messages were burned. Empty cans of K rations were dumped into a ravine. The radio, code books, and viable equipment were rushed into the bushes. Guerrillas from Paluan and the areas of Abra de Ilog and Mamburao rushed to the camp when they heard the raid.

KAZ believed that the many violations of security codes, frequency, and longer time in the air might have enabled the Japanese to break several messages. "The same code keys were used to send multiple messages instead of varying them, thus allowing the Japanese to track the station." A pro-American operative wrote that the men attended fiestas in Paluan and distributed propaganda material, chocolate bars, and leaflets of MacArthur's promise to return that lifted the people's spirits but advertised their presence in the island. Someone may have leaked information to the Japanese in resentment of Phillips' choice of Ruffy as the guerrilla leader of Mindoro. He was not Peralta's (Panay) choice.

KAZ warned all to use extreme caution. Members of the Manila net had been captured. The Japanese arrested around one hundred agents believed to be associated with the guerrillas in Manila. Some had worked with ISRM radio. A spy in Mindoro had been observing the contacts arriving at ISRM station. Mindanao was warned that codes were compromised since some members from Manila communicated with their station. They were told to use private codes instead.

A submarine was arriving, and Phillips was determined to meet it to get the reports out. He bought two sailboats to meet the submarine. His last secret message to KAZ was, "If we lose our radio equipment and cannot contact you, we will make every effort to meet vessel (submarine) on 28th February to give all data on hand." KAZ sent one more message asking if the weather situation and security of the sub was favorable for a rendezvous. There was no reply.

Phillips picked out the most vital information of maps, sketches of airfields and placed it in his mountain pack. The mailbag full of publications and other documents were left in the hiding place until the next sub. No *bolos* meant hard moving through forests to Pinuela's radio station at Aro Point, a two-night hike. Each man carried a pack and some emergency ration. A *Mangyan* who was to guide them to Pinuela's station, did not show up.

Loud machine gun bursts scattered Phillips, Galang, and the three American escapees as they hiked down the hill. The Americans ran back up and took what they needed and Songco's Carbine. Songco wanted it back, but Kramer said it was deserted property. Galang ran behind a Lauan tree, emptied two clips of his pistol while shouting for everybody to disperse and meet at Abra de Ilog. He hurled two hand grenades which silenced the enemy machine guns. He saw Phillips throw away his pack into the bushes with the intelligence reports when machine guns burst again. It was the last time he saw Phillips. Later, a Japanese prisoner in Hollandia testified that the radio code texts were recovered and decoded. The buried equipment and radio were found.

Meanwhile at Vitorio' station, Maj. Edwin Ramsey, a guerrilla commander operating out of Luzon arrived after a month-long journey across the Verde Passage. He asked for a radio so as not to continue relying on an undependable courier system. Vitorio told him to get one at ISRM. He was too late. Station ISRM had been raided and off the air.

When KAZ heard of the surprise raid, Vitorio and Harder were ordered to close their station. The Japanese knew of their presence from captured operatives in Manila and guerrillas in Mindoro who had worked for Phillips. They buried their equipment and made their way to Manila to lie low. Since they had run out of money, Vitorio went back to Balayan with two guides to dig up the money and radio they had buried. He was captured and brought to Fort Santiago in Manila for interrogation. He was last heard from in March. The need for utmost secrecy was drilled into the men in their training and again before departure as it posed a grave danger to hundreds of agents and mission parties. KAZ ordered them not to keep a diary, but Vitorio did. A Filipino prisoner interrogated after the war said that most of Phillips' activities were in his diary.

With the destruction of ISRM, old differences between the guerrilla leaders, Maj. Jose Ruffy and Capt. Esteban Beloncio, an early guerrilla organizer, flared up again.

Meanwhile Galang was lost in the forest for three days. He jumped into a *matong* (large basket for harvest carried on backs) in an abandoned hut as a Japanese patrol with sniffing dogs passed. He crept out at daylight and followed a stream to a village. The locals recognized and fed him. He had not eaten for five days. Deciding not to meet at Abra de Ilog, he disguised as a merchant selling coconuts and *buri* (palm leaves) hats and crossed the sea to Batangas on southern Luzon to reach Vitorio's radio station and report to KAZ. He found the station raided but was unaware of Vitorio's capture. Proceeding to Manila, he laid low in Tarlac observing enemy disposition in the area. He changed to civilian clothes and avoided heavy Japanese patrols. People gave him food but were afraid to let him sleep in their houses, so he slept in the fields. In Manila, the Japanese military police were looking for him. His parents in Pampanga were questioned, and the Japanese nearly took his children from their house in the city.

Using an assumed name, Galang contacted the Director of Prisons, Eriberto Misa, who gave him a job as a prison guard. For two months, he watched who were brought in as labor prisoners to maintain Nichols Airfield. He was allowed to use a short-wave radio set in the office as he received intelligence about enemy movements in the city. Warned of a Japanese investigation of a prison break, he left and joined agents to blow up the ammo dump at the North Cemetery and sabotage anything that affected the Japanese Army's supply line. Working as a rig driver, he collected newspapers, magazines, journals, and maps to send to Australia. The Kempeitai combed Manila to look for him while he worked as a janitor at the Palace Theatre and living in the backstage dressing room. Agents of Luzon guerrilla leaders Edwin Ramsey and Bernard Anderson exchanged intelligence reports with him. He misled the enemy in any way he could. He changed the position of a road sign leading truck convoys into a narrow dead-end road bordered by deep barriers on both sides. Strafing U.S. planes caught the trucks as they were trying to back out of the road. He tried to get Vitorio out from the dungeons of Ft Santiago. "Vitorio swam across the Pasig River under a rain of machine gun bullets, swimming below rapid fire. He rushed him to a (U.S. Army) hospital in Lingayen, but Vitorio died in his jeep." Ramon Vitorio's name is on the Wall of the Missing at the Manila American Cemetery.

When Phillips departed to meet the submarine to deliver intelligence documents, Wise, Songco, and Alberto were left at the camp with instructions to wait for his message from Pinuela's station. If the contact did not arrive, they were ordered to pack and go to Lt. Mosquera's camp. As they watched Phillips, Galang, and the three Americans disappear into the ravine, they heard shots fired, Wise said. A hand grenade exploded. They ran down the ravine into a hole under a tree that protected them from the mortar shells raining dirt on them. With only their pistols and a hand grenade each, they ran towards the Mananao River while hearing a dozen more explosions behind them. They used a small waterproof match box with a compass on top to reach the river and cross to the small island in the middle. They dare not build a fire that could be seen at night. Wet and exhausted, they fell asleep without blankets. None of the three spoke Tagalog that the *Mangyans* could understand. After two more days of trekking, a *Mangyan* guided them to a guerrilla's house and here they gorged on chicken and rice. All the guerrillas had gone west to Abra de Ilog and the guns and supplies given to them were buried.

At the villages they passed, Wise and Alberto waited outside for Songco to get food and find places where they might sleep. The people were scared if they saw Americans and did not give food for fear of the Japanese. The rumor was the Japanese were searching all over for them.

It was cold in the mountains and a *Mangyan* hid Alberto and Wise. Songco went ahead to look for Lt. Mosquera's men. A courier brought word that the Japanese were going to search the area. Using lighted torches, they walked in the dark holding each other's hand with the *Mangyan* leading the way until dawn. The *Mangyan* made a distinctive sound, and someone answered. Songco, Mosquera and some of his men stepped out of the bushes. All were elated to see each other.

They arrived at the camp of Lt. Nietura on the other side of the mountain, five km from Abra de Ilog, which was occupied by 70 Japanese. Nearby Mamburao had 40 Japanese. Two of the Americans, Kramer and Melody who were with Phillips were at the camp. So was Pinuela, his radio operators and a guerrilla aide. Guerrillas Lt. Jamilla with his platoon and Lt. Lim with his intelligence group from Abra de Ilog were present.

Songco related that Velona, the intelligence man from Manila was forced by the Japanese to point out Pedro Tadalan, the guide Phillips trusted and had not shown up. Tadalan revealed Pinuela and Phillip's camps. At the raid of Phillip's camp, the Japanese had forced the civilians to carry all the equipment down the mountains and dig up the six executed Japanese that Pinuela turned over to Capt. Umali. People were tortured to find out who cut off the ears. The guerrillas pointed out as responsible were taken to Batangas by boat but jumped overboard.

A wounded guerrilla escapee hid in Abra de Ilog. He carried a letter for Galang that he turned over to Pinuela. The letter was from Phillips saying that he had escaped from the raid and to return to carry on the mission with the equipment they left. But Galang never read the letter since he was in Manila.

USS *Narwhal* – March 1944

With the party and radio communication destroyed, intelligence gathering on Mindoro and communication with guerrilla groups stopped.

The *Narwhal* aborted the landing when there was no reply from Phillips and no security signal was seen on the empty shoreline or approaching sailboats for two nights. The skipper did not know of the raid. The sub quickly slipped away and proceeded to Mindanao landing all Mindoro supplies there.

By this time, Japanese activities intensified on Mindoro. G-2 Security and KAZ disputed on the next steps. G-2 Intelligence argued for Peralta of Panay to continue Phillips' mission, while PRS wanted to keep it under close GHQ control and direction.

A one-thousand-peso reward for any American was posted in the nearby town. The *Mangyans* became afraid to hide and feed them. The group split up to make it easier to find food. Lt. Mosquera gave money he received from Phillips to the men and the two Americans.

Songco and Alberto went with Lt. Nietura. They fought a Japanese patrol of fourteen men in the village of Camorong. Two of the guerrilla guards were caught gathering food in the cornfield. Alberto ran through a bamboo grove into the interior, following the general direction the rest took. He slept on the ground in the jungle for three nights and ate only roots of *talahib* (wild grass). He met up with the rest in the mountains behind Puerto Galera and proceeded to the guerrilla headquarters in Naujan located in the hills. He related that he learned the survival skills of the guerrillas, their activities, and a good deal of the enemy's tactics and weaknesses. He wanted to contact KAZ, but the local radios were registered with the Japanese. From wrecked radios, he collected parts to build a small transmitter. Power output was so low that it could reach only Mindanao station WAT. They risked charging the batteries at the rice mill in the town filled with Japanese. He trained boys of high school education as radio operators, which proved to be of much help to the following missions. Suspicious of Panay's offer to set up the radio, Alberto sent some of the trained boys to set up radios in the Sibuyan Island.

Wise, Pinuela and the rest went with Lt. Jamilla to a place with no people around but where there was some rice and food stored. They were to stay in contact by courier. They slept in *Mangyan* villages while the boys got food and told the people that there were no Americans around. The Japanese had warned they would cut off everyone's head if Americans were found. Wise said "'For the first time in my life I wished I was not a white man." A Filipino smoking a Lucky Strike cigarette was captured, and they thought it was Galang, who was a smoker. All civilian men were sent to Abra de Ilog while the Japanese searched the area. When their camp was found, Wise and Pinuela grabbed their personal equipment and hid in a ravine. Without a guide or *bolos,* and the people too scared to help, they blindly followed fingers pointed in the direction of Mamburao and further south to Sablayan.

It took 10 days to reach the house of guerrilla Lt. Unson. A cow was killed, and all ate well and rested. Wise's boots were worn out and a guerrilla gave him a Japanese officer's shoes.

Maj. Jurado arrived and brought Wise and Pinuela to his camp in San Jose in Naujan to take charge of the radio. The officer in charge had deserted back to Panay. The radio was a small weather type with a weak signal. Wise tried to send messages in his personal code to FRS in Mindanao but there was no reply. With a guide, they travelled east going through a cattle ranch, two days up the Busuanga River, across a mountain ridge, and two days down the other side to a guerrilla camp at Bongabong to get good radio reception. They kept trying but the signal was still weak.

Returning to Jurado's camp, Ramsey and USAFFE soldiers from Bataan had arrived. Ramsey opened a can of fruit cocktail to celebrate. All his intelligence he gave to Jurado. "The maps were not drawn to scale," said Wise, and he thought Ramsey "had difficulty coordinating his troops because his only means of communication was couriers taking a month to get reports from his various units. He had no finances except some from rich people in Manila, causing his men to desert to other groups."

The news of a submarine coming to Panay had Americans Kramer and Melody quickly departing to get on the sub. They thought Wise was crazy to stay to find out what really happened to Phillips.

Wise went with Ramsey passing the Japanese garrisons at Pinamalayan and Pola to get Songco. He got sick and stayed to rest while Ramsey continued. When able to travel, he hiked with guerrilla Capt. Beloncio to the town of Sta Rosa where he rested in a Chinese store. He was still weak.

Later, Ramsey related that the locals at Cabasingan told that three Filipinos and two *Mangyans* had come upon Phillips and Jorgensen bathing in the river, 4 km from Abra de Ilog. They saw Phillips shot in the back as he walked out of the river. Jorgensen escaped without clothes and was captured with some *abaca* clothes on. The Japanese had offered a 10,000-peso reward for Phillips. Jorgensen was forced to lead the Japanese to three places on the west coast which had helped the men. People pointed to as having helped were abused and fled their burning homes. Ramsey sent a guerrilla officer to Abra de Ilog to get proof from Phillips buried body. The body had completely deteriorated and his gold teeth were extracted as was the practice on Bataan.

From Jurado's camp, Alberto, Pinuela, Wise and Lt. Dodson sailed to Luzon and brought a radio for Ramsey to set up in the mountains of Cavite overlooking the entrance to Manila. Songco stayed to survey the area. Ship sightings were transmitted to Jurado's station. Another radio was set up at Lt. Dodson's camp on a hill overlooking Verde Island. For the first three days contact could not be made as a Japanese station on Verde Island jammed their signal completely.

Ramsey asked Wise to be his communications officer as he covered a large intelligence net in Manila, units around Bataan, and six regiments under guerrilla Capt. John Boone formerly with the 31st Infantry. He replied that his mission was to be in Mindoro and carry on the coastwatcher station. Before returning to Mindoro, he gave his jungle hammock to Ramsey and a few things left from Australia to his men.

The group sailed back and hiked to Sta Rosa to the same Chinese store. The Chinese owner held a dance for them. "It had been a long time since they relaxed and smoked a few Japanese cigarettes," Wise said. Using Japanese money, he bought a blanket for Alberto whose clothes were in tatters.

Everyone quickly took positions on the other side of the river when told of a Japanese patrol coming down the road into town. It turned out to be a funeral. From a lookout in a building, they saw Japanese in trucks or hiking on the highway every night. They left in 4 carts changing several times whenever one broke down.

Back at Jurado's camp, the radios were not working and there were no parts to fix them. Wise left with Jurado to the town of Bansud to get a teleradio they were told about. But it had an old receiver that could hardly hear anything and there were no spare parts to fix it. Anxious to contact KAZ about what was left of Phillips' mission, they crossed Tablas Strait to Panay but was told at Mt Baloy, Peralta's headquearters, that there was no available radio.

Jurado returned to his camp and Wise stayed on Panay to wait for the next submarine to return to Australia. He planned to return with a radio and supplies that he saw were severely needed. While waiting for the sub, he was assigned to the net control station of Lt. Irineo Ames (Panay August 1943 mission). He tried all of Ramsey's contacts without success. He met Lt. Toribio Crespo (Panay April 1943 mission), acting as the adjutant to Garcia, the combat unit commander overseeing the security of the submarines' rendezvous area. Crespo strongly expressed resentment over his position that he had no work to do. Now Wise worried about Crespo.

At dusk, the signals went up and the sub roared to the surface. Wise with guerrillas went out on a sailboat to meet the *Narwhal*. He boarded the sub without clear orders. He filled two sacks with radio parts to leave behind for the network stations before leaving. On board, a Navy officer found a letter for him from his mother. At Darwin, Wise met Parsons, who said KAZ orders were for him to stay on Panay, but he never got the orders. In Brisbane, a Pvt. Hitchcock laughed with others and said Peralta told him that "Wise will go whether he wants to or not." In a later mission, Wise was able to be included to return to Mindoro but was not as fortunate the second time.

By mid-1944, command of the Pacific theatre was split between the Navy taking islands in the central Pacific and the Army island-hopping to cut off Japanese supply sources along the northern coast of New Guinea towards the west. The Third Fleet's clearing of Japanese shipping in the southern Philippines produced minimal resistance influencing the change of the first invasion to Leyte and earlier in October. The overwhelming task of changing logistics and remobilizing hundreds of thousands of men, supplies, ships, and whatever support structures was done impressively.

The new date increased the desire for more intelligence and sooner. More missions to bring signalmen to watch the coasts and skilled men to train guerrillas in combat, demolition, weather, and aircraft warning were needed. Submarine runs doubled in the coming months. The *Narwhal* and the *Nautilus* continued their runs without yard overhaul for another year. The Navy authorized four operational submarines, the *Seawolf, Stingray, Ray,* and the *Cero* for Philippine service.

USS *Nautilus* – July 1944

The mission to set up more sub-stations on the island's west coast where Japanese ships passed from their homeland was approved. It was the first cruise of a submarine wherein three trips were made to the Philippines without refitting of the vessel and shore leave for the crew.

The party's mission was the same as the doomed first party. Communication nets were to be developed into central Luzon and Manila. A network of trained personnel conducting intelligence communication, reliable weather data for the Air Force, and radar detection in the significant areas from Luzon to the islands of Palawan in the west was to be developed. GHQ wanted naval movements in the Verde Island Passage, Mindoro Strait, and Manila Bay area. Weather information and air warning reports were highly expected. Party leader, Commander George Rowe wanted the call letter for his radio to be IHRR "I have returned – Rowe." The old ISRM remained as the call letter.

GHQ was hopeful and considered the group as "more thoroughly trained in the operation of technical equipment and better equipped than any other party previously dispatched." The party brought high-powered cameras and photo development equipment for capturing enemy shipping. After the war, in Rowe's letter to Whitney, "The crew of the sub had the highest praise for the men... have changed their idea on them after watching the (Filipino) boys. Quiet, well-behaved, and spending most of their time preparing themselves and equipment for the job to be done." They had volunteered for the mission risking the consequences to go to their home country and drive the invaders away. They had not heard from their families since the war began.

The men had the full benefit of complete physical hardening in jungle warfare and commando tactics. Not a combat unit, they can still protect themselves against the enemy, elements, and the treacherous jungle. When the orders arrived, the men had completed the first half of an intensive 3-month training. Rowe, Executive Officer of KAZ, received a roster of twelve signal men and seven reconnaissance men for the mission, including himself. He was chosen for his knowledge of the Philippines as he was a sugar broker in Manila before the war. He used a cover name fearing for his wife and sons in Manila. The recon men were Lt. Al Hernandez, Sgt, Gerard Berg, Sgt. German V. Reyes, Sgt. Peter B. Aguilar, Sgt. Vidal F. Alvarez, Sgt. Inocensio V. Pascua, Pvt. Urbano P. Delfin, and Pvt. Pedro

B. Savellano. The 978th men were Sgt. Julio V. Balleras, Sgt. Daniel B. Begonia, Sgt. Pio B. Borillo, Cpl. Urbano B. Barcenas, Cpl. Enrico M. Bugarin, Cpl. Felix T. Reyes, T4 Domingo L. Logan, T4 Jose T. Gonzales, T4 Patrocinio A. Carillo, T5 Alipio C. Toribio, T5 Angelo R. Pascual, and T5 Hermenegildo Villalon. Weathermen T5 Nicholas E. Sandoval, Pvt. John T. Tolentino, Sgt. Henry I. Chambless, and Pvt. Robert H. Gamertsfelder joined them. By the time they left, they were in the final stages of jungle warfare training.

Whitney gave each man a gold watch from General Macarthur. The policy not to bring anything that would reveal their identity had changed to "bring at your own risk". He then proceeded to give a short talk on the successes of the forces in the Pacific.

To avoid what happened to Phillips's party, Rowe was given strict instructions not to meddle with internal disputes of the guerrillas. But when his party arrived, the fight for leadership and men was still ongoing between Jose Ruffy, Esteban Beloncio, and Enrique Jurado. Mindoro was not to be under the control of Panay by supporting Jurado who was the representative. He was not to be distracted and to limit his activities to intelligence gathering.

At dawn, the men boarded a C-47 from Archerfield in Brisbane to Darwin and boarded the submarine as soon as they arrived. The thrill of the coming adventure was visible on Bugarin's face and those of the men. "None of us knew where we were going… We only knew that we were going to establish radio stations somewhere in the Philippines. Where? We had no idea."

Berg and Reyes were assembling the team's gears when Reyes broke his hand during the final loading and spent the trip in the sub's sick bay.

Hernandez, a champion dancer before the war, was seen "in a faded tropical shirt and shapeless cotton pants, heavy leather brown boots, and a gold wristwatch" while Rowe was in civilian clothes.

The *Nautilus* was proud of the number of enemy ships she had sent to the bottom. Torpedo racks had rows of the rising sun painted on the bulkhead, one for every ship sunk. She was packed with men and supplies. It was the men's first trip in a sub, and they passed the time reading, playing cards, chess but always on alert. The sub ran through enemy blockades in the Seas of Timor, Celebes, and Sulu to get to Pandan Island on the western side of Mindoro. The Japanese were aware of the submarines plying a regular route to the Philippines.

Balleras recounted that their mission was so secret that a U.S. patrol plane "gave us a taste of depth charges on the third day of our trip when it mistook us for an enemy submarine." Men dropped down hatches and ladders to man battle stations. On the Fourth of July, a Japanese cruiser chased the sub as orders blasted "take her down." Reverberating depth charges ensued. At three hundred feet underwater, batteries shut off as enemy attack boats circled overhead, dropping depth charges. "We did not eat our lunch as they could not cook. After eight hours, we all looked yellow because of the heat and lack of fresh air," related Bugarin. The sub remained submerged for thirteen hours enduring vibrations with each blast. No air conditioning with silence. Anxiety levels rose with the sound of the creaking hull under the sea pressure. When all was clear, the engines would not start. A wave of relief passed when the hum of motors came back, turning the fans on.

The men were summoned to the dining room and warned about their firearms. Reyes had been shot in the leg. Close to their destination, the island where they were to land was pointed on the map. They were relieved to know that their mission leader knew the place well and it was not far from where the late Major Phillips' station was located.

The sub arrived at the northern end of Pandan Island. In the dark, Hernandez, Aguilar, and Balleras, wearing faded civilian clothes, boarded a small yellow life raft to determine if fifteen tons of radio, weather, and photographic equipment could be landed. No one else was allowed on deck. The sub continued to observe around Corregidor under darkness.

Anxieties rose as they paddled towards the shore not knowing what awaited them. There were guerrillas in Mindoro but the newcomers were unsure of their loyalty. The feeling intensified when a small boat appeared in their direction halfway to shore and passed. Unknown to them, the sub crew had manned their guns ready to shell the boat if it did not change its course. A strong wave rose from the strong current and overturned the raft. Hernandez held on to the walkie-talkie and camera. Aguilar grabbed the raft avoiding the rocks. Balleras held the second camera, the metal box of film, and rations. They crawled onto the beach and into the shadows of the coconut trees and thick undergrowth. Balleras felt "wonderful to land on Philippine soil after a long absence, and I cried with joy." They camped at the edge of a deserted copra plant that still emitted a strong odor of coconut oil. That whole day they lived on coconut meat. Aguilar and Balleras surveyed around and declared the island inhabited and safe. They heard the familiar crows of the roosters at dawn. A quick call to the sub was made to avoid Japanese interception and an immediate short reply returned.

They moved to a higher area in the morning to place a white signal cloth, taking care not to be seen. They continued taking photos of the coastline, stopping to hide whenever a Japanese patrol flew overhead or passed. To their relief, the signal flag was not seen.

At dusk, the sub sighted the white flag and rose close to the beach. Hatches opened, and the crew loaded the cargo into the rubber boats. Part of the crew was on the lookout for any sign of danger. Bugarin remembered the kindness and humor of the skipper who said that "we could get anything from the sub except the sub itself as he needed it to go home." The sub then set out for Leyte to deliver more recon men, weathermen, and supplies.

The rest of the men paddled to shore with twelve tons of supplies in five sizeable black rubber boats lashed together. Included in the cargo were half of the 4,500 *bolos* made for the 1st and 2nd Filipino Infantry Regiment in California from Col Offley.

Halfway to shore, fear gripped the men when a small boat was sighted moving towards them. Heavy waves snapped the ropes and stacks of cargo fell into the sea. The rubber boats drifted down-shore, and some crashed on the rocks breaking crates. The practice of unloading at Camp Tabragalba paid off, avoiding unnecessary exposure to the enemy, but there was no escape from the heavy waves. It took three hours for them to beach and salvage the cargo, with their machine guns ready.

Hernandez related that the cargo boxes were hidden in holes and covered with logs and tree limbs until they could be transported across the sea to Mindoro. The town of Pandan had suffered from the Japanese with people being hanged upside down and their organs slashed. It was a reprisal for helping three American POW escapees, the three who were brought to Phillips' camp. The town mayor provided papers to pass Japanese sentries. Some of the men posed as "collaborators acting as purchasing agents for the emperor's forces buying cattle and rice inspectors to visit production areas and warehouses where harvest was stored." A week after landing, a tent was set up for the radio, generator, and bicycle for power. Carillo and 'Doc' Pascual, who doubled as a dentist, radio man, medic, and in any capacity requiring skill and ingenuity, along with Bugarin, inspected the unpacking of the radio gear. With Barcenas and Reyes taking turns pumping the bicycle to generate electrical power, they heard broadcasts from various cities and languages. Tuning to the assigned frequency, they needed three tries before KAZ replied. "We heard them every time we opened up. We broke in at every opportunity, and after a few days, they challenged us," said Bugarin. Carillo excitedly yanked off his headset when he realized that the first message got through. A reply from KAZ four days later commended the men for completing "this phase of your important mission."

In the early morning, Balleras, Aguilar, Alvares, Barcenas, Carillo, Logan, Pascua, Villalon, and Pascual with Rowe and Hernandez rowed in rubber boats against heavy winds to Mindoro as the outboard motor they brought would not start. Civilians in the town of Sablayan, right across Pandan Island, had heard of the landing and provided a sailboat and a banca. Bugarin and Savellano crossed with four civilians in the sailboat. A platoon of guerrillas was left on Pandan to guard the rest of the equipment. Men paddled as a storm broke out raising waves that rolled almost on top of their boats. Fifty feet from shore, the boats capsized. Bugarin saw Savellano, who was not a strong swimmer, thrown off and noticed he landed in shallow water. Overboard with the men went some of their guns and *bolos,* the walkie-talkie, and the precious compass.

Civilians helped retrieve the heavier boxes from the floor of the sloping shoreline. Some washed back to Pandan. The weather equipment was lost with some radio parts, and rations. The civilians brought their carabaos to carry the heavy crates while they carried what they could on their backs.

The people of Sablayan provided some dry clothes. Food was scarce, especially meat, as the Japanese took most of their animals long ago. Balleras could not help wishing that most of the cargo they brought were rations. Now they "had to live on an ear of corn each meal or the starch from the *Buri* palms edible only with syrup which unfortunately we did not have. I remember Commander Rowe saying: 'Can you tell me what makes a guy volunteer for this kind of job anyway?'"

Japanese reconnaissance planes flew low but did not see them. Bugarin credited his infantry training for keeping the men still. Several guerrillas under Lt. Pedro Nietura of Capt. Umali's organization arrived. After identifying themselves as having provided support to Phillips' mission party, they were given Carbines.

A radio was to be set up on Mount Baco (Mt Halcon), the highest peak in the central interior of Mindoro. Pascua's group, guerrilla security, and the carriers of equipment slashed trails through deep ravines wide enough for the carabao and sleds to pass. They rowed on the river they thought was the Asis River using a twenty-year-old map. At the farthest point, they found a spot to camp when sudden winds announced a storm. Boxes were piled high and covered with canvas. The river overflowed, soaking their jungle boots. Runners from guerrilla outposts brought Maj. Jose Ruffy, who travelled day and night with his men over almost impossible terrain. That boosted the men's morale as these guerrillas were disciplined and familiar with local conditions. Reserve supplies were buried in a small, wooded island midway between the banks of the river with swift current on either side. A deserted barrio became the forward base. Japanese aircrafts passed twice, making the men dig in, but no bombs fell.

A security outpost was set up at a bend in the river under Logan 's station DAK to guard the entrance to the main camp twelve kilometers uphill. Logan saw his post as the "reception center for visiting military and civilian officials." He built two fixed sentry posts with shacks and barracks for his guerrilla guards. Food was scarce but 30 hungry guerrillas he trained in strict military discipline stayed with him. He kept a registry of visitors, including undercover agents from Manila who made the perilous trip. His specific duty "as per the order of the commander was to lead the party and clear the way with a submachine gun, if and when necessary, upon encountering the enemy."

Climbing towards Mt Halcon Balleras pondered "what kept us going thru waist deep mud and unceasing rain" for three weeks until they reached Lamintao hill. They stopped for only 15 minutes every hour to rest with their equipment. The resurrected "tower of ISRM' was set up on Lamintao hill with a full circle vision of the island and Dongon Bay by the coastal town of Sablayan where ships frequently took refuge during typhoons. Hidden under large trees and boulders the men called it Malaria Hill. The *Mangyans* climbed the soaring trees and set up an adjustable antenna out of sight of the enemy air patrols. The saltwater had soaked the radios, battery chargers, and almost everything they brought. It took several days of cleaning, drying, and servicing before the radios worked.

The enemy would have difficulty reaching the tower as swampland, deep jungles, and sheer cliffs surrounded it. Whenever the surrounding valley flooded from the torrential rains, large crocodiles swam in limiting the procurement of food, but the peaceful *Mangyans* almost always found some kind of food and saved the men from starvation. Below the camp were four guard stations and quarters. At the bottom of the mountain were four villages of guerrilla fighters and *Mangyans* who spoke the same dialect.

The antenna was on Malaria hill and the main station JRH was set up on the shore in the village of Matabang. The first broadcast was heard in July but whenever they heard KAZ and tried to contact, there was no reply. KAZ may have believed that they were Japanese jamming the lines and challenged them. After some exchanges, contact was established.

The station DNS operated beside JRH to relay ship sightings, aircraft warnings, and intelligence reports to KAZ. Villalon coded and decoded messages. It was on 24-hour duty receiving a stream of agents bringing information or departing with new orders. Men took turns pedalling the bicycle contraption to produce enough electrical power for the radio to carry their beam to KAZ. A strict warning from KAZ arrived, saying that reports were to be sent on the prescribed "frequency and must reach this headquarters within ten minutes of the detection time to be of operational value."

In August, Alberto (Mindoro December 43 mission) reported for duty to Rowe. He came with Ruffy and his guerrilla unit, Doctor Felix Alegre, and two "women first- aiders." Alberto's knowledge of radio quickly cleaned the soaked radios and got them to work. Rowe gave him complete control of the communication facilities.

Several weeks later, Harder and Pinuela showed up, hungry and filthy. The three rejoiced to learn they had escaped the raid that destroyed their party.

Pinuela, eager to get back in action, was soon trekking back with a mobile transmitter to his old outpost at Aro Point by Paluan Bay, the area he knew so well. In August, he reported that his small party of guerrillas killed twenty-five Japanese, captured two lifeboats, machine guns, and supplies. His group burned an abandoned sub-chaser.

Harder reached Luzon and hid in the mountains of Montalban before joining guerrilla leader Edwin Ramsey's units in the vicinity of Manila. He had gone to Mindanao and brought back to Luzon propaganda materials from the submarine. In Manila, he was glad to find Galang. Together, they reported on Japanese troop movements, atrocities, airfield data, movements of POW, mine fields in Manila waters, rescues of fliers, and other important information they received to send to Negros for relay to KAZ. Galang reported sending enemy observations through Ramsey's and Anderson's radios until the end of November and later through Cardenas' and Herreria's radio (Samar December 43 mission) in Manila until December.

Harder moved to the combat unit of President Quezon's Own Guerrillas (PQOG) that published *The Liberator,* a counter-Japanese propaganda and trained agents. PQOG operated in central Laguna, Batangas and western central Tayabas. The Japanese were in an uproar looking for the publishers of *The Liberator.* The printing press could not be found and continued to print and distribute thousands of copies.

PQOG was warned of a raid. All radio equipment were destroyed but a notebook was found that had Galang and Harder's names on top of the list of people helping them. Harder escaped to ISRM but was told to return and stay with the PQOG unit.

A streetcar station in the city had a wanted sign with a photo of Galang from his university yearbook that showed his mole. But the mole was removed in Brisbane. The reward was for 1,000 Mickey Mouse money (Japanese currency). He escaped to a house in Pampanga posing as a vegetable and salt vendor cutting Japanese communication wires along the way. He sent intelligence to Bartolome Cabangbang (Luzon September 44 mission). Enrique, his brother who had escaped the Death March, provided information on movements of the Kamikaze bases on Luzon. Whitney commented that there were no reports from Galang since the raid at Phillips' camp. He was unaware that they were sent from Ramsey and Andersons' radios. Galang may not have had a personal secret code to send messages like Phillips did.

A half-starved and sick Songco reached Rowe's camp. He was assigned to guard the station JRH. The Japanese located the station and sent 200 troops. "Flanking the enemy, Songco and armed guerrillas headed by Capt. Dodson set bombs around the camp and shot scampering enemy soldiers," said Villalon. Some were followed to their garrison which was attacked, killing several dozen. Songco had an infected leg from a gunshot wound from the raid and suffering from malaria. He recognized a guerrilla called "Ortega" who was a traitor and informed Rowe and Hernandez. The prisoner was given a choice to stand public trial or go into combat with the enemy. Having recuperated, Songco moved to the security outpost in Naujan and set up a troop training school for the guerrillas. He joined guerrilla ambushes of the enemy. When KAZ recognized Major Jose Ruffy as the guerrilla leader of Mindoro, Songco taught the troops how to load a Carbine and military training to be ready to clear Paluan, Mamburao, and the islands of Lubang and Romblon that were heavily defended by the Japanese. The enemy was looking for him, and he kept evading them. Brave to the end, he was reported shot in the back while in action in Panay. His body was brought to

Guagua on Luzon, his hometown and the townspeople erected a monument in his memory. According to Galang, Songco had fallen in love with a girl in the neighboring town of Calapan. After the war, Col Offley heard from some of the men that they went after the people who killed Songco, destroying them and their families for betraying him and his men.

A submarine was arriving in September. With Malaria Hill discovered, ISRM had to move to another location. The radio equipment was dismantled, buried ones unearthed, and packed on carabaos for the two-day trip to Barahan, on the island's west coast. A final message was sent to KAZ to cancel codes and issue new ones at the next contact. A storm slowed them down with the heavy mud. The food was *camote* if found, edible plants, or thin rice soup. They carried two heavy metal canisters loaded with documents to turn over to the submarine's captain, stopping only to rest and eat some food. At the rendezvous site, men jumped into small boats with demolition equipment under the canisters pretending to be fishermen sailing back and forth, waiting for the periscope that did not show up. Japanese patrols kept passing by, and one shot at the boat destroying it. Men swam to shore. One floating canister reflected light and was picked up by the patrol vessel. Runners warned the farmers nearby to hide to avoid possible enemy retaliation. Rowe immediately ordered "stations to change locations. Every man who sent mail was a hunted man." Two weeks later, Tokyo Rose broadcasted the canister's contents - mail, signed receipts, and a map of the ISRM network.

The *Nautilus* did not surface because it had run out of fuel from jettisoning cargo and fuel when it ran aground in Cebu. The two air warning personnel, Sgt. Robert B. Caswell and Sgt. Robert R. George, on board had disembarked in Panay. They arrived at Rowe's camp weeks later after hiking across Panay, sailing across Tablas Strait, and travelling overland across Mindoro to reach ISRM station.

With the canister found, headquarters moved north to Abra de Ilog. It took a month over jungles and mountains to carry their equipment and supplies. It took time to recruit dozens of carriers, gather carabaos, and food for the party. Logan and Alberto directed the building of the tower transmitter at the top of a mountain overlooking Verde Island passage. *Mangyans* built huts for the radio and the men.

Rowe, a Navy man, named his new headquarters Camp Nimitz after Admiral Chester Nimitz, and set it up on the beach for better observation and security. Although the camp was close to the water's edge, it was well-covered from the sea. A battalion of guerrillas set up their huts around the camp. Not all were armed. The guerrillas had rifles but "practically out of order." They were to fight an enemy attack to give time for Rowe's men to burn and destroy what they had built up or bury the equipment and records to return to later. The frequent heavy rains and lighting drowned out the noises the men listened to as signals of pending enemy presence.

The new tower's call sign PDQ went on the air in October. The station covered the 4th Military District (Mindoro and Palawan) that included the Manila area. Villalon and Bugarin operated the station. Borillo and Pascual operated station DNS besides PDQ receiving reports on intelligence and sightings of ships and aircrafts from the sub-stations. Aguilar and Alvares were observers. Berg was head of security with a platoon of guerrilla troops.

From the new tower, the men could monitor the Verde Passage, the entrance of Manila Bay, and the airfield in Lipa, Batangas. All sea traffic from Mindanao and Leyte going to central and southern Luzon had to pass close to Mindoro at Cape Calavite, Pinuela's station.

A blockade of sailboats ran between Abra de Ilog and Naujan, evading many Japanese patrols to obtain food and supplies. Couriers plied the route with intelligence reports and rescued downed pilots. At times, Filipino boys rowed out to meet boatloads of Japanese with friendly waves to get closer. From under a bunch of bananas, they threw a stick of lighted dynamite at them. Before the smoke cleared, the boys dove to retrieve the guns, ammo, and even the engines of their boats.

At Abra de Ilog, the people elected Alberto as the "military mayor" to oversee the rice mill, coconut oil production, and the procurement and shipment of rice to headquarters.

With the camp established, men in small parties quickly departed to set up sub-stations as KAZ wanted immediate information on the enemy. Each was under the leadership of the signalmen. The guerrilla leader, Ruffy, had provided twenty men with at least a high school education to attend Villalon's training in radio communication, producing a reserve of trained radio operators. Two and sometimes three accompanied each mission to speed up the coding and encoding of messages. When the mission men left to set up other radio outposts, the guerrilla operators continued to operate the station.

Logan was assigned to Naujan to procure food for headquarters with the help of guerrilla Capt. Francisco. Receipts were issued but with the risk that reimbursement might not be possible after the war. Logan said he was unanimously elected military mayor. He was co-instructor in the "San Agustin Troop School," where Songco trained guerrillas.

The first sub-station was on the other side of Mindoro at Bogio Point near Abra de Ilog with the call sign JMAC under Balleras and Felix Reyes. Aguilar and Delfin were at Calavite Point. At Puerto Galera, Logan operated sub-station MDY which communicated with Edwin Ramsey, a guerrilla leader in the Manila area. They reported ship sighting, aircraft warning, and whatever information gathered on the enemy. Whenever a report from the operatives in Manila confirmed that the ships which they reported in the Verde Passage had sunk before entering Manila Bay, they hooted in deep satisfaction. A bit of quiet dancing around would follow.

Toribio with guerrilla guards crossed the Verde Passage to Lubang Island and set up station SKP, a remote outpost but critical for the Luzon landing as it had a clear view of Manila Bay. Enemy vessels in Manila Bay and as far as Lingayen Bay were sighted. Planes taking off and landing in an airfield were reported. Flashes to the Navy control station sunk vessels sighted off nearby Ambil Island. The Japanese radar station on the island was to be located and photographed because the U.S. bombers in September had missed the radar. With two men, Toribio dressed as a farmer joined a caravan of civilians delivering food to the Japanese in the mountains guarding the radar. They found and sabotaged the radar. Toribio's men continued to blow up ammo dumps and hundreds of drums of gasoline around the island.

To isolate Lubang due to its strategic location, native boats used by the Japanese for transporting gasoline from its port to Batangas on southern Luzon were destroyed. The men and a dozen civilians dove and salvaged several cans of ammunition, a rifle, and two Carbines that added to the five Carbines and one pistol given to Toribio's group. The engines were salvaged and given to the civilians and guerrillas of Looc village on Ambil Island. Japanese launches patrolling the waters were found and sunk except for one they used for travelling. They sunk the Japanese launch after a U.S. destroyer gave chase assuming they were Japanese soldiers in the launch. The guerrillas killed a dozen Japanese who hid in nearby Ambil Island when U.S. planes bombed their garrison in Nasugbu on Luzon. A Japanese civilian interpreted a letter found in the launch of an officer to his superior concerning their desperate need for food and ammunition. Toribio was amused with the news of the Japanese "looking for 'Skippy,' a tall white American, or 'SKP'" which was his station's call sign. He did not look at all like a tall white American.

To maintain order and command the respect of the people, Toribio related that he had to use force representing the law, as there was no civil government, to save the life of a valuable guerrilla agent who was accused as a collaborator. He "was forced to lay (his) hands on a guerrilla officer and enlisted man for firing several shots just for the fun of it while two U.S. PT boats passed by." The PT boats turned, shelled the village and his observation post, causing a civilian's death and wounding several others. Three P-38 fighter planes on reconnaissance patrol shelled the village and outpost again. Not wanting to die from friendly fire, Toribio radioed an urgent message to "advise the Air Corps and the motor torpedo squadron to avoid shelling me."

When he began to doubt some of his agents' reports or agents did not show up at agreed dates, Toribio took Mori, a civilian Japanese carpenter, to locate the enemy gun emplacements at Tilik port, Mt Ambulong, and gasoline dumps in the island. Mori did not like the Japanese although he was forced to work with them. The civilians did not like Mori because of his duties with the Japanese. He begged Toribio to take his family into hiding in the mountains, as the guerrillas would kill them because he was Japanese. The family willingly offered their lives if he did anything wrong.

Gonzales crossed Tablas Strait to operate station JSG on the mountainous island of Marinduque. He saw sick and dirty people covering only their vital parts with rotten *Sinamay* woven from the abaca plant. Only root crops kept them alive in the hills hiding from the Japanese.

As the Allied forces drew near, the need for intelligence greatly increased. ISRM was ordered ready to support the landing. Hernandez said, "ISRM radio net covered all Mindoro, city of Manila, as far as Lingayen in northern Luzon, and the important Verde Passage between Mindoro and Luzon frequently used by the enemy." Balleras and Begonia trekked to the southwestern tip of the island and set up station BEG in San Jose.

Pascua crossed the Verde Passage to Luzon with great difficulty, avoiding Japanese patrols, to bring a radio and supplies to Ramsey in the vicinity of Manila. Barcenas set up radio CRM in the Manila area and later changed to FER covering the central plains of Luzon. Radio station FGQ was set up in Lipa, Batangas to support Manila guerrilla leader, Dan Barrion. American pilots were rescued, brought to Mindoro and evacuated by submarine because submarines had still not penetrated Luzon.

An urgent request to replenish dwindling supplies was transmitted to KAZ in order of priority "weather kits, radios and peripherals, Philippine pesos ... medicines, rations, and softball and volleyball equipment" at the end. Were the sports equipment for the guerrillas' goodwill?

Civilians who arrived from Batangas told of atrocities committed by Japanese soldiers, destruction of all bridges between towns, and the dynamiting of the docks in Batangas Bay, making them impossible for landing purposes. There were Japanese Q-boats camouflaged along the beach and under the houses in Anilao coast. The villagers asked the soldiers to destroy their homes as the Japanese had driven them away and robbed them of their food. Flashes to fighter planes bombed the wooden houses to the ground. Civilians forced to work in the area under the Japanese were killed.

G-2 Security had to remind Rowe that his mission was one of secret intelligence and not the internal affairs of the locals. He was living in the Philippines with his family when the war broke out and he knew many of the influential and government officials. His request for an evacuation of the wife of new Philippine President Osmena was denied. She was better off in the mountains than risk trying to meet a submarine.

With the Leyte invasion in a couple of months, training of guerrilla volunteers to be secret agents intensified. They had to be trained "to make observations of military operations and installations, translate those observations into maps drawn to sufficient accuracy to guide an air attack, strafing, or bombing." Bombing missions could not be accomplished without those maps. New communication channels were built to accommodate the new agents' reports. Under Berg's supervision, a complete photo lab developed photographs taken of southern Luzon, the Verde Passage, northern Mindoro, and Manila Bay. Water from a nearby mountain stream was so clean that the photo lab used it without filtering. Savellano and Delfin oversaw the warehouse, storing of equipment, supplies, and food.

The camp was "hard-pressed to feed themselves and the agents who sometimes marched a hundred miles and schemed their way through a dozen sentry posts to get there." Fish from the sea helped. Escapees found their way to the camp. Japanese prisoners were interrogated, and many were shot when they tried to escape. American downed pilots from the bombing raids in August and September were also brought to the camp. All the heavy movements risked discovery of the camp.

Rowe conferred with Ramsey on the gathering of accurate reports of enemy activities in the vicinity of Manila and as far as Lingayen on Luzon. Four hundred fifty agents were doing various jobs as part of Rowe's intelligence net on Luzon. They were paid for information they produced. Many of the agents were farmers and civilians eager to be of assistance. A simple script of questions was drafted for them to use to get answers as accurately as possible. The answers were either Yes or No. In the city, new agents had to be trained in receiving intelligence from established agents. Their identity remained secret to each other as protection if caught.

An operative, Johnny Ysmael, who was providing information on the Japanese in Manila knew Rowe and went to Mindoro to see him. Ysmael would dress up in a Japanese uniform and drive officials to Baguio and back all the while obtaining information on their activities. With him was guerrilla officer Dan Barrion for whom they had set up a radio in Batangas. Their intelligence was sent directly to ISRM.

Orders to Rowe's men were to stay out of Manila, so volunteers in and around the city had to obtain the information. But KAZ demanded more specific data on so many areas and installations in the city. Ignoring orders, Hernandez went to Manila with Barcenas and Pascua. A bodyguard of well-trained guerrilla fighters accompanied them. The guide was an older Filipino man. Malaria struck Hernandez before the river crossing. He had to be carried in a blanket slung between two poles until a carabao could be found. It was a 6-day trek through the jungle to Abra de Ilog, then a thirty- six-hour trip crossing the Verde Passage to the southern tip of Luzon. They cautiously sailed past Verde Island, occupied by a strong enemy force. Reaching Balayan Bay, the people gave food and shelter, but the Japanese knew of their landing. They changed to civilian clothes to enter the city. Reports described their arrival in a truck with their equipment at the house of a guerrilla outside Manila. It was dangerous passing the menacing Japanese checkpoints. Agents they contacted who knew Rowe helped to move around and find the safest location for their equipment.

The camera was set up on the fifth floor of an office building overlooking Manila. A short distance from there, the transmitter was set up on high grounds and reported daily. Berg and Delfin with their team took photos of individuals and military installations. A *carretela* (cart pulled by several horses) with a camera in the back went around photographing essential installations. A truck with a radio went around the city gathering information and sending it off to the transmitter. The men passed a sentry successfully except for Hernandez, who was interrogated for hours in a dark, dirt-floored room with a small window high above his head. They knew who he was. In the shoot-out to rescue him, a guerrilla was killed, and another died of injuries. The worry now was the enemy's recognition of Hernandez.

After they had spent five days in Manila, a city map with all the public utilities and military installations was still not completed. Hernandez related that the Manila Hotel, the former residence of General MacArthur and his family, was now filled with Japanese generals. Most of the hotel staff were agents on the alert for any revelation of military secrets. He was introduced as a new waiter and dance instructor to a Japanese major who wanted to learn the latest Latin dances so he could dance with pretty girls and surprise friends. The major was known to hold office with a locked metal cabinet at the city hall. In a few sessions, Hernandez timed his step to keep bumping into the officer whose jacket contained the key. Encouraging the major to continue practicing the steps, he ran to give the jacket with the key to Pascua who was already in a *calesa* (horse-drawn buggy). Conflicting reports state Inocensio Pascua and Angelo Pascual. An agent working at the city hall for the puppet government was waiting to open the metal cabinet. Inside was a city map and documents of Japanese positions in the Southwest Pacific Area. So critical was the discovery that the officer killed himself when he learned what had happened, Hernandez found out later.

Many residents of the city provided information on the enemy. Hernandez met with many Filipina women who routinely provided military secrets in many details from Japanese officers who had a weakness for them. Undoubtedly, they are some of the many unsung heroes of the war.

Another malaria attack weakened Hernandez for a couple of days before the arduous trip back to Mindoro. The squadron of guerrillas was ready in the first and third rigs to shoot out for an escape. A nerve-wracking inspection of their three carts at a bridge did not reveal the map that was filled with comments from agents. They met Ramsey on their return, who requested another radio to monitor the Manila area and communicate with KAZ. He wanted arms to fight the Huks and funds. He was living on borrowed war notes. The *Hukbalahap* (Huks) were mostly Luzon farmers who initially opposed the Japanese, but some had turned to terrorize the locals and guerrillas to get whatever they wanted.

Back at camp Nimitz, traffic increased tremendously as the landing was imminent. Uninterrupted communication with KAZ was imperative. KAZ ordered all intelligence from Luzon to go directly to ISRM and no longer through Panay and Mindanao networks. ISRM sent a list describing each of the photos taken in Manila, enabling KAZ to pinpoint the targets on the maps, to avoid civilian casualties.

Days before the Allied forces arrived, radio messages flowed continuously to the U.S. 7th Fleet on the enemy's disposition, strength, gun emplacements, radio stations, and tanks on Leyte.

Leyte Island, Philippines Landing - October 1944

The coastwatchers were ready to flash Japanese shipping in the waters of the islands surrounding Leyte. The Imperial Japanese High Command would not have the Allies land without some resistance from them.

Bugarin reported that "During the operations in Leyte, we reported six big Japanese convoys and received messages of commendation from General MacArthur. Lt. Hernandez has all the records."

USS *Ray* – October 1944

On the last day of October, Bugarin with a guerrilla group, two American escapees from Corregidor and two downed Navy pilots crossed the island to meet the submarine *Ray* at Mamburao. A member of the Philippine government, Maximo M. Kalaw, was also with them. Hopefully, the radio sets needed for the Mindoro landing were in the cargo. A Japanese convoy was sighted, and Borillo at the ISRM tower reported that "it was taken care of" the next morning.

The sub encountered a convoy of five ships. She sank the cargo ship *Horai Maru* No. 7, and twice torpedoed a small tanker off Mindoro's west coast. She destroyed the cargo ship *Toko Maru* with two direct hits and escaped depth charges in another report. She landed Lt. Richmond, two men listed only as weathermen, and 2 tons of cargo. After boarding the evacuees, she disappeared into the open sea. She fired two torpedoes and sank the seaplane tender *Kagu Maru*. In a quick dive to escape aircraft detection, the conning tower flooded because the upper hatch was improperly closed. That caused damage but was brought under control before reaching eighty-five feet below. She was able to return to Mios Woendi for repairs.

There were no radio sets in the cargo. Only new weather equipment that could record upper wind levels at 30,000 feet, rations, aspirin, cigarettes, and two salvaged radio parts that were sent to KAZ three months ago arrived. Monitoring the air strips of San Jose was necessary for the coming Luzon landing operations but there was nothing that could be done without radios to send information. An urgent message concerning the missing radios was sent. The reply was that there were no requests for radios before the submarine left. Upset that his supply request did not arrive, Bugarin sent Rowe's angry message, "Your supply sub as far as I am concerned are nothing but cook tours, rpt (repeat) cook tours." KAZ sternly replied: "Highly improper tone to your commander in chief. Desired that in future dispatches, you will govern yourself with more regards for military courtesy."

Rowe felt ignored by KAZ and replied that since Whitney, head of PRS, had left KAZ to join the planning of the Leyte landing, his supplies "have been ridiculous." A shipment of supplies would enable him to "assume full responsibility for supplying GHQ with the information needed from this area." Radio traffic had greatly increased and continued between his sub-stations and KAZ. There was no reply from KAZ. In desperation,

Hernandez, with a team, raided a Japanese storehouse for anything they could use. The signalmen examined the Japanese radios and rearranged procedures to make them usable.

After Leyte, KAZ turned its sight to Luzon, wanting confirmation on the locations of all existing radios, call signs, and disposition of cipher systems. Information on the enemy troops on Luzon was of utmost importance, especially in the Lingayen area where they planned to land next. It was the same spot three years ago that General Homma came ashore in the second week of the Japanese invasion of the Philippines.

A team with a squad of guerrillas departed from Sablayan and managed to cross the vast Mindoro Strait and set up a station on Apo Island to monitor enemy ships coming from the southern islands in the Strait.

In November, a coastwatcher signaled the Japanese sub-chaser *Higu Maru* docked ten km from Camp Nimitz. The whole crew was seen landing unarmed except for one who had a saber with him. They pestered the people in the village for food. Guerrillas from camp departed in the night with a storm covering their tracks to arrive the following day to ambush the crew. The *Mangyans* went to work when a sack of salt was offered for each captured Japanese soldier. Seeing no firing from the boat, Alberto, Balleras, Savellano, Pinuela with Rowe and a team of guerrillas boarded to take whatever they could find. They were desperate for supplies especially fuel and gasoline to keep the station on the air. The galley was bare, so the crew was still onshore. There was much gasoline, oil, and fully charged batteries for the radios they needed. While the others took "a chronometer and plenty of clothing (uniforms)," Alberto and Balleras stripped the radio room of all the radio sets, logbooks, papers, and the cryptographic system of the Japanese navy. The guerrillas burned and sunk the vessel. "It was the first time in the war that the cryptographic system of the (Japanese) Navy has been captured," Rowe informed KAZ. The enemy documents were of great value, and KAZ commended their actions.

In November, the submarine USS *GAR* brought 5 tons of supplies before proceeding to Luzon.

Mindoro, Philippines Landing – December 1944

KAZ urgently requested for intelligence on San Jose, Mindoro, where the American forces planned to land. The northern part of the island would have been a better choice, but it was accessible by Japanese forces from Southern Luzon. More radio sets were requested to be sent in a submarine. The reply "was impossible." They could only hope that some radios would come in the next rendezvous of a submarine mission.

ISRM sent out the long-awaited message to Maj. Ruffy, the guerrilla commander of the island, to openly destroy enemy installations, highways, and bridges to trap them several days before the Task Force landed.

The decision to invade Mindoro Island was, at first, considered to be "audacious" since the island was within easy reach by Japanese airdromes around Manila. Even Washington D.C. advised General MacArthur that the plan was "too daring in scope, too risky in execution." Admiral Kinkaid of the U.S. Seventh Fleet objected firmly to MacArthur's plan, pointing out that to get to Mindoro, his fleet must pass through Surigao Strait and the Sulu Sea, where the vessels would be "clay pigeons for Jap land-based planes." MacArthur, however, assured him that General Tomoyuki Yamashita already "had no stomach for another Leyte" and that he was "husbanding his strength for the coming struggle before Manila." MacArthur was proven right when intelligence revealed the presence of only a couple of hundred Japanese troops on the island.

After the Battle of the Philippine Sea that captured Saipan, General MacArthur was convinced of air superiority. Mindoro was not in the path of typhoons, so airfields could be built unlike on Leyte's muddy ground. Mindoro was close enough to Luzon from which aircraft could fly, bomb targets and protect landing forces on the way to Lingayen Bay. As protection, coast watchers on Mindoro ensured that shipping in the Japanese-infested waters in and out of Manila Bay, around Mindoro, and between Leyte and the nearby islands was not hampered by the enemy.

Alberto tested the radios that arrived with the USS *GAR* and found one unusable as it did not have a power supply. They left for San Jose to prepare for the landing of the U.S. forces. but they were delayed in Naujan and could not contact ISRM. The Western Visayas Task Force landed in San Jose without them. Pinuela was reported to have led the landing party at Naujan when the area was liberated.

When they arrived in San Jose, news reached them of several guerrilla officers and men killed when over 100 Japanese troops raided the school. Balleras said they took cover with the radio equipment in a crocodile-infested swamp with "only our noses sticking above the surface." They slipped away to Mount Halcon and found the radio severely soaked and inoperable. Drying over a fire did not work. They cleaned their muddy weapons and loaded them.

ISRM's camp and guerrillas attached to the U.S. Sixth Army. Radio PDQ transferred to the American command, as reports continued to come in. A message to KAZ informed of three hundred partially armed men, waiting for orders to attack an enemy's headquarters 150 km away. Many had *bolos,* and some homemade guns made from one-inch pipes they could find. Mines floating loose in the sea were disarmed and the explosives made into bullets. The guerrillas had rescued eleven pilots and turned them over to the Navy.

With the heavy equipment brought in, two airstrips were built in less than two weeks. Many civilians and *Mangyans* worked day and night. The Japanese retreated to Ambil Island.

All personnel at Camp Nimitz and men at their outposts were brought to San Jose, the Sixth Army's headquarters by P.T. boats. Bugarin described an exciting event of a Japanese plane on a bombing run that was "either crazy or he had no idea how guns could work at short range. He was about 100 yards high... all twenty machine guns from the two PT boats converged on the plane as if they were going on a pre-arranged signal," totally obliterating it.

At the new headquarters, Alberto requisitioned for radio parts and Bugarin connected the new message center to their network of stations.

The men heartily ate their first decent meal in six months. The cooks continuously filled up their plates when they found that they had been in Mindoro that long.

Unidentified Transport – December 1944

Preparations for the Luzon landing intensified. Heavy rains turned the airfields in Leyte into muddy swamps delaying the completion needed for the Luzon landing. Mindoro was ready with airfields and the surrounding waters were monitored for Japanese attacks that would cause any delay to the Luzon landing. An "enemy garrison of 1,000 Japanese soldiers and around 200 sailors marooned from sunken ships" were surprised and eliminated. But more personnel skilled in reporting weather observations, and aircraft, were needed.

Orders arrived for a mission headed by Sgt. Cleto S. Bello. Under his charge were Sgt. Rafael Licera, Sgt. Mero L. de Ocampo, Sgt. Walter B. Lagrimas, Sgt. Felicisimo D. Magbual, Sgt. Apolinar G. Salvador, Cpl. Frank S. Leanio, Cpl. Maximo V. Maglangit, Cpl. Harry T. Tucay, Cpl. Elpidio A. Rivera, Pfc. Lacaro C. Arquero, Pvt. Pedro S. Azcueta, Pfc. Julian F. Caduang, Pvt. Benny T. Velayo, Pvt. Ricardo D. Cayabyab, Pvt. Andres D. Darang, Pvt. Victor V. Marquez, Pvt. Robert C. Minder, Pvt. Al J. Granados, Pfc. Philip Martin, Pfc. Gerardo C. Alfafara, Pfc. Tom P. Respicio, Pfc. Perfecto N. Trocio, Pfc. Alfonso L. Antonio, T5 Francisco M. Urbano, T5 Gene D. Costales and T5 Urbano M. Francisco.

Maj. Brown saw them off before daylight in Australia before they boarded a C-47 cargo plane to Hollandia, staying overnight and continuing to Biak. They picked up their supplies, including radio equipment, ammunition, medical supplies, clothing, tommy guns, automatic pistols, Carbines, and all necessary equipment for combat. They were attached to the 597th Signal Aircraft Warning Battalion (SAW), Company D.

No record was found of their type of transportation. The party joined a group of American soldiers with a SAW Battalion and likely in a convoy bound for Leyte. Ten Landing Craft Infantry (LCI) were used for transporting supplies and equipment to guerrilla forces within the Visayas and Northern Mindanao areas. The submarines were no longer available.

Sgts. Licera and Salvador disembarked and reported to the Dulag net control station in Leyte while the rest went with the convoy of invasion forces to land in Mindoro. A PBY bringing them with a radio to Abra de Ilog missed their landing and returned to Dulag. In the meantime, Japanese paratroopers had infiltrated Dulag and cut the communication lines to capture the airstrip. Japanese planes caught the men on the beach as a floating oil tank was bombed and shot flames into the sky. In between strafing, Licera and Salvador frantically dug fox holes that kept caving in on them. A week later, they boarded a PT boat with White soldiers to Mindoro and were shelled by a Japanese patrol boat nearly capsizing them. Shells passed overhead. The skilful boat commander drove through the blinding searchlight to avoid making silhouettes, making them difficult to be spotted. Torpedoes were released as they escaped to the PT boat base near San Jose.

The convoy to San Jose travelled slowly to be cautious of Japanese subs or surface crafts. Enemy aircrafts spotted the convoy and dogfights ensued for the next two days as ack-ack guns (anti-aircraft) repulsed them. Two ships were hit. The shelling of San Jose beach destroyed an enemy tanker. Men were relieved when a burning Japanese plane heading towards their ship landed 15 yards away.

The directive was to set up more radio stations on Mindoro with soldiers from the Sixth Army. Caduang, Velayo, Trocio, and Granado each with a team of guerrilla security set out to establish signal aircraft warning stations to monitor enemy aircrafts. One station operated by Salvador and Licera, was ordered closed after a week by a Sgt. Ash who wanted to go with sailors in a PT boat to raid some Japanese launches in Batangas Bay. Francisco remembered Ash with them at Camp Tabragalba eating with the Filipinos and was treated well but openly criticized the Philippines, not wanting to have anything to do with it. Salvador refused his request and kept the radio open for valuable information regarding the movements of enemy aircraft and surface vessels. He singlehandedly handled communication 24 hours a day for a week while Ash and Licera "went out romancing through the town's *dalagas* (young women)," Salvador related. A civilian technician, Modaferri, helped him. He became sick. Disillusioned, he thought of the men as "excess baggage who abandoned the ultimate goal of our mission ..." Weak from lack of sleep and rest, he kept his thoughts to himself.

Arquero and Minder, with four White soldiers who were also with them at Camp Tabragalba, were a radar group stationed at Abra de Ilog. After a week, they moved north to Puerto Galera closer to the coast and sent verified reports on aircraft and enemy surface vessels nonstop. After two months, orders were to proceeds to Palawan Island heavily garrisoned by Japanese troops. They were fired upon while attempting to land in their rubber boats, destroying their equipment. Their food supplies went overboard. An American private suffered a shattered leg. The party was sick and hungry for a week until an American patrol plane spotted them and sent an amphibious aircraft to their rescue.

With two American weathermen, Maglangit and Urbano boarded a medium landing craft to sail to Sablayan and set up an outpost. Civilians in two small boats got them over the reef to shore. It looked like the entire village was on shore to greet them and help carry their load and equipment to a small hut. They were relieved as it was the first time, they had a decent place to sleep. They set up a dispensary to cure malaria, dysentery, and tropical ulcers. Urbano expressed his despair upon seeing the destitute especially the undernourished and sick babies. The surrounding villages heard of their arrival and soon more came for treatment and medical supplies. The dispensary was open all day. In return, they were fed with whatever the people could share with them.

The radio was on the air 24 hours a day to relay enemy presence and activities to the net control station at San Jose. It was Christmas Eve, so they shared their rations and food supplies with the people for their first real Christmas in three years. The grateful locals held a dance in their honor at the schoolhouse "where Japanese tortured the civilians including children and old ones... making them stay under the sun and letting (them) stand the whole night 'till their limbs cracked" related Urbano.

The civilians told of a Japanese plane that had crashed three days earlier in their area, and the pilot had still not been found. An enemy patrol would come and torture them if they were found helping the men. With four guerrillas, Manglapit and Urbano went to look for the pilot. They heard the crashing of leaves three miles out and saw the Japanese pilot, an officer, all burned up and still alive His pistol was ready in his hand. Tackled by the brave guerrilla boys, he was taken prisoner and brought to headquarters in San Jose for questioning. The pilot did not follow "their zero Ace pilot Saburo Sakai who had long resisted wearing a parachute. They left them unbuckled because their presence might otherwise have suggested a willingness to be captured ... No fighter pilot of any courage would ever permit himself to be captured ... It was completely unthinkable."

Japanese bombers flying low just above the trees shelled the shores and strafed the village two nights later. The people rushed and took whatever they could and evacuated. Lt. Eagle from the U.S. Task Force and some of the guerrillas stayed to guard the outpost while Maglangit and Urbano frantically punched a message of a Japanese convoy on its way south to their Sablayan outpost. They packed whatever equipment they could carry and went to higher grounds to see the passing convoy. They returned the following day and continued radio operation for the next three months. Orders arrived to proceed to Palawan to support the Thirteenth Air Force landing. After a month, they returned to Leyte.

Continuous and heavy radio communications took a toll on the worn-out sets and equipment running 24-hrs at camp Nimitz. Radios had to be serviced daily to keep them running. Alberto said that "I even had to convert Japanese radios and other equipment to suit our needs." Each sub-station had to have a new cipher system set up. The men worked long shifts receiving the sub-stations reports enduring only two meals a day.

Rowe radioed a request to close his camp but was told to "hold position throughout the Luzon invasions at any cost. You and your channels of communications are the eyes and ears of GHQ."

A radar signal in Camp Nimitz was set up by the 583rd Signal Battalion in 24-hour contact with the American PT base in San Jose. PT boat crews with ISRM agents prowled the inter-island waterways to destroy enemy aircraft, gun installations, and other targets pointed out by agents. The boats brought the reports to the Sixth Army headquarters in San Jose. When Bugarin saw six sailboats of fishermen around the camp and a group of U.S. fighter planes flying low towards them thinking they were Japanese, he thought the camp would be strafed. He contacted the PT boat base to determine the plane's frequency and contacted them. What a relief when the planes gave them the wing signal and left.

Carillo had set up station PSC on Maricaban Island to monitor the activities of two Japanese garrisons The station was able to see into Batangas Bay and the towns deeper into Luzon. Three guerrillas with Pascual replaced him with station WAP receiving operatives' reports on enemy activities, locations of anti-aircraft guns, machine guns, and warehouses of food supplies and equipment in Batangas. An old man was summoned to form civilian guard units to protect the station from attacks coming from the north while guerrilla Capt. Tibay's men protected from the south.

It was imperative that the Japanese garrison on Maricaban Island in the middle of the Verde Passage be eliminated as it would impede the convoy of American forces from Leyte to Luzon. A task force to take the stronghold was shipwrecked on the way to the island killing guerrillas on board. U.S. planes did not bomb the garrison in the town of Tingloy because most of the Japanese were living in private house to avoid attacks. Two PT boats arrived and shelled other towns on the shores killing many civilians, mostly children. Pascual reported on the American bombing and the destruction of enemy vessels with Filipino laborers forced to work for the Japanese. Two PT boats were to evacuate the wounded but never came, "thus leaving them to die and suffer without any medicine or medical treatment," Pascual reported.

Verde Island in the middle of the Verde Passage had a large installation of Japanese guns that posed a threat. Posing as a German trader with papers provided by the mayor of Pandan Island, Rowe, with guerrilla security, crossed to the island and photographed the Japanese fortifications to transmit back to KAZ.

Alfafara, Martin, and two American weathermen were attached to the 3rd Battalion of the guerrilla 21st Infantry to handle their communications at Bongabong. The civilians warmly received them giving food, cooking their meals, providing lodging, and even washing their clothes. They did not ask anything in return until a streak of dysentery and malaria went through their area. Knowing the outpost had some medical supplies, they asked for help. Martin and Alfafara insisted on helping them, but the two Americans "were too selfish to share with the people who have (had) helped them, respected them, with (even) only some medicine to check the spread of the sickness...Dismayed and eyes cast to the ground, they left the outpost without a word of complaint. They buried six civilians that week..."

The 583rd SAW Battalion in San Jose received Martin's urgent message that a Japanese convoy had landed some troops. Elements of the 21st Infantry repulsed them and captured a soldier. With his limited knowledge of the Japanese language, Martin was able to take valuable information from him. He and Alfafara with the Americans were called back in January.

The American forces continued to amass and mobilize thousands of men and hundreds of warships for the Luzon landing. A separate amphibious group crossed the inter-island seas and landed in San Jose. Locals met the troops with some waving American flags.

It was Christmas time, and a box of American rations were sent from Camp Nimitz to the men at their remote outposts boosting their spirit and making the loneliness bearable.

At a radio outpost in Abra de Ilog monitoring the eight-mile-wide Verde Island Passage, a Christmas Mass was held at a church built by the locals and *Mangyans* from jungle growth. It was peaceful with no disruptive sounds of drilling, clicking rifles, and aircraft raids. It was a fiesta with a gift of Kraft cheese, canned shrimps, and Philip Morris cigarettes from the Sixth Army commanding general. Everyone danced. The following day, the men were in PT boats on their way back to Camp Nimitz. A Japanese task force was shelling San Jose when they arrived. Rowe had gone to the Army field hospital for treatment.

Luzon Island, Philippines Landing – January 1945

The New Year opened with a crackling broadcast from *The Voice of America* of Japanese planes attacking American positions and being shot down over Mindoro.

Requests for radio operators sent signalmen to support the landings. Balleras with Bugarin, Villalon, and two other signalmen were sent to Calapan to operate the radio station for two months then to guerrilla 46th Infantry Regiment which later became the 51st Infantry Regiment under guerrilla leader Maj. Ruffy. Bugarin stayed at Calapan and departed in October. Villalon was proud of "the guerrillas we trained in (nearby) Abra de Ilog ... became the keymen who held responsible positions in the communications platoons of this unit." The guerrillas showed their pride publishing the daily paper *Hilltop Calling* with Villalon as editor.

Two PT boats of the Army radar team arrived at Rowe's headquarters with sealed orders. The guerrillas were to destroy a Japanese radio outpost in Paluan that was reporting American bombers coming from San Jose airfields in the south bound north to Luzon. Planes of the 5th Air force were launching strikes from the two Mindoro airfields to destroy the enemy planes at Clark Field on Luzon. The enemy outpost was destroyed four days before the Lingayen landing.

With two technicians from the U.S. 24th Signal Service Company, Pascual sailed to the island of Marinduque to reinforce Gonzales' JSG radio. When the U.S. Task Force landed in Palawan, he sailed with a company of Philippine Army soldiers to operate the 51st Infantry Regiment (PA) communication in Busuanga, a small island in Palawan.

As the hundreds of Allied ships streamed towards Lingayen Bay passing across the central Philippines, the coastwatchers on the islands of the Visayan islands and Mindoro watched the seas surrounding their islands for Japanese attacks. At his outpost on Lubang Island, Toribio was the last to watch the convoys before they crossed the long stretch of the West Philippine Sea into Lingayen Gulf.

Reports of leaflets dropped from U.S. planes urged the people to clear the road and give way to the troops rushing from Lingayen to Manila. News of civilian atrocities abounded in the towns the forces passed. In Balanga, Bataan, locals were abused and killed by the enemy. The "Kill all" orders of prisoners from Tokyo heightened the urgency to reach Manila.

At Nimitz camp, station BM5 was ordered to receive all reports for several weeks relieving KAZ of heavy radio traffic. Rowe, sick with malaria, was ordered to report to the commanding general of the U.S. division in San Jose. Hernandez stayed behind as second in command with Berg.

Aside from taking KAZ's radio traffic, the Mindoro signalmen filled ISRM's frequency channel with Japanese atrocities. Couriers from Manila continued to arrive. Agents said trucks filled with Japanese troops left Manila for two days heading north from midnight to dawn several days before the Americans arrived in the city. The four bridges crossing the Pasig River were mined. The piers were destroyed. Japanese marines concentrated in concrete buildings along the business center. Around a dozen pillboxes blocked Jones bridge, major intersections, and buildings with one connected by trenches to the buildings of Far Eastern University. Tank barricades made of logs, rails and barbed wires ran through major streets. Plane engines and equipment were stored in residential houses and a thousand gasoline drums under the trees along Padre Burgos Street, the location of the City Hall. Reports of abused women surfaced. People were starving as all harvest in central Luzon was confiscated by the Japanese Army. Water was cut off. Thick black smoke from fuel dumps blown up polluted the air. Agents reported camouflaged patrol boats plying the Pasig River, Laguna Bay, and Pampanga River.

Thousands of gasoline drums were reported under the trees around Lubang airfield at the entrance to Manila Bay. Agents informed of Japanese soldiers willing to surrender to the American forces but not to Filipino forces. They believed that the Americans were their real enemy or that they would not live another minute if they surrendered to the Filipinos.

Ramsey who had received a radio had Harder operating it and reporting on the enemy fortifications of Intramuros and the barricading of streets. Manila did not look like an 'Open City.' Numerous artillery guns were emplaced in the Montalban and Marikina hills facing Manila Bay. The Bay was mined.

On the road leading to Manila, Galang sick with malaria and beri-beri was picked up by the 1st Cavalry on their way to Manila. He left Pampanga as his parents were being questioned by the Japanese. He guided the troops wherever he could. "(He was) still alive because he could run faster than them (Japanese)," he wryly declared.

"The center of Manila was barricaded by around five thousand Japanese marines," Galang recounted. The Japanese had constructed wooden dummy tanks and numerous anti-tank barricades of buried two-hundred-gallon fuel drums, possibly with explosives. Machine gun emplacements were at intersections in the city. A single row of wood stumps was at the Victoria Street entrance that dead-ended into Calle Victoria, once the USAFFE headquarters of General Macarthur. Eight machine gun nests facing in all directions transformed the ruins of the Binondo Church into a fortress. Rewards were offered for information on spies, paratrooper landings, and concealed American pilots. The fanatical Japanese resistance from house to house and room to room in buildings went on for several weeks. Galang could not help crying when he learned that the Japanese broke into the library of the University of the Philippines, piled up the books and set them on fire because they were in English. "Pictures of American leaders, like George Washington, were cut off from textbooks," he wrote.

He walked to headquarters in Tarlac following orders. At the Quartermaster, he changed to a complete uniform (American size) from his ragged civilian clothes. His camouflaged pants were several times bigger. The matching shirt with sleeves hung below his fingertips and he had to roll them up. He found a seamstress to fit his uniform properly. A pair of captain bars and a couple of cross rifle pins were pinned on him. Whitney unpinned the bars and replaced them with a pair of gold oak leaf. Unknowingly, he was promoted to Major on the day he left on the *Narwhal*. There was no headgear except for an oversized helmet. His size 8-1/2 (shoes) no longer fitted. Constant walking barefoot spread out his toes.

Manila was still burning when Galang was directed to clear Malacañan, the official residence and office of the Philippine president, of booby traps. General MacArthur was to turn over to President Osmena the city departments, police, health and welfare, public works, and fire department in a simple ceremony. He recalled finding guns and many dead bodies in the building. Malacanan was untouched with its stained-glass windows, tapestries, and the crystal chandeliers were intact. He wrote, "Simple and charged with emotion. Only a small crowd was present: General MacArthur and his staff, President Osmena, presidential aides, and a few members of U.S. Signal Corps." A photo showed Commander Parsons with Mrs. Osmena. When General MacArthur was delivering his speech, Galang witnessed that at a "certain point, his (MacArthur) face grew tomato red, and he faltered. He walked behind the drapes, and we saw him wipe his eyes. Everyone was silent. Then he returned to finish his speech. He saluted President Osmena and embraced and kissed Mrs. Osmena." Maritess Galang, his daughter, recalled her father telling her that General MacArthur said to President Osmena: "The Philippines should have been liberated earlier if not for the U.S. Congress. I wanted to return to Manila and liberate her because I made a promise. It took so long, and many died." The rare poignant feeling showing through the usually stoic face of the General appeared again six months later when Lucien Campeau (Mindanao mission June 44), a weatherman, returned to Manila from a mission to Mindanao. He wrote of General MacArthur "thanking him profusely in a makeshift headquarters in downtown Manila in a large room like a ballroom, (behind) a large desk on a raised platform, and nothing else in the room. General MacArthur's voice quivered, and tears came into his eyes."

Camp Nimitz with station PDQ was ordered to close in February. All signalmen at their remote outposts were to report back to Camp Nimitz then to the Signal Company of the U.S. 24th Division in San Jose. The men packed all their heavy equipment and closed Nimitz for carriers to deliver by PT boat but not all the men returned at the same time. Some stayed at their outposts for another six months. When they left, the guerrillas they trained at Abra de Ilog continued radio operations under the Philippine Army.

The PT boat streaked through the Verde Passage to Nasugbu harbor, passing through a convoy of various U.S. warships and aircrafts roaring overhead. Black squares painted on each side of their sails identified them. Hernandez recalled they were "challenged at

Nasugbu harbor by an American patrol boat carrying two menacing guns because twenty weary and seasick men did not look like any American fighting force they had ever seen." They walked to Manila, passing numerous dead bodies. Every bridge over the Pasig River was destroyed. Snipers continued picking on targets. They arrived in the chaos of Manila heightened with President Osmena's decree to drive on the right-hand side of the road.

As an officer, Galang ate at the Officers Mess with White officers, who did not want to sit with him. They passed his table and "threw cigarettes and chewing gum at him or just dropped them on his table," he related. That insulting action was usually done to Filipino civilians and guerrillas. The food at the officers' mess was too much to resist but was a shock to his shrunken stomach. He wound up in a dispensary and needed to have his stomach pumped out.

Pinuela surfaced in June in Manila at Malacanan Palace driving for a U.S. Colonel. Bugarin, Toribio, Pascual, and Alberto arrived two months later.

The strategic location of Mindoro and her extensive radio networks covering the surrounding islands, passages, and seas, significantly secured the water routes of the liberation forces to Leyte and Luzon saving many Filipino and American lives.

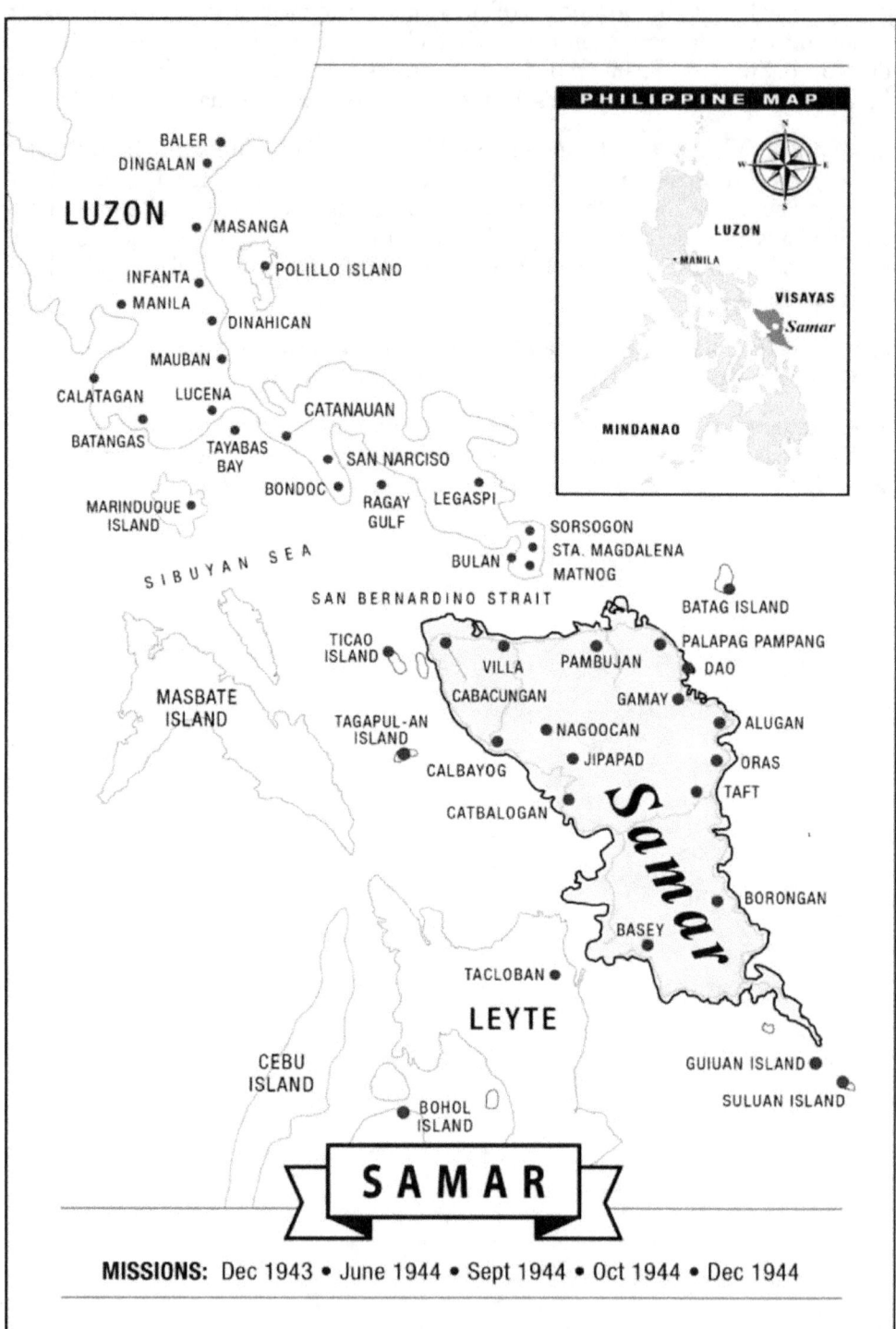

PHILIPPINE MAP

LUZON

BALER
DINGALAN
MASANGA
POLILLO ISLAND
INFANTA
MANILA
DINAHICAN
MAUBAN
CALATAGAN
LUCENA
CATANAUAN
BATANGAS
TAYABAS BAY
SAN NARCISO
BONDOC
RAGAY GULF
LEGASPI
MARINDUQUE ISLAND
SORSOGON
STA. MAGDALENA
BULAN
MATNOG
SIBUYAN SEA
SAN BERNARDINO STRAIT
BATAG ISLAND
TICAO ISLAND
PALAPAG PAMPANG
VILLA
PAMBUJAN
DAO
MASBATE ISLAND
CABACUNGAN
GAMAY
TAGAPUL-AN ISLAND
NAGOOCAN
ALUGAN
CALBAYOG
JIPAPAD
ORAS
CATBALOGAN
TAFT
Samar
BORONGAN
BASEY
TACLOBAN
LEYTE
CEBU ISLAND
GUIUAN ISLAND
SULUAN ISLAND
BOHOL ISLAND

LUZON
MANILA
VISAYAS
Samar
MINDANAO

SAMAR

MISSIONS: Dec 1943 • June 1944 • Sept 1944 • Oct 1944 • Dec 1944

XIV

Samar Island

GHQ believed that the Japanese would not expect another submarine landing in Mindanao following the USS *Narwhal's* landing the month before. The mission would then proceed to Samar to gather intelligence and set up radio networks in the island and the Bicol region along the east coast of Luzon. The sea lanes running between Samar and the southern tip of Luzon were to be monitored for enemy shipping.

Samar was not a strategic place for the Japanese. No airfields were built. At one time, around 2500 Japanese soldiers occupied the island. Less than 400 remained with their main activity looking for radio transmitters.

USS *Narwhal* – December 1943

When the orders arrived, the men had passed several months of grueling physical and technical training. They were ready. The reconnaissance men were Sgt. Restituto J. Besid, Sgt. Aniceto C. Manzano, Sgt. Gerardo A. Sanchez, T4 David D. Cardenas, and T4 Crispolo G. Robles. With them were men from the 978th Signal Service Company: T3 Robert E. Stahl, T4 David B. Sabado, T5 George R. Herreria, Pfc. Querubin B. Bargo, and Pfc. Andres S. Savellano. The Filipino men were fluent in Visayan and Tagalog, the most common languages on Samar and Luzon. The Americans were Maj. Charles M. Smith, the mission leader, and Capt. James L. Evans.

They were from the 1st and 2nd Filipino Infantry Regiment in California and the first men from the 978th Signal Co to arrive in Mindanao.

Smith recommended that the network he was to set up in Luzon be separate from that of Mindanao's FRS net control station because "by this method, we can get information which can be check(ed) against other sources." And it would "also eliminate the possibility of the capture of agents causing the stoppage of all sources of information." He was avoiding the same fate that Phillips' party encountered on Mindoro.

Most of the Filipino men were older and probably did not have 20/20 vision, but Stahl said that was not a requirement because he wore glasses throughout the war. He was a volunteer from the PIMC group of the 978th Signal Service Co in Brisbane that handled the station KAZ. His knowledge of crypt was to prevent the same misfortune that broke up the Phillips mission. The Mindoro mission earlier had numerous violations in the transmitted messages which greatly contributed to their discovery and destruction. He was to ensure that security codes were effectively used. He also prepared new codes for the Negros and Panay stations.

Ten tons of cargo were packed with propaganda, currency, and heavy weapons. Thousands of cigarettes, matches, pencils and chocolate bars with wrappers were emblazoned 'I Shall Return-MacArthur'. Torpedoes were carried in their tubes, and extras removed to leave room for more cargo and sleeping cots for the men. The two aft torpedo tubes accessible from the deck stowed the anti-aircraft gun barrels too big for the hatch. Other parts were packed in cosmoline, a brown wax that inhibits corrosion, and wedged into the sub's structure. Mortars with heavy plates which required specialized training were disassembled to fit through the submarine hatch.

Ninety tons of cargo were for Mindanao and distribution to other islands. Mindanao was expecting an earlier request for batteries and sturdier radio transmitters with a range of 100 miles, powered by a hand-driven generator, would arrive in this cargo. The set should be able to maintain a coastwatcher station longer than the 5 hours with batteries. Currently, three stations were inoperable from the lack of batteries.

After two weeks on Mindanao, the men departed with packed gear and equipment for Samar.

Doc Evans stayed behind as a signal officer and badly-needed surgeon for the 10th MD command post. In exchange, guerrilla Capt. Arrieta and 20 Muslim guerrillas from the large Naval transmitter located in Lanao at Bowler's headquarters for security and Lt. Elwood Royer formerly with the 19th Bomb Group who could operate a 50-cal machine gun joined the Samar mission. Each group moved out with a unit of guerrillas for security, help with transportation, and procurement of food. Each man carried a portable sub-station radio on his back with batteries each weighing five pounds, smaller than the radio, and with a life of five hours. A six-month supply was ninety pounds.

Smith wrote to Whitney about the Filipino men with him. He was "against bringing Filipinos from the States up here (Philippines) for a long time. Now I am glad I brought them along. Proud of how the men left in separate boats on their missions." He described the groups of men and their assignment to establish radio networks and sub-stations. His letter ended with "I am proud of the men with me so far, they have left a very good impression here and left here full of enthusiasm and confident that they can carry their missions thru to a successful conclusion. I would like to bring out one point regarding the sending of Filipinos from the States that we have not considered. These men that came up with me displayed no hesitancy when I proposed to send them out on a job for several months. They have no family ties here and are willing to go out in the bush on a watcher station while it is almost impossible for Fertig to get men to leave their home ties and go to a new location particularly to a place where they cannot take their family. Fertig has a lot of radio equipment here but does not have men willing to go out to lonely watcher stations with it. (Fertig has) neither men who are willing nor men who have training to handle the equipment."

Parsons was also convinced. "After observing the men whom Maj. Smith brought with him, I was convinced . . . They have excellent training in codework and radio operations. They have been away from the islands long enough that family ties are not as strong as with our local Filipinos. Fertig had stated earlier that family ties cause refusals to go on assignments for fear of enemy reprisal or death leaving their family to starve." Beside the printed text of Smith's message that went out to General MacArthur, Whitney wrote "Evidence is accumulating to justify the policy that brought U.S. Filipinos here for ultimate Philippine service. (Even Cmdr. Parsons is now convinced.) It is cause for personal satisfaction to me to be able to report that PRS is not caught unprepared. We have a great number of highly trained Filipino operators available for such service – men capable of sending and receiving 20 – 30 words per minute, rather than the 10 words recently specified by Fertig –- men who could give their right arm for assignment to the service indicated. We will have to substitute Filipino officers (for requests for American officers), but there too we have good material developing as evidenced recent tests at the Jungle Training School where sixteen are presently enrolled."

In mid-December, KAZ received the first message of the party's safe arrival. The main station's call sign was MACA, just like the way General MacArthur hand-signed his papers with 'MacA'. Ten stations were eventually set up. The sub-stations sent reports to the net control station MACA that relayed to FRS net control station in Mindanao for relay to KAZ. Enemy ships sighted were flashed to the Navy station, KAZ in Australia, and prowling submarines nearby.

Smith reminded KAZ of Fertig's request for trained radio operators but not officers because "he has plenty already. (It was easy to get information from Manila but) hard for an American to attempt to operate there. Do more good by having Filipinos trained in Australia in both operations of radio equipment and methods of obtaining military intelligence information. I was against bringing Filipinos here for a long time, but glad I did."

After receiving the encouraging message, PRS wasted no time requesting to land two mission parties on Luzon. But that would not happen until nine months later.

KAZ had no visibility on the guerrilla situation on Luzon. Col. Primitivo San Agustin, an officer from Bataan and with the President Quezon's Own Guerrillas (PQOG) operating in the Batangas area, arrived to get a radio transmitter and codes for Luzon. He provided guidance and information about the Luzon guerrillas. He informed KAZ about Bernard Anderson 's group with American and Philippine Scouts operating in the Tayabas area and agents operating in Manila and surrounding areas. When the use of radio was restricted by the Japanese, intelligence gathering on Luzon was through the southern island's networks until a mission could safely land on Luzon.

The men went into enemy territory in civilian clothes with sidearms and ammunition. They carried an emergency medical kit, watches, cigarettes, American magazines, and propaganda supplies. Although they could pass as local Filipinos, bringing these items with them increased their chance of death or being tortured to reveal their mission if captured.

Being able to blend in with the local people and speak their language helped them immensely in accomplishing their job. But some disadvantages had to be surmounted.

They had all the technical training as soldiers, but they had been away so long that they had forgotten the local ways and were not as resourceful as the local guerrillas. Looking well-fed and healthy when they arrived caused some grumbling from the locals. There were times when they could not command respect as easily as a White officer. It was a carry-over from the period when the islands were under American rule.

Establishing a secure code system was difficult as few guerrilla operators knew anything about cryptography. Some used "pocket-size encoding devices that the Army had used in the field to yield a substitution cipher." An interception by the enemy could decipher the message. "The only way a top-secret code system can be established between two parties without being compromised was to hand them over physically in a face-to-face meeting. Someone, a signal man, had to penetrate the Japanese lines to establish the cipher systems."

Smith began sending out groups of men to establish radio outposts while he sailed into the Philippine Sea to set up the central control station MACA.

Manzano and Robles with San Agustin as a guide travelled from Mindanao on a month-long journey overland through Samar's jungles and mountains, crossing rough seas with their heavy equipment to set up station MAA on the Bondoc Peninsula, the southern tip of Luzon. The Bondoc Peninsula was a strategic location to monitor the Ragay Gulf with Camarines Sur across and the Sibuyan Sea, which was a central passage of Japanese shipping between the islands. By January 26, the radio was on the air. They reported on the sinking of the freighter *Celebes Maru* that was meant to be Japanese reinforcement to Leyte, Japanese launches in Catanauan Bay, and several minor vessels in Tayabas Bay from their vantage point. Later, a commendation described their station as "was one of the best-operated coastwatchers in the Philippines." Robles proved to be an excellent operator who maintained his station for over a year, reporting ships and sending out military intelligence. MAA was transmitting an average of six reports per day to MACA in addition to receiving information from northern Luzon. During the months of the Allied landings on Leyte and Luzon, the station received the most radio traffic from Luzon for relay to MACA. A recommendation for their promotion was sent to KAZ.

The second team of Besid, Bargo, and Savellano crossed the San Bernardino Strait to Ticao Island. Guerrilla commander Licerio Lapus whose operatives were on the island was given money and instructions on the type of intelligence to send through their station MAB. The men asked to go to Masbate Island, concerned for the station when Lapus and his men began using propaganda material given to them to show that they were a recognized force with a radio that had direct contact with General MacArthur. The Japanese were already in Ticao looking for the radio. Besid returned to Samar while Bargo and Savellano stayed until they were discovered by the Japanese or spies two months later. Both were captured, possibly tortured to reveal their mission, and killed several days later. The gruesome news shook the men with the reality of their suicidal mission. They could be next. The radio was reported recovered later and brought back to MACA station for repair.

A Japanese headquarter garrisoned Cebu Island. It was a transportation center, site of a major airfield, and a port of call for cargo ships of the enemy. Sanchez and Sabado with a guide sailed to Mactan Island beside Cebu. Aside from radio equipment, they brought some propaganda of cigarettes, chocolates, and chewing gum. The first contact of station MAD with MACA failed. They kept trying day and night until they heard a call in January. Exchanging their uniform for civilian clothes and taking assumed names to enter the city, they gathered information on troop strength, movements, gun emplacements, and military activities in Cebu City. They reported on the waterways between the islands of Cebu, Bohol, and Southern Leyte. Their messages had to be brief to avoid interception by the enemy. Going through enemy-infested territory "was no joke," Sabado said.

After two months, Sanchez returned to station MACA with five operatives bringing all the information on Cebu while Sabado departed for Tagapul-an, a small island beside Samar to observe ship movements going north and south. Sailing in a sailboat, "If a Japanese patrol boat found us with our equipment, it would be death," Sabado recalled. Again, it took a while to contact MACA. His station was raided with the guerrilla guards killing six of the enemy. Taking a sailboat, the group avoided Japanese patrols and landed in the town of Cabacungan on Northern Samar. With a guerrilla officer as a guide, they took another sailboat through the San Bernardino Strait to MACA camp. Nearing Batag Islands, a large party of 14 Japanese sailboats made them turn back. For the next two weeks, they travelled overland. Food and money were running low. They hired civilians armed with *bolos* to carry their heavy equipment. Barefooted and in rags the group managed to cross mountains and villages. Eating coconut meat or going without a meal for a day made them begin to lose hope. They kept moving whenever they heard the Japanese were coming. They were almost caught when they camped overnight in a house with the Japanese only 30 minutes away. Leeches and mosquitoes feasted on them. They continued to send messages on the enemy whenever they were able to get the radio on the air.

Orders arrived to return to Cebu as Japanese presence in the city had increased. MACA reported that Cebu had an enemy strength of over sixteen thousand with fresh recruits from Taiwan and Tokyo. Filipino males between sixteen and sixty were rounded up to determine whether they were guerrillas. Some were tortured and killed. The group sailed on the lookout for Japanese patrols and docked on small islands for the night avoiding Japanese-occupied villages until they reached Bohol Island next door to Cebu. Unknown to them, several hundred Japanese troops had invaded Bohol the day before. The civilians told of being kicked, slapped, or hanged. The Japanese had taken everything worthwhile from them. Inside a house, the men kept deathly still as the Japanese passed back and forth in front. They could not fight as the guerrillas did not have enough arms. They escaped in the dark night, sleeping under the trees soaking wet from crossing rivers and swamps. When Japanese patrols appeared, they hid and picked them off one by one until they arrived in Cebu. "When we left (Cebu), the Japs were hunting for soldiers and relatives of the governor of Cebu. They caught one of those relatives, a young lady, and kept her in the municipal building for investigation. She was tortured and hanged. Her beheaded body was seen along the road."

1944

Smith was so impressed by the men that he wrote a memo to Whitney on New Year's Day: 'Recommendation for Promotion of Officers and Men sent from SWPA for Duty with station MACA.' He listed the names of the men and continued to state that they had been largely responsible for the success of the MACA radio net. He requested a jump in grade for rendering outstanding service.

A squad of armed guerrillas led by Capt. Arrieta and several White soldiers from Mindanao headquarters had joined Smith's group to set up station MACA. Robert V.

Ball, who served as a radio operator with the 5th Airbase Squadron and with Mindanao headquarters joined the group.

Herreria accompanied a cargo of supplies and operated station MAV in southern Samar. The load was buried in case Japanese patrols showed up. In less than a week, the Japanese arrived, "looted property, burned and shot innocent people, and raped young women." Herreria was warned and escaped to the guerrilla's command post. His guide, Alejandro Capuli, was shot in the neck and died in his arms soaking his clothes with Capuli's blood. Determination mingled with anger and fear as he hid in a log until two Japanese came into close range. He shot them both dead and rolled down a hill as more Japanese appeared. He wrote "I was lost for a week. Being unfamiliar with the terrain, I was confused about the direction I should take. There was no more food, my clothing was in tattered rags, and my jungle boots were worn out. I had crossed lagoons and deep rivers infested with blood-sucking leeches, thickets where entangling vines greatly impeded my progress. Weak, tired, hungry, my feet bleeding, what more could I do than relegate myself to the mercy of fate? I had not eaten for days which meant biting hunger, the worst enemy of man."

A group of people did not recognize Herreria with the long beard but did know his voice and brought him to their homes to feed and rest. In the meantime, the camp had wondered what happened to him and even dismissed news of his body seen floating on the Llorente river. After a month of healing, Herreria recruited forty men to dig up the buried cargo. Carabao carts pulled the heaviest loads on muddy trails. Small groups of soldiers were sent on patrol around for protection. The rainy season made the crossing of swollen streams dangerous. Cargadores (carriers) backpacked heavy loads for miles on the one-lane muddy road closest to the coast. When the small generator broke down and could not operate a more powerful radio, a diesel engine was found that could provide stronger power. A piece had to be carried by four men suspended from two bamboo poles. Six men carried the large drums of coconut oil to fuel the engine also suspended over bamboo poles. Constant recruitment of men to replace 20 exhausted carriers working in daily shifts delayed the trek. After two weeks, the group met up with the MACA station group who was looking for a good camp location. Herreria felt proud when Smith praised him for a job well done.

Station MACA's tower was on top of a mountain peak in the northeast side of the island surrounded by steep mountains. Bringing up the equipment was a nightmare on a hacked trail of switchbacks. Thick jungles stretched into the horizon. The tower had a hundred-foot antenna wire strung between two tall trees. Stahl was in charge and "hoped for the best." If compromised, "no time to yank them down. It was worth the miserable journey," he thought. Small groups of soldiers patrolled for protection.

MACA radio contact with KAZ was maintained in the town of Borongan. The townspeople rejoiced upon seeing the men and held a dance in their honor with a band playing patriotic hits. They asked the people for their full cooperation to liberate their homeland. The signalmen sat through hours of listening on the air to various interfering signals on their frequency trying to pick out the weaker MACA frequency. A schedule to maintain a continuous watch on the assigned frequency was followed. MACA station eventually connected with four sub-stations on Samar, one in Cebu, one in Masbate, and two in Southern Luzon at Sorsogon and the Bondoc Peninsula. One of the radios penetrated Manila and Baler in central Luzon. An alternate radio site was set up at Palapag Mesa with stores of fuel and money in case of a Japanese raid.

Smith faced having to unify the strongest guerrilla organizations under one leader. Pedro Merritt, a Philippine Army officer in the north and Manuel Vally an escapee from Bataan in the south vied for men, civilian support, territory, and supplies, confusing the guerrillas as to which leader to follow. Civilians were afraid of Merritt's army which killed anyone charged with cooperating with the Japanese. Vally discouraged food production so that the enemy would not get their harvest. Juan Causing, a USAFFE officer from Leyte tried unsuccessfully to unite the guerrillas. Peralta on Panay and Fertig on Mindanao had operatives on Samar also laying claim. The Japanese watched and listened amusedly increasing patrols against the guerrillas who were distracted by their infighting and disagreements.

Cardenas was sent to the west coast to set up station MAE. He received $1500 for his subsistence. Guerrilla officer Maj. Antonio Sabarre received 60 Carbines and 6 grenades for his guerrilla unit to act as a guide and to supply guards for the new station. MAE went on the air and contacted KAZ at the end of January for only a week as the radio was betrayed by two officers of Merritt. Cardenas narrowly escaped the Japanese. He returned to find the radio discovered by the enemy where he had hidden it and the guards killed.

Three months later, Cardenas with guerrilla guide, Maj. P. Cabrera sailed with transmitters and joined Herreria. An urgent request from KAZ wanted to know the Japanese dispositions in Manila. They were to penetrate the city and set up stations MAF and MAJ. Feeling that they needed someone who had some strong political contact, Lt. Leopoldo Flores, Philippine Army, and the son of an ex-governor of Sorsogon province was willing to take the risk. In the Japanese-occupied town of Matnog, Sorsogon, they barely passed unseen. Travelling mostly at night to avoid Japanese patrol, they crossed the sea to Masbate Island and the Sibuyan Sea to the Bondoc Peninsula dropping operational money to the stations they passed. There were days without food, avoiding Japanese patrols, going through malaria-infested forests, and crossing leech-filled rivers. As they saw Japanese activities, they radioed them to MACA.

Lt. Flores was captured when he bought a railroad ticket to Lucena and a suspicious Japanese ordered him to open the suitcase that held the transmitter. He was taken away leaving the men with only one radio set. They never heard from Flores again. The Japanese had registered and reconditioned all the radios to receive only local news and none from the outside world. With one transmitter left, they had to be very cautious of being captured as that meant death.

They reached Manila in August. They were the second group of U.S. Army men to enter the city. Jorge with the Negros mission had entered Manila earlier to smuggle out the Japanese propaganda movie *Dawn of Freedom* for General MacArthur. They posed as civilians and had the proper identification obtained from the city mayor. Herreria said that he had all identifications tagging him as "a goord numbeh one civirian," sounding like a Japanese.

Malaria had broken out in the city and the Japanese built defenses despite a shortage of weapons and supplies that had not arrived. The receiver of the other set they brought did not work. MACA could only hear their station calls. They left the inoperable radio set in the open so if they were raided the enemy would think that they had successfully eliminated the station and abandoned the chase.

No radio did not deter them from making several trips on their own into the city and penetrating Japanese headquarters. Filipinos who worked for the Japanese puppet government officials obtained information of Japanese homeland activities. They transmitted enemy strength, location of food storage, ammunition, and fuel dumps by courier to Robles' station MAA in the Bondoc Peninsula for relay to MACA, Mindanao and KAZ. One disturbing message was the enemy's plan for a punitive drive against the 10th MD headquarters on Mindanao which they immediately forwarded.

Herreria saw chewing gum with wrappers 'I Shall Return MacArthur' selling for 10 pesos each in the city. Moving pictures of all kinds were shown. Filipino and Japanese films were allowed but the Filipino films were censored. If not complied with, the films were burned, and the producers sent to the dungeons of Fort Santiago. Dramas were allowed but in the Tagalog language. Using the English language meant being pro-American. Although American and British films were prohibited, those showing American pictures were packed to capacity with Japanese soldiers. They were curious about an enemy they hardly knew. Films of anti-American propaganda were shown but theatres were empty.

Herreria's penetration of Manila made Smith express in a message to KAZ that the "U.S. Filipinos were able to assimilate among their people even though some have been gone for many years. Proof that sending more will be successful infiltration into other enemy-occupied centers." Smith recommended expanding the Luzon net requesting fifteen Filipino operators and three White officers "capable of hard life and not too much to eat." He reported that his party was "now scattered from Bondoc Peninsula on Southern Luzon to southern Samar and have demonstrated that they have what it takes one hundred percent." He wanted more Filipino radio operators trained in operations, radio repair and maintenance aside from the receiving and transmitting of messages. The White officers were to have cryptographic or radio experience. He requested American officers as he was doubtful of Filipino officers. Whitney replied that he "was sending outstanding ones in the next parties," but he had difficulty securing American officers who would fit into the situation.

G-2 Intelligence in Australia commented that bringing U.S. Filipinos to Australia for "ultimate Philippine service" was justified. "(White) Americans cannot be used for this type of work in enemy occupied areas since their presence is a matter of conversation among the Filipinos and with the large number of undercover agents employed by the Japanese the American would be betrayed." Whitney and Parsons were convinced and expressed satisfaction with the men's preparation at Camp Tabragalba. There were now many highly trained Filipino radio operators available capable of sending and receiving 20-30 words per minute.

By March, station ISRM of the doomed mission in Mindoro was "just another coastwatcher station that no longer existed." KAZ had warned all stations many times to avoid security violations when sending messages.

At this time, intelligence from Samar reported that the Japanese troops on the island were from New Guinea. Recuperated casualties were being reassigned to Samar. Their main activity of finding the radios relaying to Mindanao hounded MACA and the sub-stations. Messages were jammed. MACA shut down in April when Japanese aircrafts painted with American insignias flew over the camp looking for them to drop their bombs. All coastwatchers and sub-stations were warned.

The Japanese relentlessly tracked the radio network. The *Watari Group Intelligence Report from June 11 – 20, 1944* published news concerning the "many small-type wireless sets which are used for communication within the islands... (the guerrillas) are skillfully concealing their stations and are successfully preventing interruption of communication by our punitive units. Station MACA alone is equipped with 10 to 20 wireless sets. More accurate information must be collected to destroy these stations."

The Japanese used direction-finding equipment to find the stations. The men were constantly on the alert for Japanese boats along the coast and in rivers carrying the strange-shaped antennas and wires. That caused enough concern to shut down the radio in the camp and outposts' vicinity for several days and transfer relay to other net stations. Couriers were used for critical reports. Guerrilla guards were posted at each station to give time to pack up the radio and documents to avoid capture. Some stations were never heard again.

Smith bombarded KAZ for more signalmen, weapons, and morale items like cigarettes and magazines. The supplies they brought six months earlier were used up. Mindanao had not sent any. Watcher stations were almost out of batteries for their radio sets. Four watcher stations were in danger of going off the air. The unforgiving tropical weather slowly ate into radio parts and shorted out critical parts. He demanded KAZ send equipment and supplies to stay on the air. "Supplies were used up, several radios captured, generators wearing out ... need more arms and ammunitions, medicine, and above all shoes." Everyone suffered from jungle rot eating into the bones of their feet.

PRS approved the next mission and added 500,000 counterfeit Japanese notes to inflate the currency, real $50,000 and 100,000 Filipino pesos "newly printed in Washington D.C. from engraving plates taken from the Philippine treasury when the islands fell." A rendezvous was arranged for the next mission's arrival. "Finally, someone in a cushioned chair in Brisbane decided we were due for a supply run," Smith said.

Weather observation was added to the training program of the signalmen at Camp Tabragalba. Twenty signalmen with weather training were to be selected for the next missions. The 5th Air Force, Bomber command, and Navy Task Forces depended on weather forecasting especially in the tropics with the unexpected rainstorms and typhoons, for accurate bombing of enemy installations and avoiding certain targets. Schedules for transporting equipment and supplies depended on the weather reports. In clear weather, fighter planes stripped of armament and fitted with aerial cameras flew to map target areas for bombers to drop their load using the clear weather forecasts.

GHQ's decision to suddenly switch the Philippine landing from Mindanao to Leyte and two months earlier affected radio network plans. KAZ informed the guerrilla leaders and radio networks of the Leyte invasion but not of the actual date. Maps of enemy dispositions were to be transmitted daily. Smith radioed KAZ to send complete jungle warfare equipment for twelve men to send out as more agents with radios.

KAZ turned to the Bicol region in southern Luzon wanting intelligence coverage of enemy disposition around Legaspi and Sorsogon. The seas between Leyte and Luzon were heavy with Japanese shipping. KAZ wanted equipment and personnel installed on the islands surrounding Leyte. Where could supplies be landed?

USS *Narwhal* - June 1944

The *Narwhal* was ferrying missions and cargo northward every month and attacking Japanese shipping as sighted. Without stopping for servicing in between missions, her condition showed increasing risk to the men.

It was the largest group to leave on a mission in a single submarine. Thirty-nine signalmen in two groups were to reinforce Smith's party and Mindanao with supplies, weather observers, reconnaissance, and signal men.

Smith had asked for American (White) officers. Instead, five well-qualified Filipino officers arrived for reinforcement. They were Lt. Vicente Labrador, Lt. Louis P. Padilla, Lt. Carlos F. Ancheta, Lt. Paul F. Mauricio, Lt. Jose T. Mendoza, Lt. Epifanio Ibay, and Sgt. Cipriano L. Miguel. The 978th Signalmen were Sgt. Gerardo B. Nery, Sgt. Raymundo Agcaoili, Sgt. Gaudencio Guyot, Sgt. Jack Montero, T4 Pete Luz, Cpl. Agripino J. Duran, T5 Robert O. Angcas, T5 Eddie C. Holgado, T5 Leodegario O. Nuevo, T5 Rodolph H. Santos, T5 Julius C. Advincula, and Pvt. George Bulatao. The weathermen were Pfc. Jerry D. Pascua, Pvt. Isaac Aguila, Sgt. William Richardson, and Cpl. William Becker, carrying portable meteorological equipment. The weather station they would establish on Polillio Island extended weather observations to central Luzon, but the enemy discovered it, and the equipment was abandoned.

Orders arrived at Tabragalba in mid-May. It was a moonlit night when the men boarded a plane to Darwin in a closed hangar at Amberley Field. From there they were to go back to the Philippines on various missions. While climbing a wooden ramp into the plane, "MacArthur stood at the bottom shook our hands, and gave a wristwatch to each of the thirty-nine of us. MacArthur had tears in his eyes and his voice quivered as he wished each one of us God's speed and good luck. It was very dramatic as only he could be," Lucien Campeau with the party to Mindanao said. They were passengers with "various missions and 100 tons of supplies, including guns, ammo, weather gear, medicine, clothes, radio sets, a printing press, but no food. They were to live off the land.

Darwin was in ruins, from the continuous bombing raids of the Japanese. Pascua recalled the party staying at the Lugger Maintenance Section, east of Darwin for two days before being taken to Melville Island on a landing craft across from the City. They stood in a row passing their duffel bags trying to dodge jellyfish and manta rays to reach the island. For 3 days they waited. Purifying tablets made water brought in from the city fit for drinking. Sun flies infested the island. They feasted on boiled fish and crabs caught in Guyot's improvised fish trap.

As they boarded the sub, Commander Parsons handed Campeau a five-gallon can soldered shut containing 25,000 real pesos made in the U.S. and soaked in salt water to make them look old and used. A gunny sack was sewn around it so it would not reflect and be seen from the air. The money was to be used for intelligence purposes. Onboard, the cargo was dispersed on the entire subfloor and in the engine room. Torpedo room floors were covered with dunnage of a few feet of loose clothes. Some of the 16 torpedoes mainly for defense were removed to accommodate the men and cargo. The Filipinos were dispersed throughout the ship. Sleeping beside the cargo they brought, the men sweated in the rising temperature and humidity. Palmeto felt it was like a prison. The walls were so close and cramped with supplies and equipment. "There was no bathing, but the food was good," he said. There were one hundred drums of airplane gas in the deck wells where torpedoes were normally stored. It was only at sea when the men learned of their destination and mission to find out whatever they could on the enemy.

The sub sailed for two weeks submerged during the day running on batteries to avoid Japanese air patrols and surfaced at night recharging the batteries. When the weather was rough, large powerful waves slapped against the sub keeping the men awake. On the first day, thousands of gallons of fresh water were accidentally blown out instead of sea water to surface causing short rations for the rest of the trip. Cruising the Japanese-occupied east side of Mindanao, the sub's starboard main motor burned out, leaking oil slicks. After a week at sea, the sub saw a convoy of eight freighters with three escorts. The crew was ordered to battalion stations. The men kept quiet and sweat began to pour. The sub trailed the convoy for two hours before changing to top speed and encircled past the protective destroyers. The men felt the sub turning as she fired four torpedoes. She creaked as two were direct hits, sinking a troop ship and damaging another. The sub shook and chunks of cork plasters fell. She increased her speed to avoid the escorts closing in to drop depth charges. She crash-dived to 280 feet with everyone gripping anything they could hold on to. The *Narwhal* was tested for only two hundred fifty ft when commissioned in 1929.

When clear, she submerged to repair one of her main engines. Palmeto thought the sub was hit. There were "90 drums of high-octane gasoline rigged up outside the sub. The slightest hit would have blown us to kingdom come." Sgt. Fortun wished he was somewhere else. He swore that he would never set foot in a submarine again if he ever got on land.

The sub surfaced in Gamay Bay in the northwest of Samar and delivered the men with 30 tons of cargo. Nery was afraid that the Japanese would hear the roar of the motor, and everyone "would not even have time to say his last prayer."

There were "electric lamps, radio parts, and flour for the priests." Part of the cargo was lightweight Carbine rifles that GHQ wanted the guerrillas to use replacing old Enfields, Springfields and captured Japanese rifles. A flotilla of sailboats met the sub in the early evening to help unload the cargo and men to shore. Having learned efficiency from past mission parties, the unloading took only one hour. Smith and Stahl boarded the sub. The sky was so dark that hardly anything could be seen on the deck of the submarine. The men packed their barrack bags to board one of the sailboats, not knowing where they were. Nery said that the Filipinos in the sailboat told them they were in Samar. Guyot could hardly suppress his feelings. "I cried like a small boy who has gone away so long from his mother's fold." He saw the men in the sailboats. "Some were guerrillas whose long hair had not been shorn for quite some time ... the clothing they had were either taken away or burned with their houses by the Japanese."

The *Narwhal* silently submerged towards Southern Mindanao to deliver her second group of passengers and cargo to reinforce FRS, the net control station.

Once all men and cargo were accounted for, Bulatao related that he, Mauricio, Mendoza, Padilla, Guyot, Agcaoili, Aguila, Pascua, Angcos sailed to the mouth of the Kalaw River and transferred to smaller boats to reach further inland. Mangroves and nipa palms crowded both sides of the river allowing only a single sailboat line. Along the bank was the net control station MACA well camouflaged and open all day and night. There were no huts or shacks built. It was bare. Cooking was done in the boats, and they ate under the coconut trees by the river.

Mauricio, Padilla, Jerry Pascua with 5 recon men and several guerrillas hiked for three days to Nagoocan to set up outposts on hillsides that overlooked the river below the camp. That was to give time for the camp with the radio to move to another location if the enemy appeared. The Japanese were in Catubig, a town along the river around 20 km away. Guerrillas and civilians carried some of the supplies and equipment through winding muddy trails to the village. The civilians were paid one peso per day with meals. "Civilians were so happy to see them and thrilled that they came from the U.S. that they brought rice, *camotes*, and boiled fish for lunch," said Pascua. Chicken and eggs arrived with *tuba*, a fermented coconut drink they had not tasted in many years. Girls sang for them. They gave soaps, combs, candies and chewing gums in return.

Nery said they were all told that they were going north but not precisely where. They got rid of all military identities and changed to civilian clothing. "One hundred genuine pesos were given to each man for emergencies, dope about the mission, the place to hit, stations and personnel to contact with ... and everything we needed." Guyot was in charge of the carriers and guerrillas to bring the supplies over the muddy trails and forests from the river to the camp. Agcaoili oversaw another thirty-two carriers on another route to the new camp. The carriers were paid one peso per day with meals. After camping overnight with a supper of half-cooked rice, roasted dried octopus, and salt it was another day of hiking. Guyot thought of Americans Richardson and Baker eating the same food with hungry stomachs. Heavy rain fell, and the trails became very slippery. It was a long hike, and more carriers were added. At the village of Osmena the locals brought them rice, camotes, and fish in exchange for soaps, combs, and clothing. "They (people) fashioned a toothbrush with bamboo for the handle and pig's hair for the bristles." Smith arrived and went ahead to MACA camp up the mountains.

Agcaoili recalled that when he arrived at camp, everybody was tense and excited. A Japanese patrol was heading towards the camp. Taking a chance that it was false news, they ate dinner and piled their hammocks inside the bamboo barracks which was also the kitchen of the guerrillas. Agcaoili did not eat breakfast and reported to Capt Arrieta, the guerrilla security officer. A forward delaying outpost some two km from camp was set up. They waited for five days but no Japanese showed up.

News of the Japanese asking people for the whereabouts of the station in the three towns around the camp arrived. Some had been tortured to get information. Orders were to hide everything, radio, medical supplies, weather equipment in the forest away from the camp. Pascua wrote "Smith ordered us to hide and bury supplies around the camp at least one hundred meters apart, gasoline here, medical supplies there... we carried the supplies without shoes. Shoe prints to Japs meant U.S. soldier." No civilians were employed to help. Men had to carry the load with the guerrillas' help.

By midyear, Nery traveled with signalmen Holgado, Montero, and Advincula with three guerrilla bodyguards led by Stahl to set up sub-net station S3L on the Bondoc Peninsula in Southern Luzon as an alternate should MACA be lost to the enemy. They were to link up with Robles and Manzano's station MAA (December 1943 mission). They were all signal men and had no recon men with them for intelligence gathering and security. However, their training was almost identical so as to back up each other as needed. Five tons of supplies and five radio sets were loaded in a sailboat. A 1933 edition of the *U.S. Coast and Geodetic Survey Chart of the Philippine Islands* and a 1940 *Socony Oil Vacuum Company Road* map that did not show rivers, inlets, and major waterways were their guides. The group departed as the sun was setting to avoid the Japanese patrols in the San Bernardino Strait. Passing the lighthouse on Capul Island, two Japanese patrol boats with guns on deck tried to envelop and shoot them. They passed Japanese garrisons on Ticao, and Burias islands. On the 4th day, Japanese planes flew in circles around them. Finding a high position, they anchored offshore as they could not get any closer and landed at the village of Pagbilap. It was not a safe place to hide their equipment. They backtracked to a river they noticed earlier to camp for the night. Going through mangroves, they unloaded their equipment throughout the following night and hid them separately under thick foliage, trees, and up hillsides. The antenna was hung at the top of the first mountain they encountered and went on the air. MACA and the stations of guerrilla leader Bernard Anderson, an Army Air Corps officer at Infanta on Luzon and Robert Ball, an Air Corp radio operator who had been operating a radio on Samar and now in Robert Lapham's camp in Baler on Luzon were contacted. From Baler, the radio coverage extended into central Luzon. Robert Lapham, an Army officer was a guerrilla commander on Luzon and sent his area's intelligence through Ball's radio to MACA.

KAZ emphasized station S3L's purpose in a message that it was to be a backup to MACA, and act as a relay station for those stations that could not reach Mindanao. It was also to establish a sub-net and obtain all possible military intelligence in the southern Luzon sector, including Manila. The station moved four more times before settling in a secure location.

Some of the party's cargo was to go to Anderson and Ball with Lapham but Anderson's men arrived and took the cargo with money and two radio sets to their headquarters at Infanta. Smith reported that Anderson took over the cargo and only one set was sent to Lapham whose radio had died. That greatly delayed Lapham's direct contact with KAZ and radio equipment to Russell Volckmann, a guerrilla leader on Northern Luzon. Volckmann had reorganized scattered guerrillas from Nakar's unit into the United States Forces in Philippines, Northern Luzon (USFIP-NL) with three partially armed regiments totalling around ten thousand men.

A fisherman arrived in his banca and said he knew of a party like S3L who had worked in the area for quite some time. The guerrilla bodyguards did not speak the same language of the fisherman except for Nery, who translated. His wife was cooking three meals a day and bringing them to Robles who had been on the Bondoc peninsula for the last 4 months. They found Robles shoeless and worn out. Manzano who was with him had gone farther north. Nery and the signalmen took over the station's incoming messages to let him rest. The outpost was to keep an eye on ship movements in Mogpog Pass (Tayabas Bay), which lead to the Verde Passage towards Manila Bay.

They had to fend for themselves as no supplies came. The battery charger began dying. A drum of regular gasoline for the charger was gotten from another radio station and carried back to camp by four carriers. A condenser part was severely needed, and they got it from a Japanese truck after attacking the Japanese officer and men on board. Nery related that "we got their arms too… a colonel's sabre and several Japanese flags. On one flag was inscribed the victory of Bataan in Japanese characters."

Barefooted men armed with *bolos,* suddenly appeared without warning. The men grabbed their guns and were ready to defend the camp. Three weeks after their camp was set up, people had reported their presence to the guerrilla commander of the area who wanted to meet them. Nery went with Stahl to the guerrillas' camp and met Gaudencio Vera, commander of the Tayabas Guerrillas. Montero carried a sub- machine gun. The guerrillas were armed only with bows and arrows, shot guns, a few 1903 Rifles, and some captured Japanese rifles and machine guns. Vera had allowed Robles to operate the radio on Luzon and Manzano to leave for Manila in exchange for a machine gun and a rifle. Manzano had only a pistol and a radio into Manila. A record of Manzano reaching Manila was not found. Vera led about one hundred men, women, and children of his extended kin to escape the Japanese. He and guerrilla groups in the north were destroying each other for men and terrain to survive but distracting them from fighting the Japanese. Some civilians were forced to give them food and other useful materials they could find. Vera had three American escapees, Chester Konka, George McGowan, and Eldred Sattem who were beaten up by the Japanese. They were treated well, but not let go. Vera wanted direct contact with Australia and threatened to wipe out the station if he did not get supplies and contact with General MacArthur.

The meeting was disrupted with news that the Japanese were close by looking for them. About a hundred guerrillas quickly showed up to carry all the cargo over hills, rivers, and jungles in three days to the guerrillas' camp on the Bondoc Peninsula. A high position that continuously picked up a good signal was chosen. Not long after, a Filipino spy reported their location, and over a hundred Japanese entered the jungles after them. A guerrilla scout was captured but managed to escape and report back to the camp. The antenna went down, and all equipment packed. A firing line on top of a hill was established. The firing lasted for an hour, and the Japs retreated with their dead. Another Filipino spy was caught. Two carabaos of the guerrillas were taken and devoured by the Japanese. Nery recounted saving the radio, "Since our stay with the guerrillas on the Bondoc Peninsula, we had retreated about 15 different times. This meant that we had also moved all our radio set, equipment, and all the loads. Despite the hard retreats across steep mountains and thick jungles... managed to save all of our equipment in perfect condition."

The new camp of station S3L was set up north of Robles' station. Two guards were posted. They could see Robles' station at the southern tip of the peninsula. The importance of S3L was highlighted when it was the only radio strong enough and available to receive Luzon radio traffic to forward to KAZ. Time lags were eliminated bypassing the Mindanao station. Vital information on airfields in the Manila area, Batangas, Pampanga, and Tayabas provinces on Luzon was transmitted.

S3L took over when MACA radio was on the run from Japanese patrol while trying to connect sub-stations in Samar, Cebu, Masbate, Sorsogon, Baler and Manila on Luzon. Agents and radio stations were being captured. Transmissions were being jammed. Relaying through FRS on Mindanao was not possible because the Japanese were also after them and they had to go deeper into the remote jungles of Agusan. Stahl wrote, "Sgt. Nery manned the radio with two other operators and spent endless days sending and receiving message traffic, sometimes operating two radios side by side to clear the backlog of messages. Nery's knowledge of the Visayan language spoken in the area was helpful. When we thought we were caught up on the work, a piece of equipment would break down and cause another logjam." He recommended Nery and Agcaoili for promotion.

The camp worked with Vera so as not to be killed and keep a radio on the air on the Bondoc Peninsula. The American escapees were used to code, decode messages, and keep the hand-powered generator operating and batteries charged. Agents were recruited and given assignments to spy on the enemy.

A newsletter sent out to surrounding villages and towns promoted goodwill to the people. In return, they received alcoholic drinks, beer, and Japanese cigarettes.

When a submarine landed several months later, Vera received arms, and his forces protected the station. He was reminded not to fight and to gather information on the Japanese. A radio and an operator were established at his camp to send back reports. But when the guerrillas were not nearby, the men in the station had to set up an ambush to defend themselves. They knew what to do to protect the radio and codes. Their intense training at Tabragalba was helpful. In their favor was the Japanese' reluctance to go deep into the jungles. "Many of the enemy soldiers were conscripts and not regular army troops. A few bursts of rifle and machine-gun fire from a concealed position would disperse them quickly."

By July, S3L was suffering from dwindling food, medicine, clothing, money, and everything needed to keep a radio station and an intelligence network operating. Exhausted operatives had to be fed as they arrived and went out. The radio operators could not be spared from their tasks to find food. A submarine with supplies marked for Samar, landed in Suluan island at the tip of Southern Samar. The cargo did not have what they had requested as priority. They were not supposed to go into combat but felt their mission of gathering information on Japanese movements and pinpointing military installations were valuable enough to receive at least non-combative supplies they desperately needed. They wore *bakya*, the local wooden clogs the civilians gave to avoid walking barefoot.

From S3L, Montero and Advincula brought the remaining radio sets with men from Anderson's unit to watch the airfield in Lipa, Batangas. Unable to reach the airfield, they moved to a more secure remote location north of Anderson's camp and set up a station. The forest hid them as they operated the radio. Infected skin rashes, leech bites and thorn scratches turned into ulcers. From the locals, they learned to apply coconut oil to heal quickly. Always hunted by Japanese observation planes, they finished cooking before sunrise to avoid the smoke being seen. After two months of operation, the heavy humidity under the foliage made transmission unreliable. They split up and Montero left with guerrilla guards and carriers of the radio equipment to a Japanese garrison in Mauban, south of Manila to check on a rumor of the arrival of some 3000 thousand Japanese reinforcements to bolster their shore defense. He set up radio station AOR overlooking the town and the Japanese garrison. The antenna was just set up when a boy appeared and excitedly told him that a Japanese patrol of around twenty-five men was headed his way. He quickly packed his set and, with the guidance of an old civilian, penetrated deep into the jungle until they came to a clearing. A family of evacuees gave them shelter and food. They thought that they had lost the enemy and returned to their camp. The following day, more Japanese appeared. Quickly packing and disappearing again into the jungle, one of the carriers was hit by a bullet. Knee mortar shells were fired at them. For three days, they were on the run and had no way of informing Anderson. When scouts determined that the enemy patrols had given up the chase, Montero set up the radio to contact Anderson. After four minutes, the call was recognized. He sent a 'Q' signal meaning "I am on the move due to enemy raid". The batteries were weakening, and the charger would not work. It had been damaged in the constant handling while evading the Japanese. A runner he sent to Anderson returned. Two weeks later, another runner showed up from Anderson with a message to lie low until further orders.

A sealed order arrived to "report to Masanga beach (Luzon)...by boat with all equipment." The party traveled overland, loaded all equipment into a sailboat crewed by seven men, and sailed north. They procured food and some supplies at Cagbalete and Alabat Islands. Two men continually bailed water out of the boat. They sailed at night avoiding Dinahican Point, a peninsula jutting out with a Japanese garrison and a harbor for their Q boats. A violent storm stranded them on one of the tiny islands of Polillo where they stayed for a week. Montero was sick. They ate what they could find and "caught many land crabs as big as saucers and cooked them with coconut meat." They crossed the Polillo Strait to Anderson's headquarters chilled and sick from the choppy waters. Anderson was glad to see them. He thought they were gone. After resting for several days, Montero went to work again as an operator relieving the others who carried the load while he was sick. "They showed signs of suffering from loss of sleep." They found sailboats to move their supplies and equipment back to Polillo Island to set up a radio. The return of the U.S. aircrafts in September brought daily airdrops of food and clothing by large C-47s escorted by several fighter planes. They did not starve at Christmas. After the Luzon landing in January, Montero was picked up by a PBY to Lingayen. From there, he rode a truck to Calasiao in Pangasinan, the Sixth Army's communications headquarters. Due to his weakened condition, he was treated at the 51st General Hospital for malnutrition and jungle rot. Back to health after two weeks, he reported to G-2 Intelligence and worked as an operator in the Sixth Army radio room. He met up with Advincula whom he had left at Lipa. After a month, they boarded a cargo plane to Hollandia. But after another month, he was shipped back with twenty men to work in the GHQ message center in Manila. In July, they were ordered to San Miguel in Tarlac to help start a new camp for the 978th Signal Service Company.

The invasion of Guam was scheduled for July 1944, but GHQ had no visibility of the location of the Japanese task force. When Vice Admiral Jisaburo Ozawa received news of the U.S. Task Force heading to the Marianas Islands, he ordered his 1st Mobile Fleet stationed in Tawi-Tawi to meet them. Coastwatchers on Palawan (June 1943 mission) and Panay (April 1943 mission) watched the large Japanese task force pass in the vicinity of the Sulu Sea and Mindoro Strait. Priority messages were sent to FRS on Mindanao for relay to KAZ and the Navy as they tracked the convoy.

On June 15, 1944, Gerald Chapman, formerly with the 19th Bomb Group from Mindanao, was desperately sending numerous priority messages to FRS from his station MAG in Sorsogon, Southern Luzon. He would not stop as his low 4-watt radio had difficulty transmitting the information that Japanese vessels were entering the San Bernardino Strait in a single file going towards the Pacific Ocean. "Going east northeast . . . Jap Naval Fleet . . . two small patrol boats, eleven destroyers, ten cruisers, three battleships, and nine aircraft carriers."

Nery and Stahl on the Bondoc Peninsula nearby heard the persistent stream of Chapman's jammed messages. It must be urgent. The Japanese were deliberately preventing his messages from getting through to either MACA or FRS stations to KAZ. Nery was concerned that Chapman's post was in danger of being discovered by being on the air too long. Many coastwatchers were never heard of again after capture. They pieced together bits of his coded messages and moved to a higher frequency on their stronger 12-watt radio to forward to FRS on Mindanao and station KFS in San Francisco. There was no response for hours while Chapman bravely continued to transmit. When a faint call to change to a rarely used frequency came from station KFS, they realized that the location of S3L close to the coast was affecting transmission to Australia but not across the Pacific Ocean. The warning messages reached the U.S. Fifth Fleet supporting the Marine landings four days earlier of the Japanese task forces' arrival. The surprised Japanese Navy was decimated in the Battle of the Philippine Sea. Several hundred enemy planes were shot down destroying Japan's air superiority. The loss of Saipan reverberated at Imperial Tokyo and the resignation of Prime Minister Hideki Tojo's cabinet.

Interrogated after the war, Admiral Ozawa expressed his concern about the insufficient training of his air group due to the airfield in Tawi-Tawi being under construction when they departed. The pilots could land in the daytime but not at night. Their land-based planes at Palau, Yap, and Guam Islands were under the separate command of the Japanese Combined Fleet which made coordination of aerial attacks on the U.S. fleets complex.

Station S3L was moved further north with Vera guerrillas' help between the towns of Catanuan and San Narciso in the central part of the Bondoc Peninsula for a better transmission atmosphere. A 3-hr hike to San Narciso for a much-needed bath in the stream after weeks in the station was a welcome treat. They shared their supply of soap with the civilians.

With a guerrilla officer, Signalman Holgado left station S3L and sailed across Tayabas Bay to reach Batangas to report on the Lipa airfield in the area. Enemy patrols blocked the passage, so he traveled north overland to Anderson's camp before going south to Batangas. Stahl and Nery were the only two left at station S3L with three guerrilla guards – Madeja, Ochigue, and Doming.

Back at Camp MACA, recon men Labrador and Ancheta with signalmen Miguel, Luz, Duran, and Santos departed under the cover of a dark night in May to join Robert Ball, head of Lapham's radio network at Baler. They traded most of their unform clothes and shoes for civilian clothes. A Philippine Army guerrilla officer and three civilian radio men joined them. Sailing from Samar, they brought five tons of supplies in two sailboats at night. A 1928 relief map of the Philippines was their guide. They stayed far out at sea up the east coast of Luzon as Japanese garrisons were near the beach. Light storms rocked the boat and floated still in dead wind areas. They became worried when the current brought them close to Batag and Catanduanes Islands where enemy garrisons and naval guns were located. Low on water, they sailed close to the coast to find a safe spot to dock. On the fifth day, a big Japanese patrol boat appeared. They raised a Japanese flag hoping to be bypassed. It worked. Another week passed sailing between two big enemy garrisons on Polillio and Alabat Islands until the rendezvous location was sighted. Sighting the rugged coastline, they beached between the two garrisoned towns of Mauban and Infanta.

At sundown, four wooden boats approached the shore. They were ready for battle clutching their weapons and relieved when there were only refugees running from the enemy. The civilians told of enemy patrols along the coastline and an American guerrilla leader (Bernard Anderson) only ten miles to the south. Three civilians were held hostage while Ancheta surveyed the area and returned telling of an old man who could not read or write claiming he was a member of Anderson's command. The following day, they unloaded the cargo worried that their sailboat anchored away from the shore would be seen. Over fifty armed men of the communist *Hukbalahap* (Huks) sprung from the bushes prepared to open fire. The Huks saw the machine guns and rifles in the boat. Four men left the boats to talk to the Huks ashore. If anything happened to them, the men in the boat were instructed to sail away.

By sheer luck, men of Anderson happened to come by and brought Ancheta and Luz to their headquarters. The Huks held three of the civilians in the group hostage to ensure that the men would return. Alerted by his men, Anderson arrived later in the day. Their story was verified through "slang talk and Brooklyn accent." An agreement was made with the Huks to give them arms at a predestined date and location, a promise which was not kept. They had doubts about Anderson but since he was an American, they thought maybe they could find out what happened to Robert Ball who was supposed to meet them.

Because it was a starless night when they sailed for Anderson's camp, they missed the river entrance where they could hide the boats. With the water too shallow, they stayed awake all night waiting for the high tide. They were forced to turn over sealed letters for Ball intended for Lapham. Anderson and Lapham were not cooperating with each other. Their sailboats were sent back to Samar while they took a wooden boat at night to a high spot north of the Anderson's camp for a good radio connection. The rugged coastline allowed the only access by the sea. Station GRT was set up and contacted station UU2 operated by Ball in the north. It took three days to contact station MACA with the short-range transceiver. Messages piled up as other guerrilla commands in Luzon sent their messages to KAZ through their radio.

Two weeks after setting up the radio, Ancheta, Luz, and Santos were ordered to Maj. Russell Volckmann, guerrilla leader of Northern Luzon. Labrador, Miguel and Duran were to report to Ball at Lapham's camp. Anderson refused to let the men go even with the insistence of Smith. Miguel expressed his disappointment to his officer as they continued to operate the radio at various places. After three months with Anderson's headquarters, food was becoming scarce, the dry batteries almost consumed, and reports on Japanese patrols increased near their area. Miguel said they stayed out in open fields to get a better signal as the batteries ran low. Wracked with hunger, they argued with the officers. In desperation, they sent a message in their code to Ball and the almost immediate reply was to "either beg, borrow, or steal anything we can ride on and report to his place right away." The three men took a wooden boat leaving a note for Anderson with the guerrilla radio operators they trained. They brought a radio set passing enemy garrisons by sea and overland while crisscrossing the Sierra Madre mountains.

Labrador and his men did not have much time to rest when they arrived at Lapham's camp. They were to bring cargo to Anderson's camp then proceed to MACA headquarters for instructions about a job in Leyte before the U.S. Task Forces' arrival. The submarines with their load usually landed in Dibut bay, by Lapham's headquarters. He had *Dumagats,* indigenous people of the area, carry the cargo for Anderson. Some of the *Dumagats* were ambushed by Japanese patrols. Continuing to Samar, Labrador moved to different places, dodging enemy patrols and pro-Japanese Filipinos to get to a radio station. He reached the Bondoc Peninsula as the American troops were landing on Leyte. He reported his status and was ordered to proceed to Sorsogon to investigate the marine cable line from Guam to the Philippines. But halfway, he received another order to return to the Bondoc Peninsula to pick up some captured enemy documents. His boat was almost destroyed crossing the Ragay Gulf to return. For two days, he was held up by Japanese patrol boats blocking San Narciso Bay. He finally reached Bulan in Sorsogon with the documents but was grounded when a typhoon partially wrecked his boat. He took the mountain trail across Sorsogon to the coastal town of Sta Magdalena. Almost captured in the town of Matnog, he tried to sail from Sta Magdalena but another typhoon hit the island. His first attempt to cross the San Bernardino Strait to Samar failed, carrying the boat almost to Capul Island, where the

Japanese had big guns. A second attempt the following day capsized his boat. All were lost except for his weapon and the captured documents. He went through on his third attempt. Another seven days overland to Leyte and he reported to Whitney with the documents. Smith was there too.

Miguel and Duran met Maj. Lapham, commander of guerrilla forces on Central Luzon. They discussed an intelligence net in his area subdivided with reliable men from his squadron to report on Japanese activities. The sub-stations were set up high up in the jungles, accessible only by *Negritos,* indigenous people of the area, who acted as guides and guards. Runners brought the reports to the radio stations. For security, was an organization of influential men in the villages deployed by the Japanese to censor the people and had switched support. They spread radio broadcasts to the people through underground means to keep their spirits up and told them to lie low to prevent attacks from the Japanese. Everything possible was done to keep the battery charger running from hand operation to water operation. But the burned-out radio sets had rundown batteries which worked intermittently. Miguel found a battery charger while he was sailing through mined areas and had men carry it up.

A week after they were established, news of a submarine landing (October 44) near Lapham's headquarters at Dibut Bay on Luzon arrived. With a month to go before the Leyte landing, GHQ was sending multiple submarines to the islands laden with personnel and equipment. Miguel and Duran sailed for two days through mined waters and brought back the battery charger they badly needed and a more powerful radio set with spare parts that were wet. The additional signalmen they hoped for did not materialize as the new arrivals were weathermen and were to operate in central Luzon closer to strategic points. After the batteries were recharged and some parts replaced, the radio still did not work. The humid climate in the jungle and continuous rain had set mold in the radio parts. In desperation, they heated the elements until they heard "oscillation on the receiver." More careful heating increased the fluctuation until intelligence reports began coming in fast. Miguel consolidated the reports while Duran furiously tapped the keys to send them out to station S3L. Messages between the Luzon guerrilla commanders were extremely urgent and had priority. With only the two of them, Miguel and Duran depended on guerrilla operatives that did not have training on gathering intelligence. They trained some in between urgent radio messages. Their station ANA alias GNC functioned despite many mechanical problems that they improvised like using toothpaste and fine sand for grinding valves and having guerrilla laborers stealing gas and oil from Japanese-operated rice mills. Soon orders arrived to move to the central part of Luzon. Squadrons of guerrillas armed with weapons from the submarine cargo and reliable scouts as guides brought them and the radio equipment to the village of Baroy on the side of Mount Amorong in the middle of central Luzon. They set up their radio.

Meanwhile at MACA station in Samar, Guyot, Pascua, Angcos, Mendoza, and Mauricio with a guerrilla combat team for protection had no interruptions sending messages to KAZ. They followed Smith closely to maintain secrecy and facilitate escape as Smith constantly moved with his guerrilla guards whenever a Japanese patrol neared. Bulatao was head of net control station MACA almost the entire time of the mission that directly contacted KAZ. He trained a guerrilla officer and civilian operators to relieve him in the job of encoding and decoding messages to and from KAZ. KAZ received messages on downed pilots or pilots forced to land due to engine trouble and soldiers from Bataan and Corregidor who joined the guerrillas. Agcaoili encoded messages sent to the sub-stations and decoded received messages aside from his guard duty at night. Radio MRM stayed on the air all day handling air warning messages, ship flashes, and military information from the various sub-stations. Messages from the coast watcher stations sunk enemy convoys by U.S. Navy planes. Sometimes a decoy station was set up in another area to divert Japanese interference on their frequency. Guyot was assigned to that outpost for a while.

Firing shots shattered the early morning in June as guerrilla guards burst in warning of three fully armed Japanese patrols trying to encircle the camp. Japanese in the nearby town of San Vicente were moving closer to camp. Another enemy unit was coming up the trail the guards took.

With the men divided into three groups, Capt. Arrieta, the guerrilla security leader with Guyot and 20 guerrillas led the first group to occupy the hill by the camp covering the trail from the village of San Vicente to meet any Japanese. Guyot thought that he was now going to use his Army training unlike his earlier overseas assignment to the Aleutian Islands where he stood Cossack guard to protect the borders every night for eleven months. Pascua with four guerrillas were also on top of the hill by the camp overlooking a trail. Weathermen Richardson and Becker were with them. They dug into a semi-circle at 25 yards apart waiting "for some creeping devel (sic) to jump at us." The recon men were the advance party while signalman Ibay assisted Padilla in the rear. Angcos was at the end of the line with two guerrillas. They waited for three days as more volunteers arrived and provided them with Carbines and rounds of ammunition. On the fourth day, the men moved closer to camp to be able to move faster. They slept in the village below the camp, dug foxholes and offense lines around. Guyot posted guards at night. All the mission men were armed with Carbines, Enfields, BARs, grenades, and machine guns. After two days, when the Japanese did not come, the weathermen returned to camp MACA.

After several days of waiting, Aranas and Pascua saw 5 men in civilian clothes and hats coming up the trail. He called out in the Samar language to identify themselves and 'Halt' in English. They ran into the bushes as they were fired upon. The rifle shots cracked the silence. "That Carbine shot must have been fired by Jerry," said Mauricio in a low tone. Two younger guerrillas, holding Carbines, ran away scared. A BAR man went to the hills and joined his wife leaving only three men with automatic rifles. A runner arrived bleeding from a slug in his knee, shouting that around a hundred Japanese were in the village only ten kilometers away from the outposts with various arms and a radio set.

Throwing their packs into the bushes, recon men Mauricio and Padilla with signalmen Guyot, Mendoza, Angcos, Nuevo, Aguila, Pascua, and Agcaoili took up positions along the stream and along the hill parallel to the steam at "15 yards apart to cover the bluff. A BAR man was on either end and in the middle supported by Carbines and machine guns. They were less than thirty men guarding the bluff and side of the hill that was the only access to the village against two hundred fifty Japanese. Agcaoili prayed as he was assigned to guard a forty-yard stretch of tall grass. He had no cover. It was the men's first experience in battle and "their lips trembled as they tried to say a word," Agcaoili related. The men decided on 'Lapu Lapu' as the password because the Japanese could not pronounce the 'L.' They fired their automatic rifles and kept the Japanese from charging them.

Guyot and Mauricio ran to take up positions by the riverbank but suddenly fell into a big drainage trench covered with tall grass. Peering through the grass, they could see the enemy on the opposite hill. Above them was Agcaoili waiting for the sight of the

Japanese. In between the continuous barrage of bullets, there was a lull from both sides for an agonizing two hours. They could hear the Japanese moving. The guerrilla, Max, operating the BAR fired whenever he heard them moving. The Japanese openly charged in the open field, carrying rifles, and shouting fanatically. Capt. Arrieta ordered his men to open fire and they did so for five minutes. They ran short on ammo. They ducked from the Japanese reply of a hail of bullets. In the middle of the fight, Pascua heard a Japanese officer from the other side of the stream shout somewhat like, "Filipinos, surrender if you do not want to die. You are our friends. We want the white monkeys." A burst of their guns replied. The Japanese responded with machine gun and rifle fire. The BAR man, Max, began desperately digging a foxhole with his bare hands when he saw a Japanese officer pointing him out to his men. But he shot them first.

They quickly ate their one meal for the day of rice and small pieces of carabao meat wrapped in banana leaves. "Fried banana skins tasted like potato chips." Runners came gasping with news of 200 hundred Japanese reinforcements arriving from the other villages at night from across the river with knives for a silent kill. Pascua frantically looked around for a trail that the Japanese could come from at any time. It began to rain, and the guerrilla commander ordered a withdrawal back to the main camp, as they were outnumbered and surrounded. Pascua did not hear the retreat. When he saw no one around, he crawled back to his post. Footsteps became louder and a man kept repeating 'Lapu Lapu' and his name correctly. The man explained the retreat in Samar language that Pascua did not understand. He thought the man needed help. He heard from a distant "Come here" and remembered at Canungra Jungle School that Japanese were good imitators. A guerrilla appeared and told him of the withdrawal. Both ran as fast as they could. Although Pascua was limping from a boil in his leg, he "felt as light as a feather." He was so grateful that Capt. Arrieta sent back the guerrilla to get him.

The Japanese broke their main line of defense crawling up the hill. They exchanged fire. The men could hardly see each other in the night but they could not stop as the Japanese knew where they were. A BAR man covered their retreat. But the Japanese caught up with him and wrestled away his weapon. Shouting for help, he fell off a cliff. Pascua wondered why the Japanese did not stab him with his bayonet. He learned that some Japanese did not carry arms and got their arms from the enemy or dead comrades. The Japanese with rifles were followed by those who did not have arms. If killed, their weapons were taken. The guerrilla boys got scared and ran leaving the mission men and a dozen guerrillas on the hill. Swarms of mosquitoes feasted on them while they looked for edible food. Too weak to fight, guerrilla scouts went ahead armed only with *bolos* before moving into a village to find food. "*Bolos* were silent," wrote Pascua. They had thought Max, the BAR man was dead but he caught up with them. Pascua thought, "he was an owl and a dog to smell trail."

Withdrawal continued as the Japanese crawled closer. They walked single file and held hands on very dark trails to avoid losing comrades. At dawn, they could still hear the Japanese shooting believing that that they were still surrounding the guerrillas on the hill. The gunfire continued until the rains came, and civilians told them of hearing the Japanese moaning. They smelled the dead bodies as they passed through and saw many dead Japanese including their commanding officer. The stench persisted for many days.

The very steep and slippery trail up was slowing them down. Agacaoili was having a hard time keeping up. He tossed away his boots and Guyot did the same. A sharp stone cut into Agcaoili's foot, and it became painful to walk even with a stick. Walking on muddy trails and streams necessitated him borrowing another pair of boots. He limped for two months keeping up.

Guyot and Agcaoili became separated from the others in the night. They went hungry for three days except for coconuts until some people gladly fed them. They slept in the house of a family of evacuees. The people became frightened with their presence when news of around 70 Japanese crossing the river only three km below arrived. If they were captured, it meant death to the civilians. They hurriedly left when a volunteer arrived to guide them to Capt. Arrieta and the rest of the group. It was a fast whole day march.

The rendezvous point was in the jungles of Jipapad. They crossed the Kalaw River passing occupied towns. Thirty men reunited as more guerrillas volunteered to join. Capt. Arrieta, Mendoza, Padilla, Mauricio, Angcos and Guyot were in one hut. The rest were posted up the hill and down towards the rice fields. Some deserters were allowed to rejoin but a guerrilla officer and his boys were sent back to Sorsogon. They were running low on ammunition and money. Pascua found out that the guerrilla officers spent their own money to feed them.

The night passed without an attack. They camped for a week with trails guarded. A runner informed them of eighty Japanese in the village of Jipapad with mortars, machine guns, and a radio set. Orders were to shoot any moving object in sight from the direction of the village. After two days of observing enemy movements, Arrieta, his guerrilla officers, Mauricio, Mendoza, Ibay, and Padilla decided to raid the Japanese. Thirty-two men were divided into 3 squads except for Ibay, the radio man to assist Padilla in the rear. Pascua joined the squad although he continued to limp. Mendoza led the advance party. They went single file at night with a new password. One time the rear squad lost sight of those in front. They imitated the sound of frogs but stopped when too many were heard in coordination. Agcaoili and a guerrilla fell into waist-deep mud through a broken bamboo bridge. All stopped, holding their breath in case the Japanese heard the cracking and splashing, but the rushing water muffled the sounds. The first squad saw Japanese soldiers eating with the civilians, so they decided not to attack to avoid killing the civilians. They slept on the wet ground until dawn. Returning to camp, they took a different trail in case they were followed. The town mayor, who the enemy wanted, brought a carabao for their rations. His two sons were guerrillas, and he joined after he evacuated his family to a safer place.

The following day, a Japanese patrol of 100 men was 300 yards away on a trail across an open camote field by the side of a hill. They saw a civilian guiding them with his arms tied behind his back and to a Japanese soldier behind him. He was the cook of a wealthy family who had been sending food to the officers. But he was smart to take the trail where the observation post of Guyot, Arrieta, and BAR man Max were deployed along the hill so they would see them from a distance. To the outpost's right were Aguila, Bulatao, Padilla and others. Angcos was out in front. Left at camp were Pascua suffering from a boil and Agcaoili who had a deep cut in his foot.

The Japanese charged fanatically shouting, making them easy targets. When the Japanese began using mortars, the order was to withdraw into the night to the top of the hill. Pascual said he thought of the adobo and rice that they left behind. They had killed over 40 Japanese and wounded many more without a casualty on their side. It was raining and cold as they climbed a high hill for the night and tried to get some sleep amidst the swarms of mosquitoes. At daylight, they heard the Japanese using tracking dogs and they ran deeper into the forest. A bunch of wild pigs attracted the dogs and saved them. Several days passed and with only young rattan shoots to eat and coconut water, they weakened. One cup of boiled rice was for 25 men. At the edge of the forest, they climbed down a steep hill to reach the river below. They watched as the Japanese moved thru villages searching for them. They were too weak to fight and run. They had to get to an evacuation site where there would be food. Scouts armed with only *bolos* and in civilian clothes checked villages for Japanese and warned the residents of guerrillas arriving or they would run away seeing armed men. After walking for hours, they came to a house shielded by vegetation and there was a pot of rice cooking on the stove and a plate with some roasted corn. The guide began grabbing the food but stopped to speak in the native language but no one answered. Guyot spoke in English "If no one shows up, I'll burn the house." In the dark, Max and the guide spoke up in their language that they were not Japanese but very tired and hungry friends. A boy came out of the bushes and an older couple appeared. It was a good chance they took the trail that led them to Catalino, a guerrilla that Smith had told to hide important documents and equipment.

At camp, when Smith heard the automatic rifle going off warning of Japanese infiltration of the defense line, he packed the equipment of the entire camp with the White American soldiers, guerrilla radio operators and officers. He withdrew to an unknown location with the radio MACA. But much of the supplies were left behind and were looted by the enemy. Arms, ammo, gas, food, clothing, weather equipment, some radio sets and about fifty thousand pesos were lost. All the men lost their bags and field packs. They had only the clothes they wore and the arms they carried. Later, Smith said he was "glad that they did not lose the most valuable things...," Pascua related. He wondered if those were the codes.

The men did not know where Smith had moved to. A courier arrived and guided them to the new camp near Nagoocan where station MACA was first set up. Agcaoili limped while he went with the men to dig up hidden boxes of medicine, equipment, and supplies that he had hidden in the mountains. The men were ordered not to wear shoes to avoid being tracked by the Japanese. "That was tough for those who had wounds and boils like him," thought Pascua. He took Atabrine for the malaria that kept him down for many days. Two soldier guards and 24 carriers armed with only 4 Carbines were with them. They passed villages frequented by the enemy and travelled through the night in the rain. As they begged for food, civilians boiled some sweet potatoes for them and would not accept payment. They camped at a guerrilla commander's hide-out in the mountains near the town of Jipapad. The guerrillas ambushed a coming Japanese column until the enemy began shooting mortars. Retreating, they slept on the side of the mountains without eating. Shelter from the rain was large banana leaves. The ambuscades with the Japanese continued for several weeks on the way back with the cargo to the camp.

At the Camp, Guyot was put to work training radio men with a bureau of post operator because the guerrilla stations were using a commercial procedure different from what was taught to the Army signal men. The Japanese kept looking for the station until they came close again in August. MACA moved to the top of the Dao Peninsula by Gamay. Nearly a hundred carriers brought the equipment. Guyot was put in charge of four carriers that carried the radios. It was eleven hours on a rugged trail. The camp of Lt. Lentijas of the 93rd Samar guerrilla division was at the village of Dao. "He appeared to be rather young, around the age of 16." Station MACA was immediately installed, and couriers began arriving with reports for transmission. But Japanese patrols continued to follow and approach the camp, so the station moved inland again twice, with guards remaining for cover.

Radio signals were erratic as they moved. The new camp was only 2 km away from a garrison of eighty Japanese soldiers at Gamay keeping the men on the alert. A shot broke the stillness of the night and the smell of gunpower filled the air. Men rushed to get rifles and grenades. The tall grasses moved, and the cows tied nearby pulled at their ropes. A Carbine was missing, and an empty shell was found. A guard had accidentally fired. The Japanese would have heard the shots so the camp immediately packed and moved out again. The following day, the Japanese raided the camp.

The Japanese were at Dao overlooking the trail they took. To avoid being discovered, they climbed rocks and cliffs with the Pacific Ocean below. A dozen men fell. Guyot lost his wallet, and "a very good-looking *bolo*," he related. Reaching the shore, they loaded into boats and traveled along the northern coast to the village of Pangpang. The radio MACA was set up in a house on top of a hill belonging to the village leader and had a view far into the Pacific. The signals were much better. A Japanese destroyer was sighted near the mouth of the Bay, and in less than an hour that KAZ received the message, the vessel was not seen again. Using the new transmitter and receiver that the submarine brought, air warnings from around a dozen stations on the island were received in time for relaying them to KAZ at the appointed hour and frequency. But some weather reports could not be received due to atmospheric, or weather conditions or the set used for weather was not functioning well. It was challenging to keep the radio secret when dozens of curious civilians would arrive. They moved a short distance away to a larger building when they found out that the next-door neighbor was sick with a form of contagious leprosy.

With only a couple of months before the Leyte landing, all the stations connected to MACA were "hurting for many of the essentials of our existence: food, clothing, medicine money, not to mention everything needed to keep a radio station operating and an espionage network going. Supplies had not arrived, and rice was low. Food had to be found for a large number of people at camp and couriers arriving and departing." American escapees and civilians were sent to nearby towns to be fed by the local people so the radio operators could be provided with at least a meal a day with the food at camp. Birds, wild cattle, and monkeys were shot or caught then roasted. In desperation, George McGowan, an American escapee, went out and returned after three months with a pig and rice. He contracted a man to continue to send food. By this time, the men were starving and weak.

After a few weeks, the camp was awakened to the hum of engines. As the planes flew closer, they saw the white stars on the fuselage. They went into hysterics, especially Mauricio, as they counted eighty-two bombers wave after wave. Airdrops landed in the bay. Guerrillas salvaged a huge gasoline tank with the trademark of 'Grumman' makers of heavy-duty engines. Other drops brought magazines and periodicals. The sight of the planes was a cause for celebration, and the men were invited as special guests by a Bataan veteran, Sgt. Murallos, to a dance with much Filipino food. But not all the aircrafts were friendly. The radio devoted to air warnings sent messages of two low- flying Japanese planes over their station. The men had almost waved American flags to these planes, believing they were U.S. fighter planes. They were shocked at the thought if they had not noticed the red sun.

The relentless Japanese were again sighted a few hours from the station. Patrols entered the village next to Pangpang making the civilians leave to hide in the forest. An urgent message from Smith ordered a move. Packed in half an hour, they took another trail to lose the Japanese. The move at night, without sleep, followed a trail, a creek, and a dangerous deep ravine. In the morning, the guide said they had one more mountain range to climb.

They traveled as fast as they could, knowing that the Japanese were following them, and they did not have enough arms and ammunition. They found a house in a clearing in the middle of the forest. Useless Japanese equipment was scattered around. On the air the following day, messages were received and transmitted until night.

All stations were now to send aircraft warnings and weather reports as the Allied forces were nearing Leyte Island. Mauricio was the chief of MACA exclusively for transmission of vessel sightings and guerrilla communication. He plotted the grid coordinates of Japanese plane flights and friendly planes. Guyot, a signal man with infantry skills, was happy to relate that, Smith praised him for his work as a radio operator because "He seldom spoke to me except to bawl me out if I made a mistake." Earning his rank in the field, Guyot was recommended for a promotion.

Heavy rain fell when Guyot and guerrilla operators went farther inland through mangroves and set up station MRM three km from MACA exclusively for air warnings and in a separate frequency from the weather reports. All radio stations in the southern Luzon area, Samar, and as far north as Pangasinan reported all plane flights to MRM. The station worked so well that it picked up messages that MACA could not receive from remote sub-stations and relayed them to KAZ.

To ensure the safety of the radios, another move was ordered. Mendoza with his group was on the way to MACA to provide protection. Missing the rendezvous point, they landed in a town occupied by the Japanese. Two guards were captured and severely beaten but they did not talk and escaped on the eve of their execution. One was hit by a bullet that passed through his nose. All had to swim for cover in the mangroves.

The rendezvous point was a village along the coast north of Pangpang. Catalino, the guerrilla that Smith had entrusted to hide a big black rubber bag with documents was missing. The post office operator, Barba, was distraught because he was the companion of Catalino. Smith angrily ordered the men to disperse and find him immediately. He had to have the bag as he was being picked up by boat at the coast of Alugan several hours away to meet the American forces at Leyte.

In the meantime, agents discovered a sizeable Japanese garrison in the interior jungle near the town of Borongan. Recon man Ibay, and signalman Pascua with two guerrillas, volunteered to set up station KRP to watch the garrison. Richardson joined them with his equipment to set up a weather station. The weather equipment was bulky and difficult to move. Smith told Pascua that with a boil and jungle itches all over his body, he might not be able to take the running with many Japanese in the area. The group sailed for three days and hiked overland for two days with one hundred pesos for emergency expenses. Japanese patrols delayed them, but they were not stopped as they wore civilian clothes over their GI clothes. Along the way, an old quack doctor woman treated Pascua's itch with a holy smoker, prayers, plus his faith in her. He gave her two dollars, and three days later, his itches disappeared. He said he was "not superstitious but ... was able to sleep when my itch disappeared." With the guerrillas and civilians' help Pascua found a good site for the radio and had a small hut built for protection. The radio went on the air, and messages were sent in code to MACA station, while Richardson handled the weather radio and codes. An American guerrilla, Gordon Smith, joined them taking charge of security with guerrilla guards and food. Station KRP worked for a short time until the charging engine of the batteries died. It took Ibay and a mechanic two weeks to fix the engine. The Australian radio set worked.

As the Leyte landing neared, Pascua and Ibay continued operating KRP encoding and decoding the increasing number of messages and preparing intelligence reports from operatives and sub-stations to send to MACA. Orders were to send all aircraft and naval craft warnings or movements in the area besides enemy strength, location, ammunition dumps, and airfields. Sometimes call signs changed for security measures and to deceive the Japanese. Pilots who survived from their downed planes and ships were rescued and reported. They remained on the alert for Japanese discovery of the hydrogen-filled balloons released daily into the air to determine wind currents from various locations. Richardson sent four weather observation reports daily as soon as they were ready increasing to six reports when the Americans landed. A directive from GHQ sent a group south along the coast evading Japanese shore patrol to plot the locations of minefields in the Surigao Strait before the U.S. Task Forces' arrival.

Gordon Smith was called back to MACA and Richardson left to deliver a message to guerrilla commander Juan Causing in Cebu about meeting a cruiser carrying arms for his men. Pascua was left in charge. The radio was under the U.S. Army even though higher-ranking guerrilla officers were in the camp. Richardson did not return but instead sent some arms for the station. With the new rifles, Pascua taught the twenty men at his station to maintain the weapons and how to use them in jungle fighting. He taught signals and a little radio work. The ever-appreciative civilians of the basic medical service he provided gave bananas, rice cakes, and chicken.

Before year end, Pascua and Ibay moved station KRP into the town of Borongan. After a day of hiking, they found the constabulary barracks burned by the Japanese. Trenches and air raid shelters had been dug up under or near several big houses that had been looted of their furniture. Those who stayed were lacking food and clothing. Most of the people had evacuated to the hills but many slowly returned when they saw the radio and personnel. They were thankful to read the latest news written on a blackboard posted outside the station. The radio was installed in the large house of a Juan Bocar, a former Philippine official who was an operative in the area. Intelligence and weather reporting continued as the people shared with them what they had. Feeling safe, the locals celebrated the New Year with a dance. Pascua realized that he "was back in civilization when I saw nice Filipinas dressed in regular evening dress and formal Filipino costumes." He had not seen "pretty and pure Filipina girls for the last twelve years." The following month, they moved the radio overland to the capital city of Catbalogan. A guerrilla commander arrived to help with the move of men and equipment. Pascua went to Tolosa, Leyte, for a few days to report to the Sixth Army but was told to continue with radio KRP. Ibay operated the radio while he was gone until January when the guerrilla Maj. Sabarre arrived to take charge of the radio.

Recon man Besid accompanied Aguila and Becker to set up their weather station. They sailed along the northern coast of Samar for three days avoiding the Japanese naval station at Batag Island. They stopped at a small island to get food and hid in position with their weapons as an enemy launch passed by. After three days on the water, they hiked through the night to the town of Villa where radio KRM operated by Elwood Royer was located. They had seen Royer earlier at Camp MACA. KRM was an alternate net control station and a weather station. Agcaoili had earlier brought radios, but they did not work and were sent back. He stayed to operate the radio stations KRM and OBA.

Becker, Aguila and Besid trekked to the hills of nearby Cabacungan and set up a weather station. Unknown to them, the Japanese had been watching their flashlights at night at camp while taking weather observations. They were followed and constantly moved to get away from the Japanese while providing weather conditions. Frustration set in when there was no reply to their weather reports as they risked being located with each transmission. Becker sent an angry message that they were doing their job and asking for cooperation. Finally, reassurance arrived that their reports were of value in the execution of operations. As the months passed, food became scarce with the difficulty of transporting it from the lowlands. As the Leyte landing neared, they continued sending weather reports although the rice was gone, and they ate sweet potatoes. "Aguila and a guerrilla went out to hunt some wild pigs. They came back with a big monkey which we ate." Ammo was running low as their guerrilla guards sniped at enemy patrols that got too close to the camp. Their station could not close for it was in the strategic path of aircrafts entering Luzon and the Visayan islands from the Pacific Ocean. An airdrop of Carbines, ammunition and 2,800 pounds of food was ordered for Aguila.

Lt. Luis Padilla, described as an aggressive and resourceful officer, arrived at Villa to investigate the organization of two guerrilla commanders. There was a dance in the house of a town official to celebrate the baptism of his child. Padilla, Agcaoili, and Lt. Royer attended. The group of Becker, Aguila and Besid were still at Villa before their departure to set up a weather station. There was much drinking of *tuba,* an alcoholic drink, through the night. Royer, who had been drinking heavily, fired shots into the air. Padilla warned him so as not to attract Japanese patrols and ordered his revolver taken away. Two sailboats showed up, and the dance was halted. After waiting for some time for the return of the men who volunteered to check out the sailboats, Padilla went with two civilians to the beach to check for himself thinking that the arrivals might be the contact men of the two guerrilla commanders, Maj. Licerio Lapus and Gov. Salvador Escudero, from the Albay, Bicol area. Escudero passed intelligence to the Samar net control station while Lapus was a Philippine Constabulary officer who led the guerrilla 54fth Regiment. They fought for leadership. Padilla shouted out questions to the people in the sailboat until shots were fired. Everyone hid in town. The Japanese sailboat was on the way to another town when they heard the shots of Royer and decided to stop and find out. They had captured the volunteers, hid by the town's graveyard and ambushed, killing Padilla.

The Japanese knew much about the stations set up in Samar. They asked the people about Padilla and the girl he wanted to marry. They had observed the flashlights of Aguila's station with Becker in the mountains watching the weather. They heard the frequent target practices and dances near the camp.

The men dispersed to hide the radio. Agcaoili suggested to Royer to move the station to the interior. Later, KRM was closed down and the equipment transferred to the weather station in the hills until orders arrived to close. With a guerrilla as a guide, Agcaoili brought the radios back to station S3L. He sailed to Sorsogon and trekked overland to cross the treacherous Ragay Gulf with the radio, two drums of gasoline, and his Carbine as the only weapon. The strong current and wind drifted the boat in front of Bulan (Burias) Island for a whole day, where there were six hundred Japanese guarding an airstrip. They were relieved that there were no patrolling Japanese launches. A storm ripped the main mast, and the cargo soaked. The boat swiped the rocky shore. Only the radio and gasoline were saved. The cargo of precious rice was lost. Reaching shore, they were towed to Sorsogon Bay to obtain permission for their presence from the guerrilla commander Lapus and get a boat.

They returned to cross Ragay Gulf with a small sail and their main mast was torn again but they were able to arrive with a small forward sail. They reached the town of San Narciso, north of station S3L. The guerrillas questioned them about the radio before leading them to the town's mayor. The mayor fed them and suggested they not travel at night because it was a rugged country. Agcaoili remembered the "taste for some real coffee. I must have looked like a beggar with my old fatigue suit and my beard unshaven." The mayor provided transportation and carriers the following day. They travelled first by boat to the closest town where the Tayabas Guerrillas were stationed. The camp was deep in the forest and very muddy. Food was scarce. After 4 days of plain boiled corn Agcaoili's stomach was upset. At station S3L, he rested for several days, and did cryptography work between assignments. He delivered radio parts overland to Robles' station at Patabog through rough trails and jungles, avoiding Japanese patrols barefooted so as not to be tracked. He was relieved to know that there was only one Japanese garrison on the Bondoc peninsula.

Large flights of Japanese planes were sighted searching U.S. Navy ships. The stations were to report those flights as high priority in a brief code which did not require encoding and had to be passed to GHQ within ten minutes of the sighting, or they were useless. They reported on six dive-bombers and pilot Lt. Davis who crash-landed in the Ragay Gulf. Nery recounted that he and the Tayabas Guerrilla commander, Vera, had the remains of Edwin Robinson, the son of a U.S. Senator, turned over to a submarine landing. The Tayabas guerrillas had grown to around 1,000 men patrolling southern Tayabas and part of Camarines Norte on southern Luzon after 400 Carbine, rifles, including a 50 cal machine gun were given to them. Medicine was also given. In return, they provided security to the station. The station was inundated with reports on Japanese troop movements and defenses when the radio ONG and a radio operator was sent to their camp in Camarines Norte. The radio was later returned for repairs. The guerrillas destroyed two Japanese radars near Boac in Marinduque Island in support of the landing. At Atimonan south of Manila, the Tayabas guerrillas stopped the Japanese railroad operations by blowing up bridges and preventing enemy troops from reaching the city.

Once radio ONG was repaired, signalman Angcos and recon man Mendoza with a guerrilla combat team brought it to the northwest coast of Samar to report on the Japanese ships in the surrounding waters and the San Bernardino Strait that could reach Leyte. Mendoza engaged with the enemy whenever they neared.

September was a harrowing time for the civilians as the Japanese increased their patrols and brutal atrocities. They were closing in on station MACA. They may have intercepted news of the October landing as the radio traffic dramatically increased. Agents were losing radios or were themselves captured. Signals were being jammed. The signalmen knew something important was going to happen with the big jump in radio traffic but were not told of landing dates and other sensitive information in case they were captured. Radio frequencies were changed several times to protect the invasion plans. MACA spent six hours daily trying to keep up with the changes, reports from the sub-stations, and operatives from Luzon to relay to KAZ. As the Japanese increased pressure to find station MACA, KAZ was urgently informed that it was on the run and closing down. S3L in the Bondoc Peninsula took over for the next two months sending reports directly to KAZ.

Sabado reported around one hundred American bombers in the early morning of September dropping their loads on Cebu City. The bombing raid had begun in August over Mindanao and spread to the other major islands. The Japanese retreated allowing the gathering of the latest information on their ammunition dumps and quarters. Guerrillas went after the remaining Japanese. Surprised enemy planes were damaged and could not get off the ground. The second wave of bombers arrived the following day guided by their reports to destroy the missed targets in the first wave. Enemy escapees from the city holding their arms and a few rounds of ammunition surrendered, looking "skinny and in rags." After interrogation, they were turned over to the guerrilla commanding officer and probably killed. Those who escaped into the mountains were hunted down.

Reports transmitted to KAZ had to be analyzed for accuracy and differentiated from rumor. As the Japanese atrocities increased with the pending liberation landing, the civilians moved out of the area. The accuracy of reports decreased. A shipment of 1,000 Japanese troops that passed through Merritt's area by road and boat to Luzon was missed. In Bicol, Lapus could not get information on a large airfield in his area because of no dependable operatives.

USS *Seawolf* – September 1944

The Japanese and collaborators were not the only danger the men faced. The *Seawolf*, an obsolete sub converted for better use to cargo, carried a party to reinforce station MACA's extension of the radio network to Luzon. Nine tons of supplies were onboard with some going to Leyte. On the way out, guerrilla officer Maj. Sabarre was to be brought with his men to Batan Island north of Luzon to set up a weather station and two coast watcher stations capable of direct communication with KAZ.

Braynard Wise, a weatherman, volunteered again to return to the Philippines with fourteen recon and 978th Signalmen from Tabragalba. He had survived the Japanese raid that killed his leader, captured two of his comrades, and scattered the rest of his party on Mindoro nine months earlier. He was eager to return. Orders were cut for him and 978th Signalmen Sgt. Emiliano A. Almero, Sgt. George B. Bueno, T5 Amadeo C. Cendania, Pfc. Aquilino B. Ramos, and Pfc. Juan F. Rimando. Reconnaissance men were Sgt. Artemio Ibea, Sgt. George Peralta, Sgt. Alberto Francisco, Sgt. Emil Pugosa, Cpl. Antonio Tria, T5 Ruperto Ruiz, and Sgt. Irineo R. Rodriguez. Capt. Howell S. Kopp, an Alamo Scout, was onboard to an undisclosed destination in the islands to support the Leyte landing.

The personal packs of the men were marked 'J' in a diamond to differentiate them from the other cargo. For Smith, there were boxes of subversive equipment, signal equipment, medicine, rations including a case of whiskey, rifles and ammunition, and personal items marked 'SH.' A cargo of signal equipment, Carbines, ammunition rations, medicine, and personal items were for Maj. Sabarre and his men.

The *Seawolf* reached Manus Island off New Guinea and exchanged radar signals with Maj. Sabarre. Before it could discharge stores and personnel in Samar, the sub was sunk off Morotai with all men on board. Earlier a Japanese submarine had sunk a U.S. destroyer. The destroyer escort *Rowell* hunted the Japanese submarine. Four U.S. submarines were in a safety area for American subs and gave their position except for the *Seawolf*. The *Seawolf* may have maintained radio silence in avoiding the enemy ships from locating it. Other sources said that the *Rowell* did not have knowledge of any friendly subs in the area. With no contact for two days from the *Seawolf* and receiving unrecognizable sonar signals, *Rowell* attacked the spot marked with dye by a Navy aircraft from the carrier *Midway*. *Rowell* fired Hedgehog, an anti-submarine weapon more deadly than depth charges and sunk the sub.

Five days later, a memo to the Chief of Staff regretfully advised that the submarine *Seawolf* "failed to reach her first objective and must be presumed lost."

USS *Stingray* – October 1944

KAZ was unaware of the *Seawolf's* fate and sent the *Stingray* several days later with three men on board to bring supplies. Alamo Scouts Lt. John. C. S. Hall, Lt. William Rommel, and Lt. James O. Johnson landed on the small island of Suluan in the southern tip of Samar to reconnoiter the area in preparation for the Leyte landing. Minesweepers combed the area. They crossed the water to the mainland of Samar with a party of guerrillas and carriers of their supplies to reach MACA camp. Bulatao related that he provided radio support to the Scouts in an operation to dislodge an enemy garrison in the town of Taft on Eastern Samar. "Johnson had difficulty using the NEI-3 radio" to send a message on enemy locations but "Hall said all targets were wrecked." The Scouts stayed at camp reviewing landing plans with Smith before proceeding to Leyte to continue checking the area for enemy presence. They hiked overland, crossed Leyte Gulf to Tacloban arriving four days before the Allied forces landed. Kangleon provided information on the enemy for the Leyte operation. The Scout's information on Suluan Island was invaluable to the 6th Rangers when they landed in Suluan, Dinagat and Homonhon Islands to clear the entry of the invasion convoy into Leyte Gulf. Lt. Rommel was killed in the Leyte landing.

At station S3L, the radio crackled with traffic of the U.S. convoys approaching Leyte. Numerous Japanese planes overhead were seen from the Manila area flying in the direction of Leyte. Radios were switched to air warning frequency as the men banged out messages to let everyone know that there were flights of Japanese planes on their way. The Navy sent a message of appreciation on a report of a Japanese convoy that passed in MogPog Pass, a waterway towards Leyte and Manila which did not escape destruction.

As the Allied forces streamed towards Leyte, General MacArthur "alerted all radio stations and coastwatchers for any detection … Report immediately any naval or air movement during present operations." To increase the confidence and morale of the guerrillas, MacArthur broadcasted instructions not in code but plain English "to all PI commander" in the Philippines. "The Japanese were reminded of a large and potent secret army in their midst."

Preparations for the Luzon landing were ongoing as the Leyte landing was unfolding. A schedule of the landings and liberation of the major islands was to be met. Christmas found Labrador returning to Samar to pick up Bulatao and another signal man with three radio sets in preparation for the Sixth Army's landing on Luzon. He used the river route for three days to avoid enemy forces and another two days to trek to MACA. The group took the mountain trail to the northern coastal village of Pambujan to avoid enemy patrols. The trek took 2 days and three more were spent looking for a sailboat to cross the treacherous San Bernardino Strait to Luzon.

Labrador requested all guerrilla units to send their radio operators because he had only two operators. All reports in the Bicol region were to go to his station for relay to KAZ. Guerrilla officer Maj. Sabarre arrived from Leyte with a radio operator to reinforce the Albay-Sorsogon net, up the transmission on enemy depots, ammunition dumps, supply depots, big gun emplacements, and other enemy military installations to the Sixth Army. A radio net was established to cover the Bicol region of southern Luzon, monitoring the surrounding water passages. Intelligence reports and flashing signals to the 15th Weather Regional FEAF and the Air Force, were sent. Rescue parties were sent to downed pilots over the Bicol skies.

Bulatao set up a station close to the town of Matnog in Bicol where about three hundred Japanese garrisoned. He trained more operatives to send out and spy on the enemy's activities. Operatives were sent to the Bulan airfield. The guerrillas were given arms and medicine to monitor the Ticao Pass and San Bernardino Strait, looking out for Japanese shipping sailing south towards the beaches of Tacloban, which would soon be filled with landing equipment and men.

Leyte Island, Philippines Landing – October 1944

Guyot related that "the air was so full of traffic that station MACA could hardly get a message through. Every station in our net wanted to come in with urgent messages and AWAWs (air and weather,) but the interference was so bad that we had to stand by almost all day. Bulatao managed the message center, which had direct access to KAZ. In the afternoon, we heard it announced through our receiver that the landing of our Forces in Tacloban was a successful one. We were wild with joy. Civilians came pouring in for complete information. The next morning, we raised the U. S. flag high above the coconut trees. In the air, we became busier than ever. Two radio sets were used ... separating one for KAZ. In this way, we managed to function perfectly." With Samar right beside Leyte, continuous formation of planes were sighted and reported immediately without trying to identify it just to keep up.

The Americans had landed after three years of waiting. As the task forces were unloading on the beaches, station S3L on the Bondoc Peninsula and northern Samar had an excellent view. Around fifty Filipino families arrived at the Bondoc station and established a temporary campsite in the surrounding woods. For three days, they celebrated with the men with lots of dancing and *tuba*. A Catholic priest from nearby San Narciso blessed the festivities. In several days, they saw the blazing skies from the bombardment of Japanese convoys trying to prevent the landing.

Samar was next to be invaded within a month after Leyte. KAZ wanted more information on the enemy situation, but the guerrillas continued to squabble amongst each other. Several guerrilla officers were captured. Several radio station's locations were betrayed and captured. Juan Causing, a representative of Kangleon, Leyte guerrilla leader, tried unsuccessfully to unite the two prominent leaders, Capt. Pedro Merritt who commanded guerrillas in Northwestern Samar, and Maj. Manuel Vally who organized the guerrillas in the eastern part of the island. Vally allied with Kangleon while Merritt refused. Needing Samar guerrillas unified, General MacArthur appointed Charles Smith as commander of the Samar area to unite the warring groups. Smith declared himself as a coordinator, thus keeping the respect of the men. They agreed to cooperate under Smith, send rosters of their men, and expand the radio networks on Luzon and Bicol Islands. Smith promoted Merritt and appointed Vally as Executive Officer. Causing commanded northern Samar, guerrilla Capt Arrieta with station MACA commanded central Samar, and southern Samar was under guerrilla Luciano Abia.

Given the authority to induct civilians, Smith inducted men with the promise of payment from induction and recognition after the war which was not entirely fulfilled. Strict orders from KAZ kept the guerrillas from inciting the Japanese. But the unified force of around eight thousand men continued guerrilla warfare against the enemy and several rival factions continued to cause trouble. Armed with *bolos* and whatever weapons they could use, they harassed the enemy until the U.S. Sixth Army landed on Samar.

The radio crackled with messages from Manila. American planes suddenly appeared over the city from all directions dropping bombs on military objectives, installations, and ammunition dumps. Ships in the Bay were sunk or damaged. Still, the Japs said in their papers "No damage on our side." Residents did not dare show their elation to avoid Japanese harassment or being killed by a desperate Japanese soldier with a machine gun.

Smith returned to camp from Leyte bringing arms and supplies. Several newspaper men and technicians were with him, including a Mr. Brannon, a war correspondent of *The Sydney Times*. When Guyot was called to man the mortars, around two hundred guerrillas joined him to attack a Japanese garrison in the town of Taft. Brannon, was allowed to fire in the first round but did not follow instructions, almost killing himself. The Japanese radar station, command post, hospital and other targets were destroyed. Fighting continued in the town with the Japanese entrenched in the school and municipal buildings. The Japanese had dug "a tunnel in swastika fashion around the hill as storage for their radio sets, arms, ammunition, food, and medical supplies. They had set up trunks of coconut trees mounted to look like anti-aircraft guns."

Samar Island Landing – October 1944

The sub-stations pounded the keys day and night with the pending landing of the U.S. Task Forces in Basey, Samar. With Samar practically attached to Leyte, the surrounding agents and sub-stations became "search-and-rescue mission hunting for Allied pilots who had ditched their planes in their area." Guerrillas and civilians went down to the shore to help Navy fliers and the wounded. Some of the "fliers thought the Filipinos were Japanese and if they had any weapon would attempt to use them when the people would come out to help them."

The guerrillas' ferocious attacks to wipe out the Japanese presence filled the airwaves of MACA and the sub-stations. Capt. Arrieta with forty men cleared the town of Oras. Sixty guerrillas led by Maj. Sabarre cut off the Taft-Wright Road and proceeded to take the town of Taft. In the north, Merritt organized the USAFFE men and destroyed Japanese emplacements. Capt. Abia received five thousand pesos for food and 400 rifles and ammo to clear out the large Japanese garrison in the Balangiga area. Causing and Arrieta's men sent reinforcements. The town of Borongan was cleared of the enemy by Causing. With reinforcements of American troops and guerrillas, Abia cleared the islands of Guiuan.

When Japanese reinforcement arrived at Basey, the entire town fled to the mountains. Guyot went with a guerrilla officer to a guerrilla camp by the town of Basey. He reported the town's strong resistance to the Japanese stemmed from "every non-disabled male above twelve years old from the town and thirty-one other villages on Samar joining the guerrilla resistance." They fought in the hills while women and young children hunted for food. "Many of them were barefooted, naked, and others in rags and leaves," Guyot noticed. They stole rifles from the Japanese, and when their ammunition was gone, they used *bolos,* knives, and anything that could be used as a weapon. Guyot was picked up by a signal officer of the 1st Cavalry Division. Upon seeing him barefooted, Chaplain Charlies R. Loss gave a pair of jungle boots, socks and some toilet articles. He left to set up his radio along the shore near the church but moved again on Christmas Day to Lawaan, a nearby coastal village.

MACA's radio networks were turned over to the Sixth Army. Nery coordinated a meeting in the mountains on the mopping up operation between Lt. Thompson, of the Sixth Army and Vera's Tayabas Guerrillas that went well.

Three hundred Garand rifles were sent to the guerrilla commander in the north to keep the Japanese from crossing a particular road into the hills to an area which would not be accessible to the U.S. troops to go after them. In an interview after the war, Smith reported that the guns were not used.

In November, the Sixth Army commander ordered USAFFE men to active duty to clean up the Japanese. One thousand Garand rifles were sent for Samar troops.

Agcaoili was with Nery operating station S3L encoding messages between assignments. He continued to train guerrillas to encode, decode and evaluate intelligence from the Tayabas Guerrillas to ease the workload. Twenty-four Japanese survivors were reported to have arrived at San Narciso from a sunken Japanese launch and freighter "bombed by our dive bombers and B-25s along the coast of Aurora in Tayabas. The grounded freighter there could still be seen. Japanese soldiers, civilians, women, trucks, mules, and plenty of supplies...inside the freighter." Stahl took Sattem, the American POW escapee and a couple of guerrillas to meet the Japanese survivors. They were too late. Trying to get to shore, the survivors were met by the locals and hacked with *bolos*. There was no mercy.

Nery was ill when an urgent message came through wanting all available information about the Japanese within their area. Mindoro was to be approached in December with U.S. Task Forces sailing from Leyte. The waterways from Leyte to Mindoro were to be kept clear of the enemy. Agcaoili immediately left to inform Stahl at San Narciso and Robles in Patabog, who had difficulty contacting S3L.

Smith suddenly left station MACA before the Luzon landing. Pascua found out from Herreria who was the lead operator that Smith had left Samar with Richardson and the other Americans several days before the Luzon landing. They may have thought that weather observation was no longer needed now that the Sixth Army Task Forces had landed. Their departure was not told to the men out in the sub-stations, nor to whom they should report to. Stahl at station S3L expressed his disappointment at the abandonment of the men who risked their lives at their outposts. There were no provisions made for food and operational money. They went hungry begging the civilians for food. Capt Arrieta sent Guyot to Tacloban and was informed that Smith was relieved. He was to report to Juan Causing, the new commanding officer as his signal officer. Guyot related that he had no money and was not given food and quarters. He was grateful to have met some of his relatives who let him stay with them.

Unidentified Vessel – December 1944

The last submarine mission scheduled to Samar for supplies to the Bicol Provinces listed on the *Calendar of Submarine Missions to the Philippines* had Lt. Carbonell, Sgt. Senen R. Ramos, and T4 Miguel Q. Paraiso. 978th Signalmen Sgts. Mariano V. Arbis and Manuel F. Faragan joined the team. The party departed from Hollandia, and Ibay (June 44 mission) was to join them upon arrival to aid the radio station. Problems in scheduling and coordination caused the party to return without landing. Ibay continued operating a radio sending sightings of Japanese launches hidden in camouflage and nightly patrols along the coast of Albay. Intelligence from Legaspi City reported Japanese planes flying over a village by the city dropping gas and making the inhabitants tear and cough.

GHQ was elated over the more current visibility of Japanese activities from radio networks covering Luzon – MACA station on Samar, S3L on the Bondoc Peninsula, in Southern Luzon, and at Lapham's camp in Baler Bay on central Luzon. In addition, intelligence continued to arrive at KAZ from Mindanao in the south and Panay, Negros, Cebu, and Leyte in the central Visayas.

1945

The Sixth Army wanted an emergency crash landing strip built before landing in Luzon. Station S3L moved to Abuyon to monitor the construction progress of the airfield. Hundreds of local folks from neighboring villages, guerrillas, and volunteers worked in shifts. They answered the call to be paid for three meals a day. Bamboo was used as culverts to drain a swamp and covered over with dirt carried in baskets on workers' backs. Forty shovels were airdropped to spread the earth evenly. Carabaos paraded back and forth to tamp it down. Throughout construction, Agcaoili continued coding messages, copying the news, and typing them for distribution to different stations. When Stahl needed rest, Agcaoili was placed in charge, so he began making messages for encoding too. When the airfield was completed, he kept records of airdrops and all the arms, medicine, and clothing provided to the Tayabas Guerrillas. The airfield was completed in time and two hundred feet longer than required. There was not enough money to pay the laborers, so parachutes were given to make clothes, and the cord unwound into threads. The sewing machines came out with the new packets of sewing needles that arrived with the submarines. The heavy burlap material used for making clothes quickly wore out the needles.

Maj. Russel Barros, a Philippine Scout, was one of six agent stations in the Manila area of Markings Guerrillas led by Marcos Villa Agustin that provided intelligence on Japanese troop movements in the city and rescued American pilots.

A guerrilla operative on Southern Luzon was reported to have been operating a hand-cranked low-wattage radio for several months to keep in contact with S3L. Nuevo reported crossing the Ragay Gulf and setting up radio ARL at Gov. Escudero's camp to cover the Bicol region and the San Bernardino Strait. Information was relayed to station MACA. Two more stations were set up with one operating near Legaspi City. Nuevo was unfamiliar with the hand-cranked radio since the radio he knew to operate was mainly the sturdy Australian type. "It took some tinkering with it for a short while to find that the set would drift off frequency and lose contact." Once a steady speed was maintained, the frequency did not fade. Codes and security procedures were coordinated, and a message was sent to S3L and Anderson's radio at Infanta as tests.

Luzon Island, Philippines Landing – January 1945

After six months from their first request to send supplies, and three scheduled airdrops that did not arrive, cargo planes dropped about a hundred and thirty parachute loads of supplies in the first two months of the year by the Bondoc station. They had been subsisting on food provided by the guerrillas. They hoped that some boots would be sent for them to replace their rotted ones. They had resorted to wearing *bakya* (wooden clogs) with their tattered GI undershirts. The airdrops mainly had arms and ammunition, and explosives. There was a case of radio batteries, much-needed medicine, sugar, cigarettes, shoes sizes 5 - 8, assorted clothing, and forty-eight cases of rations joyously received. The explosives were to be used in Manila. Each pack weighed less than a hundred pounds and could thus be carried by one or two men.

With so many supplies, Nery stored them in the jungle "rigged with explosives so that it could be blown up in a hurry if necessary." He knew that he would oversee the station with Agcaoili once the Sixth Army called Stahl to Leyte.

Three days before the Luzon landing, Nery and all radio operators on Luzon received the following message from GHQ "Starting immediately upon receipt of this message is the time all patriotic movements of the people have thus, under local leadership, the opportunity to unleash maximum possible violence against the enemy to assist our liberation campaign. Employ all means available to restrict the movements of enemy forces on Luzon."

Mauricio was recalled to the Eighth Army in Leyte and given the assignment to reconnoiter the area of the capital town of Puerto Princesa in Palawan. In February, he landed at Brooke's Pt to survey enemy troops, anti-aircraft gun emplacements, the condition of the airstrip, and location of enemy radio stations. He sent his reports to the Eighth Army headquarters in Leyte and the XXIV Corps in Mindoro until April when he was called back to his outfit.

When the 1st Cavalry entered Manila in February, Labrador turned over his network in the Bicol area to the Sixth Army. Message traffic had lessened. Stahl related that "Labrador's skills as an undercover agent were wasted as he was sent around like a courier and a radio operator. He spent much of his time crossing waterways and islands with significant risk to his life and the men he was with." With his assuring messages of the enemy's weakening, the Sixth Army launched an amphibious landing in Legazpi, the capital of the Bicol province, in April. Continuous American air raids rained on the capital city to remove around five hundred Japanese in the nearby village of Salvacion. Demoralized Japanese deserting and committing suicide from the constant bombings were reported.

Nery took charge of the station, sub-stations, guerrilla guards, equipment, and the Abuyon landing field when Stahl was recalled to GHQ in March. Agcaoili was in charge of the radio operation and wrote the intelligence reports. Work was eased up a bit when Agcaoili trained guerrillas on cryptography and evaluating intelligence reports coming from Vera's Tayabas Guerrillas. Before he left, Stahl expressed his appreciation of Agcaoili's work and the cooperation and security provided by the Tayabas Guerrillas that enabled the men to continue the mission. The station was never off the air, he claimed and credited the American POW and "high-class Filipinos" working with him. Whenever a Japanese patrol put ashore to find them or neared the camp, Vera's guerrillas were radioed to repulse them. He had mixed emotions when he found out that the men in their stations on Luzon and outlying islands were "intentionally omitted from the supply operations to keep our activities low key in the eyes of the Japanese ... purposely been left high and dry."

"No one knew of the 978th Signal Service Company, First Reconnaissance Battalion (Special), whose headquarters were in New Guinea," related Stahl. Without proof he was in the Army, he could not get a small advance in his pay. He ate and slept wherever he could in Leyte. Without a commanding officer, he did not exist in any table of organization. That made it difficult to go home. He found Smith, who vouched for him, so he was able to collect back pay. He remembered asking Smith why he had departed without informing and leaving provisions for the men.

The Filipino recon and signalmen encountered an experience like Stahl's when they individually reported back. They looked for Whitney or anyone who could vouch for them so they could find a place to sleep, eat, given medicine for their tropical diseases, and provided with their back pay.

The Press announced Stahl as the "first American soldier from Australia" to establish a radio position on Luzon. They did not know that the earliest men to reach Luzon with a radio were Ignacio (Negros April 1943) who set up a radio in Manila and Rodriguez (Negros July 1943 mission) a radio in Tayabas, Quezon.

The Press were looking only at White American soldiers. Others who reached Luzon earlier were most likely naturalized American citizens at the February 1943 naturalization ceremony at Camp Beale. Harder and Vitorio (Mindoro November 1943 mission) set up a radio in Calatagan in January 44. Robles and Manzano (Mindoro December 1943 mission) reached the Bondoc Peninsula in January 44. Cardenas and Herreria (Mindoro December 1943 mission) set up a radio in Lucena on April 4 and reached Manila in August 44.

Station S3L was ordered to close down three months later. Nery and the men reported to G-2 Intelligence at San Fernando on Luzon. From there, Agcaoili reported that he was sent to Mindoro for crypto work and guard duty.

After the liberation of Manila, G-2 Intelligence of the Eighth Army and Causing, guerrilla commander of Samar moved station MACA from Borongan to Catbalogan closer to Japanese emplacements. It was a big push to eliminate the Japanese in the island. Pascua was directed to get more radios from Leyte. In Calbayog, Causing's guerrilla 93rd Division was attached to the 1st Filipino Infantry Regiment that had arrived in February, in the clearing out of the Japanese in the area. Guyot operated the radio and was glad to see some of the men he trained with in California.

Herreria was already in Manila operating for many months. His radio crackled with news of a purge. Four of his agents, Dr. Enriquez, his wife, son, and a Salud Siojo, were caught with intelligence papers as far as Aurora on central Luzon. They were never heard of again.

With the news of the coming Allied forces, Herreria saw Japanese troops in the City continuing to march and sing day and night in the streets. They must have been trying to keep their spirits up. Agents reported truckloads of Japanese troops moving southward and then north, bringing all the rice from the civilians, the stores, and the supplies they stored in churches, theatres, and government buildings. All means of transportation were prohibited, and entry of foodstuff was forbidden. It did not matter if the people starved. Many who tried to leave the city were stopped and caught in the crossfire. Brutal atrocities heaped on the civilians heightened as desperation set in on the enemy. Nevertheless, Herreria observed the residents trying to stifle their wild elation over the news despite the Japanese's denial through their controlled newspaper, *The Tribune*, expounding that the Americans could never again return to the Philippines because all their ships were sunk, and planes shot down in the South Pacific area.

Herreria was able to steal a Japanese map of the Manila area and immediately sent a courier, Narciso Garcia, to MACA. An American aircraft strafed the officer entrusted with the documents. He was mistaken for the enemy. With the radio set that still worked, Herreria contacted Galang outside of Manila and sent messages through his radio network. He may have also connected with Manzano, who had left Robles on the Bondoc Peninsula earlier to enter Manila. They knew that their messages were getting through from seeing planes destroy enemy installations, enemy planes hidden under bamboo groves, and ammunition and gasoline dumps with such accuracy, sending balls of fire hundreds of feet high. A memo dated September 29, 1944 verified that a radio set was sent to Col. Paciano Tangco, Chief Signal Officer, Philippine Army, but was not in operation.

Knowing that the American forces would be close to Manila within days after their landing, Herreria hiked to Bigaa, in Bulacan north of Manila, passing through rice fields and under cover of bamboo groves. He turned over information on enemy troops' location of land mines, tank traps and barricades in the city area. While on the road, he saw American planes dropping leaflets warning civilians to stay away from the roads, beaches, and near Japanese concentration areas. He observed U.S. planes flying low in a line along highways, roads, and trails leading to Manila gun down Japanese vehicles, positions, ammunition, and oil dumps.

Returning to Manila, he passed the town of Obando in Bulacan, where he spied on a garrison of 20,000 Japanese dressed in civilian women's clothes to escape and not be strafed. "They knew about our policy of not hurting the civilians unless under extreme necessity ... another evidence of Japanese indigenous cruelties and acts of barbarism," Herreria wrote. In Manila, he wrote "meeting a number of GI Joes like himself and had them camped for a while in his hiding place until they left to eliminate the Japanese at Obando."

During the battle to retake Manila, Herreria reported to G-2 Intelligence, Sixth Army on the concentration of enemy troops in the southern part of the city. "Since then, artillery operations against the enemy began. Before our forces crossed the Pasig River to the southern part of the city, the Japanese ordered the people to draw rations in the cathedral in Intramuros. When the cathedral was packed, the whole building was set on fire. Other Filipinos who escaped this holocaust were barricaded at strategic points where the Japanese were shooting (from) behind them at the advancing American troops because they knew that the Yanks would not hurt the civilians. Machine guns mowed down those who attempted to escape."

Arriving from Cebu, Sabado and Sanchez moved to another net control station when the Sixth Army landed. In March 1945, their mission was over, and they reported to the Eighth Army returning all their arms and equipment. A landing barge took them to the Army headquarters on Leyte where they were greeted and congratulated by a G-2 Intelligence officer. After a week, they were ordered to report to their unit, 978th Signal Company in Luzon.

Robles, Pascua, Guyot and Manzano reported to the Eighth Army in Tolosa, Leyte. Mauricio in Palawan was also recalled. By April, they left for the 978th headquarters in Hollandia. Herreria was called to the Sixth Army at Calasiao, Pangasinan and flew back to headquarters with Balino, burning from a slap by a Japanese sentry as a priceless souvenir because he had forgotten to bow. They felt proud when they were congratulated for their work in the Philippines. Cardenas had survived the mission and proudly accomplished his assignments only to be killed in a truck accident in Pampanga on Luzon. He was returning from a mission with the Alamo Scouts. His body was embalmed and sent to his parents' home in San Juan, La Union on Luzon for burial.

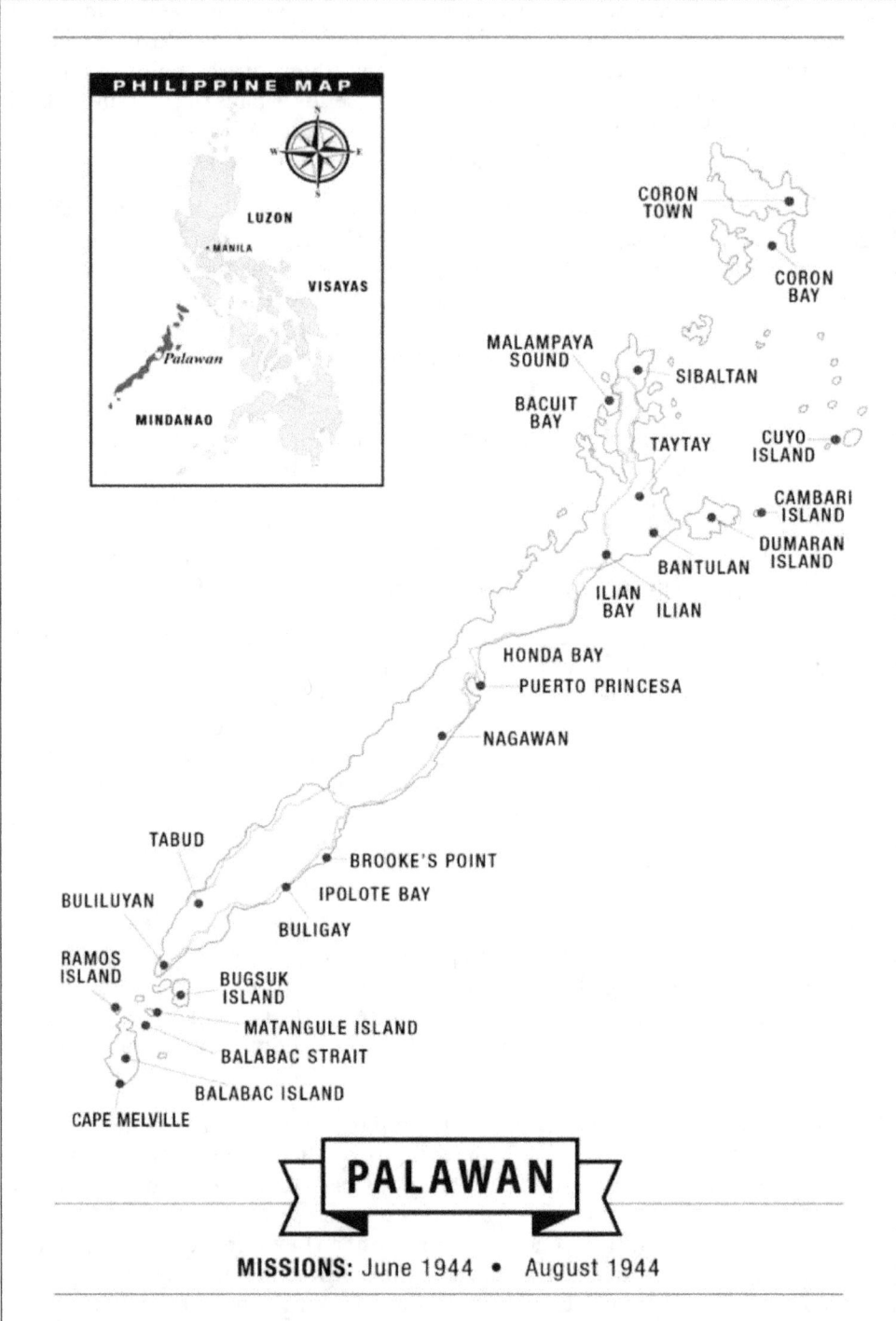

PHILIPPINE MAP

LUZON

• MANILA

VISAYAS

Palawan

MINDANAO

CORON TOWN

CORON BAY

MALAMPAYA SOUND

SIBALTAN

BACUIT BAY

TAYTAY

CUYO ISLAND

CAMBARI ISLAND

DUMARAN ISLAND

BANTULAN

ILIAN BAY   ILIAN

HONDA BAY

PUERTO PRINCESA

NAGAWAN

TABUD

BROOKE'S POINT

BULILUYAN

IPOLOTE BAY

BULIGAY

RAMOS ISLAND

BUGSUK ISLAND

MATANGULE ISLAND

BALABAC STRAIT

BALABAC ISLAND

CAPE MELVILLE

PALAWAN

MISSIONS: June 1944 • August 1944

Palawan Island

The waters around Palawan were crowded with Japanese and U.S. vessels prowling for each other. The Japanese and GHQ in Australia knew that the island's strategic location provided coverage of the South China Sea to the west and Sulu Sea in the east fronting the Visayan islands, the center of the archipelago. The Japanese created task forces and established a large naval base that protected essential waterways around Palawan. On the other hand, GHQ saw Palawan as a future aircraft base for sorties farther west that would cut off Japanese sea lanes in the South China Sea to support the planned reconquest of Borneo and the East Indies.

The primary objectives for troops in Palawan were to block the central Philippine islands of Panay, Negros, Cebu and Bohol from the enemy and expand the range of Allied air operations.

For six months, PRS had been planning a mission to land men and radio in order to establish a communication network and train secret agents to send out and obtain information on the enemy for transmission to Australia. But there was difficulty getting a submarine. The *Narwhal* and *Nautilus* were on continuous missions to other islands. The Navy had extended their dedicated service without yard overhaul. risking the crews' and mission parties' lives.

Planning for more future missions, a memo to the Chief of Staff dated July 15, 1944 and marked 'Top Secret' requested four additional submarines to be made available by the Navy department for Philippine service. Only two were allowed.

USS *Redfin* – June 1944

G-2 intelligence granted the request to use the USS *Redfin*, a war patrol submarine. Members of the small party originally came from different units of the Army - Sgt. Amado Corpus, Sgt. Carlos S. Placido, Sgt. Ramon F. Cortez, and T4 Richie D. Dacquel were from the 1st Filipino Infantry Regiment. Sgt. Jaime R. Reynoso was from the 2nd Filipino

Battalion, and T4 Teodoro J. Rallojay was from the 63rd Training Battalion at Camp Robinson in Arkansas. Except for Sgt. Corpus, the men were 978th Signalmen.

Although informed that the mission was exceedingly dangerous and knowing they would have to be on constant alert for their lives, they were dauntless and still volunteered. They were packed and ready waiting at Camp Tabragalba for a truck to take them to Brisbane. Maj. Croell and Lt. Bones, officers of the 978th Signal Service Company with Sgt. Villarta, and Sgt. Tanato saw them off. Each carried only one barrack bag. They were not told where they were going as it was 'Top Secret'.

Arriving in Brisbane at dark, "and since we knew we were going to leave soon and maybe forever," they went for a nice dinner of steak, eggs, pies, and coffee. Overnight was at an Australian army camp. They were not allowed out because they had to get ready for breakfast at two in the morning to be trucked to the airport. Whitney said, "God bless you, boys, and I will see you in Manila." Everyone shook hands, including the truck driver, before boarding the plane for a two-day trip to Adelaide and Fremantle.

A Navy truck waited for them. They were under strict instructions to tell no one that they were Army until they were inside the submarine. Maj. Brown, the commanding officer at Tabragalba, saw them board the submarine at Fremantle. They loaded radio sets, personal gear, medicine, and rations enough for a year into the boat first. They had operational money. After changing into Navy uniforms as a disguise they boarded the USS *Harder* and had supper onboard. All the pork chops they could eat. A change order arrived to transfer to the USS *Redfin*. Close to midnight, the submarine departed Fremantle with orders to stop at Exmouth Bay to practice rowing in rubber boats and making night landings on a beach under an Australian officer. Was this a refresher to the amphibious training at Fraser Island? Water was also picked up. Corpus had discovered that the gasoline to start the radio generator was missing so that was obtained at Exmouth Bay, stored in the ammo locker and aired daily because the fumes were hazardous. The captain was concerned that the fumes might ignite if depth charges come close. All were concerned and hoped for the best.

The sub stayed submerged in the daytime, running the engines that quickly heated up the inside. Temperatures soared as around one hundred men sweated while working and breathing. When it surfaced at dark in the choppy waters, fresh air rushed into the opened hatch lifting the heavy foul air.

In the moonlit night, the outlines of the islands were clear. Their destination, Ramos Island was the biggest of the islands off the south coast of Palawan Island. Corpus' orders were to "establish control in the south with close observation over the Balabac Strait and north channel, report all enemy naval movements, and to extend his net north into Palawan as far as Brooke's Pt."

They could see another island that had two big lights, and they guessed these were Japanese outposts. The sub commander and Corpus picked out a landing place through the periscope. It was a risky landing without security signals and a backup site. It was unknown whether there was an enemy presence on the shore. The landing was reminiscent of the first mission party to Negros. There was no coordination with a guerrilla party for security signals.

Their rubber boats were visible up to three hundred yards in the moonlight for any enemy patrol. Some of the sub's crew came with the men to help unload the cargo and pile them on the beach. Rallojay, Corpus, Placido and Cortez were in the first rubber boat loaded with a one-year equipment supply in cans wrapped in gunny sacks. Reynoso and Dacquel were in the second boat. The current was swift and the boat filled with water. Some of the boxes drifted away. Placido lost his Carbine and Cortez his bag. The chosen spot turned out to be barricaded by an eight to ten-foot coral bank that had to be crossed before low tide. "What we learned in training on being silent during landing we never followed this time." It took over two hours of shuttling back and forth from the sub. Before the crew left them on the beach, Miller, a gunner, gave Placido his 45 cal pistol. He may have felt that the men might not come back.

The *Redfin* disappeared into the Sulu Sea. She scouted the Japanese naval base at Tawi-Tawi and sank the Japanese tanker *Asanagi Maru* despite it having an escort ship. She patrolled the vicinity of Tawi-Tawi and warned the American forces in the Marianas of sighting the Japanese task force going into the Battle of the Philippine Sea a week later. Before returning to Fremantle, she sunk the Japanese passenger-cargo ship *Aso Maru*.

Watching the submarine fade into the horizon was difficult. There was no friendly guerrilla group to welcome and help with the cargo scattered on the beach. Some of the boxes still had to be pulled over the bank to shore. They tied the boxes to a rope, pulled them over the bank, and dragged 1 ton of supplies into the jungle before morning. It was raining and they slept for a couple of hours to get their strength back. The constant noise of the jungle creatures did not bother them as they fell asleep exhausted. If the jungle noises abruptly quieted, that meant a coming foot patrol. After their first breakfast on Philippine soil, they oriented themselves to locate what part of the island they were on. Sailboats passed not noticing them. It was rainy season and tents were set up. They had to get those boxes into the jungle or buried.

Rallojay began a diary in a small ringed notebook. GHQ had warned them not to write and keep journals to avoid compromising the mission. Being caught would mean death. In another small notebook that fitted in his palm, Rallojay wrote notes on demolition, fuses, and types of booby traps. He took a big risk in doing this, but the details of events that befell the group might never have been known without his notes. He wrote that they had "J rations," ready-to-eat dry foods of beef, peaches, apricots, and whole milk. A jungle ration was compact and fed four men in one day. In each 10-ration waterproof box, a can opener, matches, toilet paper, and cigarettes were packed. Water purification tablets were included.

Ramos Island was in a most remote place to watch Japanese ships in the Balabac Strait, a prominent shipping lane of Japanese vessels. Looking out at the vast waters of the Sulu Sea separating Palawan from the rest of the Philippines, the men could see the faint outline of Mindanao.

Orders were for the reports to be transmitted to FRS station on Mindanao, for relay to the 7th Fleet. Radio and equipment had to always be ready for packing, moving quickly, and quietly without a trace. Aware of indiscriminate Japanese atrocities, they were to use their judgement and means to accomplish the mission.

The rainy season had started, and the men welcomed it, filling their canteens and a ten-gallon milk can with good drinking water. Each man was rationed one canteen per day. Burying the boxes, including their clothing, in scattered places was slow. A big tree provided a guide to their location. Two men's items were in one box, so at least one would remember the site. Rallojay shared a box with Dacquel.

They avoided the locals as they planned to reach the highest point of Ramos Island to set up a station. If they did not speak the language of the island, they would be exposed as foreigners. But they had to trust some of the locals to find out that 180 Japanese Marines occupied nearby Balabac Island. The people and *Moro* (Muslim) inhabitants in the Balabac area and Ramos Island were pro-Japanese and formed a *bolo* battalion to help the enemy. Despite this, the *Moros* informed the group that the Japanese occupied the island.

Two weeks after they arrived, Corpus left to look for more locations to bury boxes. The men began to worry when he did not return in the agreed four hours. He finally showed up with a *Moro* with him revealing that there were three hundred *Moros* on Ramos Islands. They felt uneasy with his presence. The *Moro* was an escapee from a Japanese garrison in Puerto Princesa, the capital city of Palawan Island in the north. He revealed the location and the crippling physical condition of three hundred American prisoners forced to build airfields.

They were unable to meet the head of the *Moros* and the owner of the island, who was sick. The mayor of the town of Balabac sent word that it was dangerous to set up a station in Ramos. He said not to trust any of the non-Christians, for they might unknowingly reveal having seen them. With this warning, the *Moro* was told to go home because their rations were getting low. It became a greater risk when more people from Balabac came wanting food. They would not leave unless given clothing, medicines, cigarettes, needles, and thread. They departed with blankets, a whole roll of khaki cloth, and salt-water soap.

The realization of the riskiness of their mission hit the men. "We became religious all of a sudden. On Sundays, before we did anything else, we read a chapter or two of the bible. Butch (Rallojay) passed his bible around, and each one would read."

On June 21, thirty Japanese cargo ships escorted by a heavy cruiser and several destroyers passed Balabac Strait and crossed the Sulu Sea towards mainland Philippines. They could not send the sighting because the radios were still packed in the cans. They were anxious to establish a station.

Placido and Dacquel scouted the northeast side of Matangule Island for a new camp. All the buried supplies were dug up. In the night, the men loaded all their equipment in several large *Moro* boats and sailed across the deep Balabac Strait to Matangule Island, uninhabited, with a coral reef encircling the western tip. From there, shipping lanes and the Japanese encampment at the foot of the Balabac mountains could be seen. It rained all night and the following day. That was when they discovered that the cans containing money were soaked the night of unloading from the sub. The cans were not sealed well inside the gunny sack wrapper. Termites had entered into one of the cans. Placido was concerned that "only 1746 pesos out of how much we didn't know" was salvaged. They were paying the locals for their labor, but now cash was low. "This was really bad because the natives knowing we came from America believed we had plenty of money. The payment for their services was exorbitant." After the war, Placido felt that his party was "more fortunate because the boys in the other groups stationed on other islands were paying three, even four, times as much as we did, not only for services of the men they hired but also for food."

Three locals were hired to guide them into the jungles to set up camp. Rallojay wrote in his small spiral notebook of his uneasiness when the money box was opened in plain sight of the local help. After a trail was hacked, more locals were hired to help carry sixty-pound loads into the jungle. Without help, it would have taken them a week to haul all the supplies themselves inland.

The remoteness of the outpost affected Corpus. He had become depressed with the thought that he was unable to operate or fix a radio. He was an infantry soldier while the others were Signal men trained in radio operations. He had grown more discouraged and did not believe he was capable of doing his job although he was an older soldier with many years of service. He expressed that he "is just an excess baggage." Whitney later commented that his party was the "most trained of any who left here and yet never received a direct report from Corpus." Corpus felt that the men had lost faith in him and told Placido to take over, but he refused. Rallojay thought their situation was not good if the leader thought that way. They began to argue on how to transfer the equipment. Rallojay wrote his concerns as he laid in his hammock at night. "Everyone had their opinions, and no one agreed... argued what work they should be doing. They were no longer a close team." They decided that Placido should take over.

The important radio sets were carried on their backs to a small clearing in the jungle and local boys brought the battery and charger. Unpacked, they found the equipment all wet. Rust had settled on the parts and charging devices. A radio bulb had to be replaced. A motor charger had to be dried to start. It took a week to dry the parts before being repaired, hoping that they would still function. They could hear Morse code coming in but could not reply. With the radio not operatonal, Rallojay worried that it had been over a month and Parsons coordinating the submarine missions and PRS must be worried with the lack of contact. Food was scarce because although they had money, there was nothing to buy. Two civilians were hired to look for food. Coconuts and papayas were the only food readily available around their camp.

Desperate, Corpus with a local as a guide crossed the seas to Brooke's Pt on Palawan Island to contact the guerrilla commander of the sector for help. Placido offered to go with Corpus but Rallojay said he would go because Placido was most important in the group. The guerrilla leader was known to prefer not wasting precious bullets so no prisoners were kept alive. A *bolo* battalion headed by a *datu* collected supplies and weapons from the civilians. Sometimes force was used, but the people had no one else to turn to for protection against the Japanese.

Corpus' orders were to contact the guerrillas and an American in Palawan, but it was risky since "a few of them (Palawan locals) were patronizing with the Japanese." The rural residents of Christians and Muslims avoided each other. He brought medicines, cigarettes, and magazines for the people and Capt. Nazario Mayor, the guerrilla leader. Although it was only 10 miles from Matangule Island to the southern tip of Palawan, the trip was still dangerous due to a jungle and a maze of coral reef separating the two. Then it was a hazardous 65-mile journey along the eastern coast to the guerrilla headquarters at Brooke's Pt. Crocodiles, large lizards, snakes, and malaria-carrying mosquitos were everywhere.

With the camp in the care of Placido, the men continued to work on the radio set and rust-eaten battery charger as the rain poured through the jungle foliage for the next two days. The humidity did not help at all. With the set working, a weak signal was barely transmitted telling KAZ of their location and condition. KAZ's garbled reply could not be read before the radio went dead again. Not giving up, they managed to charge a battery. More boxes were hauled from the shore to their camp inland. The hauling delayed their communication with KAZ at the assigned time of three in the afternoon.

Three men surprised them as they worked on the radio and battery. Sgt. dela Cruz of the guerrillas from the southern Palawan area, chief of police of Balabac island Vicente Aizo, and an escapee from Bataan, George Marquez of the 48th Material Squadron appeared. They had heard of the presence of Australian soldiers. Concerned for their safety, they strongly suggested that the radio move to Palawan Island. The battery charger worked and they were able to listen to *The Philippine Hour* from Australia causing excitement to the visitors. It was their first radio broadcast since the fall of the Philippines.

Reynoso and Cortez hurriedly buried the last of the boxes that were not needed near the beach where the soil was soft and sandy. All along they were on the alert listening to the different engine sounds of Japanese and American planes. They stopped for lunch of biscuits, jam, and coconuts. A civilian, Cirilo Sunson brought cooked chickens, rice, and eggplants for the best dinner they had in weeks. They slept by the beach to be able to see the Strait.

Corpus returned from Brooke's Pt with the guerrilla commander. He was relieved to see his men still alive and the camp intact. The Japanese regularly patrolled Matangule Island and would not have taken long to find the camp. The family compound of the guerrilla leader on nearby Bugsuk Island was raided and houses burned.

Dacquel and Rallojay covered up any signs of their camp in preparation for their departure. The deputy governor of southern Palawan arrived with two large *bancas* (wooden boats) and some locals to help with the move. Placido at first wanted to stay, believing that he could get a better signal and continue contact with KAZ, but was overruled. An outboard motor towed one of the boats as they maneuvered around coral reefs that only a local would know, heading north.

It took two days for the group to arrive at the village of Buligay at Brooke's Pt. They met Americans Charlie Watkins, a Navy man and 'Red' Wakefield. The guerrilla base camp was a couple of miles south of the town of Brooke's Pt. The area was deserted after over a dozen Japanese bombings. There were small craters everywhere. The guerrilla commanders' wife fed them in her home in a coconut grove inland from the beach. The men were quartered in the houses of the guerrilla commander and Thomas Edwards, the American they were to contact. Edwards was a 'Thomasite' (One of 500 teachers who arrived on the SS *Thomas*) who came to the Philippines under the U.S. government to teach after the Spanish-American war and became a plantation owner. His place was like a small village in the mountains busy with cooking, men coming and going, and children playing.

A radio shack was built near the shore. The receiver did not work. Parts were wet and had to be dried again. They took turns sending a message at the chosen hour but there was no reply. A spare radio, ration, and extra equipment were hauled to nearby Macagua village. There Rallojay tried to work the radio, and it connected again with the *Philippine Hour* from Brisbane. A pleasant feeling of hope swept through the listeners.

After ten days of trying, the first exchange of signals and readability of messages with KAZ excited everyone. The messages were read in between Japanese jamming signals. Placido was on the air when Buck Ergas (Eugene Ergas), a Filipino signalman at the other end asked for Butch. "Operator at KAZ keyed in my initial and in plain text said, 'Hi Butch' (Rallojay). We could not believe it so asked to repeat and he sent my initials again and Hello Butch right then. I replied 'Ok Buck' which is what I called him. We all felt better after this exchange." Short messages on sightings and enemy disposition were sent and received with good reception with Australia. It had been almost two months since the sub landed the men. KAZ confirmed later that this was the first time they received a report.

USS *Robalo*

In the first two years of the war, GHQ knew that the enemy mined south of Palawan. Almost a dozen submarine patrols had made successful trips through the passage. Storms and currents can move mines, so in December of 1943, the Japanese refurbished the mine fields. Allied Intelligence did not know of the new mine fields near Balabac Strait. The path assigned to USS *Robalo* for her patrol area had her travel through Balabac Strait, a narrow passage between Ramos Island and Matangule Island.

The men had just arrived at Brooke's Pt eager to find a suitable location for radio transmissions and set up their radio when the submarine hit a Japanese mine. It sank in less than two minutes on the east coast of Balabac Island on July 26. The men did not hear of the sinking for weeks. Before she sank, her last message was about sighting a Japanese battleship with escorts in the South China Sea. She was not considered lost for days as submarines did not transmit when close to enemy installations to avoid enemy detection. An unknown number of the crew struggled in the water. Four swam ashore, and were captured, and held at the Puerto Princesa Prison Camp. A former clerk of Balabac's Treasurer under the Japanese saw the questioning of the captured survivors. Guerrillas were able to bring two survivors to the camp.

The new location at Brooke's Pt allowed the group to cover the entire southern Palawan and the waterways of the Balabac Strait. The main station TOP had direct contact with the net control stations of the other islands' military districts and the Navy's powerful transmitter in Mindanao.

The locals of the villages of Buligay and Tobtob were always glad to see Reynoso, Cortez, and Rallojay as they felt safe seeing them. Rallojay wrote of visiting the village of Tobtob under the rain to visit Judge Rodriguez. He had met the judge's wife and daughter, Connie, whom he found "very interesting and charming."

A Japanese gunboat anchored in the Bay and tried to land a patrol party, but the guerrillas held them off with rifles and shotguns provided by the men. Quickly all equipment was moved inland to a hill, missing the three o'clock afternoon usual call from KAZ. A reply was not sent because the antenna was still at the shore camp in the village of Buligay. It was placed so high that it would take some time to remove it. Placido stayed behind with a portable transceiver. In the temporary camp, the men kept busy trying to fix the spare radios. The only good receiver and transmitter left were rigged together to work.

There was nothing to report to KAZ for the next four days. They planted onions, cabbage, and mustard with the seeds they brought for food when there was nothing to report. But without enough water and the insects feasting on the stems, there was nothing to harvest. For food, they started to buy root plants and what little rice could be found at exorbitantly high prices. When the people in the nearby villages arrived starving, some of the buried boxes were opened and they were relieved to see them untouched.

Riding the pedal generators to run the radio kept them exhausted. Guerrillas took over. They were able to attend a Mass in the nearby village, uplifting them. Rallojay learned a lesson when he fired a 45 cal pistol while holding it in sighting like a rifle. Next time, he held it farther out.

In mid-August, four Japanese ships were sighted passing Brooke's Pt. Two days later, an unidentified ship was sighted far out in the Sulu Sea. A long message from KAZ was coming in when the receiver died. The oscillator tube was immediately replaced. The weather radio station did not have an anemometer (which measures wind speed, wind chill, and temperature), but reports were still sent four times per day. When Dacquel completed setting up all the weather equipment, weather reports were sent every six hours, which earned Col. Whitney's commendation of "excellent reports."

Placido left with a radio for Cape Buliluyan to watch shipping in the north Balabac Strait and the South China Sea. He set up station UPG and began sending reports to Brooke's Pt twice a day or FRS in Mindanao if the signal was not strong enough. Contact was weak, with the signal fading out until completely gone. Rallojay's notes confirmed Placido's return from the south on August 13 and brought a horse, a couple of cows, and five sacks of rice donated by the civilians for the 'Australians.' It was unfortunate that Placido was on his way back to camp when the submarine USS *Flier* sank in the area of Balabac Strait.

USS *Flier*

U.S. patrol submarines continued to stalk the passageways around Palawan for Japanese transports. The *Flier's* order was to pursue a convoy of Japanese ships heading towards the western side of Palawan. The sub was headed into the path of the approaching convoy, ready to go to battle. Radar showed proper bearings, a lighthouse ahead, reefs, and small islands as the sub headed into the Strait. The Nasubata channel, the deepest water route, was taken to avoid mines. She took the same last part of the route that the *Robalo* had been ordered to take earlier. In minutes, the *Flier* was gone. Operatives by the coast sighted the ship sailing the Balabac Strait by the Lombucan Channel south of Comiran Island and saw the explosion. The sighting did not reach the camp so their reports to KAZ did not include the *Flier's* sinking.

Eight survivors made it to shore, cared for by the guerrillas who had hardly enough food for themselves. Onboard a boat owned by a *Moro,* the survivors were brought across several islands until the guerrillas guided them to Edwards' house. They suffered from burning exposure and running sores. Any doubts and pain from their wounds dissipated when they saw the men in uniform and the radio contact with Australia. Mentioned were "Capt. Crowley, Lt. James Lidel, and Jacobson, G. Howell, D. Tremaine, E Baumgartner, Miller, and J. Russo." Corpus thought of not having enough food to feed more men. But Edwards, a planter who had access to large tracts of land was not as worried.

KAZ was immediately informed of the *Flier* having "struck a mine off Balabac and sank in thirty seconds. Eight known survivors now at Brooke's Pt. *Robalo* same fate July 3. Two Americans (from the *Robalo*) probably only known survivors and reported prisoners. Need the earliest evacuation of *Flier* survivors. Need arms, ammo, meds, supplies badly to work further." The survivors' names and addresses were noted in Rallojay's notebook. The submarine captain survivor immediately followed up with a message that the *Flier* was lost and likely hit a mine in the Balabac Strait. Although the Strait was known to be mined, it was heavily used in the narrow middle paths. He emphasized avoiding the Strait at all costs.

A scolding from the Navy arrived that the men did not properly do their job of watching the Straits for mines. The men deeply felt the admonishment. Placido said "they could not do much at the time about sending men (to reconnoiter) because Japanese planes and gunboats boarded and bombed anything floating in those waters. The survivors themselves admitted they knew the place might have been mined, but they had chanced it, nonetheless." The continuous cleaning of their rusted radio equipment from the debilitating rains and humidity also limited contacting Australia.

KAZ replied, "Advise immediately disposition of your party and radio units with exact locations." KAZ wanted the exact status of the party and equipment for non- receipt of intelligence which was a requirement of the mission. Corpus was asked "Why have not sent information on enemy dispositions in southern Palawan and Balabac Strait? Why did not advise the presence of mines in Strait. Disappointing. Immediate improvement in your intelligence coverage is desired and expected." KAZ became cautious and uncertain of what happened to the team. Placido was instructed to carry on as party leader in a message by private code which he memorized before departure. He was asked for an inventory of significant items and exact locations.

Receiving the scathing reprimand was a great embarrassment to Corpus. He sent one more report on six Japanese loaded trawlers going slowly northeast and unescorted. That was his last message.

The guerrilla Women's Auxiliary Service (WAS) gave a welcome party for the men and survivors. After several days of rest, the "survivors were well on their feet except for two who were not accustomed to our (Filipino) food." On the way to the celebration, Placido and Dacquel waited for the guerrilla commander and his wife. Suddenly a muffled cracking shot was heard, and the guerrilla commander's son yelled that Corpus was lying on the ground. He had gone behind their house and shot himself through the heart with his 45 cal pistol. Placido reached him first and tried to make him talk. He felt no pulse. The celebration did not go through. Instead, a burial was held for him in a small wooden coffin that day. Rallojay sadly wrote, "At 7.45 in the morning on August 27, Corpus took the easiest way out. Shot himself to death. He got discourage(d). His moral(e) was very low. May God have pity on his soul."

In his report, Placido said that Corpus was reprimanded for not "reporting the presence of mines in the Balabac Strait, forwarding information on enemy dispositions in south Palawan and Balabac and naval movements through Balabac Strait. Master Sgt. Corpus died at Brooke's Pt from a self-inflicted pistol wound just above the heart. Unknown on cause for committing such an act, no note left behind. "

After advising KAZ of Corpus death, Placido messaged that he "will take charge of the party until further orders from you." General MacArthur, replied, "regret over Corpus' death. Placido to carry on as party leader. Send inventory of equipment and cash as responsibility, and exact location." Whitney sent his regrets. Later, KAZ replied with an appreciation for the "written report and manner you and men took over after Corpus' death."

A Protestant missionary and his family were at the camp with other evacuees and the *Flier's* survivors, waiting for the submarine's arrival. Alexander M. Sutherland, although weak from malaria attacks, heard Corpus tell his men to take care of their share of the money because he felt he was going crazy. He was anxious to go out on reconnaissance to get more information for transmission, but the problems with his feet did not allow him to move around much. Jungle boots rotted from the humidity and muddy trails, making the men walk barefoot. He observed that Corpus had difficulty leading the men with their questions on his plans to get anything done. At times, the municipal mayor did not cooperate. The guerrilla commander of Southern Palawan actively gathered intelligence reports but was not in a position to order Corpus' men. There was a shortage of men to handle all the work and equipment. The missionary noticed unauthorized individuals were allowed in the radio room during transmission, revealing its location and mission. The arrival of the next submarine was known to many. The missionary said that security became lax when he received a "typewritten copy of the news from the radio room for the public. On the reverse side of the paper was a lengthy message in code that was not erased." He learned of Corpus' death as he was leaving to meet the sub to be evacuated. He believed that Corpus' "death might have been caused by blaming him for the *Flier* disaster." A GHQ August personnel report on men missing in action listed Corpus as having committed suicide.

KAZ became cautious of Japanese patrols after the depressing news of the *Flier's* sinking. A plan was developed with maps of a safe rendezvous point for a submarine. Only the survivors and the coastwatchers knew the details. An urgent list of radio parts was sent to bring with the submarine. The radio sets they had for communicating at close range were unreliable. Boxes were improperly marked revealing a transmitter instead of a receiver as marked. With only one receiver, the two extra transmitters were not used. The sub-station badly needed another receiver, and also asked KAZ to send the rest of the quota of khaki cloth to barter for food.

The *Flier's* radio operator, who survived, worked on the short-wave radios to connect with KAZ on the evacuation. Three *Flier* men took turns pedalling a stationary bicycle to produce electricity for the radio. Once the plan was complete, Rallojay messaged KAZ to rendezvous five miles east of Brooke's Pt lighthouse on August 30 or any day after to pick up the survivors and evacuees. Placido reminded of his request to bring the "maximum possible supplies list sent." The *Flier's* skipper requested a gallon of lubricating oil for the sailboat of the coastwatchers to return to their camp.

The USS *Redfin* prowling the area was dispatched to pick up the evacuees. A *datu* provided two boats to bring the survivors to the pickup point. Signal lights using Coleman lamps they could borrow with empty packing tins for reflectors "were (to be) hung one above the other from the lighthouse on the northern corner of Ipolote Bay, near the beach at Brooke's Pt. It was the signal that the coast was clear of Japanese for twenty miles on either side, and survivors were ready."

An hour before 8:00 pm, the rendezvous time, the *Redfin* skipper received the message of two small boats off Brooke's Pt carrying the eight survivors of the *Flier,* evacuees, and some coastwatchers. But their reply was not received as one of the radio's resistor burned out.

A Japanese cargo vessel anchored near the spot on the night of the rendezvous. The signal lights were not turned on for fear of discovery. The evacuees sailed out to meet the sub but there was no sign of them. Discouraged, the group was on their way back when the skipper decided to look for them. Out in the open sea, Placido pounded the keys of the functioning radio while others were signaling with a flashlight. It was a moonlit night, and the sea was calm, but the current was drifting their boat away from the designated spot. Everyone got so excited when the sub heard their call that the antenna pole on the mast broke down. The men scrambled to hold it up, almost tipping the boat. The waters churned as the submarine surfaced an hour after midnight.

The submarine commander was glad to see Placido. He had dropped off the party three months earlier on Ramos Island. He must have been shocked at the skeletal look of Placido and his men. In turn, they were glad to see the crew they had gotten to know. When the men asked for a few donations, the crew gave them what they did not need for the seven days in their return to Darwin. The men were so glad to receive automatic rifles, machine guns, rifles, pistols, thousands of rounds of ammo, office supplies needed to code and decode, medicine to fight diseases, flour, yeast, coffee, canned fruits and vegetables, two hundred cartons of cigarettes, diesel oil, radio parts, toilet paper, soap, and even playing cards. The two small boats dipped lower into the water with all the load. The crew explained that it was not much because they were on their way back to their base from a patrol mission when they received the secret orders to pick up evacuees. Some of the crew gave their clothes. The signalmen left feeling that Christmas had arrived early. They sadly watched the sub turn south to scout the Japanese naval base in Tawi-Tawi.

Japanese activities increased in Palawan as the target date for the Leyte landing neared. Unknown to the men, the War Ministry of Tokyo had issued a 'Kill-All Order' directive to the commandants of the various POW camps. Palawan had a POW camp at Puerto Princesa.

KAZ wanted the "number of motorboats in a given area, the numbers painted on the crafts, dimensions, armament, mounted torpedoes on patrols, depth charges, radio, and escorts of enemy ship convoys." Agents operating in Puerto Princesa, the capital city, provided the intelligence on Japanese aircraft, airfields, defenses, and military objectives.

The submarines USS *Seawolf* and USS *Stingray* were added as dozens of men and tons of cargo were leaving for the Philippines almost every week.

USS *Seawolf* – August 1944

The submarine surfaced at Pirata Head on Cambari Island off the northeast cost of Palawan. The rendezvous point was far from Brooke's Pt and security signals of three fixed white vertical lights were turned on at Brooke's Point lighthouse as a guide to Palawan Island. Placido sent a courier to bring reports and a sketch of the Japanese installation at Puerto Princesa to the sub's skipper for KAZ. A force of fifteen hundred Japanese with assorted planes, anti-aircraft guns, cannons, tanks, heavy weapons, and ammunition had unloaded at the Puerto Princesa wharf. Half remained in the city while the rest went to various points of the island. An additional eight thousand were expected for various parts of the island, including Brooke's Pt.

In the night, the sub slowly surfaced and closed in silently on the beach on battery power. A small boat identified by flashing code name met the sub and the party headed by Sgt. Eustaquio B. Cabais, a veteran of WWI in France and Belgium. With him were five Signal men: Sgt. Jacinto Cutaran, Sgt. Vincent C. Goluyugo, Sgt. Fortunato C. Dagandan, Sgt. Leo Marquina, and Sgt. Thomas C. Vergara. Their mission was to set up a radio station that would communicate with FRS and the Navy transmitter in Mindanao sending intelligence on enemy activities on Palawan up to Cuyo islands in the middle of the Sulu Sea.

The men's rubber boats and the sub's two-man rubber boats were inflated. By midnight, the last five tons of cargo were safely ashore. With her two rubber boats stowed away, the sub headed out to sea to continue her patrol.

The landing was undetected by the enemy. A good site for an observation post was found but was out in the open so the hunt for a better site continued. The constant rain added to the difficulty of moving around and hiding supplies. Three weeks after landing, the first message was sent. A request for a receiver was sent to replace a damaged one. KAZ replied with a commendation "for initiative and rapid installation of radio station."

The guerrilla commander of the area was organized and cooperated. The locals helped but were afraid of some guerrillas who continued to intimidate and abuse them. The group tried to stop the harassment wherever they could. Sickness was rampant amongst the people. They informed KAZ that the delay in transmission was mainly due to Japanese landings on the island, causing the antenna to be taken down and hiding the equipment until the Japanese finished ransacking the locals' homes. Already starving and in tattered clothes, the locals were in added danger with the party's presence.

In a short time, Cabais requested arms, ammunition, grenade launchers, clothing, and food for the guerrillas and civilians. His message ended with "hoping for shipment of relief and supplies." The reply emphasized that the "present job is the development of communication and intelligence net in your area of responsibility... move vigorously towards accomplishing this mission, so flow of reliable information on the enemy in that area best serve the primary interest of the people. Their material need for food and clothing must wait (for) more favorable opportunities for supply within military operations."

To obtain respect and support for his men, Cabais explained the higher ranks he gave to his men. It was to "combat indolent inclinations of the guerrillas and natives. Rank meant nothing to them but produced results in maintaining the cooperation of the people." Whitney commented that enlisted men fall into the guerrilla practice of self-bestowed ranks but was alright if it helps them do their job.

Peralta, guerrilla leader of Panay, convinced Cabais that their present location was too exposed to the enemy. With his help, a move to the mainland of Palawan began with one-third of foodstuff rowed to the town of Araceli on Dumaran Island before transfer to Palawan. Cabais, with two of his men arrived at Capayas, a town in Palawan. Instructing his men to contact KAZ, he left to hunt for a possible control station location. After nine hours of hacking through the jungle, he found a hill that suited his requirements. The guerrillas and civilians supplied the labor and local materials in building the camp. For two days, locals were paid one peso per day to bring all the cargo and equipment to the new location. The locals were so happy to see them that they held a party with dancing to celebrate. Much tuba was served.

An enemy convoy of five small cargo vessels was sighted but was not flashed to KAZ as the antenna was still under construction. The roof of the new station was unfinished from lack of nipa processed into strips. A motor launch to get the rest of the men at Dumaran was out of order, so runners were sent out to hire a sailboat. It took two weeks for the rest of the men to arrive at the town of Ilian about four kilometers from the hillside camp. A week later, contact with KAZ was made and a reply complimenting the tactical move arrived.

After the security of the net control station was established and supplies stored, civilians were allowed to come and receive medical treatment. Nearly every family had a case of malaria and tropical ulcers. A guerrilla officer was convinced to allow his wife, a nurse, to minister to the sick. After four weeks, most of the locals were cured of tropical ulcers, and more men could be hired. The people showed their gratefulness with whatever they could with native foodstuff.

Maj. Pablo Muyco, the area's guerrilla commander greatly assisted in the first three months. Cabais commented that Muyco's battalion was well-organized and surprisingly harassed the enemy with whatever equipment, supplies, and arms they had. Medical supplies, clothing and ammunition were turned over to them. At Muyco's headquarters, Cabais met the provincial governor for help to construct another station as an alternate in case of enemy action. A suitable place up the Ilian River around four miles from the beach was found. Hired civilians and guerrillas set up the shelter and transferred the reserve supplies inland. After two months, the relationship began to strain due to petty thieving by a few of the guerrilla officers and men. Cabais confronted Muyco to stop the intimidation and abuses of the civilians by some of his men. The commanding officer was "a fine soldier, a gentleman but too young for his command and had no control on the discipline of his scattered units."

From Brooke's Pt, Dacquel arrived at Cabais' camp to establish contact and ask for a spare battery charger. A woman offered her service as an agent inside Puerto Princesa, the capital city. The Japanese concentrated in the city and had over five thousand troops at one time commanded by a general. Convinced of her loyalty, Cabais gave her two hundred pesos, five thousand pesos of Japanese currency, and propaganda material. The detailed layout of enemy installations she brought back was sent to KAZ. Cabais did not state her name as protection and reported that she "rendered us valuable service with her accurate reports of the enemy." It was reported that Cabais did give two hundred pesos and ten sacks of rice to a widow who was bound for Cuyo Island to live with relatives. She may have been the agent, and the rice was thanking her for her services.

At camp, the unforgiving rain was steady, keeping the grounds muddy and the men's feet suffering from jungle rot. Once the radio station at the hillside camp was in full operation, Goloyugo hiked overland with several carriers to carry his equipment to Malampaya Sound. By September, there was still no report from him. Two runners were sent to find out. He returned reporting the failure of his radio to reach the camp's control station due to the "mineral contents of intervening mountains." Observations of the west coast of Palawan had now to be relayed by runners that took several days to reach.

Feeling rested, Goloyugo was ready for his next assignment to re-establish a sub-station overlooking Bacuit Bay in the north. Enemy shipping was active in the area. He carried the reserve radio accompanied by the required guerrilla escorts and paddlers. A Russian, Jack Bermont who could speak Japanese offered his services. Widely known as pro-Japanese, he was accepted after his loyalty was determined. He asked only for expense money and clothes. The deputy governor of northern Palawan, Jacinto Alli, departed with them for Bantulan to verify channels newly mined by the enemy. Earlier he had arrived from his hideout onboard his sailboat, bringing fruits and food for the men. He had escaped the Japanese when they landed in Taytay, the capital city at that time. Forty-five pesos were paid for the cow and a pig he brought to maintain the men's proper nourishment. The governor spoke little, worked hard, and was well-liked by his people. He was commended for his courage, leadership, and sacrifice after the war. Along the way, Goloyugo's group distributed medicine to village officials for the civilians.

Dagandan with a radio crossed the Sulu Sea to Cuyo Island. His wooden boat capsized while crossing the Dumaran Channel, losing his firearms and two hundred fifty pesos out of the fifteen hundred he carried. He saved the radio set and beached at the village of Araceli on Dumaran Island. Japanese patrols were raiding adjoining villages of foodstuff. His group anxiously waited on the beach for a sailboat to take them to Cuyo Island when an enemy boat docked. They killed four enemy soldiers and captured a Filipino with the Philippine constabulary. In the village at Cuyo Island, ammo was running short. A runner was sent with a request to bomb the village especially the big house with galvanized roofing where the enemy quartered. Unorganized guerrillas with firearms intimidated the civilians. When no report was heard from Dagandan for a month, a runner was dispatched. He was sick with malaria. After recuperating, he returned to camp and told of civilians being abused and intimidated by the Japanese.

Runners arrived daily reporting enemy shipping sighted in the West Cuyo passage by the Sulu Sea. It was crowded with almost daily sightings of Japanese ships. Messages flashed to KAZ and the Navy control station for the submarines patrolling the area. Information coming in gathered by the guerrillas from the civilians had to be confirmed because many were exaggerated. "It caused more trouble than the enemy," Cabais commented.

Cabais reprimanded a guerrilla security officer, Lt. Concepcion, for "neglect of duty and lack of courage" by depending on the civilians for information "as he spent most of his times sleeping with his young wife" instead of going out into occupied areas.

In September, a large convoy consisting of one heavy cruiser, four destroyers, one ten-thousand-ton cargo transport, and five medium tankers were flashed to KAZ. Two days later, they received an unconfirmed report that several enemy ships were sunk a few miles northeast of Dumaran in the West Cuyo Passage. The men were elated to feel that they "drew blood for the first time," especially when they heard the broadcast of the Office of War Information (OWI) mention the news.

Cabais left for Danlig south of Malampaya Sound to survey the area. A Japanese plane strafing killed the paddler. He saved himself by diving under the boat but lost firearms and seven hundred fifty pesos. He returned to camp and fell ill in bed for ten days with a severe bout of malaria. Vergara and Cutaran maintained radio contact as the runners continued to arrive with information. Later, when another spell hit him, a guerrilla medic gave him an injection to help with the recovery.

When he was well enough to move, Cabais canceled his plan to survey the northern part of the island. An urgent order from KAZ was to rescue and care for the many pilots being shot down before the pending Leyte landing. All guerrillas and civilians were requested to guide the American fliers to their camp. The names of pilots shot down in the bombing of Coron were sent in reports marked 'Secret'. Navy pilot officer R. H. Beatle and his gunner R. A. Johnson were reported recuperating well and entertained by the family of Deputy Governor Alli. They had gotten used "to eating native fashion, that is rice and dried meat or fish." A message was sent that "both fliers now in our station. No injury. Awaiting orders. We are taking good care of them. Native USAFFE soldiers here badly in need of arms, ammunitions. Fighting bravely."

More rescued pilots arrived. Two from the Marine's air branch happily reunited with two others from the same squadron of the Third Fleet. Social gatherings livened up the camp. One pilot officer enjoyed singing classics. When not on duty, the signalmen joined to release the pressure of work. The locals hardly had any food, but they seemed able to put up tasty native cooking, Cabais noticed. Later, one officer "had trouble with a young lady for using a Tagalog phrase taught him by someone and the wrong words to use on such timid souls." KAZ gave no date for a submarine pickup as the focus was on the Leyte landing. The pilots were getting restless. *Tuba* was readily available, and Cabais had to reprimand two of the drunken pilots.

Despite internal conflicts, sickness, evacuation of downed pilots, and radio troubles the men encountered, KAZ received weather reports four times per day from the weather observations and sub-stations established in Coron Bay, Bacuit Bay, and Cuyo Islands, as gladly reported by Whitney.

Radio traffic tremendously increased as the Leyte landing neared. Signalmen at Cabais' camp were weak with malaria and could not go out on reconnaissance with radios to report on the enemy. But three signalmen were on the air 24 hours. In their undernourished and weakened state, they were determined to compile and transmit messages and instructions without delay. Runners, guerrillas from the civilians continued to bring reports on the enemy. Enemy air and water activities were reported as soon as identified. Bombing targets were confirmed before transmission to prevent unnecessary deaths. Messages from Brooke's Pt and Buliluyan for other parties in the island were routed to their station for decoding and acknowledgement. "Perhaps the failure of equipment of the other parties was the cause for heavy traffic on our radios." Their signals were strong making four other stations pass through them to KAZ.

The camp reported heavy enemy losses as twenty-four large ships were sunk and several damaged. The coast north of Cabais' camp was getting rich in supplies salvaged from the sunken vessels but their food supply was becoming a problem. Governor Alli contracted the owner of an island who had cattle at ten pesos per head keeping the camp well-fed and allowing them to focus on their radio duties. Provincial officials arrived almost daily for medical supplies and conferences with Cabais and the stranded pilots.

After the American bombers' return in September, fewer enemy planes and ships were sighted. Morale quickly rose and Cabais ordered the flag to be raised daily noting that " our flag was the first to be raised in Palawan since the Japanese landed on the island." He did not mention if it was a Philippine or a U.S. flag.

The euphoria was short lived. Men from 16- 60 years were now being forced to work for the Japanese. The people feared for their lives, not knowing what would come next.

At Brooke's Pt, Placido and Cortez departed south to Buliluyan with three radio sets for the south to set up station UPN a month before the Leyte landing. Capt. Narizidad Mayor with his guerrillas accompanied them as security. At Mt Wangle, they could see Japanese shipping in the Balabac Strait. The location provided a strong enough signal to send four messages daily to KAZ and the Navy transmitter in Mindanao.

Placido returned to Brooke's Pt, leaving Cortez with two hired men to continue operating the station. One radio communicated with Brooke's Point and the other with KAZ. Cortez sent a message of a Japanese plane bringing troops to nearby Pandanan Island, a haven for escort enemy gunboats. A detachment of 60 Japanese marines was garrisoned in a big, corrugated iron house in the middle of a large coconut grove camouflaged with coconut palms. Three machine guns and rifles armed them in trenches. In the daytime, their seaplane patrolled around the island and attacked Cape Buliluyan. Cortez moved the radio inland and continued operation until April 1945.

Enemy actions began disrupting the transmission of weather reports from Brooke's Pt. The radio equipment was evacuated up the mountain when the Japanese launched an attack. The station NEI receiver was heavy and bulky, slowing movement when the camp had to move out quickly. It did not work for long-distance reception and contact was lost in the mountains.

Radio traffic at Brooke's Pt was at its peak in the weeks before the landing, depleting much of the fuel oil running the rice mill engine charging the radio batteries. The last drum of oil was used up. To keep the radio on the air, they risked using coconut oil supplied by Edwards. But it worked. They promised to replace it when their supplies from Australia arrived. Placido directed a message to KAZ and the commanding general of the Eighth Army, and copied Rallojay, asking to send crystals to cover the new frequency field to contact KAZ at the Leyte landing.

Once a week for the past five weeks, Brooke's Pt was an evening rendezvous of enemy vessels overnight. The waters around were busy with the traffic of unidentified ships and motorboats. They slept ready to go at any moment. The radio station was only one mile from the shore, so equipment was packed and hidden "every time a Japanese ship parks on our front door." A Japanese gunboat anchored in the Bay and tried to land men, but the guerrilla guards watching the shore fired at them. KAZ was informed and equipment packed and hidden. The gunboat sailed away, leaving Japanese troops around a mile and a half away from their location. A patrol of guerrillas went out and killed a dozen Japanese. In retaliation, Japanese planes returned and strafed the area.

Protection for the Leyte landing began. KAZ ordered Placido to "alert all radio stations and coastwatcher stations to maximum vigilance. Report any immediate naval or air movement or concentration during present operations. Instruct coastwatchers (to) report sighting of cargo vessels to give size and type." Placido was to instruct the coast watchers to show the size and type of auxiliary cargo vessels larger than a native fishing vessel, of sixty to one hundred feet in length and fifty to one hundred gross tons that could be fuel barges. Motor torpedo boats and guns on the fore deck should be reported.

Rallojay, with a portable radio and two hired men, left for Inagawan to monitor the Japanese garrison one kilometer away from Puerto Princesa city. He listed the men's equipment brought down from the mountains: a BBZ transceiver, a box of spare equipment, batteries, gas, engine charger, wind charger, and two boxes of J ration. At Honda Bay, they observed as much as fourteen man-of-war anchored at one time. Puerto Princesa was completely occupied. The civilians were ordered to move out. After two weeks, Rallojay returned to camp and went with Reynoso to relieve Cortez at Buliluyan. They quartered in the house of the deputy governor, who was a *datu*. The outpost was up in the mountains of Tabud where ships passing the entire southern part of the island, the western side of the Philippines and the Balabac Strait could be seen.

Leyte Island, Philippines Landing – October 1944

Japanese convoys with cruisers and destroyer escorts in the Balabac Strait were sighted going north and south. Two hundred Japanese marines garrisoned Balabac island overlooking the Strait at Cape Melville. The cemetery was planted with cannons, anti-aircraft, and machine guns.

The guerrillas united under the Palawan Special Battalion were now actively hunting the enemy. They were starving for supplies. Officers arrived at camp and were given a radio set and communication procedures to send intelligence including the narrow waterways around nearby islands. One of the boats the men used for traveling to survey around the island was lost to Japanese strafing, killing two of the crew. Placido compensated the families with 1500 pesos and bought another boat for their use.

At Brooke's Pt, Japanese transports continued to be sighted and reported. One Japanese convoy of four ships was seen torpedoed and shelled by a submarine in broad daylight. The biggest ship gave out "a big black pall of smoke and disappeared." Explosions were seen and heard at night. A "great blazing fire lighted the sky... Japanese convoys never travelled nights again unless heavily protected by strong naval aircraft. They sought shelter before dark, or in deep waters very close to shore. They camouflaged their masts to look like trees."

Refugees from the north and Puerto Princesa poured into the Brooke's Pt area. Placido donated one thousand pesos to the community emergency relief association to help. Rallojay was sent north to find out why the refugees were fleeing.

An interesting report from Cabais' camp was of Japanese and enlisted men in the town of Aborlan that was one of the earliest to surrender. They were located at an airfield constructed by American POWs that could hold sixty planes at a time. Another bigger airfield next to it was almost finished. Aviation fuel continuously arrived.

The Russian Bermont returned from Bacuit with guerrilla officer Capt. Carlos Amores who brought a Japanese warrant officer, Hioe Konno of the Japanese Air Force, captured by civilians. Lacking fuel, his plane crash-landed. He was found with valuable maps of the Japanese military and naval installation, covering the entire empire of Japan, China, Manchuria, Siberia, the Philippines, Indo China, Malaya, and the Dutch East Indies. Included was a detailed map of the Puerto Princesa airfield.

American planes continued to be shot down outside Puerto Princesa. Some of the crew were severely wounded. Civilians and local doctors cared for them. Commander Justine Miller and six fliers survived the crash of his twin-engine bomber a few miles north of the city. His squadron of bombers were based in Morotai and Palawan was his territory. Cabais and Marquina, with two civilians, hiked for two days to meet the survivors. Housed in Maj. Muyco's headquarters, their wounds were treated. "His staff, and wives were very helpful to the point of being a nuisance," Cabais said. The uninjured followed Marquina and Cabais to their camp. Governor Gaudencio Abordo of Palawan provided food with a large cow and one chicken. Marquina eagerly caught fish. Once the wounded could travel, KAZ sent a rendezvous date for a submarine and instructions on security signals. Two different dates were set, but no contact was made causing much disappointment. A plane scheduled to arrive was canceled at the last minute by GHQ, citing heavy military operations in Leyte. Thanksgiving Day came, and a small pig was roasted. The men were thankful to be alive. Cabais could not join as he was hit again with a bout of malaria and a severe cold.

GHQ picked December 2nd as the rendezvous date with a combat sub patrol. Marquina took charge and contacted the sub to inform of the meeting spot about twelve miles offshore. Miller brought the documents and the submarine's crew eagerly gave the little supplies they had as they were returning from a combat patrol. Cabais wrote that he was sickened by the chaos that ensued in the division of supplies as the guerrillas' wives helped themselves to candies, sugar, and some of the badly-needed items that were for his camp. The following day, he prepared to move his camp north to Bantulan to be away from the trouble with the guerrillas. He left ahead with an escort of one soldier and two boatmen. He gave the Russian Bermont 150 pesos to help the rest of the men and equipment to the new camp. The new radios were set up and functioned well. Bermont stayed with the party throughout their mission.

Barely two weeks after the pickup of the seven pilots, word arrived of eleven American pilots shot down in the vicinity of Mindoro Island. Civilians in the village of Sibaltan cared for them. Cabais immediately left through stormy weather to meet Lt. J. V. Fallon, E. V. Rinner, R. L. Harper, J. S. Kuper, E. O. Enloe, F. Thierer, R.R. Govia, R. P. Duplantis, H. wells, M. J. Frank and J. E. Walters. Some were sick and needed rest before moving. The sailboat they took needed repair.

Two more pilots were forced down in Ilian Bay. KAZ's instructions were to hold the men for twenty-four hours until the seaplane was loaded with supplies for the camp. But Lt. Fallon's insistence on leaving forced Cabais to agree and sacrifice the loss of badly-needed supplies. The empty seaplane arrived and the pilots with three Japanese prisoners departed in the first week of January.

A Japanese pilot who gave his name as Capt. Hazai Ichiro, arrived escorted by two guerrillas. He had run out of gas and was captured near Malampaya Sound. No information of value was obtained from him except his sword presented to Cabais by Capt. Amores. He was observed eating more than his ration of rice and trying to escape. Cabais wrote, "to keep him overnight as I did not relish killing him in front of women." But Bermont questioned him and his lies and claims of Japan ruling the world raised the men's anger so much that he was decapitated with his sword for his arrogance.

The watchers could see enemy ships and supply launches docked in the bay, some camouflaged with tree branches and coconut leaves. They vitally needed radio equipment, office supplies for coding and encoding, medicine, and some personal items to continue operating and hoping that they would receive them soon. The radio sets needed to be replaced. Medical supplies were almost gone, so issues to civilians were stopped. The civilians hid in the forest as Japanese combat patrols stole food from them. Instead, KAZ asked for information on suitable places for a radar station in their area, with which they complied the same day. Supplies were almost gone when a cargo plane parachuted boxes, raising the men's morale. It was a relief just in time for the increasing traffic caused by the pending landing of the U.S. Task Forces in Mindoro. From their camp, they could monitor the surrounding seas of Mindoro, the South China Sea, Sulu Sea, and Mindoro Strait.

Puerto Princesa, Philippines Massacre – December 1944

Couriers arrived telling of American prison escapees from Puerto Princesa making their way to Placido's camp at Brooke's Pt. Food supplies were meager, so four men were sent out to procure foodstuff of all sorts. The men's shelter by the station hut was enlarged to make room for the escapees. Six American POWs reached their camp. They told of their harrowing escape from a massacre at the camp garrisoned by around two thousand Japanese troops. The prisoners had built the airstrip and filled up bomb craters from air raids. When American aircrafts raided and destroyed the planes on the airfield, the Japanese thought the Americans were coming to Puerto Princesa. Their naval planes had sighted a U.S. convoy heading for Palawan but was actually heading for Mindoro. The Japanese armed and carried out the orders to kill the prisoners. Those who tried to escape were shot. Those who were able to escape had split up, swam and went through the jungles for three weeks. Local families found and tended to the prisoners' wounds. They prepared meals for the holidays that helped the men regain some of their lost weight. Some were helped by Santa Lucia Penal Colony members, a section of the Iwahig Penal Colony. Brought to the guerrilla forces of Maj. Nazario Mayor, they reached the camp at Brooke's Pt. Placido made doughnuts with some flour left.

Placido related what he heard from the escapees. "They told us of the horrible massacre of the 150 American POWs... There was an air raid alarm one afternoon, and all of them were hurried up to their respective air raid shelters, underground and opening only to the sea. They did not know that this was to be their death trap ... the exits of the shelters were locked, gasoline poured over them, and they were set ablaze. A few blazing human torches were able to force their way out of the inferno with their hands up in the air, pleading to stop shooting. In their unbearable pain, they were mowed down by machine guns. One hundred forty of these men were burned and shot to death."

After the war, one of the escapees recalled that he was able to escape because their air-raid shelter had an exit in the back that they had kept hidden. In digging up the trench, a large chunk of coal fell out and they continued to dig a tunnel that they camouflaged with brushes. The Japanese doused the men with gasoline before torching them but only threw a torch in their direction, allowing them to escape. Cpl. Elmo 'Mo' Deal said he was burned, shot twice, and had over twenty wounds from bayonet stubs or blunt instruments. Using his last strength, he crawled through the jungle until guerrillas found him.

The signalmen sent the names of the six and requested a PBY to pick them up. KAZ replied to "display security signals each day until evacuation completed. The signal consists of three smudge fires on the beach, one hundred yards apart. (They should) have sufficient boats to unload 4,000 pounds supplies. Boats with evacuees should stand offshore as soon as the plane is sighted."

After two weeks, a large bonfire on the beach guided the PBY as it inspected the area and boarded the escapees. Their anticipation to receive the 4 tons worth of requested supplies and equipment turned to disappointment. Only half of the cargo arrived with some radio equipment and medicine but no guns or ammo. How much longer could they operate at Brooke's Pt, starving, sick, and with broken down equipment, they asked themselves. One of the POW, Ensign Hagen, promised that the next plane would bring the rest. It never came.

A week later, three Marine escapees from Corregidor arrived, as an urgent request from KAZ marked 'Top Secret' also arrived, stating the "desire to set up suitable landing for flying boat. Can you have two men familiar with Honda Bay area at pickup? Security signal consist of two panels on the beach. Have all available sketches on Puerto Princesa immediately." With barely three weeks and the travel time needed to reach Honda Bay, Placido scrambled to have the sketches made. The important sketches showed Japanese beach defenses from the harbor of Puerto Princesa to Canigaran beach. Maps of the City showed bridges that could handle three tons, the location of Japanese units, guns, gasoline dumps, radio station, and their unit names. The latest intelligence reports were included. Placido requested that KAZ send a plane to pick up the sketches and the Marine escapees at Brooke's Pt.

Placido made it clear to the officer who received the reports and sketches that the charger could not run without lubricating oil. They did not want to risk continuing to use coconut oil. More guns and ammo were needed to protect the radio station. They needed shoes, clothing, and lubricating oil.

Before the year ended, they donated seventeen hundred pesos to the Women's Auxiliary Service to help the guerrilla soldiers and civilians in desperate need of clothing. None could be purchased anywhere. Instead, the men gave all their packed clothes except for what they had been using.

The prayers of the locals' religious members were answered when the USS *Gunnel* landed arms, ammunition, office supplies, many cases of medical supplies, and a variety of cases of food including 8 quarts of Worcestershire sauce and 60 cans of mustard pickles. The sympathetic crew donated toilet articles, clothing, candy, 100 cartons of cigarettes and 90 cans of nuts. But the all-important gasoline and lubricating oil were not included.

Despite all the difficulties and sickness, communication with KAZ, between Placido and Cabais' main network stations, two guerrilla stations 9JT and 14A, were well established by year end. Reports were coming in strong. Brooke's Pt and Buliluyan stations' communication with KAZ and the Mindanao stations continued to fill the airwaves.

In the spirit of Christmas, the men sent a greeting to their old unit at Tabragalba in Australia: "To the commanding officer and personnel of the 5217th Reconnaissance Battalion. Merry Christmas, and may peace be ours in the coming New Year." The men may not have known that the battalion had been replaced with the 1st Reconnaissance Battalion (Special) in Hollandia. On Christmas Eve, the men attended midnight services in a nearby town.

Luzon Island, Philippines Landing – January 1945

The year opened with the Japanese reinforcing the northern area of Palawan. Around 2,000 troops of their army and navy units garrisoned the areas of Puerto Princesa, Taytay, and the islands of Coron. More troops were added to the 4,000 in their Air Corps unit. Two hundred planes and more were assembled at the landing field. The Japanese commander quartered in the City's church house with two anti- aircraft guns. The main bulk was in the high school and elementary school buildings located at the road junction north of the city. The church was reported as the gasoline supply dump, and two light field guns were under coconut groves. Mortars and machine guns were seen. Anti-aircraft guns from ships sunk in the bay were emplaced along the beach harbor as a defense. Their five landing barges were hidden. Reports abounded from surviving victims of the Japanese' increasing atrocities of the civilians.

Placido insisted in his messages that they were following GHQ's radio instructions but were still looking for a plane to drop gasoline fuel and lubricating oil. The constant guard of a certain frequency required by KAZ was quickly depleting the oil. None could be found where they were camped. The airdrop several months earlier had mostly radio equipment and only one unbroken bottle of electrolyte. Five hundred guerrillas in an area covering one-third of Palawan were restless and eager to be equipped with mortars, automatic weapons, and grenades. They constantly begged Placido to inform Australia, "even only arms and clothing to rid isolated Japanese garrisons. They had been waiting so long for the time when they can rid Palawan of the Japanese." Again, he sent coordinates for a suitable landing of a PBY and the location of three smudge fires on the beach in front of the pier as security signals.

Palawan Island, Landing – February 1945

A PBY had landed with more supplies and orders for Cabais' station to provide logistical information of Puerto Princesa. An air and naval base were planned for Palawan. All data was consolidated and a message sent for pickup but there was no reply. Cabais left for the south and met the provincial governor at a probable pickup point. Cabais reprimanded the governor for spreading the possible date of American landing in Palawan. He said he was guessing and expressed his willingness to help the people but refused to lower commodity prices which he had set and agreed to by some guerrillas who probably got a cut on the prices. A new order arrived to bring the information to the landing site of the Task Forces. Meanwhile, KAZ sent a message to inform civilians to stay away from Coron Bay as it would be bombed before the landing. Cutaran and Vergara immediately sent out the priority messages to all stations to inform the civilians in their area. As Cabais entered Puerto Princesa, the city and nearby airfield were being shelled in preparation for the landing. Over the city, an unidentified four engine plane was shot down according to an underground report. Five of the crew were found dead. None were captured. Boats of the locals that the U.S. aircrafts could not identify were bombed and strafed as they sailed along the coast of Eastern Palawan.

The signalmen were kept on the air with the heavy radio traffic of the Eighth Army, Palawan Task Force, and guerrilla stations. The 186th Regimental Combat Team stormed ashore at Puerto Princesa "without a single shot." The only two casualties were due to booby traps. Two good guerrilla men guided the Army landing. Not until three days after the landing did the American troops catch up with the Japanese who were rapidly seeking shelter in the hills. Still carrying the logistical report, Cabais was ordered to report to the commanding general of the Palawan Task Force. He remained in the city for several days assisting the assembly of civilian officials to establish a local government as directed by G-2 Intelligence. He rode a PT boat back to Bantulan.

A PBY arrived with Maj. Hervey, G-2 Intelligence of the Eighth Army and Mauricio (Samar mission June 44) to survey for enemy troops, anti-aircraft gun emplacements, condition of the airstrip, and location of enemy radio stations. The airfields were mostly destroyed due to the Navy's pre-landing bombardments. Mauricio transmitted enemy defense emplacements around their quarters in the elementary school building and artillery along the shores. The landing force was reminded of the importance of the Balabac Strait and its danger from mines. He sent a message to the XXIV Corps to send a plane to Brooke's Pt to pick up important sketches, intelligence reports, and maps. He requested 3 machine guns and ammo to protect the main stations of Palawan. Most importantly, he pointed out the need for lubricating oil to keep the station on the air. Send clothing and size 6 shoes as the men were in tatters.

Control of Cabais and Placido's main station radios, sub-stations, and their codes were turned over to the Eighth Army for direct contact with the guerrilla forces. The men continued to send weather observations for another month until instructions arrived to close the stations, including the sub-stations in Bacuit and Cuyo, and report to the Palawan Task Force headquarters in the city. Cabais, Goloyugo, and Marquina were hospitalized while waiting for the other mission men to arrive.

Further north in Coron, a civilian operative, Tony Parmelee, reported having completed negotiation for the surrender of the puppet officials. They would not surrender to the guerrillas who were known to execute pro-Japanese and captured Japanese. The governor requested a guarantee of their lives. Cabais requested KAZ' approval to guarantee the safety of the entire puppet officials of Palawan as they surrendered to him.

The Japanese marine garrison on Balabac Island had evacuated. Placido, Mauricio, the guerrilla commander of Brooke's Pt and an American ground observer of the 13th Air Force went to Balabac. What looked like a hasty evacuation was confirmed. Boatloads of arms, ammo, rice, and a lot of new radio equipment left behind were sent to Brooke's Pt.

It took half a year until they received the much-needed supplies to support the liberation. While doing all they could to keep the radio on the air, Placido received a bewildering message: "funds given to Corpus for use in obtaining intelligence were turned over to Placido upon death. Acknowledge the amount of funds. Will require a detailed account of your expenses." Did GHQ think of the risk of documenting names, transactions and anything that might be used by the enemy?

With the radio turned over to the Task force, Placido thought it was the right time to send a request for Ramon Cortez and Teodoro Rallojay to marry their sweethearts. They had met their future wives while on the island. A separate message was sent to Eichelberger, commanding general of the Eighth Army.

Placido was still at Balabac when the orders arrived to close down the outposts. While he was waiting for a boat from Brooke's Pt, a Japanese gunboat came into the Bay. The guerrillas thought it was an American PT boat or a landing barge. When Placido woke up, he looked out and recognized a Japanese gunboat docked at the pier. He ran with only his clothes, important papers, and 'grease' gun' (submachine gun). A guerrilla partly destroyed the good portable radio set that could transmit 140 miles away. The Japanese captured everything else. A captured guerrilla was taken to Banguey Island (Banggi Island), north of Borneo, where the gunboat was presumed to have come from.

Placido arrived back at camp with the men packed and waiting for the PT boat to pick them up. He had lost his equipment and the rest of the arms left behind to the Japanese gunboat. The day before their departure, the men gave a farewell party to the people who helped them to express their appreciation of their cooperation. Two bulls and fifty chickens were butchered.

The PT boat arrived, and all equipment and men were loaded. "Butch (Rallojay) and Ray (Cortez) said their goodbyes to their wives" for now. Many goodbyes were said, and the waving continued from shore as the boat moved away. They had left behind four good head of cattle, about sixty chickens, their "pretty nifty nipa shack," and a fish corral with more fish than a "water buffalo could haul." The catches were distributed to the people. There was enough rice to last for another six months.

Wistfully, Placido wrote in his after-action report. "We all hoped that whatever little we accomplished had somehow or in a small way contributed to the success of the Philippine campaign. We had tried our best with what equipment was furnished to us. We had to live in the land, expensive as it was. Yet, we thought that it was worth it. We thank the officers and men who trained us for their patience, interest and tireless efforts. We feel deeply indebted to their ability in bringing and developing the best in their men. Believe us, we are glad and proud to belong to such a very fine organization as the 978th Signal Service Company."

Cabais expressed a commendation for Deputy Governor Jacinto Alli, an old soldier, for his devotion to duty, courage, leadership, and sacrifice. He did not talk much and worked hard. The people idolized him.

The two mission parties were in Palawan for ten months. They flew to Leyte and two days later to the 978th Headquarters in Hollandia. Undernourished, many times sick, and sometimes starving, they had been dauntless in pursuing their mission.

The Eighth Army's Combat Wing, Fifth Air Force, complimented them for their continuous operation of weather and radio control stations. The Commanding Officer of the 15th Fighting Command Weather Detachment praised their work for accurate weather reports. More than forty sunk enemy ships were reported from the party's radio stations.

Commander Justin Miller, USN, 7006, Aircraft 7th Fleet, who was shot down in Puerto Princesa and evacuated, did not forget. He cited exceptionally meritorious performance in executing great responsibility. A total of twenty-three officers and enlisted men of the Navy and one enlisted of the Army, namely Corporal Elmo Deal, one of the survivors of the massacred POWs in Puerto Princesa were evacuated by the Party. The evacuation was made possible through the loyal cooperation of the civilians and the guerrilla medical officers.

GHQ highly commended the Party after the war for their intelligence reports. Its most outstanding performance was the capture of the Japanese air force pilot that had a complete map of the Japanese military and naval installation in the entire Japanese empire, including China, Manchuria, the Philippines, Malaya, and the Dutch East Indies.

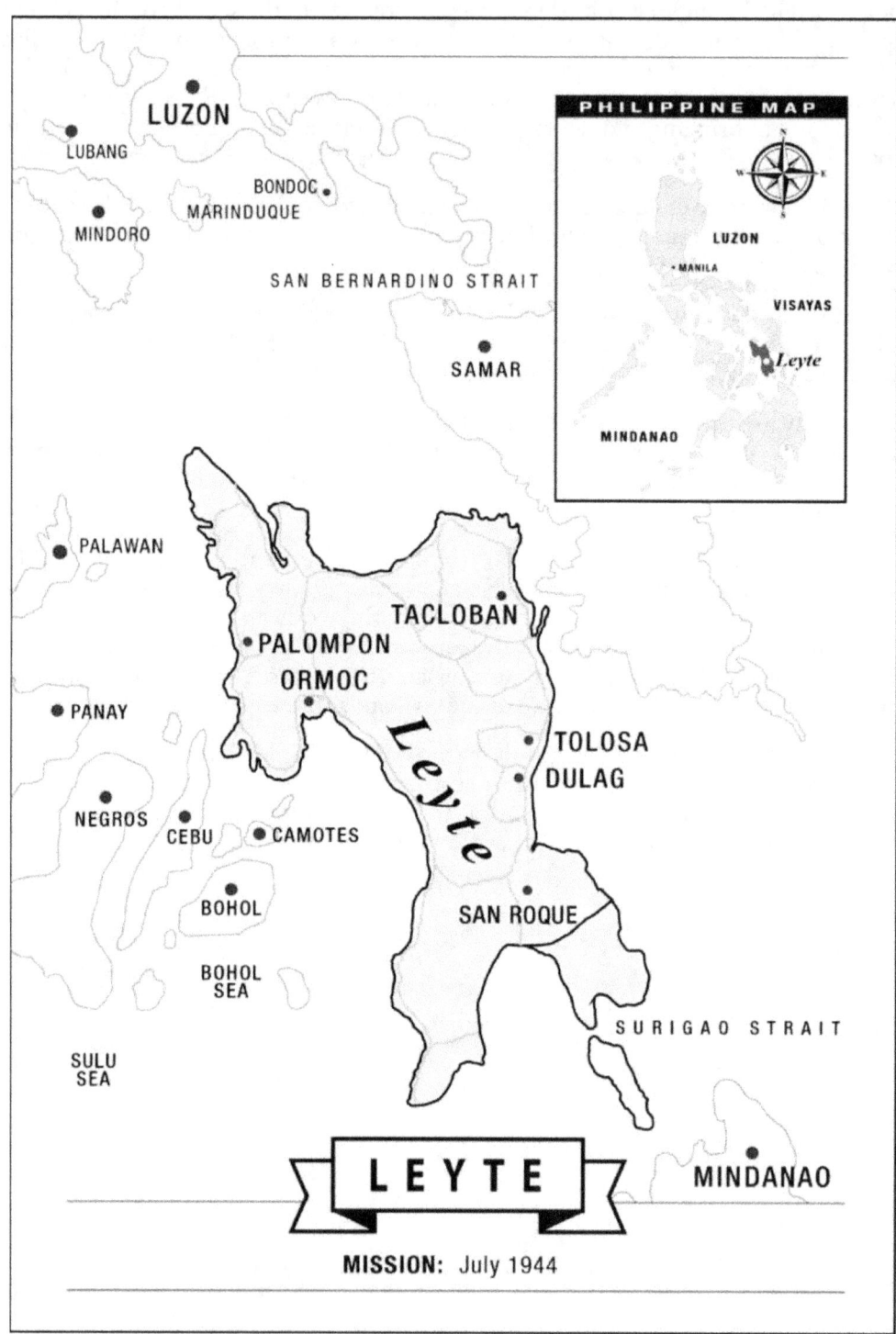

LEYTE

MISSION: July 1944

XVI

Leyte Island

Tacloban and Ormoc were garrisoned by an enemy force of 4000 Japanese soldiers a week after the USAFFE surrender in May 1942. The resistance retreated to the mountains where food was not as plentiful as in the lowlands. They ambushed enemy convoys and pressured the fearful civilians to give them food or draw Japanese retaliation against them if they refused. The guerrillas also prevented them from growing and harvesting crops that would go only to the Japanese if confiscated. People were soon starving.

Hundreds of guerrillas had been killed in the island not only because of the enemy but also because of feuds among the resistance fighters. Guerrillas in the Ormoc area led by Lt. Blas Miranda, a former constabulary officer, fought for control of the area until many of his men were killed by the Japanese. Capt. Pedro Merritt in the north, who was supported by Peralta from Panay, refused help from outsiders thus allowing Japanese ships to travel the San Bernardino Strait at the backside of Leyte without fear of attacks. The infighting stopped when Col. Ruperto Kangleon was recognized as the leader of the 9th M.D., upon the recommendation of Commander Parsons. Kangleon had earlier escaped from a Japanese prison camp in Mindanao and established a guerrilla force in Leyte.

In October 1943, Lt. Truman Heminway, a former aerial photographer turned radio operator, with Lt. Joseph St. John and another American were sent to Kangleon's headquarters with portable radios. They were American soldiers stranded in Mindanao after the Pearl Harbor attack. Their directive was to establish a radio network and provide information on guerrilla and Japanese activities. The short ranges of their radios required passing intelligence reports through the more powerful FRS Mindanao radio network. That overburdened the radio transmissions of the 10th M.D. Two months later, a larger and more powerful transmitter that could directly contact Australia was brought to Kangleon from Mindanao. But it was reported to have been buried, for being caught in possession of a radio was very dangerous. It was too heavy to move around. It took several months for Heminway to locate the radio, dig it up, and bring it to Kangleon's camp. St. John operated the radio. During the war, the large transmitter was never shut down regardless of any danger of it being located by the enemy.

The station TUM sent its first message to KAZ in February 1944, reporting an enemy of division strength plus around four thousand Marines garrisoned in the capital city of Tacloban. One unexpected message was "on the use of gas dropped by enemy planes on civilians."

Leyte received much of its supplies from the submarine landings in Mindanao, Negros, and Panay. But they would take weeks, sometimes months over waterways and land to arrive, and by the time they did, the supplies were damaged or stolen. Sometimes they did not arrive at all.

USS *Nautilus* – July 1944

The submarine *Nautilus* proceeded to Leyte after delivering men and supplies to Mindoro. She proceeded with caution at the rendezvous point as it was heavily mined and being watched by the enemy. After waiting for three days, she moved closer to shore. At dusk, she approached San Roque in the northeastern part of the island. Security signals were sighted. A small boat flying a "clean set of American colors" with three men approached. She surfaced and Kangleon came onboard. Recon men T5 Nicholas Sandoval and Pfc. John Tolentino disembarked to reconnoiter and provide protection for weathermen Sgt. Henry Chambless and Pfc. Robert H. Gamerstfelder. Their mission was to gather reliable weather data for the Air Force bombing targets. The guerrillas and coastwatchers would provide reports on enemy activities that they radioed to KAZ.

Seventy tons of equipment were unloaded in surprising record time since Kangleon and his men had no previous experience in such operations. The cargo included "two hundred fifty pounds of corned beef, one hundred fifty pounds of cured ham, three hundred fifty pounds of cured bacon, fifty pounds of powdered eggs, and two hundred pounds of white sugar," food they had not tasted in the last three years. The sub departed towards the Sulu Sea and established contact with Abcede in Negros to receive important enemy documents.

Gamerstfelder recalled that the four tons of weather equipment loaded on the backs of sixty Filipino boys were brought to a suitable site for a weather station.

By now large Japanese forces from the major islands were sent to Leyte to meet the coming Allied Task Forces. The enemy had recognized their desperate situation. After the war, the Japanese conceded: "It is impossible to fight the (U.S Forces) enemy and at the same time suppress the activities of the guerrillas." (Japanese Monograph No. 3, Philippines Operations Record, December 1942 - June 1944 Army)

The radio crackled with the message, "Tacloban, Leyte to be invaded 20 October 1944. Evacuate all Filipinos week before." The Navy was to shell the beaches before the landing. Parsons and Frank Rawolle of the Sixth Army Intelligence were flown from Mios Woendi to Leyte to survey landing beach conditions, Japanese troop strength, and locations. They met with Kangleon's guerrillas to evacuate the people from the beaches. A message to KAZ reported that the Japanese did not allow the people of Tacloban to leave. The Japanese had left the city and set a new line of defense about five kilometers outside of the city.

"Spare the City from bombing," was the urgent message. Civilians hid wherever they could to avoid any encounter with the Japanese who were indiscriminately killing with the pending American landings. Fannie Sumaoang, future wife of Federico Sumaoang, a member of the Filipino Infantry Regiment told of whole families like her sister's including her baby, being bayoneted to death by the Japanese in the town of Palompon.

The long-awaited message to attack the enemy and limit his activity arrived. The guerrillas eagerly began liberating towns and destroying Japanese garrisons. They had waited and endured three long years of suffering and watched their families killed. Now they had the arms and ammunition brought in by the submarines to exact revenge. In one report, Kangleon remarked "Caught them at dinner, used our *bolos*" which were quieter than guns.

Leyte Island, Philippines Landing – October 1944

The time had come for General MacArthur to fulfill the promise he made two and a half years earlier at Townesville, Australia. He had successfully convinced President Roosevelt that a strong Philippine resistance backed him up in planning a successful liberation of the islands. As he looked out from his ship's bridge in the early morning, he thought how familiar Tacloban looked. Forty years earlier, it had been his first assignment after leaving West Point.

Five old battleships damaged at Pearl Harbor were repaired in time to arrive with the convoy and shell the 18-mile-long Leyte beachheads of enemy presence. Three Filipino Boy Scouts who knew semaphore saved lives by redirecting the ship's artillery away from Tolosa to the location of the Japanese troops. But civilians in the areas that were not evacuated suffered heavily from the naval bombings. Dulag suffered the most casualties. From reports, it seemed that the warning did not reach the people in time. Japanese troops were reported dug in on the beach waiting. U.S. dive bombers continued to take out Japanese gun emplacements in the hills behind the beaches.

The US 3rd Fleet surrounded Leyte Gulf to protect the invasion landing against attacks by Japanese naval forces. The 7th Fleet was moved closer to shore to protect the men, equipment, and supplies piling up on the beaches.

The landing on the invasion beaches was unopposed with minimum casualties. Radios dispatched guerrillas' reports of enemy's location and strength. The Japanese now had to fight a conventional army and unconventional bands of guerrillas. The Task Force met heavy resistance about five kilometers inland as the guerrillas ambushed retreating Japanese troops.

But the Japanese High Command did not give up Leyte easily. They saw the open beaches of Leyte filled with over a hundred thousand men unloading an unimaginable amount of equipment. It was an opportunity to strike.

The coastwatchers' outposts had visibility to the sea lanes of the islands surrounding Leyte. Three days after the Allies touched Leyte soil, the coastwatchers pounded the keys for two days to alert the Air Force and Navy of the position and number of enemy ships streaming towards the area. They saw a sizeable convoy of three attack forces made up of cruisers, carriers, transports, and escorts of the Imperial Japanese Navy on the way to attack the beaches that were still unloading landing forces and equipment.

The Japanese center force of Admiral Takeo Kurita from Singapore would have been first sighted in the West China Sea by Cortez (Palawan June 44 mission) at his outpost at Cape Buliluyan in Palawan. Toribio's (Mindoro July 44 mission) outpost on Lubang island had an unobstructed view as the convoy traveled along Mindoro's west coast. Pinuela's (November 43 mission) outpost on the northwestern tip of Mindoro would have sighted them. Aguilar and Delfin (Mindoro July 44 mission) at their outpost on Calavite Point sighted them. When the force turned eastward around the southern tip of Mindoro, Balleras and Begonia saw them (Mindoro July 44 mission.). Begonia from his outpost high in the mountains could not believe the number of ships he saw as he thought the Japanese fleet had been demolished earlier in the Battle of the Philippine Sea. His fingers flew on his radio transmitting the message of what he believed was the "whole combatant strength of the Japanese Navy" – sixty-five ships going by to the Sibuyan Sea and heading toward the American Fleet anchored off Leyte. Gonzales' (Mindoro July 44 mission) outpost in Marinduque sighted the ships. Robles' (Mindoro December 43 mission) outpost on the Bondoc Peninsula in southern Luzon and outposts on Northern Samar had clear views of Kurita's force streaming into the Sibuyan Sea and entering the San Bernardino Strait.

The U.S. 3rd Fleet was lured away from the position of protecting the San Bernardino Strait to the north. The 7th Fleet was left alone to face the enemy convoy's barrage without the mightier firepower of the 3rd Fleet. Taffy 3 Naval Task Force valiantly faced Kurita's force.

Desperate to strike Leyte, the Japanese sent two more convoys three days later. A convoy led by Admiral Kiyohide Shima streamed southward passing Lubang and Mindoro. The convoy passed Panay and Negros islands. The earlier AIB mission parties' stations on the western coast of Panay at Pucio Point and Negros at Kabankalan, Cartagena, and Culipapa would have sighted the convoy. Turning left eastward around the southern tip of Negros towards Leyte, the convoy entered the Bohol Sea. The fleet of Admiral Shoji Nishimura from the west side of Palawan would have been sighted by Cortez (June 1944 mission) at Cape Buliluyan crossing the Sulu Sea into Bohol Sea towards Surigao Strait. Shima and Nishimura's forces rendezvoused in the Sulu Sea. Sarmiento (Mindanao March 44 mission) at his outpost in Northern Surigao would have sighted the convoys. Nishimura's ships faced the might of the 7th Fleet and were destroyed. Shima's attack force behind Nishimura entered Surigao Strait but unexpectedly withdrew.

Davao coastwatchers reported many Japanese bombers and fighters leaving Davao airfield and heading towards Leyte. Japanese planes attacked the cargo ships and airstrip preventing Navy fighter planes from taking off. Anti-aircraft fire filled the air as enemy aircrafts were sighted and reported. Army Air Corp planes intercepted the enemy before they could attack the landing beaches. Agents reported fighter planes that ran out of fuel and landed at the Dulag airstrip as there was not enough fuel to return to their ships.

The radio centers and communication equipment at General MacArthur's headquarters at the Price House in Tacloban were damaged from repeated aerial attacks. Big vital electric generators were sandbagged for protection. One transmitter was moved twenty miles down the coast but was bombed the first time it went on the air. Messages piled up. A radio transmitter was inoperable due to its aim off target but was hurriedly fixed when a technician moved the radio center south to Tolosa.

A typhoon struck Leyte bringing strong winds and heavy downpour that slowed down fighting with the enemy. The airfield was soaked, reducing the runway to a muddy ground. KAZ went off the air. Signalmen scrambled to fix it.

Radio codes of the guerrillas' stations were turned over to the Sixth Army upon landing. The valuable intelligence they provided enabled the advancing American forces to pinpoint enemy artillery and airstrike targets. The guerrillas were reorganized and provided tactical intelligence, security, and guides to the Army assault units. They patrolled vulnerable supply depots and lines of communication in the rear areas from Japanese infiltration.

A month after the landing, Tolentino informed KAZ that the weather station would close due to enemy operations only two miles away and that they did not have enough supplies to continue to operate. He requested an airdrop of supplies and rations. KAZ replied that they should order one of the men to go to Tacloban to get the supplies they requested. That would mean the assigned officer, Sandoval, who provided the station's security, would be away from the camp for a week of travel overland through jungles, swamps, and encounters with the enemy. The station decided to just relocate to a new outpost further away from the Japanese camp.

Imperial Tokyo would not give up. Troops already amassed on Luzon to fight the primary ground battle in the region were ordered to reinforce Ormoc and "annihilate the enemy invading Leyte". The convoy with over 30,000 Japanese troops did not escape the coastwatchers and sightings were flashed to KAZ and the Navy. Some of the Japanese units were diverted to eliminate the guerrillas in Camotes Island guarding the entrance to Ormoc Bay. Not all the ships reached Ormoc under the U.S. aircrafts' barrage.

The U.S. Task Forces of the 77th Division and guerrillas broke the Japanese resistance taking Ormoc City by December and the Japanese retreated to the mountains. Elements of the 1st Filipino Infantry Regiment took over and destroyed remaining pockets of the enemy when they arrived in February 1945.

By the year's end, General MacArthur issued a communique: "The Leyte campaign can now be regarded as closed except for minor mopping up operations." But the mopping-up operations continued until April of the following year as large Japanese contingents still held parts of Leyte. The Sixth Army was now tasked to prepare for the invasion of Luzon as the Eighth Army took over the mopping up of Leyte and the liberation of the Visayas and Mindanao.

Luzon Island, Philippines Landing – January 1945

After three months of continuing to operate, their remote outpost was almost out of food and supplies. Tolentino sent another message asking KAZ to airdrop supplies and rations to continue the weather outpost. He also asked for some time to withdraw to allow for equipment maintenance. The reply was, again, for one of them to report to camp ISRM on Mindoro and work out a satisfactory solution. Unknown to KAZ, Mindoro had closed earlier. Unable to continue the radio with their wasted bodies, starving stomachs and depleted supplies, Tolentino and Sandoval closed their station. They slowly hiked to Tacloban with equipment they could carry on their backs.

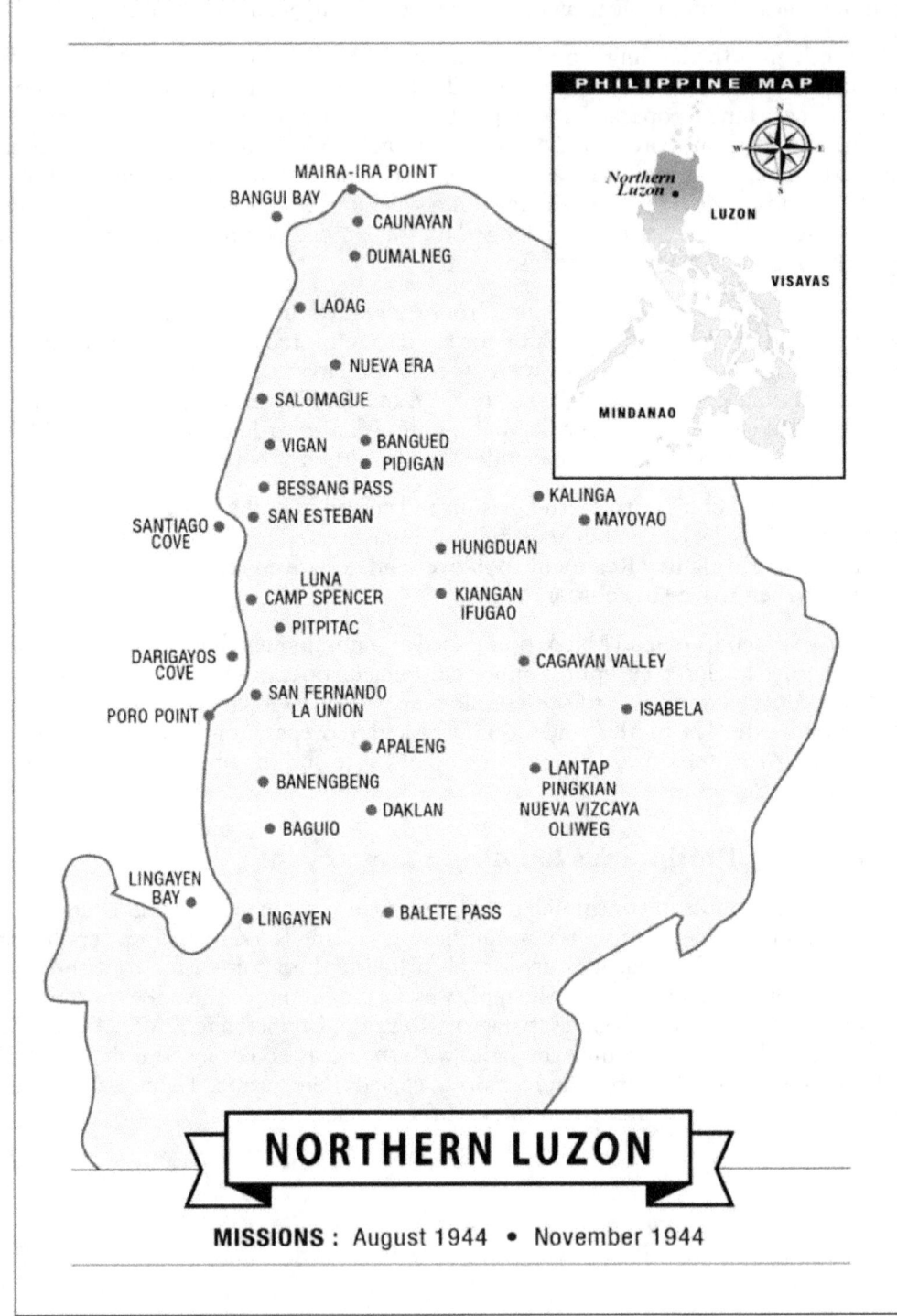

PHILIPPINE MAP

Northern Luzon

LUZON

VISAYAS

MINDANAO

MAIRA-IRA POINT

BANGUI BAY

CAUNAYAN

DUMALNEG

LAOAG

NUEVA ERA

SALOMAGUE

VIGAN • BANGUED
• PIDIGAN

BESSANG PASS

SANTIAGO COVE • SAN ESTEBAN

KALINGA

MAYOYAO

HUNGDUAN

LUNA
CAMP SPENCER • KIANGAN
IFUGAO

PITPITAC

DARIGAYOS COVE

CAGAYAN VALLEY

SAN FERNANDO
LA UNION

ISABELA

PORO POINT

APALENG

LANTAP
PINGKIAN
NUEVA VIZCAYA
OLIWEG

BANENGBENG

DAKLAN

BAGUIO

LINGAYEN BAY

LINGAYEN • BALETE PASS

NORTHERN LUZON

MISSIONS : August 1944 • November 1944

XVII

Luzon Island, Northern Area

Bataan and Corregidor stood silent, drenched with the blood of brave men. Morale had sunk for the ones who refused to surrender and had retreated to the mountains. They banded together turning to guerrilla warfare. But without arms, weapons, and supplies, guerrilla leaders were captured one after the other. Many civilians were afraid to help for fear of torture, burning of their homes, destruction of their crops, and death. The situation seemed hopeless as there was no way to contact Australia.

Then the weak short-lived messages of Lt. Col. General Guillermo Nakar and Major Ralph Praeger reached General MacArthur's headquarters. It revealed glimmers of a resistance movement that could support his desire to return.

Before Corregidor fell, Major Russell Volckmann trekked to Northern Luzon from Bataan. He united the Northern Luzon guerrillas into the United States Army Forces in the Philippines, Northern Luzon (USAFIP- NL) and organized them into regiments to secure the provinces. The 121st Infantry Regiment commanded by George Barnett trained the guerrillas in hidden camps on weapons, combat principles, scouting, patrolling, and security measures despite the difficulties of a language problem. Many of the officers from Central Luzon spoke Tagalog, a different dialect from what their men understood. Ilocano was the prevalent dialect of the north. The 11th Infantry Regiment commanded by Donald Blackburn also included *Igorots,* the indigenous tribes of the Mountain Province, and in the end the regiment had almost two thousand men speaking eleven different dialects. The support of the *Igorots* was crucial to the organization's living existence. Eventually, the difficulties of understanding each other was overcome and Volckmann's organization grew to a strength of over twenty thousand men.

The guerrilla organizations in the southern islands of Negros, Panay and Mindanao had planted agents in Luzon and were bringing back intelligence on Japanese dispositions to send to KAZ. When a radio was set up in the Bondoc Peninsula in Southern Luzon by the Samar mission party, Volckmann used a relay of couriers to send intelligence to the station for transmission to KAZ. Couriers ran between message centers every four to six hours depending on the terrain. They used miners' helmets from the gold mines in the mountains of Luzon, with carbide lamps as they ran at night. Tattered clothes, however, did not defend them much against leeches or mosquitos.

Volckmann received his own radio from Robert Lapham, guerrilla commander of the Luzon Guerrilla Army Force (LGAF) headquartered at Baler on the coastal area of Central Luzon. The radio was from a submarine landing in Mindanao that arrived with Ancheta and signalman Luz (Samar mission June 44). The radio functioned for a week but was able to contact KAZ and station MACA on Samar before its batteries died. Their first reports explained the guerrilla commands with a list of units operating in Northern Luzon from June 1943 to the present. KAZ replied with a promise of arms, supplies, and medicine in a submarine mission in September. Two radio technicians were found in Manila to improve the power pack. After a few days, the transmitter was on the air powered by a generator raided from a rice mill in a Japanese garrison and was hooked up to a water wheel in a waterfall near Volckmann's mountain hideout. But he needed supplies and more radios for his guerrilla units.

As men were chosen from the 1st and 2nd Filipino Regiments in California for further training in Australia for missions, they were restricted to an area until departure. Twelve of them were at Camp Cooke. "M.P. followed us everywhere," recalled Lorenzo Pimentel. Eleven men boarded a plane. An American colonel told Pimentel that the Army takes the boat. He confidently boarded, following orders, and was the only Filipino on the aircraft. He was given two envelopes and ordered not to open them until told to do so in the middle of the Pacific Ocean. The orders were to go to Australia. He was picked up by a female member of the Australian Army and driven from the airport. Pimentel thought that this might be her first time to see and hear a Filipino. He introduced himself as one of 'General MacArthur boys'" and she nodded and replied "Philippine." She asked, "Did you come to die?" and he said, "No. I did not come to die." She meant 'today.'

Shipped to Camp Tabragalba outside of Brisbane, the group underwent more interviews, orientation, and intelligence training. Orientation was done in a room where they were taught how to fix a radio, paddle a rowboat, and, in case they were down in the water, make signals with a mirror among other skills. Several months were spent in the jungle and amphibious survival training until the order came for a mission.

An Immigration officer roamed the various American bases across the globe, naturalizing men to bear arms. Pimentel thought about his future and decided to be naturalized as an American citizen. He recalled the unfair treatment and disrespect he received before joining the Army. But when the war brought Filipinos on Bataan fighting side-by-side with Americans, the attitudes changed. He was convinced to take citizenship if he wanted to go back to the Philippines. Jobs in the U.S. would be more accessible, he thought, when he returned and he would be able to "purchase a firearm as a citizen." He took the oath and was naturalized a week before his mission party left.

Fifteen men met Col. Whitney at the office of the AIB, on the 4th floor of Heindorf House in Brisbane. He began his talk by complimenting the men, saying that "each new group that comes in here seems to be better looking than the last one. They would be the first to land in that part of the Philippines. Not much is known about the place. No one will meet you. For this reason, I am sending instructions with your party leader. You are going to be on your own. If for any reason any of you would want to go back out now, don't be reluctant to say so. It wouldn't be a disgrace." No one did. "Not after all those months of hard physical training in the Canungra Jungle Warfare school," Sgt. William Sammons thought. All were dauntless to start on their mission. Whitney shook their hands and wished them good luck. It was raining when they left Heindorf to wait for further instructions. It was a suicidal mission.

USS *Stingray* – August 1944

Sgt. Leovigildo Giron recalled a briefing at General MacArthur's headquarters "Boys, I selected you to do a job that a general can't do. You have the training to do a job that no one else can do. You are going home to our country, the Philippines — yours and my homeland. You'll serve as my eyes, my ears, and my fingers, and you'll keep me informed of what the enemy is doing. You are going to find the Japanese for me so that I know how to win the war. Good luck, and there will be shining bars waiting for you in Manila."

In a large warehouse of equipment for the missions, they were given a list to pack and told to be ready to go. Giron packed a Carbine, bayonet, *bolo* knife, pistol, and brass knuckles. The following day the party departed in a transport for Darwin, where they stayed in a secret camp before boarding the submarine USS *Stingray*. Their equipment was loaded "by needs of equipment… least will go to the bottom and what we need most stays on top. We have to arrange the boxes so we will feel comfortable when we sleep. They did not have a bed or compartment for us in the sub." They brought eight tons of supplies and slept on their cargo boxes. Giron and another soldier slept under the torpedo racks.

Jose Valera returned to lead the party. A member of a British intelligence group in Singapore, Borneo, and East Indies, Valera had involuntarily joined the guerrillas in Sulu when he was forced to dive inside the submarine as he was removing cargo when a Japanese ship appeared during the unloading. Onboard were 978th Signal men Sgt Telesforo T. Asuncion, Sgt Leovigildo M. Giron, Pvt. Don Liberato, Pvt. Anselmo S. Moscoso, Pvt. Selmo O. Saladino, Pvt. Estanislao C. Peralta, and Pvt. Williams H. Sammons. Accompanying them were men trained in sabotage and demolition: Sgt. Victorio C. Goloyugo, Sgt. Alfredo R. Agron, Sgt. Teodoro Taluban, Cpl. Lorenzo U. Pimentel, T5 Larry G. Soliven and Pvt. Patricio Ollada. Pvt. Ralph N. Balunes was the weatherman.

They were travelling with sealed orders. Taluban said nobody knew where they were going and when they were going to get there. Ollada was excited to leave after a year and a half of training. Valera, the mission leader, and most of the men were from Northern Luzon, spoke the different languages and were familiar with the terrain. That helped in the decision of a landing site. They were the first to bring intelligence, signal, demolition, and sabotage men to Northern Luzon. Their mission was to contact the guerrillas of Governor Roque Ablan who were believed to have joined Volckmann after the capture of their leader.

The submarine crew was sworn to secrecy. That was the standard procedure whenever there was a mission party of Filipino soldiers onboard. Ed Dolinar, a torpedo man on the *Stingray*, related that the crew did not know the destination. Only the captain and executive officer knew. Secrecy was maintained, and they hardly talked to the Filipinos. He saw the Filipinos sleeping on the torpedo room floor next to, behind, or under equipment. He saw Commander Parsons writing in shorthand to keep the Japanese from knowing what he was doing if caught. Dolinar remembered that he and three other crewmen were almost left behind in Tawi-Tawi. They were offloading cargo and, when paddling back to the sub saw it turn to leave. If the executive officer did not come up to light a cigarette, they would have been left behind.

The sub travelled submerged most of the time and surfaced at night which was eagerly welcomed by the men for the fresh air that surged into the open hatches. Sammons recalled that the crew was friendly, and the food was good. Movies were occasionally shown during the ten days of travel. On the third day at sea, an enemy plane was spotted, and the sub stayed underwater, resuming course at night.

The sub reached Batanes Island north of Luzon and went through the mine- infested islands towards Bangui Bay. At dark, the security signals never came. Tension in the sub rose. A torpedo shot was detected on the starboard side. Through the periscope, they saw two sections of a Japanese transport ashore. Another Allied sub must have hit it. Japanese vessels surrounded the damaged ship and people stood around onshore to watch. Submerged, the *Stingray* went around the northwestern tip to Pasuquin Bay but it had shallow waters and rough seas. Further south at Poro Point was not a good spot either. Caunayan Bay was recommended to the skipper although Japanese convoys frequented the area. The submarine continued north to Dalupiri Island and circled it. Headlights of Japanese vehicles were seen moving along the inland road. The captain wanted to drop off the party at Dalupiri, but Valera said it would be difficult to bring the men and cargo across the forty-five miles of open sea to the northern coast of Luzon. The captain retorted that the men could just steal a sailboat to sail across as he was more concerned for the sub's safety.

The skipper and the men decided that Mayraira Point (Maira-ira) in Caunayan by Bangui Bay in the northern coast was possibly safe. Japanese vessels and aircrafts were detected but the landing had to be made even if there were no friendly guerrillas to welcome them. The sub submerged until nightfall and prepared for the landing.

At dusk, a Japanese plane spotted the sub and dropped depth charges. All were ordered to be silent, and electricity was switched off. The gyroscope was stopped. Giron recalled the trembling voice of the man on the sonar. The sub shook so severely that many of the cargo boxes fell. There was some confusion, but they tried to keep calm. The sound of motors passed overhead. An enemy boat was looking for them. A third depth charge exploded close enough to spring a leak in the galley. It was repaired quickly. Sammons said that "he never prayed so hard in my life." The sonar followed the sound of the moving boat until the echoes could no longer be heard. The Japanese knew they were in the Bay.

Caunayan was at the northernmost tip of Luzon. The enemy and their spies gathered in the area. KAZ knew of Roque Ablan's guerrilla organization and believed they were in the vicinity. The guerrillas guarded the landing sites but, without a pre- arranged contact with Ablan's unit, it was a dangerous landing. Without the civilians' presence, it was difficult to transport supplies.

Two hours before midnight, the sub surfaced in rough waters. As big waves slammed against its sides, the sub running on battery power, headed towards the beach and stopped at a safe distance. The helmeted crew manned the deck guns, while others carried the cargo up on deck. Pimentel felt ice over his back getting out of the sub. "This is my country, I am back," he thought. The light on the beach was not a predetermined security signal. Three rubber boats were inflated. The code name was 'Surawar Tabucol,' Pimentel recalled. Reconnaissance men Agron, Balunes, Taluban, Giron and Valera paddled ashore in one rubber boat armed with walkie-talkies to survey the beach. Upon landing, Taluban and

Giron were to tie a rope to a tree that was connected to the sub to ease the unloading, but the current was so strong that it started to pull their rubber boat away from the shore. Dropping the rope, they paddled until they reached land. It was a blind and "suicidal landing...under the nose of the enemy – with no guerrillas to meet us" Valera recounted.

Ashore, they went in the direction of the small light. It was a place to make salt and no one was around. Two houses were found inland where the village official and his three nephews were willing to help. A Japanese garrison was only one hundred yards away. Taluban and Giron remained while Agron and Balunes returned to the submarine to unload more of the cargo.

The waves continued to pound the sub, drenching the crew and cargo on the desk. Without waiting for word from the Recon group, Sammons and Pimentel began loading the boats with the rest of the equipment. At close to midnight, Valera radioed the sub that a Japanese convoy of twelve ships passed close to shore, and another convoy stopped between the sub and their landing site. They saw barges lowered and soldiers getting into them. Sammons remembered that it was an "airplane carrier escorted by two Japanese destroyers and the escorts passed right in between the sub and the shore." One of the rubber boats nearly collided with an enemy barge. Unloading continued as they watched two sailboats loaded with Japanese infantry soldiers and Filipino military police go ashore. That was too much risk for the concerned skipper. He stopped the unloading and ordered the rest of the cargo still on the deck below. The rest of the men and loaded cargo made it to shore, but half was left behind. The skipper was aware of the Japanese garrison close to the landing site, so he was most anxious to leave. He refused to continue unloading the following night with the Japanese presence and submerged to return to Australia.

They did a complete count onshore so as not to miss anyone in the hurried departure of the sub. Agron later expressed his pride "to be the first to land in Northern Luzon ahead of American troops, returning to the Philippines." Exhausted and hungry, they dragged their equipment into the forest.

At dawn, they wiped out signs of their footprints to avoid detection from Japanese patrols. Old ladies brought an official of Caunayan who warned them not to fight the Japanese or everyone in the area would be killed if the enemy retaliated. They dug a perimeter, taking turns as sentries while burying some of their equipment. Rubber boats were cut up and buried. They covered the pit with dried branches. A Japanese column of more than sixty men passed by as they quickly hid and watched from the bushes. Giron recalled, "My heart beat faster. I couldn't remember my prayers anymore. No doubt all of us had prayed. I held my breath as they passed by. They smelled as if they had their last bath many weeks before." He continued that he "had my Carbine down on my side and the walkie-talkie with me." The decision to ambush the column was quickly passed down. The men opened fire with their submachine guns and twenty-three Japanese soldiers fell in the fusillade. The survivors were dealt with by cutting their throats. There were no casualties among the Filipino soldiers. No equipment was lost. They retreated inland in case reinforcements arrived.

The people in the coastal towns of Bangui and Caunayan were fearful of the Japanese and hardly helped in carrying the equipment. Word had spread of their arrival and alerted the Japanese. Fear gripped the locals as their towns soon became heavily infested with Japanese troops and spies. Suspected guerrillas and their supporters were picked up and tortured to get information. Their situation was aggravated by the scarcity of food and medicine. But when the guerrillas in the vicinity saw the men, they stepped up their ambuscades targeting Japanese supply and communications. It was proof that the Americans were returning.

The group hid for a week, reconnoitering the area while continuing to ask for Roque Ablan and his guerrilla force. No one knew where Ablan's forces were until a Caunayan village official contacted Miguel Garvida, a former guerrilla under Ablan. While waiting for the contact to arrive, they recruited civilians to carry the equipment, set up an outpost in the jungle and armed guerrillas to carry out security patrols. They ate only one meal a day and slept in wet clothes and shoes for faster escape. Tents were not set up in case they had to get away quickly, so mosquitoes feasted on them. They camped close to a Japanese site, taking turns watching. They talked to people and gathered information on the enemy, eventually learning that the Japanese patrolled only the beaches. Food was hard to find even if it was harvest time because what rice the people hid were all taken by the Japanese. Food sellers wanted the correct currency and charged extremely high prices. They learned that civilians were being tortured to reveal their location.

Not wanting to waste time, the party recruited *Apayaos*, indigenous people of Northern Luzon, to guide and carry their supplies and equipment up the mountains, hoping to find Ablan or Volckmann's men. As more *Apayaos* arrived, they served as security. The group set out into the mountains and crossed the Caunayan River with food supplies and three radio sets on their backs. Runners sent in advance informed them of Japanese patrols to give them time to hide. Sometimes, the advance guards were unable to warn them and they were fired upon. It was only three kilometers but the trek became a two-day walk through rain and mud. One guide climbed a tree and announced there were three more mountain ranges. They marched on following streams, making detours and sometimes doubling back to elude the enemy. At night, they froze in their hiding places as Japanese patrols passed them. Climbing the mountains, "their jungle training in Australia came in handy," said Sammons. "We moved by sounds and hit the bush in seeking cover, and we never fired back at the Japs. Thus, we saved ourselves from detection."

Rifle shots broke through the noise of the jungle. Valera said that a "native advance guard was unaware of their (Japanese) presence." They prepared their defenses but learned that the Japanese were firing at a sailboat nearby, which they suspected of aiding the party. As they set up camp, Giron wrote of "preparing a vantage point on a riverbank overlooking the trail from the river. We could not take chances since we were not supposed to fight if we could help it... and fired on the downward incline as the Japanese started to ascend from the river... Japanese bullets whizzed overhead. Some hit rocks, some hit branches of the trees." They moved on when the enemy withdrew at dawn because the locals of Caunayan told them that the enemy always came back in greater numbers.

It took five days of travel through the remaining nine miles, hacking a trail to reach a mountain at 2500 feet above sea level and overlooking the town of Bangui. They rested for three days waiting for the rest of the cargo. In a heavily forested area, they set up camp and built a radio shack. It was an area the *Apayaos* said that no Japanese had ever entered. Huts were made to shelter them from the daily rains. Outposts were set up as Japanese numbering as many as a hundred patrolled the areas around the camp, making procurement of food difficult.

Medical supplies were dwindling, and food was almost gone. Men were becoming sick of malaria or dysentery. They also had to take care of the sick carriers and guerrilla guards at the outposts. The locals gave food, but they did not have much. It was harvest time, and the Japanese had taken their rice. Some were afraid of selling food for fear of the Japanese. The civilians did not want to accept their Australian rations as they were different from food they knew.

It was now a month since they landed. They were concerned that KAZ was waiting for a transmission from them. Three boxes were opened. One had a hand-crank generator for the radio. A cold chill went through them when they discovered that the most powerful radio set they packed that could reach Australia was left in the submarine. It was now impossible to contact KAZ directly. The sulphuric acid for the battery and gasoline for the power unit was also missing. One radio had a transmitter wet with saltwater. The other radios of smaller wattage were working but suitable only for short distances and communication. With the help of agents, they kidnapped a mechanic to fix the motor charger and sent out men to find sulfuric acid and gasoline for the radio. While waiting for the radio to be fixed, recon men went out to ask the locals what they knew of the Japanese in their area. Fifteen men who fought in Bataan soon joined to help transport equipment, secure food, and find the people on the list KAZ gave to contact.

Several guerrilla units arrived expecting to be given arms. Valera interrogated them for information and to determine their loyalty before turning them away. Medical supplies and food were almost gone. In desperation, Agron thought many times "of going out into the open to meet the enemy, but our duty was to secure intelligence reports."

They formed groups with dozens of guerrillas and *Apayaos* to gather information on the Japanese as far as La Union to the east and the Cagayan Valley in the west. They walked all day and night over mountain ranges. Giron wrote of going into villages in Bangui and Dumalneg to learn of the Japanese strength and whereabouts.

Leyte Island, Philippines Landing – October 1944

Radio traffic jumped days before the Leyte landing. The Ilocos area of Luzon had around 5000 Japanese troops with 800 of them stationed at the Gabu airfield (Laoag airport) at the mouth of the Laoag River. Signal men punched out a report to KAZ of Japanese planes parked at the airfield that resulted in their destruction by six waves of Allied aircraft bombing and strafing. The enemy's eight anti-aircraft guns did not have a chance to fire.

The guerrillas continued to ambush even with their limited weapons and training, driving the Japanese south. With the help of the demolition men, they destroyed bridges, blocked roads, and captured arms from the Japanese convoys. Some of the dynamite used had been found by the guerrillas in the gold mines of Northern Luzon.

The radio was able to tune in to many Japanese stations. But no reply from KAZ came from their calls as the sets were not strong enough to reach Australia. After four days of trying to contact KAZ, they heard a familiar signal from another radio station on Luzon. Australia was talking to the station about their landing and looking for them. After almost two months of silence, morale shot up with the knowledge that contact could be made. A message was quickly sent with the assurance that the party, codes, and radio crystals were safe. Enemy disposition was dispatched. KAZ was elated and PRS sent a congratulatory message on 30 Oct 44 to Valera. "We congratulate you and your men on the establishment of your communications and on your splendid feat in evading the enemy pd be alert to all enemy development cma movements of nip forces including aircraft and general intelligence matters pd nr two pd we will be standing for your message pd. Signed: Colonel Whitney."

An American guerrilla officer, Captain John O'Day, of the 15th Infantry, USFIP-NL, arrived and recruited locals to help transport the supplies to Volckmann's headquarters near Baguio City. It was just in time because 500 Japanese were reportedly conducting mopping up operations in the area, moving towards Valera's hideout. Their buried equipment was carried over a hundred and fifty km through trails, mountain ranges, villages, and Japanese lines. They avoided attacks as they travelled on the main road. With over 500 enemy soldiers, the guerrillas were obviously outnumbered, thought Sammons.

At each village, they gathered information and transmitted it to station S3L on the Bondoc peninsula. The locals were happy to see them and provided food. Pimentel said they were "like heroes... gave us flowers, fruits, everything." In return, they gave propaganda speeches, saying "Americans never forget Filipinos." They turned on the radio to the San Francisco station KGEI and created newsletters suited for the guerrillas and civilians to boost morale. Along with an audience numbering in the hundreds and composed of guerrillas, carriers, and civilians, they listened to a Japanese station at the correct hour with a Filipino announcer called "Guernero" who was forced to extol Japanese military accomplishments. There was a Leon Ma. Guerrero who served in the Intelligence Section of the Philippine Army before the war. After the fall of Bataan, Guerrero continued in propaganda work but on behalf of the Japanese.

Valera delayed transfer to Volckmann's camp because he wanted to have a successful contact first with KAZ. Agron was sent ahead to meet Volckmann at the end of October with a letter from Valera which stated that "he (Valera)will stay for several days on the coast to execute his mission, to report as much intelligence work as can be gathered from reliable sources from places within the perimeter of assignment." He was in a strategic location to provide intelligence for the Luzon landing.

The party passed the villages of Dingras and Banna to the camp of the 3rd Battalion, USFIP-NL at Nueva Era. The radio was again fixed and operated to send reports to Samar and Mindanao for relay to KAZ. Nueva Era was suspected by the Japanese of hiding guerrillas with arms and radios and there might be reprisals against the civilians. Saladino and Moscoso were ordered to stay behind with a radio set and the necessary codes as a sub-station. The rest trekked through mountain passes into the province of Abra in the south. Asuncion became ill but refused to be carried. The people in the villages they passed received them with kindness and food. With not much information coming from the hilly area, they moved on to the coastal area.

In the town of Vigan, the commanding officer, Capt. Laudi, of a guerrilla unit under the 121st Infantry Regiment USFIP-NL patrolling the area had to be convinced first that they were Americans as he only saw Filipino men. He supplied carriers, couriers, and someone to prepare food as they travelled. To keep moving, runners advanced to villages to prepare carriers and food before the group arrived. People were happy to see them and became hopeful again as they turned on the radio to listen to the Filipino broadcaster in San Francisco.

Pimentel recalled interviewing a well-dressed woman who was not from the village. She smoked and declared that American cigarettes were better than Japanese cigarettes. She rode in Japanese trucks selling merchandise. Giron became suspicious of her because she knew their route and found a *United Nippon* book in her pocket during interrogation. Concluding that she was a spy, they hurriedly packed their equipment after a courier reported several hundred Japanese following them. Pimentel said that they "were only fifteen men, but with the guerrillas, they were a battalion." They climbed mountains to reach the town of Abra inland then turned south when the Japanese were reported going north.

Taluban found out that one of the guides in the search party was from his hometown. He was allowed to go with the guide to an evacuation center near Baguio City. They found his mother and sister, who were very sick. At first, they were unsure if he was a family member. But when the guide showed them the chocolate candy bar from Taluban, they were so happy and convinced that it was indeed him. Taluban found out that his brother was a prisoner. That short meeting with his family was a precious moment he cherished before he continued to his mission.

While climbing the mountains of Northern Luzon, an unsuccessful effort was made to contact KAZ. The radio central communication station DKI was eventually set up in the new camp. Giron operated the radio receiving messages from different relay stations to forward to KAZ. The messages were marked with a 'P' for priority.

In the meantime, Volckmann had gotten impatient and radioed KAZ for the whereabouts of Valera's party. He justified that the party report to him "to operate more efficiently (to cover Northern Luzon) than if operated independently of USFIP- NL." He asked KAZ to order Valera to attach him and his party to his command for duty. KAZ replied that "Valera landed Mayraira Pt (Maira-ira) Aug 27 with radio and equipment. He was instructed to contact you and Gov. Ablan. No radio contact with him since arrival. Send agent to find out."

GHQ ordered Valera to "place you and your party under the command of Volckmann, use all codes, call signs, and equipment directed by him." Valera received an "'arrest and equipment into custody from Volckmann because station DKI interfered with the communication network in Luzon and the persons operating DKI set are incompetent." The fact was the 3BZ radio of station DKI had stronger signals than Volckmann's set. The radiomen were on the air 24 hours, acting as relay station per orders from KAZ to establish communication from the northernmost point of the Philippines.

With hundreds of Japanese troops mopping up the area heading towards his hideout, Valera followed the arrest orders and trekked the seventy-kilometer distance to Volckmann's camp. Giron attributed "the thirteen weeks of hard training to my ability to climb the high mountains... and survive in the jungles." He remembered having been given only three days of C-rations and needing to survive on his own for the other four days.

The men were sick but forced themselves to walk until they reached the coastal town of Capangan. Whenever they saw Japanese bonfires on the beach, they backtracked until they reached Volckmann's camp with their guides. Before entering the camp, they sent a message ahead of their arrival.

A commendation arrived before Christmas that boosted the men's morale. "Repeat of your trip to Colonel Volckmann's headquarters just received pd aye commend you all your men for resolute devotion to duty that permitted you to get through (trip to Volckmann) and preserved your equipment pd it was special achievement pd. Signed: General MacArthur".

Volckmann was not at his headquarters when they arrived. G-2, Maj. Murphy ordered all their radio sets to be given to his radio men. The men were not allowed to send a message to KAZ using their code and call sign.

Volckmann was anxious for the list of supplies he had requested earlier. His message ended to "omit nonessentials like cigarettes from the shipment." He sent instructions to GHQ to pack the arms, ammo, medical supplies, communication, and demolition equipment in containers that one or two men could carry. Offloading from the sub to the shore would quicken. But the six tons of supplies came in heavy crates. When the boxes were opened, 'I Shall Return' propaganda material took up a large part of the expected supplies. In an ironic reply, he crafted a small brass pipe with a bamboo tag attached. One side of the tag printed USFIP-NL and the names of the five regiments. On the other side were the words 'We Remained' and sent it back in the next sub arrival.

Valera requested KAZ to inform Volckmann to withdraw his arrest order, wanting no friction to happen as they were all were fighting for the same cause. When Volckmann received the instructions and the issue was cleared, the men began teaching Army radio procedures and assisting the guerrilla radio men to communicate with the rest of the radio networks on the islands. The weatherman started sending four observation reports daily.

The party was split into groups of three to four men to the five combat regiments and were set up with sub-stations covering the entire Northern Luzon from Nueva Vizcaya and La Union provinces to Cagayan and Ilocos Norte provinces. Reports were sent to USFIP-NL main station for relay to Australia. Each group had a signalman carrying a set of transmitters and receivers. Recon men were attached to each regiment's S-2 intelligence section. A demolition man in each team destroyed bridges, and the weathermen with guerrilla guards transmitted the weather conditions.

The 15th Infantry Regiment USFIP-NL guarded the western half of Northern Luzon. Valera was the intelligence officer attached to the unit together with signalmen. It was reported he was in a court-martial board composed of all Filipino radio operators from Australia created by Bernard Anderson, the guerrilla commander at Infanta.

Ollada, a recon man with radio knowledge, trained personnel of the guerrilla 121st Infantry Regiment and monitored shipping and airplane flights over the area of Poro Point, La Union. He set up a command post and trained men in ship and airplane recognition as all kinds of Japanese vessels anchored off the town capital. He stationed men inland at the village of Apaleng outside the town to gather intelligence. Enemy strength was concentrated at Poro Point with their depots of supplies, arms, and ammo. The accuracy of the reports sent to KAZ on the supply locations, hidden artillery and anti-aircraft guns, and troops movements on Poro Point destroyed the enemy garrison. The Japanese retaliated by increasing patrols combing the area of the command post. They exchanged fire with the enemy as they packed the radio to move north. When a sub landed at Darigayos Cove, north of Poro Point (November 1944), Ollada accompanied the guerrilla regiment's executive officer and remained until San Fernando, La Union was liberated.

Sammons, a signalman was assigned to Nueva Vizcaya which included the Cagayan Valley, the area of the 14th Infantry Regiment (guerrilla). Giron was left at the main camp to work as assistant chief radio operator, in charge of the signalmen, and at the same time as an instructor of radio army procedures. All Japanese installations, movements, atrocities, rescue of American pilots, guerrilla activities, and results of American raids were reported. Later, Giron and Saladino led companies and platoons.

Six Japanese divisions, including a new division from Korea, disembarked at San Fernando, La Union in Lingayen Bay. The stronghold of the Japanese Imperial Forces in the North was being reinforced. Agents reported 72,000 Japanese troops in the USFIP-NL area suggesting that Southern Luzon positions were also being reinforced.

The runner-relay system was maintained in case the radio went down. The runners were creative in concealing messages in the mouth, anus, or sewed into their tattered clothes. Eventually, forty-four runners were recognized for their dedication.

With the radio network set up, Volckmann wanted to further strengthen his command in Northern Luzon. He reported to KAZ that without a centrally recognized leader, his area was disorganized. He offered to consolidate all commands, especially the strong Huk group. Reports of less attacks by the Huks on Japanese sites were allegedly due to the Japanese providing them arms to combat the guerrillas. Upon receiving the message, Whitney commented to GHQ that "three radio nets were being set up without letting them (Volckmann) know. It was to provide security against loss through compromise and check on the information given." Luzon was not destined to be under one guerrilla leader as in the other islands.

A Japanese plane crashed returning from Manila to their post in the area of Northern Luzon. A briefcase full of documents was recovered from the dead pilots. An elderly Japanese man who had lived for thirty-five years in Baguio was detained by the Japanese as an American collaborator but was released. He translated the documents, and the signalmen coded the message to KAZ. The maps showed the location of eighty-one airfields in the Philippines and their landing strips from Kyushu to Taiwan. The documents revealed that they knew the Allies were expected to arrive at Lingayen, the same place the Japanese forces landed in December 1941.

USS *Gar* - November 1944

The submarine left Brisbane for Pidaido Lagoon in Mios Woendi for her final war patrol to board men and cargo. The skipper was warned that the rendezvous site was swarming with Japanese troops. Whitney commented in a message that "to supply Volckmann had been our greatest headache but must get in." After dropping off five tons at Mindoro for the Rowe party, the sub proceeded to Northern Luzon. Her arrival significantly boosted the confidence of the guerrillas to hunt for General Yamashita, who had retreated to the north by December. The heavy fortifications that the Japanese had built in the towering mountains with artillery reaching twenty miles had kept them away.

Volckmann waited for the submarine at Darigayos. The guerrilla 121st Infantry Regiment set up security perimeters flanking the landing site on both sides at less than five miles each. Enemy patrol boats regularly sailed the coast. Hundreds of carriers were stationed at each primary and backup site to transport the cargo to USFIP-NL headquarters. The guerrillas were ordered to stop activities in the area to avoid enemy attention and allow only transporters. But someone had tipped off the Japanese.

Two days before the rendezvous date at Darigayos, communication between Volckmann's headquarters and KAZ went down when a carabao fell into the river and was washed over the waterfall, landing on the waterwheel powerhouse of the radio. There was a scramble to fix the powerhouse. Once they got the radio functioning again, KAZ sent the security signals for the shore party – two white discs two meters in diameter with a smoke smudge fire in the middle at the designated time. Men in a small sailing boat with three vertical lights and flying a Chinese flag were to sail out as identification to the sub.

The day before the rendezvous date, a Japanese patrol boat appeared, beached, inspected some of the native huts, and left. An American airstrike on passing ships brought Japanese patrols swarming around the Darigayos area, looking for survivors. The guerrillas left them alone so as not to attract attention. On the day of rendezvous, a Japanese patrol docked and ate their meal on the beach. By nighttime they were still on the designated landing site. Darigayos was compromised. KAZ was informed, and the sub continued north to the second landing site at Santiago Cove.

The sub surfaced at Santiago Cove, the backup location at San Esteban but a Japanese destroyer, two light cruisers, and two 50 cal guns were sighted. Now both sites were swarming with Japanese. Whitney commented on his printed radio message, "Both locations compromised, but we must go ahead."

At Santiago Cove, the skipper thought he had landed at Darigayos Cove. With the Japanese presence, it was unsafe to unload. The guerrilla commander of the 121st Infantry Regiment boarded the sub to look for Volckmann. Spotting Japanese fires on the beach, he remained in the sub to return to Darigayos Cove. En route, he received Volckmann's message that Darigayos was compromised. Volckmann saw the U.S. airstrikes on the Japanese port to the south of Darigayos Cove and the Japanese patrols entering the landing site.

At night, the submarine GAR landed 17 weathermen, aircraft observers, including ten Filipino radio operators and demolition men. Twenty-five tons of supplies were unloaded. Two hundred thousand pesos was turned over to Volckmann for operational expenses. It was close to Christmas, and the crew of the GAR generously added to the cargo brought by the men. They gave five hundred pounds more of food from the ship's pantry, loaves of fresh bread, and Christmas decorations.

The party with some of the cargo was unloaded into rubber boats and paddled to shore. Wooden sailboats trying to keep afloat against the strong current and high waves brought the rest ashore. The submarine recharged its batteries while unloading to be able to leave once done. By early morning, the sub with intelligence papers and the Japanese pilot's documents submerged to depart. She met and transferred the documents to the destroyer USS Cassin on her way back to Mios Woendi.

Disembarked were Capts. Fred H. Behan and Donald V. Jamison from Coastal Artillery Corps. Capts. William A. Farrell, and William D. Vaughn along with Sgts. J. C. Bierley and Larry O. Guzman who were weathermen. Filipino soldiers trained in demolition work were Sgts. Clement B. Mapile and Larry O. Guzman along with Cpls. Leo C. Clemente, and Felix Bermudes. Sgts. Arsenio B. De la Pena, Angelo F. Villanueva, Martin Q. Arellano, and Beato C. Alves, with Cpl. Demetrio G. Medrano were recon men with radio training. The lone 978th signalman was Cpl. Ambrosio Pabros. The men shared with the guerrillas the instructions they carried for the coming Luzon invasion. Lingayen Gulf was chosen after Leyte. Morale quickly shot up.

On Christmas Eve, Giron was sent to the coastal town of Pitpitac in Luna north of Lingayen Bay to transmit enemy movements. He recounted hiking for two weeks to bring a receiver and transmitter set to Company M of the 66th Infantry Regiment USFIP-NL at Daklan in the Cordillera mountains. Along the way, "...something burst as if a shot was fired. It jerked my belt at the back. I released my finger from the trigger to feel what it was. I was already wet, and fear had crammed me to the bone. But it was not blood when I looked at my hand. My canteen was punctured. I cursed the hell to the Jap who fired the hole in it for I could not get another canteen anywhere."

Vaughn and Farrell remained at Volckmann's headquarters. Farrell was assigned to G-2 Intelligence. Vaughn reported on the efficiency and scope of Volckmann's command. In carrying out GHQ's orders, the guerrilla regiments prepared weekly intelligence reports and assessed enemy capabilities in their area. Ready to mobilize, reservist guerrillas were called from their homes to join the regulars and undergo combined drills and military lectures. The reservists ensured that there was food at the camp for the regulars. While waiting for the message to attack the enemy, the guerrillas were skilfully improvising. They were distilling alcohol for fuel, making bullets from curtain rods, and printing currency on the back of wallpaper, but they desperately needed a regular source of supplies.

Behan, assigned to the 121st Infantry Regiment, and Jameson to the 14th Infantry Regiment USFIP-NL carried out demolition orders with Guzman, Mapile, Clemente, and Bermudes.

Signalman Pabros set up a radio net that covered the provinces of Cagayan, Kalinga, Ifugao, and part of Isabela. Weathermen Lowe and Bierley requested codes and ciphers for three weather stations in Northern Luzon. The weather reports for successful aerial bombings began coming in to KAZ four times a day.

The Allied landing operations in Lingayen Gulf began in December with the bombardment of San Fernando, La Union, to destroy enemy ships. Pimentel and his group were up in the mountains and identified Allied ships in the bay by their silhouette. From his lookout, he saw American ships going north then south to bomb the Japanese's San Fernando base gasoline tank. The enemy was pushed out of the town. He went down to the beach, passing guerrillas shooting at the Japanese on the highways, capturing their ammo boxes, and burning them. His family was from San Fernando and some of the guerrillas were his classmates in school. Fearful of being recognized and his family detained, he followed orders. He reported Japanese boats and ships' movements and communicated it to the radio in Lapham's camp in Baler for relay to KAZ. He told "the small kids, if Japanese asked for the trail going to Baguio, tell them to follow the trail where guerrilla companies were waiting for them." Female cousins were surprised and happy to see him. They called him *Manong* and insisted that he go to the evacuation center packed with people from the mountains. He became emotional seeing his family after over a decade. His parent and relatives were speechless, shaking his hands. His mother cried. He found out that his younger brother was a guerrilla. He was proud that there were "no traitors amongst the guerrillas because some of the officers ... were my classmates at school. They were surprised to see me." All night the highway lit up with the popping of ammo. When the Japanese returned, his group was almost ambushed when the guerrillas at his flanks retreated to the coastal town of Bacnotan without warning.

Luzon Island, Philippines Landing – January 1945

Radio traffic significantly increased with the pending landing on Luzon. Signalmen's fingers flew punching lists of specific targets - bridges, ammo and fuel dumps, communication sites, concealed airplanes, airfields, especially beach defenses. The Japanese propaganda unit had been seen burning documents since mid-December. More than 120 radio stations and some 25 weather stations were operating twenty-four hours daily feeding intelligence to Australia.

A week before D-Day in Luzon, Taluban and Soliven successfully penetrated enemy areas and demolished roads blocking enemy movement. Their orders were to bomb the coastal village of Sulvec and the mountain caves of Tangadan. Valera in a cub plane directed the bombing of enemy installations and garrisons in the interior villages of Pidigan, Bangued, and the coastal village of Salomague. Later, Soliven was wounded on the face, arms, and legs by a hand grenade while leading his company.

The long-awaited return was now a reality. There was rejoicing at the orders to attack the enemy. Bolstered by the news, the guerrillas rose and ferociously ambushed small enemy garrisons and set up roadblocks trapping the enemy. Couriers were sent out to the people and remote locations to revolt against the enemy with whatever weapon they had. And the people responded, increasing the USFIP-NL's five infantry regiments' strength to around 20,000 men, "over one-third the size of a regular U.S. infantry division." Many of the units were armed only with *bolos* but were eager to strike back. After the war, a guerrilla in Northern Luzon replied to Peter Parsons, that they had no arms, some *bolos,* no BAR (Browning Automatic Rifle). But they had 'rebar' to fight.

When the Sixth Army landed, the main station handled by Giron was changed to LH5. Codes and radio networks with the guerrilla stations were turned over to the regular army. The signalmen continued reporting on Japanese movements, guerrilla ambushes with new weapons, disruption of communication lines, and Japanese troop movements in and out of the Cagayan Valley. Although they were not to go into combat, some of the recon men joined the fight. Reports arrived of Japanese survivors who were killed as they waded to shore from a sunken ship or crashed aircraft. More than a hundred-armed pro-Japanese civilians were killed. Japanese soldiers hiding in burning cane fields were shot as they ran. Bridges were blasted, killing crossing Japanese troops and enemy trucks like the ones going north on Baroro bridge. But they had to be alert to friendly fire. "Advise the Allied airmen not to attack our forces (as they did today at Baroro) who will display U.S. flag upon approach of planes" was an urgent message to GHQ.

As the American troops waded ashore on the beaches of Lingayen, some civilians were heard singing the 'Star-Spangled Banner'. Coordinated guerrilla operations with the landing army limited any opposition by the Japanese. Low-flying Kamikaze planes appeared, damaging, and sinking many ships in the convoy. The Sixth Army's order was to reach Manila upon landing, but the Japanese blew up all the bridges leading to Manila to delay the Task Forces.

The U.S. Sixth Army's three divisions were joined by the USFIP-NL, forming the fourth division to break the defenses of the Japanese 19th Division. The guerrillas had already struck Japanese outposts scattering enemy troops to the north and destroyed their supply and communication lines. As the more than 250,000 troops of the Japanese Army withdrew inland lacking air power, artillery, or tanks, much of the west coast of Luzon north of San Fernando by Lingayen was reported cleared of the enemy.

Large quantities of Sixth Army supplies were stored in a large building at Camp Spencer in the coastal town of Luna. Ollada was assigned to the Signal Supply and Maintenance Company at the camp until he was recalled. The men received brand new uniforms, discarding the rotting ones they had been wearing since they arrived. Although the uniforms were American size and too large, they felt like real soldiers again.

Finding General Tomoyuki Yamashita

Fifteen days before the Leyte landing on Oct 20, 1944, the radio crackled with reports from Manila that a General Yamashita flew in to assume command of the Japanese 14th Army. Reports said he was so new that he asked where Leyte was.

Messages poured of General Tomoyuki Yamashita having left Manila for Japan in December to confirm orders from Imperial Tokyo. Upon his return, he planned to make Baguio City his headquarters. Tunnels were built around the city and the residents began leaving. Reports confirmed the largest concentration of Japanese troops in the city. With Baguio as the official headquarters of General Yamashita, it became a target for American aircrafts. It also had the 2nd largest supply depot of the Japanese in Northern Luzon.

An estimate of over 250,000 troops were under General Yamashita's command, with over half on Luzon alone. A third of the forces was under his direct control, and the rest under the Air Force and Navy. The food supply was sparse. Tokyo's orders to send reinforcements to prevent the Leyte landing had diminished his forces on Luzon.

Messages from agents streamed of increasing atrocities and massacre of Manila residents as a great flow of supplies and troops were leaving the city going north day and night. The troops were observed to be hungry, weary, and unarmed with many using only bayonets. Many begged for food in the villages. They were reported to be refusing to surrender to the Filipinos for fear of revenge.

The U.S. Task Force fought their way from the coastal region to Baguio up the heavily used and guarded Naguilian Road with the guerrilla 66th Battalion. As elements of the Sixth Army neared the city, General Yamashita retreated in February with around 10,000 soldiers of his elite Shobu Group to the rugged Mountain Province marking the beginning of the liberation of Baguio. It was a stronghold of defense in the mountains. Uphill valleys, hills and ridges of Balete Pass, the Villa Verde Trail, and Bessang Pass converging towards his hideout were fortified with thousands of his men along the cliffs, caves, and tunnels to block the path to his headquarters. But the very same defenses also trapped him in the area.

Demolition men Guzman, Mapile, Clemente, and Bermudes with Jamison and Behan destroyed roads and two main bridges – Lamut bridge and Beretbet bridge towards Kiangan, the location of the U.S. headquarter in the mountains. They had observed that the Japanese usually attacked at night. The destruction prevented Japanese reinforcements from the south from reaching the main body of soldiers defending Yamashita's bastion.

Months of ferocious fighting passed following General Yamashita's trail. GHQ set the month of June as the deadline to capture Bessang Pass, the backdoor to General Yamashita's stronghold. Down south at Balete Pass on the Villa Verde Trail, the Japanese had dug up the area with caves to stop the flow of U.S. military vehicles and supplies to the men fighting in the upper end of the Cagayan Valley. A group of demolition men from four submarine missions helped widen the Villa Verde trail under torrents of rain that broke down retaining walls. Soldiers working on the road suffered heavy casualties from Japanese snipers in the caves looking down from ridges and hills.

Giron was in the mountains of Banengbeng, lying on his back covered by bushy undergrowth while a Japanese patrol passed close by. He heard a shot fired and his body shook as the compass attached to his belt was hit. "That was a close call, too close for anybody to admit it," he thought. He had taken charge of a particular station to operate north of Balete Pass with the U.S. 25th Division. Near the town of Pingkian, his group reported engaging the enemy at several villages to clear the way to the Japanese bastion. Continuing north, they joined the U.S. camp at Lantap to the U.S. headquarters at Kiangan.

Another close call was a shot from a Japanese sniper in the village of Oliweg where the 11th Infantry Regiment USFIP-NL was in combat. Giron quickly hit the dirt, saving himself when the next shot hit the magazine of his submachine gun. The bullet lodged on the side of his weapon and tore off a piece of it. Along the way, a Japanese position neared his group's location occupying a lone knoll surrounding a ravine and a small stream. They were outnumbered with only nine signalmen and 15 guerrillas armed with captured Japanese rifles and five Army-issued rifles. They sniped taking down the enemy one by one.

Giron used a 1903 Springfield rifle and added it with two tracer bullets which he fired at targets some 500 yards away. He hit a Japanese climbing an old fence. Knowing the Japanese retrieved their wounded, he waited and shot the others. Later adding to his report, he recounted that "Just like any Banzai charge, the enemy was always noisy. Yelling and shouting, they were not afraid to die. The Filipino guerrillas, on the other hand, chew their tobacco, grit their teeth, and wave their *bolos,* chop here, jab there, short daggers, pointed bamboo, pulverized chili peppers with sand deposited in bamboo tubes to spray so the enemy cannot see." He deflected a bayonet stab with his rifle and hit the Japanese bringing his sword down on him. As his group fought Japanese patrols, he wrote that he kept repeating in his mind that "our job was not to be detected by the enemy. Our mission was to send back vital information on the enemy headquarters."

The 3rd Battalion, 121st Infantry, was ordered to march inland towards Cervantes at Bessang Pass at the end of January. Pimentel's group moved to the area where the Japanese were reinforcing the pass. Around 40,000 enemy troops faced the 121st Infantry. Aside from captured Japanese weapons, supporting artilleries were two infantry guns and two antitank weapons that lacked ammunition. Through extended frontal battles, hand-to-hand combat, artillery fire, and snipers, they tenaciously held every inch of the ground from April to June 1945. The 15th and 66th Infantry Regiments USFIP-NL arrived and joined them for the final assault.

In Balangbang, a town on a ridge, the guerrillas were pushed back because of the lack of ground cover. Signalmen reported another ridge higher at five hundred yards away that machine gun fire could reach and where the enemy had caves. They precariously tried to hold their position on the ridge waiting for reinforcements that did not arrive as the U.S. Army was also fighting in another area. The guerrillas were on their own. After some continuous firing, the *Igorots* scaled the steep cliffs they grew up in carrying hand grenades they threw at every guns in caves and tunnels through the thick fog. The town of Mayoyao, closer to General Yamashita's hideout, was captured.

The capture of Cervantes drove the Japanese from Bessang Pass and opened the back door to General Yamashita's stronghold. At the narrow Villa Verde trail through the Caraballo mountains, the 32nd with the USFIP-NL guerrillas trudged through cold rain and heavy fog that shielded the ridges dotted with the enemy. Supplies were on the backs of *Igorot* men and women. A Signal Corps photo captioned *Igorot women are employed by the 32nd Division to carry supplies over the 12 1/2 miles mountainous trail between Valdez and Imugen (Imugan), Caraballo range in Northern Luzon,* showed an *Igorot* escort armed with a BAR longer than half of his body and *Igorot* women with packs on their backs looking stoic and determined. The regiments, tired, shivering, and some sick, continued their advance climbing the slippery ridges to meet more Japanese troops entrenched in the hillsides.

U.S. airplanes could only strike from 0700-1100 feet when the fog cleared at Bessang Pass. The weathermen's reports several times a day and the intelligence from the captured Japanese documents (turned over to the captain of the submarine *GAR* at Santiago Cove last November) to locate their targets and other enemy locations helped pinpoint bombing strikes. Unfortunately, guerrillas reported that many of the bombs dropped by the planes rolled downhill and exploded on them, killing many. "The ones in the field were all Filipino soldiers. We were the ones fighting the Japanese," related a soldier of the USFIP-NL.

Remnants of the Japanese units launched desperate suicide attacks or were buried alive in their caves. Many were starving and found digging for sweet potatoes or hiding between artillery attacks. Maj. Gen. Fortunato U. Abat, who was with the 14th Infantry Regiment (USFIP-NL) as a private, recalled that the Japanese broke through their defense line during a period of night fighting. They ran straight to their makeshift kitchen and voraciously ate food before being shot. Their mouths were stuffed, and their hands clutched food as they died. They could not turn on their dead comrades and eat their flesh for fear of execution for disobeying orders. A Japanese POW captured by Australian troops recounted that they were allowed to eat the enemy's flesh as a desperate move to stave off severe hunger. In December 1944 an order was issued from the Japanese 18th Army Headquarters that troops were permitted to cannibalize the dead enemy but not the bodies of their comrades. "The troops must fight the Allies even to the extent of eating them." Four men were executed for disobeying the order.

It had taken six months to reach the lair of the 'Tiger of Malaya'. *Igorot* fighters and guerrillas of the USFIP-NL hounded the trail and surrounded General Yamashita with his elite security force at Mt Napuluan in Hungduan. Holding a white cloth as he emerged, the general was taken into custody. It was reported that the "Japanese admiral who was in command of the special submarine flotilla which attacked Pearl Harbor" and Vice Admiral Okochi, who ordered Admiral Iwabuchi to defend Manila, were with him.

Giron wrote that Kiangan was his last operation before being recalled to his former outfit. His group was at the U.S. headquarters at Kiangan where they saw General Yamashita arriving to surrender. Recalled to Manila, he was able to go home. He saw the poles supporting his parent's house and had the creaky floor replaced because it might hit the pig under the house that was for sale. He had been away 20 years and everything was the same. Banana plants were everywhere. He returned to camp, got his pay and another pass to return. He gave the money to his mother who wrapped it in a big handkerchief on top of her head. The U.S. was his home now. If he does not gamble and learn to save, he can get by. He saw that the higher-class Japanese in California no longer had their farms and businesses. After working and saving for three years, he was able to buy three pieces of land.

When news of Japan's surrender arrived, the men continued at their outposts until they were no longer needed and reported back at various times. Some returned to the 978th headquarters in mid-June and August 1945.

Pimentel and his group were fighting at Bessang Pass when orders came that all Filipinos dropped by submarine were recalled to Manila. There, he quickly changed to a new uniform and arranged for a truck to take him and the men to visit their families who were in La Union. After a week, they were back in Manila when Japan surrendered. His unit, the 1st Reconnaissance Battalion, had deactivated, and the men dispersed. He was sent to Leyte to join the 1st Filipino Infantry Regiment to help with the mopping up operation in Leyte and Samar. Before the end of the year, he shipped back to California and waited for a boat to go to Seattle. At Fort Lewis on New Year's Eve, he was given a new uniform and he hitchhiked to downtown Seattle. After having been away for so long, he did not know what bus to take so he walked home. He received his discharge papers at Fort Lewis and did not re-enlist.

U.S. Army historian Robert Ross Smith wrote that the 121st Infantry Regiment, with a strength of fewer than three thousand troops were the best equipped, best trained, and the most experienced regiment of the USFIP-NL. Thousands of Japanese civilians died when General Yamashita forced them to follow him into the mountains. Large quantities of food supplies and equipment were found. "The USFIP-NL accomplished far more than GHQ SWPA, Sixth Army, or I Corps had expected or hoped." wrote Smith.

After the war, General Yamashita's chief of staff, Gen Akira Muto, observed that the "Philippines was the only conquered nation in Southeast Asia that fiercely resisted the Japanese." In his memoir and report, General Krueger, commanding general of the U.S. Sixth Army, described the USFIP-NL at the Battle of Bessang Pass as "one whose magnitude and decisiveness far surpasses the U.S. Army 32nd and 25th Divisions' battles for the Villa Verde Trail and the Balete Pass, respectively." General MacArthur commended the men: "The work of the Northern Luzon guerrillas alone was equal to a front-line Division."

PHILIPPINE MAP

LUZON

Central Luzon

VISAYAS

MINDANAO

- VIGAN
- **MT. NAPULAUAN**
- LA UNION
- CABAROAN
- SAN FERNANDO
- BAGUIO
- BALOY
- PANTABANGAN
- CARRANGLAN
- BALETE PASS
- MUÑOZ
- **MASIWAY**
- SICLONG
- SAN FABIAN
- CALASIAO
- BITULOK
- SABANI
- BALER BAY
- DIBUT BAY
- BALER
- **CABANATUAN**
- **TARLAC**
- **SAN MIGUEL**
- **DIKAPANIKIAN**
- DINGALAN BAY
- PAMPANGA
- BALIUAG
- SAN RAFAEL
- **MALOLOS**
- BATAAN
- **BULACAN**
- **RIZAL**
- POLILLO ISLAND
- LINGAYEN BAY
- MANILA
- INFANTA
- MARIVELES
- LINGAYEN BAY
- CUYABAY
- CABCABEN
- CORREGIDOR
- MAUBAN
- BICOL
- LUBANG ISLAND
- CAMALIGAN
- BATANGAS
- RAGAY
- CALAUAG
- MARICABAN ISLAND
- LEGAZPI CITY
- MINDORO

LINGAYEN BAY

PANGASINAN

CENTRAL LUZON

MISSIONS : September 1944 • October 1944

XVIII

Luzon Island, Central Area

Manila, the capital of the Philippines, was the prize and situated in the middle of Luzon. It became imperative to liberate the city when urgent reports from Manila agents arrived about the new 'Operation Plan Number 11' issued by the Imperial Japanese Army headquarters. "The Army must suppress the guerrillas with as small a group of men as possible and defend the Philippines with the greater part of its force." Imperial Headquarters set the Japanese Fourteenth Area Army Command in Manila to command Luzon and the Thirty-Fifth Army Command in Cebu to command the Visayas and Mindanao. Radio messages streamed to KAZ of airfields being built in the cities of Manila, Clark, Lipa on Luzon, Bacolod on Negros, and in Davao and Malaybalay on Mindanao. The Japanese demanded more men for their army after granting Philippine independence.

Meanwhile, in Manila, the Japanese were training Filipinos for their army but wanted more men to defend the Philippines. President Laurel of the 2nd Republic refused the Japanese overtures to declare war against the United States. Instead, he declared 'a state of war' to avoid conscription of Filipino men into Japanese combat service.

From the landing site in the Gulf of Lingayen, the U.S. liberation forces had to cross a broad central plain crisscrossed by many large rivers and controlled by the Japanese. They would need the help of the guerrillas and locals who knew the terrain and enemy locations.

The guerrillas had to be coordinated, if not united, to ease the way to Manila. Attempts to unify under a single leader had failed. The guerrillas' hostilities to each other over goals, conflicting loyalties, territory, and resources caused unnecessary deaths. The Filipino culture's emphasis on class and age caused friction. The war gave opportunities to strong men from the lower classes over the old landed elite

By July 1944, over one hundred men had been sent on submarine missions to the Visayan Islands and Mindanao. The majority were Filipino 978th signalmen. By the time of the Leyte landing, GHQ realized that there was more than one strong guerrilla leader with a large following to unite Luzon, unlike in the Visayan islands and Mindanao. Several guerrilla bands stood out led by American officers who had not surrendered. They valued their independence fighting the Japanese and amassed men who voluntarily joined from civilians, former USAFFE, and who fought in Bataan. Subordinates were American and Filipino officers leading fighting regiments and companies.

To ensure support for a successful return to Luzon, another one hundred men in missions carrying signalmen, radios, arms and supplies were sent separately to each guerrilla leader to communicate directly with KAZ.

Until the latter part of the year, MACA station on Samar and station S3L on the Bondoc Peninsula were sending their reports to the powerful FRS net control station on Mindanao powered by a robust Navy transmitter. More powerful radio sets brought into Luzon, would directly contact KAZ. Mindanao would then be backup if radios were down, the enemy interfered, or the weather disallowed successful transmission to KAZ.

Maj. Robert Lapham, an American Army officer led the Luzon Guerrilla Armed Forces (LGAF) and Maj. Bernard Anderson a U.S. Army Air Corps led the Bulacan Military Area (BMA) of over ten thousand guerrillas in the wide central plains of Luzon. In January 1944, MACA headquarters on Samar sent signalmen with Capt. Robert Ball to Lapham's camp in Baler with a radio, canned goods and cigarettes. Before Ball arrived, Lapham had relied on couriers to exchange intelligence with nearby guerrilla groups, Manila, and islands to the south of Luzon. His radio network relayed information on the enemy from Northern Luzon, Batangas-Cavite area and the Bulacan province. Maj. Edwin Ramsey a former Philippine Scout, initially in central Luzon moved his headquarters to the Manila area and produced more information on the enemy from the city than elsewhere. Most of the guerrillas, formerly USAFFE, and men who fought in Bataan reported to them.

Lapham was aware that submarines in the southern islands were bringing in men and supplies. His men were armed with weapons salvaged from battles with the Japanese and *bolos*. He requested a submarine to land on the east coast of Dibut Bay which was sheltered by mountains. The nearest Japanese garrison was five miles away at the end of a heavily forested mountain range.

USS *Narwhal* and *Nautilus* – September 1944

While General MacArthur's forces were at Morotai building an airbase to launch aircrafts into the Philippines, only five hundred miles away, the submarines *Narwhal* and *Nautilus* were harnessed for multiple missions to central Luzon. At times they landed in the same month bringing twice as many men and cargo.

Bartolome Cabangbang returned, leading a mission one year after he and Enrique Torres evacuated from Negros back to Australia. They were pilots and had fought on Bataan and Corregidor. They were the first submarine mission to land in central Luzon with radios and signal men. Before them, 978th Signal men from the Visayan islands and Samar had crossed over to set up radios in the Bondoc peninsula and other parts of Southern Luzon. As a special agent of GHQ, Cabangbang reported on the guerrilla situation on Luzon and tried to convince the several leaders (Lapham, Volckmann, Anderson, and Ramsey) to cooperate and be more effective in fighting the enemy.

Vivid memories of the doomed Corregidor stayed with Cabangbang as he returned, wanting to do his share in the liberation of his country. He wrote of the desperate last days that led to the surrender. "The Japanese placed their artillery batteries along the coast of Cabcaben and Mariveles in the southern coast of Bataan facing Corregidor. All fired towards Corregidor on April 29 (1942). Concentrated a few of the batteries on a single point for several hours. Hit and destroyed all gun batteries, brought all dirt (into) tunnels, hit ammo dumps distributed along roads of Corregidor... destroyed gun emplacements and trenches along beaches. Fired at Malinta (tunnel), Navy and hospital tunnels, and all other concrete tunnels. Air bombardment (was) in groups of 27 planes. Continuous bombing April 29 – May 6. Not a tree standing without a scratch nor lizards and birds. No more sound of our ack-ack (anti-aircraft) guns because all our batteries gone. Our forces seek shelter inside tunnels day and night except for those who had to stay on the beaches, gun batteries, outposts, and we seldom saw their faces again. Those in the tunnels not strong enough were buried alive. Those who went out for sunlight and fresh air (were) hit by shells."

After receiving their orders, the selected men were flown to an isolated island off the coast of Darwin to acclimate and "waited for two weeks because the submarine was under repair," wrote Ronald De Torres, a member of the mission. The large party of forty men and supplies boarded the *Narwhal* and departed Fremantle to Dibut Bay in the east coast of Luzon. Parsons and the son of Whitney joined the mission. Parsons was to allay Whitney's concern over whom to send supplies and give orders to build roads and bridges in central Luzon.

A source reported that the *Narwhal* and *Nautilus* were used, but the men wrote only of the Narwhal. It was reported that the Nautilus brought another one thousand assorted weapons including machine guns, Carbines, automatic rifles, and grenades, with medicine, and dynamite.

On the sixth day at sea, the sealed orders were opened, and the landing spot was revealed. Parsons warned the men that threats to their mission were from the Japanese, pro-Japanese Filipinos, and guerrilla bandits who forced whatever they wanted from the civilians. Starvation could cause betrayal.

The trip was smooth sailing even with so many men. Quarters were tight. The sub dove and quickly heated up from the diesel engines spreading to the rest of the sub. The oxygen in the air was quickly depleted from the sweating and breathing. Vapors from diesel fuel, cooking, and sewage, compounded the foulness. Water was limited for bathing and there was no laundry. All eagerly awaited their destination to be able to open the hatch and breathe a rush of fresh air.

An urgent message was received enroute that "certain sabotage materials onboard manufactured in England had been found in danger of spontaneous combustion and should be destroyed immediately." Parsons refused the order emphasizing that the guerrillas on Luzon needed the explosives. For the rest of the three weeks, the crew and passengers were on edge.

Submarines were under strict orders not to surface during the day when they were within 500 miles of a Japanese airfield to avoid aerial observation and attack. With the Japanese controlling the coastlines of the islands, this was almost everywhere. Cargo for Lapham and Anderson consisted of Carbines, ammo, grenades, medicine, CW tins, rations, candy, cigarettes, and vitamins while the mission party's supplies had a smaller amount of ammo, office supplies, submachine guns, medicine, signal equipment, and no ration. The assumption was that they would be fed by the guerrillas. To speed up the unloading, Parsons sent a message ahead to the welcome party to build several bamboo rafts.

Lapham was ready at Dibut Bay, checking all the recognition signals and making sure the recognition panels were spaced properly. A force of guerrillas commanded by Juan Pajota, a Philippine Army officer, provided security. Lapham got out to the Bay in a small boat so the sub could see him and waited all day. By early evening, the *Narwhal* sighted the security signals and suddenly rose out of the sea "like a gigantic whale," he said between him and the shore. "Huge." The sub was close to the shore in the deep Bay. Lapham ate a hearty meal onboard.

Parsons went ashore to meet with the *Dumagats,* an indigenous tribe, and civilians to check the bamboo rafts to unload the sub. The sub provided one rubber boat. A line was tied from the sub to a big tree on the shore. The large bamboo rafts loaded with fifteen tons of supplies were tied to the line and quickly reeled in by men onshore. The unloading took four hours which was the same time as earlier missions with men and supplies half the size.

Twenty-one reconn and 978th signal men were to go to Baler for Lapham and the same number to Infanta for Anderson. Each group had one weatherman. Torres' team was assigned to Lapham around 130 kilometers north of Manila, to drive communication and intelligence channels through north central Luzon. Cabangbang was assigned to Anderson south of Lapham's camp to drive communication and intelligence channels through south central Luzon. Four weather observation stations were to be set up.

After Torres' group disembarked at Baler, the sub checked for stowaways, a lesson learned from earlier missions. Parsons left a case of good scotch with Lapham. The sub slipped into the sea to bring Cabangbang's group south to Infanta.

Torres's group with the assistance of Lt. Jose B. Malan landed by Masanga River. In their party were 978th men: Sgt. Carl Wright, Sgt. Henry E. Runtal, Sgt. Marcelo B. Umipeg, Cpl. Porfirio Balino, Cpl. Lucas D. Runes, Cpl. Rosendo M. Palafox, Pfc. Andres C. Silvia, Pvt. Robert B. Corpuz, Pvt. Nicano G. Ambrosio, Pvt. Cledonaldo J. Isaac, Pvt. Juan C. Marcelino, and T4 Angelo A. Calmorin. Recon men and technicians were Sgt. Faustino B. Bacus, Sgt. Maximino C. Cacas, Cpl. Vicente Anacleto Jr., Cpl. Edilberto S. Bibat, Cpl. Estanislao S. Ibarra, and Cpl. Luis C. Quindiagan. Weatherman, Sgt. Chester H. Moore joined them.

Bibat saw the half-clothed and half-starved guerrillas who came to help. They believed that aid had come. Seeing the amount of supplies, more civilians were recruited to help carry the load through mountains and jungles to Lapham's headquarters.

The shipment was distributed amongst the larger LGAF (Luzon Guerrilla Army Forces) organizations. The supplies had radio sets, Carbines, automatic weapons, ammunitions, including candy, toothpaste, sewing needles, cigarettes, and American magazines. Transceiver radio sets found in England small enough to be carried by one man in a backpack were sent instead of the bulky sets that needed several men to carry. That made it easier to set up more radio stations to monitor a wider area and to move quickly to avoid triangulation.

Spirits of the guerrillas soared at the sight of the long-hoped-for 'aid' before them. Lapham reported that one of his men described that even with only a few packs of cigarettes that arrived, the men were satisfied with just smelling them at first because of the risk of bringing them in. There were kisses on MacArthur's picture on the 'I shall return' matches with tears flowing profusely. As soon as they received their new rifles, men began practicing. Trees with missing chunks of bark could be seen everywhere. Before, any supplies the Luzon guerrilla received were brought by carriers from the Visayas or Mindanao and took months to reach them. By the time they arrived, the load would be less than half of what was initially sent due to pilferage and loss from weather and hazardous trails.

Lapham wrote that one crate of mislabeled Thompson submachine gun turned out to be full of old U.S. cavalry sabers. The guerrillas who were comfortable using their *bolos* distributed them anyway and charged at the enemy, grateful to have a weapon.

"Improper cargo handling before departure and mix-up inside the sub caused much radio equipment to be missing," Torres reported. Half of the supplies that landed with him were damaged. Some dropped into the waters. He felt that the "native population and guerrillas refused to extend assistance" with not enough men to help transport the cargo that was mostly for them. Unknown to him was the people's fear of Japanese revenge. Some wanted to be paid to be able to buy their food. The cargo included tented hammocks and mosquito nettings zippered to protect them from malaria and dengue while on patrol, but they could also be death traps when caught in a surprise night attack. Several tin boxes soldered shut contained 1 million counterfeit Japanese currency printed in 1943. Lapham declined the fake enemy money, fifty thousand genuine Philippine pesos, and real American money because Parson told him he would have to account for it. To get receipts for money given out was difficult and, at times, impossible. Paper was scarce and if the enemy captured records with names of guerrillas or civilians, they would be branded as helping the Americans and be good as dead. Many would not write their names anywhere for fear of being known and found. For the men's operating expenses, Bacus kept track of the ninety-seven thousand pesos Torres' party was given for intelligence work. He turned over the remaining funds to Torres' when he was recalled back to Hollandia.

Torres' mission was to join Lapham's guerrillas and extend the intelligence network through the north-central area to cover Lingayen Gulf, the landing site in Luzon. Torres split his men into 4-man groups. Malan oversaw the net control station F29 with Umipeg, Palafox, Balino, Corpuz, and Sivila. The others operated the sub-stations 9GM, PVG, LOD, and IBO at their assigned outposts. Each group had money for expenses to run their organization and intelligence net.

The Japanese were seen amassing forces to prepare for jungle warfare in a line extending from Dingalan Bay in the east to Lingayen Bay in the west. Radio stations were set up from Lingayen Bay traversing the island to two stations at Baler Bay and Dingalan Bay and as far south as Manila, to transmit intelligence reports to KAZ. Weather observation posts were set up. The radios crackled with reports on a division from Japan landing in San Fernando, La Union. In the east coast, three thousand Japanese troops arrived by Baler Bay. The Japanese buried electric wires that connected explosives along the main highway. Cabanatuan in the central part had hundreds of light and medium tanks.

There were also the Huks, *Hukbong Bayan Laban sa Hapon* (People's Army Against the Japanese) that posed a serious obstacle in accomplishing their mission. The guerrilla guards of the sub-stations often clashed with the enemy and as much with the Huks who forced other guerrilla units to join them and civilians to give food. Ambrosio was killed in Tayabas and Sivila was wounded in action.

Torres stayed in Lapham's headquarters with five local radio operators to manage the radio networks. Malan operated the net control station near the camp. Messages began pouring in on Japanese fortifications, airfields, and troop strength. The Japanese defense line in the Dingalan-Lingayen Gulf sector close to the Luzon landing and fortifications in Bataan and Baler Bay were monitored. Ship sightings in Manila were flashed. Japanese veterans mostly from Tokyo and Korea were seen in the Nueva Ecija area training without arms or using arms of the Philippine Army. The men felt helpless as thousands of Huks armed with thousands of rifles obtained after the Bataan surrender continued their extortion and looting of the civilians. Whenever enemy patrols neared the net control station, the radio was moved and temporarily established in a new location.

Torres left headquarters to go into the interior and set up more stations but warned KAZ of slow progress due to lack of transportation and aid. He got sick with pneumonia but persevered when he felt better to manage his group's obtaining of intelligence data from the north central areas. Over two hundred radio messages were sent. By the end of November, the net had nine sub-stations.

He regretfully reported that Cacas contracted malaria and died. They brought medicine with them, but that did not last long helping the civilians rampant with the disease.

A month after arrival, KAZ was informed that the Baler net control station was detected. Malan reported Japanese locations with fifty pillboxes along the beaches of Eastern Luzon at fifty to one hundred yards apart camouflaged with plants. To prevent a surprise attack, Malan posted ten men around the camp. For protection, he asked KAZ for "two machine guns with five thousand rounds ammo, two light mortars with sights and 200 rounds. Also, radio communication equipment, extra batteries, binoculars, walkie-talkie, radio spare parts, gas, oil" and many more were on the list. Torres added to the list sugar, coffee, salt, and salmon. Malan reported that the Japanese left in his area were aged fourteen to seventeen years and believed that the war was lost. They were armed with only bayonets without rifles conducting extensive day patrols and confiscating all *bancas*. Agents said that the Japanese were suffering from a lack of food. Later, older Japanese troops arrived wearing green suits with canvas armed with rifles and four automatic rifles. The surrounding areas were being reinforced. But the civilian morale remained high.

Runtal's move to the lowlands to set up more stations was delayed due to lack of transportation and help in carrying supplies. Civilians and guerrillas had to be recruited and paid for their food and transportation. He reported that he did not get cooperation and help moving to the lowlands. His men did everything from carrying supplies to procuring food. The group crowded in a sailboat with supplies and equipment passing Japanese mines and shore defenses. A storm broke out, and the sailboat began to fill with water, nearly capsizing. After fifteen hours of crossing a six-kilometer stretch of water, they docked with only two of the five radios in working condition. Seven days of rain damaged the cargo, some of which was looted along the way. Taking matters into his own hands, Torres radioed KAZ for authority to form his own security and supply procurement detail to provide transportation of men and equipment under his command. Without the support, it was difficult for the men to concentrate on their mission. He saw guerrillas unable to even take care of themselves. KAZ replied in the affirmative, allowing him to organize his own band to protect intelligence operations.

Civilians paid more attention, and guerrillas responded quicker to officers' orders than enlisted men's, observed Torres. He informed KAZ that "merely for security reasons," they would assume different names with higher ranks. He became Major Jaime Flor and Malan became Captain Jose B. Miel. The others in his party assumed names and the rank of 2nd Lieutenant. With higher ranks, people would believe that they would receive protection from the Japanese and in turn provide food. No laws, jails, or judges bound the guerrillas' activities, although a guerrilla commander on Luzon decided that looting, rape, and giving aid and comfort to the enemy deserved a death sentence.

Bibat trekked to Dikapanikian Point overlooking Dingalan Bay to watch for Japanese ships. It was a harrowing trip with a Japanese patrol trailing his group for days. The group leader was American signalman, Wright, but after a week, he left and was not seen again. They sent observations for two months until the radio broke down, and couriers had to be used to deliver the reports to Lapham's headquarters. Orders arrived to return to headquarters and report to Malan, in charge of the net control station. As they were leaving the outpost, a Japanese patrol of twenty men surprised them, hitting one of the guerrillas and captured another. They exchanged fire and rapidly retreated as there were only four of them. With the help of a civilian, they hid at a guerrilla officer's camp for several days while the area was swarming with thousands of Japanese in the Bitulak and Dingalan areas. They talked to as many civilians as they could on what to do when the Allies land on Luzon. They were to hide from the retaliating enemy or revolt against them. The Japanese presence doubled in the area and Juan Pajota's guerrillas in the Cabanatuan area helped them to continue safely. Malan was relieved to see Bibat, as radio traffic had tremendously increased in the Pangasinan and Nueva Ecija area with news of the landings in Leyte and Luzon. The radio was on the air day and night for months until orders arrived to report to the Sixth Army headquarters at Calasiao in Pangasinan where they operated with the signal group until March 45.

The second mission group of Lt. Bartolome Cabangbang landed at Infanta, the camp of Bernard Anderson who had been unsuccessful in contacting KAZ. Lt. Claudio Tamayo assisted him. Two men were first sent to shore to reconnoiter and contact the guerrillas. When they gave a clear signal, the rest of the party disembarked. Under a storm of heavy rain, it took longer to unload than Torres' party using bancas and rafts that Anderson's men prepared. The 978th Signal men were Sgt. Ronald A. De Torres, Sgt. Fred Balcena, Sgt. Mariano Arce, Cpl. Astor Parong, Cpl. Cipriano Fernandez, Cpl. Cirilo Lopez, Pvt. Segundo Q. Albalos, Pvt. Carl M. Mateo, and Pvt. Juan Palac. Recon men were Sgt. Isay Real, Sgt. Florencio Baldos, and Sgt. Vidal Garcia. The technicians were Cpl. Eustacio Corpuz, Pvt. Marcelo Gonzales, Cpl. Angelo Oandasan, Cpl. Henry Sabado, and T5 Paciano Racho. Weatherman Sgt. Muri Hendrickson joined the group.

Corpuz disembarked remembering that he had sworn to return onboard the *Mactan* as it left Manila in flames with his wounded legs encased in a plaster cast. The clenched fist in the air was now tightly grasping his rifle.

Four evacuees boarded the sub after fifteen tons of medicine, arms, ammo, demolition equipment, propaganda material, radio sets, and food stuff were unloaded under heavy rain. One cargador fell into the water and drowned. The sub submerged with Parsons to deliver the rest of the cargo to Mindanao.

The men worked all night with the help of the guerrillas to clear the beaches of the cargo and carry them five miles inland to Anderson's headquarters. The arrival of twenty men with more powerful radios for direct contact with KAZ exhilarated the camp. With contact established, the message was for the party to "remain under the operational control of GHQ, charged with the development of a radio net designed to reach as many enemy positions as possible of major strategic importance to your area, and to implement plans to provide GHQ with weather and air warning information. Its facilities are available to you and should be located to serve best intelligence collection requirements of your area."

Cabangbang noticed that the camp looked organized as the supplies of medicine, other items, and money for intelligence work were turned over. Anderson burned most of the magazines after reading them and the propaganda materials of candies and matchboxes for fear of harming the Filipinos who supported his organization. Materials with a date after the fall of Bataan was proof that submarines with Americans had returned.

Miguel and Duran, 978th Signal men (Samar mission June 1944) had arrived earlier from Samar to help with contacting GHQ because KAZ was not easily convinced that they were not the enemy as they attempted to make radio contact. Even Anderson's best foul language to persuade the KAZ operators failed. The radio burned out and a generator fueled with coconut oil was used to try and keep it working. When that failed, couriers were used to dispatch messages to station S3L in the Bondoc Peninsula and to FRS on Mindanao to relay to KAZ.

Cabangbang was G-2 Intelligence for Anderson, and all reports were sent in his code. The more powerful radio brought in established contact with proper security procedures by the end of September, barely weeks before the Leyte landing. KAZ was elated as they did not expect communication that early. The signal men had the radio on the air 24 hours as reports poured in preparation for the Leyte landing.

While preparations for the Leyte landing was ongoing, plans for the Luzon landing were being laid out to establish radio stations and proper contacts with major guerrilla bands on Luzon. Agents provided intelligence from the planned Luzon landing site down to the Pandacan Railroad station in Manila, the distance of around 300 kms that the Sixth Army would have to travel.

Agents reported on the Manila area, American air raids, and Japanese defenses. At the end of September, KAZ was informed of four Japanese transports loaded with POW sunk off Bataan, the east coast of Zambales. Three months later KAZ received a list of some of the rescued prisoners that joined the guerrillas in the Zambales area.

"Messages were pouring in as fast as we could send them and continued for a long time." Transmissions with three hundred letters per message were sent daily except for aircraft warning messages that were flashed immediately. Cryptographers and radio operators were needed to avoid messages piling up, but food supply and maintaining secrecy limited recruiting more men. A school to train men in intelligence, demolition, weather, and aircraft observation was set up. Money was needed when agents were sent out to outlying provinces. Going through enemy-controlled territories was a constant hazard to their safety and security. It was becoming difficult to recognize the enemy as a majority were beginning to wear civilian clothes and appear barefooted to blend in with the locals.

Although Anderson did not get along with Ramsey, he tried to improve the dispute in his area of Rizal over territory and men between the Hunters guerrillas led by Eleuterio Adevoso, Markings guerrillas led by Marcos Villa Agustin in the mountains of Rizal north of Manila, and the communist Huks. He convinced Marking that General MacArthur's orders meant weapons and bullets. He brought a radio and three signal men, Corpuz, Parong, and Gonzales with supplies to operate with Yay Panlilio who headed the Intelligence Unit proving his commitment to fighting the Japanese, not simply establishing his own power. The radio men were needed as Japanese troops garrisoned Manila and surrounding areas. Panlilio described Anderson as pale and barefooted. She wondered at the "New Guinea boys, so called because they had been trained for Signal Corps work in New Guinea... why would they leave America, if not to give their lives for the Philippines? To us, they were heroes," she wrote. The signalmen arrived with medicine and clothing aside from helmets, boots, and Carbines to be in constant contact with Anderson's camp. They distributed cigarettes, tiny paper kits with needles neatly arranged in red flannel folders with threads of many colors, toothbrushes, toothpaste, and safety razors. KAZ's orders were definite: No hostilities against the enemy; simply gather intelligence.

Operatives rushed to camp bringing news on the bombing of Manila and Clark Field by carrier-based planes and the sinking of some 60 enemy ships in the Manila harbor. Needing a high site for better reception and to look down into the Manila harbor, the signalmen went upstream thru Cuyambay to reach higher grounds. Gonzales drowned while crossing the treacherous Agos River. The men took turns sending messages to Cabangbang, who had moved from Anderson's camp.

Edwin Ramsey's camp was located at Rizal, only 20 km from Manila, and provided valuable information on the enemy's disposition around and in the city. They reported using homemade bombs, hitting a Japanese fuel depot, railroad tanker cars, oil tanks, and starting fires in the piers. The reports were sent by couriers traveling for days and sometimes weeks on sailboats and overland to the radio net in Panay and Mindanao islands to transmit to KAZ before a radio was set up in the Bondoc Peninsula. There was no guarantee that the messages would immediately be sent and received by KAZ if the radio used was weak and powered by men pedaling furiously on a bicycle for three hours to store electricity to transmit fifteen minutes at a time. Impatient to contact KAZ directly, Ramsey risked traveling to Mindoro with four bodyguards to get a radio. It was a futile trip as the party of the Mindoro November 1943 mission had been discovered, scattered, and some of the men captured and killed. Eventually, Negros provided a radio and a radio operator if reports were relayed through Negros to forward to KAZ.

Cabangbang messaged KAZ that ten weather stations were set up on Luzon using Philippine government weathermen connected to the radio net control station at Anderson's headquarters. Intelligence reports from Northern Luzon under Volckmann, Filipino-led Hunters and Markings guerrillas, operatives under Capt. Alejo Santos, a former USAFFE officer in Bulacan, and Manila were gathered and relayed to KAZ. Santos who evaded capture on Bataan, joined the guerrillas under Anderson and formed the Bulacan Military Area (BMA) expanding to over 20,000 men to protect the women from enemy molestation and the civilians from confiscation of their harvest.

Radio, intelligence and demolition teams sent out to various guerrilla units were successful in enlisting cooperation in the coming liberation. Guerrillas exhibited much pride when Cabangbang asked the guerrilla commanders' help for his radio operators to train agents on collecting more accurate data, sabotage, and radio. Included in the training was the importance of obtaining the number and type of anti-aircraft guns by making pencil-and-paper rubbings of the gun serial numbers on the identification plates for accuracy. Sketches of enemy positions showing details were invaluable.

To obtain guerrilla support, mission leaders issued commissions at risk of not being recognized by KAZ. This was proven after the war.

Anderson reported that some guerrillas interfered with radio commands, harassed civilians for food, and left them to torture and death by the enemy when they retreated. He warned KAZ not to furnish arms to the Huks. He listed guerrilla groups that were good potential for intelligence and combat. The Chinese anti-Japanese force's patriotism stood out because they did not want to receive a salary if part of the U.S. Army.

The Bulacan Military Area (BMA) guerrilla organization agreed to have Cabangbang's group established in Victory Hills, Bulacan. After three weeks, at Anderson's camp, all equipment and personnel moved across the Sierra Madre mountains to Zambales, closer to the Lingayen landing site with the help of the BMA guerrillas. De Torres wrote of a disagreement with the guerrilla commander as a reason for the move. Tamayo oversaw moving the supplies and equipment across the mountains. A storm and flood delayed the trip. The food supply dwindled, and all suffered from exhaustion and lack of sleep as they carried their supplies across the mountains. The rainy weather and mountain obstructions prevented contact with KAZ. Iron deposits in the mountains obstructed transmission and persistent calls were unsuccessful. They had to find a higher place for the radio. A Japanese patrol found them but was repulsed.

Tamayo split the group into 3-man teams of signal and recon men with a radio set to set up eleven sub-stations to collect and transmit information to the net control station without delay. The rest of the men stayed at headquarters in Malolos, Bulacan to operate the net control station. Cabangbang oversaw the station and used the underground name of J. C. Cabrera to sign his memos. The first contact with KAZ was in October, barely a month after their sub landing.

With all the men out in the field, Tamayo decided to survey the landing forces' path. He dressed in civilian clothes and crisscrossed Luzon taking the national roads posing as a traveling salesman with a pistol in his boot. For two weeks, he noted the beach defenses of Lingayen Bay and the coastal town of San Fabian. By the end of December, a courier sent his reports on the planned site of the landing, road conditions, and the volume of Japanese traffic. He went inland to the town of San Nicolas as it was in the path of the American Task Force heading for Manila. He warned the inhabitants about the beach bombardments before the landing, but Japanese soldiers were rounding up the civilians, so he hid among the houses until it was safe. Tamayo proudly wrote that he "evacuated his nephew's village to an open rice field and wave to American planes. That way, they know you are Filipinos. I saved the whole population of six hundred people from being bombed and strafed." A week later, after the Luzon landing, he crossed the Japanese line and reported to the Sixth army at Calasiao and worked for a month in the G-2 Intelligence office until he was called back to his unit's headquarters in Malolos, Bulacan.

De Torres with a radio maintenance man set up a sub-station at San Rafael in Bulacan. He was able to penetrate the enemy lines but Japanese and pro-Japanese Huks surrounded the station. "There was only a handful of us, but we struck and hit them hard. We had a few guerrilla casualties." The station was raided, but all equipment was evacuated. He reached the net control station camp by mid-January. Two more net control stations were set up in Malolos, with him as chief radio operator directing and supervising the transmission of reports to KAZ.

Balcena handled the relay stations JWR1 and JWR2 that received all information from the field. Balcena said that "it took a week or more for couriers from operatives to reach them with information but in Bulacan, it took only three days to get the message through." They listened to important radio traffic and anti-Japanese talk working more than eighteen hours daily. Eventually 15 radio stations fed into JWR. It was crucial when a radio station was set up in a Huks headquarters in San Luis, Pampanga. Operated by guerrillas Maj. Pelagio A. Cruz and Capt. David Pelayo, the messages provided vital information on around 12 landing fields in Pampanga and Tarlac provinces. Japanese troop movements were monitored passing through Pampanga to Northern Luzon, Bataan, and Zambales. For six months, over 100 messages were sent daily to KAZ.

Baldoz, the demolition man in the group, took charge of all instructions on making booby traps and blasting railroads and bridges.

BMA G-2 Intelligence provided instructions to operatives on the type of intelligence work needed by KAZ. But the result was sporadic, done by a few at indefinite intervals and individual initiatives. Cabangbang persistently informed the agents to be more thorough and faster in obtaining information on the enemy. It was essential to assess sources from unknown locations. "Our operation covered information concerning the enemy's installations, ammunition and gasoline dumps, shipping, and troop movements from Tarlac, to and around the City of Manila within a radius of not less than 50 miles. These were done through the aid of the guerrilla men who managed very handsomely in obtaining the necessary information which we relayed to KAZ," Balcena wrote. Cabangbang prepared all the messages for transmission while guerrilla men they trained encoded and decoded.

The men crisscrossed to various locations in Luzon to set up radio outposts. The team of Arce, Palac, and Racho were in Batangas, south of Manila, and worked with the Hunters Group headed by Adevoso. In November, they were raided by a Japanese patrol and lost the radio set. The men scattered and reported back to headquarters at different times. Reports claimed that recon man Palac killed seven Japanese and captured two others. Arce was then assigned to Pampanga to operate station 2DW.

Oandasan, Machon, and Sabado were in Bulacan collecting information and training civilians on demolition, radio procedures, and intelligence gathering. Sabado's radio covered Polillo Islands where there was a large Japanese garrison. Oandasan was raided by a hostile group of Huks while teaching a class. Machon was captured but released.

Signalmen Lopez and Real succeeded in smuggling their radio set YQZ9 into the town of Aglao in Zambales after previous attempts to have one sent failed. For their efforts, they were commended by GHQ.

In Tayabas, Signalman Mateo met up with Holgado and Advincula who were from station MACA on Samar. Another source said it was Montero and Advincula. Their camp was raided, and they withdrew to the mountains of Mauban. Although the Japanese heavily patrolled the area by air and water, they decided to stay and continue observation. Mateo changed to civilian clothes and found the town of Mauban as a staging area of the Japanese. Fresh troops came and left. One morning, they saw a Japanese plane crash into the sea nearby. They captured the pilot and radio operator for questioning at Anderson's headquarters, where they tried to escape and probably shot.

Mateo returned to Malolos before proceeding to Bataan to set up station BZ9. News of the pending landing made people restless and desirous of striking back. On the way to Bataan, he feared passing through the town of Guagua that was teeming with Japanese soldiers. He recalled a young woman who asked him to sit in a *carretela* (horse cart) with her and his radio while she passed by many sentries. He was awed by her unstoppable courage and had hoped to thank her after the war someday.

Exasperated with his unsuccessful attempts to unify the Luzon guerrillas, Cabangbang issued a message to all the commanders to stop the squabbling as it gave a bad impression and amused the Japanese. GHQ had ignored their requests for command over each other. A week later, Cabangbang had to backtrack and send out another memo that General Macarthur did not want to establish a unified command on Luzon at this (late) stage of the war. GHQ directives were addressed to Bernard Anderson, Russell Volckmann, and Robert Lapham like each one was an island commander. Orders to each guerrilla leader were to monitor their own location, men, and objectives ensuring that the entire Luzon was covered. The independent guerrilla commands were to work together when the liberation forces land in Luzon. Lapham wrote that "MacArthur did not take sides but replied to each one's message directly without references to other guerrilla leaders."

Cabangbang needed more men and requested KAZ to send another party. De Torres reported that four groups of men arrived in Bulacan in late December. The arrivals might be the party that landed in Dibut Bay in October.

USS *Nautilus* – October 1944

The urgent orders for signal, demolition and aircraft warning personnel went out for the next mission. The men were trucked from Camp Tabragalba to Milton Camp. Cpl. Amrafel Pascual was not glad to be pulled out of paratrooper school as his dream was to be a paratrooper. In the U.S. his age prevented him and Sgt. Jerry Regalado, from doing so.

A seaplane brought them to Cairns for refueling and overnight at Port Moresby. A Salvation Army truck served coffee and doughnuts while refueling at Madang before arriving at Hollandia. Whitney met them and "talked nicely as a politician... at Biak we will get all we need...his last words 'will meet in Manila." Sgt. Juan Pilar thought it was a "dramatic and inspiring talk that gave him more determination and feeling never felt before." Cpl. Mack Pulido asked for their GI watches and cigarette lighters as previous parties were given. Sgt. Thomas Amante recalled, "no watches available."

The party flew to Woendi Island opposite Biak. They sweltered in their uniforms as they spent the night at the warehouse to get their equipment, arms, ammo, food, clothing, and other needed items. Amante noticed that the Army warehouse was not organized. They were told to help themselves but keep an inventory of all they took. No watches. Pascual said they were supposed to get arms, but all they got were "pistols without holsters and ammo for their Carbines." They waited for the rest of the equipment to arrive. Watches and lighters were to be with their equipment.

When the lieutenant officers found out that there were no regular orders to the Philippines, there was some soul-searching on whether to continue or go back. Without a specific destination, the submarine could bring them anywhere possible to the islands. Amante recalled "that only Lt. Campbell and us Filipinos (were) determined to go. We Filipinos decided to go on whatever situation." They talked about the slogan of their Battalion 'Bahala Na' (Come what may).

The loaded *Nautilus* departed for an unknown destination in the early morning. Ear pressures stabilized underwater. Inside the sub, they listened to music, read, watched movies, and thought chow was the best. It was their last taste of American food for the next many months. The men were in high spirits and treated well by the crew. Amante noticed the officers avoided the men and counted the money allotted to the party, dividing it. Agustin was uncomfortable with the cramped quarters. According to Pilar, there was no space to move around and they were "forced to stay in our individual spot where we slept on top of boxes of explosives." Capistrano slept with men cramped in the forward torpedo room. But the men were in high spirits. Before landing, the party leader informed the recon men of their objective to destroy bridges and communications.

On the fourth day, the sub received news flashes about naval battles around Formosa (October 12-16) between Japanese naval aircraft and the U.S. 3rd Fleet's aircraft. The sub hid during the time. The skipper was more afraid of being attacked by their own planes than by Japanese. It took fourteen days to arrive on the eastern side of Luzon as the area was heavily mined. Capistrano recalled the sub laid on the bottom for forty-eight hours to evade depth charge attacks by Japanese destroyers. The sub's radar was on the alert.

The Leyte landing was happening while the sub approached Dibut Bay with three dozen mission men bringing 40 tons of supplies, personal equipment, medicine, ration, weapons and ammunition, demolition equipment, and "cigarettes for stevedores." Lapham's supplies were marked 'LM' and 'AN' for Anderson.

A group of two teams was to support Lapham covering the central Luzon area of Nueva Ecija, Pangasinan, and Tarlac. Their mission was to penetrate enemy lines and destroy roads, railroads, and bridges leading to Manila from the north and east when the liberation forces land. The areas were in the path of the Sixth Army to Manila and had to be cleared of the enemy. The men were Sgt. Eleno Ammaguey, Sgt. Procopio M. Adia, Sgt. Julian A. Pilar, and Pvt. Jaime A. Bernal. Coastal Artillery men Capt. Frank Skundale and Lt. John Bove joined. A second team had Lt. Henry L. Baker and Lt. Sidney S. Tison from Coastal Artillery with demolition men Sgt. Jerry D. Regalado, Sgt. Paulino A. Rosales, Sgt. Modesto P. de los Santos, and Pvt. Narciso C. Generales. The two demolition teams were accompanied by Signalmen Sgt. Buenaventura L. Dolendo, Sgt. Segundo C. Bucol, Cpl. Amrafel D. Pascual, Cpl. Mack H. Pulido, Cpl. Loren E. Almojera, and Pvt. Baltazar C. Agbunag. Lapham also had coastwatchers at Baler Bay on the east coast of Luzon and in the areas of Pantabangan and Caranglan.

The second group went to Anderson's guerrilla camps in Bulacan and areas close to Manila: Sgt. Thomas T. Amante, Sgt. Alfred E. Lim, Cpl. Pedro P. Agustin, and Pvt. Emil D. Arciaga were the third team with Lt. Andrew P. Bahr and Lt. Merle E. Campbell from Coastal Artillery. Demolition men Sgt. John R. Constantino, Sgt. Gregorio F. Basug, Sgt. Marcelo Garcia, and Cpl. Romulo D. Robles were the fourth team led by Lt. Richard A. Ensor and Lt. Rhys C. Wood from Coastal Artillery. Both demolition teams were accompanied by Signalmen Sgt. Alberto N. Capistrano, Sgt. Marcelo P. Fontanoz, Sgt. Gregorio F. Bareng, Cpl. Thomas V. Rabang, Pfc. Thomas D. Gaduang, and Pfc. Barney A. Quinagon.

Skundale's group for Lapham could not be unloaded at Baler on Dibut Bay as the shore security signals were not seen. Pascual recalled that the sub used a "new pre-arranged signal that the boys on the shore challenged with an old pre-arranged signal." They were ready to go on deck with their packs but were ordered back to their compartments. The sub submerged, thinking that the guerrillas were surprised by the enemy. They waited for two days surfacing at night and did not see signals from the shore.

Not to waste time, KAZ ordered the sub to proceed south and land at Masanga Point halfway between Lapham's camp at Baler and Anderson's camp at Infanta. Clear signals from shore arrived in the evening, and small *bancas* appeared alongside the sub. Guerrillas and some mission men from earlier missions met them. The unloading took two days with not enough boats and the sub remained underwater during the day. As the waves raised the sub so high, men clutched the rope climbing down to the sailboats to avoid falling into the sea. The sea was so rough that some of the men and cargo were thrown overboard but they kept moving since a Japanese garrison was only several km away. Rocks jutted from the sea slowing the rowing to shore. A big wave hit one of the sailboats stopping the motor. The men furiously paddled to get away from the rocks. They were ready to swim to shore. Hundreds of guerrillas waited on the beach to help unload the cargo and hide them before the Japanese patrols came in the morning.

After the *Nautilus* unloaded the large party of 36 men, evacuated Americans and Gottleib Neigum boarded. Neigum was an enlisted man who survived the Bataan Death March and escaped to join the guerrillas. At first with the Markings guerrillas, he joined Anderson's camp located in the highest mountain with difficult trails to bring food supplies. He saw messages encoded by trained Filipino civilians. He evacuated when he could not accept the unequal way the rations that came in with the submarine and the food given to the soldiers and radio men who continuously transmitted intelligence was distributed. In contrast, Panlilio related that many American soldiers joined Anderson as couriers upon learning that orders came from Australia. As the *Nautilus* entered Palawan waters, she received orders to destroy the submarine USS *Darter*, which had run aground on Bombay Shoal. Stuck high up on the reef, torpedoes could not reach it. After the crew was rescued, the *Nautilus* blasted the sub's conning tower and control room with her 6-inch deck guns.

Agustin choked as his feet landed on the shore. He had been gone eighteen years. Brought inland to a nipa house, hungry looking civilians dressed in rags served them black coffee. They were crowded in the shack and slept on the floor for three days. They suffered headaches as they adjusted their eyes to light after 2 weeks in the almost darkness of the sub. There was hardly any food while civilians were being found to carry the equipment. The rice was full of stones with some husk and corn mixed in.

At Anderson's camp, Amante wrote that "Anderson and a subordinate officer had their women along who thought what the sub brought were all their assets."

On the fourth day after they arrived, demolition men Amante, Arciaga, Lim, and signalman Agustin with Bahr and Campbell, were instructed by Anderson to join the Bulacan Military Area (BMA). That meant crossing the treacherous

Alfred Lim (Luzon mission October 1944) with guerrillas.

Sierra Madre mountains. Twenty *Dumagats,* indigenous people of the area guided them to follow the shoreline and cross the mountains carrying their heavy load of supplies and equipment. Each one received five hundred pesos as operational money. Agustin commented, "I had gone through some tough training in the Army, but nothing compared to this. If you miss your grip on those vines and rocks along the cliff, you fall fifty to one hundred feet below. The tough training we had really helped. Must be in tough physical condition in this mission." The trek took nine days, skinning his feet. They were so swollen that he could not put his shoes on for two days and could hardly walk. Lim was a smoker and hid American cigarettes inside the package of a local brand. He was almost caught when a Japanese grabbed the pack but did not immediately light one to tell the difference.

At the mouth of the Umari River, they waited with the rest of the cargo several days for the boat that was delayed by a typhoon. The *Dumagats* built huts and a guerrilla roasted a pig. Amante was ordered to wait for a representative from BMA to help with transporting the dangerous cargo of dynamite to the lowlands. When the skies cleared, the group proceeded except for Amante. He angrily waited for the rest of the cargo and the guerrilla who stole his pistol. But he could not find the culprit. The ground was slippery and muddy as it rained most of the time. It was a surprise when the *Dumagat's* family joined later as there was not enough to feed everyone. Amante knew they were hungry and could not refuse but he worried about having enough food for everyone. He had six lbs of TNT with cups and fuses aside from rice, six cans of corned beef, and ammo. Every time they camped, he sent guides to find food, and they came back with birds, fish, and vegetable leaves. When the food was gone, TNT was used to blow up the deep part of a river killing many fish. Broiled fish fed everyone including the *Dumagat* children. Along the way, they exchanged fish for rice.

They met guerrilla units who lost men to hunger, malaria, fatigue, falling, or drowning. Another storm arrived while they were waiting for an officer from Victory Hill who was to bring more carriers for the equipment. They spent much time drying out clothes and equipment. Long beards had to be shaved according to one of the men. Food was short. With more *Dumagats* carrying the heavy loads, the party stopped occasionally for a 10-minute break.

Reaching Victory Hill in the Bulacan area, the Japanese presence worried Amante. There was a guerrilla sentry at every curve of the trail asking for a pass, and this slowed them down. Amante did not miss noticing "fair-looking first-aiders" aside from the countryside. But he had to reach the guerrilla camp on a hill to meet Campbell, Bahr, and the rest of the group. The equipment arrived after several weeks. Boxes that were too heavy for one man to carry were opened. Most cans were punctured, and moisture had seeped in causing damage. Personal items were lost, and demolition equipment was wet and missing parts. They were "mad as hell but knew the boys did their best" through rain and floods.

The camp impressed the men except for the food. The Stars and Stripes and Philippine flag waved together in a tall tree. There was even Reveille, Retreat, and all sorts of Army regulations. "At last, this is 'Free Philippines,'" as they slept and ate with the guerrillas learning each other's ways. More guerrillas arrived with stories of their experiences with the Japanese, communist Huks, Ganap (Filipino political party), collaborators, and the Makapili (Patriotic Association of Filipinos), who joined the Japanese. But they were more concerned with the Filipinos who worked for the enemy as spies.

Meanwhile, the second team for Lapham waited four days for the weather to clear up. Demolition men Bernal and Regalado with radiomen Dolendo, Bucol, Pulido, Agbunag, Pascual, and Almojera were ready. They had one meal from the town residents. The stormy weather discouraged going by sea, so some men took the mountain trails going north with only eight packs on their backs, carrying clothing and the demolition equipment. Guerrilla officer Lt. Bernardo and his unit who were returning to the town of Laur that was halfway to Baler guided them. The rest of the party with the supplies took the sea route when the weather cleared.

Pascual thought the one hundred pesos they were each given for subsistence was too small because the Philippines' prices were sky high and had no face value at the time. Japanese war notes were the only means of buying their food from starving civilians. He wrote, "Philippine money and Japanese war notes should have been sent through the party leader, but Skundale claimed that there was no money for us. Maybe he (Skundale) thought we did not know all previous parties were sent with enough money to be split among members ... trusted party leader to have money he heard but never gave me a single centavo." Only six cans of salmon and a little rice for the six men, six guerrillas and two civilians as guides and runners were given. They asked for more because they had the guerrillas with them. Regalado felt lucky that he had 100 pesos of his own. Their duffel bags were left behind with the rest of the cargo to Baler taking only a change of clothes, and a bottle each of quinine and Atabrine. They walked along the beach and crawled on the edges of rocky cliffs. One can of salmon was broiled for 14 men with the cooked rice they brought. Green bananas, papayas, and sweet potato leaves were added. The *Dumagats* gathered sea snails. Another can of boiled salmon was for breakfast. Rain poured soaking them and they rested as needed. Regalado was thankful for his jungle training that made him physically fit to survive.

They ate coconuts at the *Dumagat* village. They sent scouts ahead overland and evaded a force of one thousand Japanese soldiers along the beach of Dingalan Bay. Halfway to Lapham's camp, a captured Japanese soldier patrolling the mouth of the Umiray river was interrogated by the guerrillas and probably shot after. Three men at a time were brought by the *Dumagats* to a small island in the middle of the river and to the other side, evading crocodiles. In a hut, another can of salmon was cooked for all. After eating supper, the moon was behind clouds indicating a good time to move on, but the *Dumagats* said they were sick and did not want to go on, afraid of the Japanese. A guerrilla fired a shot that scared them into moving. One scampered into the bushes. They forced march along the beach in the rainy night passing a Japanese outpost at the next town of Dingalan. So tired, they drank water from streams to fill up their stomachs. Lights flashing ahead made them hit the ground and hide. A guide returned with a sort of flashlight that might have been from a Japanese life craft wrapped in his cap and handkerchief. They buried it to avoid the Japanese sending a patrol.

At daybreak, breakfast was a can of corned beef with lots of water to make soup. With no map and relying on the guide, they waded into the Amutan River waist high and sometimes followed it. Agbunag went missing from the group. They climbed the guerrilla trail up the Sierra Mountain for several seemingly endless hours. Boiled water was all they had for the whole day, and they were getting weak. Pascual thanked his stamina from cutting asparagus and topping beets, both the most demanding jobs in the labor fields of California. Guerrillas stood guard while they slept in a deserted hut looking down at a river. The last two cans of salmon were shared with Lt. Bernardo and his guerrillas before they departed for their camp with two *Dumagats*. They gave cigarettes and some toilet articles to him. A pair of underwear and tobacco went to the *Dumagats*. While waiting for another guide, they cleaned their weapons of 5 Carbines and pistols. The guerrillas had only *bolos*.

At night the cook came running from the river, as he had seen a Japanese patrol which almost caught him. They could hear the Japanese talking and shouting from below. A Japanese scout was heard climbing up but he slipped and rolled downhill. A guerrilla was caught while others escaped. Some equipment was lost in the dark as they escaped except for their Carbines and pistols. In the dark the guides took a very narrow steep trail for security. It was so dark and raining so heavily that they had to walk in a line with their hands on the backs of the men in front of them. Regalado was the last man and rear guard. Going downhill, he slipped, rolled down, bouncing on rocks with his left hand gripping his Carbine. The muzzle dug in the ground stopping his fall to the river. His left thigh was gashed and both legs scraped.

They were in the mountains for two weeks until they got to Bitulok, and Sabani Valley, which were infested with Japanese. At a guerrilla camp on the highest mountain close to the Sabani Estate overlooking the Japanese garrison, they stayed for ten days gathering information. They had to steal or beg food from the civilians. Stalks of young palm were all they ate. From guerrilla camp 180 Squadron in the hills, they could see Dingalan Bay heavily reinforced and mined with a Japanese garrison waiting for a U.S. landing on the east coast of Luzon. At night, men and guerrillas reconnoitered the garrison, gathering information on the enemy. An estimate of four thousand pro-Japanese and eighteen thousand Japanese troops were reported prepared to launch a major counterattack.

After a big storm, they left Sabani Estate, to report to Harry Mckenzie, executive officer of Lapham. Six guerrillas were instructed to provide protection, food, and all necessities for the group to reach Lapham's camp by the town of Kadaklan. A runner arrived, saying that they were to proceed to the town of Laur, 20 miles away to get identity cards from the mayor to pass sentries. Fear gripped them when dogs incessantly barked as they crossed rice paddies and main roads. Three scouts ahead were unable to warn of a Japanese company with fixed bayonets at closed formation. It was too late by the time they heard the hobnailed Japanese shoes. The first scout was caught unable to warn the second and third scouts. They all landed on their bellies by the roadside except for Bucol in the rear. The guerrillas waited for the signal to open, but Dolendo allowed the patrol to pass by. He knew that they would be wiped out. Bucol quietly winced as his foot was stepped on by a Japanese who did not see him. The party reformed and sent three scouts ahead.

They crossed the Maligaya River with dangerous boulders, in waist deep water since the bridge was destroyed. They avoided a Japanese garrison on the other side. They undressed and rolled their clothes, pistol, and Carbine tightly in their ponchos to keep dry. Stripped to shorts they tied their pants around their neck holding hands. Regalado slipped and lost his pants. A guerrilla radio operator floundered in the swift river current weighted by his pack. He did not know how to swim. Regalado crossed and pulled him out. One carabao with equipment was turned back by the strong current. The chronograph watch useful in measuring and recording in precise increments of demolition tasks and weather observations was lost.

Everyone marched with only shorts and undershirts. Marching with wet pants made noises. They became aware of a Japanese trick of placing weapons in a cot covered by a blanket looking like a wounded man and carried by two Japanese soldiers. When followed by guerrillas, they picked up their weapons and fired. Nearing a radio station, they sent all the intelligence they gathered over the two weeks of their trek by runner. Food continued to be a problem. Prices were sky high. Stomachs were hungry. In a village where one of the guerrilla's lived, a chicken dinner was most welcomed. Exhausted, they slept for 2 days. Civilians came and asked for malaria medicine.

Leyte Island, Philippines Landing – October 1944

The news of the long-awaited landing spread quickly reaching the men as they passed villages near Lapham's camp. By the end of November, they were in a guerrilla camp in the village of Siclong outside the town of Laur which was occupied by 20 Japanese with a medical group. A runner was sent ahead to contact the mayor of Laur. He returned the next day with six pairs of civilian clothes and shoes. Until this time, they were still wearing their clothes from the submarine, which made them stand out although the clothes were beginning to rot. Dressed in civilian clothes, they entered the town passing a Japanese sentry at a road junction near the marketplace. They noticed around twenty Japanese soldiers and heavy weapon emplacements. A runner was sent to report at the nearest radio outpost. Dolendo recalled seeing the guerrillas passing through Japanese sentries bowing and gesturing. They saw an old lady who greeted the sentry in Japanese but did not bow. She was slapped and told to bow three times to the sentry. Dolendo felt scared at the sight but neared "the Japanese sentry just to feel the reaction of his nerves to be face to face with the enemy." At the guerrilla commander's house, they ate with the mayor, who gave them residence certificates as identification to pass the sentries. Pascual disguised as a merchant from Manila and dressed as a civilian freely went around surveying the area. He trained three of the guerrillas on radio work.

Two more days of hiking through the jungles brought them to another guerrilla camp for the night. They were thankful for the beans and rice. Pulido recalled that they "starved most of the way. Not a single penny was given to us for our operations which made it difficult to procure food." McKenzie's men were glad to see them as their services were badly needed. Lapham was informed of their location and that Agbunag was missing. A feast of *sinigang,* a vegetable stew, boiled chicken, and broiled dried carabao meat was prepared for them for lunch. The hike continued to the radio station in the mountains by the village of Lublub, closer to Lapham's headquarters. It was safe with no Japanese in the area. Civilians they passed could not believe that they came from the U.S. because of their civilian clothes. At night they put on what was left of their uniforms and arms to prove to them. Pascual said that the civilians cried with joy, knowing that the Americans and Filipinos from the U.S. were coming. They kissed the one-hundred-peso bill he received at Camp Tabragalba when he showed it to them. They were eager to share the little food that they had with the men.

At Lublub, they were glad to see signalmen Cpl. Duran and Sgt. Miguel, whose party arrived four months earlier in Samar and were now operating the radio. The radio was busy transmitting reports on Japanese movement, activities, and atrocities committed on the civilians. Reporting on the Huks was difficult because you could not tell from the guerrilla guards who carried rifles, pistols, and machine guns.

A week later, Lapham instructed Dolendo and Duran to come to his headquarters in Baler. Pascual hiked with them for three days to the village of Baloy. Miguel and the rest of the men stayed to man the net control station at Lublub with Pulido who looked forward to using his Army training working the radios on the steady stream of intelligence reports from operatives. He was busy punching on graph papers encoding and decoding messages. He knew the results of their messages were the planes flying daily to bomb all those reported targets of Japanese concentrations and installations. The big problem was still getting food. Money was still not issued. They had to manage living off the jungle – young stalks of rattan, wild palm trees, birds, and *lugaw* (rice porridge) were the chief foods. They forced their way through Japanese lines blocking their supply route. Pulido related they were "half asleep and starved most of the time (but) we got the messages through thinking of the liberation of the Philippines and destruction of the Japanese empire."

Reaching Lapham's headquarters, Dolendo and Duran were surprised to see Agbunag alive with Skundale, Bove and the other officers. Agbunag explained that he stayed behind when the weather cleared to load the medicine and clothing into a sailboat. He had terrible cramps on both legs, weakened with only coconuts and ripe papayas for food. He laid back wet with sweat taking the pain and fell asleep exhausted. When he woke up in the dark, he realized that he was separated from his group and kept quiet to avoid warning a Japanese garrison only two hundred yards away. He backtracked to the village and found the demolition party still there, gathering enough laborers to carry the supplies. The officers had taken a *banca* to go north. It was a harrowing week of waiting for the weather to clear up since a Japanese garrison was nearby. With Bernal, Aquino and two guerrillas, Agbunag took a *banca* with the medicine, clothing, and supplies sailing for a day without food. At Lapham's camp, Bernal worked on the radio with the help of two guerrilla radio men. The radio was wet with saltwater but still worked.

When malaria struck Dolendo, Duran operated the radio until he recovered. Pulido and Dolendo were then ordered to set up headquarter at Kadaklan, McKenzie's camp, to form a ten-man net with the call sign ANA. The radio was to connect with the rusty and wet radio that Bernal fixed and established as network GNC at Lapham's camp. Dolendo hired *Aetas* (indigenous people) who lived in the jungles as guides. He said, "They were the best guides in the thick jungles. Could smell a man approaching if Japanese, guerrilla or wild pig." Along the way, they continued to send weather reports and receive messages from other stations. At Kadaklan, the radio was on the air the same day operating 24 hours waiting for the 10 radio stations to contact. Plenty of weather and intelligence reports were sent to KAZ, and messages from the ten sub-stations. To completely send all the reports, they had to persist against weather or atmospheric conditions that continually affected transmission to Australia. Sometimes the connection with California that was six times as far away as Australia got through to an Army station in San Francisco to relay intelligence reports to KAZ. After a week, the rusty radio went dead. A runner with Almojera was sent to find a radio technician in Cabanatuan and get filters from any Japanese captured American air warning sets. With the radio fixed, messages were transmitted for a month before the technicians moved to the lowlands.

The men reported how the poor farmers and pro-American people gave money, food, and clothing, gathered intelligence reports, and provided men to carry supplies. They warned of Japanese patrols and some of them died whenever some guerrilla was captured and tortured to talk. They never knew when the Japanese police would pick them up. They were open in the villages while the guerrillas were in the mountains.

With the weapons brought by the September and October parties, Lapham rapidly created new squadrons. A Chinese unit of seventy was organized from the Chinese military organization in Manila to reinforce central Luzon. "The brutality of the Japanese in China inspired the Chinese immigrants in the Philippines to resist." Additional troops now armed with automatic weapons were sent to Tarlac, a heavily garrisoned Japanese area, to clear the way to Manila.

In November, Russell Barros, a guerrilla operating in the Bicol region, reported that Garcia, Fernandez, and Albalos from Cabangbang's headquarters brought arms and set up station LRC4 to cover the Bicol province on Japanese activities. No longer would messages have to be sent to the nets of Samar and Mindanao for relay but instead they could be transmitted directly to KAZ. Barros recommended Garcia for a promotion. The first airdrop of radio equipment, weapons, clothing, and medicine arrived from Leyte and were distributed. Suffering from tropical ulcers from his ankles to his knees, Barros worked on the expansion to six radio outposts before April 1945 when the Task Forces landed in Legaspi City. He sent a message to the Sixth Army to spare the Pena Francia Church built in 1750, the mother shrine of the Bicol province at the request of the Catholic bishop of the diocese.

The demolition team of Ensor, Wood, Robles, Constantino, Basug, and Garcia reported to Barros in Camarines Sur in the Bicol province. The team blew up a 4-span railroad bridge between Ragay and Calauag towns in Camarines Sur. Three kilometers of tracks were destroyed. The bridge near the town of Sipocot was destroyed. Working with the Bicol guerrillas, they damaged one hundred drums of gasoline and oil and nine Japanese trucks in the town of Camaligan. The radio sent messages of the guerrillas improvising coconut bombs out of coconut shells with scraps of iron from old frying pans, dynamite, and detonated by a fuse that blasted Japanese trucks and enemy garrisons. The tons of dynamite were found in nearby gold mines and made into hundreds of bombs. But aside from fighting the Japanese, the Bicol guerrillas continued their infighting for recognition, men, and support from the civilians.

In the same month, demolition man Bernal, led twenty-three men to carry explosives, rations, and ammo to Mckenzie's camp in Kadaklan. Bernal did not receive ration or money from his officer. There was no time to stop to find food even though the group was starving. They simply soldiered on. Crossing an enemy patrolled highway at night, they ambushed six Japanese with a machine gun, with one of two weapons they had. It was Bernal's first encounter with the enemy, and he knew that they were not supposed to fight. After two weeks of hiking all day under continuous rain through jungles and the Sierra Madre mountains, they finally arrived.

Bernal's group gathered information on abuse of civilians in the villages of San Juan and Mariquit. They moved to San Jose in Nueva Ecija, a town crossed by two highways to observe Japanese activities in the area and to get food. In the nearby towns, the enemy had punished the civilians and taken the rifles of the guerrillas. He sent a courier to the guerrilla leader in nearby Pantabangan town to warn of Japanese presence and burning of Canaan, the next village, and the presence of other suspected guerrillas who were armed. Using his combat training, he reported disarming a Huk who tried to take his weapon when he passed through the village of Sampaloc. All intelligence was turned over to McKenzie's radio.

A patrol of three hundred Japanese raided Cabangbang's camp in December. They fought the enemy and closed the station for four days. Information came in fast when they resumed operations, keeping the radio on the air 24 hours. Japanese planes returned and bombed the camp, causing them to move to Mt. Lumot, another mountain in Bulacan.

More men with radios were sent out to areas north of Manila to report on the situation in the city. Additional sub-stations were set up in the south, east, and north of the city. Despite Luzon having the highest presence of Japanese soldiers, agents and mission men bravely penetrated Manila and observed enemy movements. Two radio stations were established in the City and agents reached Malacanan Palace and other high government offices undetected. Agents reported a scene of confusion as trainloads of haggard and hungry Japanese soldiers moved north to the mountains. Japanese civilians with children were forced to go with them. The agents' bravery and willingness to risk death if caught were reported to KAZ. Many were captured and killed.

Alarmed with the Leyte landing, the Japanese had begun strengthening defenses in the city while the residents starved, evacuated, or prepared for safety when the American Task Forces enter Manila. According to Peter Parsons, Cabangbang wrote that "two months before the Americans entered Manila, the Japanese were busy killing civilians in the Manila Districts and Bulacan towns just north of Manila by gathering men, women, and children and machine-gunning them. Town officials were being hanged and beaten." A central Luzon agent reported seven thousand American POWs, Japanese women, and children were evacuated to Formosa at the end of the year.

Agents reported the increasing recruitment of Filipinos into the patriotic Filipino organizations of the Makapili and Makabayan. They believed that the Japanese will win the war or that they were fighting for the Philippines. Most were employed as spies, and many were caught and killed by the guerrillas.

In December, KAZ wanted an investigation of a Japanese plan to use midget submarines or suicide torpedo groups. Where was the location of a powerhouse and its connections to a controlled minefield in the Manila harbor area? KAZ wanted information on the Pangasinan area in Lingayen Gulf before the landing. Japanese infantrymen were reported in the villages of Balungao, Rosales, and Tayug, inner areas of Pangasinan blocking the route to Manila. The troops were from Kobe and looked healthy, disciplined, and hardened by battle experiences in Manchuria and China. Demolition men placed booby traps in the villages and fired mortar shells into buildings occupied by the enemy. Guerrillas threw grenades at passing Japanese trucks.

"What effect would it have upon your operations if action were taken to destroy the remaining value of the enemy military currency?" was the question sent to the guerrilla commanders: Peralta, Abcede, Fertig, Cabangbang, Anderson, Lapham, and Volckmann. GHQ was looking into the distribution of Victory pesos to devalue the Japanese notes.

Radio messages crackled with reports of a network in Manila composed of three thousand men. Many were saboteurs, ready to rise against the Japanese when given the signal. They were ready to destroy anything of military value, cut telephone lines, and cables. Most of the police in Manila were members of Markings Guerrillas and ready to turn against the enemy. Horse carts could be used to block traffic. Powerhouses, gas storage tanks and bridges were to be blown up with dynamite found in mineral mines or brought from MACA camp in Samar.

Guerrillas and civilians brought around 30 pilots shot down in the central and Southern Luzon area to BMA headquarters. To facilitate evacuation, an airstrip at Acle was constructed and the pilots taken out before the Luzon landing.

Alamo Scouts, a special ad hoc group of General Krueger, commander of the Sixth Army that reconnoitered areas before any landing of U.S. Task Forces arrived at camp and took men who knew the area and could speak the language to provide radio and combat support.

Luzon Island, Philippines Landing – January 1945

As the Allied forces were streaming towards Lingayen, the signalmen had established a vast radio network operated by the guerrillas they trained in radio. The network covered the provinces of the Lingayen landing site down to the central area of the island, on both sides of the coast, and around Manila. The network connected with the radio stations set up by earlier mission parties in Camarines Sur, the islands of Lubang and Maricaban, and Southern Luzon that covered shipping routes from Leyte to Luzon.

Hearing of the coming Luzon landing, the men expected many upcoming directives. Orders were to destroy usable bridges in Pampanga and Bulacan provinces. Bridges inside Manila were not to be touched. As days passed, more guerrillas arrived to fight the Japanese. In contrast, the Huks and Makapilis swelled collaborating with the Japanese. More a problem were the Filipino civilians who worked as spies for the enemy.

Gyles Merrill, a former Philippine Scout and a guerrilla leader in the Zambales area with 17, 000 men covered an area into Manila and provided intelligence to the Sixth Army. Three thousand *Negritos,* the indigenous people of the area, had aligned with him. A Huks unit numbering five thousand men with fifteen hundred arms, had signed an agreement of cooperation and accepted tactical control. He confidently reported to KAZ that all the towns and airfields in Zambales including Olongapo could be taken at once. He would need 1,000 Carbines.

Demolition equipment to destroy bridges was dropped at Cabangbang's headquarters in the mountains of Bulacan. The Japanese saw the parachutes and bombed the camp. No damage was reported because it was the wrong mountain. The demolition groups scrambled to prepare and depart as they had only 36 hours to destroy their bridges after the drop of explosives.

Pandemonium broke out as everyone went wild with excitement when the urgent message from KAZ to all guerrilla leaders to rise and strike at the enemy arrived. Fight at last! Orders were to send all information, seen, heard, or rumored. Runners waiting for days took off to all units confirming in person. Civilians were advised to keep calm, provide support to the guerrillas, and move away from Japanese garrisons. With the guidance of the mission men, the guerrillas began destroying communication lines, railroad tracks, bridges, enemy signal stations, transports, and garrisons. Flying boats were reported landing in San Manuel, an artificial lake the Japanese created by damming the Tarlac River. Retreating Japanese passed cautiously as the guerrillas did not allow them to pass alive.

With the 'lie low' order lifted; signalmen were on the air 24 hours. The net control station in Anderson and Lapham's headquarters flooded with enemy disposition, sabotage, combat, downed pilots, and deployment of guerrilla troops in their areas and surrounding provinces. Japanese troops were on the move day and night. The first wave of Allied aircrafts and the bombing of Cabanatuan City was reported. Carabao carts were pulled by American POWs under strafing of U.S. planes. Over seven thousand civilians were evacuated from the Polillo islands to Infanta to avoid massacre by the Japanese.

Agbunag and Dolendo with eight guerrillas trekked to Ray Hunt's camp, a guerrilla leader under Lapham to establish a radio station in Pangasinan by Lingayen Bay. Without any money for food, they ate what the civilians were willing to give them. They evaded Japanese guarded trails that the guerrillas had warned them of. At the fourth town, they stopped and stayed for three days as the Japanese blocked their path. There were no spare parts to fix the radio at Hunt's camp. The battery and charger took a week to arrive from Tarlac before the radio could be set up. But the radio did not work so runners were organized to deliver messages to the Sixth Army headquarters at Leyte that took days crossing the seas between islands. As backup, couriers brought the same documents to station S3L in the Bondoc Peninsula for relay to KAZ.

Agbunag, Santos (Samar mission June 44), about a dozen guerrillas, and Ray Hunt raided the Japanese garrison in San Quintin, Pangasinan to clear the Army's path to Manila. Machine guns blazed as Santos' guerrilla security team cut all telephone wires in town. Another team sneaked into town to blow up the bridge but had to return the next night to finish the job. Two men were wounded.

While the men ran from bombings and raids to keep the radio in the air, they continued to send the enemy's disposition. Efforts now doubled having to send the same messages to the Sixth Army and KAZ. The Japanese were surprised and unable to send reinforcements because the guerrillas had blocked the roads and destroyed bridges. With the Japanese retreating to the mountains, the path to Manila was cleared for the Sixth Army Task Forces.

The Sixth Army landed on Luzon unopposed. All the mission men assigned to Lapham and Anderson's organizations, including Ramsey's in the Manila area were ordered to report to the Sixth Army at Calasiao, Pangasinan upon landing. Their network, radio codes, and cipher systems were turned over to directly coordinate with the guerrilla attacks on the enemy. Promotions were recommended for the 978th Signal Servicemen for the excellent work they performed.

Bahr formed demolition teams and sent Amante to blast the sugar central and the Calumpit bridge between the provinces of Bulacan and Pampanga. He was given three hundred pesos and told that the guerrillas would take care of them. Amante had recruited twenty-two volunteers, all high school graduates, to learn demolition. Three of them were officers in chemical warfare during the Bataan stand. Classes were held once a day, with only one-third attending at a time to avoid suspicion. The trained men were ready to set up booby traps for the Japanese using hand grenades, TNT, and captured Japanese mines.

Amante and Agustin with guerrillas left for the Calumpit bridge with demolition equipment. BMA commander, Capt. Alejo Santos, provided men to carry their equipment and for security. Twenty armed guerrillas with machine guns and two Carbines with 5,000 rounds of ammunition were on guard. Almost everyone had a pistol. They traveled only at night for four days passing many Japanese patrols. Every night, they hid their equipment. Dogs barked as they passed Japanese garrisons and occupied villages. Everyone sweated and held their breath so as not to make any noise. Everyone they passed gave them information on the Japanese nearby. The 'static guerrillas' who had no arms but carried out guerrilla duties provided intelligence. Wherever they told of where they came from, the civilians were so happy, wanting to help and share whatever little food they had. Anyone was willing to help kill a Japanese. Zigzagging through different routes, the guerrillas brought them to Col. Maclang, commander of the Republic Regiment and wanted by the Japanese near the town of Baliuag. A Japanese wanted to borrow a horse cart in the group and was nicely convinced by a guerrilla officer to find another one. Amante recalled his intense nervousness, but he learned how to talk and act in front of the enemy.

Outside the town of Malolos, they hid their equipment at guerrilla Capt. Capara's house and at a different house each night. The Malolos former chief of police had the party meet the engineer who built the Calumpit bridge. The Huks were nearby. Amante gave 50 pesos to each of the 20 guerrilla boys to go back to their camp. Dressed in civilian clothes, he surveyed the area and discovered that the Huks were following him. The Huks shot the guerrilla captain guarding him when he went to meet the engineer of the bridge. He was grateful to Helen Jones, an American lady who worked in the office of U.S. Commissioner Francis Sayre before the war, for putting him in jail for safety. She posed as pro-Japanese to obtain information and was receiving intelligence from her brother Frank Jones in Intramuros.

A message was flashed to KAZ of the Japanese defense outside Malolos. The recon men with guerrilla scouts crawled on the rice fields to intercept the Americans before reaching Malolos. Caught in between, as they radioed artillery fire of Japanese locations in the city, they were almost mistaken as Japanese by the Americans.

Agustin dressed in civilian clothes with a guerrilla entered Malolos town to survey and hide the demolition charges in one of the cemetery tombs ready for the signal. It took 4 nights to reach the heavily guarded Calumpit railway bridge. Fear of ambush enveloped them as they passed sentries and barking dogs. The few people left in the villages they passed were forced to provide food to the Japanese. People told of Japanese and Huks' locations to avoid.

All demolition parties timed to blast the bridges assigned to them at 0200 on January 5. Demolition parties along rivers attached the clamps (magnetic demolition devices) to the bridges. Detonators which looked like pencils were used. The charge at the Calumpit bridge was set with four fuses to ensure that one would go off. The explosions lit up the sky waking people up kilometers away. Quickly the group slipped away and swam through the irrigation canal to avoid a Huk unit. Cabangbang reported that the steel bridge was too strong and large to be destroyed with the limited supply of explosives. A runner was sent to get more TNT from an airdrop, and they went back the following night to finish blowing up the bridge. As they hiked back to camp through rice fields, they heard planes flying low strafing the road and nearby towns. They were mistaken as Japanese dressed in civilian clothes. They noticed that Japanese troops were running in the direction of the planes that gave out machine-gun bursts. That was how they communicated with their communication lines cut.

A Japanese truck with 4 armed Japanese regularly passed Sumapa road usually when dark. Amante reported that the truck was ambushed by the guerrillas.

The Huks gave the guerrillas more trouble than the Japanese and many times interfered with their best work in the Bulacan province. The Republic Regiment battled the Huks who were better armed and had plenty of ammo for several days. The amused Japanese watched and asked who had the most casualties. With the guerrilla's ammo running low, Amante suggested using booby traps on advancing Huks. At night, trained guerrilla boys set up traps of hand grenade, TNT, and some Japanese captured mines. When the Regiment withdrew at night, the Huks ran right into the traps killing dozens of their men. The Huks retreated believing that the guerrillas had mortars and artillery. The Japanese heard the explosion and sent troops to fight the guerrillas. Amante was so proud of his boys that he gave them a carton of cigarettes in place of a medal he wanted to give.

In Malolos City, confusion reigned. The Japanese wanted to enter the city. The Huks were battling the guerrillas to remove them from the city so they would be able to fight the American forces when they arrived. They were looking for Amante as his unit fought with the guerrillas to defend the city, killing one of his men. The civilians were fired upon by the Huks because they provided food to the guerrillas. Col. Adonis Maclang was not worried because he had 7,000 rounds of ammunition from the BMA headquarters to battle the Huks.

The American forces arrived and riddled the Malolos high school and capitol building with artillery fire, ending the Japanese presence. The jail was crowded with over 200 Huks captured by the guerrillas. The people rejoiced and brought out food they had hidden. Music blared out, and the people came out dancing. When GHQ asked for prisoners, they "were gone," Amante said. One district had backyards converted into graveyards.

KAZ was informed that the Atlag airfield was built by the whole town without the guerrillas' help, as they were defending their sector. An emergency landing field for observation planes was now available for landings. After three days, a Grasshopper plane made a landing on the airstrip that could hold ten planes at a time for observation. A message was sent via courier to the BMA camp.

By this time, Manila, the capital city, was in chaos. Streets were crowded with people begging for food. Ramsey wrote that "In Manila (an) average of 100 persons (were) dying daily due (to) starvation. Manila (is) doomed with widespread starvation." Internees and POWs were not given food. Guerrillas reported that the Japanese planned to take the entire new harvest of rice for their military use and even supervised the harvesting to ensure this.

The Sixth Army and KAZ received the message that "As of January 7, (Japanese troops) have constructed foxholes and pillboxes on practically all street corners." That did not sound like a plan for an 'Open City' to avoid destruction. Searchlights flooded the night skies of the city. The Japanese had recruited around fifty-eight thousand civilian laborers, with six thousand from Manila to build defenses. Pro-Japanese groups united and trained in Bulacan, adjoining Manila.

Mateo cited the invaluable support of the Philippine Scouts. John Boone, an Army corporal and guerrilla leader of the Bataan Military District, was particularly vigilant in harassing the enemy. Mateo operated not more than two miles from a concentration of six thousand Japanese troops entrenched in caves and concealed positions on a zigzag road where they were expected to make a last stand. Sometimes the enemy raided his camp, but he thought that "was not as bad as being strafed by our planes." As the Sixth Army moved south to the central plains, his group attached and established three radio stations throughout the Bataan peninsula. He recalled one evening while walking on the seashore with three guerrillas, they spotted a *banca* coming to shore. They knew many of the Japanese crossed in these *bancas* from Manila to escape into the mountains of Bataan. They hid the machine guns and grenades they had with them. The Japanese docked and asked them for food, thinking that they were civilians. They gave them something to eat and waited until they were asleep to "send them away to join their ancestors." A shrilling scream of a woman woke them up the following day. Soldiers had raped and brutally mutilated her 19-year-old daughter. "Her stomach was split in two with a bayonet."

As the U.S. Task Forces charged towards Manila, Mateo and his group were assigned to capture Japanese soldiers leaving Manila. "We caught four Japanese soldiers, tied their hands and feet, and laid them on a stand in the street for the civilians to see." He was disappointed to find out that they were Marines and not Signalmen. He remembered "when I was driven to the point of insanity by their constant jamming of our radio communications. I swore that if ever I lay my hands on any Japanese signalman, Hirohito had better help him." It did not take long for the civilians to dismember them.

Torres' entire group reported to the Sixth Army as it took control and directed operations of all forces in the interior northwest of Manila. The net control station at Lapham's headquarters was turned over. He and Quindiagan were in the city of Tarlac providing support in quickly identifying collaborators and appointing temporary civil officials to employ the hundreds of civilians for military projects.

When the U.S. forces reached their area in the mountains of Rizal, outside Manila, Corpuz who was in the Markings camp was instructed to take his radio and operate it behind the retreating enemy until his position became dangerous. Only then was he ordered to leave. He was gravely wounded on both legs by a Japanese machine gun while trying to return.

Demolition men Regalado, Rosales, Ammaguey, and de los Santos with Tison departed Infanta to coordinate with a guerrilla camp, Squadron 180, located north of Dingalan Bay on liberation activities. A former American mine shift boss in Northern Luzon, Alvin J. Farretta joined them. The commando and jungle training of the men showed their usefulness when they engaged in skirmishes with the Japanese while deploying demolition work. For the next two months, they ambushed and captured weapons from the enemy and destroyed bridges and roads around Dingalan Bay. The guerrillas were elated to take captured Japanese materials, especially food. Messages to Anderson's headquarters identified the men by their names, as "Australian soldiers" and commended their "magnificent work." As fighting subsided, the group joined the American troops at Cabanatuan City.

If the courier had arrived with the message on time, Pilar could have started operations a week before the landing. His group immediately moved to destroy bridges and communication lines while the guerrillas did the ambushing and raiding. He picked up materials distributed by the enemy that celebrated Filipino holidays but not the American ones celebrated before the war. Propaganda pictures of Americans deserting the Filipinos in the battles on Bataan were distributed in the towns. He discovered that most Japanese were just young kids and presumably trained while in the Philippines. On patrol, they had the habit of making lots of noises, maybe to scare guerrillas away who were told to lie low. The group with the most noise was a smaller force and vice versa. They were reported as poor shots and feared strafing. Japanese machine gun teams guarding bridges ran and left their position every time an American plane strafed them. Their anti-aircraft guns were unused. They had no aircraft to intercept and protect.

Orders arrived for the men to send all radio and equipment to the Sixth Army at Pangasinan. But a guide did not arrive as requested. After 3 days, Dolendo left with some of the equipment and six guerrillas while the rest of the men and equipment stayed not to take chances with the many Japanese patrols on the way to Pangasinan. Overland in Nueva Ecija, runners warned of Japanese with machine guns at the gates and light tanks fronting the school building in the town of Munoz. They camped at nearby Cabaroan observing

Japanese reinforcing Munoz at night. He radioed the targets and the bombing happened soon after. He reported the ammo and gasoline depot at Camp 4 (Little Bataan) and they were bombed. At the camp of recon man Anacleto and signal man Isaac, he stayed to recover from another malaria attack and later operated the radio with Isaac until the liberating forces captured the town of Rizal in Nueva Ecija. Before the battle to retake Manila began, he was flown back to Hollandia.

Finding General Tomoyuki Yamashita

Valera's men operating in Northern Luzon and Cabangbang's men in Central Luzon coordinated with the Sixth Army in the hunt for General Yamashita. The Japanese faced the guerrillas, U.S. Army troops, and the area's head-hunting *Igorots*.

Balete Pass was a delaying stand of Yamashita's forces along Highway 5 from San Jose City and Santa Fe. The U.S. 25th Division fought for four months in muddy and mountainous terrain, with inadequate supplies, and heavy casualties from an enemy that refused to surrender. In parallel, the 32nd Division battled the Japanese at Salacsac Pass along the Villa Verde trail that joined the towns of San Nicolas and Santa Fe. The trails were the only access between Central Luzon and the Cagayan Valley towards the Caraballo mountains and General Yamashita's bastion beyond.

The 32nd Infantry Division in Tayug, Pangasinan required 1500 new laborers per day. A labor camp operated on the Villa Verde trail to feed, house, and account for all civilians widening the passes and trails through the Cagayan Valley towards the Caraballo mountains.

Men who survived the battles in New Guinea, Buna and Aitape, fell on the steep slopes of the Pass. The encounters with the Japanese guarding the trail exhausted the men that units from the 126th and 128th Regimental Combat Teams were brought in as replacement. The Japanese defense broke at Salacsac Pass (Villa Verde Trail), a narrow trail that limited the supply line. After much death on both sides, a path was cleared to the mountains often shrouded in fog. Elite troops of the Japanese Fourteenth Area Army burrowed in the mountains, raining bullets, and shells on them.

The combat teams were a quarter understrength and exhausted after two months of fierce fighting. There were no additional replacements or reinforcements. The allocated forces were cut, sending troops to retake Manila and two Divisions to aid the Eighth Army's operations in the rest of the islands. Even with the support of the guerrillas, who faced frontal assaults alongside them and knew the terrain, it still took over three months to advance and break the Japanese resistance. The enemy was sealed in their caves. In recognition of their heroic actions, 1st Filipino Infantry Regiment men Pfc. Modesto A. de Samito, Pvt. Manuel B. Alcoy, Pvt. Salvador M. Abata, and Pvt. Sammy O. Espero were promoted to T5.

Bernal, a recon man trained in demolition, proudly pointed out his participation at Balete Pass. In December 1944, he was charged to reconnoiter a six-kilometer area around the Pass. It took four days to go through the Caraballo mountains under heavy rain and two days to cover the area. He sketched the entire area noting Japanese gun emplacements and observation outposts around the garrisons at Balete Pass and Santa Fe. A 50 mm gun was on a hill pointing towards the national road (Highway 5). The Pass was the only main road to the Cagayan Valley, and the Japanese were using it as their supply line. Bernal also looked for a possible camp for when his group moved in.

Bernal led Skundale, Bove, radioman Bucol, and fifteen guerrillas with radio, explosives, and equipment to Balete Pass. He showed them the observation posts and defenses around the garrisons. Runners came in announcing the Japanese nearing the camp, so they moved to a village, a four-hour hike from the Pass. They waited for the orders to strike while sending information on Japanese movements and other intelligence to Lapham's headquarters.

On the day to strike, Bernal and Bucol led the fifteen guerrillas and *Igorots* to Balete Pass with demolition cargo. Skundale and Bove stayed behind in the village. After hiking through hills to reach the Pass, they observed the road all day without food and sleep. They ate *camote* after two days of no food. Four Japanese guards were in a guardhouse, two stood on the road, and a houseful of Japanese troops was across the guardhouse near the road. Bernal prepared the charge and spent all day studying where to place it based on the enemy troops' movements. He decided to plant explosives at a "curve with a perpendicular ridge that was ten yards deep." After some Japanese tanks and foot troops passed by, the charge was planted on the saddle of the road. *Igorots* went out to cut all telephone and telegraph lines. "The explosion was so terrific that I could feel the ground tremble one hundred yards away," Bernal said. Japanese patrols immediately responded with machine guns cutting the air as the group took cover on a small hill over the road. They saw trucks coming up, and with the blackout, did not see the demolished road. Two trucks filled with Japanese troops fell down the ridge. Bernal recounted that their "systematic work of demolition on this gateway caused such a terrible jam and confusion on a long Japanese convoy. They became easy prey to the U.S. bombers who dropped their eggs and strafed the convoy." He said that the river basin below was "the sorriest sight one could hope to see. It was dammed with wrecked Japanese trucks, equipment, and dead bodies." Without food for the last three days, the group double-marched back to camp and reported a successful mission.

Barely given a chance to rest, the group went northeast to Burgos two days later, along the Villa Verde trail, which the Japanese used to retreat towards the Caraballo mountains. They surveyed the area and reported on enemy activities and locations. They heard of Japanese stealing clothes from the civilians, so they ambushed them and gave the weapons and clothing to the guerrillas. U.S. fighter planes that could not identify them strafed almost killing them. They crossed the Japanese-infested 'Spanish Trail' to the other side of the mountains and stayed in the village of Bonga observing the bombings along the trail and nearby town of Carranglan. When they heard that the liberating force was in San Jose, they proceeded to the 161st Infantry Combat team's command post and gave all gathered information in their sectors. By the end of the month, Bernal reported to the Sixth Army at Calasiao in Pangasinan. A week later, he boarded a plane with other recalled mission men to Hollandia. The success of their missions was foremost in their minds. Bernal remembered his stomach gnawing from hunger that he wrote at the end of his after-action report to send operational money for the missions.

Pulido and Pascual attached to Alamo Scout Chanley's team with the order to reconnoiter the old Spanish Trail (Villa Verde) that the Japanese were using in the area of Pantabangan and Carranglan at the foot of the Caraballo mountains. They were to monitor 4,000 Japanese reported garrisoning Salacsac Pass along the trail. Guerrilla commander of the sector, Capt. Aquino oriented them on the terrain and the enemy outposts. Two guerrilla platoons were their security. They joined the U.S. 20th and 21st Infantry to take the town of Munoz. A radio was temporarily set up at Masiway, a safer location in the same area while waiting for Scout Chanley to report to the Sixth Army. Two platoons of guerrillas returned fire when a Japanese patrol fired at them attempting to penetrate their perimeter at the Carranglan River. The Japanese retreated after half an hour and abandoned their food supplies. Another Japanese patrol showed up, and tracer bullets lit up the dark sky. The following day, they saw the area bloody and all food in sacks left behind. They reported the surrounding trails, mountains, and the junction of the Pantabangan and Carranglan Rivers teeming with Japanese troops. An artillery barrage shelled the junction. When Chanley returned, the group moved to the 25th Infantry headquarters in Rizal.

Cabangbang reported a conference between President Laurel and General Yamashita in December. The report stated that Yamashita admitted the indefensibility of Manila. Yamashita did not accept the proposal of declaring Manila an "Open City", reasoning that that would reflect on the might and reputation of the Japanese forces by doing so. The complete demilitarization of the city would lay it open to a possible U.S. paratrooper invasion from the island of Mindoro. Yamashita explained that the best option would be the immediate transfer of the Republic government to Baguio city. He would remove his troops from the city. It was up to the Americans to take notice that the Japanese left Manila. Military police would remain and enough troops to cope with a surprise attack by paratroopers. Baguio became the political and military center of the Philippines when the government was ordered to transfer.

General Yamashita was seen moving his headquarters and elite Imperial Japanese troops north to the city of Baguio and eventually to his hideout in Mt. Napulauan in the Cordilleras mountains. As of January 7, only a few Japanese troops were left in Manila. But around four thousand of the Shobu Force were left to defend North Manila, the center of the city. Retreating to the south crossing the Pasig River, they torched the city. Radio messages to KAZ contained locations of pillboxes on practically all street corners, ammo dumps, fortifications, troops, and information on buildings and bridges prepared for demolition. Recommendations were sent for U.S. planes to bomb a location on Escolta, downtown Manila where the Japanese stored weapons and explosives. The U.S. Army historian Robert Ross Smith estimated ten thousand soldiers under Admiral Iwabuchi stayed to defend the city.

In a postwar interview, Col. Hideo Ishiharu, a judicial police officer for court martials and military tribunals in the Japanese Army's Judge Advocate Section, revealed the execution of six hundred guerrillas in December without the General's signature on the death warrants. General Yamashita was pressed for time busy with his departure to Baguio.

Manila Liberation – February 1945

Agbunag was behind enemy lines when he and Santos received orders to bring their radio to the Sixth Army at Calasiao before the entry to Manila. They worked at the message center as reports poured in during the retaking of the city.

Some men were assigned as guides and interpreters for the 1st Cavalry on their way to Manila. They were the first to enter the city and liberate the internees of Sto. Tomas before a massacre could happen. *The Washington Evening Star's* article on the internees listed Jose Limon and his brother Alfred Limon with Company I of the 1st Filipino Infantry Regiment at the rescue. Messages from Anderson's station reported the BMA guerrillas and operatives guiding the 7th Cavalry to the Balara Filters, which was part of the City's water system to survey of enemy presence. It was not until May that Ipo Dam was cleared of the enemy.

Truckloads of Japanese troops moved north and south always at night, reported Robles after Manila was retaken. Guerrillas proudly displayed the American flag as they took control of surrounding towns. The Tayabas guerrillas continued to battle with the Japanese, reporting that no Japanese was seen in the open-air day or night. They were in dugouts, foxholes, and trenches. The only sign to hunt them down was the rise of smoke while cooking their food.

By March, Baler Bay was packed with a convoy of anchored U.S. ships. Pascual and Pulido joined six Alamo Scouts to reconnoiter the beach for the Navy between Baler Bay and the town of Casiguran. A truck brought them to Baler Bay for a Navy pickup but had to hike the rest of the way as bridges were destroyed. They boarded a landing craft in the Bay and saw mine sweepers busy cleaning out the many floating mines, some exploding. Every fifty meters on the shore were pillboxes. They landed close to the airfield without opposition and waded to shore. Pascual noticed that the civilians onshore looked so scared of him as he approached and realized that his fatigue cap looked like the Japanese field cap. Quickly, he stuffed it inside his pocket.

A Japanese POW was brought in by a guerrilla. Pascual could not help himself and "slapped the 'yellow' rat so hard shaking tears out from his eyes. I was ready to skin him alive ...was my dream since they invaded the Philippines." The Japanese prisoner was sent to Sixth Army headquarters with eighty Filipino POWs suspected as Makapili members. Pascual learned that not all Japanese were willing to die as they did every possible way to conceal their identity when retreating. Some Japanese officers dressed in women's clothes and mingled with civilians in harvesting rice.

An advance patrol went farther inland to survey an airstrip with Pascual assigned as the first scout and an Alamo Scout as second. Two colonels and one naval captain followed. The rest and Pulido with some guerrillas guarded both flanks. The hike took almost a day to the airfield from the beach. There were no Japanese at the three parallel airstrips. The Japanese had fled leaving heavy machine guns and plenty of food supplies. The civilians said that the first strip was a good one, and the Germans were using it. The middle strip was Japanese, and the third strip close to the mountains was American. The civilians' mention of German aircraft bolstered the regular sighting of German planes observed flying over Manila. Some downed German pilots were reported to be women when burning uniform exposed breasts. Maj. William P. Fisher who had escaped Bataan to Australia described seeing "one body that was German – tall, blonde, etc ...few white men and their crew... and also, we shot down Japanese women pilots." Sketches of the airfield were made, and signs put up at the beach for possible landing sites of barges. Their intelligence confirmed the report sent by Malan five months earlier.

Returning to Baler, the group took a plane to Mangaldan airport by Lingayen Bay. Tison, an officer in their party, was there with other mission men working the NR4 radio station. Tison instructed the men to report back to Hollandia and informed them that he opened all their bags to take arms and pistols. Pascual said that he (Tison) talked to the men like dogs and was surprised because he was not like that before. Tison claimed that he went through Pascual's backpack to take some papers that would be valuable to the enemy. Pascual said that he had all the secret papers with him. Tison insinuated that he was a liar. Pascual retorted that he doubted Tison could claim the same pride of accomplishment as he did because "he had been hiding in the mountains doing nothing but a bother to guerrillas to feed (him)."

The radio station supplies in the guerrilla camps were dwindling, and guerrilla commanders in direct contact with the Sixth Army requested supplies to continue operation. "Send more arms: grenades, air-cooled machine guns, and bazookas. Desperately needed are parts for signal equipment and five large-scale maps of Camarines Sur. And add three cases of cigarettes to the shipment." Office supplies were requested to manage the messages properly. Bicol guerrillas who were conducting frontal assaults on the enemy before the liberation forces' arrival needed medicine. A guerrilla commander in the Bicol area reported the Japanese dumping gas supply into the sea, burning artillery, rifles, and trucks in Balongay, Camarines Sur. The Japanese were seen giving away their things to civilians. Were they resigned to their pending death?

With the takeover of the Sixth Army, Cabangbang felt his skills and knowledge could be put to further use. He requested GHQ for himself and his men to join the 214th Counterintelligence Corp of the XIV Corps of the U.S. Army. He provided a list of the men with him in Manila and central Luzon in the last six months on intelligence missions. With "their mission now completed… and men have no work… in combat intelligence. They are not members of any guerrilla organizations in Luzon because they continuously worked in intelligence with me. With such experience of such a long time, it can be valuable to utilize them to help apprehend Japanese spies and collaborators in the Manila and Central Luzon areas. Assurance that these men know the Japanese spies, collaborators, Makapili, and other Japanese sympathizers."

When Cabangbang was called to Manila, Tamayo ensured that he informed as many of the men he could contact to report to the Sixth Army at Calasiao, Pangasinan. Some of the men did not receive the order from their own officers. The demolition teams reported their mission of instructing the Bulacan Military personnel in demolition and leading groups to destroy enemy installations and communications as accomplished. A commendation from the Army was much appreciated.

Men were flown to Mindoro, Palau, and Biak before transferring to a ship to Hollandia. Some returned later in July and joined their outfit, the 978th Signal Company, at San Miguel in Tarlac. Tamayo and the recon men reported to Camp Murphy.

Balcena, while at San Miguel, was assigned to drive the staff car for Col. Sidney Huff, General MacArthur's aide. He sadly saw what was left of Manila. Three-fourths of the city was reduced to rubble.

Pulido successfully fulfilled his mission but commented that there was much room for improvement. "A soldier's physical stamina was important for the kind of job together with aggressiveness and leadership. Party leaders should fully orient their men from time to time about the mission. Operational money should be supplied to individual soldiers." He thanked all the officers who trained him for that kind of military training, especially S-2 Intelligence and S-3 Operations responsible for the proper training of a special unit like theirs.

A plane picked up Pascual's group and flew them to Leyte. Without paperwork on specific orders, he reached Tacloban without knowing where to go. No one knew of his outfit. The unit was so secret and not included in the Army's table of organizations. No one could properly direct him. For thirteen days, he was at the casual camp (Base K) for transient soldiers. Lee Telesco, assistant of Col. Whitney, looked for his orders to return

A close-up shot of Filipino guerrilla fighters and a radio operator (in uniform) of the 978th Sig Svc Co., that were dropped behind Japanese lines one year before the U.S. invasion of the Philippines, to give Gen. MacArthur's forces combat intelligence reports and to conduct espionage. South Tarlac Military District, Luzon, P. I. 10-8-45. SC 216527

to Hollandia and told him that he should have stayed on Luzon. He was disappointed as he wanted to wait for his unit on Luzon and try to see his parents whom he had not seen for fourteen years. He boarded USS *Gen. Hayes* to Hollandia from Leyte. He had one tip to the boys. "Don't go for missions with officers whom you did not train with if you want your tasks to be accomplished perfectly."

In Hollandia, Pascual wrote about his experience. "When Japanese move out, it is not permanent as they come back to the same area. The Japanese use most of the big houses in a town for gasoline, ammo dumps, and quarters. When the U.S. forces began bombing the big houses, the Japanese moved to small huts for quarters and placed the ammo and gas under bamboo and mango groves. They avoided being strafed by dressing as women and mingling with the civilians. There was no security when the Japanese moved their convoys, or went on a march at night, as they were successfully ambushed by the guerrillas. The element of surprise was important because it was hard for the Japanese to reorganize. The best way to raid a small Japanese garrison was to kill the sentries silently before they could put out an alarm and the best time is in the early morning. The Japanese followed the drill of exercising early in the morning only in their loincloth and without their weapons."

After the war, Pilar regaled listeners with his story of how they taught the guerrillas in the proper uses of their weapons, Carbines, grenades, and machine guns whenever they had a chance. They shared combat principles, infantry tactics, security, and a little of everything about military training to instill some discipline. Living with the guerrillas was as he expected, moving fast to anywhere to avoid a Japanese attack. He appreciated the civilians who gave food, risking their lives, and were willing to face torture and death. He felt that Filipino officers should lead missions to avoid racial discrimination of the men he leads. It was demoralizing. He observed such attitude displayed by some American officers, not the officer in charge of his group. "In fairness, under the situation we are in, (we) cannot accuse any shortcomings as done intentionally." He suggested that officers select their men for a more efficient and successful operation.

Whenever the men gathered after the war, they humorously referred to the demolished vital bridges by those who placed the explosives and lit the charges as "Sgt. So-and-So's Bridge."

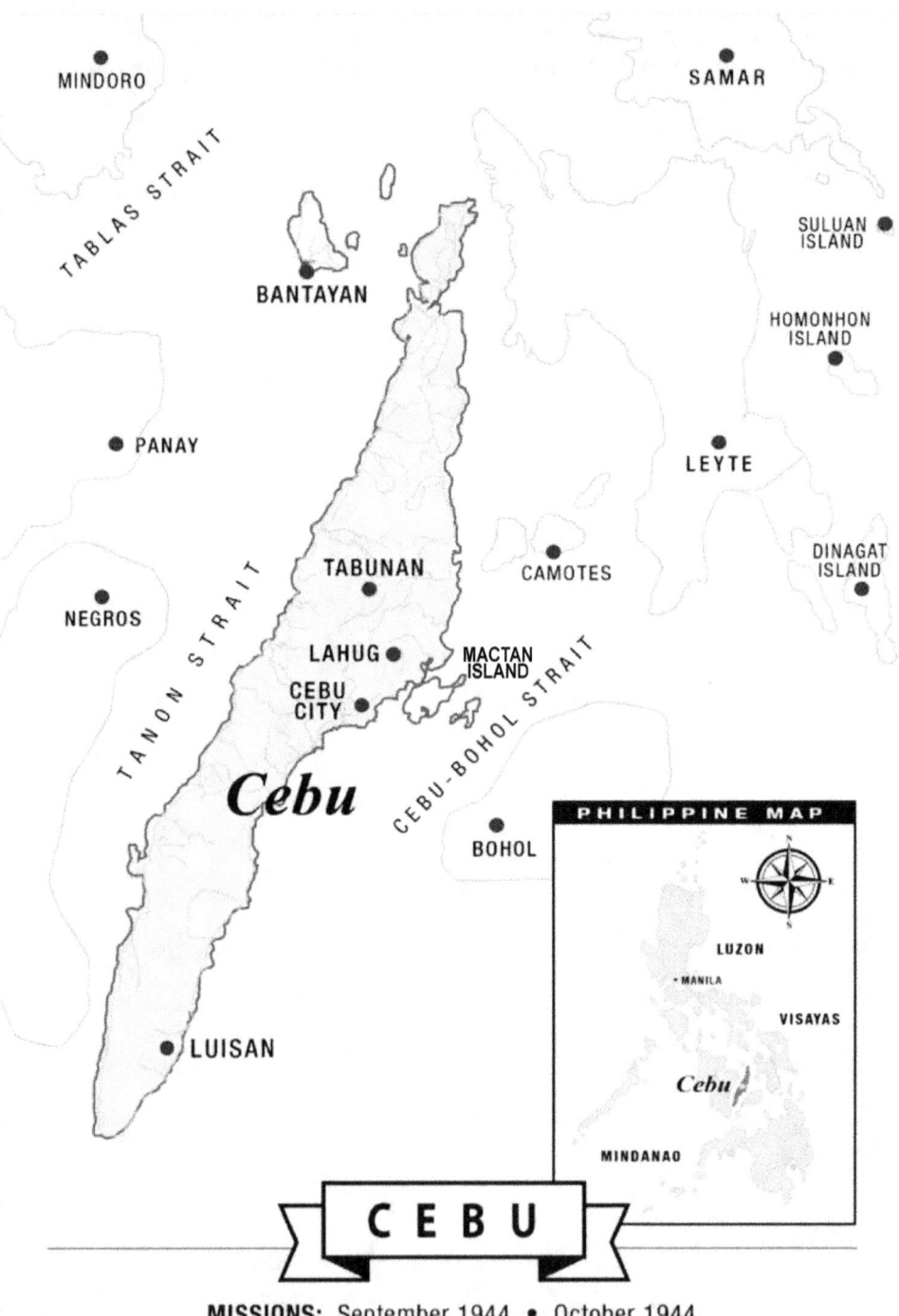

MINDORO

SAMAR

TABLAS STRAIT

SULUAN
ISLAND

BANTAYAN

HOMONHON
ISLAND

PANAY

LEYTE

TANON STRAIT

TABUNAN

CAMOTES

DINAGAT
ISLAND

NEGROS

LAHUG

MACTAN
ISLAND

CEBU
CITY

CEBU-BOHOL STRAIT

Cebu

PHILIPPINE MAP

BOHOL

LUZON

• MANILA

VISAYAS

Cebu

LUISAN

MINDANAO

CEBU

MISSIONS: September 1944 • October 1944

XIX

Cebu Island

The civilians suffered the most during the war. They were subject to constant Japanese raids and were victims of atrocities. Turning to the guerrillas for security was sometimes in vain as sometimes these guerrillas who they needed help from were the ones who stole or abused them. Some Filipinos retaliated by working for the Japanese and betraying these resistance fighters. The result was an air of mistrust constantly not knowing whom it was safe to rely on.

Harry Fenton, a former radio broadcaster, led guerrillas in the north and was feared for his brutal methods of obtaining his needs. An October 28, 1943, message of Abcede from Negros, confirmed the arrest of Fenton by Maj. Rogaciano Espiritu, guerrilla commander in the Central Cebu sector, on charges that he had ordered the execution of an Irish priest accused of espionage. The clergy was the only one left trusted to carry messages between guerrillas and amongst the people. With Fenton's death, James Cushing, a mining engineer who led the guerrillas in the south, was recognized as the commander of Cebu's Eighth Military District. Known as a fighter, Cushing developed a following that fought against a more superior enemy force.

When Mindanao became the distribution point for supplies in the early years of the war, Cushing received funds, weapons, and a radio to organize the guerrillas in Cebu. In return, Cushing radioed complete reports to FRS net control station in Mindanao on enemy activities including the islands of Bantayan and Camotes for transmission to Australia. But Cebu station PNM sent intelligence to both Negros and Mindanao to ensure it reached KAZ. Concerned for a collapse of Cushing's control, KAZ ordered Fertig (Mindanao) and Peralta (Panay) to stop influencing the Cebu guerrillas to join their forces.

A strong Japanese presence arrived in Cebu in 1944 and established a headquarter. It became imperative to strengthen the province to counter the enemy's growing presence. Twenty tons of supplies from a March submarine landing in Mindanao were hauled by carriers over water and land to Cebu. A long-range radio capable of reaching Australia was included. Signalmen Sanchez and Sabado (Samar mission December 1943) set up a radio in Mactan Island off the Cebu mainland.

At the end of March 1944, the transport plane of Japanese Admiral Mineichi Koga, commander of the Japanese Combined Fleet, was reported to have crashed before it reached its destination in Davao. There were no survivors. His chief of staff, Rear Admiral Shigeru Fukudome, with 14 staff officers, flew in another plane carrying a leather briefcase containing the top-secret Japanese Z Plan bound with a red cover in a box. The paper was entitled 'A Study of the Main Features of Decisive Air Operations in the Central Pacific.' The new strategy was to use land-based air forces as the main strength, with the support of Navy fleet units in a decisive battle to destroy the U.S. Pacific Fleet. It was committing all remaining Japanese naval power to one last major battle. Japanese surface and air strength were to be deployed after the first of April when Koga believed the American forces would be in the Philippine Sea.

Fukudome's plane ditched into the Bohol Strait between the islands of Cebu and Bohol. Fishermen captured Fukodome and ten survivors as they struggled to reach land. He did not have the briefcase. Two other survivors managed to reach land and escape in a canoe. The prisoners were brought deep into the hills of the village of Tabunan, Cushing's headquarters. Two fishermen later found a box that looked important with an embossed emblem. They buried it, afraid of the danger it might bring.

Fukudome did not reveal his rank and position during his interrogation. Cushing radioed KAZ the names of his prisoners and what actions should be taken against them. He described the documents with Fukudome covering Palau, the Philippines, and areas west of the Philippines in the South China Sea. Japanese operations maps, lists of bases, landing fields, and more were in his possession. He added that "constant enemy pressure makes this situation very precarious." The message was relayed to Mindanao, then Negros to KAZ.

Cushing moved his camp, the captives, and a small number of guerrillas across a broad ravine to Tupas Ridge. Three prisoners with serious wounds and a feverish Fukudome were carried in a litter. Japanese patrols were spotted marching toward his camp. With only 25 soldiers with him, it would be a losing fight. The camp was strafed and several of his men were killed. He received reports that Japanese soldiers had taken 100 civilians as hostages, killing some and burning villages for the return of the eleven captives. The fishermen fearing for their lives, turned over the box to him. With no directive from GHQ, he released the prisoners to stop the killing of his men, their families, and civilians.

His urgent message took more than two days to reach KAZ. Reports questioned the delay of his messages at FRS on Mindanao to KAZ. The reply was to do all possible means to keep the prisoners safe for evacuation and prevent recapture. KAZ believed that one of the prisoners was Admiral Koga. But they would know too late that the prisoners had been turned over to Cebu's Japanese commander.

Eventually, GHQ realized Cushing was forced to act while waiting for instructions on the captives. The affair was considered closed, but the Japanese continued to look for the documents and offered a reward. Finding nothing, Tokyo reported ending the search for Admiral Koga.

A POW escapee with a bodyguard of soldiers were instructed to cross Cebu and then the Tanon Strait to Negros Island to deliver the documents to Edwin Andrews, who was in charge of the radio station. They arrived at the end of April in time for a submarine landing. KAZ was informed ahead that the most important documents were in the box in Negros and the rest was on the way. The rest of the papers were tightly rolled inside two empty mortar shells and sent to Mindanao to try and catch another submarine.

By mid-May, the Allied Translator and Interpreter Section (ATIS) in Brisbane quickly translated the documents that were not in code, easily revealing Koga's secret Combined Fleet orders – Z Operations Order. It was issued from their flagship *Musashi* at Palau. The plan was to "use the combined maximum strength of all our forces to meet and destroy the enemy." Japan's crushing force was 1100 planes and 88 warships. Translated copies were quickly sent to General Macarthur, General Marshall, and Admiral Nimitz, commander of the U.S. Pacific Fleet. The documents proved their worth at the Battle of the Philippine Sea in June, when the American forces crippled the Japanese naval aviation forces before the planned Guam invasion. "Thus informed, Admiral Spruance refused to chase a Japanese decoy fleet 600 miles west of Guam and remained in place for the ensuing Battle of the Philippine Sea west of the Marianas." The Japanese Navy never recovered from their defeat to be able to deploy large-scale actions from their aircraft carriers.

While the documents were being translated, Cushing sent a message dated May 22 that the Japanese had not given up trying to find the documents. The Japanese naval commander in Cebu had planes drop leaflets warning of a deadline to return the papers or "the Japanese navy would resort to drastically severe methods against them." The camp was repeatedly attacked by the enemy unaware that the documents were no longer in Cebu but had been sent to Negros and Mindanao. Japanese patrols burned villages and continued to chase his beaten men.

As more men joined Cushing's unit, the Japanese were forced from the southern part of Cebu and the islands of Bantayan and Camotes.

USS *Nautilus* - September 1944

The submarine brought 100 tons of cargo for the islands of Cebu, Panay and Mindoro. Two weathermen, Sgts. Robert B. Caswell and Robert R. George, were also on board, bound for Mindoro Island. She proceeded using the bombing restriction lane to the Banda Sea and Luisan Point in Southern Cebu. At the rendezvous site, a small boat stood out from the coastline waving a small American flag. She surfaced at night, and 25 tons of supplies, 20 drums of gasoline, and two drums of oil were unloaded into two small boats in exchange for the captured documents, mail, and eleven evacuees.

Suddenly, the sub shook as she ran aground in eighteen feet of water amidst coral reefs and muddy sand. She tried to back up and blew her main engine ballast to no avail. The skipper was anxious for the *Nautilus* to be off before daylight to avoid being spotted by enemy planes or a patrol boat. Demolition charges were made ready to destroy the boat, and standby orders were given to burn all confidential papers in case of discovery. To lighten the sub, forty tons of cargo, the evacuees, and mail were removed and sent ashore. Her reserve fuel tank was blown dry, and reserve diesel fuel and ammunition jettisoned. The USS *Cero* was diverted from her patrol mission to assist and USS *Mingo,* another patrol submarine, left Surigao Straits ready to assist. Daylight was beginning to show, and it had been several hours of trying to get off the reef. Finally, the tide started coming in and the *Nautilus* blew her main ballast tanks again while running her engines at full speed. The sub finally came loose and cleared the reef. The cargo, evacuees, and the mail were loaded again. Assessing their situation, the skipper determined that they had only enough fuel to reach Panay Island and unload the rest of the cargo. The last stop, Mindoro Island, was cancelled. The weathermen disembarked in Panay and had to cross Tablas Strait to Mindoro.

The 85th, 86th and 88th Infantry of the Cebu Area Command guerrillas covered their battalion sectors with the propaganda materials the submarine brought boosting morale. Civilians and guerrillas wanted copies to show but there were not enough, so leaflets were passed from house to house. Some of the officers were given wristwatches and not the enlisted men. Cigarettes were passed around. They reported on the "jittery bewilderment" of enemy.

USS *Narwhal* - October 1944

From Negros, the submarine landed 50 tons of cargo for Cebu, Masbate, and Panay Islands. Radio, weather and demolition men disembarked and stayed at Cushing's headquarters. Some continued to other islands. The aircraft warning men were sent to Leyte to report to the Sixth Army. The sub slipped out to the sea for her return to Brisbane.

The signal aircraft warning men worked closely with the Army Air Corps to direct fighters and fighter-bombers to their targets. Filipino soldiers attached to the 597th Signal Aircraft Warning who disembarked were: Sgts. Jose C. Mariano, Jaime A. Ramirez, Pantaleon M. Rosete, Albert T. Talaro, Wilfred M. Regalal; Cpls. Raymond R. Fermin, Perfecto C. Quiming, Julian G. Ramos, Maurice W. Artiago, Argripino B. Flores, Emilio M. Custodio, Esteban R. Quiaoit and Ruperto Q. Quismundo; and Pvts. Benigno F. Mosquera, Sidney D. Wendam, and Hermogenes D. Oania.

Luzon Island, Philippines Landing – January 1945

On Luzon, Artiago's group set up a radar in Tuguegarao, Cagayan Valley to detect enemy aircrafts. The word was General Yamashita was in the area. Explosives were placed in the mountainside road used by the Japanese. A miscommunication asking for air support occurred that led to a bombing run on the other side of the river where Artiago and his men were hiding. Some of the men were injured while Artiago was hit by shrapnel. The locations of the correct bombing targets were immediately sent to the Air Force. Artiago spent six months in the Valley, "so scared at first, looked to God, but got used to it," he said.

Cebu Island Landing – March 1945

The radio at Cushing's headquarters crackled with the locations of heavy Japanese fortifications around Cebu City. Signalmen punched the keys transmitting the message of around 9000 guerrillas under Capt. Cushing ready for orders. The American forces expected heavy opposition when they landed at Talisay Beach but there was only slight resistance. The guerrillas had earlier received the orders to attack and secure the beach area. American Task Forces landed on both sides of the northern part of the island keeping the Japanese in retreat. Cushing's men immediately attached to the Eighth Army to enter the city.

Flanking fire from the enemy at neighboring ridges delayed the Task Force's entry to the city. The guerrillas fought and acted as guides pinpointing Japanese locations as signalmen transmitted their progress to KAZ. Fighting was so severe in the destroyed city making General Eichelberger believe that the enemy was composed of regular Japanese infantrymen. Casualties mounted as entrenched Japanese defenders in tunnels and caves in Lahug Airfield and Babag Ridge fought hard. George Fukui, a man serving in the Japanese Army as an interpreter, reported "a gruesome order issued ... Japanese children under the age of 13 would be an encumbrance and should be killed." By the end of April, the Japanese broke into small detachments to escape.

Cushing was reported demoted for his handling of the Koga documents but was reinstated after the war. Col. Whitney defended him that he did not disobey any order.

Leyte Island, Philippines: 1st Filipino Infantry soldier Pfc Nick Bachas of Reno, Nevada, and of PCAU #6 interviews native. Source: NARA

Philippine Civil Affairs Unit (PCAU) Counterintelligence Corps (CIC)

Japanese strongholds along Northern New Guinea fell as General MacArthur drove his forces towards the Philippines. After each enemy-occupied island was cleared, there were few organized and expansive civil affairs activities to establish order, alleviate the people's sufferings, and rehabilitate basic infrastructures. But for the Philippines, GHQ drew up a detailed and wide-ranging plan.

Major General Charles Willoughby, head of G-2 Intelligence, explained General MacArthur's interest in the importance of civil affairs in the Philippines. Essential to his plans was "the measure of freedom and liberty given to the Filipino people be at least comparable to that enjoyed under the Commonwealth Government before the war. Utmost care should be taken that an imperialistic attitude not be introduced into the situation under the guise of military necessity."

Edgar Crossman, a medical doctor and one of the planners of the Philippine Civil Affairs Unit (PCAU) stated that after combat, suffering civilians had to be kept from interfering with military operations. To do so, local government authority had to be re-established as soon as possible, and necessary food, clothing and medical supplies provided for civilians. The economy was to be re-established, and human suffering alleviated. Success of the operation depended heavily on the PCAUs being self-sustaining with their own transportation, mess facilities, and tent equipment.

Training of civil affairs personnel for the Far East was instituted under the War Department as early as June 1944 at Ft. Ord, California. An initial estimate of one hundred officers and four hundred enlisted men was the requirement for the Section. General MacArthur called on the War Department "to increase the number of officers to two hundred fifty leading thirty teams of enlisted men to revitalize combat-damaged areas."

GHQ looked to the Filipino soldiers stationed at Oro Bay to fill the units. A month before the Leyte landing, G-3 Operation requested a list of enlisted men from the 5217th

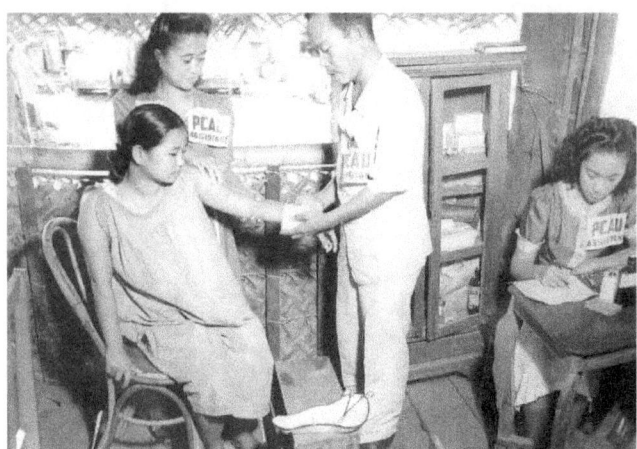

Reconnaissance Battalion (Provisional), qualified Filipinos from the 1st Filipino Infantry Regiment and the 2nd Filipino Battalion for PCAU. The officers were selected with a background in medicine, law enforcement, agriculture, labor relations, and administration. The enlisted men would provide administrative and logistical work. Men with college degrees who could type, would be authorized to appoint temporary officials in the towns they were assigned to and hire assistants.

Luzon, Philippines 1945: PCAU (Filipino Infantry men) hired assistants to provide medical treatment and economic support to the local population after military combat was over. Source: NARA

The Philippine Civil Affairs Unit (PCAU) was activated in Oro Bay, New Guinea in September 1944 and the training program began the following month in Brisbane, just weeks before the Leyte landing. The Section was "to assist in the interim period between the liberation of Philippine territory by U.S. military forces and the return of that territory to the duly authorized and recognized representatives of the Commonwealth Government of the Philippines. "

Forty-five officers were chosen for the thirty PCAU units that would follow the invasion force. The first eight PCAU for the Leyte landing were mostly composed of Filipino soldiers. *The Philippine Mail* continued to track the Filipino soldiers and suggested Capt. J. David Malig with the 1st Filipino Infantry, a physician before the war, as a possible choice for the PCAU.

PCAU Mission and Responsibilities

GHQ issued 'Standing Operating Procedure, Instructions No. 26' dated Oct 9, 1944, that spelled out the administration of the civilian population during the reconquest of the Philippines.

A PCAU unit contained ten officers and thirty-nine enlisted men. An American officer headed each section: Medical, Engineering, Transportation, Labor, Relief, Welfare and S-3 Operations, Supply and S-4 Logistics, Finance, Public Safety, S-2 Intelligence and Legal, and S-1 Personnel. The medical officer had four enlisted men as medical assistants. Not all units had a surgeon. In their designated area, they obtained information on local doctors, nurses, dentists, and pharmacists to treat the people. Midwives and qualified medical assistants were recruited in the islands to provide support. The unit was self- sufficient with its quarters, mess facilities, and transportation.

The number of PCAUs attached to a combat task force ready to take over as soon as the area was cleared of the enemy depended on the population and needs of the people. Crossman reported that the "nearer the Army units got to the fighting, the more the

Army appreciated Civil Affairs" because PCAUs relieved combat units of responsibility for civilians.

In the first months of a liberated area, temporary shelters, food, and medical relief were to be provided. Some order had to be established in the heavy destruction of the infrastructure and every effort was to be made to stabilize the civilians and not hinder military operations until turnover to the local government. PCAU men often risked their lives when they began their work, as fighting was still ongoing to remove pockets of stubborn enemy resistance. The unit's efficiency was based on the team's rapidity to get local officials to run their municipal government.

Army area commanders were charged with the responsibility of appointing temporary municipal officials in their respective areas of operations until it was possible to turn over the administration of the civil government to the Commonwealth Government. The Army and Area Commanders' power to appoint, remove and supervise the actions of temporary appointees were often delegated to the PCAU commanders.

To make the Filipinos self-sufficient as rapidly as possible in their economy, PCAU assisted in the resumption of normal civilian activities of food production (fishing, planting, harvesting, reconstruction of rice mills), education (opening of schools, hiring of teachers), health (hospitals, sanitation hygiene measures), and public utilities (repairs of water, electrical systems). Farm tools, fishing equipment, and seeds for planting were distributed. Planting was encouraged to hasten self-sustenance.

Strict laws were posted and enforced. Curfews were set at 6 pm. Manufacture and sale of alcoholic beverages was forbidden. Possession of Army equipment, food, and clothing was illegal.

The Medical Section brought 800 lbs. of medical supplies for each unit. Dispensaries and hospitals to treat casualties, conduct immunization programs, and check for venereal disease were opened. Under their supervision, sanitary inspectors checked the latrines, garbage collection, drainage ditches to control mosquitoes, and rehabilitation of cemeteries. School teachers were made PCAU assistants to instruct civilians on sanitary procedures. Garbage pits and latrines had to be used or food ration cards were confiscated. PCAU Sgt. Reynaldo R. Curva worked with a civilian hired as a sanitary inspector to ensure pigs were in their pens and the grounds cleared of trash to prevent diseases.

A Finance Section handled the cash transactions, exchange of guerrilla emergency currency and Japanese currency to Victory pesos, salaries to civilian employees, and other financial matters. The Philippine government authorized the printing of new Victory paper currency from the pre-war Treasury certificates to increase circulation and get the economy going until the government stabilized.

A commodities sales system was established to handle the cash sales to civilians. Victory pesos were used to eventually eliminate the use of printed guerrilla emergency money. Prices were set for essential commodities and medicine sold at retail stores. PCAU assistants uncovered violators of price controls. Punishment was strictly enforced in the effort to standardize costs. Soon many of the locals found good money selling souvenirs to the Americans.

The Public Safety Section supervised and assisted local chiefs of police in maintaining law and order and sponsored laws. Violators were tried by appointed Justice of the Peace recommended by PCAU. Civil government had disappeared with the war, and at times, former government officials in town continued to provide stability, but their loyalty may have become questionable. Not to take chances, PCAU coordinated with the Counterintelligence Corps (CIC) to establish laws to ensure public safety and security of warehouses, public buildings, and everywhere large supplies were stored. Judges, courts, and police were appointed. Identifying true collaborators and pro-Japanese from those accused as revenge, was difficult before arrest, interrogation, and turnover to the Philippine government. The difficulty in identifying was due to an unclear definition of a collaborationist. One was not assumed guilty simply because he worked under the Japanese government. Local leaders were approved to occupy temporary positions in the local government once loyalty was determined. For emergencies, the PCAUs had the authority to call upon the Sixth army for additional support.

The Legal Section arranged and supervised the evacuation of thousands of families from areas taken over by the armed forces. Damages to private and public property were surveyed and paid.

The Industrial and Engineering Section supervised the repair of roads, bridges, roofs of schoolhouses, churches, municipal buildings, and hospitals. The repair of churches to allow the people to attend services was important for the people's well-being. Industrial plants were revived. Equipment and materials for construction were salvaged for immediate use of the Army. An ice plant in San Juan del Monte in the Manila area was put in operation and guarded for use of the hospitals. They worked with the Medical Section, to set up or repair public latrines, hospital pits, and drainage canals. *Nipa* (a type of palm) projects created panels of *sawali* (woven split bamboo) and *nipa* shingles for home building and Army construction were established in each area. That was an object of curiosity to the White soldiers.

The Transportation Section provided trucks and hired mechanics and drivers to deliver supplies to the different districts in which the PCAU was working.

The Labor Section procured required labor for military and PCAU projects. Salaries were paid in Victory pesos for purchase in the retail stores. Thousands were hired and paid for their services such as carriers of food and ammunition to front-line troops, and litter bearers. Hired were doctors, nurses, laundry workers, nipa project workers, relief and welfare workers, census workers to register families, and emergency civilian labor for removal of houses and evacuation of families to safety outside military grounds. The police, fire department, government officials, schoolteachers and public works personnel were paid until the government took over but the salaries of government officials and schoolteachers in the liberated areas were reimbursed by the Army. A May 1945 issue of *Free Philippines* published the article *U.S. Army Repaid Php225,000 for Aid to Commonwealth*.

The Supply Section managed the flow of relief supplies. PCAU supplies were marked separately from other cargo. The War Department called the stock 'civilian relief supplies', giving the impression that they were free. In the beginning, they were given away to civilians. Some expenses were charged to the Philippine government. Supplies were supplemented from excess military stock, captured enemy material, and local produce. If the relief supplies did not arrive as soon as combat task forces cleared an area, Army rations and medical supplies were used in the interim. Relief of food and clothing included cooking utensils, machetes, needles, thread, and cloth. Trucks brought goods, fresh eggs, vegetables, and coconuts from nearby province to various relief points. Authorized by the Philippine government, rice and corn were purchased from the civilians to stock retail stores. Most of the rice came from captured Japanese warehouses. When retailers were allowed to open after careful investigation, they sold at cost and wholesalers were allowed to distribute to all retail stores. Bakeries opened and started baking with supplies from the Army base camp. Items sold were carefully inspected for any tampering. A civilian, Bubi Krohn, revealed that the Japanese drilled a hole in the bottom of the powdered milk can, drained, filled it with *gawgaw* (cassava starch), and sold it. Babies got sick and died. The unknowing Filipinos had bought them from dealers who had leftovers.

When schools reopened, schoolbooks were brought out of hiding from the Japanese who burned all related to English and American heroes. Teachers found children between six and twelve years old eager to attend classes, but many appeared without pants so children's clothes were made from captured Japanese goods. Children enrolled wearing an assortment of 'Japants.' Stealing clothes was so rampant that tombs were excavated to take the clothes of the dead.

The Relief and Welfare Section supervised the feeding of indigent and refugees. A system was established using a questionnaire to investigate disinterested parties and local officials to determine worthy cases needing aid. Periodic spot checks on each family were deployed and recorded for ration until no longer needed. To avoid double rationing from different relief distribution points, the tickets were numbered confiscating several thousand duplicate tickets. The aged, sick, blind, children, families without male members over 16 years of age, and families of war casualties received majority of the aid.

Because of the urgency for the 8 PCAU units to accompany the Leyte landing, the units had only 5 days of training in extremely limited facilities hurriedly made available in New Guinea. Aside from duties in the combat field, instructions on disaster relief and public health organization were taught. They went through lectures from the 24th Division's Civil Affairs and guest experts in their fields. Their importance was magnified with the issuance of M1 Carbine rifles to be carried and used as needed to prevent hindrance to their movement. From the time of training, a PCAU was attached to an Army or Area Command to administer civil affairs policies.

The initial assignments were the following.

PCAU #1 – 7 were attached to the 1st Cavalry Division
PCAU #1 and 2 were kept in reserve under the Sixth U.S. Army.
PCAU #3 and 4 to Army Support Command (ASCOM)
PCAU # 5 and 6 were attached to X Corps, comprising the 24th Infantry Division and the 1st Cavalry Division.
PCAU #7 and 8 to XXIV Corps, comprising the 7th and 96th Infantry Divisions. Later joined the 77th Infantry Division
PCAU #9 – 30 were assigned to combat units as they landed in various islands beyond Leyte. Duties were modified from previous PCAU landing experiences. The units continued to attach Filipino officers and enlisted men.

For the Leyte operation, an estimated twenty thousand tons of food for a population of one and a half million people with additional cargo for textiles, medical supplies, and equipment transportation were brought and administered by the 'Invasion PCAUs'.

PCAUs #1 – 8 were the 'Invasion PCAUs' for the Leyte landing supporting the task forces of the 1st Cavalry Division. The final roster had almost all the four hundred officers and enlisted men from the 2nd Filipino Battalion. The 1st Filipino Infantry Regiment was kept intact as much as possible as a combat unit for deployment. The teams embarked on the SS *J. S. Hutchinson* from Oro Bay to Finschhafen to drop off **PCAU #5 and #6** at the unit's specialized facilities. All units remained on board for three days. While unloading, Pvt. Joseph Elvena, **PCAU #1** had an accident while onshore and was sent to the station hospital. Bonner Fellers, head of Civil Affairs, members of the Philippine Commonwealth government, and Joseph R. Hayden, a former governor of the Philippines spoke to the men about the immense responsibility of the Civil Affairs Units in the coming operations of restoring the Filipino civilians to their normal way of life. High standards were expected of the men. After camping overnight on Pie Beach, **PCAU #1 and #2** joined a convoy of five hundred ships carrying X Corps Headquarters and the 24th Infantry Division to Leyte. Two days later, the convoy from the Admiralties with the XXIV Corps and the 1st Cavalry Division joined. Although the days were clear and the sea calm, there were three air raid alerts. Onboard, the PCAUs were in charge of first aid and a combat surgical station in case of enemy attacks.

Leyte Island, Philippines Landing – October 1944

The U.S. returned with a force that was the largest U.S. Army operation of the Pacific War. The armada sailing towards Leyte included reconditioned ships from the Pearl Harbor attack. Alarms went off when mines were found on the approach. All ships' portholes were shut tight for a blackout as the men sweltered all night. The men were in the "stifling holds of the ships so crowded you could not move" and heard the continuous anti-aircraft drills in expectation of attacks from Japanese medium bombers. The transport ships were not small, but half of the space was reserved for those with commissions with private dining rooms and lounges.

The transport ships arrived at San Pedro Bay four days after D-Day. PCAUs were on *LST552* escaping two directed air attacks but not the third. Unit vehicles were damaged, and money contained in one was partly burned. Sgt. Andres Ancheta, Cpl. Emiliano Tambo and T5 Larry A. Nepomuceno were missing. Sgt. Jerry Jacobs, Cpl. Gerino Catayas, and Pvt. German Sebolboro were wounded. They received Purple Heart medals for their wounds.

Crossman landed the day after and transmitted "No PCAU had arrived yet, so the combat and service troops overfed the refugees with Army rations. Civilians of all ages were coming through the lines … see soldiers carrying older people and children across carabao wallows. One baby was born on the beach that night." The Sixth Army issued a directive to guard all warehouses and public buildings until the PCAU officer and his men arrived.

He passed his observation to the signalmen to flash to KAZ. "X Corps consisting of 1st Cavalry Division and 24th Division landed on White and Red Beaches, respectively, south of Tacloban, 1st Cavalry on the right took the airstrip of Tacloban, and the 24th Division on the left circled the low hills to the east of the central Leyte Valley and proceeded up the valley. The southernmost part of this range was Hill 522 which commanded the road from

Tacloban into Palo where the road up the Leyte Valley began. Hill 522 was taken the first day. The first American flag was raised on the hill. That night the Navy kept firing star shells all night lighting it to prevent the Japs from sneaking up in the dark and recapturing it."

The PCAUs waited in their ships until combat had eased and the area cleared of the enemy. They witnessed the helplessness of the combat troops wading to shore as Kamikaze planes swarmed and picked at them. Hundreds died before they reached land. Fiery explosions and heavy black smoke filled the air as landing ships were hit. Signalmen onboard their landing crafts furiously typed sending the enemy's status and location as they received them from guerrilla stations.

Signalmen followed the landing craft carrying General MacArthur and President Osmena bumping the shallow waters of Red Beach causing them to wade ashore. General MacArthur's famous return speech filled the airwaves. He was a 2nd Lt. from West Point the last time he was on Leyte.

The Philippine flag was raised on the steps of the provincial capital of Tacloban and the government turned over to President Osmena. The radios crackled with news of President Osmena giving official status to the guerrilla fighters and incorporating them with the remnants of the old USAFFE before the war into the service of the new USAFFE and the Commonwealth Government.

Problems arose with the PCAU operations and Col. Whitney took over with a small staff in an office at Tanuan south of Red Beach. A supply base with surface vehicles and sailboats was set up, ready for the PCAU. Daily updates were provided to the military units to which they were attached but not to the War Department. That was an order from General MacArthur. PCAU was doing more for the people than required of its mission.

The Press excitedly watched as history unfolded adding to the traffic of guerrilla intelligence and internal Army communication. It was a momentous moment in history, and they were eager to publish first-hand accounts. Transmissions clogged up and signalmen scampered to keep the radios operational.

As soon as combat ended in each locality, PCAU relieved the combat troops of starving and sick civilians rushing to the American safe zones. Once the rush for food, clothing, and medical treatment eased, they recruited civilians for military work projects and for their use in carrying out their regular duties. Families were relocated to provide space for military construction. Labor was plentiful, and many were willing to work. But the lack of transportation from assembly to the place of work caused many delays. Small children living close to the camps and encouraged by the friendliness of the soldiers bartered bananas, papayas, and other fresh fruits in exchange for candy and canned goods.

Crossman observed that authorization of incumbent justice of the peace after screening for loyalty to try minor profiteers and offenders to keep the food prices down was a great help to PCAU.

The PCAUs encountered supply problems in all the assault landings without roads, docks, and buildings. Getting the relief supplies and distributing them to the people was their biggest problem. There was a lack of transportation to ship personnel and equipment from various bases. There were dead periods of waiting for additional supplies after loads were consumed. Without rest, they were relieved only when fresh PCAUs arrived or more were added to their location. The PCAU commanding officer reviewed their reports, commented on their efficiency, and moved on to the next assignment once objectives were met. Usually, the unit moved on to another location within a few weeks but stayed on depending on the problems encountered.

The people appreciated the help, but after three years of survival, it was difficult to get disciplined again to stop the looting of the warehouses and property. They did not interfere with military operations once they understood the PCAU's purposes. A local police force enforced public safety, especially against unorganized guerrilla bandits who took food and whatever they could from the civilians. In some areas, guerrillas were reluctant to relinquish their military government.

PCAU #1 to 8 known as the 'Invasion PCAU,' landed attached to the 1st Cavalry Division. The X Corps and XXIV Corps of the Sixth Army also landed with their PCAU teams. The Army Support Command (ASCOM), and Base K supply depot landed with over two hundred thousand troops in three days. Each one carried a supply for ninety days. Typhoons with heavy rains and winds slowed down the landings. Many of the soldiers experienced their first earthquake in one that rocked Leyte.

Onboard the USS *Fayette,* U.S. and Philippine flags were given to bearers who would carry them ashore in the first assault wave. Cpl. Ponciano Dacones, of **PCAU #5**, proudly received the Philippine flag from Lt. Vincent F. Zerda with Philippine Island Affairs. Earlier, he and an American soldier had won the lottery to carry their country's flag. He waited in the landing craft until all assigned were filled with troops from the transport. Unknown to him and all the other Filipino troops, the landing crafts that brought them to shore were made of Philippine mahogany bought five years earlier and manufactured in New Orleans. All the landing crafts lined up abreast approaching the beach at Palo until

Leyte Gulf: PCAU Cpl. Ponciano Dacones holding the Philippine flag on the USS Fayette ready to implant it on the shore with the invasion force. Oct 20, 1944.

the corals stopped them, forcing the men to wade knee-deep to shore with their rifles and full packs. Dacones rushed onshore and planted the Philippine flag on the beach under Japanese fire. It was reported that he had to take it down later. Only the U.S. flag remained raised. Four days later, he was killed in action at the height of the Battle of Leyte. *Free Philippines* reported that he was killed when he fired on a Japanese machine gun crew from the window of his hut and the flash of his Carbine drew their fire. He was the first PCAU from the 1st Filipino Infantry Regiment to be killed on Philippine soil. His body was shipped to his family at Paitan, Sual in the province of Pangasinan.

PCAU #1 and #2 arrived in the early morning at San Pedro Bay. Maj. Talman C. Budd headed **PCAU #1** and was temporarily replaced by Capt. Greene when he had to enter a hospital. Lt. Gregorio R. Minodin was the adjutant of **PCAU #1**. Capt. Virgil B. Ashcraft commanded **PCAU #2**. The tides rocked the landing of the LST. Kamikaze planes raided the convoy as they waited to go ashore, nearly hitting the landing crafts. Anti-aircraft guns blazed from the LSTs, but one received a direct hit, flinging bodies into the air. Men did not have enough time to hide behind a light tank the craft was carrying. Landing crafts close by began picking up survivors. The rest of the units arrived two days later, floundering ashore under sniper and machine gunfire.

Sgt. Benito S. Dalipe, Sgt. Andres Ancheta, Cpl. Emiliano Tambot, and T5 Larry A. Nepomuceno were thrown overboard and were missing. Ancheta was sitting on about $1 million worth of liberation pesos. Sgt. Benito Dalipe was killed sitting beside him. T5 Maximo P. Sebastian, Pfc. Macario A. Alicante were reported also killed. PCAUs, Sgt. Jerry G. Cabos, Cpl. Gerino C. Catavas, Lt. Igancio I. Josue, Sgt. Marcel R. Cababa, Sgt. Robert O. Guzman, Sgt. Ricardo O. Nunez, T4 Frank P. Cabrales, Cpl. Remigio A. Abalos, Cpl. Ben S. Gonzales, and Pfc. Felciano P. Obrero with Cpl. Alex de Leon Fabros were reported wounded and received Purple Heart Medals. There were others whose bodies were not found.

Disembarking at White Beach, in Palo, they worried about the supplies piled up on the beach and vulnerable to aerial attacks. The Sixth Army provided 8 tons of medical cargo for the dispensaries and hospitals to be built. Men dug foxholes on the beach under enemy strafing during the night. Water and leeches seeped into foxholes. They positioned coconut logs fallen from the bombardment on the beach around them. Pvt. German B. Sebolboro was listed as wounded on October 25. Lt. Josue received the Bronze Star for heroic achievement in military operations against the enemy in Leyte Harbor.

Maj. Budd related that they were the first team to land and wanted to sort out problems that would benefit succeeding PCAU units. Communication between the provincial officials and the Sixth Army on acute problems caused delays. They cared for the wounded civilians from the naval bombardment of the beaches and the diseases that plagued them. "The six PCAU units in the area were unclear about their authority and responsibility working with the local officials," Budd said. The training they had received focused on providing necessities and medical help to civilians and recruiting a workforce for military operations. PCAU members were transferred to support hospitals as in the case of Rodolfo M. Aquino from **PCAU #1**. They had to explain to the people that the Japanese currency would not have value and Philippine currency would be exchanged for Victory pesos. They worked with pencil and paper as they secured telephone lines because radio was subject to enemy interception. Fingers flew on manual typewriters with multiple copies for distribution to every related military unit, ship, and supply base. Some got sick from malaria or other tropical diseases like T5 Philip A Seveses.

The road from Tacloban to Palo in the south was cleared on the first day. The Navy fired star shells all night to illuminate the road and prevent recapture by the Japanese. Grenades and flame throwers did not stop the shooting from the enemy's pillboxes, an artillery major told Crossman. Bulldozers were used to bury the pillboxes. It was an intense joyous time when the Army's P-38 planes arrived several days later and circled Tacloban.

With mixed emotions, the men felt pride returning as liberators and hoped to find out if any of the family members they left behind were still alive. PCAU Sgt. Herman Brotonel arrived in Tacloban and later, went to Catbalogan in the island of Samar.

T5 Aurelio Bulosan, who had lived for almost two decades in the U.S., was eager to return and wrote of "suddenly (being) transported to the Third World" and the "admiration of the people of Leyte of these brown American fighting men." In turn, the men showed respect for the guerrillas who had fought over the years. They "saw how easily and some with a smile, the guerrillas killed any enemy they confronted after they had lost family and friends to the brutalities of the Japanese." Bulosan saw a class distinction in a guerrilla camp that vehemently disallowed *mestizos* (mixed native Filipino and Spanish blood) as they had caused the deaths of their men. He heard many of the businessmen in town "registered as Spanish citizens" were of the ruling class that often cooperated with the enemy.

The PCAU headquarters was the large house of the Price family in the outskirts of Tacloban which was also the headquarters of General MacArthur and his staff. The following day, the two units rode north on Army vehicles to the Tacloban municipal building on the alert of ongoing hostilities in the town. Near the docks, they set up camp in a warehouse, Eureka Sawmill that was still standing. A relief center was set up in the municipal building. Refugees were fed with rice from captured Japanese warehouses while enemy plane bombings dropped around the city. Many civilian workers were killed. Food stock was given to the Holy Infant Academy to continue feeding refugees from their convent kitchen and house families. Three captured Japanese trucks collected and disposed of the garbage, cleaning up the city. A Public Health Department was re-established with the aid of Capt. Frank Roman of **PCAU #1**. Captured equipment assembled a laboratory which even included live guinea pigs. PCAU Sgt. Gorgonio Quimba, with a master's degree in Bacteriology was placed in charge of the laboratory, doing commendable work. The Army and Navy requested numerous bacteriological analyses of water with the first request from two specimens of drinking water in the offices of General MacArthur and President Osmena. Medical supplies and civilian physicians were sent to other PCAU teams who could not locally obtain them. Cholera vaccines were administered to thousands of civilians living in dugouts and foxholes from the naval bombardment to prevent an outbreak.

Two days later, the Japanese were reported infiltrating the unit's service area in the city. Although they were a non-combat unit unless required to defend themselves, a small group from **PCAUs #1 and #2** located and captured the spies. Pfc. Mendoza of **PCAU #2** shot one and wounded another. Two were arrested later. Under interrogation, the spies were either Formosans or Koreans sent to obtain information for the Japanese. A deep tunnel was discovered close to the camp with Japanese rations and equipment inside.

Leyte Island: FILIPINOS GREET AMERICAN FILIPINOS.

Pvt. Demitrio Cabunoc and Pvt. Andres Lupiba, of Los Angeles, California, is welcomed home by Carlo Yanuario and Mararo Caminong, Guerrilla native partisans on Leyte Island, Philippine Group. Oct 21, 1944. Signal Corp 254864

A Signal Company photo showed PCAUs Andres Lupiba, a former bellhop at the Ambassador Hotel, Los Angeles, and Demetrio Cabunoc shaking hands with guerrillas on Leyte. The men looked like they had just arrived with their full pack, rifle, helmets, short shovel, and enclosed canteens hanging from their belt.

Another photo showed Lupiba meeting guerrilla captain Nieves Fernandez who was a schoolteacher before the war. She held Lupiba's hair at the top of his head with her left hand while her rifle was slung over her right shoulder. She held a *bolo* against his neck as she peered through her glasses to show him "how she used her long knife to kill Japanese soldiers silently."

Captain Nieves Fernandez, the only known Filipino female guerrilla leader and formerly a schoolteacher, shows U.S. Army Pvt. Andrew Lupiba how she used her long knife to silently kill Japanese soldiers during the Japanese occupation of Leyte Island. November 7, 1944

Foremost was the clearing and construction of the enemy's Tacloban airfield for air units to fly in from Morotai. U.S. air support could provide coverage for a limited time only so the 1st Cavalry had to immediately secure the airstrip. Cpl. De Leon Fabros with the 1st Cavalry Division saw the poor drainage that kept the ground muddy.

General MacArthur informed "the commanding general of the 1st Cavalry about the dangers of rape of Filipina girls by his men. As a solution, all troops were withdrawn, including military police from Tacloban that night, resulting in the capitol building's interior being wrecked and large stores of Japanese rice and other food happily carted away from warehouses by civilians using many Army trucks to their homes," related Crossman.

The carabaos (water buffalo), a beast of burden on the islands, were helpful in military operations. *Free Philippines* reported elements of the 24th Division driving a herd before them as a screen when the Japanese counterattacked outside Palo. A train of one hundred carabaos loaded with supplies trekked through miles of muddy terrain when the 96th

Division's equipment vehicles were bogged down. The commander and his aide rode on a carabao's back returning from the front.

By the end of the year, **PCAU #25** arrived to replace **PCAU #2** which moved to Tanuan where they encountered Japanese patrols as they looked for Japanese supply depots to have enough food stuff to feed the civilians. Malaria and tropical ulcers weakened the men, but they forged on, vaccinating people against smallpox, cholera, dysentery, and typhoid fever. T5 Armada was loading his vehicle with civilian laborers in the middle of a cane field when Japanese snipers opened fire on them. The snipers were those enemy soldiers who were wounded or too sick to keep up with their units and decided to climb trees and die fastened to them.

Japanese stragglers pushed inland from Tacloban roamed the island of Leyte. Thousands of Japanese troops were shipped to Ormoc on the western side of the island to attack the landing forces and supplies in Tacloban. Under heavy rains, **PCAU #2** moved to Basey, Samar adjacent to Leyte. Headquarters was at the municipal building and relief camp at the market building. Five officers and 10 enlisted men followed the U.S. 77th Infantry Division to Ormoc. Nightly machine guns and rifles broke the stillness of the nights and dead Japanese trying to infiltrate were found in the morning.

Twenty-five thousand refugees hindering the military forces had to be evacuated from the front lines. Lt. Irving, T4 Pulido, and Pfc. P. Abantao were shown on a map where to move the civilians. A crowd of around ten thousand people clutching all their personal belongings and foodstuff they could carry, with pets, rode in carts pulled by carabaos. Others walked. They observed a small group of men following the crowd from the hilltops but did not attack. As nighttime approached, a civilian was placed in charge with instructions on what to do in case of Japanese presence. Returning to camp, the men were glad to receive a verbal commendation of their day's work. Christmas was spent with the refugees before returning to Tanauan. **PCAU #2** attached to a different combat unit almost every three months.

PCAU #3 and PCAU #4 landed on Red Beach, the town of Palo south of Tacloban City. They dug into the beach for two nights with overhead fire from naval guns and artillery batteries. Quarters were in a warehouse near the docks in Tacloban City with the other PCAUs. The Japanese dropped bombs between the warehouse and docked Liberty ships, instantly killing Pvt. Moses M. Dackel and wounding T4 Gilbert Mayor, T5 Gelvacio M. Martinez and T3 Gorgonio P. Quimba who were working at the Public Health medical laboratory. The units moved their headquarters to a safer location at Bethany Hospital. Maj. Kline, head of **PCAU #3** and Capt. C. Topacio, head of **PCAU #4** located available civilian doctors and nurses to run three hospitals to care for the civilian casualties. Vaccinations began for smallpox.

Both PCAU units were attached to the Army Support Command with a finance officer who supplied funds for paying the salaries of around 25,000 recruited laborers. Wages for work encouraged independence and getting the economy started instead of just giving out free clothing and canned goods. But the American soldiers were bothered by the pitiful state of the locals who were mostly the aged, blind, war casualties, war widows, children, and families with no male member over 16 years of age. Refugees were fed from a food kitchen at the Tacloban municipal building. Civilian welfare workers were hired as assistants and investigators. Rice was given from local stores, augmented later by small stocks of rice salvaged from looted Japanese warehouses. The 1st Cavalry had opened Japanese military warehouses, and the town's inhabitants had more than what they needed, while those from surrounding villages had none. Violators of set prices in retail stores, alcohol sales, and possession of Army equipment were prosecuted. Those who prevented laborers from working with the Army were arrested.

PCAU #4 to #8 shipped to Leyte directly from Oro Bay, New Guinea to their objective area. Their training was completed in such a short time before departure that their first contact with the combat force they were supporting was when they landed on Leyte. With no orientation between PCAU and the combat unit they helped, coordination was delayed and inefficient.

PCAU #5 and PCAU #6 went ashore with the assault forces of the 24th Division and the 1st Cavalry Division on the second and third day. Their transportation and equipment did not land until nine days later. To care for the wounded civilians in the naval bombardment of the beaches immediately, they had to get medical supplies from the Army until their supplies arrived.Medical supplies were quickly exhausted in the care of wounded civilians. Captured Japanese supplies were used to augment them. From this experience, the order was issued to have future PCAU's go ashore with at least a 10-days' supply of medical items.

They moved to the town of Palo and to the surrounding municipalities. The Filipino bishop allowed the Army to put the wounded into the beautiful old Palo cathedral. Nurses were recruited. Worship continued between the cots of the injured. A little cemetery began to grow outside the cathedral.

PCAU #5 was called to Palo attached to the 24th Infantry Division and landed on the first day. They provided temporary relief to the civilians who came out after the bombardment using the Sixth Army's medical supplies until their transportation and equipment arrived nine days later. Two officers and seven enlisted men were separated to the 1st Cavalry Division. They set up relief camps in the coastal towns of Alangalang, and Carigara. The 50-bed hospital served the population of Leyte valley. Civilian laborers were recruited and housed at Capoocan, Pinamapoan and Limon areas to carry food and ammunition to the front-line troops and work with engineers on rebuilding bombed out roads. One of two men killed was Cpl. Phil P. Yuson at Capoocan near Carigara. By Christmas, they were relieved but remained attached to the Sixth Army.

PCAU #6 in the December report of the Sixth Army, was operating dispensaries in the municipalities of Leyte and Samar. They distributed posters and pamphlets to improve sanitary conditions all over towns.

PCAU #7 and PCAU #8 landed on Red Beach between Palo and Tacloban. They camped and dug in for three days on the beach under bombing raids, sniper fire, and Japanese infiltrations. They reached their assigned beaches five days later. Naval bombardment of the beach areas had been heavy, and there were many civilian casualties.

PCAU #7 moved to Blue Beach at Dulag in the south, attached to the 7th Infantry Division. The unit arrived two days after combat troops were beached. Once a sleepy town in Southern Leyte, it suffered the worst. The people were still reeling from the destruction of the town leveled by pre-invasion naval bombardment then taken over by the Army to be used as a base when a Japanese bomb scored a direct hit on an ammunition dump. Shrapnel pierced the air, killing and wounding the helpless inhabitants who were already suffering from malnutrition. The entire town and old Spanish church were destroyed. Around forty-five thousand desperate people concentrated behind the Army lines. They were afraid of the guerrillas in the outlying area. Under bombing raids, the helpless people were relocated further south to the village of Mayorga so as not to interfere with military movements. The Japanese had used starvation to control the people. The large stores of rice they left were distributed to the people. The guerrillas and civilians in the area helped locate or identify municipal officials to be checked by the Counterintelligence unit (CIC) for loyalty before an appointment. A large hospital tent was set up that had x-ray facilities and performed major surgery. The PCAU men beamed with pride at their dispensary layout until a storm wrecked it one night. A commendation from the civil affairs officer of the 7th Division endorsed by the commanding officer of the XXIV Corps boosted the men's morale. The unit covered the municipalities surrounding Dulag until **PCAU #8 and PCAU #4** took over half of the area.

After a month in Leyte, PCAU had recruited around two thousand five hundred workers, but that was insufficient. The regular Japanese air raids discouraged many from joining. Only half of the over eighteen thousand registered civilians showed up for work. Payroll was shifted from daily to weekly to increase attendance.

Officers and enlisted men of **PCAU #7** with truckloads of medical supplies accompanied the 77th Division in their capture of Ormoc City in December. Large units of the Japanese 35th Army had retreated into Ormoc Valley. The town and surrounding villages were destroyed, leaving thousands homeless, wounded, and sick. An additional two thousand people from nearby Camotes Island crossed the sea to escape the Japanese and were living along the beaches of the town of Baybay.

Civilians were not allowed in Ormoc city, but they streamed in for food, clothing, and medicine. Relief work began, and Ormoc became a restricted military base. Stores of rice and Japanese rations left in the city and in the town of Ipil were quickly depleted. Two captured Japanese trucks transported around ten thousand people to the villages of Deposito, Binulho, and Seguinon. Civilians were easily recruited to establish military cemeteries, clean up the town, and repair roads when rice was given. Public officials were authorized to enforce strict rationing to ensure no double distribution. **PCAU #8** took over municipalities in Ormoc from **PCAU #7** in November.

By the end of the year, **PCAU #7** personnel split with headquarters in Valencia and Ipil. The town of Valencia was cleared of the Japanese and became a military base. Thirty thousand people were evacuated to nearby villages. Nipa huts were erected and food distributed. The ex-mayor of Ormoc was noted to be a big help to the relief operation. When the Japanese advanced from Ormoc to the north, two hundred laborers worked as ammo and supply bearers. Roads were repaired for the military advance. Ipil was cleared. Military cemeteries were established at Ipil, Camp Downes, and Valencia. The workers were only too happy to have rice as payment.

The unit in Valencia stayed on until **PCAU #14 took over** at the end of December. By the time **PCAU #7** left Leyte, over sixty thousand civilians were fed during the initial phases of the Leyte operation and the first few days of the Ormoc operation. Thousands received clothing, and thousands of patients were treated. Around forty thousand laborers worked for the military paid with Victory pesos that they used to buy merchandise in the retail stores. On the second day of 1945, the unit loaded and returned to Tacloban to prepare for the Luzon landing.

PCAU #9 - #30

Meanwhile, in Oro Bay, Hollandia, **PCAU #9 - 30** were activated in November. Personnel came from transfers of enlisted men from the 1st Filipino Infantry Regiment, 2nd Filipino Battalion and the 5th replacement Depot. Officers were from the Civil Affairs Staging Area in Monterey, California.

The training included narrations of the problems encountered by the 'Invasion PCAUs' which was a preview of what to expect on Luzon. Working out hypothetical problems in detail proved to be instructive. Care and use of sidearms was stressed for protection. Tactical training included road marches, motor convoys and problems of defense. Each unit was completely equipped with all equipment checked and tested.

From the experience of the 'Invasion PCAUs,' civilian sanitation was declared inadequate as seen from the diseases suffered by the people. Army medical personnel and equipment would have to be increased and given priority in landing so as not to burden the Army's supply and provide immediate relief to the civilians. Civilian health services had to be established. Public health projects such as the Malaria Control Units (MCU) were needed and established. While the PCAU established hospitals and dispensaries, the MCU would engage in public health rehabilitation. The delays in supplies of clothing, shoes, farm equipment, planting seeds and the likes discouraged the civilians from working as there would be nothing to purchase with their salary. Military projects suffered from the lack of laborers.

The following month, the rest of the PCAUs departed Oro Bay with full field equipment, motor transportation, and rations for sixty days onboard the Liberty ship *David F. Barr.* They replaced some of the 'Invasion PCAUs' that landed with the Sixth Army, while others were given new assignments. **PCAU #14, 15, 17, 26, and 27** relieved **PCAU # 4, 7, and 8** to make them available for duty on Luzon. The older units became part of Base K, Services of Supply (SOS), on Leyte and continued to provide services from eight hospitals, twenty-five dispensaries, and seven sub-dispensaries until after most medical installations operated by PCAU were discontinued or handed over to the Philippine Government.

PCAU #9 landed with a convoy in San Jose, Mindoro, under Japanese attack in December. Barely a month after landing in Leyte, the Sixth Army began amassing troops to invade Mindoro, the island south of the entrance to Manila Bay. Clearing the island would seal off enemy ships from entering. But the securing of Leyte had taken longer than expected, pushing back the landing to December. The men waded to shore in 4 ft of water. With their full load, they sank. The average Filipino men was a little over five feet two inches, but most managed to get ashore with all their equipment. The unit was responsible for the entire island and the nearby islands of Marinduque, Tablas, Romblon, and Cuyo. Their office was in the former building of the Philippine Milling Company. Overhead, they watched dogfights and air battles with the enemy from the relief centers.

The unit appointed Filipino physicians and nurses to staff hospitals. Atabrine was given to civilians to control malaria. Around forty-five hundred were vaccinated against typhoid, dysentery, and cholera. Under PCAU sponsorship, a newspaper was published bi-weekly with various Filipinos contributing articles of local interest and world news. On Christmas Day, around five thousand civilians ate a holiday meal at the nearest Army kitchen. Three tons of candy were distributed to the children, their first in almost three years. On New Year's Day, the acting mayor of San Jose announced a celebration inviting everyone. Cockfights with lively betting and a lottery with a prime bull as first prize were held.

PCAU #10 and 28 were assigned to Samar. Headquarters was in the town of Guiuan. Filipino doctors and nurses staffed a small civilian hospital. Relief was provided to the residents of Guiuan, Salcedo, and the islands of Candolu and Manicani off its southern coast.

PCAU #13 disembarked at Telegrafo village in Tolosa and replaced **PCAU #3**. From December 27 to January 1, the unit assumed control of the villages in the municipality of Palo, Tanauan, and Tolosa that had around seventy thousand civilians. Guerrilla and Philippine Constabulary officer had to be reminded not to feed locals passing through but refer them to PCAU so as not to be distracted from their operations duties.

PCAU #14 was based in Ormoc in March. Two dozen Enfield rifles were issued to the municipal government for the police force who were formerly PCAU assistants. The unit was in the town of Valencia, connected by the main road to the north of Ormoc. Civilians lined up both sides of the road living under nipa palms leaning on poles. They had left their farms in fear of numerous Japanese in the area. Sgt. Mariano Angeles, a medic with **PCAU #14** who arrived at White Beach related "helping the people of Leyte who were fortunate not to die from lack of proper medical care and his other job of guarding Japanese POWs." He was glad to see his comrades when he was assigned to the 1st Filipino Regiment, that arrived in February.

PCAU #15 was at Ipil, south of Ormoc. There were many Japanese up in the hills, so they set up a perimeter defense while relieving civilians. They slept with their pistols in their hands.

A report on **PCAU #16's** activities listed White officers, thirty-three Filipino enlisted men and six native Americans. Reported were 33 Filipinos of which 27 were naturalized American citizens. Sgt. Francisco A. Floreza was appointed as 1st Sergeant.

Dr. Leo Smollar wrote of his experiences in his letters to his family. Together with the Filipinos in **PCAU #17**, he left Oro Bay in December as part of a 48-ship troop convoy that zigzagged its way to Leyte. Onboard he described an atmosphere of boredom and apprehension with daily air raids. Together with the Filipinos in his team, they saw towns pummeled by American air and artillery bombardment. They waited a week on the shore of Dulag under stifling heat, humidity, rain, and nightly air raids for their equipment to be unloaded. Although General MacArthur had declared the fighting on Leyte over, there were still pockets of Japanese resistance. Hundreds of refugees passed by, forced out by retreating Japanese. Mobilization of the invasion armada and PCAUs for Luzon was at its height. When the Japanese resistance was finally subdued, "one-third of the 80,000 Japanese army deaths on Leyte happened in the 'mopping up' period."

PCAU #17's team of ten officers and thirty-nine enlisted men went over tortuous mountain roads to set up headquarters in Palompon, Leyte, from January to May 1945. The people suffered as the town burned from a Japanese attack earlier and through Christmas Day. Dr. Smollar wrote, "The population has been underfed, in tattered clothes, and overworked by the Japanese. Many cases were of worms and parasitic infestations. Child mortality is high. Vitamin deficiencies and beriberi are widespread. Tuberculosis is high. The sanitation is very poor. No hospital but only a half-destroyed two-room structure used as a clinic. The sick are numerous, and there's a continuous stream of civilians infected and wounded, some deliberately bayoneted by Japs. It's more than enough to make your heart bleed."

Palompon was almost destroyed when the 77th Division captured it in its drive to Ormoc. Killed in action were Lt. Col. Morgan, the PCAU commander and four enlisted men when his supply convoy was ambushed on the road between Valencia and Ormoc. The enemy continued attacks on the trucks bringing supplies to the unit. Dr. Smollar recounted a "Japanese prisoner with a bullet wound in the chest was brought in by the Military Police. Without the MP protection, he (prisoner) would not have survived in the hands of the Filipinos. The usual answer in Japanese was that they had tried to escape." After four months, Dr. Smollar wrote to his son. "Civilians have filled out, no longer (have) that look of hunger."

Hundreds of Filipinos lined up the port of Palompon and, for more than two hours under a baking sun, watched the boats filled with PCAUs slowly leaving the pier. "A real poverty-stricken mother of a skeletal child I saw during the first days after fighting stopped, reckoned now as long ago in 'wartime,' handed me a dozen fresh eggs. Another former patient gave me fried chicken. The hospital men and women financed a pair of house slippers and tea cloth as a thank you for me," wrote Dr. Smollar.

PCAU #25 furnished the food and supplies for a leprosarium established in Tacloban when seven lepers were discovered living under a school building in the swamp area behind the Capitol building.

By this time, the PCAU units were attached to the Eighth Army. The Sixth Army focused on the Luzon landings while the Eighth Army was tasked to finish eliminating pockets of the enemy on Leyte and retake the Visayan and Mindanao islands. Casualties mounted in the intense fighting to clear the Japanese reinforced city of Ormoc and continued until the middle of 1945.

Luzon Island, Philippines Landing – January 1945

Initially thirteen units, Liberation PCAU, were attached to the Sixth Army for the Lingayen landing on Luzon. One additional PCAU was reserved for the Legaspi city landing on Southern Luzon. They had little time to draw supplies and equipment before moving to their loading positions. More mobile, the units also served in areas by-passed by the Army.

Many of the Filipino PCAUs were from Northern Luzon and spoke the language making it easier to recruit civilians to work for the Army on rebuilding roads, camps, or as carriers of material for forward infantrymen. Typists were in demand to handle the load of military paperwork. "The Filipino companion custom was extended to the hospitals. Shelter and food were provided for one or more members of a patient's family. Sometimes these companions would be helpful in an understaffed hospital in doing menial work, but there were disadvantages, particularly when companions would sleep at night with a patient suffering from an infectious disease."

Learning of the Filipino characteristics on Leyte, PCAU headquarters realized some gaps in the planning of supplies. The sizes of clothing they brought were the standard

U.S. sizes instead of the smaller Filipino sizes. Shoe sizes were two sizes larger than the average Filipino foot. There was more men's clothing than that for women and children. The Filipinos considered some items like four thousand rat traps or the rolls of toilet paper unnecessary when they were barely surviving. Canned meat was rarely purchased since the preference was for dried or canned fish. Some of the foods brought in were not everyday items of the Filipino diet, such as asparagus tips, and they were issued in large containers more suitable for the army than family units. But it did not take long to taste the 'new' food and satiate the hunger.

The airwaves filled with President Osmena's appeal to the Filipino people. PCAU posted leaflets in numerous places in Lingayen. "Labor is urgently required at the docks, on the airfields, in the supply dumps.... give our military forces every possible support by volunteering to work..... offer (your) services now at the nearest Philippine Civil Affairs Unit. Work in the fields must proceed. If any of you have food supplies, make them available to the nearest Philippine Civil Affairs Unit, which will buy and distribute them fairly among the people. No one will starve. Your Commonwealth Government and the United States Forces will provide you with relief supplies.... Signed Sergio Osmena. President of the Philippines."

Willard Pearson's PCAU unit was composed of a 6-man medical team, a 4-man legal team, and 4 civil affairs officers. Attached were two 4-man CIC teams. The Filipinos spoke Ilocano or Tagalog when they landed in Luzon. Those who spoke Visayan were assigned to PCAUs to be sent to the southern islands.

The ships were enveloped in thick black engine smoke that filled the sky of Tacloban as they departed for Lingayen Gulf. All types of air support left the Leyte airstrip. The convoy was a third larger than the force which invaded Leyte, and "only a portion of the Seventh Fleet" was left to defend the landing force. It surprised and must have pained General MacArthur when he was ordered "to return a portion of the Pacific fleet to Admiral Nimitz to be used in the attack on Okinawa."

The hospital ship *Mactan* returned with the landing convoy three years after her escape from Manila with wounded soldiers but this time without the danger of mines in the waters. The Navy guns' heavy shelling of the beaches extended from the low water line to four hundred yards inland. Several hundred civilians were wounded from the 3-day bombardment of the same beaches that General Homma came ashore with his Army in 1941 to take Manila.

Men in the advance convoy suffered through enemy plane raids day and night while waiting in the Gulf to land. Anti-aircraft guns opened fire hitting some of the planes. Hordes of Kamikaze planes from their bases on Luzon, sank landing crafts on their way to the beaches, killing hundreds of men before they reached the shore.

The PCAUs landed with over a quarter of a million soldiers at the Dagupan side of Lingayen Gulf. They were relieved that there was hardly any enemy resistance as the Japanese Army was widely scattered on Luzon to delay the American advance and some marching north to follow General Yamashita. News spread quickly, lifting the people's spirits.

Men were eager to do their job but were constantly on the alert for signs of the enemy. As they accompanied the fighting army and liberated villages, they heard church bells tolling, the old signal that the Japanese had fled and the village liberated. If they passed barrio after barrio emptied of all its people, pigs, and chicken, they needed to be ready with their guns, for this was an inevitable sign that Japanese were near and the first troops in would likely be shot.

When PCAU #1 arrived, the unit was assigned to the Tarlac area, then to Caloocan, Rizal and the surrounding areas around Manila to serve refugees using ration tickets. It was a large relief area serving over 40,000 people but Army priorities assigned members to other sections. Cpl. Henry A. Garcia was temporarily transferred to 116th Stations Hospital. Not to waste the relief goods for such a wide area, PCAU Lt. Anselmo P. Canapi set up 16 distribution points around Manila with 22 personnel in each to issue ration tickets and provide medical services. Main food items were rice and canned fish fixed at half a can (empty evaporated milk can) of rice per person and one can of fish for each family daily. Milk and bread were served but was discontinued when Welfareville and the Psychopathic Hospital took over the milk supply. To ease the burden, three wholesalers were established to distribute to retail stores. Bread was sold when the retail stores opened. Textile stores opened and a dozen sewing machines were provided to mass produce clothes, sheets, and pillows. **PCAU #9** was a welcome sight when it arrived to take over half of the refugees.

Welfareville in Mandaluyong, covering 20 acres, was taken over to feed and house refugee families until they returned to their province. Admission was checked by the medical staff to avoid contagious diseases. Maj. Francisco J. Roman, PCAU medical officer set up an emergency hospital as there were more battle casualties than natural sicknesses. Work took a toll on PCAUs Sgt. Manuel P. Bernardo, T4 Felipe U. Olivares, and T5 Amado O. Alaminiana, sending them to evacuation hospitals. The highest number of patients served at one time was over 125,000 persons. PCAU T5 Johnny L. Ortiz received good news for his heroic deed while in Tacloban. He received the Bronze Star Medal for rescuing a soldier from a bombed burning truck and unsuccessfully attempting to remove another man from another burning truck a few yards away. He administered first aid, risking his life during explosions and burning gasoline.

PCAU #2 was onboard LST *735* attached to the 40th Infantry Division on January 10. The LST was with the first convoy to reach Lingayen Gulf. Enemy planes flew overhead as Japanese small boats or floating logs tried to attach small time bombs to the bottom of the anchored LSTs causing minor damages. The ship's crew opened fire on the swooping planes. After three days onboard, equipment and men unloaded onshore.

The camp and headquarters were at the municipal building in the town of San Fabian. It was reported that General MacArthur was in San Fabian. A week later, they proceeded to Mangatarem and set up camp at the Mangatarem Institute building. Civilians crowded the facility for food, clothing, and medicine for their diseases. Three days later, they moved to Capas, Tarlac, and set up relief services. PCAU Lt. Ralph A. McBroom and ten men went to Bamban to guard captured Japanese foodstuff and goods. The enemy knew of their arrival and started lobbing mortar shells into the town. They remained at their post until recalled.

A signal photo dated January 28, 1945 had the caption: *Capt. Abdon Llorente of PCAU addressing the people of Tarlac on Luzon.* The town was in ruins, with only the church, town hall, and about sixty houses standing. It was the first public meeting of the people since the Japanese occupied the town. The people burst into 'God Bless America' after the meeting. Electricity was restored to operate pumps and the water supply which had been causing intestinal diseases. The market was rebuilt in a week and the ice plant, several schools and hospitals were put into operation. Relief was given to about twelve thousand people and refugees.

PCAU #2 was at New Bilibid Prison in Muntinlupa. Seeing **PCAU #8** there, they proceeded to Binan, Laguna, camping at the municipal building and providing services. A month later, they moved to nearby San Pedro where twenty Japanese soldiers had dug foxholes on the beach of Laguna Bay. A combat team of PCAU men, Sgt. Marcel Cababa and Sgt. Cardenas, guerrillas, with men from the Military Police and 1st Cavalry eliminated the enemy. Cababa killed two attacking Japanese and was wounded. An Oak Leaf Cluster was added to his Bronze Star Medal.

PCAU #3 with Army Support Command landed in the town of San Fabian. Moving to the province of Nueva Ecija, they distributed rice and canned goods at low prices while those who did not have money were given free of charge. Medical supplies were provided to the provincial hospital, and a dispensary operated in every town.

Finding General Yamashita. The unit followed the 32nd Infantry Division on the Villa Verde Trail, in the hunt for General Yamashita's hideout farther north. From March to June, the unit with **PCAU #18** provided support through months of continuous fighting and unrelenting rain and cold. They operated a labor camp to feed, house, and account for 1500 laborers daily. They carried supplies, and widened roads for military equipment to pass as the mountainside was being cleared of enemy entrenched in caves in the cliffs beside the Trail. Cpl. Alex de Leon Fabros' papers revealed his assignment with the 32nd Infantry Division, that went to the Cagayan Valley to hunt for General Yamashita. He carried the red arrow patch of the 32nd Infantry Division. "Casualties mounted and the 128th Infantry Regiment arrived as fresh troops," he related.

PCAU #6 was in La Union and saw the provincial and municipal governments functioning normally. Activities concentrated on the most rampant diseases of malaria and gastro-enteritis. Vaccinations against typhoid and cholera were ongoing. Latrines were built, and dumping pits dug up. The unit was on the alert for enemy snipers surrounding the mountains that forced the civilians to hide and not plant food crops. They withheld American goods of lima beans, sweet corn, corned beef hash, asparagus tips, and salmon to encourage the growing of local products and remove the dependency on military handouts. The nearby province of Abra had an acute food shortage, having been devastated by the Japanese.

A Japanese perimeter around Baguio prevented the city's inhabitants from leaving. The news of the Americans in Manila prompted some families who had left for Baguio City to return. A 'Refugee Trail' was set up, and hundreds of civilians petitioned the Japanese for passes to look for sweet potatoes outside the city. *Igorots,* the area's indigenous inhabitants, agreed to lead the refugees through the mountains since they could come and go without arousing suspicion. They led the evacuees down the trail to La Union. Many weakened and some died. In the following weeks, thousands got through to the American lines. Later, more evacuation trails opened. **PCAU #6** met them with food trucks. As the exodus continued, the Japanese patrols caught up with some and stripped them of valuables and food.

A Filipino family slipped through the Japanese perimeter around the city of Baguio. They subsisted on sweet potatoes and water going down the mountains to Rosario, on the western coast where the U.S. 33rd Division headquarters was located. As refugees, they were turned over to **PCAU #6**. A family member, Arellano, insisted on seeing the division commander. He broke a pillow seam he had guarded from beatings and threats of death to keep secret and pulled out an American flag. It was lowered from the flagstaff in front of the Baguio City Hall when the Japanese first occupied the city. He expressed his sincere desire that "your troops let this be the first American flag to fly over liberated Baguio." A month later, U.S. aircrafts bombed Baguio.

The Sixth Army had difficulty entering Baguio City although General Yamashita had earlier moved out to the mountainous area of Kiangan surrounded by narrow mountain ridges. His best troops defended the city that had the second biggest supply depot of the Japanese army in Northern Luzon. The city was carpet-bombed leveling much to rubble. Miraculously spared was the lone-standing cathedral saving many civilians inside during the useless destruction.

PCAU #7 was relieved by **PCAU #27** and ordered to Baguio where they camped from July to August 1945. The mountainous area was not easily accessible by carabao and cart. The unit arrived at a ruined city and prepared for the inevitable looting by starved and desperate residents. The enemy had consumed most of the local food. Thousands had fled. Around two thousand indigent families were counted consuming all the Australian food brought in. Prices of essential goods were controlled by hiring residents to sell corned beef, canned goods, California rice, and other staples. Captured Japanese medicines were used, and vaccinations against cholera, typhoid, and smallpox were ongoing.

Within the month of landing, General MacArthur moved his headquarters to Hacienda Luisita, a sugar central in San Miguel, Tarlac halfway between the beaches of Lingayen and Manila. He was ahead of the Sixth Army and closer to Manila. With the 'Kill all' mandate of the Japanese Imperial Army, it was imperative to reach the internees at Sto. Tomas. Everyone was now a guerrilla and dispensable because they were against the Japanese. Communication units and the Press were not far behind keeping up with the General. The enemy sent reinforcements to Clark Field. PCAUs and radio operators asked for more ammo to defend themselves from Japanese patrols while laying communication wires.

The order from Washington to return seventy of the transport ships to carry supplies and munitions to the Soviet forces in anticipation of the Soviets eventually joining the war against Japan shocked General MacArthur as he read the urgent message. It endangered the Philippine campaign and threatened the loss of thousands of men fighting in Luzon.

Battle of Manila: February – March 1945

On their way to Manila, the PCAU depended on the guerrillas' invaluable knowledge and support. They knew the best roads, helped rebuild bombed bridges and avoided obstructions. Filipino guides provided accurate distances since those on the American maps were inaccurate. But reports on estimates of enemy strengths had to be verified since they routinely tended to be exaggerated.

The radio outposts on Luzon flashed the destroyed bridges between the Lingayen beaches and Manila, slowing down the 1st Cavalry. Supplies could hardly keep up with rebuilding of bridges, constructing airfields, and providing landing mats for the ships due to the limited transportation from Australia and New Guinea. It was a month of bulldozing jungles to widen trails for equipment, bombarding, and keeping up constant attacks on enemy positions before arriving in Manila.

Broadcasts mixed with the sounds of combat filled the air in Manila to lower enemy morale and induce a surrender. Millions of surrender leaflets and propaganda newspapers were dropped. Members of the militant Makapili were reported supporting the enemy.

Japanese heavy artillery was aimed at Manila from positions surrounding the city. Reports of large Japanese artillery guns installed at Nichols Field and the Ft McKinley area bombarding the city before the U.S. Task Forces arrived flashed to KAZ.

General MacArthur refused General Kenney, commander of the U.S. Army Air Forces' request to bomb Intramuros because of the civilian hostages. Messages filled the airwaves following the enemy's retreat behind the walls of Intramuros, killing residents wherever they found them. The radio streamed with sightings of women and children forced to walk in the streets as Japanese defense against American planes. Many men above the age of fourteen were found murdered. With no sign of surrender, American artillery tore down buildings and heavily damaged Intramuros

As the American forces entered the city, Japanese mortars and machine guns firing from the second floor of office buildings, had to be cleared one by one. Signalmen pounded messages of Japanese naval guns mounted in major buildings and intersections. The Americans responded with howitzers raining shells on the city. The four bridges were destroyed connecting Northern Manila to the residential southern districts. Half of the twelve hundred strong police force of the city stationed on the north side of the Pasig River operated under the American Military Police and PCAU to help establish peace and order while the city burned. The Ermita and Malate districts in the south were burned scorching private food stocks and starving the people. With only three pieces of equipment left and one used as a water pump, many fire department personnel were killed or wounded as Japanese soldiers fired at them while they tried to put out fires.

The 'Invasion PCAUs,' **PCAU #1 - #8,** entered Manila with the 1st Cavalry at noon on February 7 amidst snipers and mortar fire. Battle formation was ordered. Each PCAU unit entered Manila with ten tons of relief supplies. The orders were not to interfere with the civilians who went in and out of Manila during the fighting and were caught. The Japanese had set a curfew in the city. "The scene that greeted us was very depressing," recalled a Filipino PCAU soldier. "Manila was devastated. All the buildings were down in rubbles, and as we went through the streets of Manila, we could still smell the stench of death." Hardly a major building was left intact. Debris and filth were everywhere. Street fighting and Japanese artillery shelling were ongoing and with no working hospitals in the devastated city, many civilians rushed to the soldiers for treatment. Confused civilians followed telephone lines snaking through the rubble to the American lines.

The *Frederic Galbraith*, a relief supply ship, was waiting in the Lingayen Gulf with seven thousand tons of supplies aboard. Two other vessels with fifteen hundred tons each were expected shortly in the Luzon area.

PCAU #1 reached Bilibid Prison, set up camp, and fed around thirteen hundred internees and POWs. Enlisted men of the military police were also fed for a month. PCAU and the military police set up 24-hour guard duty around the prison premises as street fighting continued around the prison grounds. Guerrillas shot snipers with hand grenades on top of the prison walls. Continuous artillery fire from both sides was heard day and night. Fires and explosions were all around, but morale continued to be high with the men. The unit served about one hundred twenty thousand people, including refugees, with rice and canned fish distributed as the primary food. "From the start, the supply was never enough to enable us to service the whole relief-seeking population in our entire area," reported a PCAU soldier. Investigators were assigned to investigate people drawing double rations from more than one relief point. Those who could afford to buy from retail stores when they opened dropped from relief, but limited supplies at the stores and difficulty getting printed ration-buying coupons, made them turn to relief again. The relief number decreased when families with non- disabled members were employed. By the middle of February, **PCAU #1** turned over duties to **PCAU #6** and proceeded to San Juan del Monte where they camped at St. Catherine's Academy. By May, the relief and refugee homes transferred to the Bureau of Public Welfare.

The Japanese razed towns south of Manila, driving fleeing refugees to the Laguna Bay area. **PCAU #2** set up a relief camp until the unit was transferred to the Cavite area. They had problems with enforcing the ceiling price of food sold as no arrests were made of those who sold above ceiling price or on the black market "The mayors were instructed to give relief supplies to the destitute only, but the mayor of Bacoor in Cavite issued foodstuff to all the people so as not to 'make anyone mad' at him resulting in the PCAU unit withdrawing from that town." He was not removed from office since there was no other CIC-cleared candidate eligible for the office.

PCAU #4 covered the slum districts of Tondo and San Nicolas in Manila. The men were shocked to see the ill-clothed and malnourished civilians. Two thousand homeless families from their burned homes were fed and temporary housing was constructed for seven thousand refugees. A group of residents set up a first-aid station with medical supplies donated by those who could afford them. Under the PCAU, it became a hospital. Patients were brought in pushcarts and on the shoulders of their relatives. The unit was in Manila until May, when their staff was reduced but they were still handling the same volume of cases, averaging two hundred dispensary treatments daily. Outside of Manila, the unit was glad that there was enough food to sustain the population. It was malaria and dysentery that wreaked havoc on people's lives.

PCAU #5 was designated the main headquarters for the eight PCAU's made available for Manila. PCAU commanders in the city held weekly meetings. The unit went straight to Sto. Tomas University with the 1st Cavalry and liberated the 3800 primarily American and British internees. The 1st Cavalry was ordered to charge ahead to secure the internment camp before the Japanese killed them.

The Red Cross worked alongside **PCAU #5**, providing relief to the internees. The Red Cross' manager Modesto Farolan arrived at his headquarters to organize his staff. The Red Cross building was hit by three shells out of the hundreds of thousands that were fired into that part of the city. With American artillery and Japanese burning the buildings around, refugees began streaming in. A week later, Japanese marines entered, closed all possible means of escape and massacred everyone found in the building including medical personnel, patients and refugees. A bullet hit Farolan's helmet. A temporary area in the education building of the UST campus was set up to feed and treat the internees. Half-starved, many of the internees overate at first with unpleasant results. Internees were then sent to the evacuation hospitals outside of Manila. Approximately one hundred Filipino civilians were hired and placed on duty with the medical team. "But three hours later, four-fifths had disappeared with shells hitting many of the buildings in the University grounds." More were hastily recruited and worked until exhaustion took over. Without electricity and running water, candles or flashlights lit many urgent surgical treatments.

With the world watching as the war unfolded in Manila, **PCAU #5** fed the internees and scores of news reporters, Red Cross workers, airmen, Filipino doctors and nurses, and visiting military personnel. To ensure the internees had enough daily food, Commander Parsons ordered cargo planes to be flown in daily especially equipped with oversized tires so they could land on Espana Ave. in front of the university. While establishing a rationing system, PCAU took a census showing over a million people leaving and arriving.

KAZ radios filled with reports of Manila liberated after a month of intense fighting, death, and destruction. But the hunt for snipers and pockets of the enemy continued. There was no victory parade with so much slaughter of civilians. The PCAU opened the first school in March, almost right after the clearing of Manila. Even with a severe shortage of books and equipment, hundreds of children showed up, eager to attend in the few buildings that were still standing. Refugees who sought shelter in the buildings were relocated.

By the end of January, **PCAU #20** and **PCAU #21** relieved **PCAU #8** in Pangasinan, which moved on to Manila.

PCAU #8 assisted and set up camp at Earnshaw St. aiding the internees at Sto. Tomas in Manila. The unit had come from Binmaley and San Carlos in Pangasinan assisting the locals. The unit hired over a thousand laborers to ensure that operations were unhampered. The water system and power plant were re-established. Hospitals and relief centers were set up in the Rosario Subdivision. Hired laborers hauled water to the hospitals. Bamboo beds for hospitals were made. They went to work relieving congested main supply routes. The police force was resuscitated to establish law and order, guarding banks, municipal buildings and records. Ceiling prices were prepared and posted throughout bus sections and marketplaces. St Joseph Academy housed the fire victims. They set up a camp at the Bureau of Animal Industry building in the Pandacan area covering the districts of Paco, Sta Ana, San Pedro Makati, Fort William McKinley and the areas to the north and west of the Pasig River. Fire victims and civilians were evacuated from Malate, the walled city of Intramuros close to the heavy combat lines and surrounding areas day and night. After a month, **PCAU #27** relieved **PCAU #8** to proceed to Lamayan St. in Sta Ana.

The 158th Regimental Combat Team sailed with **PCAU #8** from Lemery in Batangas to Legaspi, the capital city of Albay, the Bicol Peninsula in Southern Luzon. Arriving on April 1, the Task Force experienced light artillery fire in the bay and small arms fire after disembarkation at the Legaspi port. They had difficulty doing their job with only jeeps going to Bicol. They quartered at the provincial and municipal buildings in Sorsogon and Camarines Norte. Many of the people wore clothes made of *sinamay,* woven from the abaca plant that grew wild in the nearby island of Catanduanes or patches covering only their private parts. The local stock supplemented the distributed rice from plantations of corn and camote. Signs were set up in English and Tagalog to inform of the punishment for selling above the fixed prices. American and Filipino flags were raised, and guards posted at the Philippine National Bank vault. Some guerrilla leaders who had appointed officials, eliminated political competition, and looted before the liberation force arrived, were controlled. Arrests were made and guards placed at American mining companies because hidden gold bars prior to the Japanese entry were missing.

PCAU #8. Excutive Order No. 24. Any person, Firm or Corporation who shall sell any article included in the above schedule at prices in excess of the maximum selling prices herein fixed shall be punished as provided in Section 3 of Commonwealth Act numbered Six Hundred as amended. Signed: Sergio Osmeña, President of the Philippines.

An average of 1,000 laborers daily repaired sanitation facilities, water supply systems, and the power plant. The locals were informed of the discolored water in chlorinated wells so that they would use it. Two bridges were rebuilt and buildings reconditioned for hospitals, distribution points, and recruiting centers. Due to the shortage of books, pamphlets and readers were created for public schools. Pages had drawings of various local items.

A month later, **PCAU #8** moved north to Iriga in Camarines Sur. A sub-headquarters was maintained at Legaspi. Aside from relief activities and rebuilding of roads and public utilities, five rice mills were put into operation to mill the unhusked rice purchased from rice landowners. The rice was used to pay civilian laborers working in military projects, provided to hospitals, and PCAU stores to sell according to family rice quotas. Lepers were gathered, housed, and treated in the rehabilitated leprosarium that existed before the war.

James Halsema, an internee for three years at Camp Holmes in Baguio, wrote that "PCAU can't truck in enough rice to keep prices down to the controlled ceiling. One factor controlling inflation is the scarcity of money. Barter has partially replaced money as the means of trade. Food is scarce and expensive. A limited amount is rationed out by PCAU known as 'Pee-cos,' but there is not enough, not nearly enough. As an ex- internee with built-in years of scarcity, (we) fear that we will run short and starve. PCAU distributed some dress goods, and none promised by the Red Cross arrived. The headquarters of the Red Cross were stormed earlier by Japanese marines killing fifty people. Transportation crippled, burned, or carried off. Horses and carabaos were eaten up long ago. Inter-island shipping reduced to sailboats .... need a small stove, kerosene, or gasoline lantern since no electricity. Limited power supply furnished by Army portable generators and a few undamaged private units used to power hospitals, sewer pumps, and others were necessary for use. Ordinary civilians will have to wait."

Crossman wrote, "the best residential district south of the Pasig River was burned for the most part, and civilian bodies were lying all over the place. Dazed, homeless people aimlessly roam streets, picking in the rubble of burned or demolished homes to salvage, mementos and whatever they could. Clothing in tatters. The Japanese burned and mined the city, and the Americans responded to eliminate them." He related that "Between 100 and 200 thousand people were homeless in Manila after the fighting. 'Refugee homes' had to be established under PCAU supervision to accommodate about 60 thousand on March 1. The rest moved in with relatives in Manila or joined relatives in other parts of the country." Soldiers of the 32nd Infantry Division saw the slum areas outside the central city of Manila (Malate and Ermita) accept the refugees escaping from the destruction in the main city. The men were amazed that most spoke English. A PCAU warehouse fed around twenty thousand people in the slum area.

Liberty Ships of ten thousand tons capacity unloaded at Lingayen Gulf, but everything had to be trucked 120 miles to Manila over roads crowded with military traffic. PCAU had difficulty finding warehousing for the food, clothing, and medical supplies. Crossman held on to the Philippine Army Quartermaster Depot in the North Port Area as long as his (PCAU) unit was responsible for feeding Manila. A PCAU captain and two enlisted men had to quell a food riot at a warehouse near the Manila North Railroad Station that the Sixth Army was using for the scanty civilian supplies available. They were always on the lookout for pilferage. The people were starving as there was not much food coming in from the countryside.

From Manila Bay to Corregidor, thousands of ships and vessels were destroyed from the first raids in September 1944 to the time of the Luzon landing. When the island of Corregidor was captured during the height of the battle to liberate Manila, the Manila harbor was opened to Liberty ships docked at Lingayen Gulf. Supplies no longer had to be trucked overland to the city. PCAU watched out for their own medical supplies marked for them because other medical officers would try to take them for their installations.

With the hopeless physical conditions, spiritual needs had to be answered. An interesting memo marked 'Priority' to PCAU from General MacArthur requested delivery of altar supplies on troop ships on or before March, right after the Battle of Manila. Gallons of olive oil, twenty-seven cases of altar equipment and religious garb, hundreds of sealed packs of altar bread, cases of wine and sacks of flour arrived. After the chaos and destruction, the spiritual seemed the only thing left to salvage although PCAU units worked in shifts to treat the diseases that multiplied.

The 2nd Filipino Battalion arrived in Leyte in March and proceeded to Manila to help out with the clean-up. They saw bulldozers push rubble into the bay to make ramps for unloading cargo. "Swinging balls to destroy" parts of buildings were used. A powerful stench emanated from decomposing bodies that were covered with flies.

PCAU #18 had no time to rest. After three days waiting to land, the unit landed under artillery and mortar fire at White Beach. Headquarters was in the municipal building of Mangaldan, a mile inland from the coast. Thousands of starving civilians came into refugee camps from the war zones for food, clothing, and housing. Civilian physicians and nurses were hired to attend to the casualties. Fifteen towns in the Nueva Ecija area and 7 towns in the province of Tarlac were serviced for 4 months. CIC agents interviewed and appointed temporary municipal officers to restore civil government. An airstrip in Mangaldan was constructed in five days and 120 planes landed three hours after completion. The unit loaded into army trucks and proceeded to camp in an elementary school in Bayambang. By February, they were headquartered in the municipal building and the house of the Pamintuan family in Angeles, Pampanga. **PCAU #2** joined the camp for a month providing relief service. In nearby Manaoag, 30 workers were hired to repair the large church. Victims of torture were buried. In the towns of Guimba, Licab, and Munoz, **PCAU #18** encountered public safety problems with unorganized guerrilla bandits who formed their own military government, looting civilians, properties, and warehouses.

In mid-1945, **PCAU #19** was reported in Pozzurobia, Pangasinan and in the towns of Rizal by Laguna Bay providing relief to the civilians. Talim Island in the middle of the Bay had a dispensary and a relief center. Several hundred packs of rice seedlings were distributed for the next planting season.

PCAU 19 in Pozzorubio, Pangasinan. Men with PCAU armbands distributed canned goods and basic commodities to the people of Pozzurobio, Pangasinan.

PCAU #20 served the north-central area of Manila in March. They arrived, seeing the widespread destruction and wanton killings of civilians. Cpl. Toribio J. Rosal was assigned to **PCAU #20**. Old members of the USAFFE were called to active duty and Philippine Scouts received orders to work with the unit to arrest violators of retail prices and look out for counterfeit currency. Guerrillas were absorbed into the American army and paid a monthly salary aside from rations and uniforms. The rest of the male population were recruited for construction, loading and unloading of ships, and reconstruction of devastated buildings. They were paid a daily wage and rations. "PCAU was spending some half a million pesos a week for only salaries and wages in the city," wrote a Dominican priest in his diary of the Japanese occupation.

The Manila Health Department was re-established, and the water system partially restored. PCAU units administered health offices to accept permits for burial, transfer, and exhumation of human remains. Problems remained in the procurement and assignment of qualified civilians to maintain public works, develop a system of garbage and sewage disposal including fly control measures, and reopen civilian hospitals.

Many hospitals were destroyed, and civilian casualties were heavy, particularly from the devastated area south of the Pasig River. Assistance to operate and increase capacity of the civilian hospitals with medical supplies from Army stock was given by the surgeon of **PCAU #20**. Six hospitals were set up, and fifteen dispensaries opened to treat patients.

One hundred and fifty boxes of Red Cross supplies intended for internees at Bilibid Prison were taken into the medical supply depot for issue to civilians through PCAU. A shortage of Tetanus Antitoxin resulted in the loss of many civilians, internees, and guerrillas.

After the clearing of Manila and surrounding areas, morale rose amongst the civilians and the military. "The troops were 8000 miles from home, but nobody was shooting at them." Accidents and venereal disease were the biggest threats. "The venereal disease rate among troops increased steeply during February even though active fighting was in progress in the city; prostitution was widespread, and, for many troops, this was the first contact with an urban civilization for two years," wrote PCAU doctor Thomas Turner. As more prostitutes were examined, VD disease rates rose. Signal Corps photos showed Prophylactic clinics for the soldiers in the major islands. A memo dated March 1946 in a Monthly Historical Report stated that "increased number of venereal cases caused measures to be instituted to bring the rate down to an authorized rate of fifty cases per one thousand men per year."

Philippine Civil Affairs Unit No. 20. Located at Tayuman and Avenida Rizal in front of San Lazaro Hospital, Manila, Philippine Islands 1944/1945. Commanded by Col. Meader. First Filipino Infantry Regiment, U.S. Army. Source: Terry J. Rosal.

PCAU #21 arrived in Manila in March and saw what was left of the city and survivors suffering from the carnage. They set up thirty-eight refugee homes and operated a kitchen. Retail stores began selling to the public at set prices. With all the many difficulties they encountered and at great risk to their lives, they felt appreciated when store retailers and wholesalers of goods in their area honored them with a party.

The five towns in the Bulacan area on Luzon were reported to be cleared of epidemic threats. Health and sanitation conditions improved with the dispensaries and a hospital in the capital city of Malolos set up by **PCAU #22**.

PCAU #25 was on board the convoy to Lingayen on January 4. En route, the anti-aircraft guns of the ships shot at two enemy aircraft that attempted to attack. Their assignment was in the province of Isabela. The province faced a food shortage as the Japanese had taken eighty percent of all the food crops, all the work animals, poultry, and livestock before retreating to the mountains. The people went hungry as they did not plant in their fields, afraid of being killed by the Japanese raids. They formed *bolo* defense units to protect themselves but now looked to trustworthy municipal leaders to protect them. In May, they were attached to the 1st Filipino Infantry Regiment fighting in Samar. They had not seen their comrades since their camp in New Guinea.

PCAU #27 landed in Olongapo and was ordered to proceed to Manila. But ignorance of Filipino customs led to the loss of tons of ammunition and supplies sent to Subic Bay. Five hundred laborers were scheduled to work from 0800-1500, unloading the supplies, but it was a Sunday and a Filipino holiday. Only a few showed up and loaded half of the cargo in the allotted time. Twelve hundred tons were left on the beach. In the Paco, Santa Ana area of Manila, they treated the population for the most prevalent diseases of malaria and dysentery. By the end of February, **PCAU #27** had provided relief in Muntinlupa, east of the railroad, south of Calle Herran, and east of Calle Embarcadero to the Pasig River before reporting to USAFFE headquarters in Manila. Over 30,000 were fed and 6500 homeless were evacuated to refugee centers. The unit was one of the last two PCAU's in Manila to turn over responsibility to the Philippine government.

PCAU officially announced that over two hundred thousand civilians had received aid in Manila and thirty-two hospitals were in operation with a total of over twenty-three thousand patients. Law and order were enforced with the cooperation of the guerrilla units of the city, the military police, and some members of the police force. On PCAU's payroll were over seven thousand people in Manila and vicinity. A law was enacted changing driving to the right side of the road including carabao carts and horse-drawn carriages to conform with the U.S. and make use of the large number of Army jeeps left behind. That change added to the casualties to be treated as road accidents soared.

By May, the control and administration of PCAU stores in the city were taken over by the Commonwealth Emergency Control Administration (ECA) from the U.S. Army. Relieved in Manila, PCAU moved to other newly liberated areas. ECA personnel were trained to take over PCAU duties as the Army provided ration goods to control prices. ECA distributed locally produced goods to supplement the ration and enable the people to purchase basic goods at a fair and reasonable price. Army goods found in private stores were seized and restaurants were prevented from obtaining PCAU supplies to control costs. Some managers of PCAU stores who took rice and tins of sardines for their own use but listed them as distributed to relief cardholders were arrested for falsifying official documents. By June, only 6 thousand refugees were left in Manila as relatives from outside the city took them in. When the unit left Manila, six PCAU members were killed and four wounded.

Ten thousand tons of relief supplies were landed on Luzon, and six thousand eight hundred thirty tons were distributed. More than 400,000 refugees were fed and over 250,000 needed care. After providing immediate relief to the civilians, PCAU was able to open stores in some of the larger towns to sell food and textile to the public at very low prices from ships with supplies sunk by Japanese bombers.

The Army reported that "on an island of 6.5 million people, the average population served by each PCAU on Luzon was nearly seven times as large as those served on Leyte. The performance of all units was good, and that of **PCAUs Nos. 5, 7, and 8** was outstanding." **PCAU #5** was commended for its excellent work in the rehabilitation of the former Children's Hospital in Manila. The unit was one of the last teams to turn over their responsibility to the Philippine government.

Visayan Islands and Mindanao Island Landings

PCAU units were heading south with the landing forces while the Battle of Manila was going on. **PCAU #12** landed in February at Puerto Princesa on Palawan. The unit brought ten tons of rice, canned foods, miscellaneous clothing, and medical supplies with 11 trucks to haul the goods. Abandoned Japanese food and supplies, located and stored in guarded houses, added 80 tons of rice and 40 tons of canned food to feed the civilians. Evacuee camps were established for those whose homes were destroyed or taken over. Medical dispensaries opened with civilian doctors and nurses put in charge. When Filipino doctors were scarce, Filipino nurses performed much of the medical work. Malaria was the most prevalent disease reported.

Two penal colonies, Iwahig and Inagawan, had around two thousand convicts, with some having expired sentences. The old assistant director was arrested as a collaborator. Two hundred tons of relief supplies were shipped to the Leper Colony on Culion Island free of Japanese. A dispensary was set up. It was apparent the Japanese avoided the island and occupied nearby Busuanga Island. Emergency cases were sent to the government hospitals at Iwahig and Inagawan Penal colonies. After PCAU members reconnoitered Coron Island and determined that it was clear of the Japanese, half a ton of rice seedlings was brought in.

PCAU #13 followed the 40th Infantry Division into Iloilo City on Panay Island in March 1945. Col. Peralta, the commanding officer of the guerrilla forces, called on the PCAU leader and gave his assurance of cooperation, furnishing facts about the island. He said food "was generally good that the island was able to produce. The exception was the mountainous province of Antique, with a narrow coastal plain that ran along the west coast of the island." After obtaining the lay of the land, the PCAU officer reported rice shortages in a few areas, but many areas had fish, corn, root crops, *camotes,* and fruits to compensate. Lack of transportation limited the distribution of rice and other foods. Before the war, large tracts of land were planted to sugarcane but now converted to corn, rice, and vegetables. A severe problem was the guerrillas' requisitioning much of the food, supplies, and a large amount of circulated guerrilla currency. Guerrilla currency or emergency currency was approved to be printed during the war to buy and sell goods in the major islands during the war. To stop the use, the seventeen thousand guerrillas on the island were processed to return to civilian life and paid with Victory pesos.

PCAU #15 was reported to be in Cebu in May. They carried medical supplies worth ten days of relief. Emergency medical hospitals relieved the Army medical field hospitals of the burden. A dormitory of San Carlos College became a hospital. The Eversley Childs Leprosarium in Mandaue City provided relief, leading to the return of many lepers who had wandered off during the war. With Cebu as the second largest city after Manila, **PCAU #24** and **PCAU #25** arrived the following month to help with relief efforts.

The unit supervised the set-up of the customs and collection of the Internal Revenue offices by hiring staff as no one was available to represent the Philippine government. License fees on small boats and taxes were collected. PCAU had to send many of the wounded guerrillas to the 73rd hospital at Tacloban. The guerrillas destroyed the Japanese garrison entrenched in the mountains that held the reservoir, saving the waterworks but with high casualties. The people were encouraged to plant on all available land. The unit was reported in August to still be in Cebu and had expanded services to Mactan Island. Over half a million people were provided relief.

PCAU #17 was attached to the X Corps, Eighth Army. For three days, they had real meals onboard the ship that transported them to Macajalar Bay, northwest Mindanao, in August. The area was more developed than Leyte due to the Del Monte Pineapple Plantation with asphalt roads, modern buildings, and water systems though badly damaged by artillery and bombing. Hospitals were established and reopened. Many local physicians were available, but nursing help was scarce. Women hid from the Japanese. PCAU doctor Smollar wrote that in Mindanao, "one of the clinical encounters were twelve Filipino women brought in for treatment of venereal disease after being freed from an inland town. They were forced to work as prostitutes for Japanese soldiers for almost a year. 'They made us work like carabao.' ... recognized as part of a system by Japanese occupiers in Asia to force women (Comfort Women) into prostitution for their troops." Many of the civilians were still up in the hills infected with malaria and malnourished. The people badly needed food, clothing, and medical supplies. There were no Japanese supplies nearby to recapture and use to immediately feed the people. Some guerrilla groups took thirty percent of the food they produced in exchange for security against the enemy. Working animals were killed for food by the guerrillas and the Japanese. Mindanao emergency currency printed during the war for the people to buy and sell goods was the medium of exchange. It became a problem when it could not be used to purchase supplies to resell at ceiling prices.

PCAU #23 was attached to the 41st Infantry Division and landed in Zamboanga on Mindanao in March. Forty-nine PCAUs were glad to see that there was no food shortage. Clothing was most needed. Headquarters of the medical group established themselves on the grounds of the Zamboanga General Hospital and began reopening hospitals in the area, including the Sulu Archipelago. They supervised the City Health Department with its regular personnel. The people openly feasted after three and a half years under oppressive rule. "As if making up for lost time, they downed kegs of Coca-Cola, stuffed themselves with corned beef hash and K rations, and smoked themselves to black and blue with American cigarettes as they went about in PCAU- issued clothing. The locals, in turn, sold fried chicken, ripe bananas, and rice gin which the American GI gobbled up and paid for in dollars."

PCAU #26 entered Bacolod, the capital city of Negros with the 40th Infantry Division and quickly prevented the civilians from getting in the way of military operations. The Japanese still held the coastlines and towns. Public safety was enforced. After the needy were fed and clothed, medical supplies were given to the locals to prevent epidemics. Civil governments were restored, and schools reopened. A missionary wrote that the PCAU had set up a civilian post office in Bacolod to mail letters to families.

Guerrillas with the United States Forces in the Philippines (USFIP) were not allowed into the city by the Army for fear that they would shoot at known or suspected collaborators. The USFIP became angry as the American soldiers were being feted by the civilians eating and drinking American rations. Those they saw as collaborators were free to mingle and celebrate. To appease the USFIP units, a statement to the editor of the *Army News Sheet* showing appreciation of the guerrillas' support was published. The *Free Philippines* reported that the "morale of the Filipino guerrillas was boosted with the GIs sharing their rations, blankets and even clothes with the poorly equipped Filipinos many of whom are still wearing tattered civilian clothes after three years of fighting the Japanese. The commander of the USFIP pointed out to the American Task Force on a Negros Island map the eleven more places where his group cleared enemy garrisons. Accustomed to short forays of a few days, they were now fighting steadily for longer periods in the Silay mountains of Northern Negros where a large pocket of Japanese is still holding." Interesting to note is the donations of foodstuff by civilians in Dumaguete to the PCAU unit for free distribution to the impoverished citizens of Bacolod. In June, **PCAU #26** moved to Panay to reinforce **PCAU #13**.

PCAU #29 was at Santa Cruz, Digos south of Davao City on Mindanao with the initial task force of the 24th Division in April. Filipinos staffed a dispensary as starved civilians returned from the hills and needed medical attention. In May, parts of Davao city were still in combat when they entered with the liberating forces. Heavy shelling destroyed all four hospitals. It took two weeks to repair them before servicing the influx of people from the surrounding areas. The Davao General Hospital was in an old constabulary barracks. Around 100,000 locals were under the PCAU's supervision treating thousands of patients weekly in dispensaries and hospitals. A former school principal and a high school teacher were employed to investigate cases and distribute relief supplies to the needy. They insisted that relief personnel be discreet with a people who had undergone untold suffering under the savage Japanese rule for forty months. PCAU saw many civilians in "tattered clothes in months of wearing and molded on their emaciated bodies by sweat. The heavy stench and smells emanated with the curdled sourness of dried sweat in cotton." Army engineers established water points. The different religious sects in Mindanao provided enough civilian labor for military construction. But separate camps to maintain peace had to be established for the *Moros* (Muslim group) and Christians. By August, **PCAU #14** who was reported in Cotabato, Lanao, and Misamis arrived as reinforcement expanding the services north to towns beyond Davao City and Samal Island.

PCAU #30 landed in April at Parang, Cotabato, Mindanao. Few civilians were in the town. The mayor, who was a soldier of the Philippine Army, was located with several Muslim *Datus*. Thirty tons of food were offloaded.

When the PCAU team arrived in Dansalan, Lanao, in the interior of Mindanao, Dorothy Dowlen and her sister, Mercy, found work with them. The unit needed dependable workers and paid good wages. She wrote that they did inventory work, distributed food supplies to the displaced and hungry civilian evacuees, including the *Moros* who lived in the area. They helped give away tons of rice, flour, sugar, canned goods, powdered milk, and other dried goods. Each recipient had to fill out forms indicating as head of the household and how many were in the family. If approved, the appropriate amount of supplies were doled out to them. At first, distribution went smoothly with people receiving their allotted provisions, but then they noticed that *Moro* heads of household kept coming back, twice, or even three times. After prodding them with questions, officials found that they had listed all dead relatives buried in a cemetery close by. PCAU ordered that they not be given any more than their allotted rations, making the *Moros* mad, but they left the building quietly. Officials thought that they would not come back until the next time when eligible. Instead, they came back armed with machine guns and fired on officials from outside the building. The sisters quickly bolted the warehouse from the inside and dodged bullets by hiding behind the mountain of supplies. The military had to be called to prevent further trouble. Despite additional possible danger, the sisters stayed until supplies ran out and the distribution center closed.

Dr Smollar wrote that by "July 1945, five special units were still functioning, all on a reduced level as the Commonwealth government assumed most PCAU functions on the battle-scarred islands. Closing and consolidations took place slowly on every island. Some PCAU officers had cautioned early on that the Philippines' infrastructure could not sustain all the PCAU accomplishments after the teams departed."

Japan surrendered in August. The 'Invasion PCAUs' composed of around four hundred Filipino men were given leaves and with the other PCAUs returned to the 1st Filipino Infantry Regiment camp in Ormoc, Leyte, before shipping to Manila. A camp for deactivated PCAU personnel was set up at San Esteban, Ilocos Sur. Named Camp Morgan after an ambushed PCAU commander in Leyte, it was a staging area for PCAU's participation in the Japanese occupation. Thousands of Army and Navy personnel were trained by the Civil Affairs Staging Area (CASA) for the initial occupation of Japan. Medical officer, Frank Aquino went to Japan after the formal surrender.

Whitney reported to GHQ on the operations of the PCAU on October 20, 1944. "Your policy within the framework of the directive of Joint Chiefs of Staff dated Nov 11, 1944, has been to render all possible assistance to the Filipino people through the reestablishment of the national, provincial, and municipal organizations of government throughout the islands; the extension of emergency relief in the supply of essential food, clothing, and medicine to the Filipino people; the provision for hospitalization of sick and wounded and shelter for the homeless. Your policy further encompasses the transfer to appropriate government agencies of full responsibility to cover all   of these and related civil matters, as rapidly as such action could be taken without manifesting prejudice to the people's interest. As a result, starvation, epidemic, and public disorder have been avoided, and civil government under constitutional process progressively restored to the people in an orderly manner. Appropriate civil agencies of government now assumed full responsibility. This command completely withdraws with inactivation of this section August 26, 1945."

Although PCAU was deactivated, the Army continued to aid in the maintenance of public order, restoration of public utilities, and other engineering work until re-established branches of the Philippine government took over the caring of health and welfare of the people. Forty thousand Japanese prisoners were put to work cleaning up Manila from dawn to dusk. Emiliano Francisco related that his unit was in Intramuros, the walled city in Manila. They kept their Carbines pointed at the vengeful people, not at their prisoners to keep them alive.

In September, the International Military Tribunals for the Far East began hearing cases against Japanese military and government officials accused of committing war crimes and crimes against humanity. Capt. Gregorio Sese, a company commander in the 2nd Filipino Battalion who was a successful practicing attorney in Washington D.C. before the war, was assigned to review death sentences of Japanese accused of war crimes.

During the trials, Japanese Prime Minister Hideki Tojo, attempted to take his own life. Lt. Frank Aquino with Sgt. Dominic Sta Cruz were part of the medical team that administered to him. Photos taken showed Aquino administering to Tojo slumped back in a chair with a bloody bullet wound on his white shirt. Tojo was brought to the U.S. Army's 98th Evacuation Hospital in Yokohama and given transfusions of American blood. Eventually, he was declared guilty and hanged in Sugamo Prison.

Counterintelligence Corps (CIC)

All collaborationists regardless of their motives were to be arrested. That was the order from GHQ. Units of Counterintelligence Corps (CIC) coordinated with PCAU to search for possible collaborators, spies, and disloyal people when they accompanied the landing troops. The units worked with agencies in establishing security of captured enemy installations and documents regularly educating the troops on security matters. Members came from the 1st Filipino Infantry Regiment and the First Filipino Battalion. Filipinos who voluntarily gave aid to the enemy were considered collaborationists and if CIC investigations found evidence and determined by a Legal Board of Review, turned them over to the Philippine government for final disposition.

The 441st Counterintelligence Corps (CIC) Detachment of G-2 Intelligence USAFFFE was established at the Palmarosa House in Brisbane two months before arriving in the Philippines. The unit was smaller than in the European theatre but was greatly augmented by Filipino soldiers to form thirty-three Intelligence detachments. Counterintelligence was a special skill and needed training. It was so new a department that "many senior officers did not understand its role or qualifications of its personnel."

There were not enough men available at Camp Tabragalba for CIC. Those present were training and committed into selected parties scheduled for missions. The attention turned to the 1st Filipino Regiment and 2nd Filipino Battalion in New Guinea. When Offley was asked to identify qualified men who could speak or understand Japanese, he sent 20 men out on field assignments to keep the CIC from seeing them at camp. They may have been married to Japanese women or worked in Japanese farms. He wanted the Regiment to arrive at Leyte at full combat strength. But when the Regiment arrived in Leyte, men were pulled and formed into interrogation teams for each landing task force.

Trained instructors could not keep up with the increased need for CIC service hampering support of landing forces. One hundred selected Filipino men attended the USAFFE Intelligence School with classes led by instructors familiar with the Philippines. Classes on Tagalog, the national language, counter sabotage, espionage, treason, conspiracy, interrogation, evaluation of data, the law of arrests, and CIC equipment were given. The Australian Field Security Service held classes on interrogation, security, and intelligence. There were instructions on customs, character, names of locations, local laws, and secret societies. A laboratory established for analyzing fingerprinting helped identify alleged collaborators.

Fifty Filipinos from the 2nd Filipino Battalion were sent to Brisbane to attend an intensive 4-week course from August to September. Upon graduation, they returned to instruct the entire Battalion in Oro Bay. Pfc. John Bamont, who had studied auto mechanics in the Philippines, finished a six-week CIC training at Oro Bay. Upon graduation, he became an investigator and interpreter for the 11th Airborne Division. Wallace Castillo whose background was in criminology and had graduated from OCS was quickly chosen to attend two months of CIC training in Australia. Company C of the 2nd Filipino Battalion was reorganized into CIC units. A platoon landed in the islands under the command of Lt. Johnny Mendoza.

Castillo followed the 1st Cavalry into Tacloban. He contacted guerrillas to find collaborators and questioned people about the suspects. Action was taken depending on the people's consensus. He interrogated suspects to determine if they were neither pro-Japanese or pro-American but just looking out for themselves, they were coerced by the Japanese, or convinced that the Philippines was better off under Japan. Agents informed of the Kempeitai in Dulag uncovering valuable data related to Japanese operations. He discovered that Japanese officers were equipped with portable printing machines to churn out Japanese currency. After eight months in Tacloban he proceeded to Cebu, burned to the ground and a shadow of its former self.

Men like Lt. Pedro FlorCruz, who was to go on one of the submarine missions, were transferred at the last minute to CIC. He became an instructor on the local culture of New Guinea and the Philippines. "It was a court-martial offense to climb a coconut tree considered a 'mother fruit tree' in New Guinea," he wrote, which was common practice in the Philippines. He observed that trust amongst men was fragile. Serving as Defense Counsel in a court-martial, he realized that "some commanders were afraid to discipline their soldiers for mundane offenses, implying fear of retribution once they were on the battlefield." He felt the fear was unwarranted "because the men are not vengeful, but take to discipline like a duck to water".

CIC men were constantly on the alert over information leakage. Some became interpreters for the American soldiers conducting vital military intelligence-gathering missions with the front-line troops and guerrillas. They searched enemy headquarters, seized telephone exchanges, impounded and delivered to censorship teams all mail. Interrogation of enemy agents and sympathizers was in conjunction with the Allied Translator and Interpreter Section (ATIS) teams. Many times, they dressed in civilian clothes and blended in with the people to ferret out information. They were given lists of political officials and guerrilla leaders to explore.

Onboard their transport to Leyte, lectures on CIC functions, proper distribution of documents on the enemy, and security continued. More than seventy CIC American and Filipino officers landed in Dulag, Leyte with their attached combat units. The civilian situation was critical as they gathered them into a safe location. The first few days were spent interrogating persons. In the first month, denouncements of spies and collaborators by zealous civilians and guerrillas overwhelmed the agents. Interrogations that were rushed and not as thorough because invasion schedules and security of the Army's operations had to be maintained, caused many suspects to be held on barely sufficient evidence.

When the PCAUs and their supplies did not arrive for several days at the Leyte landing, CIC carried out the PCAU functions of keeping the desperate civilians from rushing to military lines. They provided medical aid and shelter for the suffering civilians using Army supplies. To enforce order, they set up a temporary police system and arranged for the distribution of food.

As the combat troops moved inland, CIC agents followed coordinating closely with PCAU. Although their missions were different and separate from the PCAUs, they worked together to obtain their objectives. General MacArthur had ordered the arrest of collaborationists and sometimes included citizens of non-enemy countries. A Legal Board of Review was created to ensure fairness and justice to the accused.

A complex problem was civilian control due to the many alleged collaborationists accused by civilians and guerrillas. They installed a police system with PCAU. Those found guilty of working with the enemy were sent to Tacloban for incarceration in provincial jail until the Commonwealth government could hear cases.

The many lectures on the importance of each scrap of written matter resulted in a large amount of intelligence captured during the operations. Agents rummaged through dead Japanese for papers or took prisoners for interrogation.

Telesforo Acosta was with the 2nd Filipino Battalion on Luzon, where he drove a jeep with a CIC inspector into the towns to conduct investigations. He was detailed as an interpreter.

The experiences gathered in Leyte made the unit better prepared for the landings on Luzon and other parts of the islands. On the third day of landing, the CIC contingent of 22 officers and more than 100 agents, mostly Filipino personnel, landed near San Fabian, Pangasinan. As in Leyte, control of the civilians required CIC roadblocks at strategic points. It was effective in screening the locals and prevented infiltration of Japanese soldiers impersonating as Filipino civilians. On Leyte, they had to protect themselves against the Japanese and unfriendly guerrillas but on Luzon, they also had to watch out for the communist Huks.

As more CIC units arrived, periodic distribution of a 'Counterintelligence Summary' listing wanted persons, suspects, information desired on individuals and organizations, and the guerrilla situation was sent out. Priests allowed to move freely carrying messages to guerrilla leaders and educated men working in Japanese emplacements told of Japanese plans. Mistresses of ranking Japanese officers knew much and passed on enemy's plans.

The landing on Luzon brought attention to the status of the 12,000 Philippine Scouts in service in 1941. As CIC identified around 500 officers and enlisted men of the unit upon entry into Manila, they ensured good treatment for them. Filipino personnel mainly from Philippine Army and Constabulary who reported into the U.S. Army units, formerly USAFFE troops were forwarded to Replacement Depots for disposition.

Jim Martinez, a pilot, was with the 2nd Filipino Infantry Regiment at Camp Cooke. From New Guinea, he was sent to CIC training in Brisbane. He was attached to the 37th Infantry Division that landed on Luzon with the mission to collect information and material that could be useful to GHQ. He interviewed locals and people who came down from the hills about the Japanese. He cleared stragglers in caves, with the help of locals who knew the hillsides. Those who resisted were shot. The hardest job, he recalled, was to "clear caves with many wires that could be connected to booby traps. Grenades were thrown in sealing the caves."

After the battle to liberate Manila, CIC established their first headquarters at the Bilibid Prison. During the vicious fighting in the city, hundreds of suspects classified as collaborationists, puppet officials, enemy nationals, and Kempeitai agents were interned as security risks. Even after many were cleared and released, there were still "1,216 arrested in the Manila area" at the end of March.

CIC activities reached their peak after the carnage in Manila. Under combat conditions, "4,140 security cases were handled by CIC. Most of the cases were possible collaboration and loyalty checks for prospective Philippine officials." The daunting task included recommending the disposition of those guilty. Thousands required investigations, but officials were also needed to run local government.

Thousands were hired to work on Army projects. Screening the employment of civilians to prevent infiltration was deployed. Agents checked lapses in unguarded conversation, morale, and potential for sabotage through covert operations. Security posters were printed, and training films were shown to the troops, loyal civilians, and guerrillas. Regular newsletters on Japanese ordnance, booby traps, and updated lists of unsuitable citizens for employment were circulated.

CIC agents were trained on general interrogation techniques but had difficulty finding more proof when they investigated government officials under Japanese rule for collaboration, subversions, and treason. They read *The Manila Tribune* left behind in buildings on what those officials had said and done during the Japanese occupations. The civilians usually waited to report suspected persons or activities after fighting ceased.

Investigations were the most significant part of the CIC agent's role to find collaborators, anti-American sentiments, and the situations of the enemy, guerrillas, and civilians. They were sworn to secrecy of their work. Al Lamata, a CIC agent, took it to heart and did not reveal details of his work when interviewed seventy years later. He said only to the people with the need to know will he talk about his work. Lamata was with the 2nd Filipino Battalion because his father was a Filipino. He was attached to the 441st CIC as a special agent and had received training from an agent from Fort Hobbard, the CIC headquarters in Maryland. It was a rush course of fifteen weeks with equipment, films, pamphlets, and books. He did not know what his assignment would be until he got to New Guinea. Aside from CIC, he underwent jungle training with the Regiment, although he did not go into combat. From Hollandia, he was shipped to the Philippines. He proudly declared that his equipment was his brain and memory.

In Manila, Lamata worked with the PCAUs conducting interrogations and identifying temporary officials to the local government. He enjoyed his work helping in the rehabilitation. He never went out alone because he did not carry a sidearm and wear a patch to show his rank, as did all CIC members to do their job without intimidation. He amusingly related that the Military Police did not know what to do with him when they found out he was a special agent.

Agents were given lists of political officials and guerrilla leaders, compiled for the past two years from the messages transmitted by the submarine parties. Commander Parsons, Maj. Joseph McMicking, and Col. Whitney, as residents of Manila before the war, were familiar with many influential names and families. Guerrillas and civilians provided names of local officials and supporters in their towns. Interrogations were carefully conducted because many of the persons accused of disloyalty were loyal to the Philippine government. Evidence of collaboration abounded, but an arrest was difficult when some men ordered to jail had political connections to the 'old hands in the Philippines.'

The Japanese destroyed almost all official records, so CIC was pressured to recommend judgment on the hundreds of Filipinos who willingly worked for the enemy or were appointed. The "Army's unclear definition of the difference between a collaborator who was pro-Japanese and one who collaborated because he saw no other way" complicated making a recommendation.

To fulfill the immediate need for temporary officials to begin a local government, they were authorized to investigate and approve the arrest and replacement of leaders disloyal to the Commonwealth government. CIC with the PCAU gathered from the people and loyal guerrillas in the area the names of those qualified to be temporary officials. They confirmed names of suspected collaborators and disloyal civilians gathered from the messages the men on the submarine missions sent in their reports from their agents operating all over the islands to KAZ. The arrests of some government officials as collaborators in some of the towns where they provided relief shocked and disillusioned the people, wrote a CIC agent. Who to trust may have been a large factor in the swearing in by President Osmena of PCAU Martin P. Lopez to the position of provincial governor of Nueva Vizcaya province. The event was published in an article *Filipino Sergeant with U.S. Army named Province Governor* in the August 8, 1945, issue of *Free Philippines*. "Lopez arrived in the Philippines last January together with the American liberation force which landed in Lingayen. He had stayed in the U.S. for more than 20 years before joining the Army." The collaborators were arrested and turned over to the Philippine government for judgment. Their arrests kept them alive from the people's pent-up revenge.

U.S. Army soldiers and not Filipino civilians or guerrillas were assigned to guard the collaborators and Japanese prisoners so that they could live another day. The *Free Philippines* published an article of a Filipino guerrilla ordered to arrest two Filipinos who worked for the Japanese as alleged spies but instead took them to a deserted place and shot them. Some Japanese prisoners wanted to be killed immediately after capture so as not to undergo torture. They were surprised by the humane treatment of the Americans. A *Free Philippines* article recounted an American soldier who spent most of his time guarding prisoners and watching their viewpoint change. "Most Japanese soldiers beg to be killed when we first see them. Often, they go into hysterics. One who begged all night to be killed was going around the next day, bowing and saying 'Yamashita no good.'"

As the liberation campaign progressed, work concerning the guerrilla movement took up most of the agents' time locating them. After questioning, some means of control over their activities had to be enforced. The guerrillas revealed valuable information on the enemy's psychological warfare and propaganda techniques, economic and political conditions, and locations of military targets. The data formed a complete and authentic picture of the Japanese Kempeitai and other enemy counterintelligence groups. A list of disloyal people was compiled but CIC formed a Legal Board of Review to investigate the cases to ensure fairness and justice.

An objective of CIC was the censorship of the enormous amount of outgoing mail from the massive build-up of troops. Incoming mail was uncensored. Some of the agents were sent to one of the six censorship offices. Many photographs were lost as the U.S. Signal Corps could not cope with censorship of the millions of rolls of films taken in the islands.

The enemy increasingly infiltrated, spied, and deployed guerrilla warfare as they pushed towards the mountains of Northern Luzon even after Japan's surrender. CIC continued to ferret out information on the location of the mountain hideouts of thousands of stragglers for combat units to destroy. For many years after the war, stragglers in the thousands would come out of hiding to take what they could, some killing civilians they encountered.

After the war, the Recovered Personnel Section, USAFFE, processed guerrilla claims of having fought and aided the return of the American Task Forces. CIC referred civilians to them for death benefits, back pay of deceased POW military personnel and civilian employees of the U.S. War Department.

After Japan's surrender, the U.S. "released over five thousand interned Filipinos pending trials of collaboration or treason ending the investigative work of CIC. The release was not an indication of either innocence or guilt. That could be determined only by the courts of the Commonwealth government."

The CIC men returned to the 1st Filipino Infantry Regiment headquarter in Leyte as they were on detached duty. Those who remained trained in a program of lectures on Japanese customs, government, and social systems in preparation for Japan. In turn, they lectured on their experience in the Philippines that was useful to the occupation troops to Japan. They broadcasted to Army radio stations on security and distributed *The Intelligence and Security Guidebook for SWPA*.

Although not in combat but in service, the men of the PCAU and CIC were proud of their significant contribution to the uplifting of the lives of their countrymen. They continued to tell stories with pride whenever they reunited after the war.

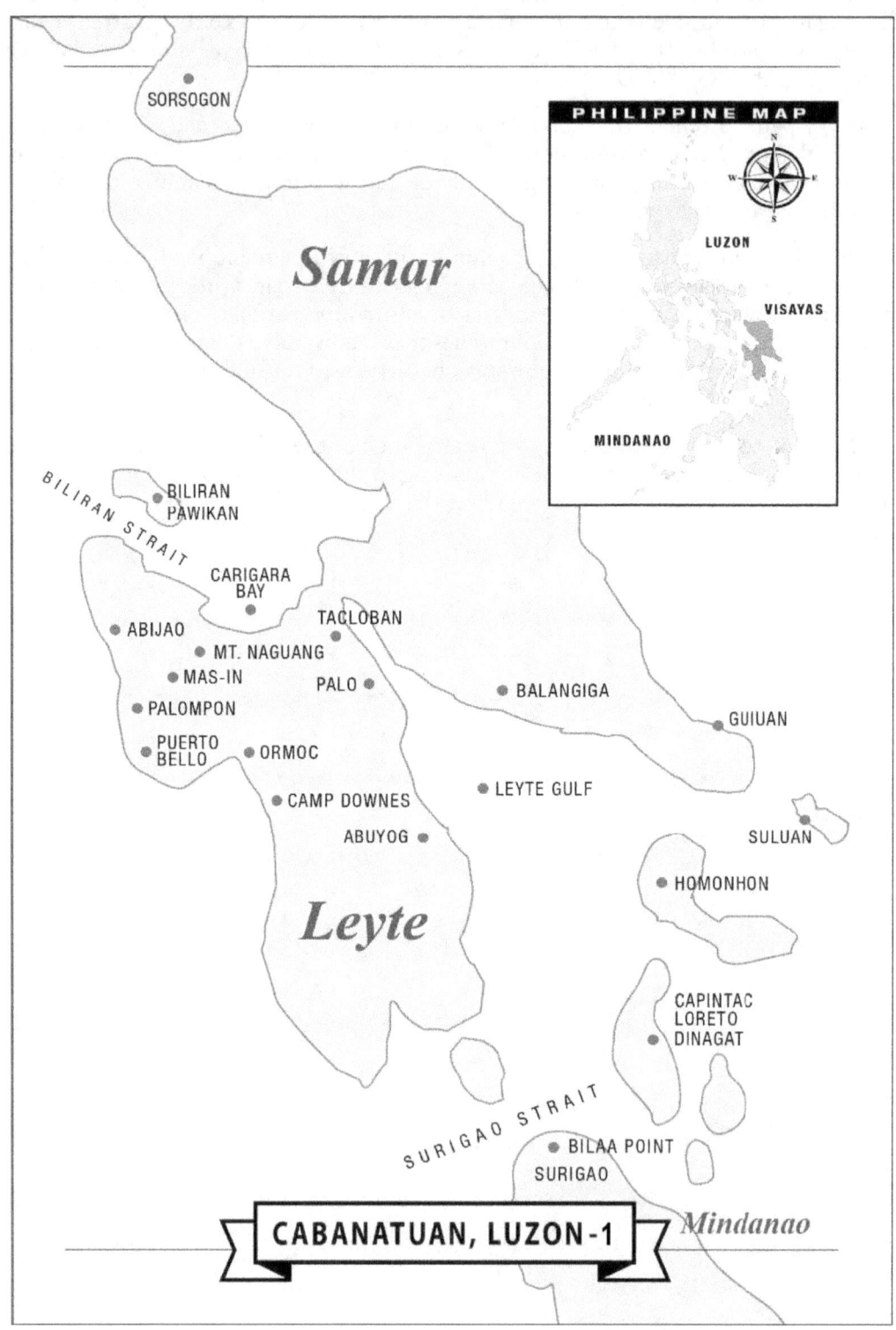

SORSOGON

Samar

PHILIPPINE MAP

LUZON

VISAYAS

MINDANAO

BILIRAN
PAWIKAN

BILIRAN STRAIT

CARIGARA
BAY

ABIJAO

TACLOBAN

MT. NAGUANG

MAS-IN

PALO

BALANGIGA

GUIUAN

PALOMPON

PUERTO
BELLO

ORMOC

CAMP DOWNES

LEYTE GULF

SULUAN

ABUYOG

HOMONHON

Leyte

CAPINTAC
LORETO
DINAGAT

SURIGAO STRAIT

BILAA POINT

SURIGAO

CABANATUAN, LUZON-1

Mindanao

Luzon Island
Cabanatuan Prisoner of War Camp

General MacArthur's forces were in Sansapor in the Vogelkop Peninsula within 400 hundred miles of Mindanao. Leyte was the new site for the first Philippine liberation landings and was scheduled two months earlier in October.

General Walter Krueger, commanding general of the Sixth Army wanted his own special force on surveillance and intelligence gathering that would be reporting directly to him. The Alamo Scouts, a Special Reconnaissance Unit, was formed in New Guinea in November 1943. They were an ad hoc unit and had no standard Army Table of Organization & Equipment similar to the Filipino 1st Reconnaissance Battalion.

The Scouts' six weeks of rugged training in the water at Fergusson Island culminated in a 26-mile jungle hike with complete equipment under sweltering heat that eliminated many aspiring members. A team of seasoned Australian special forces and local native hunters taught the trainees about jungle survival. Out of the nine training classes, over three hundred graduated. But only half qualified to join one of the twelve Alamo Scout teams that conducted over a hundred missions. The chosen men were required to be observant, to be team players, and to be disciplined enough to fight only for defense. The rest returned to their unit.

The team leaders chose men who had "individual initiative and competitive spirit, temperamentally drawn by the game of high risks, had to be a crack marksman, expert with many weapons, and willing to kill with bare hands." As a small team, they were always outnumbered, so they had to fight efficiently, quickly, and without hesitation

The small teams of highly trained Scouts operated behind enemy lines ahead of the Sixth Army's landing in the Philippines. They conducted pre-invasion reconnaissance and intelligence on enemy troops and defensive targets. And the information had to be accurate. They conducted sabotage, cleared paths of incoming task forces, and conducted rescues of hostages and POWs. They chose their weapons and equipment, wore camouflage uniforms, painted their hands and faces to blend in the shadows, and communicated via hand signals as they stealthily moved throughout the islands towards their objectives.

Many scouts with Hispanic surnames were not identified as Filipino unless specifically mentioned in a mission. Sabas A. Asis, Thomas A. Siason, and Rufo Vaquilar were with Bill Nellist's team. Alfred Alfonso also with the Nellist team was on other missions with the Rounsaville team. He was a Hawaiian-born Filipino from the 2nd Filipino Infantry Regiment. Caesar Ramirez was with John McGowen's team. Rafael Ileto, a Filipino Scout Team leader, had Cpl. Estanislao Bacat, Sgt. Juan Pacis and Sgt. Frederico Balambao. John Hidalgo was with Wilbur Littlefield's team and Juan Bergano with Herman Chanley's team. These men were from the 1st Filipino Infantry Regiment stationed in New Guinea at the time. Going behind enemy lines where the daunting chances of returning alive were almost none did not deter them. Their familiarity with the languages and terrain in the Philippines emboldened them. (A roster of Filipino Alamo Scouts is in the Appendix.)

An exception was Atanacio Chavez from the 1st Filipino Infantry Regiment, who fought with the U.S. 32nd Infantry Division in New Guinea. He was assigned to the Alamo Scouts as an instructor teaching Filipino languages, culture, people, and terrain.

Men from the Philippine Intelligence Message Center (PIMC), formerly with the 1st Filipino Infantry Regiment, were assigned to the Scouts' missions. Lt. Inocencio F. Cabrido, Pvt. Trinidad Sison, Pvt. Agapito Amano, Sgt. Leandro Reposar, Sgt. Vincent Quipo, and Ben Mones were on temporary duty as radio operators and translators.

Alamo Scout Training Center. Team leader, Rafael Ileto on the right.

Terry Santos, a graduate of the Alamo 4th training course in September 1944, underwent the maneuvers and exercises in the sea, jungle, mountains, and on the beach. He was with the 1st Filipino Infantry Regiment in New Guinea and wanted more action. He was the smallest in the class but the quickest on his feet. He recalled having essential equipment including a first- aid kit, compass, binoculars, machete, rain poncho, canteen, pistol belt, and a cartridge pouch before taken a mile offshore in a rubber boat, dumped on the side, and told to swim to shore. Anyone who was pulled out was dropped that day and sent back to their unit. Training intensified in succeeding weeks.

He learned to avoid gunfire while in the water, use different radios, and advanced first aid. He learned to plan a mission, evade the enemy, and quietly blend into dense foliage. The final test was Santos' team dropped by the New Guinea coast where the Japanese were operating. They went inland for 10 miles until they came across a Japanese platoon of twenty men chopping wood to make a fire. Knowing that the enemy was larger and had increased firepower, the team encircled and ambushed quickly. Santos walked towards the Japanese firing with his machine gun and promptly dissolved into the bushes. After graduation, impatient for an assignment, Santos decided to join the 11th Airborne Reconnaissance unit.

Filipino Sgt. Leon A. Jacobs proudly wore the patch of the 6th Ranger Battalion. When the 1st Filipino Infantry Regiment arrived in New Guinea, he volunteered for the Ranger Battalion. He was not deterred by the arduous six-week training course to be part of "the roughest, toughest outfit in the Pacific" as described by the commanding officer, Col. Henry Mucci. There may have been other Filipino men with Hispanic last names, but they were not explicitly identified as Filipino.

During a night-time mission in late 1944 to inspect a beachhead, one of the two soldiers with Jacobs triggered a mine, killing both. The forceful blast threw Jacobs onshore. Body parts of the soldiers raining on him was the last thing he remembered before blacking out. He was fortunate to have been found and shipped back to the U.S. "Months later, he woke up in DeWitt Army hospital in California with temporary amnesia and holes drilled in his head to relieve the pressure in his skull caused by the blast." For his service, he was awarded the Bronze Star and Purple Heart medals.

New Guinea

Right after graduation, the teams of Nellist and Rounsaville were transported to the Vogelkop Peninsula to free the former Dutch governor, his family, and Javanese civilians held prisoners by around 25 Japanese soldiers in the Cape Oransbari area of New Guinea. The team was composed of 13 men, including civilian guides and interpreters.

Nellist Team: Back row L-R Galen Kittleson, Thomas A. Siason, Andy Smith. Front row Wilbert Wismer, Bill Nellist. 1944 U.S. Army photo

The recollections of the men provided details of the successful rescue. Under a heavy storm at night, one of the PT boats hit a submerged log while crossing the south Pacific Ocean from Biak Island. Transferring to rubber boats, the teams paddled to shore at the mouth of the Maori River. Heavily armed, they crawled on the beach into the thick jungle. Rounsaville led the rescue while Nellist with Siason, Kittleson, and Cox stayed behind with the rubber boats on the beach. Finding four Japanese machine-gun nests, Nellist's team had to wait for the signal shot from Rounsaville before eliminating it. Rounsaville's team charged the Japanese guards' hut to free the hostages and fired the signal shot at the first sign of daylight. With a burst of his machine gun, Alfonso eliminated the sentries guarding the hostages. Several escaped into the jungle. Asis, with two other Scouts, charged the other hut with sleeping guards and freed the native chief. While seizing the enemy quarters, a Japanese officer lunged at Smith with a bayonet. Asis shot the man dead. Another source stated that Vaquilar suddenly appeared and sent a burst of lead over Smith's head, then nodded and left. Behind Smith was a dead bullet-riddled Japanese soldier with a fixed bayonet on his rifle. The governor was found, and his hut quickly searched for documents before being burned.

The team did not hear Rounsaville's first signal shot. Siason and Kittleson saw two Japanese soldiers appear, and one picked up his rifle to walk around to within a few yards where they laid hiding. When two more appeared, they decided not to wait for the signal shot to protect the landing area and shot the soldiers. Siason followed an officer into the bushes and finished him off. The Scouts led sixty-six hostages through the jungle to the pickup point at the beach. A Frenchman, his wife, and ten children who had been trying to escape unexpectedly showed up. That night, Alfonso set up a defense perimeter around the rescued prisoners. Early the following morning, the large party boarded the rubber boats to the PT boats towards Biak. The refugees were so grateful that they broke out singing the Dutch national anthem.

Leyte Island, Philippines Landing – October 1944

The Alamo Scouts' mission expanded to intelligence gathering and coordinating with guerrilla units when they entered Philippine waters ahead of the U.S. troops. The teams left Woendi Island and were inserted in the Philippines to gather information on landing beaches, water obstacles, and enemy disposition ahead of the landing convoy. Ramirez with the McGowen team was inserted at Palo, Leyte to reconnoiter future landing beaches.

The Scouts arranged with KAZ to resupply guerrillas with airdrops of arms, ammunition, radio, and medical supplies. The many radios sent in by submarine in the last three years, were wearing down or no longer operable. The guerrillas now had an assortment of portable radios, American, Australian, and "Dutch sets driven by a bicycle treadle."

The books of Larry Alexander and Lance Zedric on the Scouts gave many details of their activities. The day after the Leyte landing, Sgt. Leandro Reposar, a Filipino radio operator who arrived with the landing forces, boarded the boat of Rounsaville's team with Alfonso and Vaquilar. With a guerrilla officer as a guide, the team proceeded to Masbate Island to help the guerrillas set up a radio and observation network in preparation for the Luzon landing. Masbate's location with its surrounding waterway between Leyte and Luzon was strategic. Enemy shipping was visible. Sailing to Masbate, they were mistaken as the enemy. Both American and Japanese planes fired at them, narrowly missing them, and violently rocking their boat. Reposar later thought what a useless way to die by friendly fire as everyone jumped overboard. Rounsaville's quick thinking of "standing to risk exposing his pale torso on his tall and lanky body stopped the friendly firing". Reposar stayed behind in Masbate to set up a radio station and trained the guerrillas to provide him with the correct information. The rest patrolled and established observation posts along the coast and the nearby smaller island of Ticao. Soon, reports began to arrive from the guerrillas of heavy traffic on the short strip of water between the two islands and down towards Leyte. Six Japanese transport ships filled with troops were observed sailing from Luzon to Leyte as reinforcements for Ormoc. Reposar quickly sent a message to KAZ, and the team waited. Torpedo planes of the 5th Air Force arrived and attacked the ships. From their vantage point, the scouts and guerrillas watched smoke and flames shooting up from some of the ships as they began to sink. Fighter planes strafed survivors leaping from their ship. Another message to bomb a major airfield located in the outskirts of the city forced the Japanese to scatter.

Siason and Asis with the Nellist team left Leyte Gulf in a ship's tender during a Japanese night raid to check on the Japanese strength and defenses in Northern Mindanao that might block the Surigao Strait and jeopardize landing operations. Landing in Surigao on the northern coast of Mindanao, a guerrilla carrying an old Springfield rifle was elated to see them. Siason and Asis interpreted as the guerrilla told in his native language of a 75 mm gun in an outpost operated by eight Japanese soldiers near Bilaa Point overlooking the beach where they had landed. A guerrilla commander with 45 men armed with old Springfield rifles and assorted Japanese weapons spent several days with them reporting their findings to Siason, who had set up a radio station.

Asis and Smith went to check out the heavy Japanese machine gun on the hill. At their pickup time, there was no response to their flashlight signal from the beach. Siason heard through the radio that the Japanese navy was coming through the Surigao Strait to attack the landing forces on Leyte. Searchlights towards Surigao Straits pierced the night as they hiked over mountain ridges to the new pickup point. The following morning, dense smoke was seen from burning ships. At the pickup point, Rounsaville's team came in from their mission evading a small Japanese force in barges that landed at Bilaa Point. The teams sailed raking the coastline with

Camp Cooke: 2nd Filipino Infantry. Second from left is Alfredo Alfonso training with their bolo knives. He joined the Sixth Army's Alamo Scouts and assigned to the Nellist team who provided the reconnaissance and rescue support in the Cabanatuan Prisoner of War Raid in January 1945. Photographer: Christopher Gardner

machinegun fire and blowing up the enemy barges of ammunition. One of their PT boats ran aground, damaging its propellers. "Two engines on their rubber boats suddenly slowed down with 120 miles still to go through the Surigao Strait and across to Leyte Gulf to Tacloban but were able to sail through the high swells of the Surigao Strait. An Australian navy ship, HMAS *Hobart,* that was patrolling the sea lanes, approached and stopped them. The skipper demanded the correct password, unconvinced that they were Americans. A two- hour standoff finally ended when the PT boat skipper snapped that they would be targets for every 'Nip' plane in the Philippines come daylight if they were not allowed to move on." They beached through a screen of burning ships on fire and men scampering. It was the thick of the Battle of Leyte Gulf. Men, equipment, and supplies were strewn on the beaches open to enemy attacks. They saw the U.S. 7th Fleet bravely defending the stretch of Leyte beach as men were desperately unloading supplies, and equipment.

As a counter-offensive and to disrupt the Leyte landings, the Japanese began reinforcing their supply base in Ormoc Bay on the western side of Leyte Island with units from other islands. The guerrillas in the Ormoc area informed the Sixth Army of the growing supply base. Three Filipino radio men, Lt. Inocensio F. Cabrido, Pvt. Trinidad Sison, and Pvt. Agapito Amano, joined Sumner's team with their equipment from the Philippine Island Message Center along with two tons of weapons and ammunition. They departed Tacloban in three PT boats and traveled around the island to the western side at Matag (Matag-ob) near Ormoc. They slipped ashore in the town of Abijao north of Ormoc. The camp was set up in the largest house with security provided by the guerrilla leader. Japanese land and sea movements in the Ormoc port and surrounding areas as well as the Japanese headquarters and airfields in the town of Valencia were monitored. They were equipped for a long stay. American warplanes sunk targets as they were informed.

The townspeople were so happy to see the team that a celebration of food, music, and dancing was held. The Filipino men must have enjoyed the attention of coming back as liberators. Having been gone for many years they had become almost strangers to their homeland. The eagerness of many young Filipino women to dance with them was a welcoming contrast to the dearth of Filipino women in California. Tuning the radio to broadcasts from San Francisco made them more popular with the people and raised their morale. Popular among them was the *Voice of Liberty,* a news program aired daily through the U.S. Office of War Information. But they were careful about the radio's location, making sure that it was still well-hidden.

A radio message was sent for a location of an airdrop of weapons, ammunitions, and clothing for 200 guerrillas behind enemy lines. Muddy trails slowed down the march to the pickup point. A surprised encounter with a Japanese patrol quickly ended in a firefight. The radio broke down in the middle of all the shooting and fixing it delayed the travel to the drop zone location. To provide a signal for the arriving planes, some underbrushes were hacked out and set alight at the location. The C-47s arrived and dropped 36 large bundles that floated down under red and white parachute canopies. Garands, Carbines, BARs, ammunition, clothing, gasoline, coffee, cigarettes and bundles of *Stars and Stripes* and *Life* magazines were hauled out as the bundles were opened. The supplies lit up eyes with relief and hope. After clearing the area of any signs of their presence, the team continued to Puerto Bello, a vantage point to monitor the Japanese at Ormoc. At Mas-in, an inland area north of Puerto Bello, a radio station and an intelligence unit composed of 21 guerrillas and local constabulary was set up in a house on stilts of bamboo and palm leaves on a hillside. Cabrido, Sison, and Amano provided military and intelligence training. After several weeks scouting through the Ormoc area, a courier informed them of Japanese troops landing six miles from Puerto Bello looking for Americans. Cabrido sent a message to the Sixth Army that the radio would be down for a couple of days and, once relocated, would request another air drop. Ammunition, medical supplies, and rations were running low. The new camp was in the north on the slopes of Mt. Naguang to better protect the radio and have several escape routes in case the Japanese attacked in force. The airdrop arrived in mid-November and included sewing needles for the civilians who had been helping them to fix torn clothes. Many civilians were wearing coarse and itchy abaca fiber cloths stuck to cover their private parts. When the radio broke down again, they could not fix it due to the lack of spare parts.

Several days passed and something strange happened. A Japanese airplane from a nearby airfield flew overhead and released a cargo box attached to a parachute. Unbelievably, inside was a Japanese radio with spare tubes. The Japanese pilot must have been confused and disoriented with his instructions. The next day, the same plane dropped some spare parts. Amazed, the radio men, must have muttered a prayer of thanks before getting the radio to work again.

By this time, 23,000 Japanese troops were trapped in Ormoc Valley. Flash radio transmissions to American fighter planes eliminated any chance of supply or reinforcement to them. On the ground, elements of the 1st Filipino Infantry Regiment were hunting down and eliminating pockets of the enemy.

John Hidalgo, with Littlefield's' team reconnoitered the area of Camp Downes, one of the proposed invasion areas west of Ormoc. The Army planned to arrive on December 7 and drive north to cut the enemy in two. The team rode in a jeep to the village of Baybay, south of Camp Downes. With the help of fishermen, they sailed north not too close to the shore, taking notes. The fishermen pointed out locations of the mines, obstacles, and the slope of the beach. Information was relayed to KAZ five days before the scheduled invasion. The U.S. 77th Infantry Division linked up with other task forces and seized Ormoc City.

The Army's next target was the port of Palompon, 12 miles west of Ormoc. The Sumner team was located at Puerto Bello, a coastal village between Palompon and Ormoc, and had been sending reports to KAZ on the enemy activities in the area. The seizure of Palompon cut off reinforcements of troops, ammunition, and food to the Japanese troops. After six weeks, the Sumner team returned to their camp in Tacloban in time for Christmas dinner.

Bacat and Balambao, with their team leader, Rafael Ileto, boarded a small navy ship on December 8 from their camp in the town of Abuyog. They landed at Guiuan, on the southern coast of Samar to survey the enemy's situation. Guerrillas met and led them to their camp near the town of Salcedo. The three Filipinos communicated with the guerrillas in Tagalog and Cebuano, the language widely spoken on Leyte and nearby Samar. The locals rowed the Scouts across to Balangiga to reconnoiter the shoreline. Some 250 Japanese soldiers were in the hills five miles north of the town and patrols frequently raided the town and terrorized the inhabitants. An airdrop of weapons and ammunition raised morale to attack the estimated 90 Japanese at Bolusao west of Balangiga and in the hills north of the city.

With the impending landing on Luzon, Pacis and Berganio with team leader Herman Chanley landed near the town of Pawikan in Biliran Island that was connected to Leyte. An outpost and radio were established sending reports of enemy shipping in Carigara Bay (Coalargo Bay) and Biliran Strait heading towards Luzon.

As the Scouts gathered for a Christmas dinner at their training camp in Abuyog, they received a difficult assignment that made them uneasy. When not out on assignment, they were to be rotating bodyguards to General MacArthur and his staff whenever they inspected areas not completely cleared of the enemy. As bodyguards, they had to follow orders to ensure that the generals were not captured. It was a direct order that might end up with the generals being killed by them and not the enemy.

The U.S. 6th Ranger Battalion led the invasion of the Philippines. The islands of Suluan, Dinagat, and Homonhon guarded the entrance to Leyte Gulf. Minesweepers had removed mines in the approach to Suluan after battling heavy rains and rough seas. The urgent directive was to clear the islands of the enemy to avoid disrupting the Sixth Army's landing operations. One Ranger company stormed into Suluan preceded by naval gunfire on October 18. They cleared the lighthouse used by the Japanese for communicating with ships and aircraft.

At nightfall, four Ranger companies landed at Campintac beach nine miles north of the town of Loreto in Dinagat Island with minor opposition. They landed behind an ancestral house where a photograph commemorated the raising of an American flag with 48 stars in front of the house after three years of Japanese occupation. The U.S. Press Corps caption was *Photo taken behind the Ironwoods house of Dionesio Ompod Y Serena*. Thirty-one years later, a group of American veterans on a sentimental journey to Loreto returned to Campintac Beach, where they first landed in 1944. They met the mayor of Loreto, who was the son of the owner of the house. His father had hidden the flag. The same flag was raised a second time as tears flowed, and the veterans saluted the flag with 48 stars. A simple marker with only a helmet and a bayonet now stands where the U.S. battleships first assembled in Loreto Bay for two days before landing on the beaches of Leyte.

Luzon Island, Philippines Landing – January 1945

Four days before the landing in Lingayen Gulf, the Alamo Scouts left Leyte in two PT boats going south through Surigao Strait and Mindoro Strait. They watched the invasion convoy of over 800 ships slammed by Kamikaze planes desperate to stop the landing. Ammo ships "blew up so violently that PT boats a quarter-mile away were lifted while falling debris including unexploded shells rained down on adjacent ships causing more damage and casualties." The Scouts' PT boats rocked and were lifted violently by the ensuing waves. Unable to land, they waited until the first ships began to reach the shoreline.

As in Leyte, their missions were to establish radio stations and maintain communications between various independent guerrilla groups. They set up observation posts to watch the roads. They quickly taught guerrillas how to gather accurate information. Coordination with the various guerrilla groups corrected the inaccuracies and removed embellishments boosting the prestige of guerrilla leaders that filled the various reports to KAZ.

The landing on Luzon heightened the desperation of the Japanese. Their orders were to delay and cause as many casualties as they could on the U.S. forces.

In the town of Akle north of Manila, the Hobbs team met with Bartolome Cabangbang (Luzon mission September 1944), and a guerrilla officer of the Bulacan Military Area (BMA). A command post and a radio relay station were established. After two days, the team split up accompanied by guerrillas to reconnoiter the Bulacan area. Two scouts travelled to each of the water sources of Sibul Springs, Angat Dam and La Mesa Dam.

Cabanatuan Raid, Luzon Island, Philippines – January 30, 1945

The Cabanatuan prison camp in Nueva Ecija held American prisoners-of-war, mostly survivors of the Bataan Death March and the Corregidor surrender in 1942. They were originally held at Camp O'Donnell in Capas, Tarlac. At its peak, there were around 7,000 prisoners, but now they were down to 500 men – victims of death by starvation, diseases, and Japanese brutality. Many were shipped to Japan to work as slaves in their factories. Not having signed the Geneva Convention, the Japanese forced prisoners to do whatever they wanted done.

Robert Lapham had watched the camp for months and was concerned for the prisoners' safety when the Tokyo War Ministry's issued the 'Kill All Order' to all POW camps. All POW camp commandants who believed that their camps might fall into the hands of Americans were to kill all military prisoners and civilian internees. "Whether they are destroyed individually or in groups... with mass bombing, decapitation... dispose of them as the situation dictates... It is the aim not to allow the escape of a single one, to annihilate them all, and not to leave any traces."

When the chilling news of the massacre of POWs in Palawan circulated, Lapham pressed to meet with General Krueger of the Sixth Army about the 5,000 strong Japanese garrison around Cabanatuan City, the 1,000 enemy troops across the nearby Cabu River, and the several hundred Japanese soldiers in the POW camp. A survivor of the Palawan massacre, Pfc. Eugene Nielsen, recounted his tale to the U.S. Army Intelligence on January 7, 1945. The Japanese forced him and other prisoners to go into their earth-covered air raid shelters and there were doused with gasoline and set on fire. Many of the POWs burned to death. Others who tried to escape from the fire were shot to death by the guards.

The radio men analysed the latest enemy activities from the area of the Cabanatuan prison camp. Pascual and Pulido (Luzon mission October 44) with a recon man set up a sub-station connected to Lapham's net control station covering the central Nueva Ecija area. Guerrilla leader Juan Pajota, a Philippine Army officer who escaped from Bataan, selected the village of Santo Domingo that overlooked Cabanatuan City, two kilometers from the prison camp. The signalmen camped in the mountains of Siclong for several weeks, gathering intelligence on the Bato Ferry carrying enemy troops that crossed the Pampanga River. Reconnoitering Siclong, guerrilla troops battling the Huks were reported in the town of Talavera and at Talabutab Sur. Two hundred guerrillas clashed with 1,500 Japanese soldiers at Talavera close to Cabanatuan. Copies of Japanese maps obtained by guerrilla spies were transmitted showing the Japanese locations, concentration of tanks hidden under bamboo and mango groves along the roads, and Japanese troops in the mountains where paratrooper landings were expected. In Sto. Domingo, they radioed the nine tanks the Japanese sent to crush the guerrillas and Huks resulting in many casualties on both sides. They were saved when heavy mortar fire destroyed the city. When they moved back, they witnessed the U.S. bombing of Cabanatuan and radioed KAZ. The Japanese rushed in reinforcements as the city burned for two days.

Returning to McKenzie's camp, under Lapham, they worked all though the night until the next day under Filipino Alamo Scout officer Rafael Ileto and his team on the intelligence brought by agents. Operatives reported 10,000 Japanese in the nearby town of Muñoz and the surrounding area. Ileto had to confirm reports that were unbelievable before they were coded into messages. As Japanese patrols continued to surround the command post, the combat guerrillas repulsed them to protect the radio station. The camp moved to San Andres, a village closer to Cabanatuan in the interior when it became dangerous as the Japanese retreated into the area. Pascual said it sounded like the "world coming to an end" when friendly planes dropped bombs on the Japanese garrison.

Operatives and pro-American informants reported only 70 Japanese guards in the prison camp armed with rifles and light machine guns. But the area around the camp were swarming with Japanese, Huks, and Makapilis. The radio operated erratically and would go off the air. Runners were on standby to relay the information to the closest station.

Krueger agreed to free the prisoners, but his intelligence chief estimated that Army Task Forces would not reach Cabanatuan until the last day of January or February 1 due to the Luzon landing. By then, it might be too late to prevent the annihilation of the prisoners. The Alamo Scouts and the 6th Rangers were ordered to accomplish the rescue.

On January 27, three Alamo Scout teams with their guerrilla guides left for Guimba and Platero, two miles north of the camp. Their mission was to conduct advance reconnaissance for the Rangers to plan the raid. Asis and Siason with Nellist's team went ahead heavily armed with extra ammunition and grenades to reconnoiter the area around the camp. Messages were sent back by runners to Guimba to avoid radio interception. The guerrilla guides easily communicated in the local language with Asis and Siason while going through rice paddies, bamboo thickets, and fields of cogon grass. Villages were avoided so as not to rouse the people and alert the Japanese. Every noise they heard made them stop and drop until sure that all was safe. After wading through the Talavera River, they reached the National Highway and for the next 24 hours watched convoys of passing troops in trucks and tanks. Radio silence was maintained. They continued in a double-time march for a mile to make up for the delay. They were very much aware that a massacre could happen anytime.

At the village of Balincarin, Pajota reported about 1,000 Japanese soldiers camped across the Cabu River, around four miles from the camp, plus hundreds of troops near the Cabu River that his men could take care of. As many as 7,000 enemy troops were deployed around Cabanatuan City, further away from the camp.

Pajota and Eduardo Joson, another guerrilla leader and their men set up roadblocks in opposite directions on the main road near the camp. They were to hold traffic for the prisoners to cross over to the American lines when they hear the signal shots. Pajota, with around 100 guerrillas, some teenaged boys armed only with *bolos,* faced a Japanese battalion encamped across the wooden Cabu bridge. Mines to stop the enemy from crossing the bridge were planted in the road with the help of Filipino demolition men from the October 1944 mission. Joson and his 75 guerrillas had set up their roadblock and crouched in irrigation ditches on both sides of the road, ready to meet Japanese forces coming from the city. Both guerrilla bands were provided with new arms and were trained to use a bazooka that was part of the shipment dropped to them.

The Scouts crawled on their bellies over carabao dung to get a clear site of the camp on flat dry land 700 yards away. For the next 24 hours, they watched the camp from behind tall grasses and saw the enemy transports passing through the camp that was divided by a road – one side housed the Japanese guards and the other were barracks for the prisoners and a hospital. They saw the eight-foot-high fence surrounding the camp with three rows of barbed wire, two guard towers and four machine gun pillboxes that did not appear to be manned. The roofs of buildings were seen but the exact locations of the prisoners and the Japanese guards' barracks were unknown. More details were needed as the shootout into the camp and removal of the prisoners were to be done in 30 minutes before Japanese reinforcements at Cabu bridge could rush over.

At noon of January 30, Vaquilar decided to risk it and approached some farmers working in the fields outside the camp. He borrowed some clothes. Wearing baggy farmers' clothes concealing weapons and wide-brimmed woven hats, Nellist and Vaquilar slowly walked to a small and empty hut on stilts that overlooked the camp. They saw right into the camp and counted the number of guards at the gatehouse and tower. On an aerial photo they carried, they labeled and described the long building huts of the guards and prisoners. A Scout delivered the photos and notes to Col. Henry Mucci, the Ranger commander at Plateros. The plan to enter and exit the compound quickly with as few casualties as possible was modified with the new intelligence.

Again, at great risk, Vaquilar left the hut with pistols under his clothes, knowing that Nellist could not cover him so as not to expose the mission. He walked towards the camp along the road shoulder to get a better idea of the campsite and a description of the main gate with which way it opened. He tipped his hat at the sullen guard who ignored him. His eyes scanned the camp to see any agitated activities after he had seen a young girl pass a note to the guard earlier. Vaquilar was lucky that the guard did not slap him when he realized that he forgot to bow.

In the evening, 121 Rangers and 11 Alamo Scouts with faces smeared with mud crawled across the field among tall grass that died out as they neared the camp's fence. Since it was only two days after a full moon, their movements were easy to spot. Nellist and Vaquilar climbed from the hut and quietly joined. By daylight, Pajota's suggestion was followed to distract the guards. A fighter plane from the 547th Night Fighter Squadron roared out of the sky, twisting, and skimming the camp, backfiring the engines to distract the guards. The pilot could see the Scouts and Rangers creeping on the field. The guerrillas had cut the telephone lines from the camp to prevent sending out an alarm to the Japanese in the city. The rescue teams furiously crawled until the grass could no longer hide them and moved into their position. They were now in the open. A Ranger team arrived in the camp's rear and fired the first shot to start the raid. Rangers and Scouts rose and charged the front and rear gates. Pvt. Herbert E. Wolff recalled that after his platoon cleared out the rear gates, he moved to the front after mortar rounds to look for Dutch people as he spoke Dutch. He was of German ancestry but was assigned to the 1st Filipino Infantry Regiments at induction. Records show that after passing Alamo Scout training, he returned to the Regiment but here he was at the raid.

The rattle of automatic gunfire and blasts of grenades ripped through the morning stillness. A raider drew his pistol and shot the padlock on the main gate. A guard appeared and shot him hitting his hand. Bullets tore into the guard towers dropping the guards. Japanese soldiers scattering to get their weapons fell. Frightened prisoners hid. Nellist and Vaquilar ran for the center of the camp, firing tracer bullets at the Japanese barracks to pinpoint its location for Rangers to fire on with bazookas and machineguns. A Japanese mortar team fired three rounds near the main gate and wounded Alfonso with a shrapnel in the gut. Asis and Siason with Nellist moved quickly and directed the fire of the Rangers at the mortar, silencing it.

The Scouts stood guard at the main gate to guide the prisoners out. Some prisoners would not leave. After having been imprisoned for almost three years, they could not believe that they were being rescued. Some hid from the men who wore soft caps looking like the Japanese guard with blackened faces and weapons unknown to them. They wanted to know who they were first. In shock, prisoners shuffled towards the gate. Some had to be helped or picked up. Grenades burned the huts as they left. Once the camp was cleared, a red flare was fired, indicating all troops should fall back to the meeting point at the Pampanga River, almost two miles north of the camp. Over a hundred carabao with carts prepared by Pajota waited at the muddy Pampanga River to transport the prisoners. Civilians were ready to cover the animal and cart tracks.

Eleven Scouts stayed behind at the river and formed a firing line in case Japanese reinforcements arrived. Alfonso and another wounded Scout continued with the column of prisoners. He clutched his bloody midsection while he was carried on a stretcher. At nightfall, the Scouts in their defensive perimeter watched the burning camp. After several hours with no signs of Japanese troops, Asis, Siason and Nellist followed the column of prisoners keeping close to the edge of the forest to avoid detection. Along the way, Asis and Kittleson signaled all to get down and stay quiet as a Japanese patrol passed in front of the group.

At Platero, a makeshift hospital had been set up to care for the casualties. Alfonso was injected with painkillers as the shrapnel was removed from his lower abdomen. The radio crackled with the promise that a plane would be sent before dawn for the Ranger Battalion surgeon who was also injured in the stomach. Scouts, civilians, and some freed prisoners who stayed behind desperately hacked and flattened rice paddy dikes – many using their bare hands – to make a landing strip before daylight so a plane could land and airlift the surgeon to American lines. They watched the sky for the plane that did not arrive. A prisoner, Stephen Sitter, who was a medic, related that Nellist handed him a pistol and a magazine saying that it "might be all that was between him and the Japanese again, and if it came to that, save the last one for himself."

The Scouts and around 100 of the weakest prisoners soon encountered the Huks who blocked their path. The communist guerrillas wanted to seize the Scouts' and guerrillas' firearms. But the Huks hesitated when Asis and the Scouts cocked their weapons at ready. They calmly informed the Huk leader that the "entire U.S. Sixth Army would hunt them down and he would be the first to die." After several intense minutes of bluffing that led to a stalemate, only the Americans were allowed to pass but not Pajota's men, a rival group.

On the last day of January, the day that Sixth Army Task Forces would have been able to be in Cabanatuan, the column arrived at Talavera, a town north of the city captured by the American forces. Trucks and ambulances loaded the freed men for their last journey home. The rest of the Scouts reached Talavera, two days later with the last and weakest prisoners. Only two prisoners died during the evacuation and 21 guerrillas were wounded.

There was no time to rest between missions. The Sixth Army wanted the elimination of undetermined howitzers that fired from mountainside caves and rolled back on tracks behind camouflaged entrances. The Japanese had installed them in Leyte and Luzon. American aircraft spotters could not pinpoint their location. The guns could cause much delay in the American forces drive to Manila. The guns protected an intricate tunnel and cave network stockpiled with supplies, food, and weapons. Nellist's team was summoned to silence the guns. In a Higgins boat, Asis and Siason hugged the coastlines to the south and waded ashore towards the guns hidden four miles inland.

It was a moonless night as they stealthily passed villages. A civilian forced to work for the Japanese operating the big guns pinpointed the location and they quietly slipped away towards the guns. Shortly after nightfall, the big guns roared, and the muzzle flashes lighted the sky. The large shells roared overhead. The distance was calculated, and they began the trek to the coastal town of Cabaroan. The team boarded a wooden boat at Cabaroan but abandoned it when they saw it was leaking. They swam ashore and located the guns in the mountains of San Jose along the way towards Manila. They hiked to the American lines in Damortis to inform them of the guns' location. They did not have a radio with them because they could not quickly move around with the weight. The 158th Regimental Combat Team in Damortis wanted identification when they could not produce a password without a radio. They were "doubtful of seeing Filipino Scouts. The sight of Cox, a tall and blond-haired scout with bare chest and upper arms walking forward shouting he was an American, was convincing enough." The German-made guns' position was passed on to the artillery division in the area and soon were knocked out clearing the way to Manila.

From Cabanatuan, Alfonso and Vaquilar with their team explored a wide area of Condon, where heavy Japanese activities were seen. It was also reported that it could be one of Yamashita's hiding places.

Siason, Asis, and another Scout split from the team to the coastal town of Rawis in Albay. Encountering three Japanese by a bridge sipping tea, they shot two and took the third as prisoner. Someone must have told the locals in the town of their arrival because as they entered, the locals joyfully mobbed them. The celebration and noise became louder when they found out that General Macarthur was not far away. Asis told them to calm down and go home.

More assignments arrived in February. Ben Mones, a Filipino radio operator with the Littlefield team was ordered to set up watch stations and an intelligence network along Highway 5 from Malolos in Bulacan to Northern Manila, a span of around 45 km. Much of Luzon surrounding Manila was still swarming with the enemy. Mones was from the area and spoke Tagalog to the inhabitants of the villages they passed. The inhabitants felt safe sharing whatever food they had with him and a place to sleep at night. Coming from San Francisco, Mones may have done back-breaking work in the vegetable fields of the San Fernando Valley of California that gave him the physical stamina of a younger man. He noticed that his team leader Littlefield could hardly keep up with him. So as not to be hit by friendly fire, they coordinated with the U.S. Cavalry lines as they neared Manila. Many villages around the city were strafed by American planes to eliminate the enemy. As the team neared the village where Mones came from, Littlefield allowed him two weeks to visit his family, although he had no authority to do so. He was ecstatic as he had not been back for many years. He could not save enough to return home and, if he did, would not easily have been able to return to California because of strict immigration quotas.

From the three submarine missions that landed from September to November, the Scouts brought radio men to check a Japanese airfield two miles south of Dinalungan River by Casiguran Sound. The runway was still usable. They recruited nearby inhabitants, some *Aetas* and *Ilongots,* to clear the runway of growth and debris.

Six men with Juan Pacis, Juan Berganio, and Nicholas Enriquez, led by their leader Herman Chanley explored the wide area of Cagayan Province before hiking to the coastal town of Baler to survey the area. The 7th Fleet needed a place to anchor, and their mission was to find out if Casiguran Sound was an adequate anchorage for the ships. The mission was successful due to the accompaniment of a guerrilla unit of 500 men with only a dozen weapons of old Enfield rifles, *bolos,* spears, and anything that could kill. Around 100 Japanese troops retreated from Baler due to the continuous harassments by the guerrillas.

Sumner came down with jaundice, so McGowen's team with radio operator Quipo moved to establish a radio station on the coast of Zambales. Traveling in a wooden boat, they arrived at the coastal village of Sta. Cruz near Palauig, a few miles north of the Iba airfield. The radio was set up and information was fed from the team's patrol with the guerrillas in the area of the national highway checking out roads and bridges from the village to the airfield. The guerrillas confirmed that around 1,000 Japanese patrolled the region and around 3,000 to 6,000 Japanese were spread out along the Capiz trail. The trail led to the northern part of Manila. Quipo remained by the radio busy transmitting reports on the Japanese in the path of the advancing Americans until Manila was retaken. He did not have much to rest after his return. Reports showed him leaving Zambales after ten days and boarding a plane with Scouts Asis and Siason to Magallanes south of Manila.

Krueger was aware of the infighting among the guerrillas and the delay it may cause to the liberation timetable. Cooperation of the guerrillas was needed to obtain vital intelligence and unnecessary loss of lives. He placed Nellist in charge of the guerrillas in Albay and Sorsogon until April when the 158th Regimental Combat Team (RCT) arrived. They were to contact a poorly equipped guerrilla group that, if adequately supplied, would be of great use in the upcoming invasion of Legaspi City.

The Chanley team had been in Legaspi City for two weeks by the time Nellist's team landed on the Legaspi peninsula to survey the area for beach landings and places where enemy movements should be avoided to insert the 158th RCT. At the village of Casiguran, south of the city, the team boarded a *banca* to the guerrillas' headquarters. An unfortunate incident occurred when the generator they were unloading to power their radio fell overboard and was lost. The guerrilla commander offered them his radio to send a coded message for an airdrop of a generator provided they also received weapons and ammunition.

Siason, Nellist and Kittleson crossed Camarines Sur on foot to San Juan. The rest went to Bulan in the south. At a meeting with the guerrillas, the leaders agreed to take orders from Nellist but not from another guerrilla. To stop the bickering, Nellist assigned each group to his sector with a radio. Major Russell Barros, an American Philippine Scout, was placed in command of the guerrillas on Camarines Sur. The infighting continued with some cooperation achieved in sabotage and intelligence gathering when the American forces landed.

Meanwhile on Luzon, while the battle to retake Manila was at its height, Ileto and his team set up road watchers to the north of Cabanatuan City between Guimba and Gapan on Highway 15 and 5. Bacat and Balambao operated with the guerrillas in the nearby Carranglan and Pantabangan areas.

After several weeks, the Ileto team boarded an amphibious plane to Camarines Norte to begin a 71-day mission, the most extended of the Scout missions in the islands. Since they landed offshore, the guerrillas provided a sailboat to bring them across the Philippine Sea and Lamon Bay to Cagbelete Island to check for enemy presence. The weather diverted them to Alabat Island, south of Cagbalete. They set up a radio and organized a guerrilla unit to monitor Japanese movements in Mauban across Lamon Bay. The Scouts provided training on intelligence gathering and military tactics. Being able to give orders and commands in the local language sped up the movement. The guerrillas badly needed equipment, so an airdrop over the village of Bagasbas in Camarines Norte brought over two hundred Springfield rifles, 50 machine guns, ammunition, rice, salt, flour, cigarettes, money, and medical supplies.

The Dove team called up Bacat and Balambao from the Ileto team for a special mission ordered by General MacArthur since they had no Filipino members who could speak the language. The Dove team had scouted Fuga Island with Anderson's guerrillas from Infanta looking for downed airmen. Now they were to return and evacuate the Sycip family who had been hiding there for the last year and were now suffering from American aerial attacks. They arrived at night-time in their rubber boats and obtained the help of Filipino fishermen to lead them to the Sycip's home. Civilians informed them of around 600 enemy soldiers on the island with many of them in poor health. Transporting the large Sycip family to the beach onto rubber boats to the waiting PT boats was slow. The sky darkened further from a brewing storm. Dove carried "Mrs. Sycip in his arms like a baby... She was an elderly Chinese woman, and her feet were bound," related Vischansky, a Dove team member.

Rounsaville's team surveyed Pila in Luzon for a month before Alfonso and Vaguilar, with two other Scouts flew to Tuao to find more about General Yamashita's hideout. Rumors of the general's location abounded, and many Japanese prisoners were interrogated. Guerrillas from the 11th Infantry Unit served as guides to check out rumored locations up to the northernmost tip of Aparri. After three weeks, the men reported Japanese presence and defensive positions in the areas but no General Yamashita.

In May, the Nellist team crash-landed on an airstrip at Manaoag in Isabela Province. A command post was set up in the town. The next day, Siason and radioman Apano with Nellist and another Scout hiked through the jungles to the junction of Cagayan and Magat Rivers. A radio outpost was set up and local people were recruited to provide information on the enemy. They sought shelter in a tobacco shed under a thunderstorm and heard the heavy sloshing of the enemy's shoes passing by. Crossing the Magat River, they entered the area of the guerrilla 7th Infantry, who turned over three Japanese and one Taiwanese soldier for interrogation. Local guides brought them to the town of San Mariano, where they learned of a large presence of the enemy between the town and Palanan in the north. The prisoners willingly remained with the team. They knew that they would not stay alive in the hands of the guerrillas.

The rest of the team, Asis, with two other Scouts, set up radio posts and recruited local people of the town of Ilagan to Cabagan in the north. A sympathetic representative of the Japanese Overseas Affairs Committee used his local contacts to set up an intelligence network east of the Cagayan River. Soon reports were coming in for relay to Sixth Army headquarters. The commanding officer of the guerrilla 14th Infantry requested the bombing of several villages garrisoned by the Japanese in Isabela. The residents evacuated the night before to avoid suspicion.

Reunited at Manaoag, the team interrogated more Japanese prisoners who refused to talk. Another airdrop to resupply the guerrillas brought machine guns with clips that did not fit. There were also none of the requested mines and bazookas, as well as socks. The team continued to send reports to KAZ and the 37th Infantry Division who were eliminating Japanese forces in the northwest Luzon area. When recalled back, they met six guerrillas and helped them set up their machine gun to fire on a hut with a dozen Japanese soldiers inside. The enemy returned fire and Nellist's was hit in his right thigh ending the war for him.

In July, Ben Mones with the Littlefield team trekked deep inland to the village of Sadanga in the Mountain Province to live with the *Igorots,* the native people of the area. The *Igorots* hated the Japanese intensely and beheaded them instead of taking prisoners. Men and women proudly exposed their tattooed arms and upper bodies, especially the men whose tattoos showed how many of their enemies they had killed. Radio messages that were mostly about the exposed body parts rather than intelligence reports continuously came through the transmission forcing the Air Force, who shared the same frequency, to step in and stop the comments.

When the Eighth Army took over the liberation operations, the Sixth Army began preparing for the invasion of Japan. The Scouts began training for Operation Olympic. But when Japan formally surrendered, the training classes ceased, and the Alamo Scout flag was brought down for the last time. It had been 21 months since the first of over one hundred missions began.

The team leaders received Silver Star medals and all enlisted men received Bronze Stars. John Hidalgo received the Silver Star. Sabas Asis, Thomas Siason, and Rufo Vaquilar received the Bronze Star medal for the Mindanao mission and the Oak Leaf cluster for the Cabanatuan rescue. Alfred Alfonso who was hospitalized received the Bronze Star and Purple Heart for being wounded during the Cabanatuan Raid.

There was no formal deactivation of the Scouts. They disbanded in November and were released back to the 1st Filipino Infantry Regiment at Leyte and told not to talk about their missions when they returned home. The Scouts were lauded for their accomplishments, but their operations were kept secret by the U.S. government until they were declassified almost forty years later.

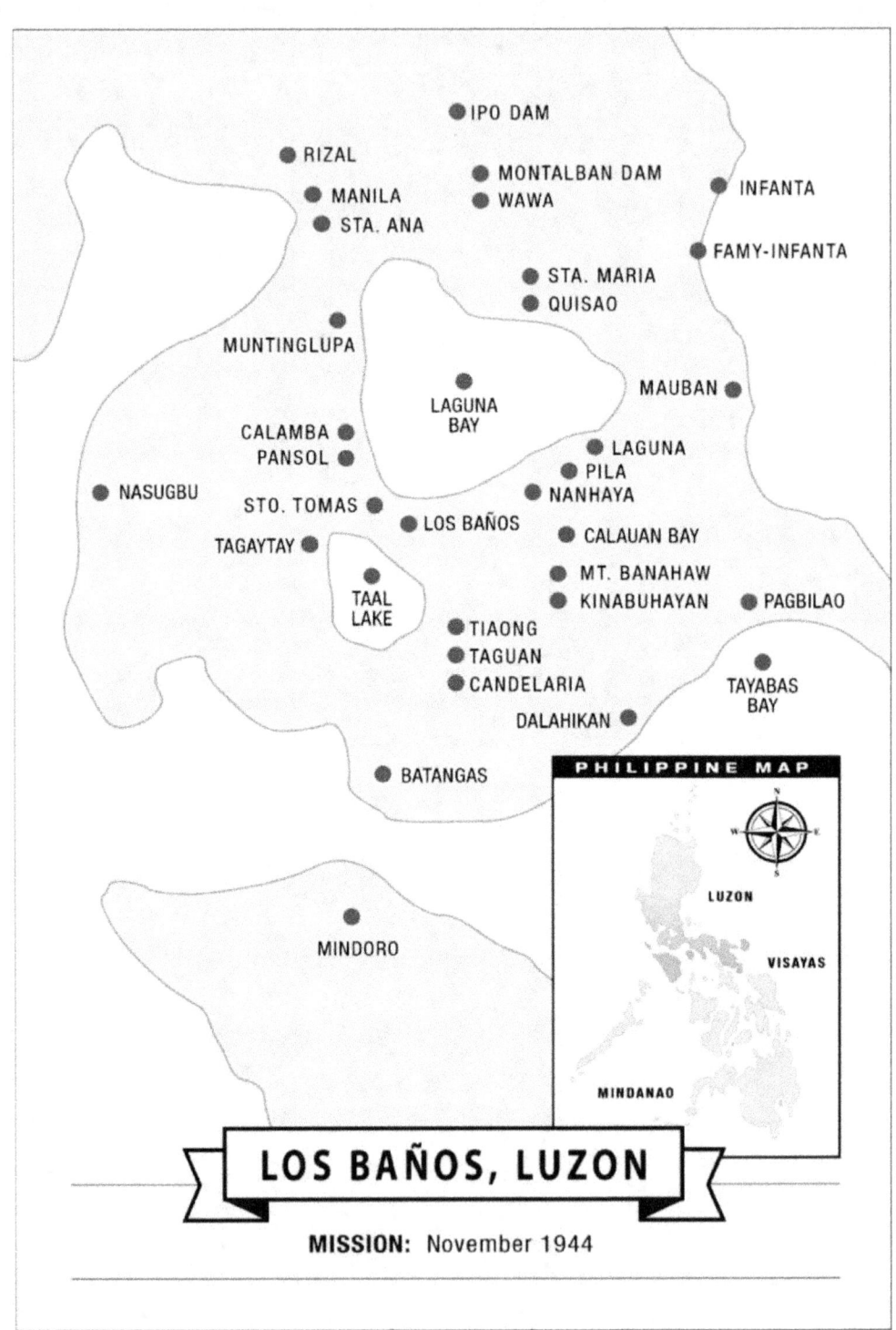

IPO DAM

RIZAL

MONTALBAN DAM

MANILA
WAWA

INFANTA

STA. ANA

FAMY-INFANTA

STA. MARIA

QUISAO

MUNTINGLUPA

MAUBAN

LAGUNA
BAY

CALAMBA
PANSOL

LAGUNA

PILA

NASUGBU

NANHAYA

STO. TOMAS

LOS BAÑOS

CALAUAN BAY

TAGAYTAY

MT. BANAHAW

KINABUHAYAN
PAGBILAO

TAAL
LAKE

TIAONG

TAGUAN

CANDELARIA

TAYABAS
BAY

DALAHIKAN

BATANGAS

PHILIPPINE MAP

N
W E
S

LUZON

VISAYAS

MINDORO

MINDANAO

LOS BAÑOS, LUZON

MISSION: November 1944

Luzon Island
Los Baños Prisoner of War Camp

The target date of the initial landing in the Philippines was pulled in from December in Mindanao to October in Leyte. The cargo submarines *Narwhal* and *Nautilus* were shuttling men, radios, demolition equipment, and supplies to various locations in different islands every month. To support the Leyte landing, two submarine missions arrived in the same location in the same month. But that was not enough. PRS added patrol submarines to bring more men and supplies.

USS *Cero* – November 1944

Sgt. Esteban Diputado welcomed the training at Canungra Jungle School after three months of 'fatigue work', including buying chickens and pigs from neighboring farms to feed the men at the camp. He said it was "mostly rugged training, hiking day and night" but admitted that "it was too strenuous for our age." Most of the men were in their mid-30s.

When orders arrived, twelve men returned to Camp Tabragalba to prepare for the Philippines. A three-day pass to Brisbane followed and amphibious training at Fraser Commando School was next. They underwent one more week of training and packing demolition equipment. Loaded into trucks, they received new gear at Milton Court in Brisbane. Diputado relived the many good memories of the camp, thinking that he might not come back.

They were given one hundred pesos each for emergency purposes only. At nightfall, they rode to the airport. Diputado was excited as it was his first plane ride. They sat close to each other as the plane's temperature inside dropped. Landing in Townsville airport to refuel, they were glad to be given coffee and doughnuts. The following morning, they arrived in Hollandia and met Whitney at GHQ on the hill. They were a demolition party first, then a reconnaissance group, Whitney reminded them. Informed of the mission, Diputado 'felt big'. After all the handshaking, a plane took them to Woendi Island, southeast of

Biak, where they stayed for three days sleeping in the same warehouse of equipment and filled their bags "'until heavy as lead." He wondered how the sub could take all the men with their packs.

Fifteen tons of supplies and sixteen Filipino and American soldiers boarded. They brought signal and demolition equipment, rations, weapons, and ammunition. Fifty cartons of cigarettes were distributed to the stevedores who loaded the cargo. Captain George Miller recalled "overnight, *Cero* was transformed into a guerrilla gunrunner. As a patrol sub, she had no beds for passengers. Off came all torpedoes save for two forward and two aft in the tubes for defense. Ships' carpenters swarmed all night aboard, building wood platforms in both torpedo rooms as a base for 17 tons of cargo consisting mostly of small-block TNT demolition charges. At daybreak, 16 army men filed onboard. ... orders were to land a hand-selected demolition squad onto Luzon, a few miles north of Manila, a hotspot." The sub cleared Darwin for her route to the Mindanao and Sulu Seas towards northern Luzon.

Maj. Jay D. Vanderpool headed the party of Filipinos and Americans. They were to reinforce earlier missions to Luzon with signal, demolition, weathermen, and to deploy specific tasks. Men from Coastal Artillery, Captain George Miller, Lt. Brooke Stoddard, and Lt. Charles P. Hope joined demolition men Sgt. Esteban M. Diputado, Sgt. Zoila A. Azcueta, and Sgt. Fortunato Santos. Weathermen were Sgt. Aton C. Chick, Sgt. William J. Males, and Cpl. Jose C. Baladad. The radio operators were Sgt. Venancio M. Mones, Pfc. Miguel B. Erana, Pvt. Hilario L. Lobendino, T5 Antonio C. Rimando and Pvt. Zoila Tolentino. Pvt. Carlos D. Udani was the lone 978th Signal in the party.

The radio broadcasted the American landing on Leyte while they passed through the Sulu Sea. The captain's log recorded a Japanese recon biplane sighted while passing between Mindoro and Palawan islands. The *Cero* stayed submerged for four days, surfacing at night off the Zambales coast, waiting for proper security signals that did not materialize.

The sub commander was ordered not to attack enemy shipping except a major warship until after the TNT and men were delivered. But the Advance Naval Operational Base gave an order to fire on a Japanese convoy with five destroyer escorts suspecting a submarine going in a southerly direction approaching them. The sub surfaced to battle stations. Deck guns fired at several small Japanese ships. One ship burned and two smaller vessels beached with the crew fleeing inland. The loud horn blared to dive, followed by hissing sounds as the sub dove. She maneuvered a steep dive to a depth of 380 feet to avoid depth charges and stayed "for six hours, which seemed like all night," according to Sgt. Santos. The men in their bunks began to slide off and roll around as the angle got steeper. Diputado recalled that they "sweated it out listening to the explosions around us." The party kept out of the way of the crew and stayed quiet. A helpless feeling swept through the men as waves of concussion from the underwater explosive charges crashed against the sub.

The crew were "nice to us," recalled Diputado. The treatment deeply contrasted from his experiences in the civilian world. They took turns sleeping in the crews' beds for several hours when they were on duty. "Real coffee 24 hours a day with toast as much as you want. Two meals a day with soup for lunch. Usually, four eggs for breakfast and turkey and ice cream on Saturdays. I helped wash dishes in the kitchen." Azcueta agreed that the food was the best in the service. To kill the monotony, Diputado started smoking, and the men sharpened their trench knives.

Proceeding north to San Esteban, the sub attempted to land for four successive nights. Shore signals were not seen. GHQ's orders were to land by the Masanga River at Infanta, Anderson's headquarters, in the east coast of Luzon. That side of the island was rough as it faced the Pacific Ocean. Diputado was seasick. When the correct security signals were exchanged from shore, the sub surfaced. Orders were given to prepare to disembark. The hatches opened and "men started coming out like worms" deeply breathing in the fresh air. "On the deck for the first time I saw the Philippine mountains after thirteen years," recalled Diputado. He could see boats coming, people waiting onshore to unload the cargo. The party and cargo transferred into the wooden boats sent by Bernard Anderson, guerrilla commander of the Infanta area. The sub recharged its batteries while the cargo was unloaded. Thirty-five tons of cargo intended for Volckmann in Northern Luzon was now diverted to Anderson in central Luzon. Another source stated fifteen tons. The *Cero* took onboard four evacuees before slipping out to sea.

Azcueta was shocked by "so much suffering, the living condition, food, and diseases." He now understood why so many kinds of supplies were loaded into the *Cero*. He wished he brought more. Amidst all the unbearable suffering, he was touched by the "undaunted spirits of the guerrillas with their dream to liberate their country". The guerrilla boys were in rags with tears in their eyes as they told him that they were the third mission to bring supplies but no clothes for them. Their clothes had rotted or eaten by termites as they may have been extra clothes buried and left behind.

Diputado's group was Hope, Santos, Azcueta, Baladad with Stoddard in charge. He was glad to meet "some of the boys ahead of us" who arrived in September at Anderson's headquarters. Their camp was set up nearby. It was crowded, so he hung his hammock in the doorway. He could hardly sleep knowing that he was back on the islands. They ate corned beef and sometimes a chicken that they traded for clothes with the locals. He saw that the people needed clothing badly. While waiting for orders, sometimes the men went to the villages with the guerrilla boys as they did not know the language of the people on Luzon. They may have been from the southern islands with different languages.

Agron (Luzon mission August 44) was at Andersons' camp with a convoy of carriers to get the supplies and weapons meant for the 11th Infantry Regiment under Volckmann. Enough rice was provided to feed the convoy for the twelve-day hike roundtrip. Traveling at night to avoid the enemy and through the mountains, the trip took a month to return because the enemy controlled the lowlands. Back at camp, he taught the guerrillas on the use of the new weapons – rocket launchers, grenade launchers, machine guns, Carbines, and cases of dynamite. One new weapon was incendiary pencils that when broken, the acid leaked out and ate through metal to cause an explosion. There was medicine for tropical diseases. Most important were the radios. But the guerrilla commander was dissatisfied that Agron did not know how to fix a radio. He was a recon man with infantry training and not in radio. Some recon men had radio operation training, but their main job was to provide support, protection, and intelligence gathering for the radio.

The arrivals were briefed on intelligence gathering, sabotage work, and enemy movements in the area. An objective of utmost importance was the cooperation to be obtained from the two warring guerrilla groups, the Hunters and Markings, in the vicinity of Manila. Their boundaries overlapped, causing continuous fighting and even killings amongst their men. Civilians suffered as they were forced to take sides. Their energy and support were to focus on fighting the Japanese instead of each other as the landing in Luzon was just weeks away. Without cooperation, there would be serious losses in material and lives at the time of landing.

A month after the party's arrival, Vanderpool sent a message by courier to Rowe's headquarters in Mindoro to relay to KAZ. "Contacted all the major guerrilla leaders– Hunters ROTC, Fil-Americans, Markings, and Castaneda. Will get radio from Rowe so can work behind lines. Established observation posts along southwestern Batangas coast. Have observation of Corregidor, Manila harbor, and all major junctions Cavite, Batangas and Western Laguna."

The party split into groups and assignments carrying equipment, medicine, and cigarettes. Some of the radiomen stayed in Anderson's headquarters, others were split with guerrillas to Hunters and Markings headquarters by the mountains located to the east and south of Manila overlooking Laguna Bay. They were to provide intelligence and tactical support.

G-2 Intelligence described the Hunters led by former Philippine Army Cadet Eleuterio Adevoso as one of the "most powerful guerrillas" in central Luzon. The organization composed of underage cadets of the Philippine Military Academy, formed in the Antipolo mountains to fight the Japanese and Huks. Philippine scouts and college students zealously increased their ranks. Although the directive from Australia was to lie low and gather intelligence until the signal to rise, the Hunters continued to harass the Japanese.

Intelligence obtained by the Hunters passed through several net control stations on Luzon via couriers until Vanderpool's group reached their headquarters in early January 1945 and set up a radio that directly reached KAZ. Sailing along the coast, hiding and avoiding Japanese patrols on Laguna de Bay, the group reached the Hunters' 44th Division headquarters south of Manila in the Laguna- Cavite-Batangas area. The only signal man, T5 Udani, was assigned to the Hunter's camp. Udani reported Vanderpool acting as an adviser forming General Guerrilla Command (GGC) with Adevoso to collect information on the enemy and relay to KAZ. Supplies were arranged from Australia and provided to the Hunters' operations in the field. For the next five months, Udani's group moved with an escort of Hunters guerrillas amongst several scattered units avoiding capture by the Japanese. The message to the guerrillas was to wait until the Americans return and to be ready to coordinate their work with them for a successful Luzon landing. Following orders meant arms and supplies brought in by the submarines.

The Hunters reported military targets in the Rizal area around the city, enemy defenses in Manila Bay, Japanese troop strength, gun emplacements, and movements of armored equipment in the city. When the Japanese bombarded the areas of Nasugbu and Wawa in Batangas, Udani punched the keys to report the results to KAZ.

In the meantime, weather observers Chick and Males from the 15th Weather Squadron moved to the weather network of Muri Hendrickson, who had arrived earlier (Luzon mission September 44) in central Luzon. They operated the weather outpost for five months and continued after Hendrickson was seriously wounded and evacuated.

Weatherman Baladad and demolition men Santos and Stoddard departed with Miller to the camp of Markings Fil-American Guerrillas led by Marcos Villa Agustin at Makantog mountains overlooking Laguna Bay. They arrived with aching bodies from the radio equipment they carried on their backs. With the pending Luzon invasion, they immediately trained men to blow up bridges. Stoddard placed dynamite and blasting caps far apart on a table to demonstrate. No smoking in classes. Do not drop or bite anything. He spoke with such a loud voice that the guerrillas agreed on whatever he said so as not to create trouble, increasing his frustration.

An estimated ten thousand trained guerrillas controlled the area of Markings with agreement from other units. They desperately needed supplies, weapons, ammo, clothing, and medicine, as only ten percent were armed. Their only source was from Anderson and that required fifteen days of carrying on the backs of men and over mountains. Miller requested an air drop of ammunition and supplies. The air drop had missing radio parts and guns that had to be cleaned from grease which took time from practice use. The parachutes were salvaged into blankets for the guerrillas.

Markings fully cooperated with the 'lie low' policy although he was tempted to retaliate when his unit of only 400 men and dozens of prisoners were relentlessly bombed and strafed for several hours in mid-1944. But they were weaponless to retaliate. Over a thousand Japanese troops, a cavalry detachment and mountain guns encircled his area while planes bombed and strafed.

KAZ received reports on the demolition work by Santos, Baladad, and Stoddard, results of American raids in Manila and rescued American fliers. Fishermen rescued pilots over Laguna Bay pushing their faces down in the banca to avoid the Japanese seeing the blond hair. They were brought to Markings camp to be cared for before proceeding to a submarine pickup at Infanta, Anderson's camp. By December, twenty- two US pilots shot down over Luzon were returned by the guerrillas.

The signalmen tapped out Miller's report of "excellent coverage of Manila, Rizal, and Laguna." Radio operators passed information from Markings guerrillas on coastal defenses in Batangas, south of Manila, and guerrilla activities in the Laguna-Rizal provinces around the city. Couriers brought maps of Manila with grids to pinpoint enemy areas to KAZ. Guerrilla fighters and farmers who knew the terrain between Anderson and Markings camps risked their lives as runners hiding in carabao wallows. When malaria attacked, they huddled beside trails, then pressed on buoyed by the thought of the coming liberation.

In mid-December, General Yamashita held a conference with generals and admirals of his Army in Sta Ana, Manila. A secretary was present to take notes. The meeting opened with "We must not expect any reinforcements from the mother country. She has to have all the forces she needs for her defense. The convoy we expected 06 December cannot cross the sea due to lack of proper convoys. This will last for about 2 months only, and it is up to us to put our brains together to stop the ever-increasing enemy . . . If retreat is imminent, we can use two routes easily, one through Rizal and the other through Cavite... we move to the north, permanently staying in Bataan. We shall have to follow in the steps taken by MacArthur when he withdrew his forces from the south... (if) the enemy strikes in the north it will be hard to move our forces from the south, and we shall find ourselves sandwiched in the city. Retreat through waters (Manila Bay) will be useless." Details of the plans and maps on the defense of Luzon were to be laid out in the next meeting.

Miller, who was in the vicinity of Manila with the Markings guerrillas, urgently sent details of the meeting to KAZ. He said that the agent who gave him the information was a Filipina assistant secretary to General Yamashita and absolutely reliable as she was formerly the secretary to Admiral Hart, commander of the Asiatic fleet before the war began. At the next meeting, she was going to try and secure operation maps. But after delivering the information, she was caught and killed, another unknown and unsung hero of the war.

A second meeting was held ten days later to furnish each of General Yamashita's staff maps of provinces showing divisions of their forces. It was reported that "a secret microphone recording of General Yamashita gave the U.S. forces details of the Japanese plan for the defense of Luzon." Signal men hiding in Manila recorded and transmitted details of the meeting. Manila was mentioned "that in case evacuated, place one infantry (Japanese) regiment..." They had not received supplies from Japan in the past five months and did not expect any in the future. They were saving their ammo. "We cannot depend on the Filipinos, although we will need 100,000 for emergency work." To enable the troops to live off the land in Mindanao, the Japanese residents in Davao were to provide support for a lasting resistance.

In early December, an order arrived for Hope, Azcueta, Diputado and Lobendino with three guerrillas as security to find a guerrilla leader near Mauban east of Manila as there was a large Japanese garrison. Their objective was the town of Pagbilao by Tayabas Bay, southeast of Manila, passing Mauban town heavily guarded by Japanese patrols and Filipino Military Police. The MPs were feared more than the Japanese as they could not identify them as guerrilla or civilian. As they waited for guerrillas stationed at Mauban to provide men, the guerrilla camp was raided. Civilians were recruited to help carry the equipment to sail south, as overland was impassable. All equipment was loaded but the large waves brought the sailboat close to the reef scaring the seasick men to return. It took days to dry the equipment and fix the torn sail. The second attempt was also unsuccessful, and more equipment had to be dried out or thrown away. After waiting a month for good weather, the party split with the demolition equipment into two sailboats in case one was captured. Hope and Azcueta, with three guerrillas were in one boat and Diputado and Lobendino, were in the other. Clearing the reef, they sailed for six hours amidst seasickness and vomiting. They passed a Japanese garrison.

The moon was bright, but the pilot lost his bearing and went in circles in the Japanese-infested waters. The boats could not see each other. They passed another Japanese garrison at Dalahican Point. Diputado was taking out his compass when a sail knocked into him, cutting his eyebrow and almost sending him overboard. Landing near Mauban, they evaded two Japanese patrol boats anchored ashore. The area was raided earlier, so there was little food to find. Japanese patrols near Mauban were raiding villages and burning houses, forcing them to take longer but safer routes. The guerrillas who met them shared whatever food they had. "C-rations did not last long. Two cans of C-ration mixed with a pot full of native potatoes fed thirty men for a few days until the canned goods were gone. Money was worthless with nothing to buy. Men ate twice a day or none at all... sometimes they ate monkey and snails." Down to eighteen men, they camped for three days until more carriers could be found in the nearby villages. People were afraid to travel farther than their village. But they willingly told of Japanese presence in their area. Still short of men, they carried whatever they could and the clothes they wore. The carriers got three pesos a day and food "if we get any. The guerrillas that joined us were not paid as regular guerrillas. I was in charge of mess and food procurement," Diputado said. They took a longer route through the mountains to avoid Japanese outposts led by a guide who was a pro-American Huk and knew the mountain trails. The soles and heels of their boots ripped off as they trekked through rugged mountains and crossed rivers over a dozen times, sometimes on their hands and feet. Many times, they were knee-deep in mud with the heavy cargo. It rained daily.

"Leeches stuck to many parts of the body, in one man's eye and on his urinal." The men continued to starve, eating only coconuts along the way. They made "stews of half a sweet potato and papaya flavored with monkey meat or Kalaw bird, eating it for 3 days" according to Diputado. Azcueta lost a can full of blasting caps along the way. Hope lost a khaki shirt hanging to dry. Lots of clothing was lost.

As they approached the lowlands, they saw Laguna Bay and coconut trees everywhere. Diputado signed a receipt for money to buy wood and supplies. He had 20 men to feed including guerrilla and Philippine Army men. They had to cross two provincial highways. Although the guards at both ends of the highway were civilians, they could not be assured that they were loyal. Three groups crossed separately in case one was detained. Diputado was ahead with twelve men and a guide. Hope's group followed fifty yards behind. Azcueta and the rest followed in the rear.

Azcueta recalled that "in the lowlands, the civilians willingly fed them and provided information on the enemy. Some of the men had stomach trouble as they were no longer used to eating pork and ate only two meals a day if there was food. They were an attraction whenever they met civilians." A guide led them to their assigned area avoiding enemy camps. At an evacuation area of civilians, they traded medicines for salt, rice, chicken, and *camotes*.

After three weeks, they reached the camp of President Quezon's Own Guerrilla Group (PQOG) at the foot of Mt Banahaw. Huks in the area ambushed the men and would have wiped them out if they did not see Hope, an American. They wanted their weapons. Hope met with the PQOG major and the Huk leader to agree in helping each other. That produced help in carrying the supplies and armed guards. In the area of Kinabuhayan, the guerrillas built three wooden buildings with rooftops of cogon grass for the party. The kitchen and mess hall were close to a creek. The officers' quarters and the guerrilla barracks were close by. The civilians were eager to provide information on the enemy and shared their rice, meat, and a few eggs. Once the radio was set up, the demolition equipment was laid out to begin training on handling the explosives.

Christmas was welcomed with a celebration by the nearby village residents. Diputado recalled, "We bought pigs, the PQOG leader provided rice, and the girls danced with us all night."

Luzon Island, Philippines Landing – January 1945

Three days before the Luzon landing, a Generational Field Operational Order from the Sixth Army headquarters commanded the start of "harassing and delaying all movements at all costs of enemy personnel and vehicles in the eastern sector of Rizal and Laguna provinces." The following day, F. Santos went to blast the embankment and road at Km 65 known as 'Pugad Lawin' to delay military movements of the enemy. A group of Hunters guerrillas was his security. Six Japanese tanks and two trucks approached as he was preparing the last charge. As the vehicles passed, they set off the charge and watched the explosion rip all the vehicles. They fled before the Japanese could retaliate. The following day, fifty Japanese soldiers and Filipino collaborators came looking for them. From their secret hideout in the village of Quisao, Pililla in Rizal, they saw two of their guerrillas cooking their noonday meal in a hut. They were captured and executed by the Japanese, and their hut was burned. Santos told his men to remain quiet unless fired upon as they heard the enemy passing their hideout.

An Alamo Scout team reported being in Laguna with around 1,000 guerrillas that had only 2 bazookas. They gave another bazooka and showed them how to use it. But the guerrillas kept asking for more ammo until they found out that the bazooka was used to shoot at individual Japanese instead of aiming at groups of the enemy. A quick training was held.

The Maytulay bridge was demolished at the end of January by Santos' team. A defensive line was set up in Sta Maria, Laguna, with a few guerrillas on the road by the hillside close to the bridge. A combat unit of the Markings attacked the enemy garrison in Sta Maria. Several hundred Japanese soldiers approached the line of defense. After an exchange of fire, the team pinned them down across in a coconut grove. The town mayor confirmed nine Japanese killed and several wounded. Santos' team withdrew back to their hideout in the mountains. A Japanese patrol of thirty-five men followed their trail and shot mortar shells to scare them.

Demolition work on the Famy-Infanta National Road between Laguna and Tayabas was Santos' team's next mission. They ambushed a truck loaded with Japanese soldiers and supplies on the Sta Maria road to Famy. All were killed except two who escaped. They set fire to the truck and dumped the supplies into the river.

In February a conference between the commanders of Markings and Hunters, the two largest Filipino led guerrilla groups in Luzon was held to cooperate and unite towards fighting the Japanese. The peace terms were not accepted, and both began planning further attacks. A Japanese punitive expedition with tanks and mortars arrived, and the guerrillas scattered, erasing any hope of cooperation. Anderson was successful for only a short time in extracting a peace agreement. Miller with Markings stepped in to unify. With an officer of Hunter's, the Rizal-Eastern Laguna command (RELCO) of the area was set up with Miller as the adviser. He combined the staff of the two guerrilla organizations. In agreement, RELCO directed the two guerrilla groups in their demolition and sabotage work in the liberation plans of Manila. A message to KAZ advised that the guerrilla forces had supplied the U.S. 43rd Division after landing with twelve hundred men to aid in the liberation. More would have been given if arms had been available.

A bridge at Pagbilao was to be destroyed but the leading roads infested with pro-Japanese were difficult to cross. The radio did not work to inform Anderson of the bridges to be destroyed so a courier was sent on a three-day hike to Holgado's radio in Batangas (Samar mission June 44) to relay the changes, but he was captured and killed. Instead, the bridge at Taguan and the railway bridge between Candelaria and Tiaong were blown up.

Diputado's and Hope's teams planted charges to destroy a steel bridge with explosives attached at both ends. Azcueta and Lobendino, with their men, planted charges under the wooden bridge close by. Three explosions destroyed the wooden bridge and loosened the last charge on the steel bridge although it failed to explode. The following night, they returned to finish the steel bridge with Japanese swarming the area. They found food near the wrecked bridges. They walked for four hours to the next village to avoid any Japanese reinforcements. From there, they climbed two thousand feet to camp. Diputado said he felt ten years older when they reached camp and collapsed in one of the three huts built by the PQOG guerrillas.

Japanese activities in nearby towns were reported. An enemy ammo dump on the hills of Burol fifty yards away was reported by guerrillas to KAZ. A week later, they saw its explosions from American plane strikes.

The Japanese did not raid their outposts at night as they preferred to stay in their garrisons. The guerrillas successfully ambushed them in the daytime as they foraged for food without an advance scout or following a tactical formation. Azcueta noted that a determined ambush party of six can destroy an enemy patrol of twenty or more men.

In early March, after five months in the mountains, Hope, Azcueta, and a guerrilla commander moved to a new headquarter near Mauban to reconnoiter Japanese activities in the area. Diputado was sent to the lowlands with a dozen guerrillas and a guide to get food supplies to last for two months because the Japanese had increased in the area with the American troops not far away. They avoided a Japanese patrol following orders not to engage as civilians would be killed and houses burned in the area if they encountered a fight.

Hope, who had reported to the Sixth Army Intelligence section in San Fernando, La Union arrived in camp. The night before, a "barbeque party... as farewell to the guerrilla boys" was held. Diputado reminisced that everybody liked Hope's management and wanted to go along with him but he could not take them all. In appreciation of the guerrilla boys support, they taught "about the U.S. Army, the only compensation to them when they return to their own organization under Anderson." At Pila in Laguna, they took a wooden boat to Binan on the other side of Laguna Bay and waited for a truck to bring them to San Fernando. They were exhausted. They slept and regained their strength for a couple of days before proceeding to the Sixth Army headquarters at Calasiao passing burning towns and wrecked bridges. Two Japanese planes raided the nearby camp as tracers lighted up the sky and anti-aircraft guns opened up. For three hours, Diputado said, they kept running between their tent and foxholes. At Lingayen, they took a plane to Leyte where they waited another week for a seven-day trip "back to god-forsaken jungle." He must have preferred the jungles of Luzon to that of Hollandia. His only regret was he did not have enough demolition charges to do more damage. Azcueta was proud of his demolition work and recognized that the guerrillas who helped them throughout the mission deserved credit.

Santos was with several men and women of Markings who were on their way to report to their headquarters at Cardona, Rizal. While approaching the mountains west of Sta Maria, Laguna, the guerrilla outposts warned of Japanese deserters in the neighboring village of Maybunga. The guide led them to the Japanese' hideout. "We captured them as prisoners of war. Later, we captured a Filipino collaborator and took him with the other prisoners to Cardona, Rizal, across Laguna Bay." At the American camp, the prisoners were questioned further. Santos was recalled to the Special Intelligence Section of the Sixth Army.

Los Baños Raid – February 23, 1945

The rescue of the POWs at the Los Baños Camp became an urgent priority after the news of the recapture of Manila and the indiscriminate killing of innocent residents. The Palawan massacre of prisoners three months earlier was not to be repeated, declared GHQ.

The chance meeting between a grower from Mindanao who passed Los Baños on his way to Manila to get medicine for his wife and Col. Henry Muller, 11th Airborne Division's G-2 Intelligence, harnessed Muller and his staff to immediately begin gathering information on the camp and enemy forces in the surrounding area from the guerrillas and civilians. The concerned grower provided what he saw of the camp.

The Los Baños internment camp had over two thousand civilians, including Navy nurses and many religious orders. The camp was in the Philippine Agricultural College and Forestry campus forty miles south of Manila by Laguna Bay. It was outside the path of the Sixth Army's advance to Manila and behind enemy lines. The prisoners' barracks, each holding 16 cubicles with 6 people in each, and huts were surrounded by three rows of barbed wire fences and four rows of Japanese sentries. Many were priests and nuns calling the place 'Vatican City'. The prisoners were dying of starvation as later autopsy showed pieces of wool blanket and leather belt were found in their stomachs. Food was purposely kept from them to starve them to death before the camp could be liberated. Catholic priest William MCarthy, an internee at the camp wrote, "The struggle for survival forced us to eat weeds, flowers, vines, salamanders, the pulpy inside of banana trees, and juicy black bugs. Deaths mounted to two a day in January 1945".

U.S. forces and around three hundred guerrillas belonging to the Hunters, sympathetic Huks, and Markings units joined forces. The planning began when Filipino Alamo scouts Alfonso and Vaquilar with Rounsaville were sent to the Los Baños area to set up radios around the village of Pila, by the coast of Laguna Bay. Several thousand Japanese were garrisoned in the area. Maps around Mt. Banahaw, southeast of Los Baños were sent to Vanderpool and the planning group. Acting as adviser to the Hunters, he coordinated plans with the Eighth Army and the 11th Airborne. The General Command Group (GCC) he formed with Adevoso coordinated with the various guerrilla groups to provide support in the rescue. The guerrillas had been watching the prison camp for some time and reported regularly to the planning group. A unit of unsung heroines were the 45th Hunters agents led by Lt. Cristina Figueroa. They smuggled intelligence hidden in their underwear as they passed through enemy checkpoints.

Gustavo Ingles, a high-ranking officer of the Hunters worked with the guerrilla groups including the President Quezon 's Own Guerrillas (PQOG), the Chinese guerrillas of Luzon, and the Huks. Guerrilla leader Romeo Espino aka Col Price of the PQOG expressed concern that if the American Task Force did not leave a garrison in the camp after the rescue, the Japanese would release vengeance on the population. The rescue plan became more critical when a disturbing report arrived of a Japanese excavation large enough for a mass grave.

Two days before the attack the people around the Los Baños campus area were warned to move to the area of Mt. Makiling. A radioman with the 511th Signal Co. sent an urgent message from guerrilla leader Espino to Vanderpool. "Have received reliable information that Japanese have Los Baños scheduled for massacres."

The first rescue plan that was to be composed of only the guerrillas from different areas was aborted. The main concern was the guerrillas might not be able to hold the camp long enough to evacuate the internees before Japanese reinforcements arrived from their garrison in Lalakay only six kilometers away. Another problem was how to evacuate the prisoners safely. A much larger force was needed for the raid and during the evacuation. At that meeting, the protection of Los Baños folks from Japanese retaliation after the raid was raised and dispelled with "an assurance of raids on Japanese concentrations after the (Los Baños) raid would be conducted." GGC was to provide arms, supplies and reinforcements.

The U.S. 11th Airborne Reconnaissance did not have enough troops assigned to the operation. The paratroops were only half a regiment division that had been depleted by casualties, illnesses, and diseases. Their other units were fighting in the battle of Manila.

Filipino Terry Santos was with the 11th Airborne reconnaissance platoon, a small highly trained group to be deployed as deemed necessary with no questions asked. On paper they were assigned other duties to maintain the secrecy of their assignments. He headed 35

The Amtrac vehicles loaded with the rescued prisoners from the Los Baños Prison camp.

volunteers to reconnoiter at night the areas surrounding Manila that would be the path to Los Baños. The guerrillas provided the reserve of troops.

Santos was the lead scout and led the assault teams. He was originally from the 1st Filipino Infantry Regiment at Fort Ord. Although born in Hawaii, he was also fluent in Tagalog which made the questioning of guerrillas they met along the way easier. He had passed the Alamo Scout class in New Guinea but was impatient to be deployed into action, making him join the 11th Airborne Reconnaissance platoon.

With his commanding officer, Santos crossed Laguna Bay to Nanhaya to reconnoiter the position of the fifty-four amphibious vehicles (Amtrac) and locate a drop zone for the 11th Airborne paratroopers. It was a ten-mile hike to the prison camp from the beach. Along the way, they reported Japanese machine guns watching Laguna Bay which they crossed, and artillery guns at the Los Baños wharf overlooking the shoreline and approach to the Lake. Enemy soldiers with guns were in the hills and could detect movement south of Manila where the raid was to take place. The Imperial Japanese Army's Eighth Infantry Division and artillery were the biggest threat on the road to Los Baños.

They identified the best route to take from the drop zone to the camp, guardhouses, two hidden machine gun pillboxes made of dirt, and location of the 7 AM Japanese exercise drill. The paratrooper drop zone was in an open rice and corn field next to the camp without railroad tracks and voltage lines that might hang the paratroopers.

The day before the raid, Terry Santos led 30 men and twenty guerrillas to cross Laguna Bay separately to avoid detection to the village of Nanhaya. Santos in one banca with five men saw a Japanese patrol boat approaching. He recalled telling the helmsman to speak only Tagalog and only what he told him to say or else he would be the first to be shot. He turned to his recon men to "stay low and he will fire the first shot if they tried to board us." One submachine gunner was to aim for the boat's waterline. If that did not stop them, throw grenades at them. The men sighed with relief when the Japanese officer accepted the helmsman's explanation of fishing late because he had a good catch. The entire mission would have been cancelled with a shootout. Then the wind died, and they had no paddles. The 3-hour crossing became 10 hours.

At Nanhaya, Santos met with his commanding officer, the Hunters ROTC officers, and two escaped internees to finalize the time of attacks on the camp sentries. The escaped internees smuggled instructions to the prisoners to prepare for the rescue. They had provided sketched maps of the camp and the surrounding areas for the recon teams and the guerrillas. The chosen time of assault was in the early morning when the Japanese were exercising in their loincloths and their weapons locked inside their barracks.

The night before the raid, Santos hiked close to the camp with three recon men and a dozen guerrillas. A local in front guided their way. Santo recalled he had a Garand rifle for more accurate shots at the sentries in the guard towers. Two bandoliers swung across his chest and six hand grenades hung from his belt. After eight hours, they arrived. He noticed that there were fewer guerrillas. It bothered him and thought they might have stayed in the forest to ambush any escaping Japanese later.

The following early morning, the recon platoons split into four assault groups. Santos led one. He crawled with his three men on their bellies to get in position close to the two pillboxes guarding the camp. Machine guns jutted out from bamboo reinforcements.

Meanwhile, the guerrillas had cut several hundred yards of Japanese communication lines in preparation for the 11th Airborne paratroopers' drop. They warned residents surrounding the area to leave securing the drop zone. The paratroopers were to be guided to the prison camp with guerrillas guarding the highway north and west of Los Baños.

Guerrilla units secured the beach landing sites for the Amtracs that were to carry the prisoners across the Laguna Bay and the evacuation routes. Guerrilla units of Markings under Col. David Estrella and the 48th Chinese Squadron under Col. Ong set up roadblocks in the towns of Calauan and Pila to delay possible Japanese reinforcements. Col. Emmanuel de Ocampo of the Hunters 47th Regiment covered the Calamba-Pansol area.

The signal to attack was when the first paratrooper jumped out of the plane, which meant hearing the planes. Prior to their roll call, a prisoner recalled seeing three planes flying so low that it almost touched the roof. That was seven in the morning. They saw six more planes dropping white sheets and realized they were parachutes. Prisoners thought they were food drops. A few claimed to see a fighter plane fly over with the word 'Rescue' painted on its side.

They laid overnight until Santos saw the parachutes over the drop zone and rushed into position. That was the start of the twenty minutes it would take for the paratroopers to reach camp and Santos' recon men to creep close to the camp entrance. Suddenly, a Japanese sentry hunting in the bushes fired at an animal. A guerrilla mistaking the shot for the signal to attack, leaped out, and hacked the man to death. Hunters' men opened fire. Santos' men rushed to the bank of the creek outside the camp.

Not quite in position, the Hunters with Santos' recon men charged the main gate, killing two front gate sentries and destroyed three stone pillboxes with a grenade and automatic weapons. Flat on his belly, Santos crawled closer, knowing that the machine gun barrels could only go down so far. As close as he dared, he threw a phosphorus grenade and a fragmentation grenade blowing the guns apart. He threw more grenades at another machine gun from a higher area that rained bullets at him. A screaming soldier ran out, and Santos finished him off. Some fled to the nearby jungles chased by the guerrillas and hacked to death. Guerrilla *bolos* sliced through the air. Grenades blew up buildings. Japanese soldiers were gunned down during their morning exercise, scattering them to get their weapons locked inside their destroyed barracks.

Upon hearing the gunfire, prisoners dropped to the floor and hid, not knowing what was going on. A prisoner recalled it was the longest thirty minutes on his belly. Some went to get their belongings. Nuns began praying.

When the paratroopers arrived, the remaining Japanese stragglers were finished off and a security perimeter was set up around the camp. Most of the Japanese sentries had been killed by recon men or guerrillas in the first few minutes of the raid. The shooting was over in 20 minutes, and no prisoner was harmed.

Proud of their participation, a guerrilla hoisted three flags on top of a barrack: "The Stars and Stripes, the Philippine national tricolor, and the Hunters 45th Regimental banner."

The Amtracs not far behind smashed over the barbed wires and through the gates to pick up the prisoners who could not walk. Some were afraid, believing they were enemy tanks. Torching their barracks convinced them to board the Amtracs. Col. Whitney with an interpreter and two crew members emerged from an Amtrac to get documents from the camp's headquarters building. Internees continued to board the Amtracs as the rest walked through the woods to the lake. A machine gun in a hut fired at one of the vehicles on the way back to the lake. A return fire blew up the hut.

Santos' recon men, Airborne infantry, and guerrillas provided flank and rear guard until the last Amtrac crossed the Bay. Engineers advanced to their roadblock positions laying mines, preparing charges, and down trees to block the roads the Amtracs did not use. A machine gun covered each roadblock in case of Japanese arrival.

The following morning Santos' men were pinned down by a machine gun by a large tree along the highway, and two were wounded. They fired back and destroyed the guns. His faithful guerrilla guide and four others were killed. Two were from Hunters. Santos with Martin Squires were the last two Recon/Ghost Platoon members to ride in the last Amtrac across the lake. Japanese bullets whizzed by from the hills. One prisoner recalled seeing the planes flying overhead as they crossed the Bay. Together with the internees, they landed at Calamba and boarded trucks to New Bilibid prison where fighting off the Japanese from the prison wall was still ongoing. That night was their first hot meal in a long time. Fresh bread with butter and bean soup was served to help their unaccustomed stomachs.

Seeing the flags on top of the barracks must have certainly angered and humiliated the Japanese. Less than 10 miles away was the Japanese Tiger Division with eight to ten thousand men.

There was a tragic and sad ending to the successful rescue. After all the Amtracs crossed the lake, the Japanese returned that same night to avenge the death of the over two hundred Japanese at the camp. Aided by the Makapili, a semi-military unit composed of rabid pro-Japanese Filipinos, civilians were bayoneted or decapitated. Houses were burned in the surrounding area. When the paratroopers returned the following day as the area was under the 11th Airborne's jurisdiction, they found "whole families tied to the posts supporting their houses then the houses burned collapsing around their former inhabitants." More than fifteen Filipinos were killed, including the murder of an American family that lived near Los Baños and were not interned in the camp. The civilians had been warned to leave before the rescue and right after. Some groups of guerrillas who were present tried to protect the civilians in their area. Around seventy civilians who ran to the Chapel to escape the rampaging Japanese were bayoneted inside before the building was set on fire. "The understanding was that the 11th Airborne was to occupy the area after liberation. But they were ordered back to Fort McKinley in Manila and the 1st Cavalry was to occupy Los Baños. But only the road from Calamba to Sto Tomas in Batangas was occupied. Los Baños and the rest of Calamba were left wide open for a Japanese operation, a free territory for massacre," wrote Gustavo Ingles, an officer of the Hunters 45th Regiment after the war.

There was an order from the 17th Infantry Regiment of the Japanese 8th Division "to kill all guerrillas, men, women, and children in Los Baños" and "to prepare to kill 100,000 men or 70 men for each of us. Also, each man must destroy one tank before he dies." This was the same Japanese division that was less than 10 miles away that was mentioned in the planning of the rescue.

Gustavo Ingles, a coordinator with the 11th Airborne, wrote about his comrades' bitterness at the loss of civilian life in the aftermath of the raid and how relatives of those massacred had not forgiven them. There was assurance that a message to the 11th Airborne headquarters confirmed the civilians' safety after the raid. To show "the validity of his intention," the U.S. radio operator with Espino was ordered to relay "any messages coming from the 11th immediately." It was assumed that the standard operating procedure occupying an area after liberating the internees would be followed, and "adequate protection would be given to the area residents after the operation." But the protection of the civilians was missed or not clearly defined during the planning of the raid. To explain to the relatives why they left right after the last Amtrac, Ingles showed orders for mopping up operations of the guerrillas out of Los Baños to another area.

The Japanese supply warrant officer, Sadaaki Konishi, who actually ran the camp, was able to escape the American raid unharmed. He was identified as having a part of the reprisal in the area. After the war, Konishi was implicated by certain Filipinos, tried for his crimes, and executed as a war criminal in June 1947.

It was a successful rescue of over two thousand civilians. But it was buried by the news and dramatic photo of the flag-raising atop the extinct volcano Mt. Suribachi in Iwo Jima.

Capture of Ipo Dam – May 1944

The Sixth Army ordered the capture of Montalban dam from the Shimbu forces at Wawa. It controlled water sources to Manila and was close enough to launch an artillery attack on the city. Alamo Scouts Berganio, Enriquez and Pacis of the Chanley team were sent to explore the area for three days before any military groups arrived. Unknown to the Army, the dam had not been providing water since 1938.

At Ipo dam, agents reported that the Japanese had planted demolition charges. An enemy strength of around five thousand was entrenched in the area. Markings guerrillas had reconnoitered both sides of the dam. Demolition men Hope, Diputado, Azcueta, and

Santos with around three thousand strong guerrillas split into two forces. On the day of the final assault, the guerrillas with the demolition men lead elements of the U.S. 43rd Division to the dam attacking the Japanese in the south and north of the dam. The men defused the huge Japanese explosives in the powerhouse ready to be activated. But "to assure the security of the dam and the aqueduct to Novaliches reservoir, remnants of the Kawashima Force west of the dam" were destroyed. They raised the American flag over the dam's powerhouse.

By July, the men were recalled as they had accomplished their objectives.

***

It had been three long years since the 1st Filipino Battalion was formed in 1942 and rapidly expanded to the 1st Filipino Infantry Regiment. The men answered orders of assignments to various units – The 1st Reconnaissance Battalion (Special), 978th Service Company, PCAU, CIC, Alamo Scouts, and the 6th Rangers. Now recalled to their units' headquarters on Luzon, or Hollandia, some continued training to join the invasion forces to Japan. Although Japan seemed defeated as she continued to lose battles by mid-1945, surrender was still not an option. The war had still to be won.

Regimental flag presented to the 1st Filipino Infantry Regiment July 1943. The same four men carried the flags throughout the war. They continued to carry the flags in parades and veterans' events after the war.

XXII

Aftermath

Although the inhabitants of the Philippines were a diverse people of various religions, classes, languages, customs, provinces, and islands, the resistance movement united them to combat the brutal invaders. The Japanese could not completely eradicate the people's loyalty and their increasing belief in their liberation as the American Forces drew near. But by the time of the landing, the people were in a most desperate situation, starving, sick, and dying. Many had turned to the enemy to survive.

Without the guerrillas, the Philippine campaign would have taken longer, required more resources, and cost more in terms of human lives. The guerrillas' morale had to be kept high to fuel and continue resistance. And the submarine missions that arrived over two years before the Americans' return did just that.

Nineteen submarines penetrated the islands bringing approximately 1627 tons of cargo and men from January 1943 to January 1945. But this was never enough to supply the needs of the resistance. Yet just the sight of the supplies gave the guerrillas and people hope and the impetus to keep resisting.

The Allied Intelligence Bureau (AIB) sent the first six secret submarine mission with Filipino soldiers to penetrate enemy lines and make the first contact with the underground resistance. These twenty-six Filipinos were recruited from people stranded or living in Australia and thus trained as the most qualified for the missions. They spent the longest time in the Philippines and were there until Japan's surrender in 1945. The Philippine Regional Section (PRS) sent two dozen more submarine missions with radios, supplies, and Filipinos from the 1st and 2nd Filipino Infantry Regiments trained in combat and radio operations until January 1945. On the submarines' return trips, 500 people were evacuated.

The radio was probably the most important equipment brought to the guerrillas. Arms and ammunitions elated the guerrillas but the "radio transmitters were far more valuable" according to Robert Lapham, Luzon guerrilla commander. Aside from being able to communicate with each other, guerrillas and their intelligence network were able to send "crucial information on the composition, disposition, and strength of Japanese forces, as well as weather and terrain data to SWPA headquarters." A report to the U.S. Congress after the war, praised the information-gathering efforts as having significantly contributed to the Philippine Resistance. GHQ had set their plans to return to the Philippines from the thousands of radio transmissions of the Japanese activities.

The Filipino mission men were behind the scenes receiving and transmitting intelligence messages and requesting supplies the guerrillas' needed to continue fighting the enemy. They provided military training to the guerrillas, advice on planning deployments, and intelligence gathering. They helped in the evacuation of prisoners of war, downed air pilots, and sunken ship survivors. In the execution of their mission, they braved rough seas, arduous mountain trails and forests, and passing through towns and areas infested with the enemy. Many a times, starving and racked with disease, they questioned their situation. They joined to free their Motherland but at such moments they just wanted to survive the war.

Two of the bloodiest campaigns of the Pacific War as to the number of U.S. troops killed in ground action were in Luzon and Leyte. More would have been killed if not for the guerrilla resistance. There is no official count of the many guerrillas who died alongside them.

Mission men who did not survive the fighting, found a final resting place in the peaceful Manila American Cemetery in the Philippines. Some bodies were never found or were sent to their parents in the provinces for burial.

Promotions and Commendations

The Filipino radio operators of the 978th Signal Service Company at their jungle outposts in the Philippines, PIMC and the KAZ radio station of General MacArthur received high praises for their dedicated work operating the guerrilla radio traffic with the Philippine islands. As the Army captured areas and islands, they turned over the codes and transmission instructions that would place the Task Forces in direct contact with the guerrillas. Many continued to operate beyond the August surrender of Japan until actual fighting in the Philippines ceased and the radio traffic was almost negligible.

Looking back, Capt. Charles Ferguson who led the signalmen in the Luzon landing had high praises for the men. Lt. James McGivney who was in charge in Leyte praised the message center and radio section: "During the period from the invasion of Leyte up to the invasion of Mindanao and the Visayas, it is doubtful whether any other comparable message center and radio station ever handled such a volume of this peculiar type of traffic with the limited personnel available, and with less services, and which contributed so much by furnishing intelligence information to the success of the Leyte, Luzon, Mindanao, and the Visayan operations." And the men handled the messages in the most treacherous locations, suffering sickness, starvation, and the constant threat of death.

Sgt. Tanato reported that the majority of 978th Signal Service Company who came to KAZ in the beginning of 1943 did not have any ratings. They did the tasks without thoughts of rank or complication. Eugene Ergas and Joe Perido stood out among the most dependable signalmen for their work equivalent to that of technical sergeants. The diligence of the signal men who trained and prepared as Army field radio operators overcame the difficulty of learning the fixed station procedures that the guerrillas used.

Earlier, mission party officers sent requests to KAZ for promotions of their men after seeing their heroic acts at their remote outposts in the Philippines. Phillips had requested promotions for his 978th men, Alfred Alberto and Ramon Vitorio, that were granted a month later. Charles Smith sent a promotion recommendation for Aniceto Manzano and Crispulo Robles, radio operators as one of the first to set up a radio station on Luzon relaying messages to KAZ. Bartolome Cabangbang requested promotions for the 978th men in his team for excellent work before turning over the radio codes and cypher systems to the Sixth Army upon their landing on Luzon. Allison Ind, head of the AIB, recommended Irineo Ames as party leader for successfully delivering supplies and personnel to Panay and operating a coastwatcher station. Ames recommended his men Jose Ramos and Felipe Ginobiagon for high quality reports to KAZ which Ind concurred.

But their headquarters and paperwork were in Hollandia as they were a secret unit, and no one knew about them. Their TOE that allowed for promotions and release of equipment was approved only in November 1944 when their unit was deactivated and replaced with the 1st Reconnaissance Battalion two years after the first party of mission men landed in the Philippines.

That caused much confusion and complication in the paperwork. Some were given conflicting answers. Jerry Pascua who manned the KRP station in Catbalogan, Samar wrote of his disappointment. He was told by Lee Telesco, Whitney's aide in Manila that he was a sergeant, and this was confirmed by his party leader. "My rank when I left Australia was Private First Class." In the Philippines he was told by Maj. Charles Smith and Telesco, the PRS Executive Officer that he was a sergeant and signed letter to family and friends as such. Returning to Hollandia, he signed letters as a corporal as he was told but he found out that he was still a Private First Class. "Friends think I am crazy promoting myself. The Army made a fool of me." Documents later stated that the men who went on missions were promoted to the next grade level.

The 1st Reconnaissance Battalion was active for 10 months from November 1944 to August 1945.

UNITED STATES ARMY FORCES, PACIFIC
Office of the Commander-in-Chief
A. P. O. 500
15 August 1945

SUBJECT: Commendation

TO: Commanding Officer, 1st Reconnaissance Battalion (SP), APO 75

1. On this day the 1st Reconnaissance Battalion (SP) is to be deactivated inasmuch as its mission has been accomplished.

2. I wish to express to all members of the Battalion my grateful appreciation for the splendid service they have rendered, both individually and collectively, toward the liberation of the Filipino people, a cause to which they have so long and so faithfully dedicated themselves. Those of them who were dispatched by me prior to the start of the Philippine campaign on secret missions within the Philippines proceeded in the path of their duty with magnificent courage and marked ability and resourcefulness. Their service in large measure inspired in the Filipino people the will to continued resistance during their darkest hours and contributed in no small degree to the success of our Philippine operations.

3. Those of them who did not have the opportunity for this type of service, through their wholehearted support of the main purposes for which the Battalion was formed, provided an esprit de corps next to none in any unit in the American Army.

4. I shall always take deep satisfaction in the accomplishments of the 1st Reconnaissance Battalion and desire to commend all of its officers and men for the high type of service which was responsible, therefore. They carry with them to their new assignments my very best wishes.

                                        /s/Douglas MacArthur
                                        /t/DOUGLAS MACARTHUR

CERTIFIED TRUE COPY

Al Hernandez
1st Lt. Infantry
Assistant Adjutant

Source: The Filipino American Experience Research Project, 5 October 1998

Battle Honors, Awards and Recognitions

(The complete U.S. Army Records, Battle Honors, and Decorations for the 1st and 2nd Filipino Infantry Regiments and the 1st Reconnaissance Battalion are in the Appendix B-33.)

The 1st Filipino Infantry Regiment earned battle honors for the New Guinea and Southern Philippines (Samar Island and Ormoc, Leyte Island)
The 2nd Filipino Battalion (Separate) earned battle honors for New Guinea
The 1st Reconnaissance Battalion (Special) earned battle honors for New Guinea.

The Philippine Presidential Unit Citation was awarded to the three organizations.

The 1st Reconnaissance Battalion was acknowledged for the campaign in New Guinea and not for their most valuable participation in the Philippine campaign - the secret missions to the Philippines Islands. The operation was Top Secret, known only to a few on General MacArthur's staff and headquartered in Hollandia which contributed to the incomplete accreditation of the men's bravery and sacrifices in the liberation of their homeland. In their mission, personal items were taken from them so as not to be identified. There were no press men or cameras when they landed behind enemy lines in the islands unlike in the Leyte and Luzon landings.

The U.S. Army Historical Report on the 1st Filipino Infantry Regiment cited men who received awards at the time of the report's completion. There were many others who received awards after completion of the report. The report listed Colonel Robert Offley, commanding officer of the 1st and 2nd Filipino Infantry Regiments having received a Bronze Star along with American regimental and battalion commanders. Bronze Star medals were awarded to Sgt. Roseno Ramirez and Pfc. Natalio C Pergis and Pvt. Romualdo B. Naval with Company C and Pvt. Gregorio B Cuaresma with Company F for meritorious service against the enemy in Leyte. For gallantry in Samar, Lt. Isidro B. Urbano, commanding officer of Company H, received the medal. Listed was Alamo Scout Sgt. John E Hidalgo as having received the Silver Star for a mission on northwest Leyte. Pvt. Sabas A. Asis, Sgt. Thomas A. Siason, and Pvt. Rufo Vaquilar who joined the Alamo Scouts were awarded the Bronze Star and an Oak Leaf Cluster for their heroic achievement in Mindanao and rescue of the prisoners at the Cabanatuan Prison camp. A Silver Star was awarded to Sgt. Agustin P. Inocencio, a squad leader in Company C for gallantry in action against the enemy at Poro Island. A Sgt. Jacinto (unreadable last name) and Pvt. Gaudencio O. Bulanguis received the Silver Star for gallantry in action in Samar.

Four hundred twenty men were sent on secret submarine missions to the Philippines. Ninety-two were White Americans and the rest Filipinos from the AIB, 1st Reconnaissance Battalion and 978th Signal Service Co. to provide intelligence, combat, and radio training to the guerrillas as listed in the 'Calendar of Submarine Shipments to Philippine Guerrillas and Agents December 1942 – December 1944'. (The Calendar is in the Appendix.)

The reports of the Battalion's disposition and activities were marked 'Secret'. Ceferino Rola, former aide to Maj. Brown with the assistance of Juan Dahilig and Apolinar Salvador was tasked to write the official 'Unit History of the 1st Reconnaissance Battalion (Special)' soon after the war. Under Maj. Croell, the history of the 978th Signal Service Company's history was written covering their general activities and commands. Only a few of the after-action reports that were invaluable in documenting the actual experiences of the men at the ground jungle level of their remote outposts were included. For dozens of years after the war, the information on their exploits remained classified at the U.S. National Archives.

Earlier, Julius Ruiz was chosen by Whitney to write the history of the 1st Reconnaissance Battalion for the U.S. Army's historical department. He interviewed the men at camp San Miguel as they sporadically arrived from their mission. Men returned and reported to camp at various times and departed for new assignments, furlough, back to Hollandia, or discharge. Not all were able to write an after-action report. Some gave verbal reports. Combat records had to be checked against other reports and radio messages.

Ruiz expressed his disappointment with the hold up of promotions for the 978th Signal men. He felt helpless when the mission men went to him for resolution. All top ratings (promotions) had been issued to the White soldiers. He considered it "unfair since majority of personnel in the 978th were Filipinos and they resent being exploited just like before the war." He strongly expressed that those resentments should not be revived when Filipino American relationship should be encouraged. He wrote a letter to Dr. Arturo B. Rotor, secretary to President Quezon in Washington D.C. to do something about the men's plight.

Overall performance was reviewed before recommendation for an appropriate medal was submitted. All the mission men were nominated for combat decorations, including the Silver Star for heroism, the Legion of Merit for exceptionally meritorious conduct and the Bronze Star for heroic achievement. They were also nominated for the Combat Infantryman's Badge for their close-in-combat with the enemy. Aside from the medals, the Battalion received numerous commendations from the Army and Navy. "For its size, the 1st Reconnaissance Battalion (Special) was one of the most decorated unit in the U.S. Army history."

The 978th Signal Service Company received a Meritorious Service Unit plaque award. The plaque did not cite two missions in November and December 1943 with 978th Signal Service men. The six AIB missions from January to August 1943 that paved the way for the success of the following parties of 978th Signal men were not mentioned.

***

COMMANDING GENERAL
HEADQUARTERS
UNITED STATES ARMY FORCES, PACIFIC
APO 50017
GENERAL ORDERS NO........115   August   1945

Meritorious Service Unit Plaque Award

Pursuant to authority contained in Circular No. 345, War Department, dated 23 August 1944, as amended by Circular No. 421, War Department, dates 26 October 1944, a Meritorious Service Unit Plaque is awarded by the Commander-in-Chief, United States Army Forces, Pacific, to the following named unit:

978th SIGNAL SERVICE COMPANY. For superior performance of duty in the execution of exceptionally difficult tasks in the Southwest Pacific Area from 1 January 1944 to 1 January 1945. Given the two-fold assignment of training and introducing radio operators behind enemy lines in the Philippine Islands and transmitting vital information from Japanese held territory for use in planning the invasion operations, the 978th Signal Service Company fulfilled each mission with the utmost zeal and competence. Over one hundred of its personnel volunteered for service in the islands to install new transmitter or to reinforce established guerrilla stations. Members of the 978th Signal Service Company established radio communications near enemy garrisons and reported convoys, troop movements, gun emplacements, supply dumps, airfields, weather conditions, and all activity of military value, enduring countless hardships in the disease-infested jungles, they took part in raids on enemy garrisons, neutralizing his patrols as well as capturing equipment, in guerrilla territory to save survivors of damaged United States aircraft and ships, they located mined waters, and on Palawan helped escaped prisoners of war. The exceptional courage, skill and devotion to duty displayed by the personnel of the 978th Signal Service Company contributed in large measure to the success of the invasion and liberation of the Philippine Islands.

AG-PA 200.6
By Command of General MacArthur

R. K. Sutherland
Lieutenant General, United States Army, Chief of Staff

A certified True Extract Copy: OFFICIAL:
Harry T. Croell   /S/ B. M. Fitch
Major, Sig. C.,   Brigadier General, U.S. Army,
Commanding 978th Signal Service Company   Adjutant General

As the mission men gathered one last time in San Miguel, on August 15, 1945, the men were anticipating the medals that would properly recognize their achievements in contributing to the liberation of their homeland. Excitement heightened as Emperor Hirohito broadcasted the surrender of his country on the same day. They stood before the flags of the two nations they fought proudly for, beaming with confidence from their mission. Long before the return, they had infiltrated the islands risking their lives and suffering through diseases, sickness, and starvation, while evading the Japanese. They recalled building a network of radio stations with the guerrillas that reported on Japanese activities and was the lifeline for the planning of the Allies' return. Silva was not present at the medal ceremony as he was undergoing the first of a dozen operations to save his leg.

Whitney presented the awards to the men in battalion formation. Awards for the mission men who had not yet returned were issued after disbandment. More than half received the Bronze Star. Some believed they deserved a better award since they were out in the field risking their lives. Many who were not out in the remote jungle outposts received the higher Legion of Merit.

On the day the 1st Reconnaissance Battalion was disbanded, Julius Ruiz was present. He had been charged to destroy the same records of the Battalion that he was to use to write the history. The records were proof of the men's exploits deserving the Legion of Merit medal instead of the Bronze Star. It is uncertain how much of the records were destroyed and how much he was able to forward to the U.S. Army's Historical Division. He fortunately brought home two cardboard boxes including the after-action reports of the men written in their own handwriting. (The list of the men's After-Action reports are in the Appendix.)

First to receive an award, the Legion of Merit, was Col Lewis Brown III, the Commanding Officer of the 1st Reconnaissance Battalion, "for his outstanding leadership in developing the Battalion to a high state of efficiency and inoculating in its members that Esprit de Corps so essential to their success in operation under most dangerous conditions."

After each soldier's name was called and battle exploits read, the Bronze Star was pinned over his heart. Questions and disappointment surfaced. Did any of the Filipinos receive the Legion of Merit? The men pondered "We went into harm's way, not those who stayed behind. We lived danger day and night, and our medals were downgraded, while those who remained in the comforts of their office and desks received more and higher decorations." Maybe no one questioned it that day. A little good news was the additional of $10.00 to their monthly pay by having been awarded the Combat Infantry Badge.

Joe San Felipe recounted that Col. Hamby who replaced Offley heard of the medals for the men of the 1st Reconnaissance Battalion and wanted them awarded in a formal presentation. He recalled many of the men received medals less than they expected but did not strongly voice their disappointment too much at the time. He noticed certain preferences in the medals that were awarded. The Legion of Merit medal for valor went to many of the White officers while most of the reconnaissance men in the jungles received the Bronze Star.

At the reunions of the men after the war, the disappointment always came up of not receiving a higher award. Were the awards fairly done? With the life-risking achievements they went through, they questioned if they received the proper medals. Maj. Jose T Mendoza wrote that he was nominated for the Legion of Merit along with other men who had gone on a mission. However, at the last minute, the decorations had been downgraded to a lesser medal. As a Veterans Administration officer assisting retired veterans after the war, Dr. Gregorio Chua who was with the Regiments, could not find the records of the 1st Reconnaissance Battalion. "There are no records," he later told Alex Fabros, Jr. "They were immediately destroyed after the war."

The disparity did not surface until Alex Fabros, Jr's research uncovered information that the majority who received higher than a Bronze Star were not the men who braved the jungles to fulfil their duties. He found the answers after 50 years. Fabros is the son of Alex de Leon Fabros, a member of the 1st Filipino Infantry Regiment. He was the first to delve into the history of the 1st Reconnaissance Battalion after finding the after-action reports that Julius Ruiz did not burn at the Filipino American National Historical Society (FANHS) in Seattle that showed proof of the men's heroic exploits. Without these records, the story of the mission men would be seriously incomplete if not unknown. Ruiz's desire to write a book after the war did not materialize and so the papers languished in his attic for over forty years. A relative who did not see the records' value had taken them to the dumps instead of fulfilling Ruiz's wish to donate them to the FANHS archives upon his wife's death. By good fortune, Fred and Dorothy Cordova saved the records from destruction and preserved them for researchers and historians.

Fabros wrote his findings as part of his master's thesis in 1997. Most men were nominated for either the Legion of Merit (LOM) or the Bronze Star (BSM) medals based on their individual combat participation. Three forms were filled out - Legion of Merit (LOM), Bronze Star Medal (BSM) and the Combat Infantrymen's Badge. The forms were filled with the soldier's name, a summary of their mission, and details of specific heroism or meritorious service. They were then forwarded to higher office for approval.

Lewis Brown, commanding officer of the 1st Reconnaissance Battalion reviewed the forms for first approval before forwarding to Courtney Whitney, head of the Philippine Regional Section. "When the files were sorted, it turned out that the records for the Filipinos who had been nominated for the Legion of Merit had been downgraded to the Bronze Star. Lt. G. Guyot, Cpl. Jerry Pascua, Capt. Petronio Huerto, Lt. Konglam Teo, and Capt. Claudio Tamayo had their medal recommendations downgraded to the Bronze Star." Most of the American soldiers' names did not have their medals downgraded to a BSM. Obviously deserving of the LOM were Jordan Hamner and Frank Young since they manned the radio themselves while risking their lives to protect it in Tawi-Tawi, the remotest islands in the Philippines. Charles Smith, leader of the Samar party, and Lawrence Phillips, leader of the doomed Mindoro party who was killed. received the Distinguished Service Cross a year later.

A Decoration & Award Information Form filled out for Maj. Saturnino Silva recommended him for the Legion of Merit for his exploits during his mission to Mindanao where he was wounded and his men for the Bronze Star. He was the only man among those sent on missions to the islands to have led a regiment of 1500 men into battle with the Japanese. A written LM for the Legion of Merit was beside his name and BS for the Bronze

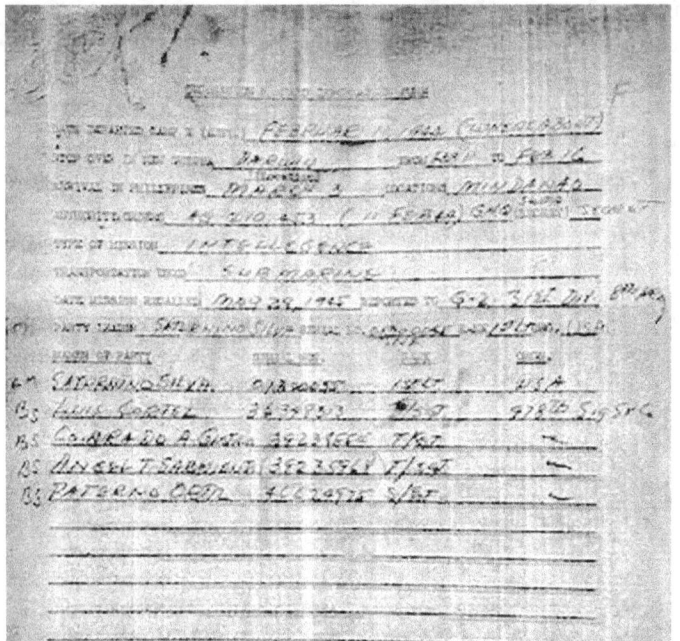

Decoration & Award Information Form filled out for Saturnino Silva's mission party (February 1944). To the left of the name is the medal each man was nominated for. LM is Legion of Merit and BS is Bronze Star.

Star beside his men's names. It was saved from the orders to destroy the Battalion's records. Instead he received the Bronze Star. The same form was also filled for Jose Valera (August 1944 mission to Luzon) and his men with LM beside his name and BS for his men. (Valera's form is in the Appendix.)

Looking for more answers, Fabros found Capt. Cecil Earl Walter, Jr. the commanding officer of the 1st Parachute platoon of the 1st Reconnaissance Battalion. Walter had reported to Maj. Brown during the war.

"Brown had nominated himself for the Distinguished Service medal, the third highest medal after the Medal of Honor, and the Distinguished Service Cross for serving as the 1st Reconnaissance Battalion commander. It was turned down. Brown was angry that his only reward was to be the Legion of Merit," Walter said. "Dismissing the exploits of the men, he downgraded LOM nominations for the Filipinos." But the Legion of Merit recommendations for many of the White American officers in the submarine missions were not downgraded to Bronze Star.

Did Whitney know and just did not question it? According to Walter, Brown said he would be "damned if any of the Filipinos were to get an equal or higher medal than him." To hide the injustice and eliminate proof of the men's activities, he ordered the records burned. GHQ's staff accepted Brown's recommendations believing that it was fair and correct.

It seemed that an injustice had denied the men their proper reward. Ricardo Galang, who was present at the awards ceremony, expressed in an interview after the war "that those who worked with the underground did not get (the proper) decorations and those who did not do anything got them."

1st Reconnaissance Battalion officers Capt. Bartolome Cabangbang and Lt. Al Hernandez avoided the injustice because they received their Legion of Merit medal earlier directly from General MacArthur. Hernandez was specially awarded for his success in obtaining a map of Manila and important Japanese locations by bravely posing as a dance instructor to a Japanese officer.

The Purple Heart was awarded to those who were wounded. Aside from the Bronze Star, Major Silva received the Purple Heart for a shattered shin from machine gun fire while leading the guerrilla 130th Infantry Regiment into a battle during the liberation of Davao City in Mindanao.

The men of the first submarine mission (Negros) to the Philippines were awarded the Philippine Legion of Honor (Officer): First Lt. Rodolfo C. Ignacio, Lt. Emilio F. Quinto, Lt. Delfin C. Yuhico, Sgt. Patricio Jorge, and Sgt. Dominador Malic. The group was cited for "setting up an intelligence network linking various parts of the Visayas and Luzon... establishing an intelligence network independent of guerrilla organizations for security and a neutral source of information... Training personnel for contacts... Paving the way for the initial assault of Allied Forces on Philippine territory in the campaign to free the islands." Silva received the Philippine Legion of Honor award after the war in June 1953.

General Krueger of the Sixth Army ordered Bronze Star medals to all the men of the Corpus/Placido party sent to Palawan. High commendations were given "for the remarkable accuracy of their reports on ship flashes and weather reports." The party was also responsible for the rescue of ten American POWs, remnants of 150 POW's held at Puerto Princesa in Palawan. The American commander and seven others who survived the sinking of their submarine USS *Robalo* in July 1944 were cared and helped evacuated by the party.

The Soldier's Medal was commended to Don S. Ruiz and Camilo Ramirez in appreciation for having volunteered for the Shangri-La rescue. They parachuted into the unknown and remote mountains of New Guinea to rescue survivors of a crashed plane. For forty-five days, they attended to the survivors until they were well enough to be taken out by glider.

Men of the Regiment assigned to **PCAU #2** were credited with participating in the Luzon and southern Philippines campaigns. They received Purple Hearts for the enemy bombing of their LST in the Leyte landing, and a Bronze Leaf Oak Cluster on their Purple Heart medals for wounds received in combat with the enemy near San Pedro, Laguna. Being in the front line while performing their mission of helping the civilians, the unit was continuously under attack. While landing with **PCAU #2**, on the beach of Luzon, Sgt. Cababa furiously dug foxholes and killed two Japanese before he was wounded. An Oak Leaf Cluster was added to his Purple Heart.

Fabros' father, Alex Fabros, Sr ended the war with a Combat Infantry Badge, Bronze Star with a V and Oak Leaf Cluster, Purple Heart with an Oak Leaf Cluster, and a Certificate of Commendation.

War Brides

With military camps set up in the rural towns retaken from the Japanese, interactions between soldiers and civilians inevitably happened. Although the military tried to suppress marriage by allowing prostitution, romantic relationships developed with the local women. In her thesis, Jeannie Magdua wrote of an official American- sponsored 'red-light district' in Tacloban to make it safe for the soldiers. Forty regularly inspected women serviced several hundred Army men daily.

Waiting for orders on the invasion of Japan, the chance to marry a Filipina, bring her back home to the U.S, and start a family overpowered the feelings of discontent and unfairness of the awards. Although, the men experienced hardship and racism in the U.S., they still believed that America offered a better quality of life after the war.

At the time, there were hardly Filipino women in the U.S., and the 1930s Miscegenation Act in California forbade interracial marriages. The Tydings McDuffie Act in 1934 changed the unlimited migration to the U.S. to only fifty per year. Marriage applications flooded the office of company commanders before the expiration of the War Brides Act in December 1945, and the Fiancée Act in June 1946 that allowed the women to enter the U.S. without visas, immigration quotas, and at the expense of the U.S.

The *Manongs,* who were already in their late thirties, found and married their wives while in the Philippines. Many of the brides were in their teens and grew up under colonial rule, learning English and developing a loyalty to the U.S. while in public school. This upbringing shaped their expectations for a better life in the U.S. Mothers reasoned that they wanted a husband for their daughters who came from the same island, spoke the same language, and had the same culture so that arguments would be resolved. With many of the *Manongs* coming from Luzon and speaking a different language, they communicated in English.

Many marriages happened during the Regiment's stay in Leyte and Samar. But before they could get married, they had to obtain permission from their commanding officer and follow certain rules. Joe San Felipe, a Hawaii boy, had the job of checking the men if they had families back in the U.S and if they followed the local tradition of making marriage announcement to the Catholic church. He was 19 years old interviewing 38- to 40-year-old men on why they wanted to marry this girl. "Because I love her. How do you know what love is? All I know is that I just want to be with her. That was good enough," San Felipe said and stamped an approval. Some of the men extended their service in the Philippines after marrying. Others married before returning to the U.S. to be discharged. They told their wives to return to their families and wait for their return.

When the Americans landed in Leyte, many starving and homeless civilians rushed to the beaches and then stayed in the camps. The Japanese had occupied the towns and taken their food stuff and livestock. Many did not have a male head of the family to earn from the military projects. Women and widows resorted to earning around military camps by providing laundry service. Sgt. Federico Sumaoang asked his future wife, Fannie, to wash his clothes and later asked her parent's permission to marry her according to Filipino custom. Her uncle acted as the interpreter speaking in English since Federico was from Luzon and spoke a different language while her family were from the Visayas. Some misinterpretations caused confusion and rejections of attention. She had thought of him like her father and refused to marry him believing she was too young. Federico was 36 years old while Fannie was 18 years old.

Dr. Leo Smollar, a PCAU doctor, wrote that several of the PCAU's Filipino-American enlisted men took local women as brides in elaborate church ceremonies as life returned to normal. His driver was married on April 30 by an American military priest. "The local priest had been charging seventy-six pesos even though the pre-war fee was 14 pesos and still is, except for soldiers. The commanding officer finally decided that the gouging had to stop and brought in our priest," he wrote.

Patricio Jorge who was with the first mission to Negros in January 1943 departed from Australia as a Filipino citizen. He reported to the Battalion's headquarters at Camp Murphy in July after the war and was assigned to the 5th Replacement depot in Alabang "for either demobilization or reinstatement to the Army of the United States (AUS) to get our citizenship." He chose not to obtain citizenship and be discharged as he was getting married. He had met his wife during his mission in Negros when he was assigned to the Bicol province where he met her. Jorge recalled in an interview that it was unforgettable when he met his wife. His commanding officer headed the Fifth Column and had brought him along. Her family were *mestizos* and her father wanted a Spaniard to marry her. With great determination, he showed his Garand to the other suitors who did not return. Mrs. Jorge sweetly said that "Patricio told her that he was going to commit suicide if she would not accept him. "It was a year later after the liberation of the province began that he returned to marry her. He requested a 45-day leave after over two years to visit Cielo, "the woman who clouded his thoughts with great anticipation." As they emerged from the church, clouds of smoke filled the horizon as U.S. bombs continued to fall on Legaspi city. The following day, the Americans entered the city without opposition.

During the Palawan mission, Sgt. Ramon Cortez and Cpl. Teodoro Rallojay met and married their wives before being recalled to GHQ. A Catholic priest was found to solemnize the marriage and Placido, their commanding officer, was best man at Cortez's wedding, the first to be married. At Conchita Marquez family's evacuation house, where some of the radio equipment was stored, the men would come for dinner. She would play the piano while Rallojay joined her with a song. In Australia, he had played the guitar and had a good singing voice. No Catholic priest was found so her father, a judge, performed the marriage ceremony. She wore her best blue dress. Several weeks later, the priest returned to their camp at Brooke's Pt and they were married again. "In their best tattered clothes and barefooted, and with much tuba, everyone had fun" wrote Placido. A bull was butchered for the happy gathering, and he was the head chef doing "Texas-style side beef BBQ". Cortez and Rallojay returned a year later to bring their wives to the U.S. The women brought only a few personal items, their birth and marriage certificates.

While operating the radio at the Bondoc Peninsula, Gerardo Nery hiked to San Narciso to bathe in the streams. There he met and married the daughter of the San Narciso mayor before reenlisting in the Army.

As PCAUs following the combat forces, many of the men met their future wives in the various islands. They were invited to the towns where they met and courted women. That was a welcome respite from the fighting. Sgt. Mamerto Ilumin met his wife in Ormoc Leyte while Alex Soria met his wife in Catbalogan, Samar. Mariano Angeles, a PCAU assigned to the Regiment at White Beach, Leyte went on weekends to the village of Alangalang where he married and had a child before returning to the U.S. He did not take citizenship earlier citing unfairness of the offer but decided to take the oath now to bring his family to the U.S.

As the men moved with their assignments to Ormoc and Tacloban, trucks brought the wives and a 'tent city' was set up for the married couples. By the time they were recalled to assemble at Dulag in September 1945, many had already married and brought their new brides.

Pascual Manipon met his wife, Espectation, in Ormoc. They arrived in San Francisco and took a bus to Pittsburgh, California for immigration processing. White women married to Filipinos helped her adjustment, especially the German wife of Pascual's brother.

Felipe Dumlao met his wife in Villaba, Leyte. He recalled that "she was buying cigarettes from us men when we put up a perimeter around her town. She was reselling them to civilians. Filipino soldiers did not smoke while in the Army, so cigarettes were spoiling. Some had several cartons. Men received one carton per week supply. She did my laundry too."

At the infirmary in Tacloban, he was treated for a wound he received in the fighting at Samar. There Pedro Visaya met his future wife while recuperating.

Allen Dawang would not have met his future wife if he did not insist that his uniform be tailored to his smaller stature. His American-sized uniform hung loosely on his body. A portion of her home had been converted to a sewing and tailoring business. He came almost every day. She was reluctant to follow procedure and meet his captain for permission to be married, so the captain went to her home to interview her. Eight weeks later, the permit came, and they were quickly married in a civil ceremony with her family present. Troop ships were ready when they went to get their marriage license at the municipal building as his unit was to leave the next day.

Arriving in Leyte as a PCAU, Mariano Angeles was a medic with the Regiment. He met his future wife in Alangalang, Leyte. Fred Ignacio also met his wife in Leyte while on assignment as a prison guard for Japanese POWs.

Teodoro Taluban told his girlfriend to stay at his mother's house until he could come back for her. When he was recalled to Leyte, she followed, and his battalion commander turned her over to him. Under a coconut tree, they were married by the regimental chaplain. His application for an extension to stay in the Philippines was approved with headquarters in the campus of the University of the Philippines. Small cottages were built around the school where he was able to stay with his new wife until orders arrived to return to the U.S.

As the Regiment divided into platoons and deployed to remote towns in Samar, men met their future wives. Jovito Liban's platoon was camped in the remote area of Hiratigan about one hundred miles away from Catarman, the capital city of Northern Samar. There he met his future wife. The platoon of Atanasio Alcala camped at Basey where he met his wife, and their first child was born. The Japanese did not bother her family since her father was the sanitary inspector of the town. A Japanese word that protected them was written on a sign attached to the front of her house.

She first saw Santiago Piscusa, a Hawaii boy, at her aunt's restaurant in Leyte as he stood out looking yellow from the malaria medicine he was taking. Her name was not mentioned in their interview after the war. He brought a friend who could speak the language to court her. She was raised with maids, so she refused to go to Hawaii unless she had a maid because she did not know how to cook and wash clothes. He told her to have the maid teach her. She lived with her parents while he re-enlisted and worked in the Manila American Cemetery registering men missing in action and supervising a crew for the burials. He travelled all over the islands working with civilians to locate bodies in villages, jungles, and provinces where Americans missing in action were buried. Today, the cemetery holds 17,000 graves and 37,000 names on the Wall of the Missing.

Before he landed in the Philippines, Philip Ordono made a promise to himself that he would "marry the first lady I met." At the end of the war, he went back to Samar and met his future wife at a store. He signed up for three years in Manila with the Ryukus Command as he did not want to leave her. He bought a Quonset hut and land for his new family. He took leave without approval from his commanding officer and "got busted back to a Private" but he was still made a platoon leader, he related.

As a PCAU with the invasion landing in Leyte, Alex Fabros, Sr. joined the invasion forces to Luzon's Cagayan Valley on the trails to General Yamashita's hideout, where he met his wife. He returned to the U.S. to get his citizenship in the Army and back to marry his fiancée in the same month that the Emperor of Japan broadcasted the surrender. *The Philippine Mail,* where Fabros used to write the column *Thorns and Roses,* published his return from the Philippines. "...with him his lovely wife and two children, a boy and a girl." He was discharged in 1946 at Camp Stoneman.

Another Filipina from the Cagayan Valley on Luzon captured the heart of Julius Ruiz, the editor of the *Bahala Na* newspaper. He had been in the U.S. for sixteen years. "He waited long enough to meet her," he recalled.

Paratrooper Sergio Solidarios with the 1st Reconnaissance Battalion was on Luzon and met his bride in Batangas City where she worked as a phone exchange operator for the U.S. Forces stationed in the city.

CIC agent Wallace Castillo who was studying Criminology before the war, met Norma Vega, a daring woman at the Temptation Bar in Manila where her sister was a singer. She worked as a secretary for the Mori Company, a manufacturer of bicycles and grenade casings for the Japanese military. Working in the main office, she passed on information to the Nakar guerrilla unit. When the manager, Shugo Ono, a former math professor at the University of Tokyo, informed her of the Americans' arrival, she quit. Newly married, they would sit in a car even when she was pregnant and watched for spies or collaborators. Castillo returned to the U.S. to get discharged and came back for her. They returned to his home in Hawaii.

The months he spent near starvation in the disease-ridden jungles and swamps of Agusan did not deter Conrado Gustilo from going back to Mindanao. It took him two years to return after he received an honorable discharge. His future wife, Araceli, did not think he would come back for her. On Mindanao, all his comrades would go to her house to visit and eat Filipino food while he shyly stayed behind at the camp. She only knew of his intentions when he wrote to her from the U.S. He took the educational benefit to finish college. Wanting to open his own radio repair shop in Mindanao, he refused many job offers. Araceli said that she wanted to put up the sign 'Gustilo Radio Shop – Experienced in the U.S.' but he refused. She remembered seeing Capt. Silva, with whom her husband arrived with in their mission, describing him as slim and quiet.

Brides told stories of accompanying their prospective spouses to personal interviews before their marriages were solemnized by the company chaplains. They felt a cultural strangeness with their Filipino American husbands as they got to know more about each other. Some of the villages became known as "The Barrio That Married the Regiment" recalled Domingo Los Banos of the 2nd Filipino Battalion.

The American Red Cross was required by the War Department to prepare the brides in the Philippines for transport. Some of the wives had never travelled outside of their town. Many did not have much education and schools were closed during the four years of the war. Classes were conducted for two days before boarding the ship on what would happen when they arrived. They were given money to buy food in case their husbands were unable to meet them.

An interesting observation was made by the brides when they arrived during the fall season while the trees were shedding their leaves. They thought that the trees were dead. Mark Cazem, whose father worked as a tractor operator in the agricultural fields before enlisting, related that his mother, Jovita, arrived aboard the U.S. Army Transport *Admiral W. S. Benson* with other war brides and their children. Expressing their dreams of living in America, the lively discussion was about adjustment to the American way of life. They had thought of the Golden Gate bridge made of gold and carefully inspected it as the ship passed under the bridge.

The arrival of the war brides began the next generation of Filipino Americans especially in the West Coast and Hawaii. According to Magdua's thesis, the total number of Filipina war brides was just over 2,200 under the War Brides Act of 1946- 1950, the largest number from Asian countries. Over 5,000 Filipino women migrated to the U.S. immediately after the war.

Accounts of war brides from other Asian countries included stories of experiencing discrimination in the U.S but Magdua stated that there were hardly any against Filipino war brides. She continued that Filipinos came from an American colony and was under U.S. protection. Racism had changed after the war especially after the Filipino soldiers' sufferings alongside American soldiers were revealed in the battles of Bataan and Corregidor.

The men brought their wives under the 1945 War Brides Act avoiding the racial laws in California. The 1934 Miscegenation Act barring interracial marriages was reversed in 1948 after the war. But the laws were not expunged from the California code until 1959 invalidating marriages before that year. Did the marriages have to be performed again? Only in 1967 was it ruled unconstitutional.

The first group of newly arrived brides adjusted to the country with the help of Caucasian women married to Filipinos. Life in the U.S. was different from what the women may have seen in the movies or magazines. Encouraged by their husbands, seven war brides began the Philippine War Brides Association in 1949 in Seattle, Washington to support the newly arrived Filipino wives of U.S. servicemen. As the group grew, their services expanded to assisting families of Filipino veterans who migrated to the U.S.

After the war, old comrades of the Regiments who returned to work in the farms of Salinas, Delano and the surrounding areas visited and reminisced. The food overflowed cooked by the men and women who attended the gathering. Many of the men were much older than their wives and conversed mostly in Ilocano which was the language that the *Manongs* spoke.

Civilian Life

At the end of the war, "over 500 soldiers returned to the U.S .... and 800 men of the Regiment remained in the Philippines." Many were discharged as they gathered enough points.

It was a somber moment for the men when their commanding officer who was like a father to them returned to the U.S. In a newspaper announcement, Col. Robert Offley, expressed his deep-seated feelings for the men. He expressed his pride and admiration of the enthusiastic efficiency displayed by the officers and men. "It is my prayer that I may be allowed to round out my service in close association with the same loyal people whom I learned to love when a young boy – the Filipino people." He returned to the US due to health problems and retired the following year.

Back on U.S. soil, anxious feelings surfaced from the men. They had proven that they were just as capable as any soldier and many times had risked more. Institutionalized laws of discrimination were slowly being repealed, expanding their personal freedom and access to public places that were once forbidden.

The G.I. Bill was the biggest advantage for the Filipino American soldiers after the war. Many attended universities and graduated to enter the political, judicial, educational fields and the once-elusive white-collar market. Some made the military their career while others became successful businessmen.

Some single men returned to work on the plantations their parents had worked on. Vincent Pinuela, a member of the doomed mission to Mindoro, bought land in Turlock, planting it to lettuce and asparagus.

Major Saturnino Silva at a Fourth of July celebration in California.
Source: Marie S. Vallejo collection

Dixon Campos, in an interview, said, "If not for the G.I Bill, I would have never gotten this far." He refused to say what was the most negative personal result of the war, dismissing the question by saying "Life goes on." Asked if he was injured, he replied, "only emotionally." It was time to begin a new life.

But discrimination laws continued to haunt the men. Sgt. Daniel Begonia graduated from the G. I. Bill as a dentist but could not practice in California. Disillusioned, he went back to working in the fields. Such an unjust reward for one who spotted the Japanese convoy streaming to attack the beaches of Leyte from his outpost in the southern coast of Mindoro and radioed Australia!

Jose San Felipe became a lawyer, but he was refused a job interview as a non-white and turned down several times to buy a house because he was the wrong color for the neighborhood. Turning to work with the Filipino community, he was handicapped because he could not speak any of the languages having grown up in Hawaii. He was told that he was not a Filipino. He finally found work in the government that did not have these quirks. Many from California moved to Hawaii where the rules were not as strict for Filipinos and there they prospered with their military experience.

Don Ruiz, decorated for his role in the rescue at Hidden Valley in New Guinea, worked in the post office as a mail carrier. He had a barbershop and grew vegetables and fruits from his garden. Still not content with the postal job's retirement, he worked as a security guard for a high-tech company well into his 90s. Others like Pablo S. Valdez went back to work with the labor unions. Manuel Buaken published a sort of memoir called *I Have Lived with the American People.*

Having gathered so many points for being overseas for four years, Lorenzo Pimentel was discharged in California. He missed Christmas with family and friends when a railroad strike forced him to take a ship that arrived on New Year's Eve. With no bus to Seattle, he walked home. Before the war, he said "Whites didn't respect Filipinos." He did not want to apply for citizenship because of what he experienced before the war. But when news of the Filipinos in Bataan fighting side by side with American soldiers were plastered in newspapers across the U.S., he enlisted. Naturalized, he was able to buy a house with the veterans' benefit.

News of the bravery of the Filipino soldiers fighting with the American armed forces in the Philippines affected the Filipinos in the U.S. After decades of treatment as second-class citizens, they were embarrassed at the new positive attitude and behavior of the American civilians towards them in the service industry. "They treat me as if I have just arrived from Bataan," a veteran commented. They had to adapt to the way they saw themselves and the way most Americans thought of them. Combat films showed positive images of Filipinos as the staunchest and bravest allies in the fight against Japan. Between 1942 and 1945, one-third of all motion pictures released by Hollywood studios focused on the war. At least eight films focused wholly or in part on the Philippines, showing the Filipino and American soldiers together in combat, death, and enduring the Death March.

As the years passed, there were hardly any publications on the men. What they underwent in the filthy trenches, disease-infested jungles, and risking their lives to evade the enemy to protect the radio were unknown. Some suffered long after the war, reliving the atrocities they witnessed, and the killing or be killed to survive. Many repressed the debilitating memories to continue life.

It was difficult for some of the men to share and have people understand what they went through that were not only victories, glory, and success but also the inhumane. Some of the men even took it to heart the orders not to reveal the secret missions they were on. Decades later, sparse records of their 'Top Secret' unit were declassified, but the men continued with the orders. By the time the children were old enough to understand and appreciate what their fathers went through, the men were in the stage of their life when memories began to fade.

A daughter asked about her father who went to New Guinea and caught "jungle rot." His hands and feet would get water blisters that persisted until he passed away. She watched her father suffer and wanted to understand what it was.

Two of the men of the first submarine mission to the islands reunited to celebrate the 50th anniversary of their pioneering assignment (January 1943 mission to Negros). They reminisced about their dauntless landing at Catmon Point beach by the USS *Gudgeon*. They felt lucky to be alive as it was a suicidal mission into an unknown. Lt. Delfin Yuhico and Sgt. Patricio Jorge stood in front of the towering boulder behind which they buried their rubber boats. After the war, Jorge went on to train the cadets at the Philippine

Military Academy (PMA) on assembling and dismantling a gun blindfolded. In an interview, he was proud that he could still do it, although he said that he would not be able to hear it as he was going deaf.

The Philippine Military Academy (PMA), with its academic curriculum and military training program, patterned after West Point was reactivated two years after the war ended. The first group of officers in the formation of the 1st Filipino Battalion returned to the Philippines to re-establish the Academy. The first post-war superintendent was Tirso Fajardo (West Point 1934). Leon Punzalan (West Point 1936) headed the study group to propose a new curriculum and select a permanent

Catmon Point, Negros: 50th Anniversary of the first submarine landing in the Philippines, January 1943. Delfin Yubico on the left and Patricio Jorge were with the mission to Negros. They are standing by the rock where they buried their rubber boats.

Source: Delfin Yubico

site of the academy. Pedro FlorCruz (West Point 1942) personally selected the permanent site at Fort Del Pilar outside Baguio City. Rafael Ileto (West Point 1943), who was an Alamo Scout team leader, became Chief of Staff of the Philippine Army, and Secretary of National Defense.

On August 10, 1959, the 1st Filipino Infantry Regiment's insignia was cancelled.

The Regiments celebrated their Silver Anniversary in San Francisco in 1967 attended by men of the Sixth Army. Another twenty-five years passed, and the men reunited at the Salinas Community Center in the summer of 1992. A dinner dance celebrated the 50th anniversary of the Regiments' activation in July 1942 at the same location. A picnic at Sherwood Park was held the following day.

The veterans' camaraderie remained strong. The original four men who marched to carry the Regimental and United States flags at their dusty camps while proudly wearing the volcano patch on their left shoulder, continued to carry the flags in the San Francisco Veterans Day Parades and Memorial Day activities at the Presidio of San Francisco for many years after the war. Thanks to the efforts of Pelagio Valdez, son of Pablo S. Valdez, a member of the Regiment, in continuing the participation and maintaining their memories.

Guerrilla Recognition Program 1945 – 1948

As the men were going on leave to find their families, finding wives, getting married, transferring to other assignments, or returning to the U.S., they thought of the guerrillas they had fought and risked their lives with. These were men who conducted espionage, acted as guides, provided security, and fought alongside them to accomplish their missions. Proof of their participation had to be submitted and reviewed before official recognition by GHQ, AFWESPAC to receive benefits.

In November 1945, the U.S. Recovered Personnel Division (RPD) deployed the Guerrilla Recognition Program to provide death and backpay of deceased military personnel and POWs who had supported U.S. objectives in the Philippines. The Headquarters Philippines Command, U.S.A. contained the details of the recognition program policy. To qualify for recognition and back benefits, proof of continuous membership in the Philippine Army or inclusion in one of the recognized guerrilla units that fought during liberation was required. One of the five-point requirements was being in an actual fighting unit or materially contributing to the war. Those who served for a time then dropped out were disqualified. Claims of providing food or manual labor to help the guerrillas were not included. Many individuals and units fought the Japanese in one way or another but were unrecognized. Providing proof of participation was difficult as rosters were lost or destroyed for fear of falling into enemy hands. Rosters were not completely maintained as guerrillas constantly moved. Others served under assumed names. Paper was scarce. Furthermore, many guerrilla units from the southern islands were unaware of the program or claims arrived too late. Originals were sent in, and some were lost by the Army Department in charge at the time. Many guerrillas quietly went home after the war without being processed and therefore dropped from the rosters. Civilians also submitted claims for compensation of labor and materials they provided to the armed forces. There were claims for what was lost to bombing action of U.S. forces or from the warehouses taken over by the Army. Those who paid someone to put their names on the list had to be ferreted out and removed before recognition. While some Filipinos no doubt submitted false claims in hopes of getting veterans' benefits, the vast majority had likely supported the guerrillas in some manner. Sometimes before a claim could be processed, choices had to be determined who would receive the death benefits of a Muslim or a Mountain Province tribesman since only one wife could be recognized to receive the benefit.

A unit roster was created after Army officers investigated each claim in the area where the guerrilla unit operated. The unit's commanding officers and political connections were investigated. It was reported that the commanding officer was then ordered to bring down the number to incorporate as a standard unit in the Philippine Army. The demobilization program was to reduce the strength of the local army to 30,000 officers and men by June 1946 as the standing peacetime army reported *The Manila Post*. The best qualified soldiers were retained with 50,000 taken into the Philippine Scouts to help garrison occupied Japan.

Only 260,000 names were approved. Out of over 1,000 guerrilla unit claims, 277 guerrilla units were recognized. No guerrilla women were recognized except for registered nurses although there were many women guerrilla leaders who led regiments into battle.

Bartolome Cabangang and Enrique Torres, officers who led the September 1944 mission to Luzon tried to help by writing letters on behalf of the guerrilla units they were with describing proof of their participation. The guerrillas worked wherever they were assigned in combat, as messengers, guards, or couriers. Cabangbang explained that the men who were transferred from the guerrilla Bulacan Military Command (BMC) unit and carried supplies and equipment across the Sierra Madre mountains were couriers and agents before the radio was set up. They trained in radio operations to help operate the radios. His request for recognition of the 500 hundred men was not approved. KAZ reasoned that he was sent to Luzon in command of an intelligence detachment composed of entirely U.S. Army personnel and had no authority to promote guerrilla personnel with whom he met. Guerrilla personnel utilized by him were supplied from their guerrilla unit.

Ben Harder, a survivor of the ill-fated mission to Mindoro, wrote a letter in support of recognition for Emilio Macabuag and his men: "On January 17, 1944, Vittorio (sic) and I sailed from Mindoro to Cape Santiago in Batangas to set up a radio station and an intelligence network covering both naval and military intelligence. Macabuag furnished the necessary security and aided in the accomplishment of mission... furnished civilian clothes and food on his personal account when Japanese were hot on our trail. He also planned and executed the evacuation of four Americans from Batangas to Mindoro in mid-January 1944."

Postwar politics may have affected the recognition program. Fertig, commanding officer of the Mindanao 10th MD wrote "It began to appear that the only recognition we would receive for our efforts was our own personal knowledge of a job well done. There were a lot of dynamite in those records for certain politicos." He reported that the records of the 10th MD were turned over to the Philippine Army headquarters and could not be found.

Rescission Act 1946

While the Guerrilla Recognition Program was ongoing, the Rescission Act was signed into law compounding the difficulty in obtaining compensation. The Guerrilla Recognition Program was now the only means of obtaining some compensation from the U.S. for services during the war. Claims ballooned to a million when the deadline neared.

When the United States needed men to defend its territory from Japanese expansion, President Roosevelt issued a presidential order in 1941 calling into service of the United States all the organized military forces of the Government of the Commonwealth of the Philippines. More than 100,000 Filipinos enlisted and were sworn in as members of the U.S. Armed Forces. And they fought alongside the American soldiers under overwhelming odds until the surrender of Bataan and Corregidor.

The Rescission Act of (February 18) 1946 (38 U.S.C. § 107) was a law of the United States that retroactively annulled full U.S. benefits that would have been payable to Filipino troops for their military service under the auspices of the United States during the time that the Philippines was a U.S. territory and Filipinos were U.S. nationals. It was reported that of the 66 countries allied with the U.S. during the war, only Filipinos were denied military benefits.

The Filipino American soldiers were not affected by the Act as they were inducted directly into the Army in the U.S. Their beginning was President's Roosevelts authorization for a formation of a U.S. Army all-Filipino Battalion.

On July 2, 1946, the Luce-Celler Act passed granting Filipinos access to citizenship. More than 10,000 Filipinos who were in the United States before 1934 were able to receive naturalization as a direct result. Was this to diffuse the Rescission Act and the forthcoming independence severing ties and responsibilities? Two days later, on July 4, the Philippines was granted independence by the United States.

As the years passed, veterans of the Philippine Army living in the U.S. did not forget. "They continued to fight to get Congress to restore benefits lost when the Philippines gained its independence," wrote *The San Jose Mercury News* in 1997. They chained themselves to a statue of General MacArthur in protest and to the fence surrounding the White House to demand full veterans benefits for Filipinos who fought in WWII.

American Recovery and Reinvestment Act 2009

As decades passed, the dwindling number of veterans continued to press the U.S. government for the benefits. To provide some closure, the American Recovery and Reinvestment Act of 2009 authorized the release of a one-time, tax-free lump sum payment to eligible World War II Philippine veterans living in the U.S. or the Philippines. A requirement was that their name had to be on the 'Missouri List' of veterans, which probably was incomplete because of a 1973 fire that destroyed millions of records, including those of many Filipino veterans. Another list that may have been used was the over 260,000 guerrilla names compiled and approved by the Guerrilla Recognition Program. The payment was a token of service and released the U.S. government from future claims. The Act did not apply to the U.S. 1st and 2nd Filipino Infantry Regiments and 1st Reconnaissance Battalion.

Sixty years and thousands of veterans have sadly passed. The registration lines of ailing veterans under the hot sun at the Manila American Embassy was long. Many were in wheelchairs, with family members holding umbrellas to protect them from the scorching sun and being fanned because of the heat. Many marked the application form with an X as they could no longer properly see or legibly write their name. The veteran had to be alive when the lump sum amount was given as it was not to go to the spouse or children.

Filipino Veterans of World War II Congressional Medal Act 2015

A Congressional Gold Medal was awarded, collectively, to the Filipino veterans in recognition of their dedicated service during World War II. The bill defined the veterans who honorably served from July 1941 to December 1946 in an active-duty status under the command of the U. S. Armed Forces in the Far East, with the Philippine Commonwealth Army, the Philippine Scouts, the Philippine Constabulary, Recognized Guerrilla Units, the New Philippine Scouts, the First Filipino Infantry Regiment, the Second Filipino Infantry Battalion (Separate), and the First Reconnaissance Battalion.

The deep faith in the Americans' return and aid was so embedded in the Filipino guerrillas' psyche for decades that it continued after the war and into their death. Robert Hudson, whose father was with the quartermaster Corps attached to the 31st Infantry Regiment, USAFFE supplying the men on Bataan, was told of a dying soldier of the Division.

Felimon Factoran had fought in Bataan, survived the Death March and subsequent imprisonment. Hudson recalled, "his daughter cried when she saw me. Her father was dying of cancer and was holding on because he knew an American was coming to see him. She showed me his prized possession, a copy of the history of the 31st Infantry Division. He could not see as he was partially blind and could not speak. When she told him an American was there to see him, he sat up in bed and saluted me. His daughter was shocked as he had not sat up in bed for weeks." Hudson was touched and "told his daughter to tell him that it was not for him to salute me but for me to salute him." He had finally met his American. Factoran soon died after at peace in his silence.

John Silva and Marie Silva Vallejo. John wears a specially made Army khaki uniform. Pinned and sewn were Dad's decorations and patches.

XXIV

My Journey

My journey was complete when the scattered bits on the Regiments came together to form their own history and simultaneously that of my father.

The war had ended. My parents returned to the Philippines while my mother was pregnant with me. They lived in Iloilo City on the island of Panay in the house of my mother's family. Her family was from the sugar-rich plantations of Negros and had a name, resources, and connections. My father worked for the various family agricultural enterprises. He soon had difficulty getting along with members of her family; it was a generation in which class, family, and island distinctions were prevalent and important. Outsiders were not readily welcomed. Dad was considered an outsider. I did not know much about his side of the family.

Saturnino 'Tony' Ramos Silva was born in Bautista Pangasinan, on Luzon, the Philippines' largest island. His parents were Tiburcio Silva and Maria Ramos from Batangas. He had three brothers and one sister. A cousin, noted painter Jun 'Bueno' Silva related that their grandfather was Juan de Silva, a *cabeza de barangay* (head of a village), and a tariff officer of Bautista, Pangasinan, during the Spanish colonial period in the Philippines. Tiburcio became the family's breadwinner after their father and younger brother were brutally murdered by thieves. I remember Dad telling us that he would earn extra money by hauling water from the river, hooking pails to both ends of a bamboo pole, and bending his knee to lift the heavy pole over his shoulder when he was a boy. He was the scholarly one, graduated as a teacher and went to Tawi-Tawi in Mindanao to teach before the war. He was in the U.S. to better his life when the Japanese bombed Pearl Harbor, and like so many other U.S. based Filipinos, joined the military. His older brother, Jose, enlisted in the Philippines, became a sergeant with the machine gun troop of the 26th Cavalry, Philippine Scouts and went missing in action during the war.

I was the first of four children born five years after his honorable discharge from the Army. Whenever I catch the tune, *The Caissons Go Rolling Along,* an Army marching song, I remember my brother John and myself waking up every morning to the tune. A portable record player was cranked up by our nanny as instructed by Dad. It was loud enough to wake him up as well in the next room. It must have reminded him of his Army training days in California.

I watched Dad dressed in a tan-colored uniform receive the Philippine Legion of Honor, the Philippine government's highest award bestowed on an American officer. My favorite photo captured the silhouette of my six-year-old head in the back row of the grandstand. I remember well the Philippine and U.S. flags flapping vigorously in the wind. Later, I realized that the award was for leading a regiment of Filipino guerrillas to battle in Mindanao during WWII.

The stress of dependence on my mother's overbearing family made my parents decide to start a new life in San Francisco. He had gone ahead and bought a house in the Avenues. He picked us up from the airport, and on the way to our new home, he sang and tapped on the steering wheel to the song from the car radio. He said it was his favorite song. As I leaned over the front seat and looked at the side of his face, he beamed a broad smile that I now understood was him expressing the joy and relief of having regained his family. He found night work in the U.S. Post Office. We became a nuclear family with only each other to rely on for decisions, unlike the usual extended family that was normally the case in the Philippines.

Barely two years passed, and we had to return to Manila as my grandfather suddenly died in an accident. My mother was the eldest child and had to run a business empire that she was unprepared for. In Manila, we lived in my grandparents' house in an exclusive gated community. It was at the entrance to Fort McKinley, a former U.S. Army Camp, turned over to the Philippine Army. A decade earlier, Dad was at the 10th General Hospital in the same camp after he was wounded in Mindanao. The afternoons in this house were always quiet, and I basked in the freedom of not having to take a nap as a ten-year old. I could hear the occasional double thumping of golf balls landing in our backyard from the nearby golf course. I would walk from room to room in the afternoons, looking for something to do. One day, I peeked into an unlocked briefcase under a desk which contained papers that I did not fully comprehend. My young mind understood Dad's name, two female names, the word Australia, a marriage date, and a birth date. Was Dad ever in Australia? Who were those names? I repeated the names several times and burned them in my mind. If my parents found out, I knew that I would get in trouble. But I promised myself that when I grew up, I would look for whoever those names belonged to.

I remember the day when John and I wanted to go swimming at the exclusive Army and Navy Club in Manila. Dad was furious when we were refused entry. Filipinos were not allowed. Moreover, a Filipino officer in the U.S. Army was unheard of. Membership was only for U.S. Armed Forces personnel and "to a carefully screened group of American business and professional men residing in the Philippines." After Dad showed his military credentials, things changed, and John and I enjoyed many days of swimming. The Filipino waiters gave us extra scoops of ice cream as they marveled at the rarity of seeing Filipino children. The guards at the club gates beamed with pride as they saluted Dad every time we arrived.

It was the same experience at Camp John Hay in Baguio City. Again, only U.S. military and their guests were allowed. Dad scowled as he carried his military ID to meet with the camp manager. We soon enjoyed the American hamburgers and hot dogs while Mom and Dad shopped at the Post Exchange. The dinners at the Officers' Club were special times to dress up and have my hair styled into a French twist held by numerous hairpins. I tasted my first Shirley Temple from the extra attentive Filipino waiters since we were usually the only Filipinos dining there.

Again, my father worked for my mother's family, which was not an ideal situation. His military ways were too stringent for the sensitive nature of the Filipino. Those close to him, however, knew he was a man of few, but direct, words that dispelled any misunderstanding. He was mostly away on business while I was growing up, and when he was, he did not say much nor was he demonstrative in his affection. I did not receive any corporal punishment except for one instance I cannot forget. I was around ten years old. He was home that day and was told of my continuous misbehavior. If I wasn't making my brother cry, I was sneaking up on the household helpers to surprise and play tricks on them. That day, Dad's eyes blazed as he held me over the toilet bowl threatening to dunk my head if I did not behave. I cried and screamed and promised to be a good girl. That was the only frightful incident. I realize now, as an adult, that for him to have punished me like that, he must have been affected by the brutalities of the war.

He eventually left again for the U.S., and we were to follow later. But my mother went through an emotional crisis, and they divorced. I can still feel the intense ambivalence of split loyalty between my parents and the many tears from it.

I remember gathering enough gut-wrenching courage to tell my mother that I wanted to live with Dad and attend high school in San Francisco. She allowed me to go. After a year of living with his closest friends in the city, he was able to buy a house as a U.S. veteran without a down payment and racial discrimination. He found work in the Office of Equal Opportunity of the U. S. government and ran a side business, buying and shipping printing presses to the Philippines. I had to learn to do laundry, iron my school uniform, clean house, and cook simple dishes - duties that I never had to do in the Philippines.

On weekends we visited his friends at farms in Salinas, Stockton, and Delano. I learned that some of them were his old comrades in the Regiment who had returned to being farm workers, hopefully with more equitable pay, better living conditions, and no discrimination. I wondered why the men looked much older than their wives. We always returned home with a trunkful of fresh fruits and vegetables.

It was just the two of us in the house. He did not talk much about the war, nor did I show any interest. If he tried to tell me war stories, I did not give him the time to listen. As a teenager, I was interested in more current topics such as 'boys.' I realize and terribly regret now that I missed so much.

Occasionally, I tried to tidy up the house in the limited way I knew how. One day, my curiosity kicked in again, and I opened the top drawer of his cabinet. There were several letters with a return address of Mindanao and a black-and-white photo of a boy scout in uniform around the same age as me. There was no smile under the piercing eyes that looked away from the camera. Who was he? Was he in Mindanao? Confused, I slammed the drawer shut and burned the image into my mind. I told no one. It took several decades for me to find the answer.

1967 San Francisco, California. The author with her father, Lt. Col. Saturnino Silva, at the Silver Anniversary of the 1st and 2nd Filipino Infantry Regiments.

When the 1st and 2nd Filipino Infantry Regiment celebrated their silver anniversary in 1967 in San Francisco, Dad was the chairman of the event. They invited members of the Sixth Army that the Regiment was attached to in New

Guinea. He had the widest smile as he greeted many familiar faces. I remember him talking to the men at each table in his usual charming way and speaking in either the Ilocano or Pangasinense language they all knew. The ballroom was crowded around the dance floor as the men reminisced, saluted each other, and sang the songs they knew at camp. Their wives eagerly shared how they met their husbands.

Dad remarried while I was at San Francisco State College, the same school he graduated from thirty years earlier. After finishing college in Manila, I returned to San Francisco to experience the independence of earning my own money for a couple of years, and then returned to the Philippines. Later, with my husband and three-year old daughter, we migrated to the U.S. and the following decades focused on work and raising a family. I made it a point to call my father every month. I was afraid that if I wrote letters, they might not reach him.

Watching my daughters grow made me realize that I knew about my mother's family up to three generations before her. Her parent's families were well-known during the Spanish colonial period and were pioneers in shipping and agriculture during the American period. I knew nothing of Dad and what he was before he married Mom. With much regret, I lost that knowledge and appreciation that was coloring my relationship with him. I could have been closer to him.

Finding photos of him as a young man before he enlisted and in an Army uniform during a vacation trip to Manila began to bring together in my mind the pieces of his early life. I collected all the photos from a neglected album that was falling apart in a closet.

A chance meeting opened another door. A conference on the Filipino regiments in a hotel I happened to walk in made me linger in the doorway to find out more. Alex Fabros Jr. stepped out and began a conversation that revealed the importance of Dad's military unit. To my amazement, he knew about Dad and his exploits. It was a strange feeling to be told what Dad did during the war, but that aroused my curiosity even more. Fabros Jr.'s pioneering research work on the 1st and 2nd Filipino Infantry Regiments became the impetus towards discovering more about Dad's generation and their military experiences.

When my daughters joined the working world my thoughts returned to my father's military exploits. The names of the two women in his briefcase continued to haunt me. During a vacation to Australia, I went through two pages with the last name of 'Cunanan' in the phone book, calling with the hope that someone might know the surnames on the papers in my father's briefcase. Years later, I found out that the name was actually spelled 'Conanan.'

When the Internet came into use, I would occasionally search for his and the two women's names in between work schedules. Retirement allowed more time for searching. Then one day, a hit! Someone had posted his name on an ancestry message board. An email thread showed a woman's reply to her mother's email that was six years old—and the mother's email address had the same first name, Isabel, as one of the women's names on the paper in the briefcase. I stared at the name for a while trying to compose the right email so as not to scare the recipient. I changed the words so many times, unsure of how and what to say. We exchanged photos of our father and each other. No DNA tests were needed, only answers to so many questions. From a short distance, my husband saw Isabel's photo and thought it was me with a new hairdo.

Several months later, Isabel and her two daughters came to Manila. Saturnino Silva Jr nicknamed 'Boy,' joined us. John and I realized then why he was not the 'Junior.' Boy was the boy scout in the photo I found in Dad's drawer. He was Dad's son with Esther while he was in Mindanao. She had met Dad at the guerrilla 107th Division headquarters when she accompanied her sister Florence who was accused of collaborating with the enemy but was eventually cleared. Father Cote, the Canadian camp chaplain, soon blessed the relationship. During the liberation of Davao City in May 1945, Dad's left leg was shattered by a Japanese machine gun while leading a regiment of Filipino guerrillas to battle north of the city. He was immediately flown to Leyte, then the U. S. Their son, Saturnino Jr., was born five months later. He never met our father.

The four of us stared at each other. It must have crossed Dad's mind so many times that someday his past would be discovered, and we would all meet.

With even more questions, John and I visited Isabel and her mother, Priscilla, in Brisbane. Meeting her, was the closest we felt to Dad. Together we tried to fill in missing parts of Dad's life in Australia. She recalled they met in a Chinese restaurant on Elizabeth Street in Brisbane, not far from St. Stephen's Cathedral. Although his unit was treated as 'Top Secret', he was allowed into the city as an officer, a privilege not afforded the enlisted men. When their relationship became serious, her parents disapproved of her marrying a military man.

Months had passed without the arrival of an approval to marry. It was reported that General MacArthur did not want any Filipinos to get married, as it would weaken the men in their mission. A memo from Col. Whitney; however, approved the marriage, and on January 8, 1944, Saturnino 'Tony' Silva and Priscilla Conanan were married at Canungra Base Camp by a Catholic priest. Silva was the first of Filipino soldiers to get married in Australia, claimed Major Carlos Arguelles who was the best man at the wedding. Together, with another vehicle loaded with musicians from the *Bahala Na* orchestra of the Battalion, the newly wedded couple drove into the city of Beaudesert to celebrate.

A month after the marriage, orders marked 'Top Secret' were issued to Silva and four 978th Signalmen for the next secret mission. They were on the "third day of their honeymoon when he was called away," Priscilla recalled. That was the last time she saw him, and she was pregnant. He was sworn to secrecy and left his new wife without revealing where he was going. Four days later, he was on board the submarine USS *Narwhal* enroute to the Philippines.

It was after the war when Silva and Priscilla connected again, and she was reluctant to join him in a strange country with a sickly child. She felt her strength and independence having raised Isabel by herself. She finally received the three letters he sent from Mindanao. But it was too late. After a couple more heated calls that did not end well, they divorced. Isabel never met our father.

Boy's daughter, Techie, sent me a brochure of Carmen — *Ising* during the war— a town in Northern Davao, Mindanao. One page was devoted to a WWII battle that happened at the road junction of Ising. The veterans were proud of the fight that devastated a Japanese garrison without American ground support, preventing the enemy from entering the northern unoccupied areas. Headed into the battle was a regiment of around 1500 men led by Maj. Saturnino R. Silva. In disbelief, my brother and I immediately went to Mindanao to find out more. It was a strange feeling, again, when a high school teacher of Carmen told the story of a battle our father led, which was completely unknown to us.

After Dad's leg was shattered by a machine gun at the battle, he underwent dozens of operations over a period of two years. Some were unsuccessful and had to be repeated. After years of therapy, when he was finally free of his crutches, he walked with a slight limp.

Colonel Wendell Fertig, the Mindanao guerrilla commander Silva reported to, began the *Mindanao Guerrilla Newsletter* in December 1947. The newsletter kept the American guerrillas on Mindanao (AGOM) connected after the war. Silva was the only Filipino American included in the newsletter.

Still in crutches, he met Elena, my mother, in San Francisco. Her parents had sent her to study at the Dominican College in San Rafael, California. Her parents had strongly advised that she marry from a 'good' family of similar status—and preferably from the same island they were from. Dad was from Luzon, spoke a different language, and not much was known about him—and he was 15 years older. This was the second time parents of the women he wanted to marry disapproved of him. I wonder how he felt. Strong-willed, my mother defied her parents and drove to Yuma, Arizona, to marry him.

Compelled to find out more about my father's military exploits. I wrote the book *The Battle of Ising*, about the regiment he led into battle from the information I found in the U.S. archives and libraries. The records listed the battle as one of the decisive encounters against the Japanese in the liberation of Mindanao.

While researching at the National Archives, College Park, Maryland, I came across the Philippine Archives Collection that covered the Pacific War during WWII in the Philippines. The need to find funding and leading a team sponsored by the Philippine Veterans Affairs Office to scan the Guerrilla Recognition Program files in the Collection both proved helpful for this book but side-tracked me for over a year.

All I knew was that Dad's unit was composed of mostly Filipinos, but no information could be found in the Philippines. No one had heard of the men or their unit. The first discovery was in the MacArthur Archives website. Dad's name was on one of the three-by-five index cards that were filled out for each soldier from the Regiment who was interviewed and chosen for further training in Australia to go on secret missions. But details on the Filipino soldiers proved elusive. Even harder to find was information on what they did when they arrived in the Philippines. The difficulty laid in the scant information. The Regiment's exploits were buried under the U.S. forces they were assigned to. The submarine missions were all marked 'Top Secret'. But stories and rare photos lovingly kept and shared by the men's families emboldened me to piece together the story. I peered at each photo hoping to find his familiar face.

Even with so much discrimination heaped upon them, my father and the men volunteered to fight. Going back to their lives in the U.S. before the war began, records affirmed the cruel disparity between my father's America in 1929 when he arrived and the America I lived in after the war. His was brutal and vicious. He lived through back-breaking work in the fields of California due to lack of other opportunities. He suffered rampant discrimination in housing, restaurants, and stores. All in places where I never experienced such. I realized that my better and kinder life experiences in America could only be as a result of the agony of the Filipinos of his generation in their constant battle for equal treatment. To prove that they were just as good as any American, they enlisted at significant risks to their lives.

My journey ends with this book about the dauntless Filipino soldiers who volunteered to fight for the United States to liberate their homeland, the Philippines.

A rifle party fired three times in unison as a salute at his military burial. Every time I hear 'Taps,' I tear up. My toothy smile is just like Dad's. Sometimes I catch myself humming his favorite song…

FREIGHT TRAIN
Sung by Elizabeth Cotten

"Freight train, freight train
Going on so fast
Freight train, freight train
Going on so fast
I don't care what train I'm on
As long as it keeps going on."

Lt. Col. Saturnino "Tony" Ramos Silva

Silva served in the Central Pacific campaign, the Philippine Liberation campaign, and the Mindanao-Sulu campaign. He was awarded the U.S. Bronze Star and Purple Heart medals. From his records, he was awarded the American Campaign Medal; Asiatic-Pacific Campaign Medal with two bronze service/campaign stars; Philippine Liberation Medal with one bronze service/campaign star; World War II Victory Medal; and Combat Infantryman Badge. In addition, his unit received the U.S. Distinguished Unit Citation (patch) and the Philippine Presidential Unit Citation. As a member of the 1st Filipino Infantry Regiment and the First Reconnaissance Battalion, he received the U.S. WWII Congressional Gold Medal.

1st and 2nd Filipino Infantry Regiments
1st Reconnaissance Batallion (Special)
Source: Pauline Valdez, Pelagio Valdez Collection

APPENDIX

Appendix A
1ST AND 2ND FILIPINO INFANTRY REGIMENTS

Appendix A-1

The Filipino Bachelors' Club
of San Francisco

Holds Its

Annual Summer Gala Night Grand Ball

Saturday, August 17, 1940 at 8:30 P.M.

Mari Club, 1355 Market Street
San Francisco, California

An Annual Summer Gala Night Grand Ball was sponsored by the Filipino Bachelors' Club of San Francisco. August 17, 1940. Source: Sergio Solidarios

You are cordially invited to the **ANNUAL SUMMER GALA NIGHT GRAND BALL**

Reservations:
Members: Single or with a Lady Friend $5.00
Non-members: Single $2.50. Couple $5.00

R.S.V.P.
Strictly formal
Note: Part of proceeds will be donated to the Building Fund of the Filipino Community of S.F.

PROGRAMME
Juan M. Dulay   General Chairman   S. R. Silva   Master of Ceremonies

PART II
PROGRAM

1.	Opening Remarks	S. R. Silva
2.	Voccal Solo	Perfecto Halog
3.	Selected Spanish Dances	Misses Juanita and Mary Alice Espiritu, Lucrecia Suguitan at the piano
4.	Nomintaion, "Bachelors' Muse of 1940"	Vicente Fuentes, Presiding Officer
5.	Vocal Solo	Miss Pacita Todtod, Dolores Ingojo at the piano
6.	"Violetera" Spanish Dance	Miss Betty Paredes, Leo Olivares at the piano
7.	Short Sketch "Bachelors Meet A Spinster"	Characters: A. Raquepo, F. Marquez, J. Ingojo, S. R. Silva
8.	(a) Presentation of the Muse	Salvador Yotoko, Sophomore Dean F.B. C.
	(b) Grand March and Special Dance	Pedro Jazmin, Grand Marshall

PART III
Dancing 'til 1:30 A.M.
Motto: "Pro Cibo Omnia Nihil Pro Amore"

Appendix A- 2

1942

Company "I"
First Filipino Infantry
Fort Ord
Calif.

Merry Xmas

May the new Year bring Victory and Freedom

**COMPANY "I" HEADQUARTERS
FIRST FILIPINO INFANTRY
FORT ORD, CALIFORNIA**

Roster

**1st LIEUT. CLARENCE L. CURTIS
Company Commander**

Company Officers

1st Lieut. Paul F. Rodriguez   2nd Lt. Saturnino R. Silva
2nd Lieut. Delbert K. Schiess   2nd Lieut. William R. Stewart

First Sergeant
Francis W. Sullivan

Staff Sergeants

Paul Bandura   Manuel F. Faragan
Reynaldo R. Curva   Edilberto T. Macagba

Sergeants

Antonio M. Ancheta   Samuel E. Johnson
Deming Y. Asensi   Arsenio Dela Pena
Daniel C. Belarmino   Julian A. Pilar
Pedro T. Tejero

Technicians 4th Grade

Regino V. Uduano   Nicomedes T. Yoro

Technicians 5th Grade

Aldrico S. Acoba   Victor M. Fangonilo
Jose I. Baltazar   Santiago De Pilar

Romeo Aban
Florencio A. Alinan
Johnny J. Albano
Nicomedes C. Cabuag
Florencio Calip

Corporals

Abraham C. Dumlao
Eliseo A. Hermosa
Manuel L. Luz
Marcelino B. Mariano
Marcos M. Valenzuela

Jose Sabado
Cristobal Sana
Ray R. Somera
Jovito Unite

Privates First Class

Ramon B. Ables
Celestino D. Alayu
Pedro B. Alcantara
Anacorito T. Alicaya
Rufino C. Ancheta
Albert C. Apanay
Leonardo V. Baclig
Ambrose P. Baggao
Claro C. Bautista
Ignacio S. Bragado
Manuel B. Buaken
Vicente V. Bugayong
Eriberto M. Butac
Cornelio A. Caluya
Cuadrato C. Casanova
Julian F. Cuanang

Gregorio P. Daguimol
Alfredo Desierto
Urvano D. Doctolero
Eulalio P. Dulay
Miguel B. Erana
Teodorico M. Espejo
Alejandro E. Esperanza
Bernabe H. Estira
Jose P. Franda
Julian R. Gonzales
Basiliso D. Guzman
Marcelino O. Jurbina
Miguel L. Lawagan
Alex S. Lucas
Avelino B. Macabeo
Albert S. Naval

Francisco I. Ouano
Nick B. Paredes
Ulpiano M. Pascual
Elias Peniafiel
Aquilino B. Ramos
Marcelino B. Rania
Manuel Reyes
Ross N. Sacramento
Santiago L. Sarceda
Jose S. Serdinia
Gregoris V. Serquina
Philip C. Sison
Jimmie C. Tolentino
Robert F. Villamore
Bill Q. Viloria
Monico E. Zumel

Privates

Federico G. Abalos
Ambrosio B. Adana
Evaristo A. Agbayani
Marciano C. Agbunag
Modesto A. Agbunag
Domingo A. Aguila
Joseph H. Alabanza
Lorenzo A. Alarcio
Socorro P. Alcarion
Catalino L. Alingato
Exequiel G. Amar
Jose B. Amoguis
Chris A. Amparo
Julian B. Ancheta
Basilio Andaya
Marcelino L. Andres
Torbio G. Andres
Manuel A. Anolin
George A. Antolin
Pedro T. Antolin
Guellina R. Aquino
Nemisio S. Aradillos
Mauro N. Aragon
Esteban A. Arcelona
Fausto R. Argao
Andres N. Arellano
Telesforo T. Asuncion
Florentino F. Aquinde
Raymond B. Aviles
Greg S. Azcueta
Marcelino C. Baldios
Jaime T. Baltazar
Estapanio C. Bautista
Leon F. Bautista

Cirilo G. Canero
Thomas A. Carbonell
Luciano C. Cariaso
Balbino H. Castillo
Simeon S. Castillo
Henry C. Castro
Vicenti R. Causin
Raymond C. Cayabyab
Miguel H. Corpus
Manuel M. Crisostomo
Alejandro A. Dacanay
Bonifacio D. Dato
Ireneo D. De Felipe
Pancho P. Directo
Julian P. Diza
Anatalio M. Domingo
Lupo C. Domingo
Eusebio G. Duclayan
Wenceslao P. Dugenia
Cornelio R. Edra
Tony C. Empleo
Henry E. Enrado
Sergio C. Escalona
Anacleto Escarlos
Antonio E. Esmeralda
Carmelo Espanol
Bonifacio M. Espe
Pedro Y. Espero
Antonio A. Estigoy
Angel C. Frigillana
Semplecio G. Galope
Teodore S. Gascon
Teodoro T. Gerardo
Dominador F. Gonzales

Catalino L. Lozada
Bernard T. Lucas
Teodolfo C. Lucero
Eugenio F. Luz
Valentine A. Luza
Dioscoro M. Mabubay
Hospicio C. Macabeo
Juan M. Macahilas
Bernard F. Macalma
Agustin B. Macaoay
Gil V. Madriaga
Catalino Y. Maghiney
Bartolome C. Magracia
Catalino A. Mahor
Magno S. Markinano
Januario R. Malalis
Toribio Malisa
Numeriano S. Malong
Miguel J. Manangan
Crestino L. Manganaan
Diosdado B. Medina
Vicente G. Mendoza
Paul H. Nacional
Geronimo V. Nebres
Albertos Niduaza
Tetronilo R. Padim
Teodoro Padua
Joaquin J. Pallaya
Deogracias D. Pascua
Gregorio Penaranda
Martin Pilipina
Moises P. Porras
Simplicio Ramel
Roque P. Reynon

Alfredo Basalo	Fausto V. Gorospe, II	Urvano M. Rola
Alejo C. Baxa	Johnny O. Guzman	Gemo Rosete
Arcadio A. Benzon	Pete D. Ilar	Feguso G. Sotelo
Bennie M. Bernabe	Paulino B. Jobahib	Constancio A. Suguitan
Peter Bernardo	Ronaldo C. Juanich	Dionisio H. Suguitan
Domingo N. Boado	Teodoro C. Juanich	Joseph Z. Trinidad
Vicente D. Boado	George D. Judalena	Roman C. Torregosa
Delfin P. Bohulano	Feliciano V. Laboca	Pablo G. Valenzona
Agustus I. Bracerus	Catalino V. Lagadon, Jr.	Gregorio D. Valin
Rufino G. Buen	Vicente C. Laron	Juan P. Villa
Arsenio B. Bumanglag	Marcelino S. Laureta	Marion B. Villoria
Rufino O. Cacayorin	Francisco U. Logan	Fermin D. Visaya
Paul H. Calupig	Antonio A. Lopez	Eugene A. Webelhardt
Sergio C. Campos	Benigno Lopez	Saturno R. Yadao

Appendix A- 3

1st And 2nd Filipino Infantry Regiments, 1942 – 1945
Names In The News And Publications*

RANK	LAST NAME	FIRST NAME	M.I.	RANK	LAST NAME	FIRST NAME	M.I.
Sgt	Abrenica	T.		T4	Laureta	Marcelino	S.
Sgt	Acosta	Tony		Pvt	Laureto	Eriberto	
Pfc	Adona	Faustino		Capt	Ledesma	Serapion	B.
	Africa	Marciano	V.	Pfc	Legaspi	Alex	
	Aguinid	Alex		Pvt	Legaspi	Bobby	
	Agsalud	Nemecio	D.	Lt	Levida		
Capt	Aguirre	Mariano		Pvt	Ligutum	D.	
T5	Alcala	Rudy		T5	Lizardo	Cipriano	
Pvt	Alcoy	M.		Capt	Llorente	Abdon	
Pvt	Alesna	Johnny		Pvt	Lomague	Florencio	O.
Cpl	Alfonso	Valli		Pfc	Lomo	Rizal	
Sgt	Alfonso	Alfredo			Lopez	Julian	L.
Pfc	Alforque	Casiano			Lopez	Frank	S.
Lt	Algieri	P.		Pfc	Lopez	Stanley	
Pvt	Almalel	Eduardo	E.	Lt	Lorenzo	Sebastian	
Cpl	Almeniana	Tranquiliano		Lt	Louise	Beth	
Cpl	Almoite	Benito		Cpl	Lozada	Leo	
Cpl	Alvarez	Vidal		Sgt	Luzano	Valentino	
Lt	Ancheta	Carlos	F.	Cpl	Macabeo	Rosario	
Cpl	Andaya	Martin		Sgt	Madamba	Avelino	
T5	Andrea	Andy		Sgt	Mahusay	Joe	
Sgt	Andreas	William			Malla	Carlos	G.
Sgt	Aninao	Quintin		T5	Mallo	Benjamin	C.
	Anselmo	Bernard	C.		Martin	Mamuyac	
Pfc	Antonio	Alfonzo		Pvt	Manalo	Armand	
Cpl	Aperido	Jimmy		Lt	Mance	Julia	
Pvt	Aquino	Tranquilino	B.	Sgt	Mangrobang	Jesus	
Pfc	Aquino	George	C.	Lt	Manresa	Miguel	
Lt	Aragon	Potenciano		Sgt	Manzano	Lino	
Lt	Arca	Ricardo	T.	Cpl	Marcelo	Benjamin	R.
Cpl	Arciaga	Honorato		Sgt	Marfori	Ernesto	J.
Lt	Arellano	Arsenio	Y.	Lt	Marfori	Edmundo	J.
Lt	Arguelles	Carlos		Cpl	Mariano	Jose	
	Arollado	Robert	L.	Sgt	Martin	P. B.	

RANK	LAST NAME	FIRST NAME	M.I.
Lt	Arroyo	Alfonzo	
	Asis	Bob	
T5	Asuncion	Nicasio	
Sgt	Azcueta	Z.	
Pvt	Baer	Buddy	
T4	Balbuera	Martin	
Pvt	Balcena	Freddie	
Pvt	Baldonado	A.	A.
Lt	Balonon	Pedro	S.
T4	Baltazar	Prudencio	
Sgt	Banaires	Arcangel	
Sgt	Barangan	Victor	
Sgt	Barbero	A.	
Pvt	Barcelona	Fred	
Sgt	Barganan	Victor	
Cpl	Barroga	Jimmie	
Pvt	Bascon	Magdaleno	
Cpl	Basconsillo	Nasario	
Pvt	Bautista	Vivencio	
Pvt	Bautista	D.	
	Bautista	G.	
Pfc	Bendia	Marcelo	
Sgt	Berces	Severo	
Pfc	Berendo	C.	
Sgt	Bermudes	Micheal	
	Bibar	Gonsalo	
Cpl	Bibat	E.	
Sgt	Bisquerra	Santos	
Pvt	Bito	M.	
Cpl	Blas		
Sgt	Boncavil	Domiciano	
Sgt	Bongolan	Floyd	
Cpl	Borromeo	Jose	
Sgt	Brigandit	Joe	
Lt	Broodboaks	Lewis	
Pvt	Buaken	Manuel	
Pvt	Buenaventura	Nicanor	
Sgt	Buenavista	Aurelio	
T5	Bulosan	Aurelio	
Cpl	Bundy	Domingo	
Sgt	Cabais	Eutiquio	B.
Pvt	Cabaldon	A.	
	Caballes	Sulpicio	A.
	Cabaran		
Sgt	Cabe	Thomas	C.
Pvt	Cabebe	Donato	
Lt	Cablay	Benny	D.
Pvt	Cabral	Salvador	C.
Sgt	Cabus	Marcello	
Sgt	Cacabelos	Rufino	
Cpl	Cacas	Maximino	C.
Cpl	Cacdac	S.	
Cpl	Calacal	Carlos	
Sgt	Calderon	Victor	C.
Cpl	Cali	Frank	
Pvt	Calivo	Atanacio	
Pvt	Calivo	Vincent	
Cpl	Canapi	Anselmo	P.
Cpl	Canedo	Delfin	

RANK	LAST NAME	FIRST NAME	M.I.
	Martinez	Pedro	M.
Cpl	Marzo	Alejandro	
Sgt	Matawaran	Raymond	R.
Lt	Mendoza	Jose	
Pvt	Mendoza	Bridido	
Lt	Mendoza	John	
Lt	Mendoza	Don	
Pvt	Mendoza	Alejandro	
Lt	Mendoza	S. P.	
Sgt	Menes	Raymond	
Sgt	Mina	Jullian	
Cpl	Mina	Moises	
Sgt	Minodin	Gregorio	
Cpl	Minon	Gaudencio	
Sgt	Mogot	L.	
Pfc	Molina	Vicente	
	Mones	E. B.	
Pfc	Montenegro		
Sgt	Montero	Alfonzo	
Lt	Montoya	Eugene	
Sgt	Morales	Francisco	
Cpl	Moreno	Celso	
	Nalangan	Ray	
T4	Navarro	Mariano	
Cpl	Navea	Vincent	O.
Pvt	Nepomuceno	Alejandro	
Pvt	Nepomuceno	Alex	
Sgt	Nerry	Gerard	
	Nicolas	Hilarion	M.
Sgt	Nocon	Alejandro	
Pfc	Nosa	Felix	
Lt	Nyman	Donald	
T4	Oliveros	Felipe	
Sgt	Oller	Conrad	O.
Sgt	Oller	Jose	
	Orbeta	Stanley	
Pvt	Orillana	Fernando	
Pfc	Orpilla		
Pvt	Osoteo	Modesto	
Cpl	Ouano	Jose	
Pvt	Paardo	Guillermo	
Pvt	Pacebe	Marcelo	
Sgt	Padilla	Balbino	
Pvt	Padin	Robert	
Sgt	Padua	Miguel	
Pvt	Paduranga	Gaudencio	
	Paguirigan	Francisco	
Sgt	Pahinag	Paul	M.
Sgt	Palac	J.	
Cpl	Palmejar	Louis	
T5	Palmeto	Daniel	S.
Pvt	Panes	Salvador	
Pvt	Pani	Salvadore	
Lt	Pantangco	Ireneo	
	Paragas	Modesto	R.
Capt	Pardo	Clay	
	Pascua	M. D.	
T4	Pastor	Anselmo	
Sgt	Pastrana	Jose	R.

RANK	LAST NAME	FIRST NAME	M.I.	RANK	LAST NAME	FIRST NAME	M.I.
Pvt	Capa	Anito			Patacsil	Ambrosio	
Sgt	Capuyan	Benito	M.	Sgt	Paulo	Restie	N.
Cpl	Carbonel	Louis	H.		Paulos	Anastacio	T.
	Casanares	Cannon		Sgt	Pedrozo	Eduardo	C.
Sgt	Castillano	Fred	F.	Sgt	Penaranda	Gabriel	C.
Cpl	Castillo	Arturo		Pvt	Peralta	Alfonso	
Sgt	Castillo	Moises		Sgt	Peralta	Silverio	
Sgt	Castillo	Martin		Sgt	Peralta	George	
Sgt	Catalan	Rafael		Sgt	Peralta	Tolentino	
Pvt	Catayas	Ray		Pvt	Peralto	Casimiro	U.
Pfc	Celaya	Genaro		Pfc	Perez	Santiago	
Pvt	Cespedes	Leo		Sgt	Pimentel	Fructoso	
Lt	Chavez	Atanacio		Pfc	Pimentel	Epifnaio	
Pfc	Clarin	Angelo		Lt	Pizzuli	Nunzio	
Lt	Clausen	Joseph	E.	Pvt	Plandez	Al	
Pvt	Collado	Macario		Sgt	Pogusa	Emil	
Sgt	Collier	Eutaquio		Pfc	Ponce	Rudy	
Sgt	Conception	Dan		Pvt	Portilla	Nick	
Sgt	Conception	Benito		Capt	Punsalan	Leon	F.
Sgt	Conde	Alphonso	T.		Pylant	Julian	
Sgt	Corpuz	Eustacio		Pvt	Quides	Julian	A.
Pvt	Corpuz	Claro		Sgt	Quillopo	Larry	
Cpl	Cortez	Steve		Cpl	Quindiagan	Gregorio	
T4	Cosca	Leo		Sgt	Quintana	Mar	
Cpl	Costales	Gerardo		Pfc	Rabong	Chester	
T4	Costales	Paul		Cpl	Ramirez	Patricio	
Pvt	Crescenti	Dominic		Pvt	Ramirez	R.	
T4	Cruz	Eladio		Sgt	Ramos	Mauricio	B.
Pvt	Custodio	Emilio		Cpl	Ramos	Frank	
Pvt	Daan	Quirico		Pfc	Randall	Jack	
Sgt	Dacquel	Isodoro		Sgt	Real	Isay	
Pvt	Daguio	Art		Pfc	Reclosado	Francisco	
	Dalao	Buenaventura	B.		Regacho	Sammy	G.
Pfc	Dalompinis	Nicholas		Sgt	Regala	Wilfred	
Pfc	Dangaran	Danny		Pvt	Renegade		
Pvt	Darang	Andres		Sgt	Reyes	Eulogio	
Pvt	Daus	Cornelius		Sgt	Reyes	German	
Pvt	David	Ignacio		Pfc	Reyes	Henry	
Cpl	De Castro	Carlos		Pvt	Reynolds	Sam	
T5	De Santos	Urbano		Cpl	Robles	Rudy	
T5	Decena	Frank		Pfc	Robles	Conrado	
Sgt	Defiesta	Modesto		Pvt	Rodriguez	Roy	
Sgt	Del Mar	Don		Lt	Roman	F. J.	
Sgt	Del Rosario	Noe		Lt	Rosal	Vincent	V.
Sgt	Delenea	Benigno		Capt	Rosete	Francisco	
Lt	Demandante	Venancio		Pfc	Rublico		
	Desingano	Antonio	G.	Cpl	Ruiz	Julius	B.
Cpl	Diado	Anacleto		Pvt	Ruiz	Rafael	T.
Sgt	Difuntorum	Santiago		Capt	Rustia	Guillermo	J.
Lt	Dionolo	Tomas		Sgt	Sabio	S. M.	
Lt	Dizon	Daniel	D.	Pvt	Sacramento	Frank	
Pvt	Doctor	Eugenio		Pfc	Saguindel	Eusebio	
Sgt	Domantay	Frank	J.	T5	Sahagun	Ponce	
Pvt	Domantay	Chris		Pvt	Saladino	Selmo	
Pfc	Domingo	Rolque	S.	Cpl	Salgado	Mauricio	B.
T4	Donaldo	Bernardo		Pvt	Salvador	Tommy	
T4	Dongallo	Isidro			Samson	Phil	
T5	Duco	Licerio			San Felipe	Clemente	
Lt	Dulay	Jose	P.	Sgt	Sanchez	Policroni	

RANK	LAST NAME	FIRST NAME	M.I.	RANK	LAST NAME	FIRST NAME	M.I.
Cpl	Dumpit	Danny		T5	Sanchez	Apolinar	
Pfc	Duro	Jacinto		Sgt	Santes	Amador	
Lt	Dy	Francisco	J.	Pfc	Santos	Remegio	
Lt	Dykes	Alonzo		Pvt	Santos	Aguilino	
Sgt	Ebarle (Jr.)	Patrick			Santos	C.	
Pvt	Edralin	Alfonzo		Sgt	Sarmiento	George	
Lt	Elefano	Alberto	C.	Lt	Sarmiento	Ignacio	M.
Capt	Elizalde	A.		Sgt	Sarmiento	Robert	
Pvt	Empinado	Ferning		T5	Sarmiento	Rufino	
Cpl	Ergas	Eugene		Cpl	Segui	Salustiano	
Sgt	Escudero	Manuel		Sgt	Seniza	Segundo	
Pvt	Esguerra	Jimmy		Cpl	Seno	Conrad	
Pvt	Espinuevo	Critino		Cpl	Sensano	Daniel	G.
	Espirito	Francisco	D.	Pfc	Serna	Vincente	
Sgt	Esposito	John			Serquinia	Johnny	P.
Pvt	Estabillo	Catalino			Setias	Benjamin	M.
Sgt	Estaquio	Ernest		Pvt	Sevilla	Lorenzo	
Pfc	Estoista	Fred		Lt	Siegal	Herbert	
Sgt	Estrada	Esteban		Sgt	Simaco	Cagaanan	
Sgt	Evangelista	Candido		Sgt	Simangan	Esteban	
Pfc	Exequiel	Pepito		Sgt	Sison	Antonio	
Pfc	Fabros	Alex	D.	Pfc	Solar	Marcelino	V.
Maj	Fajardo	Tirso	J.	Sgt	Solidarios	Sergio	V.
Sgt	Fangon	Johnny		Sgt	Somera	Roy	
Pvt	Fernandez	Sergio		Lt	Songco	Ruben	
Pvt	Fernandez	Juan		T4	Soriano	Arsenio	
Cpl	Fernandez	Simplicio		Pfc	Soriano	John	
Pfc	Filomino	Ulpiano		Pvt	Soriano	Antonio	
Lt	Flor Cruz	Pedro		Cpl	Sorisantos	Torado	
Cpl	Florendo	Teodorico		T4	Sotero	Alfonso	
Cpl	Flores	Conrado	D.	Lt	Suatengco	Eduardo	T.
Pvt	Flores	Sonny		Pvt	Sular	Marcelo	
Pfc	Flores	Alfred			Surel	Juan	C.
Pvt	Fortugaleza	Ireneo		Sgt	Tabios	Va.	R.
Pfc	Fulgencio	Antonio			Taclay	Francisco	
Pvt	Furtogaliza	Ireno		T5	Tactancan	Leoncio	
Sgt	Gaboni	F.			Tallido	Silvino	
T5	Gabriel	Antonio		Pfc	Tamsi	Marcelo	
Pvt	Galavasa	Celestio		Sgt	Tangenete	Marcial	
Pfc	Galeon	Alejo		Pvt	Tapalla	Delfin	M.
T5	Gamatero	Carlos		Pvt	Taporco	Sammy	
Pfc	Gaoiran	Regino	A.	Cpl	Tecson	Ray	
	Garcia	Rosendo	M.	Pfc	Tellano	Paul	
Cpl	Gaskell	Jose		Pvt	Tenio	James	
	Gatan	Vincent	L.	T5	Terren	Joaquin	
Pfc	Gemoya	Amario		Pfc	Terro	John	
Pvt	Genaro	Frankie		Lt	Tergasa	George	T.
T5	Gevarria	Heraclio		Pfc	Tigson	Restituto	
Pvt	Gil	Alfonso		Lt	Tobias	Manuel	
Sgt	Giron	John		Sgt	Tolentino	Gavin	
Maj	Gomez	Celestino	A.	Maj	Topacio	Conrado	S.
Pvt	Gonzales			Capt	Torralba	A. F.	
Cpl	Gorospe	Georgeo		Sgt	Torres	Simplicio	
Sgt	Gorospe	Casimiro		Sgt	Tuzon	Richard	
	Gramata	Mariano	B.	Pfc	Udarbe	Jose	
Lt	Guerrero	Fred	C.	Cpl	Umipig	Salvador	
T4	Guerrero	Maximiano		T5	Unciano	Hilario	
Sgt	Guerrero	Moises		Sgt	Unciano	Isaac	
T5	Guieb	Benny		Lt	Urbano	Isidro	

RANK	LAST NAME	FIRST NAME	M.I.	RANK	LAST NAME	FIRST NAME	M.I.
Pvt	Guile	Howard		Pvt	Urbano	Francisco	
Cpl	Gumangan	Fil		Cpl	Valderama	Abraham	
T5	Guzman	Sandy		Sgt	Valdez	Julian (Joe)	
Pfc	Halabaso	Potenciano		Pvt	Valdez	B.	
Pvt	Halican	Hilario		Sgt	Valdez	Andres	
Pvt	Halog	Macario		Sgt	Valdez	Isidro	
Lt	Hernandez	Al		Lt	Valencia	Elpidio	
Pfc	Herrera	Anastacio		Pvt	Valenzuela	Martin	
	Hilario	Jimmie	V.	Sgt	Valerio	Oliver	
Sgt	Hipolito	Sonny		Pvt	Vallejos	Ariston	
Sgt	Histo	Leoncio		Sgt	Velasco	Roque	
	Holanda	Pedro	P.	Sgt	Velasco	Juanito	
Pvt	Hufana	Frank		Col	Velasquez	Jamie	
Pvt	Hugo	Santos			Ventura	Dionisio	H.
	Ibarra	Egmidio	N.	Lt	Ventura	Tranquilino	
Sgt	Ibea	Artemio		Lt	Verceles	Pedro	P.
Sgt	Ignacio	Miguel		Pvt	Verceles	Fidel	G.
Pvt	Ignaus	Phelix		Lt	Vergara	Tranquilino	E.
Capt	Ileto	Rafael	M.	Pvt	Vierenes	Albert	
Cpl	Ilustre	Ernesto	D.	Pvt	Villaflor	Nick	
Sgt	Ines	George		T4	Villanueva	Angelo	
Pvt	Isip	Pagsilang Rey		Pvt	Villarico	V.	
	Isla	Bernardo	D.	Sgt	Villarta	Frederico	R.
Capt	Iverson	Sidney		Sgt	Viloria	Estenfanio	
Sgt	Jacobs	Leon	A.	T5	Vinluan	Victor	
Sgt	Jamin	Pedro		Cpl	Viray	Don	J.
Pvt	Jamora	Carlos		Cpl	Viray	Amado	J.
Lt	Javier	Genaro	P.	Sgt	Vite	Doroteo	
	Jerry	Jacobs		Lt	Willie	Jose	A.
Lt	Ignacio	Josue	I.	Sgt	Ylagan	Marcelo	
Pfc	Julian	Leo		Pfc	Yorro	Hilarion	
T5	Lachia	Fortunato		Sgt	Yotoko	Salvador	
Sgt	Lacuata	Ariston		Pvt	Young	Al	
Pvt	Lagacan	Laureto	A.	Pvt	Yparraguirre	Rustico	
Pvt	Lalisan	Simplicio	S.	Pvt	Yurong	Albert	
	Lamata	Alfonso			Zabala	Vincent	N.
Cpl	Lancita	Benjamin			Zapata	Ricardo	E.
Cpl	Lapitan	Carlos					

*The purpose is to gather as many names as possible of the men. These are names found in news articles and various sources. Other names may be mentioned in another Appendix.

*Source: The Filipino American Experience Project November 1998
Philippine Scouts Heritage Society by Sean Conejos.*

Appendix A- 4

St Patrick's Day Dinner and Dance
Sponsored by the 2nd Filipino Infantry. Camp Cooke, California March 1943

MENU

COCKTAIL
Half Grapefruit Supreme

RELISHES
Celery Ripe Olives Radishes
Green Onions

SALAD
Chef Mixed Green

ENTREES
Barbecued Pork – Liver Gravy
Roast Young Tom Turkey
Giblet Gravy
Celery – Onion Dressing

VEGETABLES
Potatoes Steamed Rice Green Peas

DESSERT
Co. "I" Cake Special

DRINKS
Coffee Tea

COMPANY 'I', 2ND FILIPINO INFANTRY
CAMP COOKE, CALIFORNIA

ROSTER
CAPT. FRANK K. PREVOST, JR.
Company Commander

First Lt. Stephen C. Sweeney
2nd Lt. Gray A. Mashburn
2nd Lt. Joseph P. Parente

First Lt. Alvin Manuel
2nd Lt. Spencer M. Pitts
2nd Lt. James Martinez

2nd Lt. Isidro B. Samporna

FIRST SERGEANT
Don Del Mar

SUPPLY SERGEANT
Joseph B. Caces

MESS SERGEANT
Juan Narbacan

STAFF SERGEANTS
Buenaventura L. Dolendo

Esteban M. Diputado

SERGEANTS
Benedicto S. Doleon   Juanito A. Velasco   Guillermo G. Pardo

CORPORALS
Gil D. Dalit
Vicente G. Goloyogo
Jose P. Fuentes
Felimon S. Maala

Ralph B. Noel
Leo D. Lozada
Pascual Lagarde
Maximino Cacas

TECH 4TH GRADE
Alponse M. Tolentino

Danny V. Navana

TECH 5TH GRADE
Izzy L. Carboneli   Temoteo R. Obana   Nasario A. Marchand

PRIVATES FIRST CLASS
Rafael A. Acierto
Paul S. Madriaga
Pat M. Dizon
Antonio R. Martinez

Victoriano C. Venus
Joaquin L. Laconsay
Andres C. Ramos

PRIVATES

Dionesio E. Abad	Alberto C. Esmenia	James V. Perenia
Maximo Acosta	Benny D. Felicitas	Eugenio T. Piza
Juan P. Agustin	Camelo Flores	Johnny A. Queen
Dionesio Alingatong	Joaquin Gaya	Patricio O. Ramirez
Ciriaco A. Alquiza	Agapito R. Gonzales	Eustaquio B. Ramones
Licerio A. Arreola	Federico Isla	Francisco L. Rapacon
Sinforozo Baldiviso	Pedro Joven	Demetro G. Rimando
Valentin S. Balagot	John R. Lauder	Leoncio F. Rubio
Mauro F. Balisbisana	Alfredo P. Lamparas	Catalino R. Salazar
Antonio B. Bertes	Emeterio P. Lianos	Pedro B. Salvador
Danny R. Blancas	Tony D. Luarca	Juan Salaga
Pelisimo F. Cabaran	Numeriano D. Malasig	Catalino B. Sanchez
Tom Castillo	Rufino N. Manzano	Isabelo D. Sajor
Felomino S. Carreon	Pedro C. Mendoza	Anselmo A. Tabanlar
Alijadro J. Casanares	Gelacio B. Munar	Castor T. Tongko
Ciriaco N. Garaang	Frank C. Osias	Perfecto S. Villasista
Paulino Corpuz	Alfredo B. Patiko	Feleciano A. Yoro
Bernardino M. Abang	Casamero U. Peralta	Francisco T. Dao-a

Telesforo M. Acosta
Arcadio M. Agcaoili
Macario A. Alicante
Florentino A. Alianza
Diego T. Asuncion
Anastacio P. Basilio
Maximo S. Barongan
Felipe D. Barientos
Castor D. Binan
Procopio T. Briones
Saturnino J. Cabaiz
Julian A. Castillo
Sencio V. Cardenas
Bruno T. Calubag
Juan Cornelius
Melecio T. Corpuz
Guillermo M. Abellera
Julian D. Adricula
Danny T. Agbisit
Porferio Alisna
Narciso N. Arcia
Sinforozo N. Aquino
Felipe U. Batalia
Emil C. Baynana
Raymondo L. Bernido
Benito H. Bijoc
Felix B. Budiao
Vicente G. Calindas
Serafin M. Castillo
Gay C. Castillo
Marcelo T. Calubag
Donato P. Cortez
Demetrio R. Claro
Ceferino Crisologo
Rufino A. Davis
Matias A. Diego

Damian S. Pelegrino
Sofronio L. Presquito
Eustaquio S. Quinto
Macario A. Ramones
Bernabe R. Racil
Pablo R. Rangel
James R. Renozo
Sofronio M. Sandoval
Fred A. Salde
Martin B. Salvacion
Pedro Sajor
Felomino G. Serafin
Isidro Tanes
Alipio C. Toribio
Ambrosio C. Uganiza
Francisco A. Venzon
Inacio V. David
Paulino P. Damazo
Bartolome L. Dulatre
Silvestre Escudero
Joe Fernandez
Mauro B. Garcia
Thomas D. Gaduang
Modesto A. Habon
Vincente J. Allores
Josue E. Kimpo
Marcelo M. Lavracio
Lorenzo L. Lasquite
Don J. Liberato
Andres A. Magdogo
Geronimo D. Mangubat
Flavio Medrano
Benedicto G. Menor
Teofilo G. Nudalo
Moises Pagalan
John P. Pascual

Guillermo Dayo-an
Dionesio A. Duculan
Mauro A. Evangelista
Pedro V. Flores
Joseph M. Galang
Arcadio Gines
Larry P. Hilario
Bartolome Javier
Emilio N. Lobarcan
Isidro Lagod
John T. Larioza
Catalino G. Lozano
Carl P. Malacaste
Luis F. Marquez
Frank E. Mercado
Jovencio Miguel
Crispiano G. Olivar
Pedro P. Palac
Leoncio A. Palacpac
Johnny C. Peralta
Vincente R. Plaza
Casamero A. Quines
Benigno A. Ramos
Lupe A. Ramos
Salvador Raz
Monico T. Rizalvo
Segundo L. Rubio
Jerry Salde
Jose F. Salez
Lorenzo B. Sajor
Leo M. Serquina
Hermogines Soriano
Juan Tumulak
Pedro B. Tingkang
Federico L. Viscara

Appendix A- 5

First Filipino Infantry Regiment
"Sulung" Laging Una Band Members

Conrad O. Oller
Sulpicio A. Caballes
Alfonso T. Conde
Francisco D. Espirito
Phil Samson
Hilarion M. Nicholas
Antonio G. Desingano
Bernardo D. Isla
Julian L. Lopez
Vincent N. Zabala
Anastacio T. Paulos
Ricardo E. Zapata

Bernard C. Anselmo
Mariano C. Cabrales
Buenaventura B. Dalao
Jose R. Pastrana
Mariano B. Gramata
Robert L. Arrollado
Juan C. Surel
Jimmie V. Hillario
Dionisio H. Ventura
Daniel S. Palmeto
Pedro M. Martinez

Quintin A. Aninao
Victor C. Calderon
Egmidio N. Ibarra
Rosendo M. Garcia
Nemecio D. Agsalud
Johnny P. Serquinia
Vincent L. Gatan
Sammy G. Regacho
Carlos G. Malla
Frank S. Lopez
Marciano V. Africa

Source: 1st Filipino Band Good Will Builder, The Bealiner. 10 March 1943

Appendix A-6

1st Filipino Infantry Camp Roberts Trainer 1943

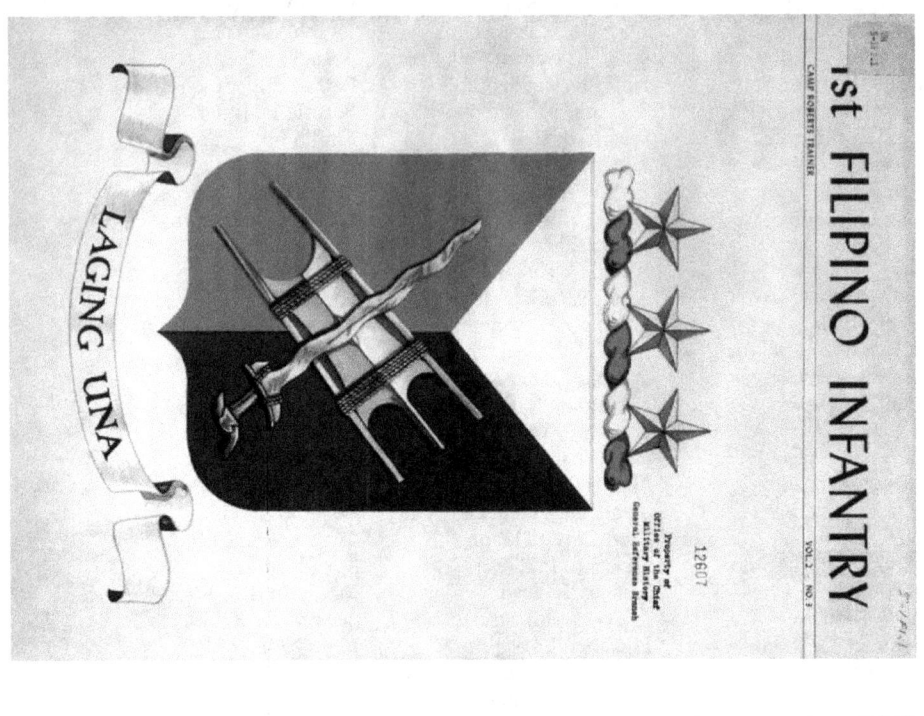

CAMP ROBERTS TRAINER

VOL.2 NO.8

1st FILIPINO INFANTRY

LAGING UNA

12607

1st FILIPINO REGIMENT

LAGING UNA

Coat of Arms

THE COLORS OF THE SHIELD REPRESENT THE NATIONAL COLORS OF THE PHILIPPINES AND THE UNITED STATES. THE CROSSED KRIS AND IGOROT WAR SHIELD REPRESENT THE TWO DOMINANT WAR-LIKE PAGAN TRIBES OF THE ISLANDS. THE THREE STARS ARE TAKEN FROM THE PHILIPPINE FLAG AND SYMBOLIZE LUZON, VISAYAS AND MINDANAO, THE THREE PRINCIPAL ISLANDS.

Shoulder Sleeve Insignia

ON A YELLOW DISC 3¼ INCHES IN DIAMETER, LEAVING ⅛ INCH EDGE, A CONVENTIONALIZED BLACK VOLCANO EMITTING SMOKE, THE LATTER CHARGED WITH THREE YELLOW MULLETS IN FESS.

THE GOLD AND BLACK COLORS REPRESENT THE BLACK OF THE VOLCANO ERRUPTING; THE GOLD BACKGROUND BEING THE GOLDEN OPPORTUNITY OF RESTORING THE COUNTRY TO ITS RIGHTFUL OWNERS. THE VOLCANO IS A REPRESENTATION OF THE NATION VOLCANO, NORMALLY INACTIVE BUT NOW SERVING WITH THE RAGE OF CONQUEST. THE THREE STARS ARE THE FILIPINO SYMBOL, TAKEN FROM THE PHILIPPINE FLAG AND SYMBOLIZE LUZON, VISAYAS AND MINDANAO, THE THREE PRINCIPAL ISLANDS.

To the Members of the Regiment....

I have considered it an honor to command this Regiment. You have made my work easy by your fine spirit and your devotion to duty. Such qualities will bring us victory... God willing, our history will fill many more books of this size and be a great source of pride to all . . . . It is my hope that this book will help to cherish the memories of days spent together during the formative period of our history. A glance at the pages will show that we have worked hard. May the work not have been in vain, but instead, help us to conduct ourselves with glory on the battlefield. -

Robert H. Offley

Robert H. Offley
COLONEL, INFANTRY
COMMANDING.

Col. Rob't H. Offley
COMMANDING

Maj. Fajardo
S-3

Capt. Ashcraft
ADJUTANT

Chaplain Noury

Lt. Col. Mann
2nd Bn.

Lt. Col. Kessner
3rd Bn.

Lt. Col. Mullinix
EXECUTIVE OFFICER

Capt. Spiotto
S-4

Maj. Iverson
1st Bn.

Maj. Mueller
S-2

CEREMONY IN HONOR OF VISIT OF VICE PRESIDENT OSMEÑA
OF THE PHILIPPINE COMMONWEALTH

COL. OFFLEY MAJ. GEN. GILLEM VICE PRES OSMEÑA COL. FITCH COL PIERCE

S-1 SECTION

OPERATIONS

SUPPLY SECTION

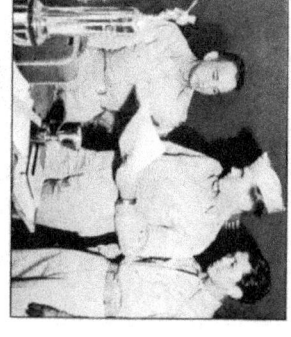

SPECIAL SERVICES

PERSONNEL SECTION

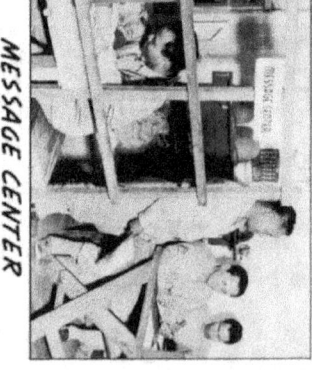

MESSAGE CENTER

FIRST FILIPINO INFANTRY BAND
CAPT. ASHCRAFT-COMMANDING

HEADQUARTERS COMPANY'S VEHICLES

HEADQUARTERS COMPANY
CAPT. SCHULPS - COMMANDING

SERVICE COMPANY
CAPT. OLLS - COMMANDING

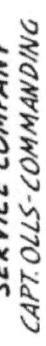

RED CROSS
FIELD DIRECTOR
ZUEGER

HEADQUARTERS COMPANY OFFICERS

SERVICE CO. OFFICERS

ANTI-TANK CO. OFFICERS

MEDICAL DET. OFFICERS

ANTI-TANK COMPANY
CAPT. PAUL - COMMANDING

MEDICAL DETACHMENT
MAJ. SPANGLER - COMMANDING

HEADQUARTERS COMPANY, 1ST BATTALION
CAPT. FIELD, COMMANDING

HEADQUARTERS STAFF, 1ST BN.
LT. COL. VELASQUEZ, COMMANDING.

COMPANY A
CAPT. DI BIANCA, COMMANDING

COMPANY B
CAPT. CHAVEZ, COMMANDING

HEADQUARTERS COMPANY, 1ST BN. OFFICERS

A COMPANY OFFICERS

B COMPANY OFFICERS

C COMPANY OFFICERS

D COMPANY OFFICERS

COMPANY D
1ST LT. DAVIS, COMMANDING

COMPANY C
1ST LT. DUNCAN, COMMANDING

A PART OF MOTOR POOL

A TYPICAL KITCHEN

PERSONNEL SECTION

PARADE - END OF BOND RALLY, SAN LUIS OBISPO, CALIFORNIA.

SWEARING IN OF 1000 MEN AS CITIZENS OF THE UNITED STATES, CAMP BEALE, CALIFORNIA.

ANTI-TANK COMPANY *in* ACTION

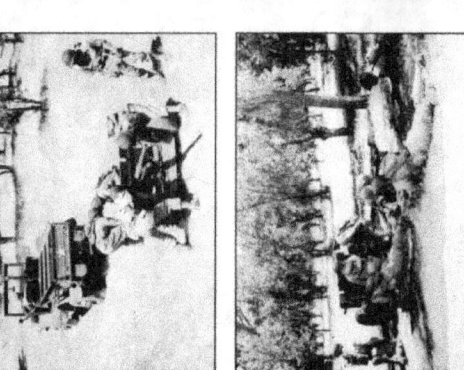

HEADQUARTERS COMPANY, 2ND BATTALION
CAPT. LLORENTE, COMMANDING

HEADQUARTERS STAFF, 2ND BATTALION
LT. COL. MANN, COMMANDING

H COMPANY OFFICERS

G COMPANY OFFICERS

F COMPANY OFFICERS

E COMPANY OFFICERS

HEADQUARTERS COMPANY, 2ND BN. OFFICERS

COMPANY G
CAPT. BERGEN, COMMANDING

COMPANY H
CAPT. PARDO, COMMANDING

COMPANY E
CAPT. PEARLMAN, COMMANDING

COMPANY F
CAPT. RUSSE, COMMANDING

HEADQUARTERS COMPANY, 3RD BN.
1ST LT. FULLER, COMMANDING

HEADQUARTERS STAFF, 3RD BN.
LT. COL. KESSNER, COMMANDING

COMPANY M OFFICERS

COMPANY L OFFICERS

COMPANY K OFFICERS

COMPANY I OFFICERS

HEADQUARTERS CO., 3RD BN. OFFICERS

COMPANY L
CAPT. FLOR CRUZ COMMANDING

COMPANY M
CAPT. WALLENFANG , COMMANDING

COMPANY I
CAPT. YANCEY-COMMANDING

COMPANY K
IST LT. KUHN-COMMANDING

CAROLE LANDIS and POWERS MODELS AT—

1st FILIPINO INFANTRY BOND RALLY

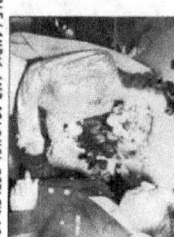
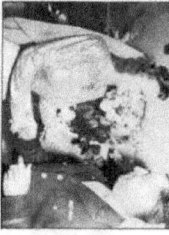
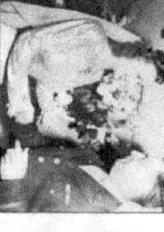

CAROLE LANDIS AND COLONEL OFFLEY AT RECEPTION BY OFFICERS

CAROLE AND CAPT. DI BIANCA

CAROLE GLADLY POSES FOR PICTURES

LECHONADO IN HONOR OF CAROLE AND POWERS GIRLS SELECTING THREE BEST DRESSED SOLDIERS

MASS—SAN ANTONIO MISSION—HUNTER LIGGETT MILITARY RESERVATION

GOOD FRIDAY MASS CAMP ANZA ROADS

1ST FILIPINO INFANTRY'S FAVORITE WEAPON "the BOLO"

CPL. ILUSTRE. EDITOR 'B' "NEWS"

MEDICAL DETACHMENT IN Action

SATURDAY MORNING INSPECTION

AUTOGRAPHS

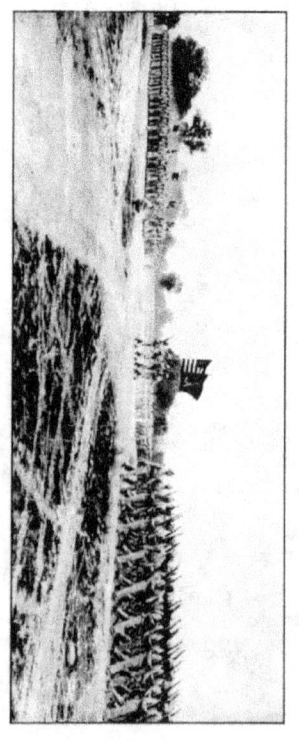

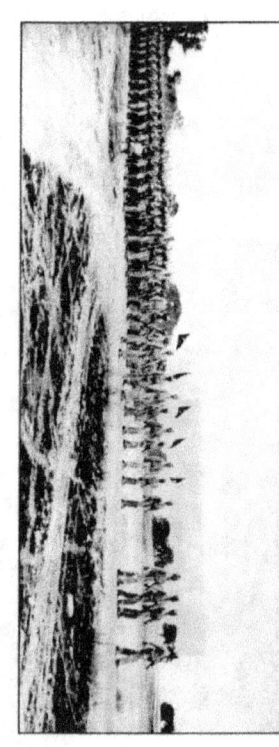

ORGANIZATION DAY REVIEW AND PRESENTATION OF REGIMENTAL COLOR
BY MAJOR GENERAL MORRIS, JULY 13, 1943

1st Filipino Infantry Regiment. The same men carried the Regimental flags from 1942 to after the war at parades and military events.

Appendix A- 7

**Summer Fiesta sponsored by the 2nd Battalion, 1st Filipino Infantry
Camp Roberts, Hunter-Liggett. August 1, 1943.**

PROGRAM
Sunday, August 1, 1943

Mass, Mission San Antonio	11:30 a. m. to 12:15 p. m.
Lechonada and Refreshments	12:30 to 2:00 p. m.
Welcome Address	S/Sgt. S. V. Solidarios
Introduction of Honored Guests	Capt. Abdon Llorente
"On To Bataan"	First Filipino Infantry Band

FEATURES BETWEEN DANCES

Solo	Miss Mary Alice Espiritu
Tap Dance Exhibition	Al Nipomoceno, Hq. Co., 2nd Bn.
Guitar Solo	Stanley Orbeta, Co. "G"
"What's The Good Word Mr. Blue Bird"	Miss Martha Braga, Stockton
Song	Pfc. Vicente Serna, Co. "E"
"That Soldier of Mine"	Miss Frances Begley, San Francisco
Tagalog Song	Pvt. Rustico Yparraguirre, Co. "F"
Hula Hula Dance	Miss Virginia Garcia, Salinas
Violin Solo	Pvt. A. Baldonedo, Co. "H"
Hit Song	Miss Consuelo Gacer, Lodi, California
Solo	Miss Dolores Catalia, Oakland
Dancing Exhibition	Miss Betty Paredes, San Francisco
Vocal Solo	Miss Pacita Todtod, San Jose
Short Talk	Major Talman C. Budd, Executive Officer, 2nd Bn
Closing Remark	Lt. Col. Geo L. Mann, Commander, 2nd Bn
National Anthem	First Filipino Infantry Band

DANCING	BALL GAMES	DRINKS
BOXING	VOLLEY BALL	SEPA (SIPA)

BN. HQ. and HQ. COMPANY

Lt. Col. George L. Mann	Commanding Officer
Major Talman C. Budd	Executive Officer
Captain Abdon Llorente	djutant
1st Lt. Wayne R. Rader	Intelligence Officer
Captain Joseph L. Grill	Operations and Training
1st Lt. Emmett C. Cochran	Supply Officer

"E" COMPANY

Capt. Chester E. Pearlman	Commanding Officer
1st Lt. Robert H. Green	Executive Officer

"F" COMPANY

Capt. Frederick R. Russe	Commanding Officer
1st Lt. Lewis C. Wingard	Executive Officer

"G" COMPANY

Capt. Edward B. Bergen	(Detach Service)
1st Lt. B. W. Williby	Commanding Officer
2nd Lt. E. H. Judge	Executive Officer

"H" COMPANY

Capt. Clay E. Pardo Jr	Commanding Officer
1st Lt. George W. Williams	Executive Officer

COMMITTEES

GENERAL CHAIRMAN — S/SGT. S. V. SOLIDARIOS

Invitation
Sgt. Dan Conception
S/Sgt. Salvador Yotoko
Cpl. J. Gaskell
Cpl. E. Bibat

Athletics
Sgt. Raymond Menes
Cpl. J. Gaskell
Pfc. Pepito Exequiel
S/Sgt. S. M. Sabio

Guides
1st Sgt. S. Hipolito
S/Sgt. Cagaanan Simaco
Sgt. Z. Azcueta
Sgt. Agcaoili, Victorio

Decoration
Cpl. Valli Alfonso
Cpl. Castillo Arturo
Sgt. Raymond Menes
Cpl. J. Gaskell

Program
S/Sgt. Joe Mahusay
Sgt. L. Alcaide
Sgt. P. B. Martin

Refreshments
Sgt. Escolastico M. Vasquez
S/Sgt. Eulalio M. Agustin
S/Sgt. Barry Barrios Pfc. Stanley Lopez

Finance
S/Sgt. Claudio M. Cabel
S/Sgt. Sergio Solidarios
1st Sgt. Sonny Hipolito

S/Sgt. Felipe R. Pobre
Sgt. Jose P. Sotelo

Appendix A- 8

Roster of 1st Filipino Infantry Regiment Officers December 31, 1943
Regimental Headquarters & Headquarters Company

Colonel Robert Offley ***
Commanding Officer
Joined 07/31/42

Lt. Colonel Earl L. Mullinix
Executive Officer
Joined 06/15/42

Major George H. Mueller
Intelligence Officer, S-2, G-2, A-2

Major Tirso C. Fajardo*
Operations and Training Officer, S-3, G-3, A-3
Joined 07/28/42

Captain Julius L. Klein
Adjutant

Captain Walter J. Schulps
HQ Company Commander

Captain Russell E. Jones

1st Lieutenant Charles H. Tennis
Motor Transportation Officer

1st Lieutenant Norman E. Goodwin

1st Lieutenant Henry M. Durham
Communications Officer

1st Lieutenant Edwin J. Johnson
Liaison Officer

1st Lieutenant Truman P. Lambert
Liaison Officer

1st Lieutenant A. V. Mossotti
Liaison Officer

1st Lieutenant Francisco Paguirigan*
Intelligence Officer, G-2, A-2, S-2

1st Lieutenant Oliver L. Rowland
Intelligence Officer, G-2, A-2, S-2

2nd Lieutenant John L. Robinson
Communications Officer

2nd Lieutenant Honorato S. Echavez*

WOJG Richard G. Snyder
Band Leader

Officers Attached from Other Organizations

1st Lieutenant Maurice B. S. Legare
Chaplain

1st Lieutenant Pedro Verceles
Chaplain

1st Lieutenant Peter O. Monleon
Chaplain

1st Battalion

Lt. Colonel Jamie C. Velasquez*
Commanding Officer
Joined 11/19/42

Major Talman C. Budd
Executive Officer

Captain Atanacio T. Chavez*
Student Officer

Captain Edwin H. Field
Student Officer

1st Lieutenant James E. Healey
Operations and Training Officer, G-3, A-3, S-3

1st Lieutenant Arthur C. Smith
Communications Officer

2nd Lieutenant William A. Simpson
Anti-Tank Platoon Officer 2/16/43

2nd Lieutenant Alfred J. Servant
Ammunition Pioneer Platoon Officer

1st Lieutenant Francis J. Clancy
HQ Company Commander

2nd Lieutenant William O. Martin
Motor Transportation Officer

2nd Lieutenant Venancio D. Demandante
Student Officer

Company A

Captain Joffre L. Gueymard
Company Commander

2nd Lieutenant Robert L. Keys
Platoon Officer

1st Lieutenant Issac F. Unciano***
Platoon Officer

2nd Lieutenant Herbert H. Schmidt, Jr.
Platoon Officer

2nd Lieutenant Alexander Dram
Platoon Officer

2nd Lieutenant Harvey H. Thomas
Platoon Officer

Company B

Captain J. Di Bianca
Company Commander

2nd Lieutenant Lloyd L. Mize
Platoon Officer

1st Lieutenant Morris H. Barasch
Executive Officer

2nd Lieutenant George E. Schlachter
Platoon Officer

2nd Lieutenant Sydney N. Friedman
Platoon Officer

2nd Lieutenant Joseph A. Price
Platoon Officer, Heavy Weapons

Company C

Captain Harry N. Duncan
Company Commander

2nd Lieutenant Herbert C. Johnson
Platoon Officer

1st Lieutenant Harvey E. Sweeney
Executive Officer

2nd Lieutenant John V. Swango
Platoon Officer

2nd Lieutenant John D. Davidson
Platoon Officer

2nd Lieutenant Gervasio G. Sese
Platoon Officer

2nd Lieutenant Frank J. Dwelle, Jr.
Platoon Officer

2nd Lieutenant Charles F. Vanderpool
Platoon Officer

2nd Lieutenant William E. Vogel
Platoon Officer

Company D

Captain Vernon L. Davis
Company Commander

1st Lieutenant Alonzo T. Dykes
Machine Gun Platoon Officer

1st Lieutenant Harrison B. Hair
Company Commander

1st Lieutenant Albert J. Eger
Platoon Officer

1st Lieutenant William E. Price
Executive Officer

2nd Lieutenant Claud M. Gregory, Jr.
Platoon Officer

2nd Battalion

Lt. Colonel George J. Mann
Commanding Officer

1st Lieutenant Lee W. Kelly
Motor Transportation Officer

Major William C. Parker
Executive Officer

2nd Lieutenant Doyce Hamilton
Intelligence Officer, G-2, A-2, S-2

Captain Fred C. Guerrero
HQ Company Commanding Officer
Joined 12/12/42

Captain Abdon Llorente
HQ Company Executive Officer
Joined 04/10/42

2nd Lieutenant Irving H. Selzer
Ammunition Pioneer Platoon Officer

2nd Lieutenant Edwin Livermore
Anti-Tank Platoon Officer

2nd Lieutenant Ellis E. Myers
Communication Officer

Company E

Captain Chester A. Pearlman
Commanding Officer

1st Lieutenant Robert H. Green
Executive Officer

1st Lieutenant Richard E. Hoyt
Executive Officer

1st Lieutenant Walter L. Crosby
Platoon Officer

1st Lieutenant Eugene H. Judge
Platoon Officer

2nd Lieutenant Henry R. Bittick
Platoon Officer

2nd Lieutenant Maurice G. Boyer
Platoon Officer

2nd Lieutenant Thomas P. Landers
Platoon Officer

2nd Lieutenant Ignacio I. Josue**
Platoon Officer, Heavy Weapons

Company F

Captain Frederick G. Russe
Student Officer

1st Lieutenant Robert D. Irving
Executive Officer

2nd Lieutenant Jasper W. Farrow
Platoon Officer

2nd Lieutenant Leonard M. Jones
Platoon Officer

2nd Lieutenant Charles F. Leonard
Platoon Officer

2nd Lieutenant Jack H. Morris
Platoon Officer

2nd Lieutenant Rafael M. Ileto**
Platoon Officer

2nd Lieutenant Frank C. Jackson
Platoon Officer, Heavy Weapons

2nd Lieutenant Kenneth I. Siegel
Platoon Officer

Company G

Captain Joseph L. Grill
Company Commander

1st Lieutenant Pedro S. Balono***
Platoon Officer

2nd Lieutenant Terrell O. B. Harrison
Platoon Officer

2nd Lieutenant Arthur R. Hines
Platoon Officer

2nd Lieutenant Richard W. Murdock
Platoon Officer

2nd Lieutenant Donald H. McBride
Platoon Officer

2nd Lieutenant Jack Palmer
Platoon Officer

2nd Lieutenant Domenick Zema
Platoon Officer

Company H

Captain Clay E. Pardo, Jr.***
Heavy Weapons Company Commander

2nd Lieutenant William F. Lee
Platoon Officer

1st Lieutenant Robert M. Plangman
Executive Officer

2nd Lieutenant Erwin A. Bellm
Platoon Officer

2nd Lieutenant Isidro D. Urbano
Platoon Officer

2nd Lieutenant Robert M. Viale
Platoon Officer

3rd Battalion

Lt. Colonel Jerome Kessner
Commanding Officer

Major Edwin E. Hopson
Executive Officer

Captain Pedro R. Flor Cruz,***
Operations and Training Officer, G-3, A-3, S-3
Joined 6/26/43

1st Lieutenant George R. Ferguson
Operations and Training Officer, G-3, A-3, S-3

2nd Lieutenant Charles E. Meyers
Anti-Tank Platoon Officer

1st Lieutenant Ross B. Hill
Intelligence Officer, G-2, A-2, S-2

1st Lieutenant William F. Jeffreys
HQ Company Commanding Officer

1st Lieutenant Dorsay V. Bias
Communication Officer

2nd Lieutenant Morris A. Frazier
Communications Officer

Company I

Captain Don A. Yancey
Company Commander

1st Lieutenant Paul F. Rodriguez***
Executive Officer

1st Lieutenant Delbert K. Scheiss
Platoon Officer, Heavy Weapons

2nd Lieutenant John J. Stockton
Platoon Officer

2nd Lieutenant Albert S. Kelly
Platoon Officer

2nd Lieutenant Jacob J. Klomp
Platoon Officer

2nd Lieutenant Eduardo T. Suatengco
Platoon Officer

1st Lieutenant Edward A. Kuhn
Company Commander

Captain Robert J. Terrening
Student Officer

1st Lieutenant Clarence C. Bernett
Executive Officer

2nd Lieutenant Martin C. Kehoe
Student Officer

Company K

1st Lieutenant Norman F. McCarthy
Executive Officer

1st Lieutenant James O. Moran
Platoon Officer, Heavy Weapons

2nd Lieutenant Fred M. Hyde
Platoon Officer

Captain Robert J. Fuller
Company Commander

1st Lieutenant Joseph M. Fahey
Executive Officer

2nd Lieutenant William G. Pounds, Jr.
Platoon Officer

Company L

2nd Lieutenant Jim E. Cherry
Platoon Officer

2nd Lieutenant Miguel Manresa, Jr.
Student Officer

Company M

Captain Lee L. Wallenfang
Heavy Weapons Company Commander

2nd Lieutenant Benny D. Cablay
Student Officer

1st Lieutenant Philip A. Pitsker
Executive Officer

2nd Lieutenant John P. Diaz
Platoon Officer

1st Lieutenant Vernon C. Anderson
Platoon Officer

2nd Lieutenant Gerald S. Marshak
Platoon Officer

1st Lieutenant Clifford D. Sivil
Machine Gun, Platoon Officer, Heavy Weapons

2nd Lieutenant Edward F. Swain
Machine Gun, Platoon Officer, Heavy Weapons

Service Company

Captain Victor Olls
Commanding Officer

1st Lieutenant John T. Gill, Jr.
Supply Officer, S-4

Captain Virgil B. Ashcraft
Personnel Officer

2nd Lieutenant Anselmo P. Canapi
Student Officer

Captain Compton J. Gault
Information Officer, Special Services

2nd Lieutenant Myron B. Haas
Platoon Officer

Captain Clyde M. Haas
Automotive Maintenance Officer

2nd Lieutenant Stuart J. Linc
Student Officer

Captain Anthony Spiotto
Supply Officer, S-4

2nd Lieutenant Vincent V. Rosal
Student Officer

1st Lieutenant Robert H. Bessey, Jr.
Supply Officer

2nd Lieutenant Tranquilino E. Vergara
Student Officer

1st Lieutenant Emmett C. Cochran
Ammunition Supply Officer

CWO Herschel Womack
Automotive Maintenance Officer

WOJG Pedro P. Holanda***
Supply Officer General, S-4

Cannon Company

Captain Clarence W. Bruce
Company Commander

2nd Lieutenant William W. Fehtis
Platoon Officer

1st Lieutenant Arthur J. Rosenlof
Cannon Howitzer Platoon Officer

2nd Lieutenant Rafael T. Ruiz
Student Officer

2nd Lieutenant Joseph E. Crum
Platoon Officer

2nd Lieutenant Jack H. Williamson
Motor Transportation Officer

2nd Lieutenant Vincent P. Yuscavitch
Platoon Officer

Anti-Tank Company

Captain Charles C. Paul, Jr.
Commanding Officer

2nd Lieutenant John A. Devine
Platoon Officer

1st Lieutenant Myer Cohen
Executive Officer

2nd Lieutenant Ralph A. McBroom
Platoon Officer

1st Lieutenant Bennett F. Pope
Platoon Officer

2nd Lieutenant Allan A. Smith
Platoon Officer

Major Frederick E. Spangler
Commanding Officer

Captain Mariano R. Aguirre
Medical Officer, General Surgery
Joined 08/08/42

Captain Modesto R. Paragas
Medical Officer, General Surgery

Captain Francisco A. Rosete
Medical Officer, General Surgery

2nd Lieutenant Eugene Montoya, Jr.
Platoon Officer

Medical Detachment
Captain Gregorio P. Chua
Medical Officer, General Surgery

Captain Conrado S. Topacio
Medical Officer, General Surgery

Captain Jose A. Willie
Medical Officer, General Surgery

Captain Benjamin M. Setias
Dental Officer General

1st Lieutenant Celestino A. Gomez
Dental Officer General

Unknown Assignment
2nd Lieutenant Alberto C. Elefano
Platoon Officer

Legend:
Bold text – Filipino officers
* Served with the Philippine Scouts before and after WWII
** Served with the Philippine Scouts after WWII.

Other officers not marked may have served with the Philippine Scouts after the war.

Source: Sean Conejos' research from the U.S. National Archives and Records Administration

Appendix A- 9

**Christmas Day 1943 sponsored by the Regimental Headquarters,
1st Filipino Infantry Regiment Camp Beale**

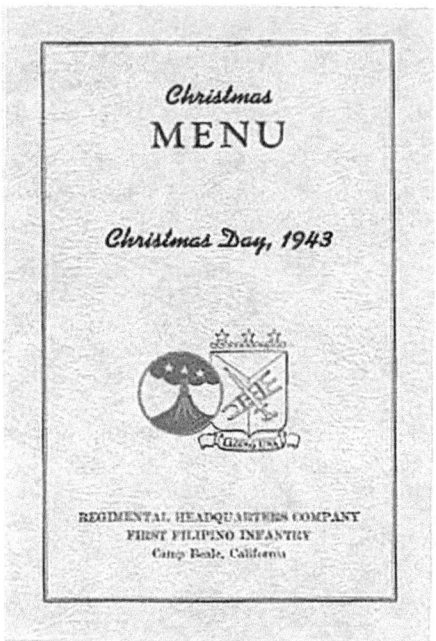

Christmas
MENU

Christmas Day, 1943

REGIMENTAL HEADQUARTERS COMPANY
FIRST FILIPINO INFANTRY
Camp Beale, California

MENU
Chicken Noodle ala Prancaise

Sweet Mixed Pickles   Olives
Assorted Vegetables With Dressing
Roast Stuffed Tom Turkey
Giblet Gravy   Sage Dressing
Butter Peas
Mashed Potatoes Fanchonian
Victory Cake
Fruit Punch ala Englaise
Assorted Nuts   Assorted Hard Candies
Coffee

Roster
REGIMENTAL HEADQUARTERS COMPANY
FIRST FILIPINO INFANTRY

COMPANY OFFICERS
CAPTAIN WALTER J. SCHULPS
Commanding

1ST LT. STANLEY W. ZEGEL
Executive Officer
1ST LT. FRANCISCO PAGUIRIGAN
Intelligence Officer
1ST LT. HENRY M. DURHAM
Communication Officer

1ST LT. OLIVER ROWLAND
2ND LT. JOHN L. ROBINSON
2ND LT. ELLIS E. MYERS
2ND LT. JULIUS ARIAIL
WO J-G RICHARD G. SNYDER
Band

ACTING FIRST SERGEANT
Senando Cuevas

MASTER SERGEANT
Wallace H. Martin

TECHNICAL SERGEANT
John O. Aspiras

STAFF SERGEANTS

Jose G. Ancheta	Emiterio Romero	Maurice A. Estioko
Mark J. De Guzman	Isidoro Refuerzo	Lawrence Aglipay

SERGEANTS

Gilbert Biloy	Anastacio C. Malia	Pablo Viduya

TECHNICIANS FOURTH GRADE

Cezar L. Mamaril	Martin Quibuyen	Valentino V. Velasco
Braulio Orate	Tranquilino Z. Soriano	Cenon Q. Viloria
	Edilberto P. Velasco	

CORPORALS

Felix Contilo	Florencio Madamba	Serviliano Y. Masinda
	Clemente Maigue	

TECHNICIANS FIFTH GRADE

Angel R. Carino	Arcadio Felix	John A. Ratliff
Severo Cres	John B. Finones	George C. Ubungen
Elpidio De Leon	Roman I. Ilar	Gabriel Valdez
Felix Della	Hilario Gravilias	Danny Ventura
Delix Dilodilo	Austin L. Quia	Teofilo O. Zabala

PRIVATES FIRST CLASS

Cleto T. Anacleto	Francisco Galvez	Glicerio M. Molina
Rufino C. Ancheta	Juan E. Garcia	Esteban Monteclaros
Johnny A. Ascuncion	Pio C. Huligana	Vincent Pojas
Juan C. Berganio	George C. Madamba	Rafael C. Porciunco
Basilio A. Calub	Jose B. Mapili	Pedro J. Rayas
Eugenio J. Casenas	Segundo J. Maratas	Andres Reyes
Juan Dacanay	Philip Martin	Pedro F. Reyes
Federico B. Dequina	Eulalio N. Mascarinas	Dioscoro M. Ureta
Joseph Estep	Paul G. Mata	Muguel C. Vinoya
Zoilo G. Eugenio	Pablo Q. Menor	Benjamin S. Yere

PRIVATES

Federico A. Antonio	Francisco Fernandez	Mike P. Ragasa
Antonio A. Arcala	Selvino F. Fernandez	Zacarias R. Rivera
Glicerio V. Ayson	Eddie Foronda	Ignacio Roque
Graciano Baclig	Vicente Fuentes	Juan A. Salindong
Iaias Balotite	Simplecio Galope	Lee L. Sapaden
Emil Bartolome	Roque Hugo	Jose Sinay
Jacinto J. Bolante	Miguel Y. Inong	Sammy Sotelo
Perfecto Cabututan	Celestino M. Labaco	Rizalino C. Tejada
Magno C. Candilinia	George P. Pablico	Primo Valledor
Zoilo Castillo	Pedro B. Padayhag	Felix Viloria
Candido M. Celestino	Mariano Pasay	Eustaquio Vinoya
Rupert V. De Castro	Patrocinio S. Pascua	Alfredo S. Villagarcia
Frederico S. Doria	Hilario T. Pertubal	Oldorico M. Vicitacion

Appendix A- 10

Headquarter Company
1st Battalion, 2nd Filipino Infantry Regiment
Camp Cooke 15 January 1944

Richard Valiente
Maurice Skelson
Deminador Bautista
Hornato Cases
Thomas Hallaway
Stephen McElroy
Jon Jones
Alfredo Nasarino
Gregorio De Lima
Antonio Ballaros
Macario Aliante
Casemero Rabanal
Ricardo Sarabio
Manuel Ruero
Luiciano Florendo
George Lalaun
Elpedio Leparzo
Felix Floresca
Celedonio Viloria
Demitrio Trenidade
Rosario Macabio
Ambrosio Lolyola
Diego Miro
Refino Garey
Diosdad Aba
Guillermo Abellera*
Tanineo Agbeinta
Manuel Garcia
Vesente Reyes
Meck Folor
Benny Ayco

Domingo Cayabyab
Paul Peralta
Lorenzo Sevillia
Phrase Rayborn
Joe Leaming
Thomas Wilkinson
Danny Dompit
Francisco Pulido
Lynn Lopes
Dominador Parcio
Luis Dulyaga
Pedro Batalla
Demetrio Laraf
Telesforo Acosta
Pedro Veray
Augosto B. Dogue
Honario Villanaga
Antonio Martinez
Sedora Mangoing
Frank Sarimento
Urbano Pentescostis
Pedro Tuazan
Santiago Baldua
Santos Nilo
Geralado Igarta
Cornelio Flores
Juan Capusan
Diego Aroncion
Alpredo Pronda
Goldalpredo Rubio

Vitaliano Sugan
Benny G. Bacnat
Remegio DeLos Santos
Antonio Manginsay
Salustino Segui
Catalino Data
Arcadio Banez
Lorengo Valdez
Benito Corpoz
Arcadio Oloroso
Francisco Santos
Vesinte Molina
Mariano Larina
Felix Abanel
Victorio Campos
Sesenado Cabel
Benito Bergano
Gregorio Tayabas
Ambrosio Isadora
Larry Osalva
Hilario Pera
Juan Balasica
Marino Rabang
Ulpiano Revera
Augasto Besguera
Marcalino Balo
Eugino Patacsel
Andres Deguzman
Richard Lupez
Gregorio Armada

Source: Lawrence Acosta from the back of a photo

Appendix A- 11

HEADQUARTERS
SECOND FILIPINO INFANTRY
Camp Cooke, California

GENERAL ORDERS)   27 March 1944
NUMBER   6)

DISBANDMENT: REORGANIZATION

1. Under the authority of Letter, War Department, File AG 322 (25 Feb 44) OB-I-GNGCT-N, Dated 6 March 1944, the 2d Filipino Regiment less the 1st Battalion, Battalion Medical Detachment and Attached Chaplain, is disbanded effective this date.

2. Concurrent with this disbandment the 1st Battalion and Battalion Medical Detachment reorganized in accordance with T/O and E 7-95, 21 July 1943, is redesignated as the 2d Filipino Battalion.

3. The surplus officers will be attached unassigned to the 2d Filipino Battalion; the surplus enlisted men will be assigned to the 2d Filipino Battalion. All surplus personnel will be formed into two (2) temporary companies, the 1st and 2d Provisional Companies, with the enlisted personnel being placed on detached service to these companies.

                                        /t/CHARLES L. CLIFFORD
                                        Colonel, Infantry Commanding
OFFICIAL:
            /s/LESLIE F. PALMER
            /t/LESLIE F. PALMER
            Captain, Infantry Adjutant
            Certified true copy:
            EDWIN L. SALMAN
            Major, Infantry Commanding

Source: Filipino American Experience Research Project. November 6, 1998

Appendix A- 12

HEADQUARTERS
2D FILIPINO BATTALION
APO 565

                                                                                7 February 1945
314.7
SUBJECT: Unit History
TO: The Adjutant General, Washington 25, D.C. (Thru: Channels)

1. Pursuant to Par 11b AR 345-105 dated 18 November 1929, the following history of the 2d Filipino Battalion is submitted for the period from the date or reorganization, 26 March 1944 to 31 December 1944.

    a. Original Unit.
        (1) Designation: 2d Filipino Battalion
        (2) Date of organization: 26 March 1944
        (3) Place of organization: Camp Cooke, California
        (4) Authority for organization: War Department Letter Order,
            File: AG 322 (25 Feb 44) OB-I-GNGCT-M, subject: "Reorganization of the 2d Filipino Regiment" dated 6 March 1944. (Incl#1)

b. Changes in organization: Effective 15 September 1944 the 2d Filipino Battalion was reorganized under T/O & E 7-95, 12 July 1944. AUTH: General Orders No 84, Headquarters USAFFE, 26 August 1944 (Incl #2) Reorganization effected by General Orders No 14, Headquarters 2d Filipino Battalion, 15 September 1944.

c. Strength commissioned and enlisted.

    (1) As of 28 March 1944 the assigned strength was 45 officers, 1 warrant officer junior grade and 1416 enlisted men.

    (2) Net increase each month: Officers: 0; warrant officers: 0; enlisted men 0.

    (3) Net decrease each month: Officers: 0; warrant officers: 0; enlisted men: 39.

    (4) At end of period: Officer: 36; warrant officer junior grade: 1; enlisted men: 1064.

d. Stations

    (1) Unit departed Camp Cooke, California 31 may 1944

    (2) Unit arrived at Camp Stoneman, California on 1 June 1944.

    (3) Unit departed Camp Stoneman, California on 13 June 1944 and traveled to Oakland, California by ferry boat. Unit boarded USAT Noordam 13 June 1944 and left Oakland pier 14 June 1944.

    (4) Unit disembarked at Oro Bay, New Guinea on 6 July 1944.

    (5) Unit boarded USNT Lurline on 3 November 1944.

    (6) Unit disembarked at Hollandia, New Guinea on 7 November 1944.

e. Marches: No battalion marches were made.

f. Campaigns: None.

g. Battles: None.

h. Former and present members who have distinguished themselves in action: None.

i. Photographs: None.

EDWIN L. SALLMAN
Lt Col, Infantry Commanding

3 Incls:

1 – WD Ltr Order, 6 Mar 44
2 – GO#84, Hq USAFFE, 26 Aug 44

Source: Filipino American Experience Research Project, November 6, 1998

Appendix A- 13

Historical Report
Operations of Northwest Leyte and Camotes Islands in the Philippine Liberation Campaign During Period 8 May – 20 September 1945
First Filipino Infantry Regiment
Colonel Robert H Offley, Commanding

#Annex 2, Strength Commissioned and Enlisted

DATE	OFFICERS	ENLISTED
1 Feb 1945	137	2425
1 Mar 1945	137	2428
1 APR 1945	136	2539
1 MAY 1945	136	2535

**Headquarters
1st Filipino Infantry**

Strength: Commissioned and Enlisted

DATE	OFFICERS	ENLISTED MEN
1 JUN 45	132	2600
1 JUL 45	113	2627
1 AUG 45	110	2737
1 SEP 45	114	3047
21 SEP 45	119	3346

Appendix A- 14

The Soldier's Medal was awarded to men who distinguished themselves by heroism not involving conflict with an enemy. Men who volunteered to rescue the survivors of the plane crash in New Guinea on May 1945 were awarded the medal. This was a citation for Tech Sgt Benjamin Bulatao.

28 August 1945

GENERAL ORDERS)
NO   140)

Soldier's Medal

By direction of the President, under the provision of the Act of Congress approved 2 July 1926 (Bulletin 8, WD, 1926) a Soldier's Medal is awarded by the Commander-in-Chief, United States Army Forces, Pacific, to Technical Sergeant Benjamin C. Bulatao, 39841434, (then Staff Sergeant), Medical Department, United States Army. For heroism in Dutch New Guinea on 19 May 1945. Three survivors of a C-47 crash had been located by aerial search on a mountain side at approximately 7,000 feet altitude. Radio contact revealed that two of them suffered from potential gangrenous infections and were in immediate need of medical attention. Although the endeavor was made extremely hazardous by the high altitude and rocky terrain dotted with sharp stumps, Sergeant Bulatao, parachute medical corpsman of the First Reconnaissance Battalion (Special), volunteered to jump to the rescue of the isolated crash survivors. With one comrade, he successfully executed the jump and landed in the rugged mountains near the scene of the crash. Proceeding to the survivors through the dense jungle growth, he rendered immediate medical assistance to the injured and remained to give aid and comfort until the subsequent rescue. By his heroism in volunteering to make the hazardous jump, Segeant Bulatao was largely instrumental in saving two fellow soldiers from their death or permanent disability.

The Purple Heart was awarded to those killed or wounded while serving. This was a citation for Sgt Andres C. Sivila.

Purple Heart

Sgt. Andres C. Sivila, 39843086, awarded at 51st General Hospital in April 1945. Diagnosis: Wound, gun shot, 25 cal rifle, perforating, severe wound right chest exiting posterior. Comminuted fracture of scapula.

Source: Filipino American Experience Research Project November 1998

Appendix A- 15

HEADQUARTERS, SPECIAL TROOPS
GENERAL HEADQUARTERS
UNITED STATES ARMY FORCES, PACIFIC

APO 500
15 August 1945

GENERAL ORDERS)
NUMBERS   13)

AWARDS OF COMBAT INFANTRYMAN BADGE

Pursuant to authority contained in War Department Circular 408, 17 October 1944, as amended by Section II, War Department Circular 93, 24 March 1945, a Combat Infantryman Badge is awarded to each of the following officers and enlisted men by reason of his satisfactory performance of duty in ground combat against the enemy while assigned as reconnaissance troops in enemy territory on the dates indicated. Additional pay is authorized. (TX 200.6 XO)

Major Ricardo C. Galang	18 February 1944	Infantry
Major Frank H. Skundale	17 November 1944	Coast Artillery Corps
Captain Irineo A. Ames	16 September 1943	Coast Artillery Corps
Captain Andrew P. Bahr	16 December 1944	Cavalry
Captain Fred H. Behan	10 December 1944	Coast Artillery Corps
Captain John E. Bove	10 January 1945	Cavalry
Captain Petronio B. Huerto	9 July 1943	Transportation Corps
Captain Rodolfo C. Ignacio	9 July 1943	Transportation Corps
Captain Donald V. Jamison	11 January 1945	Coast Artillery Corps
Captain Ruben P. Songco	18 February 1944	Infantry
Captain Enrique L. Torres, Jr.	28 September 1944	Air Corps
Captain Frank H. Young	18 December 1941	Infantry 1st
Lt. Carlos F. Ancheta	5 June 1944	Infantry
1st Lt. Henry L. Baker	15 January 1945	Coast Artillery Corps
1st Lt. Merle E. Campbell	16 December 1944	Coast Artillery Corps
1st Lt. Toribio Crespo	16 September 1943	Coast Artillery Corps
1st Lt. Richard A. Ensor	24 October 1944	Coast Artillery Corps
1st Lt. Maynard C. Hawley	19 October 1944	Army of the United States
1st Lt. Al Hernandez	12 August 1944	Infantry
1st Lt. Vicente W. Labrador	2 June 1944	Army of the United States
1st Lt. Jose B. Malan, Sr.	25 October 1944	Field Artillery
1st Lt. Paul F. Mauricio	5 June 1944	Infantry
1st Lt. Jose T. Mendoza	1 June 1944	Infantry
1st Lt. Emilio F. Quinto	9 July 1943	Signal Corps
1st Lt. Albert Schmaltz	19 October 1944	Cavalry
1st Lt. Claudio M. Tamayo	2 January 1945	Infantry
1st Lt. Konglam Teo	18 December 1941	Infantry
1st Lt. Sidney S. Tison	20 January 1945	Coast Artillery Corps
1st Lt. Jose V. Valera	28 August 1944	Army of the United States
1st Lt. Rhys C. Wood	24 October 1944	Coast Artillery Corps
M/Sgt Gerard G. Berg	15 August 1944	Infantry
M/Sgt Restituto J. Besid	1 October 1944	Infantry
M/Sgt Eutiquio B. Cabais	22 September 1944	Infantry
M/Sgt Isidoro D. Dacquel	20 June 1944	Infantry

BASIC: General Orders No. 13 Hq Sp Trps, GHQ, AFPAC, APO 500, dtd 15 August 1945

M/Sgt Victorio C. Goloyugo – 28 Aug 1944 – Infantry
M/Sgt Benjamin Harder – 18 Feb 1944 – Infantry
M/Sgt Aniceto C. Manzano – 15 June 1944 – Infantry
M/Sgt Cipriano L. Miguel – 1 Feb 1945 – Infantry
M/Sgt Wilfred E. Regala – 21 Oct 1944 – Infantry
M/Sgt Crispolo G. Robles – 3 April 1944 – Infantry
M/Sgt Gerardo A. Sanchez – 22 April 1944 – Infantry
M/Sgt Fortunato Santos – 20 Oct 1944 – Infantry
T/Sgt Alfredo R. Agron – 28 Aug 1944 – Infantry
T/Sgt Peter B. Aguilar – 5 Aug 1944 – Infantry
T/Sgt Thomas T. Amante – 9 Jan 1945 – Infantry
T/Sgt Zoil C. Azcueta – 24 Jan 1945 – Infantry
T/Sgt Faustino B. Bacus – 1 Jan 1945 – Infantry
T/Sgt Arcangel Baniares – 18 Feb 1944 – Infantry
T/Sgt Cleto S. Bello – 1 Jan 1945 – Infantry
T/Sgt Custodio P. Bolivar – 10 Aug 1944 – Infantry
T/Sgt Secundio G. Bucol – 13 Jan 1945 – Infantry
T/Sgt John R. Constantio – 24 Oct 1944 – Infantry
T/Sgt Modesto De Los Santos – 25 Dec 1944 – Infantry
T/Sgt Esteban M. Diputado – 12 Mar 1945 – Infantry
T/Sgt Buenaventura Dolendo – 18 Jan 1945 – Infantry

T/Sgt Vidal Garcia – 4 Oct 1944 – Infantry
T/Sgt Larry O. Guzman – 11 Jan 1945 – Infantry
T/Sgt Rafael C. Licera – 26 Feb 1945 – Infantry
T/Sgt Alfred E. Lim – 12 Dec 1944 – Infantry
T/Sgt Jose S. Mariano – 17 Oct 1944 – Infantry
T/Sgt Julian A. Pilar – 25 Dec 1944 – Infantry
T/Sgt Vicente A. Pinuela – 8 Feb 1944 – Infantry
T/Sgt Jaime A. Ramirez – 17 Oct 1944 – Infantry
T/Sgt Jose J. Ramos – 15 June 1944 – Infantry
T/Sgt Isay M. Real – 20 Oct 1944 – Infantry
T/Sgt Jerry v. Regalado – 15 Jan 1945 – Infantry
T/Sgt Pantaleon M. Rosete – 17 Oct 1944 – Infantry
T/Sgt Jose T. Sitjar – 28 July 1944 – Infantry
T/Sgt Val R. Tabios – 29 Mar 1945 – Infantry
S/Sgt Marcelo B. Abalos – 24 Dec 1944 – Infantry
S/Sgt Procopio M. Adia – 25 Dec 1944 – Infantry
S/Sgt Orlando A. Alfabeto – 15 June 1944 – Infantry
S/Sgt Vidal F. Alvares – 10 Oct 1944 – Infantry
S/Sgt Vicente H. Anacleto – 15 Dec 1941 – Infantry
S/Sgt Eleno D. Annaguey – 16 Jan 1945 – Infantry
S/Sgt Fred Balcena, Jr. – 3 Dec 1944 – Infantry

BASIC: General Orders No. 13 Hq Sp Trps, GHQ, AFPAC, APO 500 dtd 15 August 1945

S/Sgt Florencio A. Baldos – 1 Dec 1944 – Infantry
S/Sgt Julio V. Balleras – 15 Nov 1944 – Infantry
S/Sgt Gregorio J. Basug – 24 Oct 1944 – Infantry
S/Sgt Felino L. Bautista – 14 May 1945 – Infantry
S/Sgt Daniel B. Begonio – 5 Dec 1944 – Infantry
S/Sgt Pio B. Borrillo – 10 Oct 1944 – Infantry
S/Sgt Pascual M. Calix – 10 Feb 1945 – Infantry
S/Sgt Angelo A. Calmorin – 15 Jan 1945 – Infantry
S/Sgt Patrocinio A. Carrillo – 16 Sep 1944 – Infantry
S/Sgt Eustacio C. Corpuz – 8 Dec 1941 – Infantry
S/Sgt Jacinto Cutaran – 2 Sep 1944 – Infantry
S/Sgt Marciano R. Daelto – 8 Nov 1944 – Infantry
S/Sgt Fortunato C. Dagandan – 3 Oct 1944 – Infantry
S/Sgt Arsenio B. De La Pena – 26 Nov 1944 – Infantry
S/Sgt Mero L. De Ocampo – 1 Jan 1945 – Infantry
S/Sgt Cipriano C. Fernandez – 4 Oct 1944 – Infantry
S/Sgt Marcelo P. Fontanoz – 1 Jan 1945 – Infantry
S/Sgt Fermin M. Francisca – 20 June 1944 – Infantry
S/Sgt James R. Gallemore – 15 July 1944 – Infantry
S/Sgt Pedro L. Garces – 19 Oct 1944 – Infantry

S/Sgt Marcello Garcia – 24 Oct 1944 – Infantry
S/Sgt Felipe R. Ginobiagon – 20 June 1944 – Infantry
S/Sgt Jose T. Gonzales – 10 Oct 1944 – Infantry
S/Sgt Marcelo G. Gonzales – 28 July 1944 – Infantry
S/Sgt Patricio D. Jorge – 16 Aug 1944 – Infantry
S/Sgt Walter B. Lagrimas – 12 Apr 1945 – Infantry
S/Sgt Doming L. Logan – 10 Oct 1944 – Infantry
S/Sgt Cirilo C. Lopez – 25 Jan 1945 – Infantry
S/Sgt Felcisimo D. Magbual – 10 Jan 1945 – Infantry
S/Sgt Dominador Malic – 2 June 1944 – Infantry
S/Sgt Eusebio J. Manuta – 19 Oct 1944 – Infantry
S/Sgt Leo Marquina – 15 Feb 1945 – Infantry
S/Sgt Venancio M. Mones – 26 Dec 1944 – Infantry
S/Sgt Juan M. Palac – 11-22 Feb 1945 – Infantry
S/Sgt Rosendo M. Palafox – 15 Jan 1945 – Infantry
S/Sgt Astor C. Parong – 28 Oct 1944 – Infantry
S/Sgt Inocencio V. Pascua – 20 Aug 1944 – Infantry
S/Sgt Paulino A. Rosales 16 Jan 1945 – Infantry
S/Sgt Lucas D. Runes – 29 Nov 1944 – Infantry
S/Sgt Apolinar G. Salvador – 28 Feb 1945 – Infantry

S/Sgt Albert T. Talaro – 10 Jan 1945 – Infantry

BASIC: General Orders No. 13 Hq Sp Trps, GHQ, AFPAC, APO 500 dtd 15 August 1945

S/Sgt Teodoro M. Taluban – 28 Aug 1944 – Infantry
S/Sgt Thomas C. Vergara – 3 Sep 1944 – Infantry
S/Sgt Angelo F. Villanueva – 26 Nov 1944 – Infantry
S/Sgt Doroteo V. Vite – 29 Sep 1944 – Infantry
S/Sgt Ramon D. Vitorio – 18 Feb 1944 – Infantry
Sgt Gerardo C. Alfafara – 26 Feb 1945 – Infantry
Sgt Loren D. Almojera – 7 Feb 1945 – Infantry
Sgt Beato G. Alvez – 26 Nov 1944 – Infantry
Sgt Carlos F. Antikoll – 19 Oct 1944 – Infantry
Sgt Alfonso L. Antonio – 10 Jan 1945 – Infantry

Sgt Arthur G. Lopez – 28 Nov 1944 – Infantry
Sgt Maximo V. Maglangit – 13 Dec 1944 – Infantry
Sgt Philip Martin – 28 Feb 1945 – Infantry
Sgt Angelo P. Oandasan – 5 Oct 1944 – Infantry
Sgt Amrafael D. Pascual – 16 Feb 1945 – Infantry
Sgt Lorenzo U. Pimentel – 28 Aug 1944 – Infantry
Sgt Mack H. Pulido – 15 Jan 1945 – Infantry
Sgt Antonio M. Quesada – 19 Oct 1944 – Infantry
Sgt Luis C. Quindiagan – 3 Sep 1944 – Infantry
Sgt Paciano S. Racho – 15 Jan 1945 – Infantry

Sgt Lasaro C. Arquero – 16 Feb 1945 – Infantry
Sgt Pedro S. Azcueta – 1 Jan 1945 – Infantry
Sgt Jose C. Baladad – 7 Jan 1945 – Infantry
Sgt Edilberto S. Bibat – 28 Nov 1944 – Infantry
Sgt Serafin B. Buenavista – 6 July 1944 – Infantry
Sgt Gene D. Costales – 10 Jan 1945 – Infantry
Sgt Raymond R. Fermin – 17 Oct 1944 – Infantry
Sgt Urbano M Francisco – 15 Dec 1944 – Infantry
Sgt Estanislao S. Ibarra – 5 Feb 1945 – Infantry
Sgt Aniceto S. Kintanar – 20 June 1944 – Infantry

Sgt Tom P. Respicio – 12 Apr 1945 – Infantry
Sgt Antonio C. Rimando – 15 Jan 1945 – Infantry
Sgt Elpidio A. Rivera – 10 Jan 1945 – Infantry
Sgt Romulo D. Robles – 24 Oct 1944 – Infantry
Sgt Juan A. Rodriguez – 24 Nov 1944 – Infantry
Sgt Henry D. Sabado – 15 Feb 1945 – Infantry
Sgt William H. Sammons – 28 Aug 1944 – Infantry
Sgt Nicolas E. Sandoval – 21 Oct 1944 – Infantry
Sgt Larry G. Soliven – 28 Aug 1944 – Infantry
Sgt Zoilo Tolentino, Jr. – 3 Jan 1945 – Infantry

Sgt Perfecto M. Trocio – 22 Dec 1944 – Infantry

BASIC: General Orders No. 3 Hq Sp Trps, GHQ, AFPAC, APO 500 dtd 15 August 1945

Sgt Harry T. Tucay – 10 Jan 1945 – Infantry
Sgt Apollo P. Tullo – 25 June 1944 – Infantry
Sgt Celestino R. Valdez – 20 Oct 1944 – Infantry
Sgt Hermenegildo Villalon – 10 Oct 1944 – Infantry
Corp Baltazar C. Agbunag – 1 Jan 1945 – Infantry
Corp Isaac Z. Aguila – 5 June 1944 – Infantry
Corp Sahibad H. Aliacbar – 12 Apr 1945 – Infantry
Corp Maurice W. Artiago – 17 Oct 1944 – Infantry
Corp Martiniano S. Atad – 17 Oct 1944 – Infantry
Corp Ralph M. Balunes – 28 Aug 1944 – Infantry
Corp Jaime A. Bernal – 17 Nov 1944 – Infantry
Corp Pedro F. Cariaga – 30 Aug 1943 – Infantry
Corp Ricardo D. Cayabyab – 10 Jan 1945 – Infantry
Corp Julian F. Cuanang – 15 Dec 1944 – Infantry
Corp Emilio M. Custodio – 17 Oct 1944 – Infantry
Corp Andres D. Darang – 7 Jan 1945 – Infantry
Corp Urbano P. Delfun – 5 Nov 1944 – Infantry
Corp Miguel B. Erana – 24 Oct 1944 – Infantry
Corp Leon C. Fernandez – 15 Feb 1945 – Infantry
Corp Agripino B. Flores – 17 Oct 1944 – Infantry

Corp Thomas D. Gaduang – 25 Oct 1944 – Infantry
Corp Narciso C. Generales – 9 Jan 1945 – Infantry
Corp Dominador F. Gobaleza – 22 Aug 1943 – Infantry
Corp Ali H. Ladjahasan – 16 Sep 1943 – Infantry
Corp Jalil H. Ladjahasan – 16 Sep 1943 – Infantry
Corp Mangona H. Ladjahasan – 16 Sep 1943 – Infantry
Corp Leodegario Lascano – 27 Nov 1944 – Infantry
Corp Hilario L. Lobendino – 11 Mar 1945 – Infantry
Corp Sergio M. Machon – 17 Jan 1945 – Infantry
Corp Victor V. Marquez – 10 Jan 1945 – Infantry
Corp Robert C. Minder – 16 Feb 1945 – Infantry
Corp Benigno E. Mosquera – 17 Oct 1944 – Infantry
Corp Manuel M. Nerez – 29 Sep 1944 – Infantry
Corp Federico J. Niebres – 25 May 1945 – Infantry
Corp Hermogenes C. Oania – 17 Oct 1944 – Infantry
Corp Patricio Ollada – 28 Aug 1944 – Infantry
Corp Jerry D. Pascua – 6 June 1944 – Infantry
Corp Esteban R. Quiaoit – 21 Oct 1944 – Infantry
Corp Ruperto Q. Quismundo – 21 Oct 1944 – Infantry
Corp Camilo M. Ramirez – 9 Dec 1941 – Infantry

Corp Alfredo L. Salvacion – 26 Nov 1944 – Infantry

BASIC: General Orders No. 13 Hq Sp Trps, GHQ, AFPAC, APO 500 dtd 15 August 1945

Corp Pedro E. Savellano – 10 Oct 1944 – Infantry

Corp John T. Tolentino – 21 Oct 1944 – Infantry

Corp Benny T. Velayo – 20 Jan 1945 – Infantry

BY COMMAND OF BRIGADIER GENERAL BAIRD:
HOWARD V. JUDSON
Major, Adjutant General's
Department Adjutant General

OFFICIAL:
HOWARD V. JUDSON
Major, Adjutant General's
Department Adjutant General

Appendix A- 16
American Officers of the Philippine Civil Affairs Units (PCAU)*

DATE	PCAU	RANK-FIRST-LAST NAME	NOTE
Sept44		Col M. A. Snyder	Charged with training at Civil Affairs Staging Area, Hollandia.
Oct44	1	Maj Carl E. Erickson	CAC, Joined PCAU 1 as a Legal Officer.
Oct44	1	Maj Charles W. Conklin	Finance Dept. Later dropped from PCAU 1.
Oct44	1	Lt William I. Turnbull	Joined.
Dec44	1	Maj Talman C. Budd	CO PCAU 1, confined to hospital.
	1	Maj Carr	Replaced Budd as CO PCAU 1.
	1	Maj James R. Sorenson	Relieved Maj Carr. Later CO of PCAU 3.
	1	Capt John B. Greene	Relieved Sorenson
Feb45	1	Maj Talman C. Budd	Returned to PCAU 1 two months later.
	1	Lt Joseph S. Walker	Handled Finance Section.
	1	Capt Myron W. Reed	Relieved Lt. Joseph S. Walker.
	1	Lt Clint M. Koch	Headed Transportation Section.
	1	Capt William I. Turnbull	Supervised Supply Section.
Mar45	1	Maj Robert M. Collins	Attach to PCAU 1 for rations. Temp detail at Welfareville with Capt John B. Greene.
Mar	1	Maj Charles D. Scott	Became Exec Officer of PCAU 1. Came from PCAU 20.
Mar45	1	Lt Stanley L. Wilson, Jr.	Assigned to PCAU 1.
Mar45	1	Lt Dudley I. Krenz	Assigned to PCAU 1.
Apr45	1	Maj Talman C. Budd	Sent to 80th General Hospital, Manila.
May45	1	Lt Glen O. Allen	Relieved from PCAU 1 for duty with Allied Geographical Section.
	1	Dr. George I. Adamson	Head of Industrial and Engineering Section while at San Juan del Monte.
	2	Maj Virgil B. Ashcraft	CO PCAU 2.
	2	Lt Lionel E. Blankenship	Wounded in San Pedro, Leyte landing. Awarded Bronze Star. Burned attempting to rescue a trapped soldier.
	2	Lt Ralph A. McBroom	With 10 EM to Bambam, Tarlac to guard captured Japanese goods and foodstuffs.
Apr4,45	2	Lt Irving	With a PCAU team joined groups from the Military Police,
Apr4,45	2	Lt Blankenship	1st Cavalry Div, some guerr, to form recon party to engage the
Apr4,45	2	Lt White	enemy that landed close to PCAU 2 hqs at San Pedro, Laguna.
	5	Maj Chester A. Pearlman	CO.
	9	Maj F. L. Nemeck	CO.
	12	Maj Robert H. Overton	Departed for PCAU 9.
	12	Lt Edward A. Rhatican	Joined from PCAU 23.
	13	Col Nephi C. Christensen	CO.
	16	Maj William J. Gordon	CO.
	16	Maj Albert H. Knoll	Medical Officer.
	16	Capt Jesse S. Baskett	Engineer Officer.
	16	Capt Robert O. Jones	Transportation Officer.
	16	Capt George W. Puckett	Executive and Labor Officer.
	16	Capt Marion E. Tisdale	Relief and Welfare Officer, S-3 and CO of Enlisted Detachment.
Apr4,45	16	Capt Stanley A. Yesukiewicz	Supply Officer and S-4.
	16	Lt Leon F. Hartley	Finance Officer.
	16	Lt Ralph M. Kearney	Public Safety Officer and S-2.
	16	Lt John R. Jacobs	Legal Officer and S-1
	16	Sgt Elmer H. Trost	EM.
	20	Maj Meader	CO.

*There were more PCAU units. Only names found are listed.

Source: A History of the 1st & 2nd Filipino Infantry Regiments, 1992 by Alex S. Fabros, Jr. Volume VII; Philippine Civil Affairs Units; U.S. National Archives and Records Administration and MacArthur Archives

Appendix A- 17

EIGHTH ARMY (VICTORY OPERATIONS)
PHILIPPINE CIVIL AFFAIR UNIT (PCAU)
AREAS OF OPERATION January – May 1945*

JANUARY

During the period covered by this report, Eight Army PCAU's operated in the following localities:

PCAU #9	San Jose, Mindoro (Detachment Marinduque Island)
PCAU #10	Guiuan, Samar
PCAU #11	Mindoro
PCAU #12	San Jose, Leyte
PCAU #14	Valencia, Leyte
PCAU #15	Ipil, Leyte (Detachment Camotes Islands)
PCAU #16	Limon, Leyte (Now Luzon)
PCAU #17	Palompon, Leyte
PCAU #27	Abuyog, Leyte (Now Luzon)
PCAU #28	Catbalogan, Samar
Naval Civil Affairs Unit	Guiuan, Samar

FEBRUARY

During the period covered by this report, PCAU's under Eight Army operated in the following areas:

PCAU #10	Guiuan, Samar
PCAU #11	Calapan, Mindoro (Detachment Marinduque Island)
PCAU #12	Palawan Task Force
PCAU #13	Leyte (Staging for V-I)
PCAU #14	Ormoc-Ipil, Leyte
PCAU #15	Carigara, Leyte
PCAU #17	Palompon, Leyte
PCAU #23	V-IV Operation
PCAU #26	Baybay, Leyte
PCAU #28	Catbalogan, Samar
PCAU #29	Enroute to Leyte
PCAU #30	Enroute to Leyte

MARCH

During the period covered by this report. PCAU's under Eight Army operated in the following areas:

PCAU #9	Mindoro and Marinduque
PCAU #10	Guiuan, Samar
PCAU #12	Puerto Princesa, Palawan
PCAU #13	Arrived in Iloilo, Panay, 19 Mar 45
PCAU #14	Ipil-Ormoc-Valencia, Leyte
PCAU #15	Arrived Cebu City, Cebu, 27 Mar 45
PCAU #17	Palompon-Villaba, Leyte
PCAU #23	Zamboanga, Mindanao – Arrived 10 Mar 45
PCAU #24	Departed Leyte for V-II, 30 Mar 45
PCAU #26	Arrived Bacolod, Negros Occidental, 29 Mar 45
PCAU #28	Catbalogan, Samar
PCAU #29	San Jose, Leyte (Staging V-V)
PCAU #30	San Jose, Leyte (Staging V-V)

APRIL

Group "A" (K-2, L-3, and Operations other than Victor)

PCAU #9	Mindoro and Marinduque
PCAU #10	Guiuan, Samar
PCAU #17	Palompon, Leyte, and Masbate Island
PCAU #25	Ipil-Ormoc-Valencia, Leyte
PCAU #28	Catbalogan, Samar

Group "B" (Victor Operations)

PCAU #12	Palawan and Culion
PCAU #13	Iloilo, Panay
PCAU #14	Cotabato, Mindanao
PCAU #15	Cebu City, Cebu
PCAU #24	Cebu City, Cebu and Bohol Island
PCAU #26	Bacolod, Negros Occidental
PCAU #29	Digos, Mindanao
PCAU #30	Parang, Mindanao

MAY

Group "A" (K-2, L-3, and Operations other than Victor

PCAU #9	Mindoro and Marinduque
PCAU #10	Southern Samar
PCAU #25	Northwest Leyte and Masbate
PCAU #28	Northern Samar

Group "B" (Victor Operations)

PCAU #12	Palawan and Culion
PCAU #13	Iloilo, Capiz, and Antique
PCAU #14	Southern Cotabato
PCAU #15	Cebu
PCAU #17	Misamis Oriental and Bukidnon
PCAU #23	Zamboanga, Misamis Oriental, and Sulu
PCAU #24	Negros Oriental
PCAU #26	Negros Occidental
PCAU #29	Davao
PCAU #30	Northern Cotabato

*Covers the Luzon operations from January – May, 1945. There were PCAU's #1 – 8 who landed in Leyte, October – December, 1944.

Source: U.S. National Archives and Records Administration

Appendix A- 18
ALAMO SCOUTS*
Filipino Infantry Regiments, Philippine Army and Philippine Guerrillas
WWII

RANK/FIRST/LAST NAME	ALAMO CLASS TEAM LEADER	UNIT
TSG Agapito G. Amano	Sumner	92nd Div PA**
CPL Agrifino J. Duran		978th***
TSG Alfred Alfonso	4th Rounsaville	D Company
Pvt Alfred Pulito	6th	
Angelo A. Calmorin		978th
TSG Antonio Militar		Lapham's guerrillas
SGT Antonio Rimando		1st Rcn Bn
3Lt Asclapiades B. Kuison		
Capt Atanacio Chavez	4th	1st FIR******
CPL Barney A. Quinagon		1st Rcn Bn *****
SSG Benjamin Mones	Littlefield	Signal Service Company****
Carl Wright		978th
1Lt Clinton B. McFarland		978th
PFC Cruz C. Vega	8th Grimes	
TSG Daniel C. Belarmino	4th	1st FIR (KIA)
PFC David Bello	4th	1st FIR
David Cardenas		1st Rcn Bn*****
PFC Don J. Liberato		
PFC Estanislao S. Bacat	5th Ileto	Hq Company
1SGT Eudosio B. Edon		92nd Div PA
SSG Federico Balambao	5th Ileto	F Company
PVT Felix Viloria	4th	1st FIR
CPL Genaro L. Faelnar		
CPL George R. Urbano	4th	1st FIR
SGT Gregorio F. Bareng		1st Recn Bn
SSG Henry E. Runtal		978th
2LT Honorato S. Echaves	4th	1st FIR
SGT Ignacio C Sacmar		
2LT Inocensio F. Cabrido	Sumner	
T5 Isaac J. Cleodoalde		978th
2LT Isidro Urbano	4th	1st FIR
PFC Jose E. Estrada	4th	1st FIR
1LT Jose L. Rifareal		
PFC Joseph S. Macaluso		
PFC Juan B. Medina	6th	
SGT Juan D. Pacis	5th Chanley	A Company
PFC Juan E. Berganio	2nd Chanley	Hq Company
SGT Juan E. Hidalgo	4th Dove	1st FIR
T5 Julio C. Advincula		978th
Leandro Reposar	Rounsaville	
PFC Lino Canebaro	9th Wewit	1st FIR
TSG Lorenzo Bartholomew		Anderson's guerrillas
SSG Marcelo P. Fontanoz		1st Rcn Bn
TSG Marciano C. Alfonso		92nd Div PA

SGT Mariano T. Arso		978th
SSG Mauro Dacanay		Lapham's guerrillas
1SGT Melquiades Ventura	4th	
CPL Miguel B. Erana		1st Rcn Bn
TSG Nemesio Cabaong		
PFC Nicholas C Enriquez	2nd Chanley	1st FIR
TSG Pablo P. Julian		
SGT Paciano S. Racho		1st Rcn Bn
SGT Porfirio Ramos		
TSG Rafael C. Cuico		
1LT Rafael M. Ileto	5th Ileto	H Company
SGT Rafael Santisas		
TSG Ray Corpuz	4th	1st FIR
PFC Raymond Aguilar	3rd Farkas	1st FIR
Rosendo M. Palafox		978th
T5 Rudolph B. Santos		978th
PFC Rufo Vaquilar	4th Rounsaville	Cannon Company
SGT Sabas A. Asis	4th Nellist	G Company
TSG Samuel M. Quimbo		92nd Div PA
CPL Terry R. Santos	4th	1st FIR
PFC Thomas D. Gaduang		1st Rcn Bn
SSG Thomas A. Siason	4th Nellist	Cannon Company
TSG Trinidad S. Sison	Sumner	
SGT Venancio M. Mones		
Vincent Quispo	McGowen	

*These are names found as Alamo Scouts.   **Philippine Army (PA)
***978th Signal Company   ****Signal Service Company (PA or US)
*****1st Reconnaissance Battalion   ******1ST Filipino Infantry Regiment

Source:
Zedric, Lance and Blaise, Russ. Silence no more: the Alamo Scouts in their own words. Peoria, Illinois.
Alexander, Larry. Shadows in the Jungle: The Alamo Scouts Behind Japanese Lines in World War II. Filipino
American Experience Research Project. November 1998
Alamo Scouts Historical Foundation

Appendix A- 19

1st and 2nd Filipino Infantry Regiments*
Casualties 1944 - 1945

Last	First	M.I.	Rank	Death Date	Status
Martinez	Jose	V	SSgt	19-Aug-44	DNB
Pepito	Exequiel	S	Sgt	19-Aug-44	DNB
Virgara	Plabiano	V	Pfc	21-Dec-44	DNB
Gatan	Inocencio		Pvt	Feb-45	KIA
Remonde	Ireneo	A	Pfc	1-Mar-45	KIA
Iloreta	Edward	D	SSgt	15-Mar-45	KIA
Dumo	Eddie	A	Pvt	22-Mar-45	DNB
Crowson Jr.	William	B	Capt	23-Mar-45	KIA
Bittick	Henry	R	Lt	26-Mar-45	KIA
Pilotin	Emilio		Pvt	17-Apr-45	KIA
Pequit	Felix	D	Pvt	20-Apr-45	DOW
Floresca	Vincent	T	Pfc	1-May-45	KIA
Antonio	Federico	A	Pfc	31-May-45	KIA
Fontamillas	Venvencio	M	Pfc	22-Jul-45	KIA
Galicia	Teodorico	G	Pfc	6-Aug-45	KIA
Salomon	Juan	P	Sgt	11-Feb-45	KIA
Savella	Felix	S	T5	14-Nov-45	DNB
Butanas	Andres	S	Pfc	25-Mar-46	DNB
Sales	Esteban		Sgt	17-Dec-44	KIA

1st and 2nd Filipino Infantry Regiments

Last	First	M.I.	Rank	Death Date	Status
Belarmino	Daniel	C	Sgt	4-Mar-45	KIA
Canubag	Ecolastico	C	Pfc	7-Mar-45	KIA
Abuten	George	C	Pvt	8-Mar-45	KIA
Alngog	Eulalio	C	Pvt	13-Mar-45	KIA
Bernido	Esparidion	C	Pvt	13-Mar-45	KIA
Baladad	Simplicio	P	Pfc	23-Mar-45	KIA
Pira	Hilario	P	Pfc	30-Mar-45	KIA
Velasco	George	C	Pfc	15-Apr-45	KIA
Bandana	Feling	G	Pvt	18-Apr-45	KIA
Calamasa	Alfredo	G	Pvt	9-Mar-45	WIA
Tamanao	James	B	Cpl	28-Feb-45	WIA
Evangelista	Jacinto	B	Pfc	3-Apr-45	WIA
Regnon	Andres	P	Pfc	6-Apr-45	WIA
Pajo	Maximo	L	Pfc	20-Apr-45	WIA
Maga	Jose	P	Cpl	26-Apr-45	WIA
Cabling	Andres	C	Pfc	13-Mar-45	WIA
Saribay	Telesporo	R	Pfc	13-Mar-45	WIA
Verzosa	Lorenzo	R	Pfc	13-Mar-45	WIA
Orphilla	Alfonso	S	Pfc	13-Mar-45	WIA
Brasis	John	C	Lt	3-Mar-45	WIA
Sacramento	Ross	H	Sgt	3-Mar-45	WIA
Tejero	Pedro	T	Sgt	3-Mar-45	WIA
Balig	Leonardo	V	Pfc	4-Mar-45	WIA
Lopez	Patricio	V	Sgt	1-Mar-45	WIA
Escano	Baltazar	B	Sgt	1-Mar-45	WIA
Gundaya	Benny		Pvt	1-Mar-45	WIA
Baniqued	George	C	Pvt	7-Mar-45	WIA
Tejada	Bonifacio	B	Pfc	7-Mar-45	WIA
Malan	Toribio	T	Pfc	14-Mar-45	WIA
Calban	Jacinto	M	Sgt	14-Mar-45	WIA
Riego	Rufino	R	Pfc	14-Mar-45	WIA
Lopez	Enrique	M	Pfc	7-Mar-45	WIA
Aquino	Tranquilino	A	Pvt	17-Apr-45	WIA
Mendoza	Mariano	C	Pvt	18-Apr-45	WIA
Raneses	Grigido	C	Pvt	18-Apr-45	WIA
Cayabyab	Eustaquio	M	Pfc	18-Apr-45	WIA
Sadiarin	Jimmy	T	Pfc	26-Apr-45	WIA
Ebreo	Bienvenido		Pvt	1-Mar-45	WIA
Quinones	Esteban	S	Pfc	12-Apr-45	WIA
Ramones	Macario	A	Pfc	9-Apr-45	WIA
Peruna	Pedro	P	Pfc	18-Mar-45	WIA
Urbano	Edulfo	O	Pfc	21-Apr-45	WIA
Jiminez	Narciso		Pfc	1-Mar-44	WIA
Asuncion	Corneli	S	Pfc	14-May-45	KIA
Ami	Apolonio	A	Sgt	27-May-45	KIA
Antonio	Federico	A	pfc	31-May-45	KIA
Galicia	Teodoro	C	Pfc	5-Aug-45	KIA
Nana	Ricardo	M	Pfc	26-May-45	WIA
Dasali	Pedring	D	Pfc	18-May-45	WIA
Arcol	Manuel	V	Pfc	29-May-45	WIA
Quinones	Esteban	S	Pfc	9-Jun-45	WIA
Cuaresma	Gregorio	B	Pfc	19-Jun-45	WIA
Almerol	Paul	B	Sgt	21-Jun-45	WIA
Pagdilao	Gregorio	C	Pfc	28-May-45	WIA
Mina	Silvano		Pfc	7-Jun-45	WIA
Rondon	Paulino	T	Pfc	5-Aug-45	WIA
Garces	Luiano	T	Pfc	5-Aug-45	WIA
Inocencio	Agustin	P	Sgt	5-Aug-45	WIA
Flores	Camelo		Pfc	7-Jun-45	WIA
Collado	Jose	G	Pfc	1-Jun-45	WIA
Brono	Sajulan	L	Pvt	27-May-45	WIA
Dominisac	Paulino	R	Sgt	27-May-45	WIA
Volmonte	John		Pfc	27-May-45	WIA
Baliscao	Federico	B	Pfc	5-Aug-45	WIA
Lucas	Alejandro	S	Pfc	12-Jun-45	WIA
Gregory	Rubin		Sgt	?	DNB
Gosmo	Louis	Ros	Pfc	14-Nov-45	DNB

Philippine Civil Affairs Unit PCAU***

Last	First	M.I.	Rank	Death Date	Status
Dacones	Ponciano	S	T5	24-Oct-44	KIA
Ancheta	Andres	C	SSgt	27-Oct-44	KIA
Nepomuceno	Larry	A	T5	27-Oct-44	KIA
Tambot	Emiliano	P	Cpl	27-Oct-44	KIA
Dalipe	Benito	S	Sgt	27-Oct-44	KIA
Blanco	Eugenio		Sgt	7-Apr-44	KIA
Josue	Ignacio	I	Lt	2-Jul-44	WIA
Cababa	Marcel	R	Sgt	4-Apr-45	WIA
Guzman	Robert	O	Sgt	2-Jul-44	WIA
Nunez	Ricardo	O	Sgt	2-Jul-44	WIA
Cabrales	Frank	P	T4	2-Jul-44	WIA
Abalos	Remigio	A	Cpl	2-Jul-44	WIA

Sebastian	Maximo	P	T5	24-Oct-44	KIA
Rodriguez	Paul		Lt	2-Jul-44	KIA
Alicante	Macario	A	Pfc	2-Jul-44	KIA
Dackel	Moses	M	Pvt	25-Oct-44	KIA

Gonzales	Ban	S	Cpl	2-Jul-44	WIA
Obrero	Feliciano	P	Pfc	2-Jul-44	WIA
Gilbert	Mayor		T4	25-Oct-44	WIA
Gervasio	Martinez	M	T5	25-Oct-44	WIA
Gregorio	Quimba	P	T3	25-Oct-44	WIA

This is not a complete list.
*Buried in Manila American Cemetery
***Wall of Missing: Manila American Cemetery

DNB - Died Non-Battle: Sickness, Training, Homicide
KIA - Killed in Action
WIA - Wounded in Action
DOW - Died of Wounds
FOD - Foreign Object Damage: Torture MIA - Missing in Action
SELF - Suicide

Source: Adam Earle; National Archives and Records Administration; Filipino American Experience research Project Nov 1998; Historical Reports. Operations on Northwest Leyte, Northern Samar, Camotes, Burias, and Ticao Islands 16Feb-7May45, 8May-20 and Sep45 First Filipino Infantry Regiment. Col robert H. Offley Commanding

Appendix A- 20

CANAO NEWS

Canao News was a weekly newsletter published by members of the 1st Filipino Infantry Regiment in Leyte.

The staff of the paper in 1946 were:
1st Lt. I.M. Sarmiento - Advisor to Canao News
Technician Fourth Grade (T4) Peter A. Aduja - Associate Editor
Technician Fourth Grade (T4) Benjamin Menor – Editor
Corporal Benny Sacramento - News Reporter
Technician Fifth Grade (T5) William Pasco – Production.

The Honolulu Star-Bulletin, March 12, 1946, published names of men from Hawaii with the Regiment: Johnny Benjamin, Manuel Abergas, Henry A. Aurio, Panfilo M. Bascar, Johnnie D. Heneraiau, Joseph R. Miguel, William Monroy, Lucio Sanico, Paul Bibilone, Abel Miraflor, and Joe Asontista. Others were Abundo Repunte, and Janito De Garcia.

Appendix A- 21

<table>
<tr><td>WAR DEPARTMENT
DISPOSITION FORM</td><td>SECURITY CLASSIFICATION</td></tr>
</table>

FILE No.   SUBJECT

AGAO-I 322 (20 Sep 48)   Status of Unit

TO Organization Records Br   FROM Orgn & Dir Section   DATE 20 Sepr 1948   COMMENT No.1

Records Adm Center, AGO	Adjutant General's Office
4300 Goodfellow Blvd	Room 10 552, The Pentagon

1. In the event the records of 1st Filipino Infantry Regiment are on file in your office, it is requested that the date of activation and reference to general order or other orders, directed by 320.2/13 (Inf)-GNOPN (1 Jul 42), 1 July 1942, be furnished this office.
2. It is also requested that the dates of all movements, by element, from station of activation, Salinas, California to station of inactivation, CP Stoneman, Calif. 10 April 1946, be furnished this office.

Signed: Adjutant General

AGRS-AC-B 322 Status of Unit   OA/kka

TO: Demob Pers Records Br   FROM: Orgn Records Br   DATE: 19 Oct 48   COMMENT No. 2

Misc Records Unit	RAC, AOO	Arth/2151
Photo Process Section		
Bldg 123		

(IN TURN)

1. For necessary action and direct reply to writer of basic communication who has not been advised of this action. Morning reports of the 1st Filipino Inf prior to 1 Aug 43 are not on file in this office.
2. Morning Reports of the 1st Filipino Inf, by element, are not on file for the periods indicated:

Organization	Period
Co A, 1st Filipino Infantry	7 Jan 45 thru 3 Mar 45
Hq Co, 1st Bn, 1st Filipino Infantry	1 Jan 45 thru 16 Feb 45
Co G, 1st Filipino Infantry	1 Jun 45 thru 12 Dec 45
Anti-Tank Co, 1st Filipino Infantry	21 Mar 46 thru 10 Apr 46
Cannon Co, 1st Filipino Infantry	1 Nov 45 thru 10 Apr 46

45913

AGRS-AC-B 322 Status of Unit, COMMENT NO. 2, dated 19 Oct 48 Cont'd   OA/kka

3. Records of the 1st Filipino Inf show activation effective 13 Jul 42 per GO 24, Hq VII Army Corps, San Jose, Calif dated 11 Jul 42.
4. Retained copies of Morning Reports of the 1st Filipino Inf show all elements at the following stations during the period 1 Aug 43 thru 28 Apr 44:

STATION		ARRIVED	DEPARTED	STATION	ARRIVED	DEPARTED
Hunter Liggett Military Reservation						
Camp Roberts, Calif	Prior to	1Aug43	24Sep43	Aboard Gen John Pope at Sea	20Apr44	20Apr44
Camp Beale, Calif		24Sep43	31Mar44	Aboard SS Gen John Pope anchored New Caledonia	20Apr44	22Apr44
Camp Stoneman, Calif		31Mar44	5Apr44	Aboard SS Gen John Pope at Sea	22Apr44	25Apr44
Aboard SS Catalina Enroute to San Francisco, Calif		5Apr44	5Apr44	Aboard SS Gen John Pope anchored Milne Bay, New Guinea	25Apr44	27Apr44
Aboard SS Gen John Pope anchored San Francisco, Calif		5Apr44	6Apr44	Aboard SS Gen John Pope at Sea	27Apr44	28Apr44

5. Retained copies of Morning Reports of the 1st Filipino Inf, by element, show the following stations for the period 28 Apr 44 thru 10 Apr 46, except for periods of missing Morning Reports as shown in Par 2.

Hq Co, 1st Filipino Inf – 28 Apr 44 thru 10 Apr 46

STATION	ARRIVED	DEPARTED	STATION	ARRIVED	DEPARTED
Oro Bay, New Guinea	28Apr44	18Jan45	Arrived LSM, Catbalogan, Samar, P.I.	16Feb45	1Mar45
Aboard MS Van Heutsz at Sea	18Jan45	19Jan45	Arrived LSM, Calbayog, Samar, P.I.	1Mar45	7May45
Aboard MS Van Heutsz Finschhafen, N.G.	19Jan45	19Jan45	Arrived Barge, Ormoc, Leyte, P.I.	7May45	21Sep45
Aboard MS Van Heutsz at Sea	19Jan45	21Jan45	Dulag, Leyte, P.I.	21Sep45	18Dec45
Aboard MS Van Heutsz Hollandia, N.G.	21Jan45	31Jan45	San Antonio, Samar, P.I.	18Dec45	23Mar46
Aboard MS Van Heutsz at Sea	31Jan45	6Feb45	Aboard USS Gen Callan at Sea	23Mar46	9Mar46
Aboard MS Van Heutsz San Pedro Bay, Leyte, PI	6Feb45	7Feb45	San Francisco, Calif	9Apr46	9Apr46
Alangalang, Leyte, P.I.	7Feb45	15Feb45	Camp Stoneman, Calif (Inactivated)	9Apr46	10Apr46
Aboard LSM, White Beach, Tolosa, Leyte, P.I.	15Feb45	16Feb45			

Hq Co, 1st Bn, 1st Filipino Inf – 28 Apr4 4 thru 31 Dec 45 and 17 Feb 46 thru 10 Apr 46

STATION	ARRIVED	DEPARTED	STATION	ARRIVED	DEPARTED
Oro Bay, New Guinea	28Apr44	24Jan45	Arrived LSM, Tacloban, Leyte, P.I.	24Apr45	25Apr45
Aboard Linley M Garrison Oro Bay, N.G.	24Jan45	25Jan45	Villaba, Leyte, P.I.	25Apr45	23Jun45
Aboard Linley M Garrison at Sea	25Jan45	28Jan45	Ipil, Ormoc, Leyte, P.I.	23Jun45	17Jul45
Aboard Lindley M Garrison anchored Humboldt Bay, N.G.	28Jan45	31Jan45	Seguinon, Albuera, Leyte, P.I.	17Jul45	25Sep45
Aboard Lindley M Garrison at Sea	31Jan45	6Feb45	Dulag, Leyte, P.I.	25Sep45	No date
Aboard Lindley M Garrison anchored San Pedro Bay, Leyte, P.I.	6Feb45	7Feb45	San Antonio, Samar, P.I.	No date	23Mar46
Alangalang, Leyte, P.I.	7Feb45	12Feb45	Aboard USS Gen Callan at Sea	23Mar46	9Apr46
Arrived LSM, Guiuan, Samar, P.I.	13Feb45	24Apr45	San Francisco, Calif	9Apr46	9Apr46
			Camp Stoneman, Calif (Inactivated)	9Apr46	10Apr46

Co A., 1st Filipino Inf – 4 Mar 45 thru 10 Apr 46

STATION	ARRIVED	DEPARTED	STATION	ARRIVED	DEPARTED
Oro Bay, New Guinea	28Apr44	No date	Seguinon, Leyte, P.I.	17Jul45	21Sep45
Gen Hq Tolosa Leyte, P.I.	No date	3May45	Dulag, Leyte, P.I.	21Sep45	3Jan46
Villaba, Leyte, P.I.	3May45	4May45	Tigbao, Tacloban, Leyte, P.I.	3Jan46	11Mar46
Ormoc, Leyte, P.I.	4May45	11May45	San Anotnio, Samar, P.I.	11Mar46	23Mar46
Villaba, Leyte, P.I.	11May45	17Jun45	Aboard USS Gen Callan at Sea	23Mar46	9Apr46
Arrived LSM, Ipil, Ormoc, Leyte, P.I.	17Jun45	17Jul45	San Francisco, Calif	9Apr46	9Apr46
			Camp Stoneman, Calif (Inactivated)	9Apr46	10Apr46

Co B, 1st Filipino Inf – 28Apr 44 thru 10Apr46

Location	From	To
Oro Bay, New Guinea	28Apr44	18Jan45
Aboard MS Van Heutsz	18Jan45	7Feb45
Tacloban, Leyte, P.I.	7Feb45	7Feb45
Alangalang, Leyte, P.I.	7Feb45	12Feb45
Aboard LSM #131	12Feb45	13Feb45
Guiuan, Samar, P.I.	13Feb45	24Apr45
Arrived LCM, Tacloban, Leyte, P.I.	24Apr45	25Apr45
Villaba, Leyte, P.I.	25Apr45	1May45
Abijao, Leyte, P.I.	1May45	2May45
Arrived LCM, Villaba, Leyte, P.I.	2May45	3May45
Palompon, Leyte, P.I.	3May45	25May45
Villaba, Leyte, P.I.	25May45	19Jun45
Arrived LCM, Ipil, Ormoc, Leyte, P.I.	19Jun45	4Jul45
Palompon, Leyte, P.I.	4Jul45	18Aug45
Seguinon, Leyte, P.I.	18Aug45	23Sep45
Dulag, Leyte, P.I.	23Sep45	29Nov45
Tacloban, Leyte, P.I.	29Nov45	11Mar46
Arrived LCM, San Antonio, Samar, P.I.	11Mar46	23Mar46
Aboard USS Gen Callan at Sea	23Mar46	9Apr46
San Francisco, Calif	9Apr46	9Apr46
Camp Stoneman, Calif (Inactivated)	9Apr46	10Apr46

Co C, 1st Filipino Inf – 28Apr44 thru 10Apr46

Location	From	To
Oro Bay, New Guinea	28Apr44	24Jan45
Aboard Lindley M. Garrison anchored Oro Bay, N.G.	24Jan45	25Jan45
Aboard Lindley M. Garrison at Sea	25Jan45	28Jan45
Aboard Lindley M. Garrison anchored Humboldt Bay, N.G.	28Jan45	31Jan45
Aboard Lindley M. Garrison at Sea	31Jan45	6Feb45
Aboard Lindley M. Garrison anchored San Pedro Bay, Leyte, P.I.	6Feb45	7Feb45
Alangalang, Leyte, P.I.	7Feb45	12Feb45
Arrived LSM, Guiuan, Samar, P.I.	13Feb45	24Apr45
Arrived LCM, Tacloban, Leyte, P.I.	24Apr45	25Apr45
Villaba, Leyte, P.I.	25Apr45	23Jun45
Ipil, Ormoc, Leyte, P.I.	23Jun45	17Jul45
Seguinon, Leyte, P.I.	17Jul45	23Sep45
Dulag, Leyte, P.I.	23Sep45	9Jan46
San Antonio, Samar, P.I.	9Jan46	23Mar46
Aboard USS Gen Callan at Sea	23Mar46	9Apr46
San Francisco, Calif	9Apr46	9Apr46
Camp Stoneman, Calif (Inactivated)	9Apr46	10Apr46

Co D, 1st Filipino Inf – 28 Apr 44 thru 10 Apr 46

Location	From	To
Oro Bay, New Guinea	28Apr44	24Jan45
Aboard Lindley M. Garrison anchored Oro Bay, N.G.	24Jan45	25Jan45
Aboard Lindley at Sea	25Jan45	28Jan45
Aboard Lindley M. Garrison anchored Humboldt Bay, N.G.	28Jan45	31Jan45
Aboard Lindley M. Garrison at Sea	31Jan45	6Feb45
Aboard Lindley M. Garrison anchored San Pedro Bay, Leyte, P.I.	6Feb45	7Feb45
Alangalang, Leyte, P.I.	7Feb45	12Feb45
Arrived LSM Guiuan, Samar, P.I.	13Feb45	24Apr45
Arrived LCM, Tacloban, Leyte, P.I.	24Apr45	25Apr45
Villaba, Leyte, P.I.	25Apr45	22Jun45
Ipil, Ormoc, Leyte, P.I.	22Jun45	15Jul45
Sequinon, Leyte, P.I.	15Jul45	25Sep45
Dulag, Leyte, P.I.	25Sep45	26Nov45
Arrived LCM, San Antonio, Samar, P.I.	26Nov45	23Mar46
Aboard USS Gen Callan at Sea	23Mar46	9Apr46
San Francisco, Calif (Inactivated)	9Apr46	10Apr46

Hq Co, 2d Bn – 1st Filipino Inf – 28 Apr44 thru 10 Apr46

Location	From	To	Location	From	To
Oro Bay, New Guinea	28Apr44	18Jan45	Bobon, Samar, P.I.	23Mar45	26Mar45
Aboard MS Van Heutsz at Sea	18Jan45	19Jan45z	Catarman, Samar, P.I.	26Mar45	18Apr45
Aboard MS Van Heutsz Finschhafen, N.G.	19Jan45	19Jan45	Cahicsan, Samar, P.I.	18Apr45	21Apr45
Aboard MS Van Heitz at Sea	19Jan45	21Jan45	Nenita, Samar, P.I.	21Apr45	27Apr45
Aboard MS Van Heutsz anchored Hollandia, N.G.	21Jan45	31Jan45	Catarman, Samar, P.I.	27Apr45	28Apr45
Aboard MS Van Heutsz at Sea	31Jan45	6Feb45	Pagsanghan, Samar, P.I.	28Apr45	1May45
Aboard MS Van Heutsz anchored Leyte, Gulf, P.I.	6Feb45	7Feb45	Catbalogan, Samar, P.I.	1May45	19May45
Tacloban, Leyte, P.I.	7Feb45	7Feb45	Arrived LSM, Ormoc, Leyte, P.I.	20May45	24May45
Alangalang, Leyte, P.I.	7Feb45	12Feb45	Palompon, Leyte, P.I.	25May45	1Jul45
Tolosa, Leyte, P.I.	12Feb45	24Feb45	Ormoc, Leyte, P.I.	1Jul45	8Jul45
Aboard LCI, Tolosa Beach, Leyte, P.I.	24Feb45	25Feb45	Camp Downes, Leyte, P.I.	8Jul45	29Sep45
Arrived LCM, Catbalogan, Samar, P.I.	25Feb45	27Feb45	Dulag, Leyte, P.I.	29Sep45	23Dec45
Calbayog, Samar, P.I.	27Feb45	2Mar45	Tacloban, Leyte, P.I.	23Dec45	15Mar46
Oquendo, Samar, P.I.	2Mar45	3Mar45	San Antonio, Samar, P.I.	15Mar46	23Mar46
Tarabucan, Samar, P.I.	3Mar45	16Mar45	Aboard USS Gen Callan at Sea	23Mar46	9Apr46
Lope de Vega, Samar, P.I.	16Mar45	21Mar45	San Francisco, Calif	9Apr46	9Apr46
Calbayog, Samar, P.I.	21Mar45	22Mar45	Camp Stoneman, Calif (Inactivated)	9Apr46	10Apr46
Arrived LCM, Allen, Samar, P.I.	22Mar45	23Mar45			

Co E, 1st Filipino Inf – 28 Apr 44 thru 10 Apr 46

Location	From	To	Location	From	To
Oro Bay, New Guinea	28Apr44	18Jan45	Cervantes, Samar, P.I.	4apr45	6Apr45
Aboard MS Van Heutsz at Sea	18Jan45	19Jan45	N. Pagsanghan, Samar, P.I.	6Apr45	13Apr 45
Aboard MS Van Heutsz Finschhafen, N.G.	19Jan45	19Jan45	Guibunauan, Samar, P.I.	13Aor45	14Apr45
Aboard M Van Heutsz at Sea	19Jan45	21Jan45	Cahicsan, Samar, P.I.	14Apr45	20Apr45
Aboard MS Van Heutsz anchored Hollandia, N.G.	21Jan45	31Jan45	Pagsanghan, Samar, P.I.	20Apr45	1May45
Aboard MS Van Heutsz at Sea	31Jan45	6Feb45	Catbalogan, Samar, P.I.	1May45	21May45
Aboard MS Van Heutsz, San Pedro Bay, Leyte, P.I.	6Feb45	7Feb45	Arrived LSM Ormoc, Leyte, P.I.	22May45	28May45
Alangalang, Leyte, P.I.	7Feb45	10Feb45	Palompon, Leyte, P.I.	28May45	29May45
Telegrafo, Leyte, P.I.	10Feb45	24Feb45	Rizal, Leyte, P.I.	29May45	7Jun45
Tolosa Beach, Leyte, P.I.	24Feb45	25Feb45	Palompon, Leyte, P.I.	7Jun45	17Jun45
Catbalogan, Samar, P.I.	25Feb45	26Feb45	Arrived barges Villaba, Leyte, P.I.	17Jun45	27Jun45
Calbayog, Samar, P.I.	26Feb45	28Feb45	Ormoc, Leyte, P.I.	27Jun45	7Jul45
Oquendo, Samar, P.I.	28Feb45	2Mar45	Camp Downes, Ormoc, Leyte, P.I.	7Jul45	27Sep45
Tarabucan, Samar, P.I.	2Mar45	3Mar45	Dulag, Leyte, P.I.	27Sep45	13Dec45
Lope De Vega, Samar, P.I.	3Mar45	14Mar45	Tacloban, Leyte, P.I.	13Dec45	14Mar46
Arrived boat Allen, Samar, P.I.	14Mar45	15Mar45	Arrived LSM, San Antonio, Samar, P.I.	14Mar46	23Mar46
Arrived barges Bobon, Samar	15Mar45	29Mar45	Tacloban, Leyte, P.I.	23Mar46	23Mar46
Pagsanghan, Samar, P.I.	30Mar45	30Mar45	Aboard USS Gen Callan at Sea	23Mar46	9Apr46
Santander, Samar, P.I.	31Mar45	3Apr45	San Francisco, Calif	9Apr46	9Apr46
Pagsanghan, Samar, P.I.	3Apr45	4Apr45	Camp Stoneman, Calif (Inactivated)	9Apr46	10Apr46

Co F, 1ˢᵗ Filipino Inf – 28 Apr 44 thru 10 Apr 46

Oro Bay, New Guinea	28Apr44	18Jan45	Cahicsan, Samar, P.I.	14Apr45	21Apr45
Aboard MS Van Heutsz at Sea	18Jan45	19Jan45	Pagsanghan, Samar, P.I.	21Apr45	1May45
Aboard MS Van Heutsz Finschhafen, N.G.	19Jan45	19Jan45	Catbalogan, Samar, P.I.	1May45	19May45
Aboard MS Van Heutsz at Sea	19Jan45	21Jan45	Ormoc, Leyte, P.I.	20May45	26May45
Aboard MS Van Heutsz anchored Hollandia, N.G.	21Jan45	31Jan45	Palompon, Leyte, P.I.	26May45	29May45
Aboard MS Van Heutsz at Sea	31Jan45	6Feb45	Santa Rosa, Leyte, P.I.	29May45	21Jun45
Aboard MS Van Heutsz anchored Leyte Gulf, P.I.	6Feb45	7Feb45	Wague, Leyte, P.I.	21Jun45	27Jun45
Alangalang, Leyte, P.I.	7Feb45	11Feb45	Ormoc, Leyte, P.I.	27Jun45	7Jul45
Tanauan, Leyte, P.I.	11Feb45	13Feb45	Camp Downes, Ormoc, Leyte, P.I.	7Jul45	7Sep45
Tolosa, Leyte, P.I.	13Feb45	25Feb45	Puerto Bello, Leyte, P.I.	7Sep45	15Sep45
Catbalogan, Samar, P.I.	25Feb45	1Mar45	Camp Downes, Ormoc, Leyte, P.I.	15Sep45	27Sep45
Allen, Samar, P.I.	1Mar45	11Mar45	Dulag, Leyte, P.I.	27Sep45	13Dec45
Enriqueta, Samar, P.I.	11MaR45	27Mar45	Tacloban, Leyte, P.I.	13Dec45	15Mar46
Acereda, Samar, P.I.	27Mar45	3Apr45	Arrived LCM, San Antonio, Samar, P.I.	15Mar46	23Mar46
Cervantes, Samar, P.I.	3Apr45	5Apr45	Aboard USS Gen Callan at Sea	23Mar46	9Apr46
S. Pagsanghan, Samar, P.I.	5Apr45	13Apr45	San Francisco, Calif	9Apr46	9Apr46
Guibunawan, Samar, P.I.	13Apr45	14Apr45	Camp Stoneman, Calif (Inactivated)	9Apr46	10Apr46

Co G, 1st Filipino Inf – 28 Apr 44 thru 31 May 45 and 13 Dec thru 10 Apr 46

Oro Bay, New Guinea	28Apr44	18Jan45	Casolgan, Samar, P.I.	8Mar45	9Mar45
Aboard MS Van Heutsz at Sea	18Jan45	19Jan45	Bobon, Samar, P.I.	9Mar4	21Mar45
Aboard MS Van Heutsz Finschhafen, N.G.	19Jan45	19Jan45	Magdalig, Samar, P.I.	21Mar45	23Mar45
Aboard Ms Van Heutsz anchored Hollandia, N.G.	21Jn45	31Jan45	Bobon, Samar, P.I.	23Mar45	27Mar45
Aboard MS Van Heutsz at Sea	31Jan45	6Feb45	Polangi, Samar, P.I.	27Mar45	30Mar45
Aboard Ms Van Heutsz anchored Tacloban, Leyte, P.I.	6Feb45	7Feb45	Samoge, Samar, P.I.	30Mar45	4Apr45
Alangalang, Leyte, P.I.	7Feb45	10Feb45	Cervantes, Samar, P.I.	4Apr45	18Apr45
Kiling, Leyte, P.I.	10Feb45	24Feb45	Lope De Vega, Samar, P.I.	18Apr45	19Apr45
Aboard LCI #1008	24Feb45	25Feb45	Kayhagdan, Samar, P.I.	19Apr45	20Apr45
Catbalogan, Samar, P.I.	25Feb45	26Feb45	Lope De Vega, Samar, P.I.	20Apr45	1May45
Tarabucan, Samar, P.I.	26Feb45	27Feb45	Catbalogan, Samar, P.I.	1May45	19May45
San Rufino, Samar, P.I.	27Feb45	28Feb45	Ormoc, Leyte, P.I.	20May45	26May45
Tarabucan, Samar, P.I.	28Feb45	1Mar45	Palompon, Leyte, P.I.	26May45	No date
Lope De Vega, Samar, P.I.	1Mar45	2Mar45	Dulag, Leyte, P.I.	No date	13Dec45
Quezon, Samar, P.I.	2Mar45	3Mar45	Tacloban, Leyte, P.I.	13Dec45	14Dec45
Kagbunga, Samar, P.I.	3Mar45	4Mar45	Arrived LCM, San Antonio, Samar, P.I.	14Mar45	23Mar45
Mabini, Samar, P.I.	4Mar45	5Mar45	Tacloban Bay, Leyte, P.I.	23Mar45	23Mar45
Maraganan, Samar, P.I.	5Mar45	6Mar45	Aboard USS Gen Callan at Sea	23Mar45	9Apr46
Santander, Samar, P.I.	6Mar45	7Mar45	San Francisco, Calif	9Apr46	9Apr46
Baras, Samar, P.I.	7Mar45	8Mar45	Camp Stoneman, Calif (Inactivated)	9Apr46	10Apr46

Co H, 1ˢᵗ Filipino Inf – 28 Apr 44 thru 10 Apr 46

Location	From	To	Location	From	To
Oro Bay, New Guinea	28Apr44	18Jan45	Bobon, Samar, P.I.	15Mar45	31Mar45
Aboard MS Van Heutsz atSea	18Jan45	19Jan45	Madalig, Samar, P.I.	31Mar45	4Apr45
Aboard MS Van Heutsz Finschhafen, N.G.	19Jan45	19Jan45	Polange, Samar, P.I.	4Apr45	10Apr45
Aboard MS Van Heutsz at Sea	19Jan45	21Jan45	N. Pagsanghan, Samar, P.I.	10Apr45	20Apr45
Aboard Ms Van Heutsz anchored Hollandia, N.G.	21Jan45	31Jan45	S. Canaua, Samar, P.I.	20Apr45	21Apr45
Aboard MS Van Heutsz at Sea	31Jan45	6Feb45	Lope De Vega, Samar, P.I.	21Apr45	30Apr45
Aboard MS Van Heutsz anchored Tacloban, Leyte, P.I.	6Feb45	7Feb45	Catbalogan, Samar, P.I.	1May45	22May45
Alangalang, Leyte, P.I.	7Feb45	10Feb45	Ormoc, Leyte, P.I.	22May45	28May45
Tanauan, Leyte, P.I.	10Feb45	25Feb45	Palompon, Leyte, P.I.	28May45	19Jun45
Aboard LSM #151 to Tolosa Beach, Leyte, P.I.	25Feb45	25Feb45	Villaba, Leyte, P.I.	19Jun45	26Jun45
Catbalogan, Samar, P.I.	25Feb45	27Feb45	Arrived Barges Ormoc, Leyte, P.I.	26Jun45	7Jul45
Calbayog, Samar, P.I.	27Feb45	28Feb45	Camp Downes, Ormoc, Leyte, P.I.	7Jul45	25Sep45
Oquendo, Samar, P.I.	28Feb45	1Mar45	Dulag, Leyte, P.I.	25Sep45	13Dec45
Tarabucan, Samar, P.I.	1Mar45	2Mar45	Tacloban, Leyte, P.I.	13Dec45	14Mar46
Lope De Vega, Samar, P.I.	2Mar45	7Mar45	San Antonio, Samar, P.I.	14Mar46	23Mar46
Tarabucan, Samar, P.I.	7Mar45	13Mar45	Aboard USS Gen Callan at Sea	23Mar46	9Apr46
Calbayog, Samar, P.I.	13Mar45	14Mar45	San Francisco, Calif	9Apr46	9Apr46
Allen, Samar, P.I.	14Mar45	15Mar45	Camp Stoneman, Calif (Inactivated)	9Apr46	10Apr46

Hq Co, 3ʳᵈ Bn, 1ˢᵗ Filipino Inf – 28 Apr44 thru 10 Apr 46

Location	From	To	Location	From	To
Oro Bay, New Guinea	28Apr44	24Jan45	Mondragon, Samar, P.I.	12Apr45	18Apr45
Aboard Lindley M Garrison anchored Oro Bay, N.G.	24Jan45	25Jan45	Carania, Samar, P.I.	18Apr45	20Apr45
Aboard Lindley M Garrison at Sea	25Jan45	6Feb45	Nenita, Samar, P.I.	20Apr45	21Apr45
Aboard Lindley M Garrison anchored San Pedro Bay, Leyte, P.I.	6Feb45	7Feb45	Ginualgan, Samar, P.I.	21Apr45	23Apr45
Alangalang, Leyte, P.I.	7Feb45	15Feb45	Pagsanghan, Samar, P.I.	23Apr45	24Apr45
Aboard LSM, Tolosa, Leyte, P.I.	15Feb45	16Feb45	Nenita, Samar, P.I.	24Apr45	29Apr45
Arrived LSM, Catbalogan, Samar, P.I.	16Feb45	18Feb45	Catarman, Samar, P.I.	29Apr45	3May45
Arrived Boat Calbayog, Samar, P.I.	18Feb45	27Feb45	Calbayog, Samar, P.I.	3May45	3Sep45
Arrived Boat Allen, Samar, P.I.	27Feb45	26Mar45	Catbalogan, Samar, P.I.	3Sep45	10Oct45
Caranglan, Samar, P.I.	26Mar45	27Mar45	Arrived LSM Dulag, Leyte, P.I.	10Oct45	15Dec45
Pangubian, Samar, P.I.	27Mar45	29Mar45	White Beach, Leyte, P.I.	15Dec45	12Mar46
Bobon, Samar, P.I.	29Mar45	2Apr45	Arrived LCM, San Antonio, Samar, P.I.	12Mar46	23Mar46
Caranglan, Samar, P.I.	2Apr45	3Apr45	Aboard USS Gen Callan at Sea	23Mar46	9Apr46
Catarman, Samar, P.I.	3Apr45	12Apr45	San Francisco, Calif	9Apr46	9Apr46
			Camp Stoneman, Calif (Inactivated)	9Apr46	10Apr46

Co I, 1st Filipino Inf – 28 Apr 44 thru 10 Apr 46

Oro Bay, New Guinea	28Apr44	24Jan45	Mondragon, Samar, P.I.	12Apr45	13Apr45
Aboard Lindley M Garrison anchored Oro Bay, N.G.	24Jan45	25Jan45	Cahicsan, Samar, P.I.	14Apr45	19Apr45
Aboard Lindley M Garrison at Sea	25Jan45	6Feb45	Canaua, Samar, P.I.	19Apr45	21Apr45
Aboard Lindley M Garrison anchored San Pero Bay, Leyte, P.I.	6Feb45	7Feb45	Bantayan, Samar, P.I.	21Apr45	3May45
Alangalang, Leyte, P.I.	7Feb45	15Feb45	Catubig, Samar, P.I.	3May45	7Jul45
Aboard LCM, San Pedro Bay, Leyte, P.I.	15Feb45	16Feb45	Calbayog, Samar, P.I.	7Jul45	10Oct45
Arrived LCM, Catbalogan, Samar, P.I.	16Feb45	17Feb45	Dulag, Leyte, P.I.	10Oct45	14Dec45
Calbayog, Samar, P.I.	17Feb45	22Feb45	White Beach, Leyte, P.I.	14Dec45	12Mar46
Capul Island, P.I.	22Feb45	6Apr45	San Antonio, Samar, P.I.	12Mar46	23Mar46
Allen, Samar, P.I.	6Apr45	7Apr45	Aboard USS Gen Callan at Sea	23Mar46	9Apr46
Catarman, Samar, P.I.	7Apr45	8Apr45	San Francisco, Calif	9Apr46	9Apr46
Malijao, Samar, P.I.	8Apr45	12Apr45	Camp Stoneman, Calif (Inactivated)	9Apr46	10Apr46

Co K, 1st Filipino Inf – 28 Apr 44 thru 10 Apr 46

Oro Bay, New Guinea	28Apr44	24Jan45	Quezon, Samar, P.I.	7Apr45	9Apr45
Aboard Lindley M Garrison anchored Oro Bay, N.G.	24Jan45	25Jan45	Catarman, Samar, P.I.	9Apr45	12Apr45
Aboard Lindley M Garrison at Sea	25Jan45	6Feb45	De Maria, Samar, P.I.	12Apr45	15Apr45
Aboard Lindley M Garrison anchored San Pedro Bay, Leyte, P.I.	6Feb45	7Feb45	Cahicsan, Samar, P.I.	15Apr45	16Apr45
Alangalang, Leyte, P.I.	7Feb45	15Feb45	Gibarogwan, Samar, P.I.	16Apr45	18Apr45
Aboard LSM, Tolosa, Leyte, P.I.	15Feb45	16Feb45	Cahicsan, Samar, P.I.	18Apr45	19Apr45
Arrived LSM, Catbalogan, Samar, P.I.	16Feb45	21Feb45	De Maria, Samar, P.I.	19Apr45	20Apr45
Calbayog, Samar, P.I.	21Feb45	26Feb45	Loangan, Samar, P.I.	20Apr45	21Apr45
Arrived LCM, Allen, Samar, P.I.	26Feb45	28Feb45	Nenita, Samar, P.I.	21Apr45	22Apr45
Mauo, Samar, P.I.	1Mar45	3Mar45	Loangan, Samar, P.I.	22Apr45	25Apr45
Lavesares, Samar, P.I.	3Mar45	8Mar45	Nenita, Samar, P.I.	25Apr45	30Apr45
Caranglan, Samar, P.I.	8Mar45	15Mar45	Catarman, Samar, P.I.	30Apr45	4May45
Lavesares, Samar, P.I.	15Mar45	27Mar45	Calbayog, Samar, P.I.	4May45	2Aug45
Caranglan, Samar, P.I.	27Mar45	27Mar45	Camp Downes, Ormoc, Leyte, P.I.	3Aug45	21Sep45
Polange, Samar, P.I.	27Mar45	31Mar45	Dulag, Leyte, P.I.	21Sep45	13Dec45
Bobon, Samar, P.I.	31Mar45	31Mar45	White Beach, Leyte, P.I.	13Dec45	12Mar46
Gimanalud, Samar, P.I.	1Apr45	4Apr45	Arrived LCM, San Antonio, Samar, P.I.	12Mar46	23Mar46
Ponod, Samar, P.I.	4Apr45	6Apr45	Aboard USS Gen Callan at Sea	23Mar46	9Apr46
Poratan, Samar, P.I.	6Apr45	7Apr45	San Francisco, Calif	9Apr46	9Apr46
			Camp Stoneman, Calif (Inactivated)	9Apr46	10Apr46

Co L, 1ˢᵗ Filipino Inf – 28 Apr 44 thru 10 Apr 46

Location	From	To
Oro Bay, New Guinea	28Apr44	25Jun45
Aboard Lindley M Garrison at Sea	25Jan45	6Feb45
Aboard Lindley M Garrison anchored San Pedro Bay, Leyte, P.I.	6Feb45	7Feb45
Alangalang, Leyte, P.I.	7Feb45	15Feb45
Aboard LSM, Tolosa, Leyte, P.I.	15Feb45	16Feb45
Catbalogan, Samar, P.I.	16Feb45	19Feb45
Allen, Samar, P.I.	20Feb45	13Mar45
Catarman, Samar, P.I.	13Mar45	23Mar45
Allen, Samar, P.I.	23Mar45	26Mar45
Caranglan, Samar, P.I.	26Mar45	29Mar45
Magdalig, Samar, P.I.	29Mar45	29Mar45
Pangubian, Samar, P.I.	29Mar45	31Mar45
Bobon, Samar, P.I.	31Mar45	1Apr45
Pangubian, Samar, P.I.	2Apr45	4Apr45
Malijao, Samar, P.I.	4Apr45	6Apr45
Polange, Samar, P.I.	6Apr45	12Apr45
Manalibang, Samar, P.I.	12Apr45	14Apr45
Polange, Samar, P.I.	14Apr45	16Apr45
Cervantes, Samar, P.I.	16Apr45	30Apr45
Polange, Samar, P.I.	30Apr45	8May45
Catarman, Samar, P.I.	8May45	2Jul45
Calbayog, Samar, P.I.	2Jul45	25Aug45
Arrived LCM, Catbalogan, Samar, P.I.	25Aug45	10Oct45
Arrived LSM, Dulag, Leyte, P.I.	10Oct45	13Dec45
White Beach, Tacloban, Leyte, P.I.	13De45	12Mar46
San Antonio, Samar, P.I.	12Mar46	23Mar46
Aboard USS Gen Callan at Sea	23Mar46	9Apr46
San Francisco, Calif	9Apr46	9Apr46
Camp Stoneman, Calif (Inactivated)	9Apr46	10Apr46

Co M, 1ˢᵗ Filipino Inf – 28 Apr 44 thru 10 Apr 46

Location	From	To
Oro Bay New Guinea	28Apr44	25Jan45
Aboard Lindley M Garrison at Sea	25Jan45	6Feb45
Aboard Lindley M Garrison anchored San Pedro Bay, Leyte, P.I.	6Feb45	7Feb45
Alangalang, Leyte, P.I.	7Feb45	15Feb45
Aboard LSM, Tolosa, Leyte, P.I.	15Feb45	16Feb45
Catbalogan, Samar, P.I.	16Feb45	18Feb45
Calbayog, Samar, P.I.	18Feb45	26Feb45
Allen, Samar, P.I.	26Feb45	10Mar45
Magdalig, Samar, P.I.	10Mar45	31Mar45
Enriqueta, Samar, P.I.	31Mar45	8Apr45
Malijao, Samar, P.I.	8Apr45	10Apr45
Catarman, Samar, P.I.	10Apr45	20Apr45
Somoge, Samar, P.I.	20Apr45	23Apr45
Polange, Samar, P.I.	23Apr45	3Jul45
Catarman, Samar, P.I.	3Jul45	8Jul45
Calbayog, Samar, P.I.	8Jul45	25Aug45
Catbalogan, Samar, P.I.	25Aug45	10Oct45
Arrived LSM, Dulag, Leyte, P.I.	10Oct45	15Dec45
White Beach, Leyte, P.I..	15Dec45	12Mar46
San Antonio, Samar, P.I.	12Mar46	23Mar46
Aboard USS Gen Callan at Sea	23Mar46	9Apr46
San Francisco, Calif	9Apr46	9Apr46
Camp Stoneman, Calif (Inactivated)	9Apr46	10Apr46

Med Det, 1ˢᵗ Filipino Inf – 28 Apr 44 thru 10 Apr 46

Location	From	To
Oro Bay, New Guinea	28Apr44	18Jan45
Aboard MS Van Heutsz at Sea	18Jan45	6Feb45
Aboard MS Van Heutsz anchored Tacloban, Leyte, P.I.	6Feb45	7Feb45
Alangalang, Leyte, P.I.	7Feb45	15Feb45
Aboard LSM, White Beach, Tolosa, Leyte, P.I.	15Feb45	16Feb45
Catbalogan, Samar, P.I.	16Feb45	1Mar45
Calbayog, Samar, P.I.	1Mar45	7May45
Arrived Barge Ormoc, Leyte, P.I.	7May45	23Sep45
Dulag, Leyte, P.I.	23Sep45	19Dec45
Tacloban, Leyte, P.I.	19Dec45	19Dec45
Arrived LCT San Antonio, Samar, P.I.	19Dec45	23Mar46
Aboard USS Gen Callan at Sea	23Mar46	9Apr46
San Francisco, Calif	9Apr46	9Apr46
Camp Stoneman, Calif (Inactivated)	9Apr46	10Apr46

Sv Co, 1ˢᵗ Filipino Inf – 28 Apr 44 thru 10 Apr 46

Oro Bay, New Guinea	28Apr44	25Jun45	Ormoc, Leyte, P.I.	7May45	21Sep45
Aboard Lindley M Garrison at Sea	25Jun45	6Feb45	Dulag, Leyte, P.I.	21Sep45	19Dec45
Aboard Lindley M Garrison anchored San Pedro Bay, Leyte, P.I.	6Feb45	7Feb45	San Antonio, Samar, P.I.	19Dec45	23Mar46
Alangalang, Leyte, P.I.	7Feb45	15Feb45	Aboard USS Gen Callan at Sea	23Mar46	9Apr46
Aboard LSM, White Beach, Tolosa, Leyte,P.I.	15Feb45	16Feb45	San Francisco, Calif	9Apr46	9Apr46
Catbalogan, Samar, P.I.	16Feb45	1Mar45	Camp Stoneman, Calif (Inactivated)	9Apr46	10Apr46
Calbayog, Samar, P.I.	1Mar45	7May45			

Anti-Tank Co, 1ˢᵗ Filipino Inf – 28 Apr44 thru 20 Mar 46

Oro Bay, New Guinea	28Apr44	25Jan45	Aboard LSM, White Beach, Tolosa, Leyte, P.I.	15Feb45	16Feb45
Aboard Lindley M Garrison at Sea	25Jan45	6Feb45	Arrived LSM, Catbalogan, Samar, P.I.	16Feb45	19May45
Aboard Lindley M Garrison anchored San Pedro Bay, Leyte, P.I.	6Feb45	7Feb45	Ormoc, Leyte, P.I.	20May45	29Sep45
Alangalang, Leyte, P.I.	7Feb45	15Feb45	Dulag, Leyte, P.I.	29Sep45	28Dec45
			San Antonio, Samar, P.I.	28Dec45	No date

Cannon Co, 1ˢᵗ Filipino Inf – 28 Apr 44 thru 10 Apr 46

Oro By, New Guinea	28Apr44	24Jan45	Catbalogan, Samar, P.I.	16Feb45	18Feb45
Aboard Lindley M Garrison anchored Oro Bay, N.G.	24Jan45	25Jan25	Arrived LCM, Calbayog, Samar, P.I.	18Feb45	16May45
Aboard Lindley M Garrison at Sea	25Jan45	6Feb45	Catbalogan, Samar, P.I.	16May45	20May45
Alangalang, Leyte, P.I.	7Feb45	15Feb45	Arrived LSM, Ormoc, Leyte, P.I.	22May45	29Sep45
Aboard LSM, White Beach, Tolosa, Leyte, P.I.	15Feb45	16Feb45	Dulag, Leyte, P.I.	29Sep45	No date

Robert H. Wagner, Capt, AGD
Acctg Chief of Branch

....

AGRS-DS-R 322 Status of Unit
TO: Photo Process Sec, RAC   FROM: Misc Rec Unit   DATE: 28 Oct 48   COMMENT NO. 3
cec/786

1. For appropriate action.
2. Morning Reports (Station & Records of Events) of Hq Co, Band, Service Co, Anti-Tank Co, Hq Co 2ⁿᵈ Bn, Hq Co 3ʳᵈ Bn and Med Det, 1ˢᵗ Filipino Inf from 13 Jul 42 to 31 Jul 43 shows:

Activated Salinas Garrison California	13 Jul 42	20 Aug42
San Luis Obispo, California	20 Aug 42	14 Oct 42
Ft Ord, California	14 Oct 42	8 Jan 43
Camp Beale, California	8 Jan 43	2 May 43
Hunter Liggett Military Res., California	2 May 43	No date

3. Morning Reports (Station & Records of Events) of Co's A, B, C, D, E, F, H, I, J, K, L & M 1st Filipino Inf from 13 Jul 42 to 31 Jul 43 shows:

Salinas Garrison, California	13 Jul 42	21 Aug 42
San Luis Obispo, California	21 Aug 42	14 Oct 42
Ft. Ord, California	14 Oct 42	30 Nov 42
Hunter Liggett Military Res., California	30 Nov 42	18 Dec 42
Ft Ord, California	18 Dec 42	8 Jan 43
Camp Beale, California	8 Jan 43	2 May 43
Hunter Liggett Military Res., California	2 May 43	No date

4. Morning Reports of Cannon Co., Hq Co 1st Bn and Co G, 1st Filipino Inf Regt have not been received.

Coleman
Misc Rec Unit
Demob Pers Rec Br

....

AGRS-EP 322 (20 Sep 48) Status of Unit
TO: Chief Operations Br, AGO   FROM: Chief Photo Process Sec, AGO   DATE: 19 Nov 48
COMMENT NO. 4   fnc/2136

1. Morning reports of Co A, 1st Filipino Inf Regt for 7 Jan thru 3 Mar 45 show unit stationed at the following locations:

STATIONS	ARRIVED	DEPARTED
Oro Bay		18 Jan 45
Aboard SS VAN Heutsz	18 Jan 45	7 Feb 45
Anchored at Hollandia	21 Jan 45	31 Jan 45
Anchored at Leyte	6 Feb 45	7 Feb 45
Alangalang, Leyte	8 Feb 45	10 Feb 45
Tacloban, Leyte	10 Feb 45	23 Feb 45
GHQ, Tolosa, Leyte	23 Feb 45	No date

2. Morning reports of Hq Co, 1st Bn, 1st Filipino Infantry Tegt for 1 Jan thru 16 Feb 45 show unit stationed at the following locations:

Oro Bay (Base Camp)		24 Jan 45
Aboard Lindley M Garrison	24 Jan 45	7 Feb 45
Alangalang, Leyte	7 Feb 45	12 Feb 45
Aboard LSM	12 Feb 45	13 Feb 45
Vicinity Guiuan, Samar, P.I.	13 Feb 45	No date

3. Morning reports of Co G, 1st Filipino Inf Regt for 1 Jun thru 12 Dec 45 show unit stationed at following locations:
(CP moved from Palompon to Abijao 31 May 45)

Abijao, Leyte		27 Jun 45
Ormoc, Leyte	27 Jun 45	7 Jul 45
Camp Downes, Ormoc, Leyte, P.I.	7 Jul 45	29 Sep 45
Dulag, Leyte	29 Sep 45	No date

4. Morning reports of Cannon Co, 1st Filipino Inf Regt for 1 Nov 45 thru 21 Mar 46 (final M/R) show unit stationed at following locations:

Dulag		26 Dec 45
San Antonio, Samar, P.I.	26 Dec 45	21 Mar 46

5. Morning reports of AT Co, 1st Filipino Inf Regt for 21 Mar 46 (final M/R) shows unit stationed at San Antonio, Samar, P.I.

THOMAS M. HARFORD, Major, AGD
Chief, Photo Process Section, RAC

Appendix A-22

TIMELINE
1ST AND 2ND FILIPINO INFANTRY REGIMENTS

1942

Apr - 1st Filipino Battalion activated at Camp San Luis Obispo
Training:
> Jun - 1st FIR, Salinas CA
> Aug - 1st FIR, San Luis Obispo
> Oct - 1st FIR, Fort Ord. Attach to Second Armored Corps
> Nov - 2nd FIR, Ft. Ord. Attach to Second Armored Corps

1943

Training:
> Jan - 1st FIR, Camp Beale
> Jan - 2nd FIR, Camp Roberts, Camp Cooke
May - Men chosen for Australia to train for secret missions
Training:
> Jun - 1st FIR, Camp Roberts
> Aug - Hawaii men, Camp Roberts
> Oct - 1st FIR, Camp Beale

1944

Training and Combat:
> 1st and 2 nd FIR
> Mar - 2nd FIR disbanded. 2nd FIB activated, Camp Cooke
Mar - 1st FIR, Oro Bay, New Guinea. Units attached to U.S. Divisions in combat
Jul - 2nd FIB and Hawaii men, Oro Bay, New Guinea
Men joined Alamo Scouts and Rangers
Sep - Men transferred to PCAU, CIC

1945

Combat:
> Feb - 1st FIR in Leyte and Guiuan, Samar
> Feb - 1st FIR units in Catbalogan, Calbayog, Capul and Buri Islands, Samar
> Feb - Hawaii men in Leyte, Samar, Mindoro
> Apr - 1st FIR in Leyte and Samar
> Jun - 1st FIR units in Ipil, Ormoc, Leyte
May – 2nd FIB to Philippines
Aug - Japan surrendered
Oct - 1st FIR attached to Western Pacific Area Command (WPAC)
Dec - 2nd FIB inactivated in Luzon

1946

Combat:
> Jan - 1st FIR in San Antonio, Samar
Mar - Hawaii men attached to 86th Division, Philippine Army, Manila
Apr - 1st FIR inactivated at Camp Stoneman, California
Sep - Hawaii unit discharged in California

Legend: FIR – First Filipino Infantry Regiment; 2nd FIR – 2nd Filipino Infantry Regiment; FIB – Filipino Infantry Battalion; KAZ – General MacArthur's radio station; PIMC – Philippine Intelligence Message Center

Source: Documents on the 1st Reconnaissance Battalion (Special), 97th Signal Service Company, 1st and 2nd Filipino Infantry Regiments, 2nd Filipino Infantry Battalion, Philippine Civil Affairs Unit (PCAU), Counter Intelligence Corp (CIC), Hawaii soldiers, and from the U.S. Army Service Summary, and Abdon Llorente's 1st Filipino Infantry

Appendix A-23

1st and 2nd Filipino Infantry Regiments Silver Anniversary
San Francisco, California. July 14-16, 1967

COMMITTEE

Col. Earl L. Mullinix, USAR-Ret	Advisor	M/Sgt Rudolph F. Corpuz, USA-Ret	Heraldry
Lt. Col Saturnino R. Silva, AUS-Ret	Chairman	Lazaro Fabian	Registration
Gregorio P. Chua, M.D.	Secretary	Bernabe F. Flores	Facilities
Atonio A. Calagos	Treasurer	Frank M Pulido	Social &Hospitality
John S. Bamont	Entertainment	Maj Agustin L. Santos, USAR-Ret	Refreshments
Maj Anotnio D. Campos, USAR	Program	M/Sgt Anastacio E. Velarde, USA-Ret	Registration
M/Sgt Frank B. Corpus, USA-Ret	Publicity	Mrs.Mullinix, Mrs Satos, Mrs Silva	Hostesses

The welcome speech was given by Col Earl L. Mullinix former Executive Officer, 1st Filipino Infantry Regiment and Reunion Advisor. The guest speaker was Hawaii Assemblyman Peter A. Aduja former member of the 1st Filipino Infantry Regiment. Entertainment were a Violin solo by Gilopez Kabayao, Fiilipino folk dances by the Maligayan Troupe of City College of San Francisco Students, Vocalist Regina Arenas, and Trumpet Trio by the Delgado family. Lt. Col. Silva introduced the Committee Members before the dance music started by the "Gaylanders" with Mr. Delgado of "Bahala Na" until 1:00 a.m.

Welcome speeches were given by Col Robert H. Offley, former CO 1st Filipino Infantry Regiment, Col Charles L Clifford, former CO 2nd Filipino Infantry Regiment, and Brig Gen John C. Crowley, former officer in the 1st Filipino Infantry Regiment. The response was given by Lt. Col Saturnino R. Silva, Reunion Chairman. A review of the units by Regimental Commanders and Honorary Chairman followed. The floor reunion of members and their families followed the Retreat of Colors. In the evening, the Reunion dance was held at the Mark Hopkins Hotel. Uniforms were optional at Registration if the men could still fit in them.

1st FILIPINO INFANTRY REGIMENT

Activated – 13 July 1942, Salinas, California
Inactivated – 10 April 1946, Camp Stoneman, California
Battle Credits – New Guinea and Southern Philippines

2nd FILIPINO INFANTRY REGIMENT

Activated – 21 November 1942, Ft Ord, California
Disbanded – 27March 1944, Camp Cooke, California

1st RECONNAISSANCE BATTALION (SPECIAL)

Activated – 20 November 1944, Hollandia, New Guinea
Disbanded – 15 August 1945, Manila, Philippines
Battle Credits - Philippines

2nd FILIPINO INFANTRY BATTALION

Activated – 28 March 1944 – Camp Cooke, California
Disbanded – 31 March 1946, Manila, Philippines
Battle Credits – New Guinea and Southern Philippines

IN MEMORIAM*

Ricardo T. Arca
Edward Bergen
Vevencio M. Fortamillas
Clyde H. Haas
Jerry Kessner
Modesto R. Paragas
Francisco A. Rosete

Ruben T. Songco
Edon Zueger
Moses M. Dackel
Henry R. Bittick
Leonilo T. Digal
Genaro Frias

James Healy
Serapion Ladesma
Luis Padilla
Joe Santos
Isidro Urbano
Doyce Hamilton
Dionisio Santos

*Names are those known at that time.

1st FILIPINO INFANTRY REGIMENT

Early Registrants

Pacifico A. Abaua
Sixto C. Aduana
Peter A. Aduja
Joseph P. Agbannawabe
Vincent A. Agbayani
Pacifico D. Agpaoa
Richard O. Alvarado
Cleto I. Anacleto
Carlos F. Ancheta
Valentino P. Andrada
Arcadio S. Andres
Jose K. Angeles
George A. Antolin
Enrique J. Aquino
Isidoro D. Aranas
Antonio V. Arce
Manuel V. Arcol
Numeriano C. Armas
John O. Aspiras
Bernardo G. Asuelo
Martiniano S. Atad
Felix E. Ayson
Glecerio V. Ayson
Faustino Bacus
Mariano L. Balancio
Leo M. Balangue
Julito C. Balolong
Teodor L. Balugo
Joseph C. Bisue

Florencio P. Dizon
Eusebio G. Duclayan
Antonio C. Duldulao
Teodoro Q. Dumpit
Leocadio B. Escalana
Victor Fanganillo
Justino E. Felicitas
Louis A. Filarca
Melvin D. Fortes
Juan Francisco
William Gale
Benny Galvez
Feliciano J. Garcia
Rosendo M. Garcia
Theodore S. Gascon
Sixto S. Gaylon
Andres B. Gomez
Celestino A. Gomez
Pedro Q. Gorospe
Hyacinth E. Gorre
Jose E. Guzman
Myron B. Haas
Macario A. Halog
Doyce Hamilton (Deceased)
Graciano Z. Hortaleza
Richard Hoyt
Mamerto Ilumin
Robert D. Irving
Sidney E. Iverson

Francisco Padin
Francisco Paguirigan
Jock Palmer
Satero D. Panis
Clay E. Pardo
William C. Parker
Ramon Partolan
Jose M. Pasion
Max P. Pelayo
Johnny M. Perdido
Vincente A. Pinuela
Willie F. Polido
Antolin Quitoriano
Pedro J. Rayas
Isidore L. Refuerzo
Welfred M. Regala
Johnny Rodriguez
Vincent V. Rosal
Felix R. Rosario
Don S. Ruiz
Daniel B. Sabado
Thearareco E. Sabiniano
Santos M. Sabio
Rafael G. Sajonas
Alfredo L. Salvacion
Ruperio O. Sampayan
Chris G. Sana
Clemente J. San Felipe
Mariano O. Seroge

Luis Bosque
Clarence W. Bruce
Philip A. Caballes
Antonio A. Calagas
Theodore A. Calivo
Antonio D. Campos
Larry M. Cannonizado
Frank U. Carina
Fred V. Carino
Gregorio P. Chua
Francisco C. Clarin
Dan G. Conception
Fermin R. Consolacion
Estanislao V. Corpus
Juan B. Corpus
Ruddy F. Corpuz
Jimmy S. Cortez
John D. Crawley
Emilio L. Cubillo
Frank L. Dacpano
Leopoldo P. De Mateo
Vincent A. De Ocampo
Pastor A. De Padua
Alfredo H. Despy
Serapio S. Dionisio
Esteban M. Diputodo

Harry I. Javier
Russell Jones
George J. Josue
Thomas A. Jusa
Catalino V. Lagadan
Walter B. Lagrimas
Frank S. Leanio
Felix R. Lopez
Sebastian B. Lorenzo
Antonio M. Lucas
Pete L. Luz
Miguel Manresa Jr.
Aniceto C. Manzano
Potenciano Marapao
Serviliano G. Masinda
Vincente B. Mayor
Jose F. Mendoza
Teddy C. Mendoza
Edgar W. Morris
George H. Mueller
Earl L. Mullinix
Hilarion M. Nicholas
Robert H. Offley
Victorrio C. Orlanes
Valentino A. Pabros
Juan D. Pacis

Thomas A. Siason
Ramon M. Silva
Saturnino R. Silva
Sylvester G. Soriano
Johnny M. Soy
Fred E. Spangler
Brigida T. Taay
Anacleto R. Taborangoo
Artemio R. Tactaquin
Jimmio B. Tejado
Emerterio C. Tolentino
Gene S. Torio
Arsenio A. Turqueza
Primo B. Vallador
Frank C. Valmonte
Anastocio C. Velarde
Roque M. Velasco
Silvino D. Venenciano
Pablo Viduya
Marion B. Vilaria
Protacio G. Vilaria
Miguel C. Vinoya
Henry E. Voltz
George M. Wisham
Domenick Zoma
Paul Q. Zugama

Appendix A-24

Hawaii Personnel Assigned to
1st and 2nd Filipino Infantry Regiment U.S. Army*

Abad, Orlino
Abolos, Demetrio
Abrenica, Epitacio
Abrigas, Manuel
Academia, Bernardo
Adona, Florentino
Aduja, Peter
Afalla, John
Agai, Douglas
Agan, Maxirro
Agbulus, Louis
Aguinid, Alexander
Agricula, Manuel
Alao, Zoilo
Alcon, Emilio
Alcosiba, Larry
Alehaga, Petronilo
Alensonorin, Hipolito
Alferez, Raymond
Alidon, Geronimo
Alimboyogen, Trespolo
Almadin, Thomas
Amis, Victor
Amor, Simeon
Anaque, Aurelio
Ancheta, Fortunato
Antonio, Bobby
Antonio, Zacarias
Aquino, Elpidio
Aquino, Teodoro
Arao, Theodolfo
Arkangel, Carmelito
Asentista, Jose
Asuncion, Mariano (Edward)
Atabay, Hipolito
Atiz, William
Aurio, Alfred Sr
Aurio, Henry
Avelino, Rufino
Bacon, Ted
Bacoro, William
Bacos, Andres
Badayos, Manuel
Badua, Filomeno
Badua, Pedro
Lacara, Carlos
Lagbasf, Edward
Lagon, Cosme
Lagrimas, Manuel
Laping, Stanley
Lehano, Melanio
Lerma, Peter
Leyson, Jude
Lianos, Conrado
Liberato, Fortunato
Licayan, Patricio
Limbago, Joseph
Linogon, Frank

Baguio, Pete
Bagor, Flabio
Bailado, sotero
Baladay, Herman
Balaney, Jose
Balbirona, Joseph
Balderas, Juanito
Baldisarro, Felix
Baldonro, Blas
Baligad, Alfredo
Ballos, Thomas
Balmores, Jose
Bartolome, Leon
Bartolome, Raymond
Bambico, Anacleto
Banglos, Simplicio
Bareng, Orlino
Baroman, Raymond
Barroga, Jaime
Beralas, Alfonso
Betita, Ernest
Bedoya Vincent
Benjamin, James
Benjamin, Jhonny
Beter, Eugene
Bibilone, Pablo
Bingo, Victor
Bihag, Elicrio
Bismani, Kablayan
Bitte, Joseph
Biton, Ladjabeto
Bolivar, Benny
Bonachita, William
Bontog, Patricio
Bulawit, Faustino
Bolilan, Philip
Boltiador, Reyno
Cabacungan, Alfred
Cabael, Alejandro
Cabanatan, Ireneo
Cagalawan, Jose
Cahilig, Perfecto
Calio, Cypriano
Caluya, Frederico
Calpito, Pablo
Nasaros, Manuel
Omega, Francis
Orio, Federico
Ortez, Benedicto
Ortoyero, Thanas
Ozoa, William
Paas, Filirnino
Pacleb, Agrafino
Padilla, Elareo
Padin, Francisco
Pagala, Jose
Pagdilao, Honorable Roland
Paling, Stephen

Caminos, Henry
Canete, Estaliato
Canon, Andrew
Cantil, Clarence
Capio, Jose
Castillo, Bernardo
Castillo, Wallace
Clarabal, Manuel
Clarabal, Jeffrey
Consas, George
Cristobal, Alberto
Catli, Donald
Cauton, Victor
Celes, Maximo
Clavecillas, Richard
Corpus, Sebastian
Corpus, Sosimo
Cua, Richard
Cubillo, Emiliano
Dabin, Eulogio
Dahilig, Juan
Dana, Antonio
Daog, Consuelo
Dayacos, Alfred
DeGracia, Juanito
Dela Cruz, Salvador
Dela Cruz, Vicente
Dela Pena, Rodrigo
Delaries, Carmelo
Del Rosario, Aurelio
Desuacido, Boleriano
Diama, William
Dimalanta, Sosimo
Directo, Julius
Domdoma, Alexander
Domingo, Emilio
Duguran, Pedro
Dulumpinis, Jesus
Dulay, George
Dupio, Tiburcio
Dumaran, Ben
Dumaneus, Sarapin
Ebanez, Bonifacio
Eharis, Francis
Eharis, John
Quijano, Alejandro
Quizon, Barney
Rabago, Ernesto
Ragaso, Ermesto
Ragasa, Jaime
Ramento, Leocadio
Ramo, Alfred
Ramento, Leocadio
Ramo, Alfred
Rania, Tony
Raposas, Marcus
Reynon, Francis
Ribucan, Florencio

Erece, Alfredo
Ermitano Theopilo
Escalante, Anheal
Espalana, Salvador
Esquirra, Dionisio
Estabillo, Norman
Esteban, George
Estrella, Alvin
Evangelista, Alfred
Fernandez, Prino
Fortune, Emilio
Flores, Jose
Fragio, Antonio
Fuerte, Jaime
Fullon, Ben
Gaboya, Philip
Galban, Peniales
Galbiso, Moses
Galisa, William
Galut, Paul
Garcia, Joseph
Gersaba, George
Glariada, Gregorio
Gomez, Rufino
Gorgonio, Modesto
Gorospe, Guillermo
Guerpo, Felix
Guerzon, Jerry
Gunsalvez, Manuel
Hamile, Walter
Hatico, Crisistomo
Hesapene, Raymond
Hesapene, Tanacio
Hierro, Anthony
Himalaya, Benjamin
Himalaya, Larry
Ibanez, Angel
Icallia, Deogracias
Ihan, pekelino
Ilac, Hemernis Norman
Ingano, Socino Stanley
Jamito, Silvero
Jamil, Ottoinsani
Jimenez, Pablo
Kobalis, Sisoye
Sequito, Joseph
Sevilla, Cresencio
Siapno, Leoncio
Sibonga, Joshua
Simbre, Rufino
Soniega, Dionicio
Tejada, Moses
Tejada, Rosindo
Tiro, Loserio
Tobosa, Rufino
Togadi, Jimmy
Tolentino, Manuel
Toribio, Florentino

Lorenzo, Albert
Lorico, Tharias
Los Banos, Alfred
Los Banos, Domingo
Lutao, Domingo
Lopes, John
Magalion, Juanito
Maglinti, William
Malia, William
Manalo, Thomas
Mancao, Alfred
Mandac, Matias Michael
Maneseses, Theodore
Manibog, Benjamin
Manuel, Antonio
Mendiola, Don
Menor, Benjamin
Meral, Dominador
Mira, Jose
Miguel, Flordelino
Mirafuentes, Joseph
Moralino, Milton
Moratin, Alvin

Pascua, Segundo
Pasoquen, Marcelino
Patacsil, Anastacio Jimmy
Palacat, Joseph
Pama, Joaquin
Panerio, Candido
Pasco, Antonio
Pasco, William
Pascua, Florentino
Pascua, Santos
Patria, Nelson
Pave, Alfred
Pavo, Pranquilino
Pineda, James
Piscusa, Santiago
Plunkett, Deliner
Prosbitero, Angelino
Puma, Jose
Puzon, Quirino
Queja, Basilio
Queja, Celestino
Quempo, Ismael
Quireula, Richard

Richardson, William
Rivera, Agripino Larry
Rivera, Alfonso
Rivera, John
Rivera, Emeterio
Ronia, Joseph
Roponte, Abundeo
Ruiz, Alejo
Rupio, Felix
Sabate, John
Sacapanio, Francisco
Salinas, Apolonio
Saloricman, Vincent
Samante, Bernardo
San Felipe, Clemente
Sanico, Charles
Sanico, Ireneo
Sanico, Lucio
Santos, Vincente
Santa Aria, Isayas
Santos, Joseph
Saromines, Jose
Satot, Timoteo

Torrate, Victorino
Urmeneta, Froilan
Valentin, Orlando
Valenzuela, Daniel
Vendiola, Marcelo
Ventura, Antonio
Ventura, Domingo
Verdadero, Bonifacio
Vidad, Teofilo
Villafuerte, Pedroco
Villanueva, Albert
Villanueva, Antonio
Villaro, Johnny
Villarosa, Alfred Arro
Wabinga, Robert C.
Wakalina, Juan
Wayas, Dickilon
Yadao, Alfred
Yalca, Frederico
Yaris, Salvador
Yuson, Moses

*Identified at the time of printing by the source.

Source: An Untold Triumph: Regimental History inspired by Simeon Amor, Jr. Spearheaded by Domingo Los Banos and Noel M. Izon. Supported by the Filipino American National Historical Society, Linda Revilla and Stephanie J. Castillo.

***

Appendix B
1ST RECONNAISSANCE BATTALION

Appendix B-1
1st Reconnaissance Battalion

1st and 2nd Filipino Infantry Regiment: Interviewed for further training in
Australia as Radio Operators or Reconnaissance Men for Submarine Missions*

LAST	FIRST	SERIAL	RANK	AGE	LAST	FIRST	SERIAL	RANK	AGE
Abad	Otilio V	39104114	T5	30	Libatique	Sotero G	39236076	Sgt	34
Abalos	Marcelo B	39841113	T4	28	Liberato	Don J	39547094	Pvt	33
Abang	Bernardino B	39101198	Pvt		Licera	Rafael C	39843410	Sgt	36
Abanto	Rizalito	01301590	Lt		Lim	Alfred E	39846111	Sgt	32
Abel	Servando	39237333	Pvt	36	Llapitan	Carmel A	36601496	Sgt	32
Abellera	Antonio D	39526289	Sgt	38	Lobendino	Hilario L	39310376	Pvt	37
Abrencia	Santiago D	39233753	Sgt	33	Logan	Doming L	39825025	T4	36
Acenas	Semeniano D	39236952	Sgt	34	Loma	Magdalino S	39245633	Cpl	37
Adia	Procopio M	39238258	Sgt	35	Long	Howard D	17089737	Pfc	
Adona	Nicholas C	10645038	Pvt	18	Lopez	Marcos T	36617437	Sgt	38
Adricula	Julian D	19138366	T5	30	Lopez	Apolonio E	19112739	Pfc	29
Aduan	Feliciano C	39843095	Pfc	30	Lopez	Arthur G	39395298	Cpl	32
Advincula	Julio C	39234349	T5	37	Lopez	Cirilo G	39402580	Cpl	32
Africa	Robino	39841147	Sgt	32	Lopez	Felix R	39026351	Cpl	29
Agbayani	Vincent A.	39839127	T4		Lorenzo	Sebastian B	01301171	Lt	
Agcaoili	Raymundo E	32356031	Sgt	38	Lowe	William H	39036006	Cpl	
Agcaoili	Julian S	39174730		37	Lozano	Ray J	18155440	Sgt	
Aglubat	Al C	36332411	Pfc	33	Lucas	Benny G	39848430	T5	36
Agpaoa	Pacifico D	01323652	Lt	32	Lucas	Felix M	39103180	Pvt	36
Agron	Alfredo R	39838778	Sgt	33	Lucero	Florentino L	37199289	Sgt	31
Aguda	Benjamin C	39239307	Pvt	34	Luis	Monico B	39685498	Sgt	35
Aguila	Isaac Z	39031494	Pvt	32	Luntao	Celestino D	39025001	Sgt	36
Aguilar	Hermenegildo A	10645056	Pvt	20	Luz	Pete L	39845409	T4	38
Aguilar	Peter B	36339127	Sgt	37	Macabeo	Hospicio C	39099831	Pvt	30
Agustin	Pedro P	37156107	Cpl	35	Machon	Sergio M	33194571	Pfc	37
Alabanza	Joseph E	39004210	Pvt	33	Macilong	Ignacio	10645052	Pfc	29
Alayu	Celestino D	36600603	Pfc	25	Madali	Henry M	39243875	Pvt	36
Albadiano	Reynaldo H	10647176	Pvt		Madarang	Jose S	39174692	Sgt	38
Albalos	Segundo Q	19121481	Pvt	39	Madariaga	Fred M	39459947	Pvt	35
Alberto	Alfred A	36515325	T4		Madden	George		Sgt	
Alcala	Atanasio L	39402929	Cpl	40	Madura	Edward J	33610331	Pvt	27
Alcantara	Marcelino C	39126719	Pvt	38	Magale	Isidro P	33195320	Pvt	34
Alcaras	JuanD	39246396	Pfc	37	Magbual	Felecisimo D	39843914	Sgt	35
Alegre	Emilio A	39841451	T5	37	Magday	Primitivo V	39028846	Pvt	35
Alerta	Custodio P	39391259	Pvt	31	Maglangit	Maximo V	39397902	Cpl	36
Alfabeto	Orlando A	10645012	Sgt		Malan	Jose B	01314382	Lt	
Alfafara	Gerardo C	36342823	Pfc	35	Males	William J	16071526	Cpl	
Aliacbar	Sahibad M	10647057	Pvt		Malic	Dominador	10641050	T4	
Allen	Robert, Jr. C	02036590	Lt		Malitovsky	Sidney M	33426118	Pfc	
Almero	Emilliano A	39838277	Sgt	33	Malvar	Angelo B	39538482	Pfc	33
Almojera	Loren N	39604988	Cpl	33	Manalo	Armand	37341885	Pvt	21
Alms	Vincent P	36016673	Sgt		Manuel	Crispin S	39843911	Pvt	43
Altorino	Earle B	39842418	Pfc	32	Manuta	Eusebio J	39844969	Sgt	31
Alvarado	John R	39109140	Pfc	34	Manzano	Aniceto C	39171906	Sgt	36
Alvarez	Vidal F	39174517	Sgt	36	Mapalo	Juan D	36325840	Pfc	38
Alvez	Beato C	33202729	T5	36	Mapalo	Martin L	39402485	Pvt	32
Amante	Thomas T	39839215	Sgt	36	Mapile	Clement B	39838672	T4	36
Ambrosio	Nicanor G	36028978	Pvt	31	Marapao	Simeon P	33734261	Pfc	33
Ames	Ireneo A	0888247	Lt		Maravel	Antonio L	33595833	Pfc	21
Anacleto	Vincent, Jr.	6739381	Cpl	30	Marcelino	Juan C	39848180	Pvt	38

LAST	FIRST	SERIAL	RANK	AGE	LAST	FIRST	SERIAL	RANK	AGE
Ancheta	Alfonso	33739514	Sgt	36	Marcelo	Benjamin R	39121351	Pvt	20
Ancheta	Carlos F	01316979	Lt	34	Marfori	Edmundo J	01301660	Lt	32
Anderson	Moses A	39043724	Cpl	19	Mariano	Joe S	39097444	Sgt	35
Andres	Manuel V	37211856	Pvt	34	Marinas	Toby L	39844037	T5	35
Angcos	Robert O	39563906	Pvt	30	Marquez	Victor V	33737506	Pvt	32
Annaguey	Eleno D	39246485	Sgt	37	Marquina	Leo	39234336	T4	33
Ano	Gaudioso A	39399878	Sgt	37	Martin	Isidoro C	39844854	Pvt	36
Antikoll	Carlos F	10647180	Pvt	33	Martin	Estanislao C	36601591	Pvt	30
Antonio	Alfonso L	39120977	Pfc	19	Martin	Philip	34202018	Pfc	38
Aquino	Alejandro C	39841158	Pfc	30	Martinez	Chris M	36343941	Pfc	37
Aquino	Bernardino P	39031576	Pvt	37	Martinez	Pio M	39839788	Pfc	29
Arbis	Mariano V	10647023	Sgt		Marzan	Eleuterio E	39843932	Pfc	36
Arca	Ricardo T	01301603	Lt		Marzan	Ricarte S	39843944	Pvt	26
Arce	George U	36346801	Pfc	37	Mateo	Carl M	36629977	Pvt	34
Arce	Mariano T	36605781	Sgt	41	Mauricio	Abamile L	39235184	Sgt	31
Arcia	Narciso N	39116432	Pfc	30	Mauricio	Paul F	01319099	Lt	
Arciaga	Emiliano	36651864	Pvt	33	Maxam	Harry A	W2129706	WO	
Arellano	Arsenio Y	01301594	Lt		McAdams	Albert L	33600490	Pfc	
Arellano	Martin Q	39102486	Cpl	35	McCarrick	Donald R	11050775	Cpl	
Arguelles	Carlos D	0499852	Lt	25	McDonald	Lawrence W	39272111	Cpl	
Arias	Narciso T	39843524	Sgt	38	McFarland	Clinton B	01641561	Lt	
Arquero	Don T	36346436	Pfc	36	McGivney	James B	0-2034823	Lt	
Arquero	Lasaro C	39187222	Pfc	35	McGrath	Charles E	13081616	Cpl	
Artiago	Maurice W	39330914	Pfc	19	Medina	Mariano A	39024053	T4	36
Astronomo	Telesforo R	39402174	Pvt	33	Medrano	Bonifacio	10645022	Pvt	
Atad	Martiniano S	39393185	Pfc	36	Medrano	Demetrio G	36606417	Pvt	31
Ayson	Sammy L	39413091	Cpl	32	Mejia	Fulgencio L	39272113	Pvt	36
Azcueta	Pedro S	36350540	Pvt	35	Melanson	George A		Pfc	
Azcueta	Zoil S	39845469	Sgt	28	Mendoza	Serge M	39131199	Pvt	34
Bacus	Faustino B	39845477	Sgt	24	Mendoza	Jose T	01313662	Lt	34
Bader	Edward C	36208130	Sgt		Mendoza	Primo T	36356345	Cpl	33
Baggao	Mariano B	39692256	Cpl	20	Merca	Juan A	39553806	Cpl	34
Bahr	Andrew P	01030214	Lt		Mickelson	John C	36743700	Sgt	
Bairulla	Hadjula H	10647053	Pvt		Miguel	Matias L	39044522	Pfc	29
Bairulla	Sabtal H	10647051	Pvt		Miguel	Cipriano L	39154141	Sgt	31
Baker	Henry L	0534045	Lt		Milana	Philip G	39114222	Pvt	37
Baker	Clinton M	39309940	Pfc	37	Miller	George	0394905	Cpt	
Baker	George B	01643111	Lt		Miller	George F	0496303	Lt	
Baladad	Jose C	39702550	Cpl	42	Minder	Robert C	39229726	Pvt	36
Balcena	Fred, Jr.	39285012	Sgt	22	Modillas	Crisanto M	10645037	Pvt	18
Balderama	Eduardo G	39245632	Pvt	38	Mones	Venancio M	39842206	Sgt	36
Baldos	Florencio A	39176535	Sgt	31	Montero	Al B	39087782	Sgt	33
Balino	Porfirio	32416859	Cpl	29	Montero	Jack	36361427	Sgt	35
Balinton	Vicente	39042951	Pvt	19	Montesclaros	Melecio J	01305289	Lt	
Balintong	Flaviano	10641064	Sgt	42	Moore	Chester H	18219454	Sgt	
Balleras	Julio V	39840483	Sgt	30	Morales	Catalino U	39841687	Pfc	33
Balunes	Ralph M	39235966	Pfc	36	Morris	Joseph A	33478565	Pfc	
Basilio	Platon L	39406650	Pvt	39	Morris	Nick G	35753617	Pfc	
Bautista	Apolinario A	10645043	T5		Moscoso	Anselmo S	39244854	Pfc	34
Bautista	Felino L	39538438	Sgt	39	Mosquera	Benigno F	39402969	Pvt	43
Bautista	Mariano L	39842940	Pfc	28	Nagtalon	Alex A	37204648	Sgt	35
Bautista	Moises V	33189145	T5	31	Nalangan	Restituto I	39839845		35
Becker	William, III	16168669	Sgt		Nastail	Lakibul P	10647050	Pvt	
Beedy	Newell J	17108104	T5		Nastor	Domingo N	39014924	Sgt	28
Beeler	Richard M	39231980	T5		Nava	David U	39684950	Cpl	30
Begonia	Daniel B	38168017	Sgt	34	Navarro	Ambrosio	39041984	Cpl	29
Behan	Fred H	01081198	Lt		Neighbors	John, Jr. G	34777462	Pvt	
Bello	Cleto S	39238037	Sgt	35	Nelson	Joseph A	32893649	Pfc	
Bentillo	Venancio L	39238829	Pvt	37	Nerez	Manuel M	39044285	Pvt	34
Berg	Gerald G	6541910	Sgt		Nery	Gerardo B	39023603	Sgt	33

LAST	FIRST	SERIAL	RANK	AGE	LAST	FIRST	SERIAL	RANK	AGE
Bermudez	Felix	32350341	Pvt	38	Nichols	Victor R	15354489	T5	
Bernal	Jaime A	31318657	Pvt	32	Niebres	Federico J	36355586	Pfc	32
Besid	Restituto	32272318	Sgt	32	Nietes	Iluminado	36362331	Sgt	
Beuno	George B	35466492	Sgt	38	Noel	Ralph B	36382203	Cpl	33
Biado	Anacleto	36608611	Sgt	38	Nuevo	Leodegario O	36601310	T5	35
Bibat	Edilberto S	36346416	Cpl	34	Oandasan	Angelo P	36325336	Cpl	37
Bierley	James, Jr. G	13157755	Cpl		Oania	Hermogenes C	39839084	Pvt	38
Bill	Peter C		Pfc	37	Oates	Kenneth H	01645882	Lt	
Bolante	John A	36332733	Sgt	35	Obrero	Dick V	39246368	Pvt	31
Bolivar	Custodio F	32513810	Sgt	34	O'Connor	Maurice N	0-349057	Cpt	
Boncavil	Domiciano O	39838429	T4	33	Ogoy	Dale Y	36352596	Pvt	38
Bone	Alvin V	01641443	Lt		O'Gwin	Charles W	14185206	T5	
Borce	Dionicio G	39240147	Pfc	30	Ollada	Patricio	36345911	Pfc	33
Borillo	Pio B	39838832	Sgt	38	Olson	Clerence Y	37663674	T5	
Bove	John E	01031295	Lt		Ortiz	Paterno A	36624975	Sgt	31
Brillante	Leonardo T	36398274	Sgt	35	Ouano	Francisco I	39236026	Pfc	35
Brown	Lewis, III	0290292	Maj		Ovalles	Marcelo C	39320985	Pfc	38
Bucol	Secundio G	39112603	Sgt	23	Pablo	Cipriano B	39238786	Sgt	33
Buenaventura	Nicolas	19126906	Sgt	49	Pabros	Ambrosio A	39280569	Pvt	33
Buenavista	Serafin B	39803460	T5	37	Pacificar	Nicomedes	32443020	T5	42
Bueno	George B	35466492	Sgt		Padgett	Henry L	15319828	T5	
Bugarin	Enrico M	39119957	Cpl	35	Padilla	Bartolome M	39839177	Pfc	37
Bugayong	Vincente V	36356958	Pfc	41	Padilla	Louis P	39842719	Sgt	36
Bugh	Marion W	18003254	T3		Padilla	Balbino R	39845504	Sgt	36
Bulat	Edward	33142467	Pvt		Padilla	Luis P	01305300	Lt	
Bulatao	Benjamin C	39841434	T4	30	Paez	Artemio D	10641082	Pvt	23
Bulatao	George M	39103897	Pvt	34	Paguirigan	Francisco	0904302	Lt	33
Bumanglag	Arsenio B	39391750	Pvt	36	Pajara	George T	39032439	T4	37
Burgos	Edward M	36262428	Pvt	37	Palac	Juan M	6735755	Pvt	38
Burns	Edward P	33589084	Pvt		Palafox	Rosendo M	32325042	T5	38
Cabais	Eutiquio	R2781272	Sgt	42	Paler	Luis B	39532836	Sgt	39
Caballes	Philip A	39842714	Pfc	33	Palmejar	Louis A	39839102	Sgt	39
Cabangbang	Bartolome	0888747	Lt		Palmer	Ronald F		Pvt	
Cacabelos	Rufino F	39179223	Sgt	30	Palmeto	Daniel S	39837031	T5	36
Cacas	Maximino C	32355986	Sgt	35	Panes	Rafael F	39174931	Pfc	33
Cachapero	Frank G	39844821	Pfc	31	Paraiso	Irineo V	39236917	Sgt	33
Cacho	Thomas S	39123704	Cpl	36	Paraiso	Miguel Q	32232121	T4	38
Calaustro	Pedro A	R2291087	Pfc	48	Parcasio	Joe B	33194737	Sgt	35
Caliboso	Mariano S	39203189	Pfc	36	Parong	Astor C	39243014	Cpl	30
Calip	Aproniano E	39207135	Pvt	33	Parrish	John E	14091291	T5	
Calix	Pascual M	39244114	Sgt	32	Parsons	Charles		Cmr	40
Calmorin	Angelo A	32418740	T4	34	Pascua	Inocencio V	39844828	Sgt	36
Camp	Lowell A	15132532	T5		Pascua	Jerry De F	36359348	Pfc	35
Campbell	Merle E	01051505	Lt		Pascual	Ambrosio	39683927	Cpl	29
Campeau	Lucien V	W2114938			Pascual	Angelo E	32246630	Pfc	35
Campos	Henry A	19148020	Sgt	33	Pasquarella	Joseph R	15121431	T5	
Campos	Norberto C	39402744	Pvt	36	Patacsil	Regino T	39397341	Pvt	32
Candelario	Jose E	39246432	Sgt	35	Patton	John F	15130537	T5	
Caoili	Hermenegildo R	39390597	Sgt	34	Patubo	Gaudioso L	32502733	Pvt	24
Capis	Theodore G	19058858	Sgt		Paz	Rafael A	39410214	Pfc	38
Capistrano	Alberto N	39241197	Sgt	40	Pedro	Florendo R	39838516	Pfc	37
Cardenas	David D	39240843	T4	31	Pelgrims	Andrew	39256737	T4	
Cariaga	Pedro A	10641012	Pfc		Pena	Doroteo M	36710307	T5	37
Carpenter	Charles E	15319681	Pfc		Penalver	Pedro R	39105287	Pvt	35
Carpio	Amado G	10645054	Pfc	27	Peneyra	Cipriano C	39839100	Cpl	35
Carpio	Jose P	37290686	Pfc	38	Pens	Bernard B	34808820	Pvt	
Carrasca	Romain S	36636009	Pfc	25	Peralta	Alfonso A	39105174	Pvt	32
Carrillo	Patrocinio S	39842421	T4		Peralta	Estanislao C	39561110	Pfc	37
Carter	Loy W	39679011	Sgt		Peralta	George E	36022809	Sgt	24
Casco	Lainvencion R	39412952	Pvt	31	Peralta	Materno P	39843935	Cpl	32

LAST	FIRST	SERIAL	RANK	AGE	LAST	FIRST	SERIAL	RANK	AGE
Casia	Melanio C	39828878	Cpl	34	Peralta	Maximo	10647009	Cpl	54
Castillo	Moises C	39242766	Sgt	38	Peredo	Joe E	39325510	Pfc	35
Castro	Ciriaco	10641060	Sgt		Perez	Pedro M	36345693	Sgt	36
Castro	Ray F	39101320	Sgt	37	Phillips	David A	36669207	Cpl	
Caswell	Robert B	31283847	Sgt		Phillips	Lawrence H	0-916722		
Catalina	Eugenio S	10647055	Pvt		Piana	Victor L	39243842	Pfc	38
Cayabyab	Adolfo V	39110054	Pvt	33	Pickering	Abner K	01283130	Cpt	22
Cayabyab	Ricardo D	39255604	Pvt	34	Pilar	Julian A	36341812	Sgt	33
Chambless	Henry I	7084115	Sgt		Pilar	Mariano F	39405528	Pvt	37
Chick	Anton, Jr. C	11136180	Cpl		Pilipina	Leon P	39846122	Pfc	37
Claunch	Alphie L	W2129712	WO		Pimentol	Lorenzo V	39389581	Cpl	34
Clemente	Leo C	39174917	Pfc	33	Pinuela	Vicente A	39238489	Sgt	31
Coastales	Fermin D	37216144	Cpl	34	Placido	Carlos S	39267792	Sgt	32
Cohen	Wilfred A	01644722	Lt		Pollini	Raymond F	31267530	Cpl	
Conception	Eufrecio F	39098265	Sgt	33	Pomarez	Leon V	39413450	Cpl	28
Conception	Nemesio B	39238480	Cpl	33	Pompea	Edward T	01639505	Lt	
Consolacion	Fermin R	39842904	Pfc		Powers	Harry E		T5	
Constantino	John R	36380819	Pvt	33	Pressnall	Richard M	37551898	T5	
Cook	Gerald M	37385466	T5		Profetta	Stephen J	33489135	Pvt	
Corpus	Aquilino M	39032447	Sgt	35	Pugosa	Emil L	39837296	Sgt	36
Corpuz	Amado S	6737896	Sgt	35	Pulanco	Luis M	39845632	Pfc	38
Corpuz	Eustacio	10301905	Cpl	24	Pulido	Mack H	39840672	Cpl	
Corpuz	Robert B	39101348	Pvt	36	Punsalan	Ray M	39843430	Pfc	
Cortez	Calixto	39196262	Pvt	38	Quanico	Isidro B	10645013	T5	
Cortez	Ramon F	39237487	Sgt	33	Quejarro	Ponciano B	39844129	T4	39
Costales	Gene D	37216159	T5	40	Querubin	Patricio C	39242123	T5	36
Crespo	Torribio	0888253	Lt		Quezada	Tony M	36649313	Cpl	31
Croell	Harry T	0262144	Cpt		Quiaoit	Estoban R	39191353	Pfc	35
Cuanang	Jualian F	39095248	Pfc	37	Quiming	Perfecto C	36199212	Cpl	32
Curaming	Eluterio B	39103871	Cpl	36	Quinagon	Barney A	39391096	Pfc	29
Custodio	Emilio M	39686249	Pfc	36	Quindiagan	Luis C	39105190	Cpl	33
Cutaran	Jacinto		Sgt	37	Quinitio	Basilio A	39843138	Cpl	34
Dacquel	Isidoro D	R5274389	Sgt	40	Quinto	Emilio	0888442	Lt	
Dacquel	Richie D	39243962	T5	32	Quintoriano	Estanislao Q	39396531	Sgt	37
Daelto	Marciano R	36340286	Sgt	33	Quintua	Max T	19161005	Sgt	38
Dagandan	Fortunato C	39098413	T5	31	Quismundo	Ruperto Q	39201741	Pfc	31
Dahilig	Juan F	13043938	Sgt		Rabang	Thomas V	39526403	Cpl	36
Daley	Patrick	W2129709	WO		Racho	Paciano S	36345447	T5	36
Dalida	Igmicio P	10645050	Pvt	33	Ragupos	Geronimo R	39207501	Pvt	38
Darang	Andres D	39828848	Pvt	29	Rallojay	Teodoro J	39235964	Cpl	31
Davidson	James B	36130584	Sgt		Ramirez	Camilo	10641061	Pfc	24
Davin	Numeriano D	39536974	Pfc	32	Ramirez	Jaime A	32320293	Sgt	37
Davis	William J	02036593	Lt		Ramos	Aquilino B	33203696	Pfc	30
Dawal	Leoncio	10647060	T4		Ramos	Julian G	39836592	Cpl	39
De La Pena	Arsenio B	39604577	Sgt	31	Ramos	Pascual M	39312421	Pfc	41
De La Pena	Geronimo M	39260071	Pfc	36	Ramos	Paul L	10645019	Cpl	20
De Leon	Benedict		Pvt	32	Ramos	Senen R	39245810	Sgt	45
De Leon	Peter B	39101202	Cpl	34	Randall	Roberto A	39129170	Cpl	20
De Los Santos	Modesto P	39245941	Sgt	32	Rasmussen	Paul O	36647192	T4	
De Ocampo	Mero	39097470	Sgt	32	Real	Isay	39096439	Sgt	35
De Torres	Ronald A	39264501	Sgt	27	Reddick	Lyle V	37501417	Cpl	
DeLara	Enrico	39245627	Pfc	35	Regala	Wilfred M	39075555	Sgt	35
DeLara	Mauricio	39097479	Pvt	37	Regalado	Jerry V	36345374	Sgt	37
Delfun	Urbano P	36734497	Pvt	31	Respicio	Sergio P	39836837	Cpl	38
Despy	Alfredo	39114329	Pvt	19	Respicio	Tom P	39235454	Pfc	36
DeVera	Pas Y	39246918	Sgt	34	Rexford	Sidney S	01646058	Lt	
Diana	Serafin G	36323869	T4	32	Reyes	Eulogio	6613779	Sgt	40
Digal	John V	39571379	Pvt	37	Reyes	Felicisimo	39246393	Cpl	37
Dionolo	Thomas	01298007	Lt		Reyes	Felix T	39845120	Cpl	27
Diputado	Esteban M	36343475	Sgt	35	Reyes	Hermen	39241963	Sgt	33

LAST	FIRST	SERIAL	RANK	AGE
Dodson	Hadley W	36354468	Cpl	
Dolendo	Buenaventura	33323406	Sgt	21
Doliente	Danny D	39528392	T5	34
Domalaog	Benigno N	39846507	Sgt	39
Domingo	Baldomero N	39527280	Sgt	38
Dongallo	Fernando L	39844743	Sgt	30
Dugenia	Wenceslao P	39390905	Pvt	37
Dulay	George R	36268641	Pvt	36
Dumag	Geronimo L	39392761	Sgt	33
Dumbrigue	Santiago C	39683965	Cpl	35
Dumpit	Fred N	39200226	Cpl	37
Dunat	Eugene F	16129465	Cpl	
Dupitas	Aurelio M	39845473	Cpl	34
Duran	Agrifino J	39838815	Cpl	31
Durano	Pio B	10641025	T5	22
Eals	Orien N	35678066	Pfc	
Ebarle	Patrick, Jr. F	39109132	Sgt	23
Eccleston	Edward J	33544428	T5	
Elefano	Alberto C	0530514	Lt	33
Ellis	James E	36022164	Sgt	
Enderiz	Victor P	39105219	Pfc	33
Ensor	Richard A	01050244	Lt	
Erana	Miguel B	39837011	Pfc	37
Ergas	Eugene N	39105167	Cpl	22
Esguerra	Jimmy R	39284055	Sgt	19
Espinoza	Cesar E	38251151	T3	
Espiritu	Artemio C	39123285	Pvt	34
Espiritu	Florentino C	35099797	Pfc	35
Estavillo	Sergio B	39130649	Pfc	28
Esteban	Peter M	39109680	Pvt	34
Evangelista	Vincento E	12141207	Cpl	38
Evans	James, Jr. L	0356781	Maj	
Faragan	Manuel F	19090770	Sgt	34
Farrell	William A	0411608	Cpt	
Felipe	Ramon S	39413717	Pvt	30
Felizar	Pete G	39391936	Pfc	32
Ferguson	Charles B	045881	Lt	
Fermin	Raymond R	39039876	Cpl	23
Fernandez	Cipriano C	39258733	Cpl	29
Fernandez	Leon C	36707169	Pvt	36
Ferrell	John M	14106667	T5	
Fesler	Earl W	32668743	T5	
Filek	Joseph F	37552127	Pfc	
Finnegan	George J	12139211	Sgt	
Fitzpatrick	Bernanrd C	32617351	T5	
Flores	Agripino B	39836186	Pfc	33
Flores	Alfred T	39490596	Pfc	20
Flores	Vicente M	10645004	T5	29
Fontanoz	Marcelo P	39181438	Sgt	34
Formoso	Gabriel F	36337830	Pfc	29
Forrette	John E	16135053	T4	
Fortes	Melvin D	39838794	Sgt	35
Fortun	Salvador B	36361421	Sgt	34
Francisco	Alberto C	39103160	Sgt	33
Francisco	Urbano M	39526401	T5	37
Francisco	Fermin	39236836	Sgt	37
Frias	Arthur F	10324763	Lt	27
Frias	Generoso D	01306357	Lt	
Gabatan	Rodolfo	10645047	Pvt	19
Gatchalian	Hipolito M	39841475	Pvt	32
Reyno	Jacinto U	39245105	Pfc	37
Reynoso	Jaime, Jr. R	39547229	Sgt	42
Richmond	John R	0409767	Cpt	
Rillera	Pastor A	39838420	Sgt	30
Rimando	Juan F	37287078	Pfc	35
Rimando	Antonio F	37156057	T5	31
Rinien	Eusebio T	39244456	Sgt	36
Riotoc	Manuel	39095507	Pvt	38
Rivera	Elpidio A	36337587	Cpl	36
Robert	Louis	35571878	T5	
Robinson	Irving H	36372422	T4	
Robles	Crispolo G	39837867	T4	28
Robles	Romulo D	33190611	Cpl	33
Rodrigues	Juan A	10641048	Cpl	
Rodriguez	Mariano	10641056	Sgt	
Rodriguez	Pedro P	10645064	Pvt	
Rodriguez	Fred N	39838799	Sgt	36
Rodriguez	Feliciano S	39391220	Pfc	29
Rodriguez	Ireneo R	39244217	Sgt	36
Rola	Ceferino R	01308379	Lt	37
Romas	Jose J	10641041	Sgt	
Romero	Eleuterio Z	10641053	T4	
Rommel	William, II	01307079	Lt	
Roque	Arthur F	39294730	Pvt	31
Rosales	Doroteo A	39839480	Pfc	18
Rosales	Paulino A	39396552	Sgt	32
Rosario	Gabriel Y	39838412	Pfc	34
Rosario	Juan R	37341033	Pvt	32
Rosete	Pantaleon M	39174290	Sgt	36
Rosher	Douglas A	0473642	Maj	
Routson	Merl D		Pfc	
Rowe	George F	115529	Lt	
Rubenstein	Joseph	36857371	Pvt	
Rubio	Melchor Q	36600677	Pfc	38
Rudolph	William P	39696463	Cpl	
Ruiz	Ruperto	39838264	T5	33
Ruiz	Julius B	39198058	T5	30
Ruiz	Don Saura	39838780	Sgt	25
Rullo	Modestino		Pvt	
Rulloda	Ambrosio	36360816	Sgt	
Runes	Lucas D	33196248	Cpl	33
Runtal	Henry E	39237726	Sgt	32
Sabado	Daniel B	39841843	T4	27
Sabado	Henry D	39092481	Cpl	
Sabado	Victor R	39455357	Sgt	36
Sabala	Severino	10647059	T5	
Sabiniano	Teodoreco S	39307228	Sgt	38
Sagun	Fernando C	39105188	T5	36
Saladino	Selmo C	37198262	Pfc	29
Salamanca	Albert	37205594	Cpl	33
Salgado	Mauro B	39837635	Sgt	33
Salvacion	Alfredo L	39109014	Pvt	35
Salvador	Apolinard G	39235112	Sgt	38
Salvador	Frank A	39101922	Pvt	38
Sammons	William H	39414170	Pvt	38
Sanchez	Gerardo A	32332823	Sgt	31
Sandoval	Nicolas E	39843993	T5	30
Santorio	Ceferino D	36353496	T4	33
Santos	Alberto E	39844995	Pvt	30
Santos	Amador A	39097583	Sgt	

LAST	FIRST	SERIAL	RANK	AGE
Gaduang	Thomas D	36625135	Pfc	33
Gago	Cecilio P	39394973	Pvt	30
Galam	Benito P	36631878	Pfc	39
Galang	Richardo C	0515082	Cpt	36
Galas	Eriberto	10641037	T5	19
Galban	Primo R	39110111	Pfc	31
Galey	Raymond D	14070966	Sgt	
Gallemore	James R	6580887	Sgt	25
Gallopa	Ladislao	10641017	T5	21
Gamertsfelder	Robert H	35749997	Capl	
Garces	Pedro L	39246607	Sgt	33
Garcia	Juan	39240550	Pfc	35
Garcia	Marcello	36352933	Sgt	33
Garcia	Marcelito H	32524610	Pvt	23
Garcia	Vidal	36345798	Sgt	32
Garlitos	Anton T	36346578	Pfc	33
Gascon	Clemente Y	39118002	Pfc	33
Gaston	Meleseo M	36337268	Sgt	34
Generales	Narciso C	39541136	Pvt	34
George	Robert R	38119149	Sgt	
Ginobiagon	Felipe R	10645011	Sgt	
Ginsberg	Seymour	32648872	T4	
Girard	Stanley J	31336956	Pvt	
Giron	Leovigildo M	39536590	Sgt	32
Glenn	Bobb B	0390157	Cpt	
Gobaleza	Dominador F	10641049	Pfc	
Goloyugo	Victor	36145937	Sgt	34
Goloyugo	Vincent C	36545428	Sgt	43
Gonzales	Marcelo G	39843509	T4	37
Gonzales	Jose T	39098904	T4	31
Gonzales	Marcelo U	39308995	Pfc	38
Gonzalez	Juan L	36549488	Pfc	35
Gonzalo	Arthur V	39021504	Pvt	30
Gordon	Kenneth E	20725988	Pvt	
Gorospe	Pedro Q	39839584	Pfc	34
Goroza	Edward E	39395656	Pvt	36
Gouldner	Frederick J	32346634	Sgt	
Granados	Vicente J	39846508	Pfc	
Green	James L	36614103	Pfc	
Gregorio	Querubin V	32278916	Sgt	32
Gregory	George F	39380936	T4	
Greu	Frederick			
Guazon	Bienvenido	10641021	T5	24
Gubatan	Severo	39235982	Sgt	35
Guico	Manilio	10641043	T5	
Guieb	Lorenzo M	39852408	Pfc	
Guion	Pete S	39246024	Cpl	40
Gurrea	Presciliano B	39112642	Pvt	31
Gustilo	Conrado A	39238554	Sgt	33
Guyot	Gaudencio V	39024838	Sgt	33
Guzman	Larry O	39234838	Sgt	23
Guzman	Simeon R	36289291	Pvt	35
Hale	Edward H	01637589	Lt	
Hall	John C	0131246	Lt	
Halla	Albert F	36737954	Pfc	
Halog	Rufino D	39395667	Pvt	37
Hammill	Charles H	39034039	Sgt	
Hamner	Jordan A	0888470	Cpt	
Harder	Benjamin M	32423772	Sgt	21

LAST	FIRST	SERIAL	RANK	AGE
Santos	Fortunato	33323793	Sgt	37
Santos	Gregorio	39685483	Pfc	29
Santos	Rudolph B	16128144	Pvt	31
Sapio	Lucas J	39033374	Pfc	30
Sapon	Leoncio S	39535461	Pvt	40
Saria	Juan, Jr. A	38313180	Pvt	22
Sarmiento	Angel T	39235968	Sgt	34
Sarmiento	Florentino M	32888314	Pvt	40
Sarmiento	Ignacio M	01822465	Lt	
Savellano	Andres S	39845429	Pfc	32
Savellano	Pedro E	32355686	Pvt	36
Schmaltz	Albert	01031837	Lt	
Schraml	Robert R	16056585	Pfc	
Seares	Pascual P	39273631	Pvt	35
Segomalian	Domingo S	10641019	Pvt	21
Sensano	Daniel G	39692205	Cpl	20
Serrano	Agapito B	39394382	Pfc	29
Sibonga	Emil J	39166093	Sgt	25
Silva	Saturnino R	1300095	Lt	36
Silverio	Ben V	39263745	Pfc	36
Siose	Francisco E	39837070	Pvt	36
Sison	Carmelo	6739345	Sgt	
Sison	Agapito R	36358723	Cpl	36
Sison	Ceferino P	36601469	Cpl	
Sison	Tony C	39839111	Sgt	34
Sitjar	Jose T	39241984	Sgt	34
Sivila	Andres C	39843086	Pfc	34
Skundale	Frank J	0377668	Cpt	
Smith	Charles M	0888471		
Smith	Frank B	01324796	Lt	28
Smith	Lloyd E	39550482	T5	
Solar	Marcelino V	39046660	Pfc	37
Solidarios	Sergio V	39838417	Sgt	33
Soliven	Larry G	39390506	T5	32
Songco	Ruben P	0478156	Lt	22
Sooter	Joyce M		Pvt	
Sosion	Teofilo C	10645039	Pfc	
Sperandeo	William P	36640046	Pfc	
Spiro	Meyer L	01646026	Lt	
Springmeyer	Robert L	39118902	T5	
St. Pierre	Maurice G	31292423	Cpl	
Stahl	Robert E	33236898	T4	
Stoddard	Brooke	01030523	Lt	
Strickland	Lee A	01638609	Lt	
Supnet	Flaviano	39690876	Pvt	33
Taban	Lorenzo	39112726	Pvt	31
Tabios	Val R	32020950	Sgt	33
Tabofunda	Pedro P	39246444	Pfc	35
Tactaquin	Rodrigo R	39846544	Pfc	33
Talaro	Albert T	39390649	Sgt	34
Tallido	Silvino B	01313063	Lt	37
Taluban	Teodoro	39842888	Sgt	29
Tamayo	Claudio M	01305526	Lt	33
Tanato	Leandro P	39097588	Sgt	36
Temple	Pedro S	39099772	Pfc	28
Teo	KongLam	0888892	Lt	28
Theys	Alfred P	36039602	Sgt	
Thomas	James P	15319546	T5	
Thompson	Clifford P	35546828	T5	

LAST	FIRST	SERIAL	RANK	AGE	LAST	FIRST	SERIAL	RANK	AGE
Harris	Thurston	34682819	T5		Thompson	James C	15316411	Pfc	
Hawley	Maynard C	0168144	Lt		Tinnel	Leon C	01646058	Lt	
Hayden	Joseph R				Tison	Sidney S	0534061	Lt	
Hendrickson	Murl S	17088657	Sgt		Tobias	Manuel A	01305346	Lt	
Herbig	Robert P	35583726	Sgt		Tolentino	John T	36355508	Pfc	36
Hernandez	Al	01324774	Lt	34	Tolentino	Zoilo, Jr.	32870608	Pvt	22
Herrera	Donato E	32278953	Sgt	24	Toribio	Alipio C	36713799	Pfc	31
Herreria	George R	32230716	T5	32	Torio	Isaias T	39536985	Cpl	34
Higbee	Donald M	0121674	Cpt		Torres	Enrique Luis	0888748	Lt	
Hilario	Carl G	39241054	T3	32	Tria	Antonino B	39264160	Cpl	32
Hipol	Felipe C	36515089	T5	35	Trocio	Perfecto M	36601493	Pfc	35
Hoffman	Charles L	35678553	Cpl		Tucay	Harry T	39236723	Cpl	28
Hoke	Max E	16041689	Sgt		Tudtud	Joe L	01315532	Lt	
Holgado	Eddie C	39197647	Pfc	36	Tullo	Apollo P	39253426	Pfc	37
Hondolero	Felix S	01311893	Lt		Turqueza	Arsenio A	39181851	Sgt	36
Hope	Charles, Jr. P	01059496	Lt		Ubarro	Patricio	0-359964	Lt	34
Hosberg	Glen B	35233573	Pvt		Udani	Carlos D	36325243	Pvt	34
Huerto	Petronia B	0-888249	Lt		Udundo	Esteban C	36665915	Pvt	37
Hufana	Frank A	39564521	Pvt	36	Ulivas	Albert U	36369345	Pvt	35
Hunter	Edward T	39263063	Cpl		Umipeg	Marcello B	39080051	Sgt	25
Ibarra	Estanislao S	39100658	Cpl	31	Unciano	Isaac F	01102098	Lt	36
Ibea	Artemio I	39535788	Sgt	36	Valdez	Celestino R	39842304	Cpl	32
Igmundson	Charles B		T4		Valdivia	Enrique D	39325758	Pvt	33
Ignacio	Fred C	39827659	Sgt	33	Valera	Jose Valle	0888893	Lt	35
Ignacio	Rodolfo	0888248	Lt		Valido	Pastor W	39841139	T5	36
Ignacio	Miguel E	39842016	Sgt	37	Valin	Zemon A	39174609	Sgt	32
Ilusorio	Johnnie A	39839473	Cpl	29	Valmonte	Frank G	39237209	Pvt	38
Ilustre	Ernesto D	33189080	T5	35	Vanderpool	Jay D	0410776	Maj	
Ind	Allison W	0304920	Col		Vaughan	William D	023978	Cpt	
Interior	Iluminado	10641057	T3	29	Velasco	Roque M	39839128	Sgt	40
Isaac	Cledoaldo J	39527750	Pfc	38	Velayo	Benny T	39106003	Pvt	33
Itchon	Jay F	39240629	Pvt	37	Ventura	Dionisio H	39242076	Pfc	32
Jamison	Donald V	01043373	Lt		Ventura	Frank C	01318237	Lt	
Jaramilla	Santiago Q	36358865	Cpl	37	Vergara	Thomas C	39838398	T4	36
Javier	Bartolome E	39547130	T5	34	Vicente	Victor I	39099852	Pvt	33
Javier	Floro O	36359894	Pfc	32	Vicente	Johnny Q	39180069	Sgt	32
Javonillo	Juan O	39093222	Sgt	28	Vidal	Doroteo C	39244250	Pfc	36
Jenson	Thomas, Jr. A	36650485	Cpl		Viduya	Jose F		Pvt	32
Jesena	Carl	39096093	Sgt	34	Villalon	Hermenegildo	39242352	T5	31
Jones	Enoch G	0368362	Cpt		Villamor	Jesus A	0888172	Maj	
Jorge	Patricio	10641011	Sgt		Villanueva	Angelo F	39263304	T4	30
Jose	Teodoro M	39687460	Pvt	36	Villanueva	Conrad G	39241988	Pvt	35
Joyce	Charles J	12143918	Cpl		Villarta	Federico R	39242985	Sgt	37
Juanito	Eduardo	10645036	Pfc	25	Villote	Paul P	39113338	Cpl	37
Jubane	Valentine C	39838455	Sgt	31	Viray	Don Javier	39527346	T5	32
Jurika	Thomas W	0890348			Visaya	Vincent L	39244458	Sgt	
King	Walter E	34465533	Sgt		Vite	Doroteo V	32249269	Sgt	31
Kintanar	Aniceto S	35114187	T5	37	Vitorio	Ramon D	39097646	T5	36
Knight	Quincy S	31220205	T5		Vivit	David T	01306774	Lt	
Kopp	Howell S	0386137	Cpt		Watson	Carl T	15326834	T5	
Kunkel	John, Jr. E	16096076	Pfc		Weber	Henry J	39053889	T5	
Labrador	Vicente M	0888256	Lt		Weber	Paul		WO	
Ladjahasan	Ali H	10647054	Pvt		Weissman	Edward	32964674	Pvt	
Ladjahasan	Jalil H	10647052	Pvt		Welch	Homer, Jr. J		Pfc	
Lagpacan	Laureto A	39097394	T4	33	Wendam	Sidney D	39028642	Pvt	33
Lagrimas	Walter B	39844692	Sgt	32	Wilcox	Charles S	01574556	Lt	
Landrum	Jesse E	36449262	T5		Wise	Braynard L	W2129711	WO	
Lapuente	Guillermo N	39257431	Pvt	37	Wood	Rhys C	01081044	Lt	

LAST	FIRST	SERIAL	RANK	AGE	LAST	FIRST	SERIAL	RANK	AGE
Lascano	Leodegario	36328888	Pfc	38	Worthington	Robert P	6148193	Pvt	
Lauricio	Marcelo M	39548895	Pvt	36	Wright	Carl	36601513	Sgt	37
Laurina	Filomeno A	39393681	Pfc		Yost	Fred E	33521883	T5	
Lazo	Enoch A	39240485	Cpl	31	Yotoko	Salvador L	39845715	Sgt	39
Lazo	Ernesto L	39282541	Cpl	21	Young	FrankHopes, Jr.			23
Leake	Charlie A	6398540	Pfc		Yu Hico	Delfin		Lt	
Leanio	Frank S	39839199	Cpl	30	Yuponco	Leo R	39030177	Pfc	32
Legaspi	Alex B	39555046	Sgt	20	Zola	Edward J	16124657	Sgt	
Leger	Herbert J	0503272	Cpt			Gordon M	29193466	T5	

*Out of the total 847 men, only 95 men were under 30 years old.

Source: MacArthur Archives, Norfolk VA

Appendix B - 2

QUESTIONNAIRE FOR MEMBERS OF THE
5217TH RECONNNAISSANCE BATTALION, AUSTRALIA

1. Name
2. Serial Number
3. Place of birth: Village   Province: Island:
4. Date Left Philippines   How long in the U.S.
5. Present Rank
6. Religion:
   a. Catholic   b. Mohammedan   c. Protestant
7. What military experience have you had:
   a. ROTC   b. CMTC   c.National Guard
   d. Philippine Army Reserve   e. Army of U.S.
8. Give length of time served
   a. Duties performed   b. Ranks held
9. Do you have relatives in the Philippines
   a. Number and relationship?   b. Where do they live c. Island, province, or barrio
10. Do you have friends in the Philippines
    a. Same data as above?
11. Occupations in the Philippines
12. Occupation in United States
13. Languages spoken: English Spanish Cebuano Tagalog Bicol Other
14. Part of Philippines most familiar with
15. Physical Condition
16. What do you consider yourself best fitted for
    a. Radio Operator   b. Clerk   c. Runner with Military Forces in the Philippines
    d. Interpreter   e. Guide in the Philippines and if so, what Province
    f. Handling of Explosives   g. Mechanic   h. Other
17. Would you be willing to volunteer for service in the Philippines as an individual?

1st Indication
GENERAL HEADQUARTERS, SOUTHWEST PACIFIC AREA, A.P.O. 500
1 June 1944

TO: Company Commander_____, 5217th Philippine Battalion

1. In your opinion, does the man possess the initiative, resourcefulness, and intelligence to undertake an individual mission, or to be a member of a small party on their own.
2. In your opinion, what position is the man most fitted for, as a soldier, in the freeing of the Philippines

Source: MacArthur Archives, Norfolk VA

Appendix B - 3

**Transfer Order of Selected Personnel from
the 1st Filipino and 2nd Filipino Infantry Regiments to Australia
Depart from Camp Roberts, California**

....

HEADQUARTERS, II ARMORED CORPS
San Jose, California

SPECIAL ORDERS)
8 JULY 1943
NUMBER.   166)

EXTRACT

31. Following O are held fr further asgmt and dy with the 1st Filipino Inf Regt, Hunter Liggett Mil Res, Cp Roberts, Calif and WP without delay to Hamilton Field, Calif reporting o/a 10 July 1943, O/C Transient Personnel Section, Pacific Wing, ATC, for transportation to destination. Travel by air directed fr Hamilton Field, Calif to destination and will be charged to Appropriation 2 – 64770:

Capt Ricardo C. Galang	0515082	Inf
1st Lt Carlos D Arguelles	0499852	Inf
1st Lt Ruben P. Songco	0478156	Inf
2nd Lt Saturnino R. Silva	0130095	Inf

In lieu of subs a flat per diem of $6.00 is auth the aforementioned O while traveling by mili- tary, naval, or commercial P431-01, -02, -03, -07, -08 a 0425-24. Auth: Telg, Tag, dated 7 July 1943

By Command of Major General Morris:

/s/Basil G Thayer
/t/ BASIL G THAYER
Colonel, CSC
Chief of Staff

CERTIFIED TRUE COPY:

CEFERINO R. ROLA
1st Lt., Infantry
Adjutant

Source: The Filipino American Experience Research Project. August 27, 1996 By Alex S Fabros Jr.

Appendix B - 4
UNITED STATES ARMY SERVICES OF SUPPLY
OFFICE OF THE COMMANDING GENERAL

APO 501
GENERAL ORDERS)

NO.....................31)
SECRET

AUTH CG Unit

Date: 1 July 1943

ACTIVATION OF UNIT

1. Pursuant to authority contained letter FEGC 322, Hq. USAFFE, 22 June 1943, subject: "Activation of 978th Signal Service Company is activated, effective July 1943, with station at Hq Base Section No. 3 and with strength of 11 Officers, 5 Warrant Officers and 280 enlisted men as follows:

	OFFICERS	WO	EM
One Hq Platoon (columns 3, 4, 5 and 6, T/O 11-97, Apr 1, 2942) Five Radio	2		34
Platoons (x) One Signal Intelligence	5	5	235
Section (column 8, T/O 11-200-1 April 1st 1942	4		11
	11	5	280

2. Signal Intelligence personnel for this unit now assigned to the 832nd Signal Service Company will be transferred to constitute the Signal Intelligence Section of this company (GSXM 320.2).

By command of Brigadier General RILEA:

t/ DWIGHT F. JOHNS,
Brigadier General, GSC,
Chief of Staff

Source: Filipino American Experience Research Project. November 6, 1998

Appendix B - 5
5217TH RECONNAISSANCE BATTALION (PROVISIONAL)
OFFICERS 1943

Battalion Headquarters

Major BROWN, Lewis III	0290292	Captain LEGER, Herbert J.	0503272
Captain JONES, Enoch G.	0368363	Captain O'CONNOR, Maurice N.	0349057

1st Lt. UBARRO, Patricio (NMI)

Headquarters & Headquarters & Service Company

2nd Lt ABAYA, Rosalino A.	01305174	2nd Lt. ROLA, Ceferino R.	01308379
2nd Lt. CAPIZ, Pascual (NMI)	01320599	2nd Lt. ZERDA, Vincent M. F.	01399973

5218th Reconnaissance Company

Major PHILIPS, Lawrence H.	0916722	2nd Lt. GINNETE, Pedro A.	01317005
Major SMITH, Charles M.	0888471	2nd Lt. HERNANDEZ, Al (NMI)	01324774
Captain EVANS, James L.	0356781	2nd Lt. HONDOLERO, Felix S.	01311893
Captain GALANG, Ricardo C.	0515082	2nd Lt. LABRADOR, Vicente W.	0888256
Captain HAMNER, Jordan A.	0888470	2nd Lt. LICUDINE, Raymond B.	01313105
Captain MORGAN, Luis P.	0888702	2nd Lt. LORENZO, Sebastian B.	01301171
1st Lt. ABANTO, Rizalito (NMI)	01301590	2nd Lt. MALAN, Jose B. Sr.	01314382
1st Lt. AMES, Ireneo A.	0888247	2nd Lt. MARFORI, Edmundo J.	01301660
1st Lt. ARCA, Ricardo T.	01301603	2nd Lt.MAURICIO, Paul F.	01319099
1st Lt. ARGUELLES, Carlos D.	0499852	2nd Lt. MENDOZA, Don (NMI)	01320889
1st Lt. CABANGBANG, Bartolome C.	0888747	2nd Lt. MENDOZA, Jose T.	01313662
1st Lt. DIONOLO, Tomas (NMI)	01398007	2nd Lt. PADILLA, Luis P.	01305300
1st Lt. FRIAS, Generoso (NMI)	01306357	2nd Lt.QUINTO, Emilio F.	0888442
1st Lt. HUERTO, Petronio B.	0888249	2nd Lt. SAMPORNA, Isidro B.	01306758
1st Lt. IGNACIO, Rodolfo C.	0888248	2nd Lt. SARMIENTO, Ignacio M.	01822465
1st Lt. MONTESCLAROS, Melecio J.	01305289	2nd Lt. SILVA, Saturnino R.	01300095
1st Lt. SONGCO, Ruben P.	0478156	2nd Lt. SMITH, Frank B.	01324796
1st Lt. TORRES, Enrique L. Jr.	0888478	2nd Lt. TALLIDO, Silvino B.	01313062
2nd Lt. ANCHETA, Carlos F.	01316979	2nd Lt. TAMAYO, Claudio M.	01305526
2nd Lt. ALLEN, Robert C. Jr.	02036590	2nd Lt. TEO, Konglam	0888892
2nd Lt. AQUINO, Enrique J.	01324126	2nd Lt. TOBIAS, Manuel A.	01305346
2nd Lt. ARELLANO, Arsenio Y.	01301594	2nd Lt. TUDTUD, Joe L.	01315532
2nd Lt. CARAG, Vicente D.	01320804	2nd Lt.VALERA, Jose V.	0888893
2nd Lt. CRESPO, Torribio	0888253	2nd Lt. VENTURA, Frank C.	01318237
2nd Lt. DAVIS, William J.	02036593	2nd Lt. VIVIT, David T.	01306774
2nd Lt. EDILLON, Marcial A.	01305220	2nd Lt. YOUNG, Frank H.	0888453
2nd Lt. FRIAS, Arthur F.	01324763	2nd Lt. YU HICO, Delfin C.	0888175

Note: The men were assigned and trained in the 5218th Reconnaissance Company (Provisional) then transferred to 517th Reconnaissance Battalion to form missions to the Philippines.

Source: MacArthur Archives, Norfolk, Virginia

Appendix B - 6

PERSONNEL ASSIGNED TO ALLIED INTELLIGENCE BUREAU (AIB), AUSTRALIA 1943 - 1944

3 September 1943

RANK	NAME	RANK	NAME
Col	Roberts, C. G.	1st Lt	Wilcox, Charles S.
Col	Whitney, Courtney	1st Lt	Arguelles, Carlos D.
Lt Col	Ind, Allison W.	1st Lt	Songco, Ruben P.
Maj	Brown, Lewis III	1st Lt	Ubarro, Patricio
Maj	Cruz, Emigdio C.	Lt	McCauley, M.
Maj	Villamor, Jesus A.	2d Lt	Crespo, Toribio
Maj	Caporn, S. S.	2nd Lt	Martinez, Gregorio M.
Capt	Pickering, Abner K.	2d Lt	Quinto, Emilio F.
Capt	Galang, Ricardo C.	2d Lt	Romillo, Guillermo M.
Capt	Glenn, Bobb B.	2d Lt	Young, Frank H.
Capt	Hamner, Jordan A.	2d Lt	Yuhico, Delfin C.
Capt	Smith, Charles M.	2d Lt	Ancheta, C.
Capt	Davidson A. L.	2d Lt	Marfori, Edmundo J.
1st Lt	Ames, Ireneo A.	2d Lt	Mendoza, J.
1st Lt	Huerto, Petronio B.	2d Lt	Rola, Ceferino R.
1st Lt	Ignacio, Rodolfo C.	2d Lt	Silva, Saturnino R.
1st Lt	Jones, Enock G.	2d Lt	Tallido, Silvino B.
		2d Lt	Tamayo, Claudio

Source: MacArthur Archives, Norfolk, Virginia

Appendix B - 7

Transfer Order of Selected Personnel from the 1st Filipino and 2nd Filipino Infantry Regiments to Australia Depart from 5th Replacement Depot, California

HEADQUARTERS FIFTH REPLACEMENT DEPOT
USASOS APO 711

SPECIAL ORDERS)
NUMBER.......284)

15 November 1943

Extract

25. Fol EM br Inf SSN as shown cas atchd orgns indicated asgd in gr 5217th Rcn Bn APO 923 & WP first available Govt MT reporting for dy TDN TCNT. (VOCG Hq USAFFE) :

SSGT Peter B. Aguilar	36339127	PFC Rafael A. Paz	39410214
SSGT Modesto P. Delos Santos	39245941	PFC Luis M. Pulanco	39845632
Sgt Caudioso A. Ano	39399878	PFC Esteban R. Quiaoit	39191353
Sgt Felino L. Bautista	39438438	PFC Barney A. Quinagon	39391096
Sgt Pascual M. Calix	39244114	PFC Ruperto Q. Quismundo	39201741
Sgt Geronimo L. Dumag	29292761	PFC Pascual M. Ramos	39312421
Sgt Jimmy R. Esguerra	39284055	PFC Jacinto U. Reyno	39245105
Sgt Eusebio J. Manuta	39844969	PFC Melchor Q. Rubio	36600677
Sgt Venancio V. Mones	39842206	PFC Marcelino V. Solar	39406660
Sgt Luis B. Paler	39532836	PFC Rodrigo R. Tactaquin	39846544
Sgt Inocencio V. Pascua	39844828	PFC Pedro S. Temple	39099772
Sgt Benjamin A. Pastor	39097480	PFC John T. Tolentino	36355508

Sgt Estanisla Q. Quitorian	39396531	PFC Doroteo C. Vidal	39244250
Sgt Teodoro M. Taluban	39842888	PFC Leo R. Yuponco	39030177
Cpl Gregorio F. Bareng	39841167	Pvt Servando Abel	39237333
Cpl Fermin D. Costales	37216144	Pvt Baltazar C. Agbunag	39840188
Cpl Eleutero B. Curaming	39103871	Pvt Emiliano Arciago	36651864
Cpl Teodoro Q. Dumpit	39842830	Pvt Eduardo G. Balderama	39245632
Cpl Estanislao S. Ibarra	39100658	Pvt Venancio L. Bentillo	39238829
Cpl Ernesto L. Lazo	39282541	Pvt Felis Bermudez	32350341
Cpl Arthur G. Lopez	39395298	Pvt Lainvencion R. Casco	39412952
Cpl Juan A. Merca	39553806	Pvt Ricardo D. Cayabyab	39355604
Cpl David U. Nava	39684950	Pvt Andres D. Darang	39828848
Cpl Henry D. Sabado	39094281	Pvt Mauricio Delara	39097479
Cpl Albert S. Salamanca	37205594	Pvt Urbano P. Delfun	36734497
Cpl Antonio B. Tria	39264160	Pvt Alfredo H. Despy	39114329
Cpl Celestino R. Valdez	39842304	Pvt John V. Digal	39571379
T-5 Julian D. Adricula	19138366	Pvt Wenceslao P. Dugenia	39390905
T-5 Antonio F. Rimando	37156057	Pvt Peter M. Esteban	39109680
PFC Celestino D. Alayu	36600603	Pvt Ramon S. Felipe	39413717
PFC Gerardo C. Alfafara	36342823	Pvt Leon C. Fernandez	36707169
PFC Maurice W. Artiago	39330904	Pvt Cecilio P. Gago	39394973
PFC Dionisio G. Borce	39240147	Pvt Jacob E. Insular	36505130
PFC Mariano S. Caliboso	39203189	Pvt Hilario L. Lobendino	39310376
PFC Jose P. Carpio	37290678	Pvt Felix M. Lucas	39103180
PFC Florentino C. Espiritu	35099797	Pvt Hospicio C. Macabeo	39099831
PFC Primo R. Galban	39110111	Pvt Crispin S. Manuel	39843911
PFC Juan Garcia	39240550	Pvt Victor V. Marquez	33737506
PFC Marcos P. Genantiano	39838509	Pvt Demetrio G. Madrano	36606417
PFC Marcelo U. Gonzales	39308995	Pvt Manuel M. Nerez	39044285
PFC Vicente J. Granados	39846508	Pvt Alfonso A. Peralta	39105174
PFC Flor O. Javer	36359895	Pvt Mariano F. Pilar	39405528
PFC Leodegario Lascano	36328888	Pvt Juan R. Rosario	37341033
PFC Apolonio E. Lopes	19112739	Pvt Alfred L. Salvacion	39109014
PFC Juan D. Manalo	36325840	Pvt Alberto E. Santos	39844995
PFC Pio M. Martinez	39839788	Pvt Pedro E. Savellano	32355686
PFC Eleuterio E. Marzan	39843932	Pvt Francisco E. Siose	39837070
PFC Restituto I. Nalangan	39839845	Pvt Flaviano Supnet	39690876
PFC Fedrico J. Niebres	36355586	Pvt Enrique D. Valdivia	39325758

By Order of Col SMITH:
Lt Col JOHN S. GIBBS, Infantry
Adjutant

CERTIFIED TRUE COPY:
1st Lt CEFERINO R. ROLA, Infantry
Adjutant

Source: The Filipino American Experience Research Project, August 27, 1996 by Alex S. Fabros Jr.

Appendix B - 8

USS NARWHAL DEPARTURE TO MINDORO AND MINDANAO
FEBRUARY 16, 1944*

GENERAL HEADQUARTERS
SOUTHWEST PACIFIC AREA

~~SECRET~~

CHECK SHEET   DECLASSIFIED PER JCS LTR C.
20 AUG. 75

(Do not remove from attached sheets)

File No.:   Subject:

From: PRS   To: Chief of Staff   Date: 17 February 1944

Advise that the NARWHAL having arrived at Darwin yesterday morning was reloaded and dispatched to the Philippines on her fourth trip last night.

The missions of the vessel are as follows:
- a.   Delivery of supplies to Colonel Suarez at TAWITAWI:
- b.   Delivery of supplies to PHILLIPS (N.W. Mindoro) and collection of documentary intelligence material
- c.   Delivery of supplies to 10th M.D. – 20 tons of which is for transhipment to CEBU.

The vessel took on board Lieut. M. M. WHEELER, USNR, assigned to take charge of the Navy Control Station in the 10th M.D., and the following Filipino officer and enlisted men, assigned to the same area to strengthen signal communications:

2nd Lieut. SATURNINO R. SILVA, 0-1300095, Inf.
S/Sgt. Louis Cortez, 36398313
S/Sgt. Conrado A. Gustilo, 39238554
S/Sgt. Angel T. Sarmiento, 39235698
Sgt. Paterno A. Ortiz, 36624975

Thirty-three American evacuees and Dr. Cruz are expected back this afternoon.

Signed C.W.

DCS to Chief of P.R.S. thru G-2   18 February 1944

Noted by the Commander in Chief

Signed RJM

---

*This page was retyped for readability with the original document's header and footer.

Source: MacArthur Archives, Norfolk VA

Appendix B - 9

Mother's Day celebration sponsored by the 978th Signal Service Company. Camp Tabragalba. May 7, 1944

PROGRAMME

1ST Sgt Frederico A. Villarta	Gen Chairman
S/Sgt Leandro P. Tenato and Cpt Julius B Ruiz	Asst Chairmen
Sgt Vincent Goloyugo	M. C.

Part I
Athletics to commence at 930 – 1230
Part II
Dinner 1230 – 1400
Part III
Music Entertainment Dance
Part IV
Cakes Refreshments Tea

MAJ. CROELL GREETS YOU ALL

In behalf of the officers and men of this organization, I bid you welcome, to play with us, to eat with us, to drink with us, and to enjoy with us on this great day we all hold dear and which we are about to celebrate. This special occasion is indeed a great day when we honor the fines person, "Moth- er," in the whole world. Let's remember the care and fond hopes she lavished upon us, trying each day throughout the year to live up to the things she expected us to be.

Harry T. Croell
Major, U.S. Signal Corp
Commanding

MESSAGE FROM MAJ. BROWN

Everyone in Camp Tabragalba will be thinking of home today, the first in a foreign country for the majority of us. I wish to express the appreciation of the battalion to Major Croell and the members of 978th Signal Service Company for arranging the fine schedule of entertainment of this occasion.

Lewis Brown III
Major Commandant

MENU

Chilled Fruit Punch, (Lady Beautiful); Mixed Relishes, (a la Madame Too Loose); Fried Choice Spring Chicken (unjointed) In butter with pan gravy, (Hug it); Corn-fed Pork Lechon, Luvimin (Signal Corps Special) Ask S/Sgt Rodriguez how it is done; Arroz Valencia (Spanish Styles; Maria Montez); Baked Spa- ghetti with Giblets and Savory Sauce (As you like it); Our delicious hot rolls (Prepared by the "Bataan" Camp Bakers"; Assorted Delicious Homemade Cakes (American gal); Chilled Coca Cola (Sip it); Ice cold beer (Shoe me the way to go home)

SWEETEN THE DAY WITH CANDIES TO YOU, OUR MOTHERS. VISITORS: DON'T BRING YOUR RATION BOOKS!

A TOAST TO ALL MOTHERS

Here's to the dearest of all mothers.
Who stayed at my side while I was ill.
She is the joy and pride of her son.
Who fights this battle for freedom!
To come home to her when it's won.

Here's to the dearest of all mothers.
Who bears the heartache of us all.
The one who always dreams of you.
May she have God's blessing and good will.
And find her dream come true --- VICTORY.

Philippine National Anthem
U.S. National Anthem

COMMITTEES

Staff Officers: Maj Harry T. Croell, Commanding Officer
Lt Charles B. Ferguson; Lt. George T. Baker; Lt. Alvin V. Bone; Lt. Edward H Hale; Lt. Wilfred A. Cohen; Lt. Clinton B. McFarland; Lt. James B. McGivney; Lt. Kenneth H. Oates; Lt. Mayer J. Spiro; Lt Leo A. Strickland; WO (jg) Alphic L. Clanneh; WO (jg) Patrick Daley; WO (jg) Harry A. Maxam; WO (jg) Braynard L. Wise

GENERAL COMMITTEES
S/Sgt Manuel Faragan Finance Chairman

INVITATION	PROGRAM	FOOD COMMITTEE
S/Sgt Ibon, A.	Sgt Goloyugo, V.	S/Sgt Rodriguez, T.
Sgt Lopez, M	Sgt Candelario, J.	Sgt Cutarna, J.
Sgt Ignacio, M.	Sgt Reynoso, J.	T4 Boncavil, D.
Sgt Guyot, G.	Cpl Bugarin, E.	T4 Calmorin, A.
Sgt Vite, D.	T5 Palafox, R.	T4 Logan, D.
Cpl Torio, I.	Pfc Abadilla, R.	T4 Luz, F.
Cpl Lopez, C. G.	Pfc Ovalles, M.	Cpl Duran, A.
T5 Padgett, H.	Pfc Sperandeo, M.	Cpl Navarro, A.
Pfc Morris, J.	Pvt Burges, E.	Pvt Cortez, C.
Pvt Domantay, C.	Pvt Madariaga, F.	Pvt Cayabyab, A.
Pvt Guzman, B.	Pvt Salvador	Pvt Martin
Cpl Sheppard, E.	S/Sgt Tanato, L.	Sgt Runtal, H.
Sgt Acenas, S.	S/Sgt Cacabelos, R.	Sgt Balleras, J.
Sgt Cortez, R.	S/Sgt Fortun, S.	Sgt Biado, A.
Sgt Herrera, D.	S/Sgt Nery, G.	Sgt Bueno, G.
Sgt Montero, J.	S/Sgt Rulloda, A.	Sgt Sigonga, E.
Sgt Placido, C.	Sgt Agcaoili, R. E.	Cpl Balino, P.
Sgt Wright, C.	Sgt Begonia, D.	Cpl Randall, R.
Sgt Zola, E.	T4 Fajara, G.	Cpl Runes, L.
T4 Carillo, P.	T5 Halls, A.	T5 Ferrell, J.
Pfc Carpenter, C.	T5 Kobliska, L. V.	Pfc Saladino, S.
Pfc Toribio, A.	Pfc Peralta, E.	Pfc De la Pena, G.

ARRANGEMENT
Sgt Arce, M.
Sgt Giron, L.
Sgt Umipeg, M.
Sgt Padilla, B.
Sgt De Torres, R.
T4 Gonzales, J.
T4 Thomas, J. P.
T4 Vergara, T.
T5 Svobodny, R.
T5 Thompson, C.
Pfc Cabrera, R.
Pfc Guieb, L

ATHLETICS
Sgt Legaspi, A.
Sgt Balcena, F.
Sgt Borillo, P.
Sgt Herrera, D.
T4 Rasmussen, P.
T5 Dagandan, F.
T5 Pasquarella, J.
T5 Prosenell, R.
Pfc Long, H.
Pvt Angcos, R.
Pvt Bulatao, G.
Pvt Santos, R.

TRANSPORTATION
Sgt Sturgeon, C.
T4 Medina, M.
T5 Nichols, V.
T5 Watson, C.
T5 Watson, C.
Pfc Alcaras, J.
Pfc. Burns, E.
Pfc Juan, E.
Pfc Maravel, A.
Pfc. Thompson, J.
Pvt Madura, E.
Pvt Manalo, A.

ENTERTAINMENT
Sgt Visaya, V.
Cpl Pomarez, L.
T4 Marquina, L.
Pfc Enderiz, V.
Pvt Batara, Lope
Pvt Martin, E.

TECHNICIANS
T5 Olson, C. Y.
T5 Bendy, N.
T5 Pacificar, N.
Pfc Bals, O.
T5 Pena, D.
Pvt Viduya, J.

BAND COMMITTEE
Cpl Sensano, D.
Pfc Taddeo, C.
Pvt Hufana, F.
Pvt Roduta, P.
Cpl. Rallojay, T.
Cpl. Ilusorio, J.

PUBLICITY
Cpl Ruiz, Julus B.
Cpl Alcala, A. L.
T5 O'Gwin, C.
T5 Viray, D. J.
Pvt Espiritu, A.

DISBURSTING
S/Sgt Almero, E.
Sgt Ignacio, F.
Cpl Barcenas, U.
T5 Thompson, C. P.
Pfc Estavillo, S.

MIMEOGRAPHERS
Sgt Candelario, J.
Sgt Ignacio, M.
Cpl Alcala, A. L.
Pvt Viduya, J.

SPECIAL NOTICE
Mother's Day Mass services will be held at 0700 at new recreation hall,
with Father Leger delivering the sermon

Appendix B - 10

Salo-Salo A Get Together sponsored by
the 5218th Reconnaissance Company.
Camp Tabragalba. June 4, 1944

5218th RECONNAISSANCE COMPANY
5217th RECONNAISSANCE BATTALION
UNITED STATES ARMY FORCES IN THE FAR EAST
A. P. O. 923

GREETINGS:

IN BEHALF OF THE OFFICERS AND THE MEN OF THE 5218TH RECONNAISSANCE COMPANY, I WELCOME YOU AND THANK YOU ALL FOR ACCEPTING OUR INVITATION TO PARTAKE WITH US IN THIS HUMBLE "SALO-SALO".

TO THOSE WHO HAVE HELPED US PROMOTE THIS OCCASION, I WISH TO EXTEND THE GRATITUDE OF THIS COMMAND.

CARLOS D. ARGUELLES
1st Lieut., Infantry
COMMANDING

THE LEGEND OF "SALO-SALO"

An old Filipino custom which had come down through the ages, and which still is being observed throughout the Philippines is the "Salo-Salo" party. More than anything else, it symbolizes the Filipinos' love for fun, comradeship, feasting, and for everything that is beautiful.

To Filipinos, a "Salo-Salo" is more than a party. It is a community affair wherein everyone shares in making it a success. One contributes whatever he wishes to give. One may bring with him chickens, others may bring with them wine, rice, fish, vegetables, fruits, spices, etc. These offerings are pooled together and cooked to perfection by culinary experts. While the food simmers over the fire, the crowd makes merry with music and games. The gaiety is supplemented by the serving of succulent dishes prepared in abundance.

The highlight of the fiesta comes during the selection of a "LAKAMBINI", the fairest among the fairest present. The chosen muse — usually endowed with great charm— presides over the festival and, for that day, is the QUEEN of her people.

Our celebration today serves to perpetuate that good old Filipino custom. Today, we will have to forget, for the moment, the stark and brutal realities which have brought many of us here to this friendly and hospitable country.

Let's have fun, and revive that Filipino way of life. Let's get-together in the true spirit of a "SA- LO-SALO".

SCHEDULE OF EVENTS:

0800 – 0845	HOLY MASS
	Recreation Hall
1000	ASSEMBLY OF COMMAND
	Tent Area
1000 – 1030	RECEPTION
1030 – 1200	GAMES
1200 – 1330	LUNCHEON
1330 – 1400	THE NATIONAL ANTHEMS
	ANNOUNCEMENTS
1400 – 1600	DANCE AND PROGRAM
1600 – 1700	AFTERNOON TEA
1700 – 1800	DANCE

"Goodnight and pleasant dreams"

BAHALA NA BAND MEMBERS:

2nd Lt EDMUNDO J. MARFORI
T5 Don Viray

1/Sgt Peralta, G.	Cpl Ilusorio, J.	Cpl Curaming, E.
Sgt Ebarle, P.	Pfc Miguel M.	Pfc Cachapero, F.
Cpl Sensano, D.	Pvt Peralta, A.	Pfc Gorospe, P.
Cpl Lazo, E.	Pfc Ventura, D.	Pfc Punsalan, R.
Cpl Pomarez, L.	Pvt Hufana, F.	

MENU
Luncheon:

CHILLED FRUIT PUNCH
PEANUTS AND PRETZELS
"L E C H O N"
(Barbecued Pig)
BARBECUED HEIFER
"ARROZ CON POLLO" (Spanish Rice)
COMBINATION SALAD
ICE CREAM......REFRESHMENTS
***

Afternoon Tea:

TEA --- CAKES

A L I B I:

As much as we would like to be the perfect host, we can't beg, borrow, or steal
any silverware. So, won't you please bring your own . . . .
PU-LEEZE ! THANK YOU.

PROGRAM:
2nd Lt Marfori, E.
1st Sgt Domalaog, B.
S/Sgt Quintua, M.
S/Sgt Libatique, S.
S/Sgt Luis, M.
Sgt Adia, P.
Cpl Lopez, F. R.
Cpl Jaramilla, S.
T5 Racho, P.
Pvt Estabillo, S.

FINANCE:
2nd Lt Carag, V.
1st Sgt Buenaventura, N.
1st Sgt Regala, W.
Sgt Alvarez, V.
Sgt Pascua, I.
Pfc Diaz, J.
Pvt Ordonez, M.
Pvt Balderama, E.

ENTERTAINMENT:
2nd Lt Smith, F.
Cpl Ilustre, E.
1st Sgt Peralta, G.
S/Sgt Garcia, V.
S/Sgt Bacus, F.
Sgt Parcasio, J.
Sgt Daelto, M.
Sgt Basug, G.
Pfc Balunes, R.
Pfc Solar, M.

TRANSPORTATION:
2nd Lt Aquino, E.
M/Sgt Reyes, E.
1st Sgt Santos, F.
S/Sgt Constantino, J.
S/Sgt Rosete, P.
Sgt Bolante, J.
Sgt Bareng, M.
Cpl Quiming, P.
T5 Hipol, F.

COMMISSARY:
2nd Lt Tobias, M.
1st Sgt Ramos, S.
S/Sgt Aguilar, P.
Sgt Baldos, F.
T4 Aurelio, M.
Cpl Corpus, E.
Cpl Anacleto, V.
Cpl Pimentel, L.
Cpl Taylan, G.
Cpl Quindiagan, L.
T5 Soliven, L.
T5 Sandoval, N.
Pvt Bugayong, V.
Pvt Rosario, J.
Pvt Medrano, D.
Pvt Savellano, P.
Pvt Mundo, F.
Pvt Pilipina, L.
Pvt Azcueta, P.
T5 Abad, O.

COMMITTEES:

STAFF:

1st Lt CARLOS D. ARGUELLES   Commanding Officer
2nd Lt EDMUNDO J. MARFORI   Special Service Officer
2nd Lt SILVINO B. TALLIDO   Supply Officer
1st Sgt GERMAN V. REYES   First Sergeant
S/Sgt ANTONIO D. ABELLERA   Supply Sergeant

ARRANGEMENTS:	USHERS:	INVITATION-RECEPTION:
2nd Lt Tallido, S. B.	1st Lt Dionolo, T.	2nd Lt Arellano, V.
2nd Lt Valera, J. V.	2nd Lt Tamayo, C.	2nd Lt Davis, W. J.
1st Sgt Goloyugo, V.	M/Sgt Dacquel, I.	2nd Lt Allen, R. C.
S/Sgt Campus, H.	1st Sgt Abrenica, S.	1st Sgt Cabais, E.
S/Sgt Diputado, E.	S/Sgt Barruga, F.	S/Sgt Guzman, L.
S/Sgt Javonillo, J.	S/Sgt Malinab, L.	S/Sgt Lim, A.
Sgt Ebarle, A	Sgt Cacas, M.	S/Sgt Ruiz, D.
Sgt Quilala, P.	Sgt Fontanoz, M.	S/Sgt Agron, A.
Sgt De Ocampo, M.	Sgt Rillera, P.	S/Sgt Real, I.
T4 Mapile, C.	Cpl Agustin, P.	Cpl Anderson, M.
Cpl Evangelista, V.	Cpl Arellano	Pvt Esteban, P.
Cpl Bibat, E.	Cpl Maglangit, M.	Pvt Casco, L.
Cpl Costales, F.	Cpl Rivera, E.	Pvt Arciaga, E.
Pfc Alayu, C.	Cpl Salamanca, A.	Pvt Marte, G. S.
Pfc Artiago, M.	Pvt Bernal, J.	Pvt Darang, A.

Source: Paula Valdez, 1st Filipino Regiment, U.S. Army, 1942-1946. January 13, 2015

Appendix B - 11

1ST RECONNAISSANCE BATTALION
5217TH RECONNAISSANCE BATTALION AND 978TH SIGNAL SERVICE COMPANY
ENLISTED MEN

CPL

Cpl Agustin, Pedro P.	37156107	Cpl Mendoza, Primo T.	36356345
Cpl Almojera, Loren N.	39604988	Cpl Merca, John A.	39553806
Cpl Anacleto, Vicente H.	36739381	Cpl Oandasan, Angelo P.	36325336
Cpl Anderson, Moses A.	39043724	Cpl Pascual, Amrafel D.	39683927
Cpl Arellano, Martin Q.	39102486	Cpl Peneyra, Cipriano C.	39839100
Cpl Ayson, Sammy L.	39413091	Cpl Pimentel, Lorenzo U.	39389581
Cpl Baladad, Jose C.	36702550	Cpl Pulido, Mack H.	39840672
Cpl Bareng, Gregorio F.	39841167	Cpl Quesada, Antonio M.	36649313
Cpl Bibat, Edilberto S.	36346416	Cpl Quiming, Perfecto C.	36199212
Cpl Casia, Melano C.	39828878	Cpl Quinitio, Basilio A.	39843138
Cpl Corpuz,Eustacio C.	10301905	Cpl Quindiagan, Luis C.	39105910
Cpl Costales, Fermin D.	37216144	Cpl Rabang, Thomas V.	39526403
Cpl Curaming, Eleuterio B.	39103871	Cpl Ramos, Julian G.	39836592
Cpl De Leon, Peter B.	39101202	Cpl Respicio, Sergio F.	39836837
Cpl Dupitas, Aurelio M.	39845473	Cpl Reyes, Felicisimo	39246393
Cpl Fermin, Raymond R.	39039876	Cpl Rivera, Elipidio A.	36337587
Cpl Ibarra, Estanislao S.	39100658	Cpl Robles, Romulo D.	33190611
Cpl Jaramilla, Santiago Q.	36358865	Cpl Rodriguez, Juan A.	10641048
Cpl Lazo, Enoch A.	39240485	Cpl Sabado, Henry D.	39094281
Cpl Lazo, Ernesto L.	39282541	Cpl Salamanca, Albert S.	37205594
Cpl Leanio, Frank S.	39839199	Cpl Santos, Isaac L.	39245489
Cpl Loma, Magdaleno S.	39245633	Cpl Sison, Agapito R.	36358723
Cpl Lopez, Arthur G.	39395298	Cpl Tria, Antonino B.	39264160
Cpl Lopez, Felix R.	39026351	Cpl Tucay, Henry T.	39236723
Cpl Malangit, Maximo V.	39397902	Cpl Valdez, Celestino R.	39842304

T5

T5 Abad, Otilio V.	39104114
T5 Agcaoili, Julian S.	39174730
T5 Alvez, Beato C.	33202729
T5 Bautista, Moises V.	33189145
T5 Baylon, Alfred S.	36398604
T5 Buenavista, Serafin B.	39843064
T5 Costales, Gene D.	37216159
T5 Francisco, Urbano M.	39526401
T5 Hipol, Felipe R.	36515089
T5 Javier, Bartolome E.	39547130
T5 Kintanar, Aniceto S.	35114187
T5 Racho, Paciano S.	36345447
T5 Rimando, Antonio C.	37156057
T5 Ruiz, Ruperto	39838264
T5 Sagun, Fernando C.	39105188
T5 Sandoval, Nicolas E.	39843993
T5 Soliven, Larry G.	39390506
T5 Valido, Pastor W.	39841139

PFC

Pfc Alayu, Celestino D.	36600603
Pfc Alfafara, Gerardo G.	36343823
Pfc Antonio, Alfonso L.	39120977
Pfc Arcia, Narciso N.	39116432
Pfc Argueza, Don T.	36346436
Pfc Arqueo, Lazaro C.	39187222
Pfc Artiago, Maurice W.	39330914
Pfc Atad, Martiniano S.	39393185
Pfc Baker, Clinto M.	39309940
Pfc Balunes, Ralph M.	39235966
Pfc Barba, Conrad A.	39341805
Pfc Bugayong, Viconete V.	36356958
Pfc Cachapero, Frank G.	39844921
Pfc Calustro, Pedro A.	32291087
Pfc Cariaga, Pedro T.	10641012
Pfc Carpio, Jose P.	37290686
Pfc Clemente, Leo C.	39174917
Pfc Cuanang, Julian F.	39096248
Pfc Custodio, Emilio M.	39786249
Pfc Erana, Miguel B.	39837011
Pfc Flores, Agripino B.	39836186
Pfc Formoso, Francisco H.	36337830
Pfc Gaduang, Thomas D.	36625135
Pfc Garlitos, Anton T.	36346578
Pfc Gobaleza, Dominador F.	10641049
Pfc Gonzales, Marcelo U.	39308995
Pfc Granados, Vicente J.	39846508
Pfc Javier, Floro O.	36359894
Pfc Lascano, Leodegario	36328888
Pfc Machon, Sergio M.	33194571
Pfc Malvar, Angelo B.	39538482
Pfc Martin, Philip	34202018
Pfc Martinez, Chris M.	36343941
Pfc Nalangan, Restituto L.	3983984
Pfc Nievres, Frederico J.	36355586
Pfc Ollada, Patricio	37345911
Pfc Ouano, Francisco I.	39236026
Pfc Pascua, Jerry D.	37359348
Pfc Quiaoit, Esteban R.	39191356
Pfc Quismundo, Ruperto Q.	39201741
Pfc Quinagon, Barney A.	39391096
Pfc Respicio, Tom P.	39235454
Pfc Sapio, Lucas J.	39033374
Pfc Serrano, Agapito B.	39394382
Pfc Solar, Marcelino V.	39406660
Pfc Tolentino, John T.	36355508
Pfc Trocio, Perfecto N.	37701493
Pfc Tullo, Apollo P.	39253426

PVT

Pvt Abel, Servando A.	39237333
Pvt Agbunag, Baltazar G.	39840188
Pvt Aguila, Isaac A.	39031494
Pvt Alerta, Custodio P.	39391259
Pvt Aliacbar, Sahibad M.	10647057
Pvt Antikoll, Carlos F.	10647180
Pvt Arciaga, Emil D.	36350540
Pvt Azcueta, Pedro S.	36350540(Duplicate)
Pvt Bairulla, Sabtal M.	10647051
Pvt Bairulla, Hadjula H.	10647053
Pvt Balderama, Eduardo G.	39235632
Pvt Bentillo, Venancio L.	39238829
Pvt Bernal, Jaime A.	31318657
Pvt Bermudez, Felix	32250341
Pvt Casco, Lainvencion R.	39412952
Pvt Catalina, Eugenio S.	10647055
Pvt Cayabyab, Ricardo D.	39355604
Pvt Darang, Andres D.	39828848
Pvt De Leon, Benedict S.	39840292
Pvt Delfin, Urbano P.	36734497
Pvt Despy, Alfredo N.	39114329
Pvt Esteban, Peter M.	39109680
Pvt Fernandez, Leon C.	36707169
Pvt Generales, Narciso C.	39541136
Pvt Halog, Rugino G.	39395667
Pvt Itchon, Jay F.	39240629
Pvt Ladjahasan, Jalil H.	10647052
Pvt Ladjahasan, Mangona H.	10647056
Pvt Ladjahasan, Ali H.	10647054
Pvt Lobendino, Hilario L.	39310376
Pvt Lucas, Felix M.	39103180
Pvt Marquez, Victor V.	33737505
Pvt Medrano, Demetrio J.	36606417
Pvt Minder, Robert C.	39229726
Pvt Mosquera, Benigno F.	39402969
Pvt Nerez, Manuel M.	39044285
Pvt Oania, Hermogenes C.	39839084
Pvt Peska, Martin Y.	39846550
Pvt Pilar, Mariano F.	39405528
Pvt Rosario, John R.	37341033
Pvt Salvacion, Alfredo L.	39109014
Pvt Santos, Alberto E.	39844995
Pvt Sarmiento, Florentino M.	32888314
Pvt Savellano, Pedro E.	32355686
Pvt Tolentino, Zoilo, Jr.	32870608
Pvt Velayo, Benny T.	39196008
Pvt Wendam, Sidney D.	39028642

T/SGT

T/Sgt Hoke, Max E.   16041686

S/SGT

S/Sgt Chambless, Henry O.	17084155	S/Sgt Hammill, Charles H.	39034039
S/Sgt Finnegan, George J.	12139211	S/Sgt Moore, Chester H.	10219454
S/Sgt George, Robert R.	38119149	S/Sgt Richardson, William F.	13153656

SGT

Sgt Becker, William III	16168669	Sgt Lowe, William M.	39036006
Sgt Casell, Robert O.	31283847	Sgt MacCarrick, Donald R.	11051775
Sgt Lozano, Ray J.	18155440	Sgt McGrath, Charles E.	13091616
Sgt Hendrickson, Murl S.	17088657		

CPL

Cpl Bierly, James N. Jr.	13153755	Cpl Herbig, Robert P.	35583726
Cpl Chick, Alton C. Jr.	11136180	Cpl Males, William J.	16071525
Cpl Gamertsfelder, Robert H.	35749997		

978TH SIGNAL SERVICE COMPANY

MAJOR

Major Croell, Harry T.   0262144

CAPTAIN

Captain Ferguson, Charles B.   0454881

1ST LT

1st Lt Hale, Edward H.	01637589	1st Lt Strickland, Lee A.	01638609

2ND LT

2nd Lt Becker, George T.	01643111	2nd Lt McGivney, James B.	02034825
2nd Lt Bone, Alvin V.	01641443	2nd Lt Oates, Kenneth H.	01645882
2nd Lt Cohen, Wilfred A.	01644722	2nd Lt Spire, Meyer J.	01645026
2nd Lt McFarland, Clinton E.	01641562	2nd Lt Wootin, Harold B.	02037246

WARRANT OFFICERS

CWO Wise, Braynard L.	W-2129711	WOJG Daley, Patrick	W-2129709
WOJG Claunch, Alphie L.	W-2129712	WOJG Maxam, Harry A.	W-2129706

M/SGT

M/Sgt Alberto, Alfred A.	36515325	M/Sgt Clark, Walter G.	33202546

1/SGT

1/Sgt Villarte, Frederico B.   39242985

T/SGT

T/Sgt Almero, Emiliano A.	39838277	T/Sgt Bugh, Marion W.	18003254
T/Sgt Arbis, Mariano V.	10647023		

S/SGT

S/Sgt Cacabelos, Rufino F.	39179223	S/Sgt Ibea, Artemio I.	39535788
S/Sgt Camp, Lowell A.	15132532	S/Sgt Nagle, David A.	12180125
S/Sgt Carter, Loy W.	39679011	S/Sgt Nery, Gerardo B.	39023603
S/Sgt Corpus, Aquiliano M.	39032447	S/Sgt Palac, Juan M.	6745755
S/Sgt Ergas, Eugene N.	39105165	S/Sgt Rodriguez, Fred N.	39838799
S/Sgt Faragan, Manuel F.	19090770	S/Sgt Sarmiento, Angel T.	39235968
S/Sgt Fortun, Salvador B.	36361421	S/Sgt Tanato, Leandro F.	39097588
S/Sgt Gustilo, Conrado A.	39238554	S/Sgt Thomas, James P.	15319546

T3

T3 Africa, Robino	39841147
T3 De Torres, Ronald A.	39264501
T3 Ginsberg, Seymour	32648872
T3 Giron, Leovigildo N.	39536590
T3 Gregory, George F.	39830936
T3 Lopez, Marcos T.	36617437
T3 Robinson, Irving	36372422
T3 Stahl, Robert E.	33236898
T3 Umipeg, Marcello B.	39080051
T3 Visaya, Vincent L.	39244558

SGT

Sgt Acenas, Semeniano D.	39236952
Sgt Agcaoili, Raymundo E.	32356031
Sgt Allen, David A.	20238136
Sgt Ancheta, Alfonso	33739514
Sgt Arce, Mariano T.	36605781
Sgt Balcena, Fred	39285012
Sgt Balleras, Julio V.	39840483
Sgt Barr, Charles N.	37257753
Sgt Begonia, Daniel B.	38168017
Sgt Biado, Anacleto	36607611
Sgt Borillo, Pio B.	39838832
Sgt Brown, James R.	36617705
Sgt Bueno, George B.	35466492
Sgt Candelario, Jose B.	39246432
Sgt Castro, Ray F.	39101320
Sgt Cortez, Ramon F.	39237487
Sgt Cutaran, Jacinto	39566822
Sgt Goloyugo, Vincent C.	36545428
Sgt Douldner, Feredrick J.	32346634
Sgt Guyot, Gaudencio V.	39024838
Sgt Herreria, Donato E.	32278953
Sgt Ignacio, Fred C.	39827659
Sgt Ignacio, Miguel E.	39842016
Sgt Legaspi, Alex B.	39555046
Sgt Mauricio, Abamile L.	39235184
Sgt Montero, Jack	36361427
Sgt Nagtalon, Alex A.	37204648
Sgt Ortiz, Paterno A.	36624975
Sgt Padilla, Balbino R.	39845504
Sgt Parcasio, Joe B.	33194377
Sgt Placido, Carlos S.	39267792
Sgt Reynoso, Jaime R.	39547226
Sgt Runtal, Henry E.	39237726
Sgt Salgado, Mauro B.	39837635
Sgt Sibonga, Emil J.	39166093
Sgt Sturgeon, Charles G.	33557030
Sgt Thys, Alfred P.	36039602
Sgt Vite, Doroteo V.	32249269
Sgt Wright, Carl	36601513
Sgt Zola, Edward J.	16124657

T4

T4 Agbayani, Vincent A.	39839127
T4 Boncavil, Domiciano O.	39838429
T4 Calmorin, Angelo A.	32418740
T4 Carrillo, Patrocinio S.	39842421
T4 Dagandan, Fortunato C.	39098413
T4 Davin, Numeriano D.	39536974
T4 Fernandez, Cipriano C.	39258733
T4 Gonzales, Jose T.	39098904
T4 Logan, Domingo L.	39825025
T4 Lopez, Cirilo G.	39402580
T4 Luz, Pete L.	39845409
T4 Marquina, Leo	39234336
T4 Medina, Mariano A.	39024053
T4 Pacificar, Nicomedes	32443020
T4 Pajara, George T.	39032439
T4 Palafox, Rosendo M.	32325042
T4 Parong, Astor C.	39243014
T4 Pomarez, Leon V.	39413450
T4 Runes, Lucas D.	33196248
T4 Rusmussen, Paul D.	36647192
T4 Sabado, Daniel B.	39841843
T4 Vergara, Thomas O.	39838398
T4 Watson, Carl T.	15326834
T4 Victorio, Raymond D.	39097646
T4 Wright, Clarence L.	33418570
T4 Young, Clarence R.	33107594

CPL

Cpl Alcala, Atanancio L.	39402929
Cpl Balin, Porfirio	32416859
Cpl Barcenas, Urbano B.	39111236
Cpl Buccat, Alejandro A.	39248525
Cpl Bugarin, Enrico M.	39119957
Cpl Dodson, Hadley W.	36354468
Cpl Duran, Agrifino J.	39838815
Cpl Guico, Pete S.	39246024
Cpl Ilusorio, Johnnie A.	39838473
Cpl Navarro, Ambrosio A.	39041984
Cpl Peralta, Materno P.	39843935
Cpl Rallojay, Teodoro J.	39235964
Cpl Randall, Roberto A.	39129170
Cpl Reyes, Felix T.	39845120
Cpl Sensano, Daniel G.	39692205
Cpl Sheppard, Edward T.	33576718
Cpl Torio, Isaias T.	39536985
Cpl Villote, Paul P.	39113338

T5

T5 Advincula, Julio C.	39234349
T5 Albalos, Segundo Q.	19121481
T5 Alegre, Emilio A.	39841451
T5 Angcos, Robert O.	39563906
T5 Nuevo, Leodegario O.	36601310
T5 O'Gwin, Charles W.	14185206
T5 Olson, Clarence Y.	37663674
T5 Padgett, Henry L.	15319828

T5 Arce, George U.	36346801	T5 Padilla, Bartolome M.	39839177
T5 Beedy, Nowell J.	1718104	T5 Palmeto, Daniel S.	39837031
T5 Beeler, Richard N.	37231980	T5 Pascual, Angelo E.	32246630
T5 Burgos, Edward M.	36262428	T5 Paquarella, Joseph R.	15121431
T5 Burns, Edward P.	33589084	T5 Pena, Doroteo M.	36710307
T5 Candania, Amadeo C.	36513719	T5 Peredo, Joe E.	39325510
T5 Dacquel, Richie D.	39243962	T5 Pressnell, Richard M.	37551898
T5 Estavillo, Sergio B.	39130647	T5 Robert, Louis	35571878
T5 Ferrell, John N.	14106667	T5 Ruiz, Julius B.	39198058
T5 Filek, Joseph F.	37552127	T5 Saladino, Selmo C.	37198262
T5 Green, James L.	36614103	T5 Sammons, William H.	39414170
T5 Halla, Albert F.	36737954	T5 Santos, Rudolph B.	16128144
T5 Herreria, George R.	32230716	T5 Sivila, Andres C.	39843086
T5 Holgado, Eddie C.	39197647	T5 Esperandeo, William P.	36640046
T5 Isaac, Cledonaldo J.	39527750	T5 Svobodny, Raymond D.	17073137
T5 Kobliska, La Verne J.	37663771	T5 Tadeo, Carmine	3229820
T5 Landrum, Lesse E.	36449262	T5 Thompson, Clifford P.	35546828
T5 Long, Howard D.	17089737	T5 Toribio, Alipio C.	36713799
T5 Malitovsky, Sidney M.	33426118	T5 Ulivas, Albert U.	36369345
T5 Marinas, Toby L.	39844037	T5 Villalon, Hermenegildo	39242352
T5 Mateo, Carl M.	36629977	T5 Viray, Don J.	39527346
T5 Nichols, Victor E.	15353489	T5 Wheat, Jesse P.	34495654

PFC

Pfc Aglubat, Al C.	36332411	Pfc Maravel, Antonio L.	33595833
Pfc Alcantara, Marcelino C.	39126719	Pfc Marcelino, Juan C.	39848180
Pfc Alcaras, Juan D.	39246396	Pfc Morales, Catalino U.	39841637
Pfc Alvarado, John R.	39109140	Pfc Morris, Joseph A.	33478556
Pfc Ambrosio, Nicanor G.	36028978	Pfc Moscoso, Anselmo S.	39244854
Pfc Andres, Manuel V.	37211856	Pfc Ovalles, Marcelo C.	39320985
Pfc Aquino, Alejandro C.	39841158	Pfc Pedro, Florendo R.	39838516
Pfc Asuncion, Telesforo T.	39391748	Pfc Peralta, Estanislao C.	39561110
Pfc Bargo, Querubin B.	39024844	Pfc Petro, Frank J.	35521431
Pfc Bell, Harry W.	17123222	Pfc Piana, Victor L.	39243842
Pfc Bent, George F.	11088398	Pfc Proffetta, Stephen J.	33488135
Pfc Carrasca, Romain S.	36636009	Pfc Ramos, Aquilino R.	33203696
Pfc Corpuz, Robert B.	39101348	Pfc Rimando, Juan F.	37287078
Pfc De La Pena, Geronimo M.	39260071	Pfc Rodriguez, Feliciano S.	39391220
Pfc Eals, Orien N.	35678066	Pfc Rosales, Doroteo A.	39839490
Pfc Enderiz, Victor P.	39105219	Pfc Santos, Gregorio	39685483
Pfc Felizar, Pete G.	39391936	Pfc Sapon, Leoncio S.	39835461
Pfc Gangl, Leo J.	17157229	Pfc Savellano, Andres S.	39845429
Pfc Gonzales, Juan L.	36549488	Pfc Silverio, Ben V.	39263747
Pfc Guieb, Lorenzo M.	39852508	Pfc Tobofunda, Pedro P.	39246444
Pfc Jose, Teodoro M.	39687460	Pfc Thompson, James C.	15316411
Pfc Laurina, Filomeno A.	39393681	Pfc Ventura, Dionicio N.	39242076
Pfc Liberato, Don J.	39547094	Pfc Wurst, Nelson A.	32477867

PVT

Pvt Abang, Bernardino B.	39101198	Pvt Marcelo, Benjamin R.	3921351
Pvt Aquino, Bernardino P.	39031576	Pvt Martin, Isidro	39844854
Pvt Astronomo, Telesforo R.	39402174	Pvt Martin, Sammy C.	36601591
Pvt Balinton, Vicente	39042951	Pvt Marzan, Ricarte S.	39843944
Pvt Bulatao, George M.	39103897	Pvt Mejia, Fulgencio L.	39272113
Pvt Bumanglag, Arsenio Y.	39391750	Pvt Milana, Philip G.	39114222
Pvt Calip, Ronnie E.	39207135	Pvt Neighbors, John G.	34777462
Pvt Campos, Norberto C.	39402744	Pvt Nichols, Howard M.	38590704
Pvt Carpenter, Charles E.	15319681	Pvt Obrero, Dick V.	39246368
Pvt Cayabyab, Adolfo V.	39110054	Pvt Ogoy, Dale Y.	36352596
Pvt Cortez, Calixto T.	39196262	Pvt Pabros, Ambrosio A.	29280569

Pvt Dulay, George R.	36269641	Pvt Patacsil, Regino T.	39397341
Pvt Espiritu, Artemio C.	39123285	Pvt Patubo, Gaudioso L.	32502733
Pvt Gachallan, Hipolito M.	39841475	Pvt Penalvar, Pedro R.	39105267
Pvt Girard, Stanley J.	31336956	Pvt Riotoc, Manuel E.	39095507
Pvt Gonzalo, Arthur V.	39031504	Pvt Salvador, Frank A.	39101922
Pvt Gurrea, Preciliano B.	39113642	Pvt Saria, Juan A.	38313180
Pvt Guzman, Simeon R.	36289291	Pvt Schneider, Robert J.	37624936
Pvt Goroza, Edward E.	39395656	Pvt Seares, Pascual P.	39273631
Pvt Hendelman, Lambert J.	36767588	Pvt Smith, James J.	34779782
Pvt Higgins, Donald V.	17060313	Pvt Taban, Lorenzo F.	39112726
Pvt Hosberg, Glen B.	35233573	Pvt Udani, Carlos D.	36325243
Pvt Hufana, Frank A.	39564521	Pvt Udaundo, Eseban C.	36665915
Pvt Madariaga, Fred M.	39459947	Pvt Valmonte, Frank G.	39237209
Pvt Madura, Edward J.	33610331	Pvt Vicente, Victor L.	39099852
Pvt Magale, Isidro P.	33195320	Pvt Viduya, Jose F.	39105201
Pvt Manalo, Armand	37341885	Pvt Woods, Alwin C.	33748679
Pvt Mapalo, Martin L.	39402485		

Source: Filipino American Experience Research Project, 1998

Appendix B - 12
Bahala Na Anniversary Special Edition August 1944 celebrated the men's first year at Camp X (Tabragalba).

Source: MacArthur Archives, Norfolk VA
Filipino American Experience Research Project, Oct 1998

Appendix B - 13

RESTRICTED
HEADQUARTERS
UNITED STATES ARMY FORCES IN THE FAR EAST
3846 FEXO, APO 501

FEXO 210.31   20 Nov 44
SUBJECT: Orders

TO: Officers Concerned

Following named officers are released from Detached Service 5217th Reconnaissance Battalion and assignment Miscellaneous Group USAFFE, assigned 1st Reconnaissance Battalion Special. Officers now on further Tour Duty or Detached Service status will remain on present Tour Duty or Detached Service. No transportation involved.

Lt Col Lewis Brown III	0290292	CAV	2nd Lt Merle E. Campbell	01051505	CAC
Capt James L. Evans	0356781	MC	2nd Lt Torribio Crespo	0888253	CAC
Maj Lawrence H. Phillips	0916722	CE	2nd Lt William J. Davis	02036593	AUS
Maj Charles M. Smith	0888471	CE	2nd Lt Alberto C. Elefanio	0530514	INF
Capt Ricardo C. Galang	05105082	INF	2nd Lt Richard A. Ensor	01050244	CAC
Capt Enoch G. Jones	0368362	SIG	2nd Lt Arthur F. Frias	01324763	INF
Capt Frank J. Skundale	0377668	CAC	2nd Lt Maynard C. Hawley	01686144	AUS
Capt Frank H. Young	0888453	INF	2nd Lt Al Hernandez	01324774	INF
1st Lt Rizalito Abanto	01301590	INF	2nd Lt Felix Hondolero	01311893	INF
1st Lt Ireneo A. Ames	0888247	CAC	2nd Lt Charles P. Hope, Jr.	01059496	CAC
1st Lt Carlos D. Arguelles	0499852	INF	2nd Lt Vicente W. Labrador	0888256	AUS
1st Lt Andrew P. Baker	01030214	CAV	2nd Lt Sebastian B. Lorenzo	01301171	INF
1st Lt Fred H. Behan	01081198	CAC	2nd Lt Jose B. Malan, Sr.	01314382	INF
1st Lt John E. Bove	01031295	CAV	2nd Lt Edmundo J. Marfori	01310660	INF
1st Lt Bartolome C. Cabangbango	888747	AUS	2nd Lt Paul F. Mauricio	01319099	INF
1st Lt (CH) Edmond J. Dietzel	0541243	USA	2nd Lt Jose T. Mendoza	01313662	INF
1st Lt Thomas Dinolo	01298007	INF	2nd Lt Luis P. Padilla	01305300	INF
1st Lt Generoso D. Arias	01306357	INF	2nd Lt Emilio F. Quinto	0888442	AUS
1st Lt Peronio B. Huerto	0888249	CAC	2nd Lt Ceferino N. Rola	01308379	INF
1st Lt Rodolfo C. Ignacio	0888248	AUS	2nd Lt Ignacio M. Sarmiento	01822465	INF
1st Lt Donald V. Jamison	01043373	CAC	2nd Lt Albert Schmaltz	01031837	CAV
1st Lt Melecio J. Montesclaros	01305289	INF	2nd Lt Saturnino R. Silva	01300095	INF
1st Lt Ruben P. Sangco	0478156	INF	2nd Lt Frank B. Smith	01324796	INF
1st Lt Brooke Stoddard	01030523	CAV	2nd Lt Silvino N. Tallido	01313063	INF
1st Lt Enrique L. Torres	0888748	AUS	2nd Lt Claudio M. Tamayo	01305526	INF
1st Lt Patricio Ubarro	0359964	INF	2nd Lt Konglam Teo	0888892	AUS
1st Lt Isaac F. Unciano	01102098	INF	2nd Lt Sidney S. Tison, Jr.	0534061	INF
1st Lt Cecil E. Walter, Jr.	0131467	INF	2nd Lt Joe L. Tudtud	01315532	INF
2nd Lt Pacifico D. Agpaoa	01323652	INF	2nd Lt Manuel A. Tobias	01305346	INF
2nd Lt Robert C. Allen, Jr.	02036590	AUS	2nd Lt Jose V. Valera	0888893	AUS
2nd Lt Carlos F. Ancheta	01316979	INF	2nd Lt Frank C. Ventura	01318237	INF
2nd Lt Arsenio Y. Arellano	01301594	INF	2nd Lt David T. Vivit	01306774	INF
2nd Lt Henry L. Baker	0534045	CAC	2nd Lt Rhys C. Wood	01081044	CAC
2nd Lt Delfin C. Yuhico	0888175	TC			

By Command Of General Macarthur:
/s/ M.B. KENDRICK
Major A. GD

Legend: CAB-Cavalry; MC-Medical Corps; CE-Combat Engineer; INF-Infantry; Signal, AUS-Army of the United States; CAC-Coast Artillery Corps

Source: Filipino American Experience Research Project Oct 1998; History of the 1st Reconnaissance Battalion by Ceferino Rola

Appendix B - 14
SCOUT COMPANY AND COLUMN
FIRST RECONNAISSANCE BATTALION, SPECIAL
UNITED STATES ARMY FORCES IN THE FAR EAST
A. P. O. 565

21 December 1944

ROSTER AS OF 21 DECEMBER 1944

SCOUT COMPANY

OFFICERS:

1. Capt Arguelles, Carlos D.	0499852	INF	4. 2d Lt Frias, Arthur F.	01324673	INF**
2. 2d Lt Allen, Robert O. Jr.	02036590	AUS	5. 2d Lt Tallido, Silvino V.	01313063	INF
3. 2d Lt Davis, William J.	02036593	AUS			

ENLISTED MEN:

1. 1st Sgt Buenaventura, Nicolas	19126906		7. Tec 5 Abad, Otilio V.	39104114	
2. T/Sgt Abellera, Antonio D.	39526289		8. Tec 5 Agcaoili, Julian S.	39174730	
3. S/Sgt Lopez, Felix R.	39026351		9. Tec 5 Halog, Rufino D.	39395667	
4. S/Sgt Merca, John A.	39553806		10. Tec 5 Hipol, Felipe C.	36515089	
5. Tec 4 Lucas, Felix M.	39103180		11. Tec 5 Martinez, Chris M.	36343941	**
6. Tec 4 Sison, Agapito R.	36358723		12. Tec 5 Rosario, John R.	37341033	

COLUMN
OFFICERS:

1. Major Evans, James L.	0356781	MC*	28. 2d Lt Crespo, Torribio	0888253	CAC*
2. Major Phillips, Lawrence H.	0916722	CE*	29. 2d Lt Elefano, Alberto C.	0530514	INF
3. Major Smith, Charles H.	0888471	CE*	30. 2d Lt Ensor, Richard A.	01050244	CAC*
4. Capt Cabangbang, Bartolome C.	0888747	AC*	31. 2d Lt Hawley, Maynard C.	01686144	AUS*
5. Capt Galang, Ricardo C.	0515082	INF*	32. 2d Lt Hernandez, Al	01324774	INF*
6. Capt Skundale, Frank J.	0377668	CAC*	33. 2d Lt Hondolero, Felix S.	01311893	INF
7. Capt Young, Frank H.	0888453	INF*	34. 2d Lt Hope, Charles P. Jr.	01059496	INF*
8. 1st Lt Abanto, Rizalito	01301590	INF	35. 2d Lt Labrador, Vicente W.	0888256	AUS*
9. 1st Lt Ames, Irineo A.	0888247	CAC*	36. 2d Lt Lorenzo, Sebastian B.	01301171	INF
10. 1st Lt Bahr, Andrew P.	01030214	CAV*	37. 2d Lt Malan, Jose B. Sr.	01314382	FA*
11. 1st Lt Behan, Fred H.	01081198	CAC*	38. 1st Lt Marfori, Edmundo J.	01301660	INF
12. 1st Lt Bove, John E.	01031295	CAV*	39. 2d Lt Mauricio, Paul F.	01319099	INF*
13. 1st Lt Dionolo, Thomas	01298007	INF	40. 2d Lt Mendoza, Jose T.	01313662	INF*
14. 1st Lt Frias, Generoso D.	01306357	INF	41. 2d Lt Quinto, Emilio F.	0888442	AUS*
15. 1st Lt Huerto, Petronio B.	0888249	CAC*	42. 2d Lt Sarmiento, Ignacio M.	01822465	INF
16. 1st Lt Ignacio, Rodolfo C.	0888248	AUS*	43. 2d Lt Schmaltz, Albert	01031837	CAV*
17. 1st Lt Jamison, Donald V.	01043373	CAC*	44. 2d Lt Silva, Saturnino R.	01300095	INF*
w	01305289	INF	45. 2d Lt Smith, Frank B.	01324796	INF
19. 1st Lt Songco, Ruben P.	0478156	INF*	46. 2d Lt Tamayo, Claudio M.	01305526	INF*
20. 1st Lt Stoddard, Brooke	01030523	CAC*	47. 2d Lt Teo, Konglam	0888892	AUS*
21. 1st Lt Torres, Enrique L. Jr.	0888748	AC*	48. 2d Lt Tison, Sidney S. Jr.	0534061	CAC*
22. 1st Lt Unciano, Isaac F.	01102098	INF	49. 2d Lt Tudtud, Joe L.	01315532	INF
23. 1st Lt Walter, Cecil E. Jr.	01314697	INF	50. 2d Lt Valera, Jose V.	0888893	AUS*
24. 2d Lt Agpaoa, Pacifico O.	01323652	INF	51. 2d Lt Vivit, David T.	01306774	INF
25. 2d Lt Ancheta, Carlos F.	01316979	INF*	52. 2d Lt Wood, Rhys C.	01081044	INF*
26. 2d Lt Baker, Henry L.	0534045	CAC*	53. 2d Lt Yuhico, Delfin C.	0888175	TC*
27. 2d Lt Campbell, Merle E.	01051505	CAC*			

ENLISTED MEN:

1. M/Sgt Berg, Gerald G.	6541910 *	147. Cpl Gallemore, Jose M.	6592592
2. M/Sgt Dacquel, Isidoro D.	R5274389 *	148. Cpl Garlitos, Anton T.	36346578
3. M/Sgt Reyes, Eulogio (NMI)	6613779	149. Cpl Ibarra, Estanislao S.	39100658 *
4. 1st Sgt Abrenica, Santiago D.	39233753	150. Cpl Jaramilla, Santiago Q.	36358865 **
5. 1st Sgt Besid, Restituto J.	32272318 *	151. Cpl Javier, Floro O.	36359594 **
6. 1st Sgt Cabais, Eutiquio B.	R2781272 *	152. Cpl Lazo, Ernesto L.	39282541
7. 1st Sgt Corpus, Amado S.	6737896 *	153. Cpl Leanio, Frank S.	39839199 *
8. 1st Sgt Dumalaog, Benigno N.	39095173 *	154. Cpl Lopez, Arthur G.	39395298 ***
9. 1st Sgt Goloyugo, Victorio C.	36145937 *	155. Cpl Maglanit, Maximo V.	39397902 *
10. 1st Sgt Harder, Benjamin (NMI)	32423772 *	156. Cpl Malvar, Angelo B.	39538482
11. 1st Sgt Manzano,Aniceto C.	39171906 *	157. Cpl Mendoza, Primo T.	36356345
12. 1st Sgt Peralta, George E.	36023809 *	158. Cpl Nalangan, Restituto I.	39839845
13. 1st Sgt Ramos, Senen R.	39245810 *	159. Cpl Noel, Ralph B.	36382203
14. 1st Sgt Regala, Wilfred E.	39075555 ***	160. Cpl Oandasan, Angelo P.	36325336 *
15. 1st Sgt Santos, Fortunato (NMI)	33323790 *	161. Cpl Pascual, Amrafel D.	39683927 *
16. T/Sgt Campos, Henry A.	19148020	162. Cpl Peneyra, Cipriano C.	39839100
17. S/Sgt Agron, Alfredo R.	39838778 *	163. Cpl Pimentel, Lorenzo U.	39389581 *
18. S/Sgt Aguilar, Peter B.	36339127 *	164. Cpl Pulido, Mack H.	39840672 *
19. S/Sgt Amante, Thomas T.	39834215 *	165. Cpl Quezada, Antonio M.	36649313 *
20. S/Sgt Arias, Narciso T.	39843524	166. Cpl Quiming, Perfecto C.	36199212 ***
21. S/Sgt Azcueta, Zoil S.	39845469 *	167. Cpl Quindiagan, Luis C.	39105190 *
22. S/Sgt Bacus, Faustino B.	39845477 *	168. Cpl Quinitio, Basilio A.	39843138
23. S/Sgt Baniares, Arcangel	6735925 *	169. Cpl Rabang, Thomas V.	39526403 *
24. S/Sgt Barruga, Frank C.	39239271 **	170. Cpl Ramirez, Camilo H.	10641061
25. S/Sgt Bello, Cleto S.	39238037 *	171. Cpl Ramos, Julian G.	39836592 ***
26. S/Sgt Bolivar, Custodio P.	32531810 *	172. Cpl Ramos, Paul L.	10645019
27. S/Sgt Brillante, Leonardo T.	36398274	173. Cpl Respicio, Sergio P.	39836837
28. S/Sgt Caoili, Hermenegildo R.	39390597	174. Cpl Reyes, Felecisimo	39246393
29. S/Sgt Castillo, Moises C.	39242766	175. Cpl Rivera, Elpidio A.	36337587 *
30. S/Sgt Constantino, John R.	36380819 *	176. Cpl Robles, Romulo D.	33190611 *
31. S/Sgt Cortez, Luois (NMI)	36398313 *	177. Cpl Rodriguez, Juan A.	10641048 *
32. S/Sgt Delos Santos, Modesto P.	39245941 *	178. Cpl Sabado, Henry D.	39094281 *
33. S/Sgt Diputado, Esteban M.	32343475 *	179. Cpl Salamanca, Albert S.	37205594 **
34. S/Sgt Dongallo, Fernando L.	39844743	180. Cpl Santos, Isaac L.	39245489
35. S/Sgt Dolendo, Buenaventura L.	32233406 *	181. Cpl Solar, Marcelino V.	39406660
36. S/Sgt Francisco, Alberto C.	39103160 *	182. Cpl Tria, Antonio B.	39264160 *
37. S/Sgt Garcia, Vidal (NMI)	36345798 *	183. Cpl Tucay, Henry T.	39236923 *
38. S/Sgt Gaston, Meleseo M.	36337268	184. Cpl Valdez, Celestino R.	39842304 ***
39. S/Sgt Gregory, Rubin V.	32278916	185. Tec 5 Alvez, Beato C.	33202729 *
40. S/Sgt Gubatan, Severo D.	39235982	186. Tec 5 Bautista, Moises V.	33189145
41. S/Sgt Guzman, Larry O.	39234838 *	187. Tec 5 Baylon, Alfred S.	36398604
42. S/Sgt Javonillo, Juan O.	39093222	188. Tec 5 Buenavista, Serafin B.	39843064 *
43. S/Sgt Jubane, Valentine C.	39838455	189. Tec 5 Costales, Gene D.	37216159 *
44. S/Sgt Libatique, Sotero G.	39236076 *	190. Tec 5 Francisco, Urbano M.	39526401 *
45. S/Sgt Licera, Rafael C.	39843410 *	191. Tec 5 Javier, Bartolome E.	39537130 **
46. S/Sgt Lim, Alfred E.	39846111 *	192. Tec 5 Kintanar, Aniceto S.	35114187 *
47. S/Sgt Lucero, Florentino L.	39199289	193. Tec 5 Racho, Paciano S.	36345447 *
48. S/Sgt Luis, Monico B.	39685498	194. Tec 5 Rimando, Antonio C.	37156057 *
49. S/Sgt Mariano, Jose S.	39097444 ***	195. Tec 5 Ruiz, Ruperto (NMI)	39838264 *
50. S/Sgt Miguel, Cipriano L.	39154141 *	196. Tec 5 Sandoval, Nicolas E.	39843993 *
51. S/Sgt Pablo, Cipriano B.	39238786	197. Tec 5 Sagun, Fernando C.	39105188
52. S/Sgt Padilla, Louis P.	39842719 **	198. Tec 5 Soliven, Larry G.	39390506 *
53. S/Sgt Palmejar, Louis A.	39839102	199. Tec 5 Valido, Pastor W.	39841139 **
54. S/Sgt Paraiso, Ireneo V.	39236917	200. Pfc Abel, Servando A.	39237333
55. S/Sgt Pilar, Julian A.	36341812 *	201. Pfc Aguilar, Hermenegildo A.	10645056
56. S/Sgt Pinuela, Vicente A.	39238489 *	202. Pfc Alerta, Custodio P.	39391259
57. S/Sgt Quintua, Maximo T.	19161005 **	203. Pfc Alfafara, Gerardo C.	36342323 *
58. S/Sgt Ramirez, Jaime A.	32320293 ***	204. Pfc Antonio, Alfonso L.	39120977 *

59. S/Sgt Real, Isay M.	39095439	*
60. S/Sgt Regalado, Jerry V.	36345374	*
61. S/Sgt Rosete, Pantaleon M.	39174290	***
62. S/Sgt Ruiz, Don S.	39838780	
63. S/Sgt Sabiniano, Teodoreco S.	39307228	
64. S/Sgt Sanchez, Gerardo A.	32332823	*
65. S/Sgt Sison, Tony C.	39839111	**
66. S/Sgt Sitjar, Jose T.	39241984	*
67. S/Sgt Solidarios, Sergio V.	39838417	
68. S/Sgt Tabios, Val R.	32020950	*
69. S/Sgt Velasco, Roque M.	39839128	
70. S/Sgt Yotoko, Salvador L.	39845715	
71. Tec 3 Hilario, Carl G.	39241054	
72. Sgt Adia, Procopio M.	39238258	*
73. Sgt Alfabeto, Orlando A.	10645012	*
74. Sgt Alvarez, Vidal F.	39174517	*
75. Sgt Annaguey, Eleno D.	39246485	*
76. Sgt Ano, Gaudioso A.	39399878	
77. Sgt Baldos, Florencio A.	39176535	*
78. Sgt Basug, Gregorio J.	36306596	*
79. Sgt Bautista, Felino L.	39538438	***
80. Sgt Bolante, John A.	36332733	
81. Sgt Bucol, Segundo G.	39112603	*
82. Sgt Cacas, Maximino C.	32355986	*
83. Sgt Calix, Pascual M.	39244114	***
84. Sgt Capistrano, Alberto N.	39241197	*
85. Sgt Carpio, Jose P.	37290686	
86. Sgt Daelto, Marciano R.	36340286	*
87. Sgt Dela Pena, Arsenio B.	39604577	*
88. Sgt De Ocampo, Mero L.	39097470	*
89. Sgt Despy, Alfredo H.	39114329	
90. Sgt Ebarle, Patrick F. Jr.	39109132	
91. Sgt Esguerra, Jimmy R.	39284055	
92. Sgt Fontanos, Marcelo P.	39181438	*
93. Sgt Fortes, Melvin D.	39838794	
94. Sgt Francisca, Fermin M.	39236836	*
95. Sgt Gallemore, James R.	6580887	*
96. Sgt Garces, Pedro L.	39246607	*
97. Sgt Garcia, Marcello	36352933	*
98. Sgt Ginobiagon, Felipe R.	10645011	*
99. Sgt Jorge, Patricio D.	10641011	*
100. Sgt Lagrimas, Walter B.	39844692	*
101. Sgt Llapitan, Carmel A.	36601496	
102. Sgt Magbual, Felecisimo D.	39843914	*
103. Sgt Manuta, Eusebio J.	39844969	*
104. Sgt Mones, Venancio M.	39842206	*
105. Sgt Nastor, Domingo N.	39014924	
106. Sgt Paler, Luis P.	39532836	***
107. Sgt Pascua, Inocencio V.	39844828	*
108. Sgt Pugosa, Emil L.	39837296	*
109. Sgt Ramos, Jose J.	10641041	*
110. Sgt Rillera, Pastor A.	39838420	
111. Sgt Rinien, Eusebio T.	39244456	
112. Sgt Rodriguez, Ireneo R.	39244217	*
113. Sgt Rosales, Paulino A.	39396552	*
114. Sgt Salvador, Apolinar G.	39236112	*
115. Sgt Serrano, Agapito B.	39394362	
116. Sgt Talaro, Albert T.	39390649	***
117. Sgt Taluban, Teodoro M.	39842888	*
118. Sgt Turqueza, Arsenio A.	39181851	
119. Sgt Jesena, Carl	39096093	

205. Pfc Arcia, Narciso N.	39116432	
206. Pfc Arqueza, Don T.	36346436	
207. Pfc Arquero, Lasaro C.	39187222	*
208. Pfc Artiago, Maurice W.	39330914	***
209. Pfc Atad, Martiniano S.	39393185	***
210. Pfc Baker, Clinton M.	39429940	
211. Pfc Balunes, Ralph M.	39235966	*
212. Pfc Bugayong, Vicente V.	36356958	
213. Pfc Carriaga, Pedro A.	10641012	*
214. Pfc Casco, Lainvencion R.	39412852	
215. Pfc Clemente, Leo C.	39174917	*
216. Pfc Cuanang, Julian F.	39095248	*
217. Pfc Custodio, Emilio M.	39686249	***
218. Pfc Erana, Miguel B.	39837011	*
219. Pfc Esteban, Peter M.	39209680	
220. Pfc Flores, Agripino B.	39836185	***
221. Pfc Formoso Francisco H.	36337830	
222. Pfc Gaduang, Thomas D.	36625135	*
223. Pfc Gobleza, Dominador S.	10641049	*
224. Pfc Gonzales, Marcelo U.	39308995	*
225. Pfc Granados, Vicente J.	39846508	*
226. Pfc Itchon, Jay F.	39240629	
227. Pfc Kinanahan, Santiago	10647016	
228. Pfc Lascano, Leodegario (NMI)	36328888	***
229. Pfc Machon, Sergio M.	33194571	*
230. Pfc Martin, Philip (NMI)	34202018	*
231. Pfc NIebres, Federico J.	36355586	***
232. Pfc Ollada, Patricio (NMI)	36345911	*
233. Pfc Ouano, Francisco I.	39236026	
234. Pfc Pascua, Jerry D.	36359348	*
235. Pfc Pilar, Mariano F.	39405520	
236. Pfc Quiaoit, Esteban R.	39191353	***
237. Pfc Quinagon, Barney A.	39391096	*
238. Pfc Quismundo, Ruperto Q.	39201741	***
239. Pfc Respicio, Tom P.	39235454	*
240. Pfc Rosario, Gabriel Y.	39838412	
241. Pfc Sapio, Lucas J.	39033374	
242. Pfc Tolentino, John P.	36355508	*
243. Pfc Trocio, Perfecto N.	36601493	*
244. Pfc Tullo, Apollo P.	39253426	*
245. Pvt Altorino, Earle B.	39842418	
246. Pvt Abadiano, Reynaldo H.	10647176	
247. Pvt Adona, Nicolas C.	10645038	
248. Pvt Agbunag, Baltazar C.	39840188	*
249. Pvt Aguila, Isaac Z.	39031494	*
250. Pvt Alabanza, Joseph E.	39004210	
251. Pvt Aliacbar, Sahibad M.	10647057	*
252. Pvt Antikoll, Carlos F.	10647180	*
253. Pvt Arciaga, Emiliano D.	36651964	*
254. Pvt Azcueta, Pedro S.	36350540	*
255. Pvt Bairulla, Hadjula H.	10647053	*
256. Pvt Bairulla, Sabtal H.	10647051	*
257. Pvt Balderama, Eduardo G.	39245632	
258. Pvt Bentillo, Venancio L.	39238829	
259. Pvt Bermudez, Felix	32350341	*
260. Pvt Bernal, Jaime A.	31318657	*
261. Pvt Catalina, Eugenio S.	10647055	*
262. Pvt Cayabyab, Ricardo D.	39255604	*
263. Pvt Darang, Andres D.	39828848	*
264. Pvt De Leon, Benedict S.	39840292	
265. Pvt Delfun, Urbano P.	36734497	*

120. Tec 4 Abalos, Marcelo B.	39841113	***	266. Pvt Felipe, Ramon S.	39413717	
121. Tec 4 Bulatao, Benjamin C.	39841434		267. Pvt Fernandez, Leon C.	36707169	***
122. Tec 4 Cardenas, David (NMI)	39240845	*	268. Pvt Generales, Narciso C.	39541136	*
123. Tec 4 Gonzales, Marcelo G.	39843509	*	269. Pvt Ladjahasan, Ali H.	10647054	*
124. Tec 4 Mapile, Clemente B.	39838672	*	270. Pvt Ladjahasan, Jalil H.	10647052	*
125. Tec 4 Malic, Dominador (NMI)	10641050	*	271. Pvt Ladjahasan,Mangona H.	10647056	*
126. Tec 4 Paraiso, Miguel Q.	32232121	*	272. Pvt Lobendino, Hilario L.	39310376	*
127. Tec 4 Robles, Crispolo G.	39837857	*	273. Pvt Marquez, Victor V.	33737506	*
128. Tec 4 Santorio, Ceferino D.	36353496		274. Pvt Medrano, Demetrio G.	36606417	*
129. Tec 4 Villanueva, Angelo F.	39263304	*	275. Pvt Mendoza, Serge M.	39131199	
130. Cpl Agustin, Pedro P.	37156107	*	276. Pvt Minder, Robert C.	39229726	*
131. Cpl Alayu, Celestino D.	36600603		277. Pvt Modellas, Crisanto M.	10645037	
132. Cpl Almojera, Loren D.	39604988	*	278. Pvt Mosquera, Benigno F.	39402969	***
133. Cpl Anacleto, Vicente H.	6739381	*	279. Pvt Nerez, Manuel M.	39044285	*
134. Cpl Arellano, Martin Q.	39102486	*	280. Pvt Oania, Hermogenes C.	39839084	***
135. Cpl Ayson, Sammy L.	39413091		281. Pvt Peska, Martin Y.	39846550	
136. Cpl Baladad, Jose C.	36702550	*	282. Pvt Salvacion, Alfredo L.	39109014	*
137. Cpl Bareng, Gregorio F.	39841167	*	283. Pvt Santos, Alberto E.	39844995	Base
138. Cpl Bibat, Edilberto S.	36346416	*	284. Pvt Sarmiento, Florentino M.	32888314	
139. Cpl Cachapero, Frank G.	39844821		285. Pvt Savellano, Pedro E.	32355686	*
140. Cpl Casia, Melanio C.	39838878		286. Pvt Siose, Francisco E.	39837070	51GH
141. Cpl Corpuz, Eutacio C.	10301905	*	287. Pvt Tablante, Pedro	10644002	
142. Cpl Costales, Fermin D.	37216144	**	288. Pvt Valdivia, Enrique D.	39325758	
143. Cpl Curaming, Eleuterio B.	39103871		289. Pvt Tolentino, Zoilo Jr.	32870608	*
144. Cpl De Leon, Peter B.	39101202		290. Pvt Velayo, Benny T.	39106003	*
145. Cpl Dupitas, Aurelio M.	39845473		291. Pvt Wendam, Sidney S.	39028642	***
146. Cpl Fermin, Raymond R.	39039876	***			

---15TH WEATHER SQD (ATTACHED)---

1. T/Sgt Hoke, Max N.	16041689	*	11. Sgt Lowe, William H.	39036006	*
2. S/Sgt Chambless, Henry I.	7084115	*	12. Sgt Lozano, Ray J.	18155440	*
3. S/Sgt Finnegan, George J.	12139211	*	13. Sgt MacCarrick, Donald R.	11050775	
4. S/Sgt George, Robert R.	38119149	*	14. Sgt McGrath, Charles E.	13081616	*
5. S/Sgt Hammill, Charles H.	39034039	*	15. Cpl Bierly, James N. Jr.	13157755	*
6. S/Sgt Moore, Chester H.	19219454	*	16. Cpl Chick, Alton C. Jr.	11136180	*
7. S/Sgt Richardson, William F.	13155656	*	17. Cpl Gamertsfelder, Robert H.	35749997	*
8. Sgt Becker, William III	16168669	*	18. Cpl Herbig, Robert P.	35583726	*
9. Sgt Caswell, Robert O.	31283847	*	19. Cpl Males, William J.	16071526	*
10. Sgt Hendrickson, Murl S.	17088657	*			

For the Company Commencer: 1st Sgt N, BUENAVENTURA,

MEDICAL DETACHMENT
FIRST RECONNAISSANCE BATTALION, SPECIAL
UNITED STATES ARMY FORCES IN THE FAR EAST, APO 565

ROSTER AS OF 21 DEC 44
Capt O'Connor, Maurice N. 0349057 Det Comdr Temp dy fr BS-3 Dispensary

ENLISTED MEN
Tec 4 Diana, Serafin G. 36323869 SD fr Scout Co and Column
Pfc Barba, Conrad A. 39241805 SD fr Scout Co and Column
Pfc Calustro, Peter A. 2291087 SD fr Scout Co and Column
Pvt Dobios, Clement 32560741 Atchd to 1st Rcn Bn Sp fr 21st Ground
Obar Plat Co D 597th Sig AW Bn and plcd SD
Pvt Morgan, Charles M. 37609373 Atchd to 1st Rcn Bn Sp fr 21st Ground
Obar Plat Co D 597th Sig AW Bn and plcd SD
Pvt Salvador, Frank A. 39101922 SD fr 978th Sig Serv Company

*	Temporary duty in the Philippines
**	Medical
***	597th Signal Aircraft Warning

Source: MacArthur Archives, Norfolk Virginia

Appendix B - 15

Filipino Soldiers in Australia and New Guinea 1942 – 1945
Allied Intelligence Bureau, United States Army Forces in the Far East,
Philippine Regional Section, 5217th Reconnaissance Battalion (Provisional),
978th Signal Company, 1st Reconnaissance Battalion (Special)*

RANK-NAME	MEMO DATE**	LOCATED ON MEMO OR FROM***	DESTI-NATION	RANK-NAME	MEMO DATE**	LOCATED ON MEMO OR FROM***	DESTI-NATION
T5 Abad, Otilio V.	Aug44	Anniv Edition		Pvt Ladjahasan, Jalil H.	Jun43	PRS-5217th	Mindanao
Pvt Abadilla, Ray H.	Jun44	PRS-978th		Pvt Ladjahasan, Mangona H.	Apr43	PRS-5217th	Panay
T4 Abalos, Marcelo B.	Oct44	PRS-5217th-Demoltn	Negros-Cebu	Sgt Lagrimas, Walter B.	Dec44	PRS-5217th	Mindoro
Cpl Abalos, Remigio A.	Oct45	FilUnit-PCAU		T4 Lagpacan, Laureto A.	Mar44	PRS-5217th	
Pvt Abang, Bernardino B.	Jan45	1stRB-Paratrp-Radio		Pvt Lakibul, Nastail H.	Jun43	AIB	Mindanao
Lt Abanto, Rizalito	Sep44	PRS-5217th		Pvt Lapuente, Guillermo N.	Dec43	PRS-5217th	
T5 Abata, Salvador M.	Oct45	FilUnit-PCAU		Pvt Lascano, Leodegario	Sep44	PRS-5217th	Mindanao
Lt Abaya, Rosalino A.	May44	PRS-5217th	1stFilInf	Pvt Laurina, Filomeno A.	Jan45	PRS-978th-Signal	
Pvt Abel, Servando A.	Jan45	1stRB-Paratrp-Radio		Cpl Lazo, Enoch A.	May44	PRS-5217th	
Sgt Abellera, Antonio D.	Aug44	Anniv Edition		Cpl Lazo, Ernesto L.	Aug44	Anniv Edition	
Pvt Abraham, Ismael A.	Nov43	PRS-5217th		Cpl Leanio, Frank S.	Dec44	PRS-5217th	Mindoro
Sgt Abrenica, Santiago D.	Jan45	1stRB-Paratrp-Radio		Sgt Legaspi, Alex B.	Jan45	PRS-978th-Signal	
Cpl Abril, Godofredo	Nov43	AIB-USAFFE	5217thRB	Sgt Libatique, Sotero G.	Feb44	AIB	
Sgt Acenas, Seminiano D.	Jun44	PRS-978th	Mindanao	Pvt Liberato, Don J.	Aug44	PRS-978th-Signal	Luzon
Lt Acosta, Rafel S.	Sep42	USASOS		Sgt Licera, Rafael C.	Dec44	PRS-5217th	Mindoro
Sgt Adia, Procopio M.	Oct44	1stRB-Paratrp-Radio	Luzon	Lt Licudine, Raymond B.	May44	PRS-5217th	1stFilInf
T5 Adricula, Julian D.	Mar44	PRS-5217th	1stFilInf	Pvt Lideros, Bernardo D.	Nov42	USASOS	
Pvt Adona, Nicholas C.	Dec43	PRS-5217th		Sgt Lim, Alfred E.	Oct44	PRS-5217th-Demoltn	Luzon
Pvt Aduan, Feliciano C.	Dec43	PRS-5217th		Pvt Limbaga, Vicente	Nov42	USASOS	
T5 Advincula, Julio C.	Jun44	PRS-978th	Samar	Pvt Liunoras, Ignacio	Aug43	Hq.DetachGHQ	5217thRB
Pvt Afaro, Rufino	Nov42	USASOS		Sgt Llapitan, Carmel A.	Jan45	1stRB-Paratrp-Radio	
T3 Africa, Robino	Dec43	PRS-978th-Signal	Darwin	Pvt Lobendino, Hilario L.	Nov44	PRS-5217th-Radio	Luzon
Sgt Agacaoilo, Raymundo E.	Jun44	PRS-5217th	Samar	T4 Logan, Domingo L.	Jul44	PRS-978th	Mindoro
T4 Agbayani, Vicente A.	Nov44	PRS-978th-Signal	Leyte,Phil	Cpl Loma, Magdaleno S.	Jan45	1stRB-Paratrp-Radio	
Pvt Agbunag, Baltazar G.	Oct44	PRS-5217th-Radio	Luzon	Pvt Lopez, Apolonio E.	Dec43	PRS-5217th	
T5 Agcaoili, Julian S.	Dec43	PRS-5217th		Cpl Lopez, Arthur G.	Sep44	PRS-5217th	Mindanao
Sgt Agcaoilo, Raymundo E.	Jun44	PRS-978th	Samar	Cpl Lopez, Cirilo G.	Sep44	PRS-978th	Luzon
Pvt Aglubat, Al C.	Sep44	PRS-978th	Mindanao	Cpl Lopez, Felix R.	Aug44	Anniv Edition	
Lt Agpaoa, Padifico D.	Sep44	PRS-5217th		Sgt Lopez, Marcos T.	Jan44	1stRB-978th-Signal	
Sgt Agron, Alfredo R.	Aug44	PRS-5217th	Luzon	Pvt Loremas, Benito	Aug43	Hq.Co.USAFFE	5217thRB
Pvt Aguda, Benjamin C.	Jun44	PRS-5217th		Lt Lorenzo, Sebastian B.	Sep44	PRS-5217th	
Pvt Aguila, Isaac Z.	Jun44	PRS-5217th-Weathr	Samar	Cpl Lozano, Ray J.	Jun44	PRS-5217th-Weathr	Mindanao
Pvt Aguilar, Hermenegildo A.	Aug44	Anniv Edition		T5 Lucas, Benny G.	Jun44	PRS-5217th	
Sgt Aguilar, Peter B.	Jul44	PRS-5217th	Mindoro	Pvt Lucas, Felix M.	Dec43	PRS-5217th	
Cpl Agustin, Pedro P.	Oct44	PRS-5217th	Luzon	Sgt Lucero, Florentino L.	Nov43	AIB-USAFFE	5217thRB
Pvt Aholal, John A.	Nov44	PRS-978th-Signal	Leyte,Phil	SSgt Luis, Monico B.	Aug44	Anniv Edition	
Pvt Alabanza, Joseph E.	Jan45	1stRB-Paratrp-Radio		Sgt Luntao, Celestino D.	Mar44	PRS-5217th	
T5 Alaminiana, Amado O.	Oct45	FilUnit-PCAU		T4 Luz, Peter L.	Jun44	PRS-978th	Samar
Pvt Alayu, Celestino D.	Jan45	1stRB-Paratrp-Radio		Pvt Macabeo, Hospicio C.	Jun44	PRS-5217th	
Pvt Albalos, Segundo Q.	Sep44	PRS-978th-Signal	Luzon	Pvt Machon, Sergio M.	Sep44	PRS-978th	Luzon
Cpl Albano, Johnny J.	Nov43	PRS-5217th		Pvt Macilong, Ignacio	Nov43	AIB-USAFFE	5217thRB
T4 Alberto, Alfred A.	Nov43	PRS-978th-Signal	Mindoro	Pvt Madali, Henry M.	Dec43	PRS-5217th	
Pvt Alberto, Eligio S.	Nov43	PRS-5217th		Sgt Madarang, Jose S.	Mar44	PRS-5217th	
Cpl Alcala, Atanasio L.	Nov44	PRS-978th-Signal	Leyte,Phil	Pvt Madariaga, Fred M.	Jul44	PRS-978th	Leyte
Pvt Alcantara, Marcelino C.	Apr44	PRS-978th		Pvt Madura, Edward J.	Apr44	PRS-978th-Signal	
Pvt Alcaras, Juan D.	Jan45	PRS-978th-Signal		Pvt Magale, Isidro P.	Jun44	PRS-978th	Mindanao
T5 Alcoy, Manuel B.	Oct45	FilUnit-PCAU		Sgt Magbual, Felecisimo D.	Dec44	PRS-5217th	Mindoro
T5 Alegre, Emilio A.	Nov44	PRS-978th-Signal	Leyte,Phil	Pvt Magday, Primitivo V.	Mar44	PRS-5217th	
Pvt Alerta, Custodio P.	Jan45	1stRB-Paratrp-Radio		Pvt Magno, Ramon	Nov42	USASOS	
Sgt Alfabeto, Orlando	Apr43	PRS-5217th	Panay	Pvt Magsino, Eddie P.	Nov43	PRS-5217th	
Pfc Alfafara, Gerardo G.	Dec44	PRS-5217th	Mindoro	Lt Malan, Jose B.	Sep44	PRS-5217th	Luzon

Name	Date	Unit	Location
Pvt Alfaro, Rufino	Aug43	ARC-Base	5217thRB
Pvt Aliacbar, Sahibad M. Pfc	Jun43	AIB	Mindanao
Alicante, Macario A.	Oct45	FilUnit-PCAU	
Lt Allen, Robert C.	Aug44	Anniv Edition	
Sgt Almero, Emiliano S.	Sep44	PRS-978th	Samar
Cpl Almojera, Loren N.	Oct44	PRS-5217th-Radio	Luzon
Pvt Alohan, Disocoro R.	Nov42	USASOS	
Pvt Altorino, Earle B.	Dec43	PRS-5217th	
Pvt Alvarado, John R.	Apr44	PRS-978th	
Sgt Alvarez, Vidal F.	Jul44	PRS-5217th-Weathr	Mindoro
T5 Alvez, Beato C.	Nov44	PRS-5217th	Luzon
Sgt Amante, Thomas T.	Oct44	PRS-5217th-Demoltn	Luzon
Pvt Ambrosio, Nicanor G.	Sep44	PRS-978th-Signal	Luzon
Pvt Amen, Ladislao M.	Nov42	USASOS	
Lt Ames, Irineo A.	Aug43	AIB	Panay
Sgt Amodia, Susano B.	May43	AIB	5217thRB
Cpl Anacleto, Vicente H.	Sep44	PRS-5217th	Luzon
Sgt Ancheta, Alfonso	Dec43	PRS-978th-Signal	Darwin
Sgt Ancheta, Andres	Oct45	FilUnit-PCAU	
Lt Ancheta, Carlos F.	Jun44	PRS-5217th	Samar
Cpl Anderson, Moses A.	Jan45	1stRB-Paratrp-Radio	
Pvt Andres, Manuel V.	Nov44	PRS-978th-Signal	Leyte,Phil
T5 Angcos, Robert C.	Jun44	PRS-978th	Samar
Sgt Ano, Gaudioso A.	Dec43	PRS-5217th	
Sgt Annaguey, Eleno D.	Oct44	PRS-5217th-Demoltn	Luzon
Pvt Antikoll, Carlos F.	Oct44	PRS-5217th-Demoltn	Negros-Cebu
Pvt Antonio, Alfonso L.	Nov44	PRS-5217th	
Pvt Aquino, Alejandro C.	Aug44	PRS-978th	Mindanao
Pvt Aquino, Bernardino P.	Jun44	PRS-978th	Darwin
Lt Aquino, Enrique J.	May44	PRS-5217th	1stFilInf
Pvt Aranez, Catalino	Nov42	USASOS	
Sgt Arbis, Mariano V.	Jan45	1stRB-Paratrp-Radio	Samar-Bicol
Lt Arca, Ricardo T.	Jun44	PRS-5217th	
T5 Arce, George A.	Nov44	PRS-978th-Signal	Leyte,Phil
Sgt Arce, Mariano T.	Sep44	PRS-978th-Signal	Luzon
Pvt Arcia, Narciso N.	Jan45	1stRB-Paratrp-Radio	
Pvt Arciaga, Emiliano	Oct44	PRS-5217th	Luzon
Pvt Arcilla, Gaudencio G.	Nov42	USASOS	
Capt Arechavala, Mariano	Sep42	USASOS	
Lt Arellano, Arsenio Y.	Aug44	Anniv Edition	
Cpl Arellano, Martin Q.	Nov44	PRS-5217th	Luzon
Lt Arguelles, Carlos D.	Aug44	5217th-1stRB	
Pvt Argueza, Don T.	Jan45	1stRB-Paratrp-Radio	
Sgt Arias, Narciso T.	Jan45	1stRB-Paratrp-Radio	
Pvt Arquero, Lacaro C.	Dec44	PRS-5217th	Mindoro
Pvt Arqueza, Don T.	May44	PRS-5217th	
Cpl Arreola, Licerio A.	Jul43	PRS-5217th	
Pvt Artiago, Maurice W.	Oct44	PRS-5217th	Negros-Cebu
Pvt Astronomo, Telesforo R.	Aug44	PRS-978th	Woendi
Pvt Asuncion, Telesforo T.	Aug44	PRS-978th-Signal	Luzon
Pvt Atad, Matiniano S.	Oct44	PRS-5217th-Demoltn	Negros-Cebu
T4 Aurelio, Mariano L.	Feb44		
Pvt Awing, Sergio	Nov42	USASOS	
Cpl Ayson, Sammy L.	Jan45	1stRB-Paratrp-Radio	
Pvt Azcueta, Pedro S.	Dec44	PRS-5217th	Mindoro
Sgt Azcueta, Zoilo S.	Nov44	PRS-5217th-Demoltn	Luzon
Sgt Bacus, Faustino B.	Sep44	PRS-5217th	Luzon
Pvt Badilla, Porfirio	Aug43	Hq.Co.,GHQ	5217thRB
Cpl Baggao, Marino B.	Mar44	PRS-5217th	
Pvt Bairulla, Hadjula H.	Mar43	PRS-5217th	Mindanao
Cpl Malangit, Maximo V.	Dec44	PRS-5217th	Mindoro
T4 Malic, Dominador Sgt	Jan43	AIB AIB	Negros
Malinab, Cipriano B.	Feb44		
Sgt Malinab, Lamberto C.	Sep44	AIB-USAFFE	5217thRB
Pvt Malvar, Angelo M.	Jan45	1stRB-Paratrp-Radio	
Sgt Manalese, Hermias	May43	AIB	5217thRB
Pvt Manalo, Armand	Jan45	PRS-978th-Signal	
Pvt Manuel, Crispin S.	Dec43	PRS-5217th	
Sgt Manuta, Eusebio J.	Oct44	PRS-5217th-Demoltn	Negros-Cebu
Sgt Manzano, Aniceto C.	Dec43	PRS-5217th	Mindanao
Pfc Mapalo, Doroteo Q.	Sep44	PRS-5217th	CivilAffair
Pvt Mapalo, Juan D.	Mar44	PRS-5217th	
Pvt Mapalo, Martin L.	Sep44	PRS-978th-Signal	Mindanao
T4 Mapile, Clemente P.	Nov44	PRS-5217th-Demoltn	Luzon
Pvt Marapao, Simeon P.	Nov44	PRS-978th-Signal	Leyte,Phil
Pvt Maravel, Antonio L.	Jan45	1stRB-Paratrp-Radio	
Pvt Marcelino, Juan C.	Sep44	PRS-978th-Signal	Luzon
Pvt Marcelino, Teodoro	1942	41stMedical-Bn	
Pvt Marcelo, Benjamin R.	Sep44	PRS-978th	Mindanao
Lt Marfori, Edmundo J.	Aug44	5217th-1stRB	
Sgt Mariano, Joe S.	Oct44	PRS-5217th	Negros-Cebu
T5 Marinas, Toby L.	Nov44	PRS-978th-Signal	Leyte,Phil
Pvt Marquez, Victor V.	Dec44	PRS-5217th	Mindoro
T4 Marquina, Leo	Aug44	PRS-978th-Signal	Palawan
Pvt Marte, Gonzalo S.	Sep44	PRS-5217th	CivilAffair
Pvt Martin, Estanislao C.	Sep44	PRS-5217th	Mindanao
Pvt Martin, Isidoro C.	Jan44	PRS-978th	
Pvt Martin, Philip	Dec44	PRS-5217th	Mindoro
Pvt Martinez, Antonio	Mar42	USAFIA	
Pvt Martinez, Chris M.	Aug44	Anniv Edition	
T5 Martinez, Gervasio M.	Oct45	FilUnit-PCAU	
Lt Martinez, Gregorio M.	May43	AIB	5217thRB
Pvt Martinez, Pio M.	Mar44	PRS-5217th	
Pvt Marzan, Eleuterio E.	Mar44	PRS-5217th	
Pvt Marzan, Ricarte S.	Jan45	1stRB-Paratrp-Radio	
Pvt Mateo, Carl M.	Sep44	PRS-978th-Signal	Luzon
Sgt Mauricio, Abamilo L.	Nov44	PRS-978th-Signal	Leyte,Phil
Pvt Mauricio, Marcelo M.	Mar44	PRS-5217th	
Lt Mauricio, Paul F.	Jun44	PRS-5217th	Samar
T4 Mayor, Gilbert	Oct45	FilUnit-PCAU	
Maj McMicking, Joseph	1942	AIB	
Capt Medina, Luis	Sep42	USASOS	
T4 Medina, Mariano A.	Sep44	PRS-978th-Signal	Mindanao
Pvt Medrano, Bonifacio	Nov43	AIB-USAFFE	5217thRB
Pvt Medrano, Demetrio G.	Nov44	PRS-5217th	Luzon
Pvt Mejia, Fulgencio L.	Jan45	PRS-978th-Signal	
Lt Mendoza, Don	May44	PRS-5217th	1stFilInf
Lt Mendoza, Jose T.	Jun44	PRS-5217th	Samar
Cpl Mendoza, Primo T.	Jan45	1stRB-Paratrp-Radio	
Pvt Mendoza, Serge M.	Jan45	1stRB-Paratrp-Radio	
Cpl Mensalvas, Julio D.	Jul43	PRS-5217th	
Cpl Merca, Juan A.	Aug44	Anniv Edition	
Sgt Miguel, Cipriano L.	Jun44	PRS-978th-Signal	Samar
Pvt Miguel, Matias L.	Mar44	PRS-5217th	
Pvt Milana, Philip G.	Jan45	1stRB-Paratrp-Radio	
Pvt Minder, Robert C.	Dec44	PRS-5217th	Mindoro
Sgt Minodin, Gregorio B.	Oct45	FilUnit-PCAU	
Pvt Misa, Luis	Nov42	USASOS	
Pvt Misa, Ricardo	Nov42	USASOS	
Pvt Modellas, Crisanto	Jan45	1stRB-Paratrp-Radio	

Name	Date	Unit	Location
Pvt Bairulla, Sabtal H.	Mar43	PRS-5217th	Mindanao
Cpl Baladad, Jose C.	Nov44	PRS-5217th-Demoltn	Luzon
Pvt Balderama, Edwardo G.	Jan45	1stRB-Paratrp-Radio	
Sgt Baldos, Florencio A.	Sep44	PRS-5217th	Luzon
Sgt Balcena, Fred	Sep44	PRS-978th-Signal	Luzon
Cpl Balino, Porfirio	Sep44	PRS-978th-Signal	Luzon
Pvt Balinton, Vicente	Sep44	PRS-978th-Signal	Mindanao
Sgt Balintong, Flaviano	Jul44	PRS-978th	
Sgt Balleras, Julio V.	Jul44	PRS-978th	Mindoro
Pvt Balunes, Ralph M.	Aug44	PRS-5217th-Weathr	Luzon
Sgt Baniares, Arcangel	Nov43	PRS-5217th	Mindoro
Pvt Barba, Conrad A.	Feb44	PRS-5217th	
Cpl Bareng, Gregorio F.	Oct44	PRS-5217th-Radio	Luzon
Pvt Bargo, Querobin B.	Dec43	PRS-978th-Signal	Mindanao
Pvt Baring, Ricardo D.	Nov42	USASOS	
Sgt Barruga, Frank C.	Feb44	PRS-5217th	
Cpl Barcenas, Urbano B.	Jul44	PRS-978th	Mindoro
Pvt Basilio, Platon L.	Mar44	PRS-5217th	
Sgt Basug, Gregorio J.	Oct44	PRS-5217th-Demoltn	Luzon
Pvt Batara, Lope G.	Jun44	PRS-978th	
Pvt Bautista, Apolinario	Nov43	AIB-USAFFE	5217thRB
Sgt Bautista, Felino L.	Sep44	PRS-5217th	Mindanao
Pvt Bautista, Mariano L.	Mar44	PRS-5217th	
T5 Bautista, Moises V.	Jan45	1stRB-Paratrp-Radio	
Lt Bautista, Salvador R.	Sep42	USASOS	
T5 Baylong, Alfred S.	Jan45	1stRB-Paratrp-Radio	
Sgt Begonia, Daniel B.	Jul44	PRS-978th-Signal	Mindoro
Sgt Bello, Cleto S.	Dec44	PRS-5217th	Mindoro
Pvt Bentillo, Venancio L.	Jan45	1stRB-Paratrp-Radio	
Pvt Bermudez, Felix	Nov44	PRS-5217th-Demoltn	Luzon
Pvt Bernal, Jaime A.	Oct44	PRS-5217th-Demoltn	Luzon
Sgt Bernardo, Manuel P.	Oct45	FilUnit-PCAU	
Sgt Besid, Restituto J.	Dec43	PRS-5217th	Mindanao
Sgt Biado, Anacleto	Jul44	PRS-978th	Leyte
Cpl Bibat, Edilberto S.	Sep44	PRS-5217th	Luzon
Pvt Bibuya, Jose F.	Nov44	PRS-978th-Signal	Leyte,Phil
Sgt Bolante, John A.	Jan45	1stRB-Paratrp-Radio	
Sgt Bolivar, Custodio F.	Jun44	PRS-5217th	Negros
T4 Boncavil, Domiciano O.	Sep44	PRS-978th-Signal	Mindanao
Pvt Bongar, Fermin	Nov43	AIB-USAFFE	5217thRB
Pvt Borce, Dionicio G.	Dec43	PRS-5217th	
Sgt Borillo, Pio B.	Jul44	PRS-978th	Mindoro
Pfc Brigabriel, Ceferino	Apr44	PRS-5217th	
Sgt Brillante, Leonardo T.	Jan45	1stRB-Paratrp-Radio	
Pvt Buangsi, Kahil	Feb44	PRS-5217th	
Cpl Buccat, Alejandro A.	Jun44	PRS-978th	Darwin
Capt Buenaflor, Narciso	Sep42	USASOS	
Sgt Bucol, Segundo G.	Oct44	PRS-5217th-Radio	Luzon
1st Sgt Buenaventura, Nicolas	Aug44	Anniv Edition	
T5 Buenavista, Serafin B.	Jun44	PRS-978th-Signal	Mindanao
Pvt Buencamino, Felino D.	Aug42	Dutch-Ship	5217thRB
Sgt Bueno, George B.	Sep44	PRS-978th	Samar
Cpl Bugarin, Enrico M.	Jul44	PRS-978th	Mindoro
Pvt Bugayong, Vicente V.	Jan45	1stRB-Paratrp-Radio	
Pvt Bulat, Edward	May44	PRS-978th	
T4 Bulatao, Benjamin C.	Jan45	1stRB-Paratrp-Radio	
Pvt Bulatao, George M.	Jun44	PRS-978th-Signal	Samar
Pvt Bumanglag, Arsenio B.	Jan45	PRS-978th-Signal	
T5 Burgos, Edward M.	Jun44	PRS-978th	Mindanao
Sgt Cababa, Marcel R.	Oct45	FilUnit-PCAU	
Sgt Mones, Venancio M.	Nov44	PRS-5217th-Radio	Luzon
Sgt Montero, Al B. Sgt	Feb44	PRS-5217th	
Montero, Jack	Jun44	978th	Samar
Lt Montesclaros, Melecio J.	Sep44	PRS-5217th	
Pvt Morales, Catalino U.	Jan45	PRS-978th-Signal	
Capt Morgan, Luis P.	May44	PRS-5217th	1stFilInf
Pvt Moscoso, Anselmo S.	Aug44	PRS-978th-Signal	Luzon
Pvt Moscoso, Honorio C.	1942	2ndFilInf	
Pvt Mosquera, Benigno F.	Oct44	PRS-5217th	Negros-Cebu
Pvt Mundo, Francisco G.	Jun44	PRS-5217th	
Sgt Nagtalon, Alex A.	Nov44	PRS-978th-Signal	Leyte,Phil
Pvt Nalangan, Restituto L.	Jan45	1stRB-Paratrp-Radio	
Sgt Nastor, Domingo N.	May44	PRS-5217th	
Cpl Nava, David U.	Mar44	PRS-5217th	
Cpl Navarro, Ambrosio A.	Jun44	PRS-978th	Mindanao
Pvt Nazareno, Considenino	Nov42	USASOS	
T5 Nebre, Namaso	Apr44	PRS-978th	
T5 Nepomuceno, Larry A.	Oct45	FilUnit-PCAU	
Pvt Nerez, Manuel M.	Sep44	PRS-5217th	Mindanao
Sgt Nery, Gerardo B.	Jun44	PRS-978th	Samar
Pvt Niebres, Frederico J.	Sep44	PRS-5217th	Mindanao
Sgt Nietes, Iluminado	Apr44	PRS-978th	
Sgt Nocon, Alejandro L.	Jul43	PRS-5217th	
Cpl Noel, Ralph B.	Nov43	PRS-5217th	
T5 Nuevo, Leodegario O.	Jun44	PRS-978th	Samar
Sgt Nunez, Ricardo O.	Oct45	FilUnit-PCAU	
Cpl Oandasan, Angelo P.	Sep44	PRS-5217th	Luzon
Pvt Oania, Hermogenes C.	Oct44	PRS-5217th	Negros-Cebu
Lt Obedoza, Benito Y.	Mar42	AIB	
Pvt Obrero, Dick V.	Jul44	PRS-978th	Darwin
Pfc Obrero, Feliciano P.	Oct45	FilUnit-PCAU	
T5 Ocampo, Jack R.	Apr44	PRS-5217th	
Pvt Ogama, Donata L.	Nov42	USASOS	
Pvt Ogoy, Dale Y.	Sep44	PRS-978th	Mindanao
T4 Olivares, Felipe U.	Oct45	FilUnit-PCAU	
Pvt Ollada, Patricio	Aug44	PRS-5217th	Luzon
Pvt Ordonez, Marcelo	Sep44	PRS-5217th	CivilAffair
T5 Ortiz, Johnny L.	Oct45	FilUnit-PCAU	
Sgt Ortiz, Paterno A.	Mar44	PRS-978th-Signal	Mindanao
Pvt Ouano, Francisco I.	Jan45	1stRB-Paratrp-Radio	
Pvt Ovalles, Marcelo C.	Feb44	PRS-5217	
SSgt Pable, C. E.	Aug44	Anniv Edition	
Sgt Pablo, Cipriano B.	Jan45	1stRB-Paratrp-Radio	
Pvt Pabros, Ambrosio A.	Nov44	PRS-978th	Luzon
Pvt Pacificar, Nicomedes	Jun44	PRS-78th	Darwin
Cpl Padgett, Henry L.	Feb44	PRS-5217th	
Sgt Padilla, Balbino R.	Aug44	PRS-978th	Mindanao
Pvt Padilla, Bartolome M.	Dec43	PRS-978th-Signal	Darwin
Lt Padilla, Luis P.	Jun44	PRS-5217th	Samar
Pvt Paez, Artemio D.	Nov43	AIB-USAFFE	5217thRB
Lt Paguirigan, Francisco	Sep44	PRS-5217th	
Pvt Pagunsan, Vivencio	Nov42	USASOS	
T4 Pajara, George T.	Jun44	PRS-978th	Darwin
Pvt Palac, Juan M.	Sep44	PRS-978th-Signal	Luzon
T5 Palafox, Rosendo M.	Sep44	PRS-978th-Signal	Luzon
Pvt Palarca, Marcelino	Nov42	USASOS	
Sgt Paler, Luis B.	Sep44	PRS-5217th	Mindanao
Sgt Palmejar, Louis A.	Apr44	PRS-5217th	
T5 Palmeto, Daniel S.	Jun44	PRS-978th	Mindanao
Pvt Panes, Rafael F.	Dec43	PRS-5217th	

Name	Date	Unit	Location
1st Sgt Cabais, Eustaquio	Aug44	5217th	Palawan
Pvt Caballes, Philip A.	Dec43	PRS-5217th	
Lt Cabangbang, Bartolome C.	Sep44	PRS-5217th	Luzon
T4 Cabrales, Frank T.	Oct45	FilUnit-PCAU	
Pvt Cabrera, Remy B.	Jun44	PRS-978th-Signal	
Sgt Cacabelos, Rufino F.	Jan45	PRS-978th-Signal	
Sgt Cacas, Maximino C.	Sep44	PRS-5217th	Luzon
Pvt Cachapero, Frank G.	Aug44	Anniv Edition	
Cpl Cacho, Thomas S.	Mar44	PRS-5217th	
Pvt Caduang, Julian F.	Dec44	PRS-5217th	Mindoro
Pvt Calaustro, Pedro A.	Aug44	Anniv Edition	
Pvt Caliboso, Mariano S.	Dec43	PRS-5217th	
Pvt Calip, Aproniano E.	Jun44	PRs-978th	Darwin
T4 Calmorin, Angelo A.	Sep44	PRS-978th-Signal	Luzon
Sgt Campos, Henry A.	Apr44	PRS-978th	
Pvt Campos, Norberto C.	Jun44	PRS-978th	
Lt Canapi, Anselmo P.	Oct45	FilUnit-PCAU	
Sgt Candelario, Jose E.	Jan45	PRS-978th-Signal	
Sgt Caoili, Hermenegildo R.	Jan45	1stRB-Paratrp-Radio	
Sgt Capistrano, Alberto A.	Oct44	PRS-5217th-Radio	Luzon
Lt Capiz, Pascual M.	Sep44	PRS-5217th	Mindanao
Lt Carag, Vicente D.	May44	PRS-5217th	1stFilInf
Lt Carbonell	Jan45	PRS-5217th	Samar-Bicol
T4 Cardenas, David	Dec43	PRS-5217th	Mindanao
T4 Carillo, Patrocinio A.	Jul44	PRS-978th	Mindoro
Pvt Carpio, Amado G.	Nov43	AIB-USAFFE	5217thRB
Sgt Carpio, Jose P.	Jan45	1stRB-Paratrp-Radio	
Pvt Carrasca, Romain S.	Nov44	PRS-978th-Signal	Leyte,Phil
Pvt Carriaga, Pedro A.	Jul43	PRS-5217th	Negros
Pvt Casco, Lainvencion R.	Jan45	1stRB-Paratrp-Radio	
Cpl Casia, Melanio C.	Jan45	1stRB-Paratrp-Radio	
Sgt Castillo, Moises C.	Jan45	1stRB-Paratrp-Radio	
SSgt Castro, Ciriaco	Aug44	Anniv Edition	
Sgt Castro, Ray F.	Jan45	PRS-978th-Signal	
Pvt Catalina, Eugenio S.	Mar43	PRS-5217th	Mindanao
Cpl Catayas, Gerino C.	Oct45	FilUnit-PCAU	
Pvt Cayabyab, Adolfo V.	Jun44	PRS-978th	Darwin
Pvt Cayabyab, Ricardo D.	Dec44	PRS-5217th	Mindoro
T5 Cendania, Amado C.	Sep44	PRS-5217th	Samar
Maj Cisneros, Rafael J.	Aug44	Anniv Edition	
Pvt Clemente, Leo C.	Nov44	PRS-5217th-Demoltn	Luzon
Pvt Colmenares, Jorge A.	Nov42	USASOS	
Sgt Concepcion, Eufracio F.	Mar44	PRS-5217th	
Cpl Concepcion, Nemesio B.	Feb44	PRS-5217th	
Pvt Consolacion, Fermin R.	Nov43	AIB-Radio	5217thRB
Sgt Constantino, John R.	Oct44	PRS-5217th-Demoltn	Luzon
Sgt Corpus, Amado S.	Jun44	PRS-5217th	Palawan
Sgt Corpus, Eustacio C.	Sep44	PRS-5217th	Luzon
Cpl Corpuz, Anastacio G.	Feb44	PRS-5217th	
Sgt Corpuz, Aquilino M.	Nov44	PRS-978th-Signal	Leyte,Phil
Pvt Cortez, Robert B.	Sep44	PRS-978th-Signal	Luzon
Pvt Cortez, Calixto T.	Sep44	PRS-978th-Signal	Mindanao
Sgt Cortez, Louis	Mar44	PRS-5217th	Mindanao
Sgt Cortez, Ramon S.	Jun44	PRS-978th	Palawan
Cpl Costales, Fermin D.	Jan45	1stRB-Paratrp-Radio	
T5 Costales, Gene D.	Dec44	PRS-5217th	Mindoro
Pvt Coyoca, Albaro D.	Nov42	USASOS	
Lt Crespo, Toribio	Apr43	AIB	Panay
Pvt Cruz, Jose Y.	Nov43	AIB-USAFFE	5217thRB
Pfc Cuanang, Julian F.		PRS-5217th	
Sgt Paraiso, Irineo V.	Jan45	1stRB-Paratrp-Radio	
Sgt Parcasio, Joe B.	Feb44	PRS-5217th	
Cpl Parong, Astor C.	Sep44	PRS-978th-Signal	Luzon
Sgt Pascua, Inocencio F.	Jul44	PRS-5217th	Mindoro
Pvt Pascua, Jerry D.	Jun44	PRS-5217th-Weathr	Samar
Cpl Pascual, Ambrosio D.	Dec43	PRS-5217th	
Cpl Pascual, Amrafel D.	Oct44	PRS-5217th-Radio	Luzon
T5 Pascual, Angelo E.	Jul44	PRS-978th	Mindoro
Sgt Pastor, Benjamin A.	Mar44	PRS-5217th	
Pvt Patacsil, Regino T.	Sep44	PRS-978th	Mindanao
Pvt Patubo, Gaudioso L.	Sep44	PRS-978th	Mindanao
Pvt Paz, Rafael A.	Dec43	PRS-5217th	
Pfc Pedro, Florendo R.	Jul44	PRS-978th	Hollandia
T5 Pena, Doroteo M.	Jan45	PRS-978th-Signal	
Pvt Penalver, Pedro R.	Jan45	PRS-978th-Signal	
Cpl Peneyra, Cipriano C.	Jan45	1stRB-Paratrp-Radio	
Pvt Peralta, Alfonso A.	Mar44	PRS-5217th	
Pvt Peralta, Estanislao C.	Aug44	PRS-978th-Signal	Luzon
Sgt Peralta, George E.	Apr44	PRS-5217th	
Cpl Peralta, Materno P.	Dec43	PRS-978th-Signal	Darwin
Cpl Peralta, Maximo	Nov43	AIB-USAFFE	5217thRB
Pvt Peredo, Joe E.	Dec43	PRS-978th-Signal	Darwin
Sgt Perez, Pedro M.	May44	PRS-5217th	
Pvt Peska, Martin Y.	Dec43	PRS-5217th	
Maj Phillips, Lawrence H.	Nov43	PRS-5217th	Mindoro
Pvt Piana, Victor L.	Jun44	PRS-978th-Signal	Darwin
Sgt Pilar, Julian A.	Oct44	PRS-5217th-Demoltn	Luzon
Pvt Pilar, Mariano P.	Aug44	Anniv Edition	
Pvt Pilipina, Leon P.	Jun44	PRS-5217th	
Cpl Pimentel, Lorenzo U.	Aug44	PRS-5217th	Luzon
Sgt Pinuela, Vicente A.	Nov43	PRS-5217th	Mindoro
Pvt Pio, Sonico	Aug43	HqCo.,Base	5217thRB
Sgt Placido, Carlos S.	Jun44	PRS-978th	Palawan
Cpl Pobre, Felipe R.	Sep44	PRS-5217th	CivilAffair
Lt Pojas, Juan M.	May43	AIB	5217thRB
Cpl Pomarez, Leo	Feb44	PRS-5217th	
Lt Cpl Pompea, Edward T.	Sep44	PRS-5217th	Mindanao
Sgt Pugosa, Emil L.	Sep44	PRS-5217th	Samar
Pvt Pulanco, Luis M.	Dec43	PRS-5217th	
Cpl Pulido, Mack H.	Oct44	PRS-5217th-Radio	Luzon
Pvt Punzalan, Ray M.	Nov43	AIB-Radio	5217thRB
Pvt Quanico, Isidro B.	Nov43	AIB-USAFFE	5217thRB
T4 Quejarro, Ponciano B.	Nov43	AIB	5217thRB
Sgt Querubin, Gregorio V.	Nov43	AIB-USAFFE	5217thRB
T5 Querubin, Patricio C.	Mar44	PRS-5217th	
Cpl Quesada, Antonio M.	Oct44	PRS-5217th-Demoltn	Negros-Cebu
Pvt Quiaoit, Esteban R.	Oct44	PRS-5217th	Negros-Cebu
Sgt Quilala, Pelagio P.	Apr44	PRS-5217th	
T3 Quimba, Gregorio P.	Oct45	FilUnit-PCAU	
Cpl Quiming, Perfecto C.	Oct44	PRS-5217th	Negros-Cebu
Pvt Quinagon, Barney A.	Oct44	PRS-5217th-Radio	Luzon
Cpl Quindiagan, Luis C.	Sep44	PRS-5217th	Luzon
Cpl Quinitio, Basilio A.	Jan45	1stRB-Paratrp-Radio	
Lt Quinto, Emilio F.	Jan43	AIB-Radio	Negros
SSgt Quintua, Maximo T.	Aug44	Anniv Edition	
Sgt Quitoriano, Estanislao Q	Aug44	Anniv Edition	
Cpl Rabang, Thomas V.	Oct44	PRS-5217th-Radio	Luzon
T5 Racho, Paciano S.	Sep44	PRS-978th	Luzon
Pvt Ragupos, Geronimo R.	Mar44	PRS-5217th	
Cpl Rallojay, Teodoro J.	Jun44	PRS-978th	Palawan

Cpl Curaming, Eleuterio B.	Aug44	Anniv Edition	
Pvt Custodio, Emilio M.	Oct44	PRS-5217th	Negros-Cebu
Sgt Cutaran, Jacinto	Aug44	PRS-978th-Signal	Palawan
Pvt Dackel, Moses M.	Oct45	FilUnit-PCAU	
T5 Dacones, Ponciano S.	Oct45	FilUnit-PCAU	
MSgt Dacquel, Isidro D.	Jun44	5217th	Panay
T5 Dacquel, Richie D.	Jun44	PRS-978th	Palawan
Sgt Daelto, Marciano R.	Aug44	PRS-5217th	Mindanao
T5 Dagandan, Fortunato C.	Aug44	PRS-978th-Signal	Palawan
SSgt Dahilig, Juan F.	Aug44	Anniv Edition	
Pvt Dalida, Ignacio B.	Nov43	AIB-USAFFE	5217thRB
Sgt Dalipe, Benito S.	Oct45	FilUnit-PCAU	
Pvt Darang, Andres D.	Dec44	PRS-5217th-Weathr	Mindoro
Pvt David, Ignacio V.	May44	2ndFilInf	5217thRB
Pvt Davin, Numeriano D.	Nov44	PRS-978th-Signal	Leyte
Pvt Dawal, Leoncio	Nov43	AIB	5217thRB
Cpl De Garcia, Henry A.	Oct45	FilUnit-PCAU	
Sgt De La Pena, Arsenio B.	Nov44	PRS-5217th	Luzon
Pvt De La Pena, Geronimo M.	Sep44	PRS-978th	Mindanao
Pvt De Lara, Enrico	Mar44	PRS-5217th	
Pvt De Lara, Mauricio	Mar44	PRS-5217th	
Pvt De Leon, Benedict S.	Jan45	1stRB-Paratrp-Radio	
T4 De Leon, Benjamin C.	Jan45	1stRB-Paratrp-Radio	
Cpl De Leon, Peter B.	Jan45	1stRB-Paratrp-Radio	
Sgt De Los Santos, Modesto P.	Oct44	PRS-5217th-Demoltn	Luzon
Sgt De Ocampo, Mero L.	Dec44	PRS-5217th	Mindoro
Pvt De Peralta, Mauro G.	Jun44	PRS-978th	
T5 De Samito, Modesto A.	Oct45	FilUnit-PCAU	
Sgt De Torres, Ronald A.	Sep44	PRS-978th Signal	Luzon
Sgt De Vera, Pas Y.	Jun44	PRS-5217th	
Pvt Deang, Jose A.	1942	31stField-Artillery	
Pvt Delfin, Urbano P.	Jun44	PRS-5217th	Mindoro
Pvt Demillo, Jose A.	Nov42	USASOS	
Pvt Despy, Alfredo H.	Aug44	Anniv Edition	
T4 Diana, Serafin G.	Aug44	Anniv Edition	
Pvt Diaz, Jose C.	Jun44	PRS-5217th	
Pvt Digal, John V.	Dec43	PRS-5217th	
Pvt Dimaya, Avelino B.	Jun44	PRS-5217th	
Lt Dionolo, Thomas	Sep44	PRS-5217th	
Sgt Diputado, Esteban S.	Nov44	PRS-5217th-Demoltn	Luzon
Sgt Dolendo, Buenaventura L.	Oct44	PRS-5217th-Radio	Luzon
T5 Doliente, Danny D.	Apr44	PRS-5217th	
Pvt Domantay, Chris G.	Jun44	PRS-978th	
Sgt Domingo, Baldomero N.	Jun44	PRS-5217th	
Pvt Domingo, James R.	Jun44	PRS-5217th	
Sgt Dongallo, Fernando L.	May44	PRS-5217th	
Pvt Dorone, Constrato S.	May43	AIB	5217thRB
Pvt Dugenia, Wenceslao P.	Dec43	PRS-5217th	
Pvt Dulay, George, R.	Jun44	PRS-978th	
Sgt Dumag, Geronimo L.	Aug44	Anniv Edition	
Sgt Dumalaog, Benigno	Sep44	PRS-978th-Signal	Mindanao
Cpl Dumbrigue, Santiago C.	Dec43	PRS-5217th	
Cpl Dumpit, Fred N.	Jun44	PRS-5217th	
Cpl Dumpit, Teodoro Q.	Mar44	PRS-5217th	
Cpl Dupitas, Aurelio M.	Jan45	1stRB-Paratrp-Radio	
Cpl Duran, Agrifino J.	Jun44	PRS-978th-Signal	Samar
Pvt Duran, Leonisio O.	Aug43	Ship&GunCrew	5217thRB
Pvt Durano, Pio B.	Nov43	AIB-USAFFE	5217thRB
Sgt Ebarle, Audelon L.	Aug44	Anniv Edition	
Sgt Ebarle, Patrick F., Jr.	Aug44	Anniv Edition	
Cpl Randall, Roberto A.	Apr44	PRS-978th	Leyte
Pfc Ramirez, Camilo	Jan45	1stRB-Paratrp-Radio	
Sgt Ramirez, Jaime A.	Oct44	PRS-5217th	Negros-Cebu
Pvt Ramos, Aquilino B.	Sep44	PRS-978th-Signal	Samar
Sgt Ramos, Jose J.	Aug43	AIB	Panay
Cpl Ramos, Julian G.	Oct44	PRS-5217th	Negros-Cebu
Cpl Ramos, Paul	Aug44	Anniv Edition	
Pvt Ramos, Pascual M.	Dec43	PRS-5217th	
Sgt Ramos, Senen R.	Jan45	PRS-Paratroop-978th	Samar-Bicol
Pvt Ravelo, Tim A.	Nov43	PRS-5217th	
Sgt Real, Isay M.	Sep44	PRS-5217th	Luzon
Sgt Regala, Wilfredo M.	Oct44	PRS-5217th	Negros-Cebu
Sgt Regalado, Jerry D.	Oct44	PRS-5217th-Demoltn	Luzon
Cpl Respicio, Sergio F.	May44	PRS-5217th	
Pvt Respicio, Tom P.	Dec44	PRS-5217th	Mindoro
MSgt Reyes, Eulogio	Aug44	Anniv Edition	
Cpl Reyes, Felecisimo	May44	PRS-5217th	
Cpl Reyes, Felix T.	Jul44	PRS-978th	Mindoro
1st Sgt Reyes, German V.	Aug44	Anniv Edition	
Pvt Reyno, Jacinto U.	Dec43	PRS-5217th	
Sgt Reynoso, Jaime R.	Jun44	PRS-978th	Palawan
Sgt Rillera, Pastor A.	Jan45	1stRB-Paratrp-Radio	
T5 Rimando, Antonio C.	Nov44	PRS-5217th-Radio	Luzon
Pvt Rimando, Juan P.	Sep44	PRS-978th	Samar
Sgt Rinien, Eusebio T.	May44	PRS-5217th	
Pvt Riotoc, Manuel E.	Jan45	PRS-978th-Signal	
Cpl Rivera, Elpidio A.	Dec44	PRS-5217th	Mindoro
T4 Robles, Crispulo G.	Dec43	PRS-5217th	Mindanao
Cpl Robles, Romulo D.	Oct44	PRS-5217th-Demoltn	Luzon
Lt Roces, Marcos B.	Jul43	PRS-5217th	
Pvt Rodriguez, Feliciano S.	Aug44	PRS-978th-Signal	Mindanao
Sgt Rodriguez, Fred N.	Jan44	1stRB-978th-Signal	
Sgt Rodriguez, Irineo R.	Sep44	PRS-5217th	Samar
Cpl Rodriguez, Juan A.	Jul43	PRS-5217th	Negros
Sgt Rodriguez, Mariano	Nov43	AIB-USAFFE	5217thRB
Pvt Roduta, Peter	Feb44	PRS-5217th	
2nd Lt Rola, Ceferino R.	Aug44	Anniv Edition	
Maj Roman, Francisco J.	Oct45	FilUnit-PCAU	
Pvt Romero, Eleuterio Z.	Nov43	AIB-USAFFE	5217thRB
Lt Romillo, Guillermo M.	Sep42	USASOS	
Pvt Roque, Arthur F.	Mar44	PRS-5217th	
Pvt Rosal, Diosdado	Nov42	USASOS	
Sgt Rosal, Toribio	Oct45	FilUnit-PCAU	
Pvt Rosales, Doroteo A.	Aug44	PRS-978th	Woendi
Sgt Rosales, Paulino A.	Oct44	PRS-5217th-Demoltn	Luzon
Pvt Rosario, Gabriel Y.	Mar44	PRS-5217th	
Pvt Rosario, Juan R.	Aug44	Anniv Edition	
Sgt Rosete, Pantaleon M.	Oct44	PRS-978th-Signal	Negros-Cebu
Pvt Rubio, Melchor Q.	Dec43	PRS-5217th	
Sgt Ruiz, Don S.	Jan45	1stRB-Paratrp-Radio	
T5 Ruiz, Julius B.	Jan44	PRS-978th	
T5 Ruiz, Ruperto	Sep44	PRS-5217th	Samar
Sgt Ruiz, Saura	Feb44	PRS-5217th	
Sgt Rulloda, Ambrosio	Jan45	1stRB-Paratrp-Radio	
Cpl Runes, Lucas D.	Sep44	PRS-978th-Signal	Luzon
Sgt Runtal, Henry E.	Sep44	PRS-978th-Signal	Luzon
Pvt Ruperto, Quismundo Q.	Oct44	PRS-5217th	Negros-Cebu
T4 Sabado, Daniel B.	Dec43	PRS-978th-Signal	Mindanao
Cpl Sabado, Henry D.	Sep44	PRS-978th-Signal	Luzon
Sgt Sabado, Victor R.	Mar44	PRS-5217th	

Name	Date	Unit	Extra
Lt Edillon, Marcial A.	May44	PRS-5217th	1stFilInf
Pvt Ejurcadas, Maximo	Aug43	Ship&GunCrew	5217thRB
Lt Elefano, Alberto C.	Sep44	PRS-5217th	
Pvt Elvena, Joseph	Jul45	FilUnit-PCAU	
Pvt Enderiz, Victor P.	Apr44	PRS-978th	
Sgt Esguerra, Jimmy	Jan45	1stRB-Paratrp-Radio	
T5 Espero, Sammy O.	Oct45	FilUnit-PCAU	
Pvt Espiritu, Artemio C.	Apr44	PRS-978th	
Pvt Espiritu, Florentino C.	Dec43	PRS-5217th	
Pvt Esteban, Peter M.	Jan45	1stRB-Paratrp-Radio	
Pvt Estabilio, Silvestre	Jun44	PRS-5217th	
Pvt Estavillo, Sergio B.	Jul44	PRS-978th	Hollandia
Pvt Erana, Miguel B.	Nov44	PRS-5217th-Radio	Luzon
Sgt Ergas, Eugene N.	Dec43	PRS-978th-Signal	Darwin
Cpl Evangelista, Vicente	Apr44	PRS-5217th	
Pvt Fabon, Pedro Y.	Nov42	USASOS	
Sgt Faragan, Manuel F.	Jan45	1stRB-Paratrp-Radio	
Lt Farres, Agustin F.	Sep42	USASOS	
Pvt Felipe, Ramon S.	Dec43	PRS-5217th	
Pvt Felizar, Pete G.	Jul44	PRS-978th	Darwin
Cpl Fermin, Raymond R.	Oct44	PRS-5217th	Negros-Cebu
Lt Fernandez, Antonio P.	Sep42	USASOS	
Cpl Fernandez, Cipriano C.	Sep44	PRS-978th-Signal	Luzon
Pvt Fernandez, Leon C.	Sep44	PRS-5217th	Mindanao
Pvt Ferrer, Fortunato	1942	12thOrdnances	
Pvt Flores, Agripino B.	Oct44	PRS-5217th	Negros-Cebu
T5 Flores, Vicente M.	Apr44	PRS-5217th	
Sgt Floreza, Francisco A.	Oct45	FilUnit-PCAU	
Sgt Fontanoz, Marcelo P.	Oct44	PRS-978th-Radio	Luzon
Pvt Formoso, Francisco H.	Jan45	1stRB-Paratrp-Radio	
Pvt Formoso, Gabriel F.	Dec43	PRS-5217th	
Sgt Fortes, Melvin D.	Jan45	1stRB-Paratrp-Radio	
Sgt Fortun, Salvador B.	Jun44	PRS-978th	Mindanao
Sgt Francisco, Alberto C.	Sep44	PRS-5217th	Samar
Sgt Francisco, Fermin M.	Jun44	PRS-5217th	Panay
T5 Francisco, Urbano M.	Aug44	5217th	Mindoro
Lt Frias, Arthur F.	Sep44	PRS-5217th	
Lt Frias, Generoso D.	Sep44	PRS-5217th	
T4 Fugosa, J.	Jul43	PRS-5217th	
Pvt Gabatan, Rodolfo	Nov43	AIB-USAFFE	5217thRB
Pvt Gabuten, Elpidio A.	Nov42	USASOS	
Pvt Gachalian, Hipolito M.	Apr44	PRS-978th	
Pvt Gaduang, Thomas D.	Oct44	PRS-5217th-Radio	Luzon
Pvt Gago, Cecilio P.	Dec43	PRS-5217th	
Pvt Galam, Benito P.	Feb44	PRS-5217th	
Capt Galang, Ricardo C.	Nov43	5217th	Mindoro
Pvt Galas, Eriberto	Nov43	AIB-USAFFE	5127thRB
Pvt Galban, Primo R.	Dec43	PRS-5217th	
Sgt Gallemore, James R.	Jun44	PRS-5217th-Weathr	Negros
Pvt Gallopa, Ladislao	Nov43	AIB-USAFFE	5127thRB
Pvt Gamboa, Lorenzo	1942	32ndInfantry	
Sgt Garces, Pedro L.	Oct44	PRS-5217th	Negros-Cebu
Pvt Garcia, Francisco	1942	Coast-Artillery	
Pvt Garcia, Juan	Dec43	PRS-5217th	
Sgt Garcia, Marcelo H.	Oct44	PRS-5217th-Demoltn	Luzon
Sgt Garcia, Vidal	Sep44	PRS-5217th	Luzon
Pvt Garlitos, Anton T.	Jan45	1stRB-Paratrp-Radio	
Pvt Gascon, Clemente Y.	Mar44	PRS-5217th	
Sgt Gaston, Meleseo M.	Jan45	1stRB-978th-Paratrp	
Pvt Gatchalian, A.	Jan45	PRS-978th-Signal	
T5 Sabala, Severino	Aug44	Anniv Edition	
Sgt Sabiniano, Teodoreco S.	Jan45	1stRB-Paratrp-Radio	
T5 Sabobo, Crispin	Nov43	AIB-USAFFE	5217thRB
T5 Sagun, Fernando C.	Apr44	PRS-5217th	
Pvt Saladino, Selmo C.	Aug44	PRS-978th-Signal	Luzon
Cpl Salamanca, Albert S.	Aug44	Anniv Edition	
Pvt Salazar, Narciso	Nov42	USASOS	
Sgt Salgado, Mauro B.	Jul44	PRS-978th	Darwin
Pvt Salvacion, Alfredo L.	Jun44	PRS-5217th	Panay
Sgt Salvador, Apolinar G.	Nov44	PRS-5217th	Mindoro
Pvt Salvador, Frank A.	Mar44	PRS-978th	
Pvt Sammons, William H.	Aug44	PRS-978th-Signal	Luzon
Lt Samporna, Isidro B.	May44	PRS-5217th	1stFilInf
Sgt Sanchez, Gerardo A.	Dec43	PRS-5217th	Mindanao
T5 Sandoval, Nicolas E.	Jul44	PRS-5217th	Mindoro
T4 Santorio, Ceferino D.	Mar44	PRS-5217th	
Pvt Santos, Alberto E.	Dec43	PRS-5217th	
Sgt Santos, Amador A.	Jun44	PRS-5217th	
Sgt Santos, Fortunato	Nov44	PRS-5217th-Demoltn	Luzon
Pvt Santos, Gregorio	Nov44	PRS-978th-Signal	Leyte,Phil
Cpl Santos, Isaac L.	May44	PRS-5217th	
T5 Santos, Rudolph B.	Jun44	PRS-978th-Signal	Samar
Pvt Sapio, Lucas J.	Jan45	1stRB-Paratrp-Radio	
Pvt Sapon, Leoncio S.	Nov44	PRS-978th-Signal	Leyte,Phil
Pvt Saria, Juan A.	Jan45	PRS-978th-Signal	
Sgt Sarmiento, Angel A.	Mar44	PRS-5217th	Mindanao
Pvt Sarmiento, Florentino M.		PRS-5217th	
Lt Sarmiento, Ignacio M.	Sep44	PRS-5217th	
Pvt Savellano, Andres S.	Jul44	PRS-978th-Signal	Mindanao
Pvt Savellano, Pedro E.	Jul44	PRS-5217th	Mindoro
Pvt Seares, Pascual P.	Apr44	PRS-978th-Signal	
T5 Sebastian, Maximo P.	Oct45	FilUnit-PCAU	
Pvt Sebolboro, German B.	Oct45	FilUnit-PCAU	
Pvt Segomalian, Domingo S.	Dec43	PRS-5217th	
Pvt Senarosa, Jose H.	1942	Post-Serv-Comd	
Cpl Sensano, Daniel G.	Feb44	PRS-5217th	
Pvt Serrano, Agapito B.	Jan45	1stRB-Paratrp-Radio	
T5 Seveses, Philip A.	Oct45	FilUnit-PCAU	
Sgt Sibonga, Emil J.	Jan45	1stRB-Paratrp-Radio	
Lt Silva, Saturnino R.	Feb44	5217th	Mindanao
Pvt Silverio, Ben V.	Jan45	PRS-978th-Signal	
Pvt Siose, Francisco E.	Dec43	PRS-5217th	
Cpl Sison, Agapito R.	Aug44	Anniv Edition	
SSgt Sison, Carmelo	Aug44	Anniv Edition	
Cpl Sison, Ceferino P.	Feb44	PRS-978th-Signal	
Sgt Sison, Tony C.	Apr44	PRS-5217th	
Sgt Sitjar, Jose T.	Jun44	PRS-978th-Signal	Mindanao
Pvt Sivila, Andres C.	Sep44	PRS-978th-Signal	Luzon
Capt Smith, Charles M.	Dec43	PRS-5217th	Samar
Lt Smith, Frank B.	Sep44	PRS-5217th	
Cpl Solar, Marcelino V.	Jan45	1stRB-Paratrp-Radio	
Sgt Solidarios, Sergio V.	Jan45	1stRB-Paratrp-Radio	
T5 Soliven, Larry G.	Aug44	PRS-5217th	Luzon
Lt Songco, Ruben P.	Nov43	5217th	Mindoro
Pvt Sosa, Braulio	Aug43	Port-Detach	5217thRB
Pvt Soseon, Teofilo	Nov43	AIB-USAFFE	5217thRB
T3 Stahl, Robert E.	Dec43	PRS-5217th	Mindanao
Pvt Suansing, Celso	Nov42	USASOS	
Lt Supnet, Flaviano	Aug44	Anniv Edition	
Pvt Supnet, Santiago	Aug43	Port-Detach-D	5217thRB

Name	Date	Unit	Location
Pvt Generales, Narciso C.	Oct44	PRS-5217th-Demoltn	Luzon
Pvt Genetiano, Marcos P.	Jun44	PRS-5217th	
Pvt Giango, Porferio	Nov43	AIB-USAFFE	5217thRB
Lt Ginete, Pedro A.	May44	PRS-5217th	1stFilInf
Sgt Ginobiagon, Felipe R.	Aug43	AIB	Panay
Sgt Giron, Leovegildo M.	Aug44	PRS-978th-Signal	Luzon
Pvt Gobaleza, Dominador S.	Jul43	PRS-5217th	Negros
Sgt Goloyugo, Victorio C.	Aug44	PRS-5217th	Luzon
Sgt Goloyugo, Vincent C.	Aug44	PRS-5217th	Palawan
Pvt Gomez, Frank A.	Nov43	PRS-5217th	
Cpl Gonzales, Ban S.	Oct45	FilUnit-PCAU	
Pvt Gonzales, Juan L.	Jan45	PRS-978th-Signal	
Pvt Gonzales, Marcelo G.	Sep44	PRS-978th	Luzon
T4 Gonzalez, Jose T.	Jul44	PRS-978th	Mindoro
Pvt Gonzalo, Arthur V.	Jan45	PRS-978th-Signal	
Pvt Gorospe, Pedro Q.	Mar44	PRS-5217th	
Pvt Goroza, Edward E.	Aug44	PRS-978th	Mindanao
Pvt Granados, Vicente J.	Dec44	PRS-5217th	Mindoro
Pvt Guazon, Bienvenido	Nov43	AIB-USAFFE	5217thRB
Sgt Gubatan, Severo D.	Apr44	PRS-5217th	
Pvt Gurrea, Presciliano B.	Apr44	PRS-978th-Signal	
Pvt Guerrero, Austin B.	Sep43	Hq.Co.,GHQ	5217thRB
Pvt Guico, Menilio O.	Nov43	AIB-USAFFE	5217thRB
Cpl Guico, Pete S.	Jan45	PRS-978th-Signal	
Sgt Guinobiagon, Fellipe R.	Aug43	PRS-5217th	Panay
Sgt Gustilo, Conrado A.	Mar44	PRS-978th-Signal	Mindanao
Sgt Guyot, Gaudencio V.	Jun44	PRS-978th-Signal	Samar
Sgt Guzman, Larry O.	Nov44	PRS-5217th-Demoltn	Luzon
Sgt Guzman, Robert O.	Oct45	FilUnit-PCAU	
Pvt Guzman, Simeon R.	Feb44	PRS-978th-Signal	
Pvt Halog, Rufino D.	Dec43	PRS-5217th	
Sgt Harder, Benjamin	Nov43	PRS-5217th	Mindoro
Pvt Hawrysio, Eugene J.	Nov44	PRS-978th-Signal	Leyte,Phil
Lt Hernandez, Al	Jul44	PRS-5217th	Mindoro
Sgt Herreria, Donato E.	Jun44	PRS-978th	Mindanao
T5 Herreria, George R.	Dec43	PRS-978th-Signal	Mindanao
T3 Hilario, Carl G.	Jan45	1stRB-Paratrp-Radio	
T5 Hipol, Felipe R.	Aug44	Anniv Edition	
T5 Holgado, Eddie C.	Jun44	PRS-978th	Samar
Lt Hondolero, Felix S.	Sep44	PRS-5217th	
Lt Huerto, Petronio	Jul43	AIB	Negros
Pvt Hufana, Frank A.	Feb44	PRS-5217th	
Cpl Ibarra, Estanislao S.	Sep44	PRS-5217th	Luzon
Sgt Ibea, Artemio I.	Sep44	PRS-978th	Samar
Sgt Ignacio, Fred C.	Sep44	PRS-978th-Signal	Mindanao
Sgt Ignacio Miguel E.	Jun44	PRS-978th	
Lt Ignacio, Rodolfo C.	Jan43	AIB	Negros
T5 Ilustre, Ernesto D.	Feb44	PRS-5217th	
Cpl Ilusorio, Johnnie A.	Apr44	PRS-978th	
Pvt Insular, Jacob E.	Dec43	PRS-5217th	
T3 Interior, Iluminado	Nov43	AIB-USAFFE	5217thRB
Pvt Isaac, Cledonaldo J.	Sep44	PRS-978th-Signal	Luzon
Pvt Itchon, Jay F.	Dec43	PRS-5217th	
Sgt Jacobs, Jerry G.	Oct45	FilUnit-PCAU	
Cpl Jaramilla, Santiago Q.	Jan45	1stRB-Paratrp-Radio	
Lt Jaranilla, Ernesto J.	Sep42	USASOS	
T5 Javier, Bartolome E.	Jan45	1stRB-Paratrp-Radio	
Pvt Javier, Floro O.	Jan45	1stRB-Paratrp-Radio	
Sgt Javonillo, Juan O.	Jan45	1stRB-Paratrp-Radio	
Sgt Jesena, Carl	Mar44	PRS-5127th	
Pvt Taban, Lorenzo F.	Jun44	PRS-978th-Signal	Darwin
Sgt Tabios, Val R.	Jun44	PRS-5217th	Negros
Pvt Tabofunda, Pedro P.	Jun44	PRS-978th	Darwin
Pvt Tactaquin, Rodrigo R.	Dec43	PRS-5217th	
Sgt Talaro, Albert T.	Oct44	PRS-5217th	Negros-Cebu
Lt Tallido, Silvino B.	Aug44	Anniv Edition	
Sgt Taluban, Teodoro M.	Aug44	PRS-5217th	Luzon
Lt Tamayo, Claudio M.	Sep44	PRS-5217th	Luzon
Capt Tamayo, Julian C.	Sep42	USASOS	
Cpl Tambo, Emiliano	Oct45	FilUnit-PCAU	
Pvt Tamoria, Rubin	1942	26thCavalry	
Sgt Tanato, Leandro P.	Jan44	1stRB-978th-Signal	
Cpl Taylan, Genaro V.	Jun44	PRS-5217th	
Pvt Temple, Pedro S.	Dec43	PRS-5217th	
Lt Teo, KongLam	Aug44	PRS-5217th	Mindanao
Lt Tobias, Manuel A.	Aug44	Anniv Edition	
Pvt Tolentino, John T.	Jun44	PRS-5217th-Weathr	Mindoro
Pvt Tolentino, Zoilo, Jr.	Aug44	Anniv Edition	
Pvt Toribio, Alipio C.	Jul44	PRS-978th-Signal	Mindoro
Cpl Torio, Isaias T.	Aug44	Anniv Edition	
Lt Torres, Enriquque Jr.	Sep44	PRS-5217th	Luzon
Cpl Tria, Antonio B.	Sep44	PRS-5217th	Samar
Pfc Trocio, Perfecto M.	Dec44	PRS-5217th-Demoltn	Mindoro
Cpl Tucay, Henry T.	Dec44	PRS-5217th	Mindoro
Lt Tudtud, Joe L.	Jan45	1stRB-Paratrp-Radio	
Pvt Tugab, Jose	Aug43	5thReplDepot	5217thRB
Pvt Tullo, Apollo P.	Jun44	PRS-5217th-Weathr	Negros
Sgt Turquesa, Arsenio	Jan45	1stRB-Paratrp-Radio	
Lt Ubarro, Patricio	Aug44	Anniv Edition	
Pvt Udani, Carlos D.	Nov44	PRS-5217th-978th	Luzon
Pvt Uduando, Esteban C.	Jul44	PRS-978th	Darwin
Pvt Ulivas, Albert U.	Jun44	PRS-978th-Signal	Darwin
Sgt Umipeg, Marcelo B.	Sep44	PRS-978th-Signal	Luzon
Lt Unciano, Isaac F.	Sep44	PRS-5217th	
Cpl Valdez, Celestino R.	Sep44	PRS-5217th	Mindanao
Pvt Valdivia, Enrique D.	Dec43	PRS-5217th	
Lt Valera, Jose V.	Aug44	5217th	Luzon
T5 Valido, Pastor W.	Jan45	1stRB-Paratrp-Radio	
Sgt Valin, Zenon A.	Apr44	PRS-5217th	
Pvt Valmonte, Frank G.	Jun44	PRS-978th-Signal	Darwin
Sgt Velasco, Roque M.	Jan45	1stRB-Paratrp-Radio	
Pvt Velayo, Benny T.	Dec44	PRS-5217th	Mindoro
Pvt Ventura, Dionisio H.	Feb44	PRS-978th-Signal	
Lt Ventura, Frank L.	Jun44	PRS-5217th	
T4 Vergara, Thomas C.	Aug44	PRS-978th	Mindanao
Sgt Vicente, Johnny Q.	Dec43	PRS-5217th	
Pvt Vicente, Victor L.	Jan45	1stRB-Paratrp-Radio	
Pvt Vidal, Doroteo C.	Dec43	PRS-5217th	
Pvt Viduya, Jose F.	Jun44	PRS-978th-Signal	Darwin
T5 Villalon, Hermenegildo	Jul44	PRS-978th-Signal	Mindoro
Maj Villamor, Jesus A.	Jan43	AIB	Negros
T4 Villanueva, Angelo F.	Nov44	PRS-5217th	Luzon
Pvt Villanueva, Conrad G.	Dec43	PRS-5217th	
Sgt Villarta, Federico R.	Jul43	PRS-978th-Signal	
Cpl Villote, Paul P.	Feb44	PRS-978th-Signal	
T5 Viray, Don J.	Jan45	PRS-978th-Signal	
Sgt Visaya, Vincent L.	Feb44	PRS-5217th	
Sgt Vite, Doroteo V.	Sep44	PRS-978th-Signal	Mindanao
T5 Vitorio, Ramon D.	Nov43	PRS-978th-Signal	Mindoro
Lt Vivit, David T.	Sep44	PRS-5217th	

Sgt Jorge, Patricio D.	Jan43	AIB	Negros	Pvt Wendam, Sydney D.	Oct44	PRS-5217th	Negros-Cebu
Pvt Jose, Theodore M.	Nov44	PRS-978th-Signal	Leyte,Phil	WO Wise, Braynard L.	Sep44	PRS-978th	Samar
Lt Josue, Ignacio I.	Oct45	FilUnit-PCAU		Sgt Wright, Carl	Sep44	PRS-978th-Signal	Luzon
Pvt Juan, Eustaquio L.	Jun44	PRS-978th		Pvt Ybanez, Nemesio	Nov42	USASOS	
Pvt Juanito, Eduardo	Aug44	Anniv Edition		T4 Young, Clarence R.	Nov44	PRS-978th-Signal	
Sgt Jubante, Valentine C.	Jan45	1stRB-Paratrp-Radio		Lt Young, Frank H.	Jun43	AIB	Mindanao
Pvt Jucutan, Adriano A.	Apr44	PRS-5217th		Sgt Yotoko, Salvador L.	May44	PRS-5217th	
Pvt Kinanahan, Santiago	Jan45	1stRB-Paratrp-Radio		Lt Yuhico, Delfin C.	Jan43	AIB	Negros
T5 Kintanar, Aniceto S.	Jun44	PRS-5217th	Panay	Pvt Yuponco, Leo R.	Dec43	PRS-5217th	
Sgt Laanan, Catalino	May43	AIB-Radio	5217thRB	Cpl Yuson, Phil P.	Oct45	FilUnit-PCAU	
Lt Labrador, Vicente W.	Jun44	PRS-5217th	Samar	Lt Zeller, Emilio	Sep42	USASOS	
Pvt Ladjahasan, Ali H.	Apr43	PRS-5217th	Panay	Lt Zerda, Vincent M.	Aug44	Anniv Edition	
				Sgt Zola, Edward	Aug44	Anniv Edition	

LEGEND

PRS-5217th	Philippine Regional Section-5217th Reconnaissance Battalion	PRS-5217th-Weathr	Philippine Regional Section-5217th Reconnaissance Battalion-Weather
PRS-978th	Philippine Regional Section-978th Signal Service Company	PRS-5217th-Radio	Philippine Regional Section-5217th Reconnaissance Battalion
PRS-5217th-Demoltn	Philippine Regional Section-5217th Reconnaissance Battalion-Demolition	Hq.DetachGHQ	Headquarter Detachment-General Headquarters
FilUnit-PCAU	Filipino Unit-Philippine Civil Affairs Unit	Hq.Co.USAFFE	Headquarter Company, United States Army Forces in the Far East
1stRB-Paratrp-Radio	1st Reconnaissance Battalion-Paratroop-Radio	2ndFilInf	2nd Filipino Infantry Regiment
AIB-USAFFE	Allied Intelligence Bureau-United States Army Forces in the Far East	1stFilInf	1st Filipino Infantry Regiment
USASOS	United States Army Services of Supply	5217thRB	5217th Reconnaissance Battalion
PRS-978th-Signal	Philippine Regional Section-978th Signal Service Company-Signal	CivilAffair	Civil Affairs
Anniv Edition	1st Reconnaissance Battalion Aug44 Anniversary Edition Magazine	5thReplDepot	5th Replacement Depot

*This roster was compiled from documents found with Filipinos in Australia.
**MEMO DATE – Date on the memo that lists the latest location of Filipinos in Australia.
***LOCATED ON MEMO OR FROM – Location on the record when they arrived in Australia or New Guinea. Some men on the memo did not go to Australia.

Source: MacArthur Archives and U.S. National Archives and Records Administration.

Appendix B - 16

1st Reconnaissance Battalion (Special)
Casualties 1944 - 1945

Last	First	M.I.	Rank	Death Date	Status	Last	First	M.I.	Rank	Death Date	Status
Vitorio	Ramon	D	SSgt	1945***	FOD	Corpus	Amado		Sgt	1944	DNB
Bargo	Querubin	B	Pfc	1944	FOD	Lawrence	Philips*	H	Maj	1944	
Savellano	Andres*		Pfc	1944	FOD	Cardenas	David		Sgt	1944	
Ambrosio	Nicanor*		Pfc	1944		Cacas	Maximino		Sgt	1944	DNB
Gonzales	Marcelo	U	T4	1944	DNB	Baniares	Arcangel*		Sgt	1944	
Padilla	Luis*	P	Lt	1944	KIA	Loma	Magdaleno*		Sgt	?	
Songco	Ruben	P	Lt	1944	KIA	Anderson	Moses		Cpl	?	
Almero			TSgt			Fugosa	J		T4	?	
Gregorio	Q		SSgt			Ambrosio			Pfc		

USS Seawolf Sunk*

Last	First	M.I.	Rank	Death Date	Status	Last	First	M.I.	Rank	Death Date	Status
Peralta	George	E	Sgt	25-Sep-44	KIA	Ruiz	Ruperto	R	T5	25-Sep-44	KIA
Francisco	Alberto	C	Sgt	25-Sep-44	KIA	Wise	Braynard	L	CWO	25-Sep-44	KIA
Ibea	Artemio	I	Sgt	25-Sep-44	KIA	Bueno	George	B	Sgt	25-Sep-44	KIA
Pugosa	Emil	L	Sgt	25-Sep-44	KIA	Ramos	Aquilino	B	Pfc	25-Sep-44	KIA
Rodriguez	Irineo	R	Sgt	25-Sep-44	KIA	Rimando	Juan	F	Pfc	25-Sep-44	KIA
Tria	Antonio	B	Cpl	25-Sep-44	KIA	Almero	Emiliano	C	Sgt	25-Sep-44	KIA
						Cendana	Amado	C	T5	25-Sep-44	KIA

*Found in the Manila American Cemetery
***Wall of Missing: Manila American Cemetery

DNB - Died Non-Battle: Sickness, Training, Homicide
KIA - Killed in Action
WIA - Wounded in Action DOW - Died of Wounds
FOD - Foreign Object Damage: Torture
MIA - Missing in Action
Self - Suicide

Source: Adam Earle; National Archives and Records Administration; Filipino American Experience research Project Oct 1998; Historical Reports. Operations on Northwest Leyte, Northern Samar, Camotes, Burias, and Ticao Islands 16 Feb-7 May 45, 8 May-20 and Sep 45. First Filipino Infantry Regiment. Col robert H. Offley Commanding

Appendix B - 17

U.S. Submarine Missions to the Philippines 1942-1945*
Allied Intelligence Bureau and Philippine Regional Section

Submarine	Arrival Date/Location	Mission Men	
Allied Intelligence Bureau Missions 1942 – 1943			
Gudgeon-211 (operational sub)	January 1943 Catmon Pt. Negros Mission to Mindanao but landed on Negros. 1 ton of equipment. Two complete Radio sets and supplies	Maj Villamor, Jesus A. Lt. Ignacio, Rodolfo C. Lt. Quinto, Emilio F. – Signal	Lt. Yuhico, Delfin C. Sgt. Jorge, Patricio Sgt. Malic, Dominador
Tambor -198 (operational sub)	March 1943 Tukuran, Pagadian Bay, Zamboanga 1st guerrilla supply mission. Seven tons arms, ammunition, and 60,000 pesos.	Comdr Parsons, Charles - GHQ Lt. Col. Smith, Charles. M. Pvt. Catalina, Eugenio. S.	Pvt. Bairulla, Hadjula. H. Pvt. Bairulla, Sabtal. H.
Gudgeon-211 (operatonal sub)	April 1943 Pucio Pt. by Pandan Bay, Panay 3 tons equipment and supplies	Lt. Crespo, Toribio Sgt Alfabeto, Orlando	Pvt. Ladjahasan, Ali Pvt Ladjahasan, Mangona
Trout-202 (operational sub)	June 1943 Pagadian Bay and Tawi Tawi, Mindanao 3 tons of Radio sets, supplies, ammunition, and $10,000.	Capt. Hamner, Jordan Lt Young, Frank H Pvt Aliacbar, Sahibad H	Pvt Ladjahasan, Jalil H Pvt Nastail, Lakibul P
Thresher-200 (operational sub)	July 1943 Cansilan Point, Negros. 1st guerrilla supply mis- sion. 500 lbs supplies and ammunition.	1st Lt Huerto, Petronio B Cpl Rodriguez, Juan A	Pvt Carriaga, Pedro F Pvt Gobaleza, Dominador F
Grayling – 209 (operational sub)	Aug 1943 Libertad, Panay. Two tons of supplies and money.	Lt Ames, Irineo Sgt Ginobiagon, Felipe R	Sgt Ramos, Jose J
Philippine Regional Sections Missions 1943 - 1945			
Narwhal -167 (cargo submarine)	November 1943 Paluan Bay, Mindoro with men and 10 tons of supplies. Nasipit, Butuan Bay, Mindanao with 50 tons of supplies. 1st civilian evacuees on return trip. Cmdr Charles Parsons was onboard. Remained on Mindanao	Maj. Phillips, Lawrence H. Capt. Galang, Ricardo C. Lt. Songco, Ruben P.	WOJG Wise, Brainard L. – 978th Signal Sgt. Harder, Benjamin M. Sgt Pinuela, Vicente A. Sgt. Baniares, Arcangel T4 Alberto, Alfred A. – 978th Signal T5 Vitorio, Ramon D. – 978th Signal

Philippine Regional Sections Missions 1943 - 1945

Narwhal-167	December 1943 Cabadbaran, Butuan Bay, Mindanao 90 tons supplies for Mindanao. 10 tons supplies and men to Samar. Cmdr Parsons returned on submarine.	SAMAR Maj. Smith, Charles M Capt. Evans, James L. Jr. Sgt Besid, Restituto J. Sgt. Manzano, Aniceto C. Sgt Sanchez, Gerardo A.	T4 Cardenas, David D. T4 Robles, Crispulo G. T3 Stahl, Robert H. – 978th Signal T4 Sabado, Daniel B. – 978th Signal T5 Herreria, George R. –978th Signal Pvt Bargo, Querubin B. – 978th Signal Pvt Savellano, Andres S. – 978th Signal
Narwhal-167	February 1944 45 tons supplies to Pandan Bay, Panay 45 tons supplies to Balatong Point, Negros Cmdr Parsons may have been onboard.	Commander F. Kent Loomis USN	
Narwhal-167	March 1944 70 tons supplies landed Mindanao for Tawi Tawi, Mindanao, Mindoro, and Cebu. Men to Mindanao. 40 civilians evacuated. Cmdr Charles Parsons onboard and returned with sub.	MINDANAO Lt Montgomery Wheeler USNR Lt Silva, Saturnino R Sgt Cortez, Louis	Sgt Gustilo, Conrado J. – 978th Signal Sgt Sarmiento, Angel T – 978th Signal Sgt Ortiz, Paterno J – 978th Signal
Nautilus – 168 (cargo submarine)	June 1944 100 tons supplies to Tukuran, Pagadian Bay, Mindanao	Lt. J. D. Simmons	
Narwhal-167	June 1944 30 tons of supplies, currency, and men to Gamay Bay, Samar for Smith party 15 tons of supplies and men to Tukuran, Pagadian Bay, Mindanao for Fertig. 15th Weather Squadron Army Air Force	SAMAR Lt. Labrador, Vicente W. Lt Padilla, Louis L. Lt Ancheta, Carlos F. Lt Mauricio, Paul F. Lt. Mendoza, Jose T. Lt Ibay, Epifanio Sgt Miguel, Cipriano L. Cpl Duran, Agripino J. – 978th Signal T5 Nuevo, Leodegario O – 978th Signal T5 Advincula, Julius C. –978th Signal T5 Angcos, Robert O. – 978th Signal T5 Santos, Rudolph B – 978th Signal T5 Holgado, Eddie C. – 978th Signal Sgt Nery, Gerardo B. – 978th Signal Sgt Agcaoili, Raymundo E. – 978th Signal Sgt Guyot, Gaudencio V. – 978th Signal Sgt Montero, Jack – 978th Signal Pvt Bulatao, Geoge M. – 978th Signal T4 Luz, Pete L. – 978th Signal Pvt Pascua, Jerry D. – Weather	Pvt Aguila, Isaac Z. – Weather Cpl Richardson, William F. - Weather Cpl Becker, William III – Weather MINDANAO Maj Rosenquist, Harold WOJG Campeau, Lucien – Weather Sgt Hoke, Max E. – Weather Sgt. Finnegan, George – Weather Sgt. McGrath, Charles B – Weather Cpl Lozano, Ray - Weather T5 Burgos, Edward M. – 978th Signal T5 Palmeto, Daniel S. – 978th Signal Pvt Magale, Isidro P. –978th Signal Sgt Fortun, Salvador B. – 978th Signal Sgt Acenas, Seminiano D. – 978th Signal Sgt Herrera, Donato E. – 978th Signal Cpl Navarro, Ambrosio – 978th Signal Sgt Sitjar, Jose T. T4 Gonzales, Marcelo G. T5 Buenavista, Serafin B

Redfin-272	June 1944 Ramos Island, Palawan 1 ton of supplies Coast watch and weather watch.	Sgt Corpus, Amado S. Sgt Placido, Carlos S. – 978th Signal Sgt Cortez, Ramon F.– 978th Signal	Sgt Reynoso Jr., Jaime R. – 978th Signal Cpl Rallojay, Teodoro J. – 978th Signal T5 Dacquel, Richie D. – 978th Signal
Narwhal-167	June 1944 Lipata Point, Antique, Panay 100 tons of equipment supplies and weathermen.	Sgt Dacquel, Isidoro D Sgt Francisco, Fermin M. Pvt Salvacion, Alfredo L. T5 Kintanar, Aniceto S.	
Nautilus-168	June 1944 Balatong Point, Negros 80 tons supplies for Negros. 20 tons for Cebu.	Sgt Tabios, Val R. Sgt Bolivar, Custodio P.	Sgt Gallemore, James R.- Weather Pvt Tullo, Apollo P. – Weather
Nautilus-168	July 1944 Pandan Island, close to Mindoro Men and 12 tons of Radio, weather, photographic equipment and supplies. Weathermen and 70 tons of supplies to San Roque, Leyte.	**MINDORO** Comdr Rowe, George Lt Hernandez, Al TSgt Berg, Gerard G. Sgt Reyes, German V Sgt Aguilar, Peter B. Sgt Alvarez, Vidal F. Sgt Pascua, Inocencio V. Pvt Delfin, Urbano P Pvt Savellano, Pedro B Sgt Balleras, Julio V. – 978th Signal Sgt Begonia, Daniel B. – 978th Signal Sgt Borillo, Pio B. – 978th Signal Cpl Barcenas, Urbano B. – 978th Signal	Cpl Bugarin, Enrico M. – 978th Signal Cpl Reyes, Felix T. – 978th Signal T4 Logan, Domingo L. – 978th Signal T4 Gonzales, Jose T. – 978th Signal T4 Carillo, Patrocinio A. – 978th Signal T5 Toribio, Alipio C. – 978th Signal T5 Pascual, Angelo R. – 978th Signal T5 Villalon, Hermenegildo – 978th Signal **LEYTE - BOHOL** Sgt Chambless Henry I.- Weather Pvt Gamertsfelder, Robert H. – Weather Pvt Tolentino, John T. – Weather T5 Sandoval, Nicholas E. - Weather
Seawolf – 197	August 1944 Tongehatan Pt., Tawi Tawi, Mindanao 6 men and 10 tons of supplies Pirata Head, Cambari Island, Northern Palawan 6 men and 10 tons of Radio and supplies	**PALAWAN** Sgt Cabais, Eustaquio B. Sgt Cutaran, Jacinto – 978th Signal Sgt Goloyugo, Vincent C. T4 Marquina, Leo – 978th Signal T4 Vergara, Thomas C. – 978th Signal T5 Dagandan, Fortunato, C. – 978th Signal	**MINDANAO** Lt Konglam, Teo Sgt Daelto, Marciano R. Sgt Padilla, Balbino R. –978th Signal Pvt Aquino, Alejandro C. – 978th Signal Pvt Rodriguez, Feliciano S. – 978th Signal Pvt Goroza, Edward E. – 978th Signal
Stingray-186 (operational submarine)	August 1944 Mayraira Pt (Maira-ira), Bangui Bay, North east tip of Luzon 8 tons of radio equipment and supplies. Men for Intelligence, sabotage, demolition Cmdr Charles Parsons onboard.	Lt Valera, Jose Sgt Goloyugo, Victorio C Sgt Agron, Alfredo R. Sgt Taluban, Teodoro M Cpl Pimentel, Lorenzo U T5 Soliven, Larry G. Pvt Balunes, Ralph N. - Weather Pvt Ollada, Patricio	Pvt Liberato, Don–978th Signal Pvt Moscoso, Anselmo S. – 978th Signal Pvt Saladino, Selmo O. – 978th Signal Pvt Peralta, Estanislao C. – 978th Signal Pvt Sammons, Williams H – 978th Signal Sgt Asuncion, Telesforo T. – 978th Signal Sgt Giron, Leovegildo M. – 978th Signal

		BALER for LAPHAM	INFANTA for ANDERSON
Narwhal–167 Nautilus–168	September 1944 Dibut Bay, by Baler, east coast of Luzon Men and 30 tons of radio, equipment, and supplies Mail and supplies to eastern Mindanao. Cmdr Charles onboard to Mindanao.	Lt Torres, Enrique L. Jr. Lt Malan, Jose. B. Sgt Bacus, Faustino B. Sgt Cacas, Maximino C. Cpl Anacleto, Vicente, Jr. Cpl Bibat, Edilberto S. Cpl Ibarra, Estanislao S. Cpl Quindiagan, Luis C Sgt Wright, Carl – 978th Signal Sgt Runtal, Henry E. – 978th Signal Sgt Umipeg, Marcelo B. – 978th Signal Cpl Balino, Porfirio – 978th Signal Cpl Runes, Lucas D. – 978th Signal Cpl Palafox, Rosendo M. – 978th Signal Pvt Sivila, Andres C. –978th Signal Pvt Corpuz, Robert B. – 978th Signal Pvt Ambrosio, Nicanor G. – 978th Signal Pvt Isaac, Cledonaldo J.–978th Signal Pvt Marcelino, Juan C. – 978th Signal T4 Calmorin, Angelo A. – 978th Signal Sgt Moore, Chester H.- Weather	Lt Cabangbang, Bartolome C. Lt Tamayo, Claudio M. Sgt Garcia, Vidal Sgt Real, Isay M. Sgt Baldos, Florencio A. Sgt Balcena, Fred – 978th Signal Cpl Corpuz, Eustacio C. Cpl Oandasan, Angelo P. Cpl Sabado, Henry D Cpl Fernandez, Cipriano C. – 978th Signal Cpl Lopez, Cirilo G.–978th Signal Cpl Parong, Astor C.– 978th Signal Pvt Gonzales, Marcelo U Pvt Machon, Sergio M. Pvt Palac, Juan M. – 978th Signal T5 Racho, Paciano S. Sgt Arce, Mariano T. – 978th Signal Sgt De Torres, Ronald A. – 978th Signal Pvt Albalos, Segundo Q. – 978th Signal Pvt Mateo, Carl M – 978th Signal Sgt Hendrickson, Murl - Weather
Narwhal-167	September 1944 Illana Bay, Mindanao Macajalar Bay, Mindanao Men and 55 tons radio and weather observation equipment 597th Signal Aircraft Warning 15th Weather Squadron Army Air Force	Lt Pompea, Edward T.-Aircraft Cpl Jenson, Jr. Thomas A. –Aircraft Cpl Hoffman, Charles L.–Aircraft T5 Springmeyer, Robert L.- Plot Filter Sgt King, Walter B. –Aircraft T5 Fitzpatrick, Bernard C. –Aircraft Pvt Schram, Robert R. Aircraft Cpl Joyce, Charles J. - Aircraft Pvt McAdams, Albert. L.- Aircraft Cpl Visingardi, George A. - Aircraft T5 Patton, John F. - Aircraft T5 Fesler, Earl W.- Aircraft Sgt Alms, Vincent P.–Weather Sgt Capis, Theodore G. –Weather Cpl Mickelson, John C.- Weather Pvt Weissman, Edward –Weather	Sgt Dumalaog, Benigno Sgt Capiz, Pascual M. - 978th Signal Aircraft Sgt Paler, Luis B. - 978th Signal/Aircraft Sgt Bautista, Felino R – 978th/Signal Cpl Lopez, Arthur G. - 978th Signal Aircraft Cpl Valdez, Celestino - 978th Signal/Aircraft Pvt Lascano, Leodegario - 978th Signal/Aircraft Pvt Niebres, Federico - 978th Signal/Aircraft Pvt Fernandez, Leon C. - 978th Signal/Aircraft Pvt Nerez, Manuel M. - 978th Signal/Aircraft Pvt Martin, Estanislao C. - 978th Signal/Aircraft Sgt Vite, Doroteo V. Sgt Ignacio, Fred C. - 978th Signal/Aircraft Pvt Aglubat, Al C. - 978th Signal/Aircraft Pvt De La Pena, Geronimo – 978th Signal/Aircraft Pvt Balinton, Vicente – 978th Signal/978th Signal Pvt Cortez, Calixto, T. – 978th Signal/978th Signal Pvt Mapalo, Martin L. - 978th Signal/Aircraft Pvt Patacsil, Regino T. - 978th Signal/Aircraft Pvt Patubo, Gaudioso L. – 978th Signal/Aircraft Pvt Ogoy, Dalel Y. – 978th Signal/Aircraft Pvt Marcelo, Benjamin B. – 978th Signal/Aircraft T4 Medina, Mariano A - 978th Signal/Aircraft T4 Boncavil, Domiciano O. – 978th Signal/Aircraft
Seawolf-197	September 1944 Men and supplies to Samar, Batan Island and Leyte. Sunk in error by USS *Rowell* on Oct 1944 before men were discharged on Samar.	Capt Kopp, Howell S. – Alamo Scout Lt Miller, George F. – Alamo Scout** Sgt Peralta, George E. Sgt Francisco, Alberto C Sgt Ibea, Artemio I. –978th Signal Sgt Pugosa, Emil L. Sgt Rodriguez, Irineo R Cpl Tria, Antonio B T5 Ruiz, Ruperto R	CWO Wise, Braynard L – 978th Signal Sgt Bueno, George B.–978th Signal Pvt Ramos, Aquilino B. – 978th Signal Pvt Rimando, Juan F.–978th Signal Sgt Almero, Emiliano S.–978th Signal T5 Cendania, Amado C. Sgt Hammill, Charles H. - 15th Weather Sqd Cpl Herbig, Robert P.–15th Weather Sqd

Nautilus-168	September 1944 Men and 100 tons of supplies to Cebu, Panay, and Mindoro islands. Cargo and men landed on Panay.	**MINDORO** Sgt. Caswell, Robert B. - Weather Sgt George, Robert R. - Weather	
Stingray-186	October 1944 Suluan Island, Samar Supplies for Samar. Personnel and supplies for Leyte.	Lt. Hall, John C. – Alamo Scout Lt. Rommel, William N. Lt. Johnson, James O. – Alamo Scout	
Narwhal-167	October 1944 10 tons supplies to Tongehatan, Tawi Tawi 50 tons supplies to Negros, Cebu, Masbate and Panay Islands. Demolition and weather men sent on to Negros and Cebu. 597th Signal Aircraft Warning unit.	**NEGROS** Lt Rexford, Sidney, S. – Aircraft Sgt Ellis, James. E. - Medic T3 Espinoza, Cesar. E. – Medic Cpl. Dunst, Eugene F. Cpl Hunter, Edward T Cpl. Phillips, David A. Cpl. Reddick, Lyle V. Cpl. Rudolph, Prescott T5 Knight, Quincy S. T5 Smith, Lloyd E T5 Weber, Henry J. Pvt Kunkel, John E. Pvt Owens, Bernard B. Pvt Rubenstein, Joseph Sgt Madden, George W. Pvt Rullo, Modestino Pvt Malenson, George A. Pvt Routson, Merle D. Pvt Bill, Peter T4 Igmundson, Charels B. T5 Powers, Harry E. Pvt Welch, Homer J. T5 Nickay, Rene W T5 Grau, Frederick Pvt Sooter, Joyce M. Pvt Palmer, Donald F. Pvt Atad, Martiniano – Aircraft T4 Abalos, Marcelo B - Aircraft	Lt Hawley, Maynard C. - Demolition Lt. Schmaltz, Albert - Demolition Sgt Garces, Pedro L - Demolition Sgt Manuta, Eusebio J. - Demolition Cpl Quesada, Antonio M - Demolition Pvt Antikoll, Carlos P - Demolition **CEBU - LEYTE** Lt Tinnel, L. C. T4 Forrette, J. E. Sgt Mariano, Jose C. - Aircraft Sgt Ramirez, Jaime A. - Aircraft Sgt Rosete, Pantaleon M. - Aircraft Sgt Talaro, Albert T. - Aircraft Sgt Regala, Wilfred M. - Aircraft Cpl Fermin, Raymond R. - Aircraft Cpl Quiming, Perfecto C. - Aircraft Cpl Ramos, Julian G. - Aircraft Pfc Artiago, Maurice W. - Aircraft Pfc . Flores, Argripino B. - Aircraft Pfc Custodio, Emilio M. - Aircraft Pfc Quiaoi,t Estebran R. - Aircraft Pfc . Quismundo, Ruperto Q. - Aircraft Pvt Mosquera, Benigno F. - Aircraft Pvt Wendam, Sidney D. - Aircraft Pvt Oania, Hermogenes D.- Aircraft
Ray-271 (operational submarine)	October 44 Mamburao, Mindoro 2 tons of supplies and weathermen	Lt Richmond 2 unidentified weathermen	
Nautilus-168	October 1944 Dibut Bay, east coast of LuzonMasanga Pt, Tayabas, Luzon 40 tons of supplies Demolition and signal men.	**BALER, LUZON FOR LAPHAM** Capt Skundale, Frank J. – Aircraft 1st lt Bove, John E. - Aircraft Lt. Baker, Henry L. - Aircraft Lt. Tison, Sidney S. Aircraft Alamo Scout Sgt Pilar, Julian A – Demolition Sgt de los Santos, Modesto P. - Demolition Sgt Regalado, Jerry D. - Demolition	**INFANTA, LUZON FOR ANDERSON** Lt. Ensor, Richard A. - Aircraft Lt. Wood, Rhys C. – Alamo Scout Lt Campbell, Merle E. - Aircraft Lt Bahr, Andrew P Sgt Basug, Gregorio F. - Demolition Sgt Garcia, Marcelo – Demolition Cpl Robles, Romulo D - Demolition

		Sgt Ammaguey, Eleno. - Demolition	Sgt Constantino, John R. - Demolition
		Sgt Adia, Procopio M - Radio	Sgt Amante, Thomas T. - Demolition
		Sgt Rosales, Paulino A. - Demolition	Sgt Lim, Alfred E. - Demolition
		Pvt Generales, Narciso C. - Demolition	Pvt Arciaga, Emil D – Demolition
		Pvt Bernal, Jaime A - Demolition	Sgt Capistrano, Alberto N. - Radio
		Sgt Dolendo, Buenaventura L. - Radio	Cpl Agustin, Pedro P. – 978th Signal
		Sgt Bucol, Segundo C. - Radio	Sgt Fontanoz, Marcelo P. - Radio
		Cpl Pascual, Amrafel D. - Radio	Sgt Bareng, Gregorio F. - Radio
		Cpl Pulido, Mack H. - Radio	Cpl Rabang, Thomas V. - Radio
		Cpl Almojera, Loren E. - Radio	Pvt Gaduang, Thomas D. - Radio
		Pvt Agbunag, Baltazar C. - Radio	Pvt Quinagon, Barney A. - Radio
Cero-225	November 1944 Infanta, east coast of Luzon 35 tons supplies to Volckmann	FOR ANDERSON Maj. Vanderpool, Jay. D. Capt Miller, George F. Lt. Stoddard, Brooke - Demolition Lt Hope, Charles P Jr. - Demolition Sgt Santos, Fortunato - Demolition	Sgt Mones, Venancio M. - Radio Pvt Erana, Miguel B. - Radio Pvt Lobendino, Hilario L - Radio Pvt Tolentino, Zoilo Jr. - Radio T5 Rimando, Antonio C. – Radio
	Weather, signal and demolition men	Sgt Diputado, Esteban M. - Demolition Sgt Azcueta, Zoilo A.-Demolition Cpl Baladad, Jose C. - Demolition	Pvt Udani, Carlos D. – 978th Signal Sgt Chick, Aton C. – Weather Sgt Males, William J – Weather
Gar-206 (small cargo submarine)	November 1944 5 tons of supplies to San Jose, Mindoro	LUZON for VOLCKMANN Capt Behan, Fred H. - Aircraft Capt Jamison, Donald V. – Aircraft Sgt Guzman, Larry O. – Demolition	Sgt. De la Pena, Arsenio B. Sgt Villanueva, Angelo F. Sgt Arellano, Martin Q. Sgt Alves, Beato C.
	25 tons of supplies and personnel to Darigayos Cove, Luzon	Sgt Mapile, Clement B. - Demolition Cpl Clemente, Leo C. - Demolition Cpl Bermudes, Felix - Demolition	Cpl Medrano, Demetrio G. Cpl Pabros, Ambrosio – 978th Signal
	Demolition, weather, radio personnel, and 200,000 pesos.	Capt. Farrell, William. A. Capt Vaughn, William. D. Sgt. Lowe, William H. – Weather Sgt Bierley, James. C. – Weather	
Unidentified. Probably Landing Craft Infantry (LCI)	December 1944 2 men to Dulag, Leyte Rest of the men to Mindoro	Sgt Bello, Cleto S. Sgt Licera, Rafael C. Sgt de Ocampo, Mero L. Sgt Lagrimas, Walter B. Sgt Magbual, Felicisimo D.	Pvt Cayabyab, Ricardo D. Pvt Darang, Andres D. Pvt Marquez, Victor V. Pvt Minder, Robert C. Pvt Granados, Vicente J.
	1st Reconnaissance Battalion men attached to 597th Signal Aircraft Warning Battalion	Sgt Salvador, Apolinar G. Cpl Leanio, Frank S. Cpl Maglangit, Maximo V. Cpl Tucay, Harry T. Cpl Rivera, Elpidio A Pvt Arquero, Lacaro C. Pvt Azcueta, Pedro S Pvt Caduang, Julian F. Pvt Velayo, Benny T.	Pvt Martin, Philip Pvt Alfafara, Gerardo C. Pvt Respicio, Tom P. Pvt Trocio, Perfectio N. Pvt Antonio, Alfonso L. T5 Costales, Gene D. T5 Francisco, Urbano M.
Small cargo submarine	January 1945 Tawi Tawi, Mindanao 35 tons of supplies	Four Filipino soldiers reported on board.	
Ship or plane	December-January 1945 Samar for Bicol Province	Sgt Ramos, Senen R. – Aircraft Sgt Faragan, Manuel F. – 978th Signal T4 Paraiso, Miguel Q.	Sgt Arbis, Mariano V. – 978th Signal Lt Carbonell - Radio

* The first six submarine missions were under the AIB. From November 1943 to December 1944, the missions were under the control of PRS-SPYRON.

** Lt George F. Miller was with the USS Seawolf September 1944 mission. Confirmed by the Alamo Scout Historical Foundation.

Source: MacArthur Archives, Norfolk Va; 978th Signal Service Company, United States Army of the Far East, 21-Dec-44; Subsowespac, Special Missions, Southwest Pacific; The Filipino American Experience Research Project, Vol. 2, In Honor of Our Fathers, A History of the 1st Reconnaissance Battalion, United States Army, Edited by Alex S. Fabros, Jr., October 5, 1998; United States Submarine Operations in World War II; Theodore Roscoe, U.S. Submarine Special Missions Chronology, U.S. Naval Institute, Annapolis, Maryland, 1949, pgs 508 – 522; Alamo Scouts Historical Foundation, Inc.

Appendix B - 18

Submarine Activities Connected with Guerrilla Organizations
SEVENTH FLEET INTELLIGENCE CENTER
CONFIDENTIAL [Declassified]

SPECIAL MISSIONS
Involving Delivery of Radio and Supplies to, and Personnel to and from,
The Philippine Islands by SUBMARINES
Period covered: Approximately 1 February 1943 to 23 January 1945

Total number of missions - 41
Number of submarines – 20

NAME	NO. OF MISSIONS	Brought radio 978th Signal/Recon Men	NAME	NO. OF MISSIONS	Brought radio 978th Signal/Recon Men
Bowfin	1		Gar	2	X
Narwhal	9	X	Blackfin	1	
Angler	1		Gunnel	1	
Crevalle	1		Hake	1	
Harder	1		Ray	1	X
Redfin	2	X	Gudgeon	2	X
Nautilus	6	X	Grayling	1	X
Seawolf	2	X	Tambor	1	X
Stingray	5	X	Trout	2	X
Cero	1	X	Thresher	1	X

Source: Naval History and Heritage Command - https://www.history.navy.mil Marie Vallejo – 978th Signal Men

Appendix B - 19

Approximate Tonnages Delivered by Submarine to Guerrillas
1943 - 1944

Mindanao .... 600 Tons
(Unknown quantities of supplies landed on Mindanao were subsequently dispatched to the Sulu, Samar, Leyte, Bohol, Cebu, Negros, Panay and Luzon. These have not been sub-tracted from the Mindanao total or added to that indicated below as received by guerrillas.)

Negros .... 187 Tons
(This includes a small amount which probably was subsequently transferred to Cebu.)

Panay .... 180 Tons

Cebu .... 140 Tons

Leyte–Samar–Bohol .... 100 Tons
(This also includes supplies to Samar which were later sent to Luzon.)

Lt. Col. Bernard ANDERSON (Tayabas Province) .... 85 Tons

Sulu Area Command .... 75 Tons

Lt. Col. Russell VOLCKMAN (Northern Luzon) .... 60 Tons

Maj. Robert LAPHAM (Nueva Ecija) .... 50 Tons
(Other supplies intended for LAPHAM were landed at ANDERSON's Headquarters.)

Miscellaneous shipments to intelligence parties etc. .... 150 Tons
(Small amounts of supplies were dispatched with small intelligence parties such as VALERA to Northern Luzon, CABAIS and CORPUS–PLACIDO to Palawan, etc., and with the ROWE, PHILLIPS, etc. parties.)

TOTAL .... 1627 Tons

Note:
After December 1944, airplanes were used to land supplies at guerrilla-held airfields. Beginning in February 1945 supplies were delivered by LCI (landing craft) as well as by air. None of these latter figures are included in the above totals.

Source: MacArthur Archives, Norfolk, VA

Appendix B - 20

MONTHLY TOTALS OF RADIO MESSAGES SENT TO AUSTRALIA FROM GUERRILLA COMMANDERS AND AGENTS (1942 – 1945) (a)

Guerrilla Leader/Radio	Dec42 Jan43	Feb Mar (a)	Apr May	Jun Jul	Aug Sep	Oct Nov	Dec43 Jan44	Feb Mar	Apr May	Jun Jul	Aug Sep	Oct Nov (m) (k)	Dec44 Jan45 (n)
PRAEGER (Northern Luzon)	21	15		24	10 (c)								
PERALTA (Panay)	38	58	155	337	247	15 (d)	68	110	185	176	290	305 (g)	347
AUSEJO (Southern Negros)		8	(b)										
VILLAMOR/ANDREWS/BENEDICTO (S. Negros)		28	32	35	47	69	35	27	20	22	29	50	52
FERTIG (Mindanao)		67	66	61	235	172 (f)	288	270	577	317	395	758	1,193
HAMMER/YOUNG/SUAREZ (Tawi Tawi)			7	19	58	12	49	14	10	15	31		29
ABCEDE (Negros)				19	15	45	20	33	53	71		174	280
PHILLIPS (Mindoro)					6	75	17 (h)						
INGENIERO (Bohol)							22	8	11	7 (i)	(j) 12	20	37
SMITH (Samar)							5	42	50	41	63	210	(l)
KANGLEON (Leyte)								6	8	25	30	68	(l)
CUSHING (Cebu)									50	40	84	124	101
ANDERSON (Central Luzon)										57	52	113	198
HALL (Central Luzon)										17	50	67	55
ROWE (Mindoro)											130	131	122
VOLCKMANN (Northern Luzon)											45	255	610
CABAIS (Northern Palawan)											14	37	46
CABANGBANG (Central Luzon)											32	127	1,061
RAMSEY (Central Luzon)													60
STAHL (Bondoc Peninsula)										10	17	17	63
CORPUS – PLACIDO (Southern Palawan)											19	12	20

Totals do not include radio messages from sub-networks to the guerrilla leader main radio in each island. Aircraft sighting messages not included in totals after Jan45, Luzon landing.

a. Only a few ship sighting messages were included in totals after Dec44-Jan45
b. AUSEJO – radio taken into VILLAMOR net
c. PRAEGER – off air due to capture of radio and personnel
d. PERALTA – off air due to enemy in mountains near headquarters
e. Not used
f. FERTIG – main radio bombed and destroyed
g. PERALTA – back on air
h. PHILLIPS – killed and party dispersed
i. INGENIERO – off air due to capture of radio
j. INGENIERO – back on air with new radio from Leyte
k. Leyte landing
l. SMITH and KANGLEON – turned radio net over to Sixth Army
m. Aircraft sighting messages not included in totals after this date
n. Luzon landing

Source: MacArthur Archives, Norfolk VA

Appendix B - 21
KAZ and Philippine Intelligence Message Center (PIMC)
American Radio Operators*

RANK	NAME	NOTE	RANK	NAME	NOTE

April 27, 1943

832nd Signal Service Company at Camp Tabragalba sent to GHQ message center. T3 Marion W. Bush and T5 George F. Gregory from the former GHQ group stayed on to train the men.

RANK	NAME		RANK	NAME	
1st Lt	Charles B. Ferguson		T3	Marion W. Bugh	
2nd Lt	Kenneth F. Bry		T4	Robert E. Stahl	
2nd Lt	Edward H. Hale		T4	Seymour Ginsberg	
2nd Lt	Clinton B. McFarland		T5	Melvin S. Cohen	
T/SGT	Walter G. Clark		T5	George F. Gregory	
T/SGT	George K. Bailey		T5	Norman J. Lipman	
T3	Henry D. Ephron		T5	George K. Schweig	
T3	Arthur W. Shirley		T5	Irving H. Robinson	
T3	Byron E. Snider				

November 1943

832nd Signal Battalion working at KAZ, Brisbane Station.

RANK	NAME		RANK	NAME	
T4	Young		Sgt	Wickman	

December 1943

978th Signal Service Company sent to AIB, Brisbane.

RANK	NAME		RANK	NAME	
Lt	Charles B. Ferguson		Lt	James B. McGivney	
Lt	Edward H. Hale		Lt	Meyer J. Spiro	
Lt	Clinto B. McFarland		Lt	Lee A. Strickland	

December 1943

Sent to GHQ Message Center but returned to USAFFE Signal Section after two weeks.

RANK	NAME		RANK	NAME	
T/SGT	George K. Bailey		T4	George K. Schweig	
T3	Henry D. Ephron		T5	Melvin S. Cohen	
T3	Arthur W. Shirley		T5	Norman J. Lipman	
T3	Byron E. Snider				

August 15, 1943

Signal and Warrant Officers sent to Camp Tabragalba.

RANK	NAME	NOTE	RANK	NAME	NOTE
2nd Lt	George T. Baker	Supply Officer	WOJG	Wallace Booker	978th Sig Inst
2nd Lt	Alvin V. Bone	978th Sig Inst	WOJG	Alphie L. Claunch	Electrical and Utility Officer
2nd Lt	Kenneth H. Oates	Radio Maint – 978th Sig Inst	WOJG	Patrick Daley	Transportation
2nd Lt	Wilfred A. Cohen	PX	WOJG	Harry A. Maxam	Mess
2nd Lt	Meyer J. Spiro	Radio Maint – 978th Sig Inst	WOJG	Braynard L. Wise	978th Sig Inst

October 1943

Capt. Harry T. Croell from U.S. Signal Corps and staff at Camp Trabagalba

RANK	NAME		RANK	NAME	NOTE
Lt	Lee A. Strickland		Pvt	Sidney Malitovsky	
2nd Lt	Alvin V. Bone		Pvt	Jesse L. Landrum	
2nd Lt	Meyer J. Spiro		WOJG	Braynard L. Wise	978th Sig Inst
Pvt	Howard D. Long		WOJG	Wallace Booker	978th Sig Inst

August 21, 1943

Cadre from U.S. Camps to Camp Tabragalba to help build up the camp.

RANK	NAME		RANK	NAME	
Pvt	James R. Brown		Pvt	Jesse P. Wheat	
Pvt	Newell J. Beedy		Pvt	Joseph R. Paguarella	
Pvt	Edward P. Burns		Pvt	Frank J. Petro	
Pvt	Lowell A. Camp		Pvt	Richard M. Presnall	
Pvt	Charles E. Carpenter		Pvt	Stephen J. Profetta	
Pvt	Orien N. Eals		Pvt	Paul O. Rasmussen	
Pvt	John M. Ferrell		Pvt	Louis Robert	
Pvt	Joseph F. Filek		Pvt	Edward T. Sheppard, Jr.	

RANK	NAME	NOTE	RANK	NAME	NOTE
Pvt	James L. Green		Pvt	William P. Sperandeo	
Pvt	Albert F. Halla		Pvt	Raymond D. Svobodny	
Pvt	Laverne J. Kobliska		Pvt	Carmne Taddeo	
Pvt	Joseph A. Morris		Pvt	James P. Thomas	
Pvt	Victor R. Nichols		Pvt	Clifford P. Thompson	
Pvt	Charles W. O'Gwin		Pvt	James C. Thompson	
Pvt	Clarence Y. Olson		Pvt	Carl T. Watson	
Pvt	Henry L. Padgett		Pvt	Edward J. Zola	

December 9, 1943

Personnel appointed to new positions.

RANK	NAME	NOTE	RANK	NAME	NOTE
Lt	Alvin Bones	2nd in command of 978th	WOJG	Harry A. Maxam	Transportation Officer
2nd Lt	George T. Baker	Platoon Leader	WOJG	Patrick Daley	Property Officer
2nd Lt	Kenneth H. Oates	Platoon Leader	Lt	Harley M. Claussen	Utility Officer
T5	Melvin S. Cohen	Mess Officer			

December 21, 1943

978th Signal Men sent to Darwin on temporary duty.

RANK	NAME	NOTE	RANK	NAME	NOTE
Lt	Lee A. Strickland		Pvt	Joseph F. Filek	
Cpl	James R. Brown		Pvt	James L. Green	
Sgt	Lowell A. Camp		Pvt	Jesse P. Wheat	

March – June 1944

Sent to Brisbane. Went to Hollandia on September 1944.

RANK	NAME	NOTE	RANK	NAME	NOTE
Sgt	Loy W. Carter		T/SGT	Irving Robinson	
Sgt	Frederick J. Gouldner		T5	George F. Bent	
Sgt	Clarence Wright		T5	Seymour Ginsberg	
Sgt	David A. Nagle		Pfc	Howard Nichols	
Pvt	Lambert J. Hendleman		Pvt	Lambert Hendleman	
T5	Richard Beeler		Pvt	James Smith	
Capt	Charles Ferguson		Pvt	Alwyn Woods	

May 6, 1944

Sent to 978th Signal Service Company at Camp Tabragalba.

RANK	NAME	NOTE	RANK	NAME	NOTE
Sgt	Alfred P. Theys		Pvt	Kenneth E. Gordon	
Cpl	Hadley W. Dodson		Pvt	Glen B. Hosberg	
Cpl	George D. Osborne		Pvt	John G. Neighbors, Jr.	
Pfc	Charlie A. Leake		Pvt	John C. Plummer	
Pvt	Edward Bulat		Pvt	George T. St. Peter	
Pvt	Stanley J. Birard		Pvt	Robert P. Worthington	

No Date

Assigned to Base G in Hollandia as support to PIMC personnel

RANK	NAME	NOTE	RANK	NAME	NOTE
Sgt	David Allen		2nd Lt	James B. McGivney	
Sgt	Charles Barr		2nd Lt	Edward H. Hale	
Pfc	Harry Bell		T4	Clarence Wright	
Pfc	Leo Gangl		T/SGT	Marion Bugh	
Pvt	Robert Schneider		T5	Richard Beeler	
Pvt	Donald Higgins		2nd Lt	Harold B. Wooten	
2nd Lt	Harley M. Claussen				

August 1944

978th Signal Men sent to Biak area, Woendi Island from Brisbane.

RANK	NAME	NOTE	RANK	NAME	NOTE
Sgt	Lowell A. Camp		T5	James L. Green	
T5	Joseph F. Filek		T5	Jesse P. Wheat	

October 20, 1944

KAZ and PIMC personnel who joined the Leyte landing convoy.

RANK	NAME	NOTE	RANK	NAME	NOTE
Capt	Charles B. Ferguson		T4	Irving Robinson	
T3	Seymour Ginsberg		T5	Richard Beeler	
T4	Clarence Wright		Pfc	George Bent	

RANK	NAME	NOTE	RANK	NAME	NOTE
S/SGT	Dean W. McCleeary		Pvt	John R. Kupetz	
Sgt	Marvin O. Schuler		Pvt	Joseph E. Pino	
Pvt	John A. Ahola		Pvt	Gerald J. Schmitt	
Pvt	Eugene J. Hawrysio		Pvt	Wallace F. Strow	
Pvt	Lester D. Haymore				

No Date

125th Radio Intelligence Company Personnel attached to the 978th Signal Service Company in the Luzon landing.

Lt	Card		T5	Frentz	
Lt	Phelan		Pfc	Altomare	
Sgt	Rullie				

November 10, 1944

Departed KAZ and PIMC stations in Hollandia.

Lt	Edward H. Hale		Pvt	Joseph Pino	
Sgt	Charles Barr		Pvt	James Cahill	
Pfc	Harry Bell				

Another group arrived.

2nd Lt	Harold B. Wooten		Pvt	Howard Nichols	
T/SGT	Marion Bugh		Pvt	James Smith	
T3	George Gregory				

November 24, 1944

With PIMC in Hollandia and arrived in Tacloban, Leyte.

Lt	Clinton B. MacFarland		Pvt	John Ahola	
Pvt	Eugene J. Hawrysio		Pvt	Gerald J. Schmitt	
Pvt	Donald Higgins		Pvt	John R. Korbetz	
Pvt	Lester Haymore				

November 27, 1944

From Hollandia to Tacloban

Lt	Harley M. Claussen		Pvt	Lambert Hendleman	
Lt	James B. McGivney		Pvt	Alwyn Woods	
Sgt	Marvin Schuler		Pvt	Robert Schneider	
T4	David Allen		Pvt	Wallace Strow	
Pfc	Leo Gangl		Sgt	Loy Carter	

January 1945

With PIMC group in the convoy for Luzon landing.

Capt	Charles B. Ferguson		Pvt	Dean H. Darkow	
Sgt	Duane J. Pittsford		Pvt	George H. Lieger	
Cpl	Arthur Kilpartrick		Pvt	Walter J. Mackey	
Pfc	Alvin L. Elias		Pvt	Richard G. Sexauer	
Pvt	Luther B. Bentley, Jr.		Pvt	Robert J. Thomas	
Pvt	Herbert A. Burns				

January 2, 1945

125th Radio Intelligence Company Personnel on temporary duty with the 978th.

Sgt	Rullie		Pfc	Altomare	
T5	Frentz		Lt	Phelan	

No Date

Sent to PIMC in Hollandia to replace the advance group sent to Tacloban, Leyte.

January 5. 1945

PIMC Leyte personnel with the convoy from Luzon landing.

Capt	Charles Ferguson		Pvt	Eugene J. Hawrysio	
S/SGT	Loy Carter		Pvt	John Kupetz	
Sgt	Charles Barr		Pvt	Joseph Pino	
T3	Seymour Ginsberg		Pvt	Gerald J. Schmitt	
Pvt	George Bent		Pvt	Lester Haymore	
Pvt	Donald Higgins				

RANK	NAME	NOTE	RANK	NAME	NOTE

February 1945

317oth Signal Battalion personnel with KAZ and PIMC in Leyte to replace the advance groups to Luzon

RANK	NAME	RANK	NAME
Cpl	Bernard E. Grownstein	Pfc	Harold R. Bengson
Cpl	Edward Curtice	Pfc	Floyd S. Nelson
Pfc	Gordan Joblon	Pfc	Anthony O'Donnell
Pfc	Arthur E. Parks	Pfc	David L. Smith
Pfc	Kenneth P. Schwartz	Pfc	Arthur H. Wenners
Pfc	Robert M. Willis	Pfc	Harold E. Williams

April 1945

From Leyte to Manila.

RANK	NAME	RANK	NAME
Lt	James B. McGivney	Sgt	David Allen
2nd Lt	Harley M. Claussen	Sgt	Marvin Schuler
T/SGT	Marion Bugh	Pfc	Wallace Strow
T/SGT	David Nagle	Pfc	Howard Nichols
T4	Clarence Wright	Pfc	James Smith

May 31, 1945

From KAZ Leyte to Manila Bay

RANK	NAME	RANK	NAME
T3	Frederick Gouldner	Pfc	Lambert Hendleman
T/SGT	Dean McCleeary	Pfc	Robert Schneider
T3	Irving Robinson	Pfc	James Cahill
Pfc	Harry Bell	Pfc	John A. Ahola
Pfc	Leo Gangl		

June – August 1945

Sent to the Signal Intelligence Section (SIS).

RANK	NAME	RANK	NAME
Sgt	Marvin Schuler	Pfc	Wallace Strow
S/SGT	Dean McCleeary	Pfc	Donald Higgins
Pfc	Gerald J. Schmitt	Pfc	Eugene J. Hawrysio
Pfc	John Ahola	Pfc	Leo Gangl
Pfc	Harry Bell		

*978th Signal Service Company men were identified as such or as Radio Operators.

Source: The Filipino American Experience Project October 5, 1998

Appendix B - 22

Source: Morningstar, James Kelley. War and Resistance in the Philippines, 1942-1944. Naval Institute Press, 1 April 2021. With permission from James K. Morningstar

Notable Guerrilla Leaders and Units Tracked by SWPA.[1]
[1]Compiled from data in General Headquarters, Southwest Pacific Area, Military Intelligence Section, General Staff, "Guerrilla Resistance Movements in the Philippine," 1-81, 31 March 1945, Box 255, RG 407, Philippines Archive Collection, National Archives II, College Park, Maryland.

EXTRACT

Luzon:

Central Luzon (26 October 1944)
COL Claude Thorp organized and led the Central Luzon Guerrilla Force in Mount Pinatubo area of Pampanga until he was captured in October 1942. Lapham, Anderson and Ramsey inherited his troops.

- Anderson's Guerrillas: MAJ Bernard Anderson in Bulacan and Tayabas in early 1943 – 'pro- Japanese Filipino agent' obtained Anderson's roster leading to purge that hurt strength — had agents in Manila, Mt. Kanlaon, Bulacan, Rizal, Cavite, Bicol – by October 1944, had 20,000 members.

- East Central Luzon Guerilla Area (ECLGA): MAJ Edwin P. Ramsey (26th U.S. Cavalry) sent by Thorp to organize guerrillas in Montalban, Rizal – by September 1944, Ramsey claimed 45,000 members, 7,000 armed – fought with Hukbalahaps – lacked direct communication and relayed through Fertig and then Smith – guerrilla 4th Ordnance Detachment under Pedro Villaluz operated in Zambales, Tarlac, Pangasinan and Nueva Ecija working with Ramsey.

- Hukbalahaps (*Hukbong Bayan Laban sa Japon* -"People's Army to Fight the Japs"): General Mateo del Castillo's large independent group in Bulacan and Pampanga – used communists Chinese communist model as 'military phase of the United Party Front' — from 10,000 members in 1942 grew to 100,000 by 1944 –areas in Tarlac, Zambales, Bataan, Nueva Ecija, Pangasinan, Rizal, Manila and Laguna under their control –opposed and fought all other guerrilla groups and anyone 'pro-American.'

- The Hunters (aka The ROTC, Terry's Hunters): Terry Magtanggol Adevoso organized 300 Philippine Military Academy and ROTC cadets personnel in the Antipolo Mountains -- with COL Ramirez's 34th Division until he was captured – openly fought with Marking in early 1944 and with Huks -- Anderson intervened with loose association.

- Lapham's Guerrillas: COL Robert B Lapham (CPT, 26th US Cavalry) in North Central Luzon, Nueva Ecija and Pangasinan – well organized 13,000 man force -- sent sub-groups under CPT Albert Hendrickson to western Tarlac and Ray Hunt to San Quentin, Pangasian –SWPA sent LT Enrique Torres to assist in intelligence matters.

- Marking Guerrillas: Marcos Villa Agustin founded in April 1942 in the Sierra Madre Mountains in Rizal - spread throughout central Luzon -- - "backbone of the organization is a woman known as Yay Panillo." – by May 1944, claimed 200,000 members, reportedly 5,000 armed – supported by Anderson

- President Quezon's Own Guerrillas (PQOG): Vicente Umali, former Mayor of Tiang, Tayabas, organized this group in central Laguna, Batangas and western central Tayabas in early 1942 – claimed by Marking – contact with Anderson and Fertig– claimed 10,000 members in 11 regiments.

- Tayabas Guerrilla Vera's Army (TVG): General Guadencia Vera (PVT Philippine Scouts) - about 100 men women and children in early 1942 – ruthlessness and banditry – contact with LT Bob Stahl from SWPA in May 1944 – by July was a well organized paramilitary unit of 1,000 – secured Bondoc Peninsula.

Bicol Area of Southern Luzon (7 November 1944)

COL Montano Zabat of Albay, MAJ Lapus and Governor Escudero of Sorsogon fought for leadership in this area. None were recognized by SWPA as district commander.

- **Escudero:** with17 men from Lapus, Governor Salvador Escudero kept government going when guerrillas disbanded – supplied P20,000 to Lapus but accused him of "embezzlement and bandity" – broke with Lapus by 1943 -- built 1,500 man group in Sorsogon province – fought with Lapus' rebuilt group – March 1943 Escudero fell ill and was evacuated to Samar – Lapus convinced Merritt to drive Escudero off Samar – June had 300 armed men -- sent son Antonio to Panay for authority to organize all of Bicol – July 1943 announced Straughn had promoted him to COL and warned Peralta he would not tolerate recognition of anyone else in Bicol – gained support of wealthier citizens – March 1944 drove Lapus from province – established contact with LTC Smith on Samar – April 1944 rebuffed attempt by Lapus to secure cooperation – May 1944 rebuffed Barros liaison from Anderson.

- **Lapus' Guerillas:** MAJ Licerio P. Lapus (Provincial Inspector of Sorgoson PC) evacuated Sorgoson in December 1941 and conducted raids – built large organization around Carachayon temporarily disbanded but sent 17 men to Governor Escudero – developed conflict with Escudero – December 1942 Peralta recruits Lapus as 67th Infantry Regiment – stays with Peralta as the 54th Regiment (5th MD) - August 1943 Lapus ceded command to Sandico in 56th Regiment - Escudero charged Lapus with unlawful declaration of martial law – Peralta created bloody competition between Lapus, Zabat and Escudero – Lapus persuaded Miranda to join 54th Regiment – February 1944 made contact with LTC Smith on Samar – Escudero evicted Lapus from province – Lapus contacted Andrews on Negros, Merritt on Samar and Miranda on Camarines Sur – July 1944 Smith gave Lapus P8,000 – 19 October Lapus requested GHQ immediately recognize him as CO of 5th MD – claimed 2,600 armed men with a division in reserve (probably closer to 1,500 poorly armed men) in Albay Province, west coast of Sorsogon and part of Ticao Island – "most military" in Bicol.

Northern Luzon (16 November 1944)

No regional commander appointed.

- **14th Infantry and Coordinated Command (Nakar):** MAJ (later LTC) Guillermo Z. Nakar assumed command of remnants of the 14th Infantry Regiment mainly in Nueva Vizcaya – made radio contact with SWPA from June to September - captured near Cabanatuan, Nuevo Ecija -- executed in November and as many as 5,000 of his men surrendered or dispersed -- LTC Manuel Enriquez assumed command and coordinated 43rd, 121st and 14th Infantry units with HQ in a Nacoco Store in Baguio -- Enriquez detained by the Japanese but released in October 1943 General Amnesty -- reorganized the guerrillas in the Mountain Province -- December 1943 began radio communications from Baguio through CPT Ali Al-Raschid, Chief of Police of Baguio -- late 1943 swindler Franco Vera Reyes obtained a 14th Infantry roster -- Enriquez disbanded most of his unit -- Japanese raid on the Nacoco Store rounded up most of the remainder of Enriquez's men -- Enriquez was captured in February 1944.

- **26th Cavalry (Praeger's Guerrillas):** CPT (later MAJ) Ralph B Praeger kept elements of Troop C of the 26th Cavalry (PS) active first near Tuguegaroa, Cagayan, then east of Vigan in Ilocos Sur -- former Cagayan Province governor Marcelo Adduru worked closely with Praeger before he went to Manila in June 1943 to surrender and secure a position with the Japanese to funnel intelligence to the guerrillas -- LT Bonito Bulan worked with Adduru in Isabela -- Adduru and Praeger were captured in August 1943 and executed in November.

- **121st Infantry (Barnett Guerrillas):** MAJ Walter Cushing commanded remnants of the 121st Infantry Regiment mainly in Abra and La Union Province in late 1942. -- Japanese captured most of the guerrilla leaders in Abra -- Cushing was captured in Pangasinan in September 1942 -- CPT William Arthur succeeded him but was captured in later 1942 -- MAJ George Barnett took command northwest of San Fernando, La Union and played havoc with enemy communications in the province – joined Volckmann.

- <u>Ablan's Guerrillas:</u> Roque Ablan, former governor of Ilocos Norte, established group in the Ilocos-Abra area but under pressure from the Japanese went inactive with about 100 men around April 1943 in the hills near Carasi, Ilocos Norte.

- <u>Lapham's Guerrillas:</u> CPT Lapham met with CPT Ball near Baler, Tayabas, in May 1944 and received a radio -- maintained contact with SWPA after August 1944 -- maintained an uneasy separation from Volckmann.

- <u>Volckmann's Guerrillas:</u> After the capture of Moses and Noble, MAJ (later COL) Russell W. Volckmann assumed command -- no contact with SWPA for a long time -- coordinated with remnants of the 14th after Enriquez was captured -- Volckmann reorganized guerrillas into the United States Forces in Philippines (Northern Luzon) (USFIP-NL) with three partially armed regiments of the 11th Division totaling 10,000 men -- radio contact began in September 1944 reporting on all 1st Military District in Northern Luzon -- robust program of intelligence gathering, sabotage, ambushes and training – few supplies from SWPA reached Volckmann.

Visayas:

Bohol Island (15 December 1944)
In December 1943 SWPA recognized MAJ Ismael P. Ingeniero as the Bohol Area Commander (not including Cebu).

- <u>"Behind the Clouds":</u> 3LT Ismael P. Ingeniero formed the largest band in Bohol, the "Behind the Clouds" guerrillas, in June 1942 in the north and central part of the island -- inclined to take orders from his old bass Gador, whom Ingeniero alone recognized as commander of Negros Oriental.

- <u>Boforce:</u> Ingeniero assumed the rank of MAJ and command of all Bohol forces in November 1943 -- developed well-organized guerrilla Boforce with a HQ at Carmen at the center of Bohol -- organized several regiments of battalions and a "Women's Auxiliary Service' (WAS) that produced clothing and equipment and raised funds for the guerrillas -- circulated a newspaper, "Bolos and Bullets," edited by attorney G. Lavilles -- Ingeniero claimed command of the 8th MD including Cebu and Bohol and extended intelligence to Cebu City – strained relations with Cushing's Cebu Area Command -- SWPA delivered some supplies in late 1943 -- In December SWPA recognized Ingeniero as the Bohol Area Commander (but not Cebu) and sent him a radio -- June 1944, plot by his junior officers to kill and overthrow Ingeniero disrupted by the arrival of Senator Carlos P. Garcia -- Ingniero reportedly went to Panay -- 23 June, Japanese force landed and swept Bohol, dispersing the guerrillas -- end of July 1944, Japanese withdrew except for a small garrison -- Ingniero's deputy, CPT Esteban Bernido, reorganized the remaining men -- Ingeniero returned in August/September -- Boforce sent agents to Cebu, Negros and Leyte to request arms and ammunition and a radio -- end of September, Ingniero reported to SWPA that his force was active against the Japanese garrison -- Boforce was reorganized into the 84th, 85th, and 86th Regiments with a total of about 8,000 poorly armed men.

Cebu Island (15 November 1944)
LTC James Cushing recognized as CO of Cebu area and sent him supplies by SWPA -- established direct radio contact in March 1944

- <u>Cushing-Fenton:</u> Harry Fenton (American born Aaron Feinstein) coalesced several small bands of guerrillas in north Cebu by mid-1942 – merged with LTC James Cushing southern guerrillas -- Fenton hung all suspect collaborators and spies -- Cushing was protective of the Filipino civilians -- administrative HQ at Fenton's camp in Maslog and combat HQ seven miles away at Cushing's camp at Mangalon Heights – March 1943 Japanese campaign in Cebu forced Fenton to suspended operations – mid 1943 food and currency ran short -- Fenton-Cushing had about 9,000 men, 1/2 armed, and 1/3 civilian volunteers – popular Governor Hilario Abellana carried on government in exile – July 1943 Japanese captured Fenton's wife and children and Cushing's brother – Cushing went to Negros in August to meet Villamor, leaving orders with his executive officer LTC Richard Estrella to arrest Fenton -- 15 September, Estrella arrested and executed Fenton and his loyal officers -- Estrella then reorganized his remaining 3,500 men -- Cushing returned to Cebu in November and found Estrella in command with evidence that he accepted P60,000 from the Japanese to deliver Cushing -- through early 1944 Cushing rebuilt his unit into the 85th, 86th, 87th, and 88th Regiments with 5,687 men and some 2,700 mixed arms -- June 1944 Cushing claimed 25,000 volunteer guards and faced revived Japanese attacks in August that damaged Cushing's organization -- Cushing requested command of Bohol (also part of 8th MID). Fertig and Cushing claimed Bohol – through which Fertig sent supplies -- Fertig (Mindanao), Peralta (Panay) and Abcede (Negros) maintained agents on Bohol – April 1944 captured Japan's Z Plan.

Leyte Island (as of 7 October 1944)
COL Rupert K. Kangleon appointed Leyte Area Commander on 21 October 1943. 209 officers and 3,190 men organized as the 92nd Division, Philippine Army.

* Miranda Group: BG Blas Miranda (alias COL Briguez and former LT in USAFFE . PC) – started from Palompon to Baybay – refused to serve under Kangleon – strongly influenced by 6th MD Commander Peralta – Kangleon drove Miranda and some of his top men to Bohol and – reorganized group into 96th Regiment of Leyte Command.

Negros Island (10 December 1944)
SWPA appointed LTC Salvador Abcede Acting 7th District Commander in July 1943, confirmed in March 1944, and made official in October 1944.

* Abcede: LTC Salvador Abcede (CPT, PA), refused to surrender with COL Hillsman and turned battalion with supplies and 600 rifles into guerrilla unit in the central and southern Negros Occidental area in July 1942 – with officers from nearby plantations, maintained supply sources -- MAJ Enrique Torres formed a unit for Abcede near Binalbagan -- pushed into the mountains with many families -- early August, fight with Japanese near Buenavista -- reached out to LTC Mata in the north and Ausejo in the south – by November 1942, Abcede had 7,000 men and joined Peralta's IV Philippines Corps as LTC and commander of the 72nd Division (Negros) -- reorganized into the 73rd Provisional Division in Negros Oriental -- Japanese attack around Kabankalan-Binalbagan -- January 1943, IV Corps disbanded but Peralta continued to support Abcede -- 8 July 1943, SWPA replaced Villamor with Abcede as acting commander of 7th MD -- Japanese increased attacks on the south coast from Dumaguete to Sipalay -- food production collapsed -- Abcede dispersed his men and supplies, established early warning systems, and maintained high levels of secrecy -- developed highly productive intelligence networks and assisted with a number of downed pilot recoveries -- evacuation of American left Negros with an entirely Philippine guerrilla organization.

* Ausejo and Bell: Dr. Henry Roy Bell (American Professor organized the evacuation of the Siliman University equipment and personnel to Malabo and Lake Balinsasayao when Japanese landed on Negros -- assisted in government affairs in south Negros -- Bell set up a camp at Malabe, west of Dumaguete -- slowly students and alum formed Bolo Battalions: ROTC instructor MAJ B.N. Viloria led one battalion; an USAFFE LT from Mindanao led another at Malabo; high school teacher CPT Felix Estrada and CPT Leon Flores each led battalions near Dumaguete; Victor Jornales and SGT David Cirilo led two more -- all under the guidance of Bell -- confined the Japanese to their garrisons in Dumaguete, Bais, and Tanjay --local Chinese merchant Manuel Sy Cip obtained supplies – after Gador declined Bell's request to take command, Bell recruited MAJ Placido Ausejo in October 1942 -- renamed their 1,000-man force as the 75th Regiment with headquarters in Malabo -- Bell got food and money from Fertig on Mindanao in late 1942 in return for the 75th Regiment joining Fertig's command -- February 1943, USAFFE LT Louis Vail helped Bell make radio contact with SWPA -- May, SWPA pulled the 75th from Fertig for Villamor 's 7th MD -- Villamor appointed Bell as a MAJ under and Chairman of the 7th MD Research Board -- Vail become 73rd Division Signal Officer -- Ausejo became Abcede's G-3 in the 7th MD -- Bell served as senior Civil Administrator -- June and November 1943, Japanese raids burned Bell's camps and he was nearly captured with his family in November -- early 1944 Bell and his family were evacuated via submarine to Australia.

* Gador: LTC Gabriel Gador was the 7th MD Commander (Negros and Siquijor) when war began -- General Sharp fired Gador from 7MD and moved him to staff on Mindanao – Gador claimed Sharp ordered him to return to Negros to organize guerrillas -- returned to Negros in June 1942 and remained hidden until August/September -- emerged in central Negros Oriental and accepted Mercado's offer to command his guerrillas -- obstructed surrounding organizations with arguments and pilfering of personnel – clashed with other guerrillas -- maintained discipline, established

training schools, improved communications, and set up an apparently effective civil government -- little actual engagement with the enemy -- had perhaps 2,000 poorly armed men -- began promoting relatives to senior positions -- October 1942, Mercado broke with Gador -- Gador refused Bell and Abcede who asked him to take command of all of Negros -- when Abcede united Negros under Peralta's IV Corps, Gador declared his command over Negros on orders of Sharp -- January 1943, Fertig claims command of all Mindanao and Visayan guerrillas and asks Gador to Mindanao to discuss his role -- Gador announced himself as commander of all the Philippines as Major General -- ignored by all other guerrillas -- April 1943, Gador rebuffed Villamor's call for unity on the behalf of SWPA -- Peralta supported Mata, Fertig supported Ausejo, Villamor supported Abcede -- May 1943, SWPA appointed Villamor temporary commander of the 7MD --- Mata, Abcede and Ausejo gave support to Villamor -- Gador agreed to be Executive Officer but resigned and resumed his opposition – undermined Villamor's mission – September, Gador refused to support Abcede as acting 7th MD commander -- threatened to disrupt Ausejo's organization, intelligence collection, and communication to Cebu -- December 1943, former Chief of Staff LTC Aspilla led a coup against Gador who fled to Bohol with a handful of loyal subordinates -- Aspilla brought Gador's remaining organization under 7th MD command -- on Bohol Gador reported to MAJ Isamael P. Ingeniero -- Gador disappeared after the Japanese landed in Bohol in late June 1944 until December 1944 when he went to Leyte.

- Villamor— SWPA inserted MAJ Jesus Villamor into southern Negros in January 1943 when Abcede, Ausejo and Gador were arguing and Peralta and Fertig fought over control of Negros Villamor ordered to remain out of guerrilla affairs -- Gador exploited the lack of Villamor's authority -- appointed by SWPA as temporary 7th MD commander in April 1943, organized a District HQ on Negros with Mata, Ausejo and Abecede -- appointed Bell as Negros Civil Administrator -- set up a Research Board and civil government for free Negros under Negros Occidental pre-war governor Alfredo Montelibano -- some of Gador's officers defected to Villamor's command -- 8 July 1943, Villamor relieved and Abcede appointed acting commander of 7th MD – October 1943, SWPA evacuated Villamor to Australia and replaced him with LTC Edwin Andrews (PA) as SWPA agent for 7th MD intelligence.

Panay and Surrounding Islands (27 November 1944)
February 1943 Peralta was officially appointed as commander of the 6th Military District which included Panay, the Romblons, and Guimaras Island.

- 61st Philippine Army Division: General Christie, division commander, followed orders to surrender but allowed his Filipinos to resist. Few surrendered. Division G-3 CPT Macario Peralta took troops to the northeast, Division Engineer Leopoldo Relunia went to eastern Panay, and 3rd Battalion, 63rd Regiment, Commander Julian Chaves headed to central Panay. Officers Braulic Villasis went to Capiz and Cirilo Garcia went to the northwest. They all promised Christie they would not conduct guerrilla resistance until two months after the surrender. The officers, however, were Tagalog (central Luzon), outsiders on Panay, and held in suspect by local Visayans.

- Tomas Confessor, Governor of Iloilo: The governors of Antique and Capiz surrendered but Confessor went to the hills with guerrillas. He organized the Provincial Guards and a messenger organization. Confessor remained in Iliolo and reconstructed civil government despite occasional pressure from Peralta. He managed to maintain a radio to broadcast news to the people of Panay and protected the people from the guerrillas but agreed to collect taxes and give 3/4s to Peralta's guerrillas.

- CPT Macario Peralta Peralta – took troops to northeast Panay -- began organizing resistance in August 1942 -- proposed martial law – reunited other parts of the 61st Division by end of November – contained the Japanese in San Jose (Antique), Capiz town, and Iloilo City and established radio contact with SWPA -- promoted to LTC on 13 January 1943 and to COL in August -- tolerates but ignores Americans -- competed with Confesor -- LTC Pedro Serran created and ran effective intelligence network for two years -- attempted to unite the guerrilla movements across the Islands into a Philippine IV Corps with 61st Division under Relunia – maintained contacts in Negroes, Leyte, Northern Luzon. Bicol and Samar – expanded influence to Masbate, Marinduque, Mindoro and Palawan -- only Cebu and Mindanao defied Peralta -- February 1943 SWPA appointed Peralta commander of the 6th MD (ended the IV Corps) -- July 1943 the Japanese attacked, captured and destroyed stocks of supplies – late 1943 the Japanese resumed attacks in northwest Panay and the Romblons – Peralta reorganized into Combat Teams reported as 22,600 in October 1944 with 8,000 arms of various types and about 160 rounds of ammo per weapon -- between mid-1942 and October 1944 SWPA delivered 350 tons of supplies to Panay.

- **Marinduque:**
6th MD took control of small guerrilla force on Marinduque in 1942 as a base for penetration into Luzon.

  - LT Sofronio T. Untalan (PC commander at Boac): led his men into the hills when the Japanese landed on 7 July 1942 -- surrendered on 20 July -- paroled at end of 1942 -- visited Peralta and promoted to CPT-- January 1943 returned to Marinduque as Peralta's commander of Company M under LTC Garcia's 60th Infantry Regiment -- January 1944 declared truce with the Junior BC of Marinduque, LT Rudolpho Tescon -- March and April 1944 reportedly assumed command of the Romblons -- October 1944 claimed 400 men with 90 arms in 4 companies.

- **Romblon:**
6th MD took control of a weak and insignificant guerrilla force on Romblon in 1942 as a base for penetration into Luzon.

  - MAJ Enrique Jurado -- maintained base for radio and intelligence operations -- February 1943 organization became weak and inefficient – commandeered supplies from civilians -- executive officer CPT Mario Guarinia reported to be a politician allied with LTC Garcia in northern Panay -- inducted American refugees -- expanded to Sibuyan Island -- strength grew to about 700 by July 1943 -- November 1943 Japanese campaign in Sibuyan -- many Romblon guerrillas surrendered -- organization broken, and equipment captured or destroyed -- Jurado escaped to Mindoro and became commander there in March 1944 -- Guarinia remained in Romblon to reassemble the guerrillas as part of LTC Garcia's 1st Combat Team -- Guarnia reportedly surrendered to the puppet government officials on Sibuyan and moved to Manila leaving a vacuum in Romblon that Untalan reached out to fill.

- **Mindoro:**
6th MD claimed Mindoro without approval of SWPA.

  - Beloncio Group: CPT Esteban P. Beloncio organized 250 guerrillas with 150 arms in 1942 around Lake Naujan -- joined with Ruffy in August 1943 in the Bolo Battalion -- influenced by Peralta and supported by Jurado -- broke with Ruffy at the end of March 1944 – Beloncio created a battalion of about 600 men with 230 arms with A Company under CPT Jose L. Garcia in northern Mindoro, B Company under 2LT Gomeraindo de la Toore in central Mindoro, and C Company under 3LT Ruel G. Beloncio in northern Mindoro -- November 1944 Beloncio reported killed in a fight with local guerrillas.

  - Jurado Group: Jurado arrived from Romblon late 1943/early 1944 with men from 1st Combat Team to observe the Verde Island passage and establish a base for penetration of Luzon -- after SWPA MAJ Lawrence H Phillips' death, entered into the Mindoro guerrilla dispute – March, Peralta promoted Jurado to LTC in charge of Mindoro -- Ruffy ordered him to leave – Jurado promoted Beloncio to replace Ruffy -- Bolo Battalion collapsed -- July 1944 SWPA's CMDR George F. Rowe (USNR) arrived in western Mindoro to establish a radio net and kept out of guerrilla affairs and local politics – by October 1944 Ruffy and Jurado were actively fighting, despite SWPA pleas to unite.

  - Ruffy Group: MAJ Jose M. Ruffy organized 60 constabulary troops and about 200 local volunteers near Pinamalayan into four companies into a Bolo Battalion – clashed with Beloncio's guerrillas -- November 1942 SWPA MAJ Lawrence H Phillips arrived – December Ruffy and Beloncio asked him to mediate -- they agreed Ruffy would command the Mindoro provisional guerrillas with Beloncio as executive officer – Jurado arrives from 6th MD -- February 1944 Ruffy had 23 officers and 600 men with local governments established and volunteer Home Guard formations -- March 1944 Phillips killed by the Japanese -- orders Jurado to leave -- Beloncio and

Ruffy fell out and the Bolo Battalion collapsed – Beloncio (supported by Peralta through Jurado) took the A, C, and D company to the 6th MD -- Ruffy took B Company to central Mindoro and built his force to 300 to 400 men.

- **Palawan:**
  6th MD sent units to take control of Palawan guerrillas in October 1943.

  - Carlos Amores: secretly organized 200 men into a guerrilla group to resist the Japanese takeover of the manganese mines in Busuanga-Coron in 1942 -- September the Japanese discovered the group – uprising with few arms killed all the Japanese at the mines and many more in the city -- sealed the mine with dynamite and destroyed stocks of ore before being driven into the hills -- Amores went to Danlig to ask to the Cobb Brothers for aid -- returned to Busuanga but Japanese attacks and lack of food forced Amored to leave with 100 men for Sibaltan in northern Palawan where he joined forces with the Cobb Brothers as commander of C Company, Palawan Special Battalion.

  - Palawan Special Battalion: October 1943 LT Garcia returned from Panay to Palawan in with MAJ Pablo Muyco of the G-3 section of the 6th MD -- got all local guerrilla leaders, except Manigque, to merge into the Palawan Special Battalion of the 6th MD with Muyco as CO and Garcia as XO -- organized four companies: A Company under CPT Higino Mendoza at Malcampo, B Company under 3LT Felipe Batul at Danlig, C Company under CPT Carlos Amores at Taytay, and D Company under CPT Narizidad Mayor at Brooke's Point -- by early 1944, battalion had 57 officers and 954 men with about 300 arms -- covered Palawan and limited coverage of Balabac, Cuyo and Agutaya, Cagayancillo, Busuango, Culion and Coron. Dumaran Island -- at least two agents in every town.

  - Southern Palawan Group: Several groups developed around a number of Americans near Brooke's Point -- August 1942 three US Navy sailors and three US Marines escaped from a prison camp and joined them -- October 1942 Japanese landing force driven off by the guerrillas -- one American leader killed by a renegade Filipino and the others dispersed to Tawi Tawi -- Vens T. Kerson kept the Brooke's Point guerrillas together with the aid of constabulary SGT Tumbaga, American planter Thomas Edwards, and Datu D.M. Jolkipli Narrazid -- they built a local civil government and a Bolo Battalion -- Tumbaga, promoted to 3LT, succeeded Kerson in July 1943 – Kerson drowned in December 1943 -- LT Alegre assumed command.

Samar Island (10 October 1944)
LTC Charles Smith appointed Commander of Samar area 4 October 1944 approximately 8,500 men – 2/3 with prewar training – but lacked trained officers.

- Abia Group: CPT Luciano Abia (former constabulary officer) – at time of surrender organized 100 guerrillas at Basey – later joined with Valley under Causing.

- Causing Group: COL Juan Causing -- former Kangleon's chief of staff – sent to Samar in September 1943 to unify guerrillas but only got Vally to ally with Kangleon

- Merritt Group: COL Pedro V. Merritt in the mountains in northwest Samar minus Japanese towns of Calbayog and Catbalogan – early 1944 became 93rd Division with 322 officers and 1,408 men in four regiments — refused to work under Kangleon.

- Vally Group: MAJ Manuel Vally — organized 30 guerrillas in October 1942 in Guiuan, Salcedo, Balangiga, Basey and Pambujan Sur -- joined Kangleon – refused to join Merritt – by February 1944 group numbered 1,200 men.

Southern Islands:

Mindanao and Sulu (31 January 1945)
Fertig was named Commander of the 10th Military District (Mindanao and Sulu) and ordered to organize his forces and develop his intelligence net on Mindanao and Samar-Leyte. Sulu was later separated and placed under COL A. Suarez.

- LTC Robert Bowler: at Bukidon at the time of the surrender and went into the hills until August 1942 -- gathered several guerrilla bands formed by other American officers -- second in command, his adjutant, was LTC Ciriaco Mortera – joined Fertig in late 1943.

- LTC Clyde Childress: was given command of the 107th Division in early 1944 from elements of the 110th Division in the area south of a line through the Agusan Province down through Lianga, Surigao, and Davao City.

- LTC Wendell W. Fertig: American mining engineer in Mindanao to construct an airfield refused to surrender and took to the hills at Kolambugan, Lanao – September took over Morgan's organization -- organized Morgan's and Tate's force as the 106th Regiment -- assumed command of all guerrillas in Mindanao with headquarters at Misamis – recruited scattered bands across Mindanao – organized administrative divisions based on prewar tables of organization for a Philippine Army Reserve Division -- October/November 1942, organized a civil government -- LTC Charles W. Hedges to Lanao Province -- McClish to Misamis Oriental in November with MAJ Clyde C. Childress as his chief of staff – LT Robert Ball remained at Fertig's HQ and became district communication officer -- February 1943, Parsons brings supplies and communications from SWPA -- Quezon authorized Mindanao Currency Board to print emergency currency -- July 1943, the Japanese reoccupied Misamis Occidental and north Zamboanga and chased Fertig to Liangan, Lanao -- appointed LTC Robert Bowler (instead of Morgan) to succeed him if he was captured or killed - late 1943, Japanese pressure Fertig's District Headquarters -- Fertig moved it to the Agusan Valley in January 1944 and set up a subcommand called "A" Corps under \ Bowler for communication and control -- replaced Bowler as commander of the 109th Division with LTC James Grinstead – sent his district communications officer, Ball to Luzon to develop contacts and intelligence there.

- LTC James Grinstead: recruited by Fertig in February 1943 to organize guerrillas -- later appointed commander of the 109th Regiment and later still chief of staff of the 109th Division.

- MAJ Manuel D. Jaldon: organized the guerrilla bands around Alubijid, Misamis Oriental -- May 1943, reorganized the 121st Regiment – June, Fertig learned that Jaldon was persuaded by his brothers to meet with the Japanese in Zamboanga and sign a peace agreement -- Fertig suspended Jaldon and the regiment -- No supplies reached the unit until December when MAJ Felipe Fetalvero arrived to disbanded the regiment and form a new 1st Separate Battalion with former American enlisted man LT Donald Lecouvre in command.

- LTC Claro B. Laureta: left as commander of constabulary Camp Victor at Davao after the surrender and went into the hills with about 30 of his men -- found 4,000 civilian refugees along the Libuganon River -- provide them law and order - organized farms that supported his guerrillas – by 1943, had become the sole authority of this new community with a headquarters at Maniki -- Laureta became the acknowledged resistance leader in the vicinity of Davao -- absorbed the active guerrilla bands near Hijo and Kingking, at Saug, and Compostela -- early 1943, joined Fertig's 110th Division -- reported to Misamis Oriental in July and accepted organization as the 130th Regiment of the 110th Division -- early 1944 stood up part of the 110th Division as a separate 107th Division.

- LTC Frank McGee: was in Bukidnon in May 1942 and fled to the mountains east of Malayba -- when Pendatun's guerrillas came to the area in December, McGhee joined them – appointed by Fertig to command the Cotabato-Bukidnon guerrillas in new 106th Division.

- LTC Ernest McClish: organized guerrillas in Imbatug, Bukidnon in August 1942 -- expanded to Balingasag, Misamis Oriental over several months with fellow Americans LT Robert Ball, LT Anton Haratik, and CPT Wiliam Knortz – January 1944, relieved as commander of the 110th Division and evacuated to Leyte for health issue -- CPT Marshall assumed command of the division on 21 January.

- CPT Luis Morgan: Philippine Constabulary company commander at Kolambugan, Lanao -- organized his troops to defend the Christian civilians and fight the Moros around Baroy in the north plain of Lanao – gained reputation of ruthlessness, daring and lack of consideration for civilians -- September 1942, Morgan and Tate successfully liberated Misamis Occidental and the north coast of Zamboanga – tried to recruit Fertig as figurehead leader while Morgan would run field units as chief of staff -- January 1943, left with 80 men for Leyte and Negros to recruit guerrillas for Fertig --March led a failed attack on Butuan with LTC McClish and his guerrillas went to Leyete in April 1943 and tried to unite the 9th MD an (Leyte and Samar) under Kangleon went to Siquijor and recruited guerrilla leader MAJ Benito Cunanan to command Cebu -- went to Negros and tried to recruit LTC Gador into line with the 7th MD -- met MAJ Villamor on Negros -- returned to Mindanao in June -- Japanese landed unopposed in Misamis in June -- Morgan responded by attacking Misamis to take arms from the guerrillas who had fled – angered when Fertig appointed LTC Bowler as successor to 10th MD command -- September 1943, Morgan quit the 10th MD and set up his own command in Misamis Occidental – Fertig sends him to Australia at the end of September.

- Salipada Pendatun: with his brother-in-law Datu Matalam Udtog organized a band of Moros into an untrained Bolo Battalion in the vicinity of Catabato before Corregidor fell -- armed with knives, the battalion set out to block the Digos-Kabakan Road but it quickly fell apart when it met the Japanese -- August 1942, Pendatun led attack on the Japanese garrison at Pikit, Cotabato success attracted many Moros to join -- Pendatunn then attacked Kabakan and secured the Digos-Kabakan Road by September -- joined by Datu Aliman, Datu Mantil Dilanglan and Gumbay Piang -- December 1942, Pendatun reorganized his Moro guerrillas into the Bukidnon- Cotabato Force and appointed LTC Edwin C. Andrews as his Chief of Staff -- January 1943, attacked Malaybalay from the south without promised support from LTC Bowler and failed in costly siege -- established radio contact with Fertig and accepted an offer to join his 10th MD as the 117th Regiment with Pendatun as their COL -- friction arose as Pendatun resented Fertig's presumptions -- Bukidnon assemblyman Manuel Fortich helped settle the friction but disagreement remained until Parsons arrived in May -- Fertig then sent Andrews to Negros to join Villamor –Pendatun formed the 118th Regiment with LTC Soriano, MAJ Gabutina, Datu Aliman and Datu Udtog -- regiment kept the Digos-Kabacan Road closed to enemy traffic until early 1944.

- Gumbay Piang: captured by the Japanese who released him to house arrest in Cotabato to appease the Moros -- when the Japanese revealed they had evidence of his involvement with guerrillas, he fled to the jungle to join his men full time -- joined guerrillas in Cotabato -- joined the 10th MD with his men in September 1943 as the 119th Regiment.

- COL Alejandro Suarez: arrived in Bato Bato in January 1943 – integrated 1LT Trespeces' 30 Sulu guerrillas with those on Tawi Tawi -- consolidated LT Imao's units-- made contact with Fertig and became the 125th Regiment of the Mindanao Command in March-April 1943 with one battalion of 350 men in Tawi Tawi, another of 250 men on Siasi, and a third on Jolo of 200 men -- maintained his headquarters on the Malum River near Bato Bato -- June 1944, Japanese attacked the first battalion of the 125th Regiment on Tawi Tawi.

Appendix B – 23
Guerrilla Radio Net, June 1944

Source: Morningstar, James Kelley. War and Resistance in the Philippines, 1942-1944. Naval Institute Press, 1 April 2021. With permission from James K. Morningstar.

Source: Military intelligence Section, General Staff, General Headquarters, United States Armed Forces Pacific, Intelligence Activities in the Philippines During the Japanese Occupation (Volume II, intelligence series) (Tokyo, Japan: 10 June 1948), 47-52 and Plate 10.

NET	Call Sign	Vicinity	Island	Remarks
Abcede-Andrews	?	Manila	Luzon	Sent by Villamor in Nov 43. Contact not made until Nov 44.
Peralta	?	Manila	Luzon	Sent by Peralta, disaapears from the net)
Smith	GRT	NE Tayabas	Luzon	Operated by CPT Anderson, USAC
Fertig	UAM	Tayabas	Luzon	Sent from Mindoro for PQOG 24 Feb 44
Smith	UU2	Baler, Tayabas	Luzon	Operated by CPT Ball
Smith	MAA	Bondoc	Luzon	Sent from Samar 17 Jan 44
Smith	S3L	Bondoc	Luzon	Operated by W/O Stahl, May 44
(independent)	ISRM	Paluan	Mindoro	Operated by MAJ Phillips
Smith	MAG	Sorsogon	Luzon	Sent to Escudero, Operated by LT Chapman
Smith	MAB	San Fernando	Masbate	Coastwatcher, moved to Ticao Island, and captured in May 1944
Fertig	TUT (TUH)	San Isidro	Samar	TUT Coastwatcher Operated first by LT Chapman, over San Bernadino Straight, When Chapman goes to Sorsogon, replaced by TUH operated by MAJ T. Jurika
Smtih	MACA	Catubig Valley	Samar	Operated by Stahl, then Gordon Smth
Smith?	?	Calbayog	Samar	uncertain station, possibly TUT
(independent)	NIX	Bongabong	Mindoro	Operator LTC Jurado, Romblons
Peralta	TAR	Pucio Pt	Panay	coastwatcher sent by SWPA, May 43
Peralta	DLA	Culasi	Panay	61st Division control station
Peralta	MLR (KSA)	Dumalag	Panay	6th MD contact station, used KSA to 10th MD
	YMN	Dumalag	Panay	alternate 6th MD control station
Peralta	DXR	Carles	Panay	coastwatcher sent by SWPA, May 43
Peralta	CBY	Concepcion	Panay	coastwatcher
Peralta	ANA	Barotac Viejo	Panay	coastwatcher
Smith	MAF	Santa Rita	Samar	Guerrilla contact station
Smith	MAC	Pambujan Sur	Samar	operated by SGT Herreria
Peralta	RMM	Bugasong	Panay	contact for 65th Inf and coastwatcher at Culasi
Peralta	RKS	Maasin	Panay	Ilolio coastwatcher, intell from Ilolio City
Peralta	FAU	San Joaquin	Panay	Coastwatcher
Peralta	AGC	San Remigio	Panay	Dao coastwatcher, sent May 43
Abcede-Andrews	?	Cadiz	Negros	Guerrilla contact station
Abcede-Andrews	?	Murica	Negros	operated by MAJ F. Soliven
Abcede-Andrews	?	e. Bacolod	Negros	operator by MAJ Benedicto
Abcede-Andrews	?	San Carlos	Negros	guerrilla contact station
Abcede-Andrews	?	Himamaylan City	Negros	with Governor Montelibano HQ
Abcede-Andrews	WGN	Candoni	Negros	7th MD alternate contact station
Abcede-Andrews	WPI (WBA)	Mabinay	Negros	7th MD control station. Used WBA to 10th MD and coastwatcher reports
Abcede-Andrews	YAF	Bayawan	Negros	Villamor's intell station, early 1943. became 7th MD admin control station

NET	Call Sign	Vicinity	Island	Remarks
Abcede-Andrews	FNS	Tolong	Negros	relay station
Abcede-Andrews	WSK	Santa Catalina	Negros	coastwatcher
(independent)	PGA	Catmon	Cebu	Cushing's control station
Abcede-Andrews	PNM	Naga City	Cebu	Radio sent by Villamor, operated by LT Alvarez
Smith	MAD	Mactan Island	Mactan Island	Operated by SGT Sanchez, coastwatcher and intelligence
(independent)	BMTD	Abuyog	Leyte	SWPA sent radio to Kangleon, captued in 44
Fertig	TUM	Saint Bernard	Leyte	Coastwatcher LT Richardson, USNR
Fertig	TUF	Loreto	Dingat Island	Moved from Panaoan Island
(independent)	VIM	Danao	Bohol	MAJ Igneniero control station
Fertig	MBF	Mahinog	Camiguin Island	Coastwatcher and intell. Operated by LT Snyder
Fertig	MBR	Claver	Mindanao	Coastwatcher and intell
Fertig	MBX	Gigaquit	Mindanao	Coastwatcher and area contact station
Fertig	MBA	Butuan	Mindanao	Guerrilla contact station
Fertig	MBG	Los Arcos	Mindanao	Coastwatcher and intell from Lianga Bay
Fertig	WPP	Butuan	Mindanao	Guerrilla contact station and intell
Fertig	MBS	Las Nieves, Agusan	Mindanao	Guerrilla contact station
Fertig	MBD	Esperanza	Mindanao	Guerrilla contact station
Fertig	MBJ	Santa Fe, Agusan	Mindanao	operator CMDR Wilson, 10th MD CofS
Fertig	MBY	Tugo, Surigao	Mindanao	Coastwatcher and area contact station
Fertig	MBC	Balingasag, Misamis Oriental	Mindanao	110th Regiment contact station
Fertig	HT9	Esperanza	Mindanao	Coastwatcher (formerly Z4V)
Fertig	MLX	Waloe, Agusan	Mindanao	10th MD contact station
Fertig	CH	Liangan, Lanac	Mindanao	Guerrilla contact station
Fertig	TAB	W. Misamis Oriental	Mindanao	109th Regiment contact station
Fertig	TAX	Sumilao, Bukidnon	Mindanao	111th Regiment contact station
Fertig	TAC	Talakag	Mindanao	109th Division contact station
Fertig	MWA	Waloe, Agusan	Mindanao	Guerrilla contact station
Fertig	MWQ	Bunawan, Agusan	Mindanao	Guerrilla contact station
Fertig	TD5	Loreto, Agusan	Mindanao	Guerrilla contact station
Fertig	WUS	N.C. Agusan	Mindanao	Guerrilla contact station
Fertig	MBQ	Port Lamon	Mindanao	Coastwatcher and area contact station
Fertig	WAP	Baliangao, Misamis Occidental	Mindanao	Coastwatcher and area contact station
Fertig	WAJ	Sindangan	Mindanao	Coastwatcher and area contact station
Fertig	XBV	Bonifacio, Misamis	Mindanao	Western Mindanao net control station
Fertig	WAN	n. Bonifacio	Mindanao	105th DIV contact station
Fertig	WAM	Lala, Lanao	Mindanao	"A" Corps control station under COL Bower
Fertig	WAG	Iligan, Lanao	Mindanao	120th Regiment contact station
Fertig	WAA	Pagadian, Zamboanga	Mindanao	115th Regiment contact station
Fertig	WYZB	Liangan, Lanao	Mindanao	10th MD alternate control station
Fertig	WAQ	Zamboanga City	Mindanao	Guerrilla contact station
Fertig	WAL	Vitali, Zamboanga	Mindanao	Coastwatcher and area contact station
Fertig	WAB	Malangas, Zamboanga	Mindanao	Coastwatcher and area contact station, operated by McCarthy
Fertig	WOO	Kapalong, Davao	Mindanao	Guerrilla contact station, operated by LT Laureta
Fertig	MWL	Moncayo, Davao	Mindanao	Guerrilla contact station

NET	Call Sign	Vicinity	Island	Remarks
Fertig	MBN	Caraga, Davao	Mindanao	Guerrilla contact station, inteligence collection, operated by LT Wilson
Fertig	REG	Lebak	Mindanao	Coastwatcher and intell
Fertig	ARC	Caburan	Mindanao	Coastwatcher and intell
Fertig	REC	Kiamba, Cotabato	Mindanao	coastwatcher
Fertig	REB	Baliton, Cotabato	Mindanao	Coastwatcher and intell
(independent)	FGQ	Languyan	Tawi Tawi	Operators LT Young and LTC Suarez

unlocated coastwatcher stations:
Panay: SIN, NIX (was UAE), CBY (was LEC), VKI (was NLR)
Negros: KNS, REA, JER, AFD, KLY, KSA, KRT (was RSW), DSR (was CBQ)

Appendix B – 24

RESTRICTED

HEADQUARTERS
FIRST RECONNAISSANCE BATTALION, SPECIAL
UNITED STATES ARMY FORCES IN THE FAR EAST

A. P. O. 565
1 Mar 1945

SPECIAL ORDERS)
NUMBER... 16)

1. Under the provisions of AR 615-5, the following-named EM Scout Co & Column, are appointed to the grade indicated:

TO BE TECHNICAL SERGEANT (TEMP)

S/Sgt	Cleto S. BELLO	39238037	S/Sgt	Rafael C. LICERA	39843410

TO BE STAFF SERGEANT (TEMP)

Sgt	Mero L. DE OCAMPO	39097470	Sgt	Felecisimo D. MAGBUAL	39843914
Sgt	Walter B. LAGRIMAS	39844692	Sgt	Apolinar G. SALVADOR	39235112

TO BE SERGEANT (TEMP)

Cpl	Gene D. COSTALES	37216159	Pfc	Gerardo C. ALFAFARA	36342823
Cpl	Urbano M. FRANCISCO	39526401	Pfc	Alfonso L. ANTONIO	39120977
Cpl	Frank S. LEANIO	39839199	Pfc	Lasaro C. ARQUERO	39187222
Cpl	Maximo V. MAGLANGIT	39397902	Pfc	Pedro S. AZCUETA	36350540
Cpl	Elpidio A. RIVERA	36337587	Pfc	Philip MARTIN	34202018
Cpl	Harry T. TUCAY	39236723	Pfc	Tom P. RESPICIO	39235454
Pfc	Perfecto M. TROCIO	36601493			

TO BE CORPORAL (TEMP)

Pfc	Ricardo D. CAYABYAB	39255604	Pfc	Victor V. MARQUEZ	33737506
Pfc	Julian F. CUANANG	39096248	Pfc	Robert C. MINDER	39229726
Pfc	Andres D. DARANG	39828848	Pfc	Benny T. VELAYO	39106003
Pfc	Peter B. DE LEON	39101202	Pvt	Fermin D. COSTALES	37216144
Pfc	Vicente J. GRANADOS	39846508	Pvt	Paul L. RAMOS	10645019
Pvt	Sergio P. RESPICIO	39836837			

2. Under the provisions of AR 615-5, following-named EM Hq & Hq & Service Co, are appointed to grades indicated:

TO BE SERGEANT (TEMP)

T/4th	Gr Thomas S. CACHO	39123704

TO TECHNICIAN FIFTH GRADE (TEMP)

Pfc   Luis M. PULANCO   39845632   Pfc   Mariano S. CALIBOSO   39203189

Under the provisions of AR 615-5 following-named EM, Medical Detachment, is appointed to grade indicated:

TO BE TECHNICIAN FIFTH GRADE (TEMP)

Pvt   Clinton M. BAKER   39309940

By order of Lieutenant Colonel BROWN:

OFFICIAL:
CEFERINO R. ROLA
1st Lt., Infantry
Adjutant

CEFERINO R. ROLA
1st Lt., Infantry
Adjutant

Source:
U.S. National Archives and Records Administration
MacArthur Archives, Norfolk VA

Appendix B - 25

SECRET
HEADQUARTERS
A.P.O 501

GENERAL ORDERS)
NO....................72)

10 April 1945

REORGANIZATION OF THE 978TH SIGNAL SERVICE COMPANY

1.   Effective 20 April 1945, the Headquarters Platoon of the 978th Signal Service Company is reorganized under Column six (6) of T/O & E 11-97 19 May 1944, with an authorized strength of two (2) officers and thirty-three (33) enlisted men, without change in station or assignment.

2.   Upon reorganization the authorized strength of the 978th Signa Service Company will be eleven (11) officers, five (5) warrant officers, two hundred seventy-nine (279) enlisted men and an aggregate of two hundred ninety-five (295).

3.   a. No enlisted man will be reduced in grade as a result of this action.
b. The company commander will report all personnel rendered surplus by this action to this headquarters by name, rank, serial number and specification of serial number.
c. A requisition for personnel required and not available will be submitted to this headquarters.

4.   Using 20 April 1945 as the effective date of reorganization, an appropriate entry will be made in the unit morning report in accordance with paragraph 34, AR, 345-400, 3 January 1945.

5.   Equipment
a. The only equipment authorized this unit is that provided by Enclosure No.1 to War Department letter, SPX 400 (9Sep 44) OB-S-SPMOO-M 11 September 1944, subject: "Special List of Equipment for the 978th Signal Service Company," with the quantities of items therein, whose basis of distribution is on (1) per individual reduced in accordance with the reduced strength.
b. Equipment rendered excess will be returned to appropriate supply agencies. Equipment required will be requisitioned in the normal manner.

FEGC 322

By command of General MacARTHUR:

RICHARD J. MARSHALL
Major General, General
Staff Corps, Chief of Staff

OFFICIAL: R.E. FRAILE
Colonel, A.G.C., Adjutant General

Source: Filipino American Experience Research Project. November 1998.

Appendix B - 26

CONSOLIDATED ROSTER
10 June 1945
FIRST RECONNAISSANCE BATTALION, SPECIAL
UNITED STATES ARMY FORCES IN THE FAR EAST
APO 565

TABLE OF CONTENTS

HEADQUARTERS
FIRST RECONNAISSANCE BATTALION, SPECIAL
APO 565
ROSTER
10 June 1945

Lt Col	Lewis Brown III	0292092 CAV BN COMDR
Capt	Carlos D. Arguelles	0499852 INF BN BX OFF
Capt	Cecil E. Walter, Jr.	01314697 INF S-2 S-3
Capt	Aleysius Y. Torralba	0499682 GH BN CHAPLAIN
1st Lt	Cefefrino R. Rola	01308379 INF BN ADJUTANT
Sgt	Diala, Vicente C.	6739474

MEDICAL DETACHMENT
FIRST RECONNAISSANCE BATTALION, SPECIAL

Capt	Mark A. Yessian	0502633	NC RN MED OFF		
S Sgt	Diana, Serafin G.	36323869	Cpl	Barba, Conrad G.	39241805
Tec 4	Calustre, Peter A.	R2291087	Tec 5	Baker, Clinton M.	39399940
Tec 5	Salvador, Frank A.	39101922			

SCOUT COMPANY AND COLUMN
FIRST RECONNAISSANCE BATTALION, SPECIAL
APO 565

OFFICERS' ROSTER AS OF 10 JUNE 1945

Capt	MONTESCLAROS, Melecio J.	01305289 INF	1 Lt	FRIAS, Arthur F.	01324673	INF
1 Lt	TALLIDO, Silvino V.	01313063 INF				

COLUMN**

Major	GALANG, Ricardo C.	05150082 INF	1 Lt	ENSOR, Richard A.	01050244 CAC
Capt	ABANTO, Rizalito	01301590 INF	1 Lt	HAWLEY, Maynard C.	01686144 AUS
Capt	AMES, Ireneo A.	0888247 CAC	1 Lt	HERNANDEZ, Al	01324774 INF
Capt	BAHR, Andrew P.	01030214 CAV	1 Lt	HOPE, Charles P., Jr.	01059496 INF
Capt	BEHAN, Fred H.	01081198 CAC	1 Lt	LABRADOR, Vicente W.	0888256 AUS
Capt	BOVE, John E.	01031285 CAV	1 Lt	MALAN, Jose B., Sr.	01314382 FA
Capt	CABANGBANG, Bartolome C.	0888747 INF	1 Lt	MARFORI, Edmundo J.	01301660
Capt	DIONOLO, Thomas	01298007 INF	1 Lt	MAURICIO, Paul F.	01319099 INF
Capt	FRIAS, Gene	01306357 INF	1 Lt	MENDOZA, Jose T.	01313662 INF
Capt	HUERTO, Petronio B.	0888249 CAC	1 Lt	OATES, Kenneth H.	01645882 SIG C
Capt	IGNACIO, Rodolfo C.	0888248 AUS	1 Lt	SARMIENTO, Ignacio M.	01822465 INF

Capt	JAMISON, Donald V.	01043373 CAC	1 Lt	SILVA, Saturnino R.	01300095 INF
Capt	SONGCO, Ruben P.	0478156 INF	1 Lt	SCHMALTZ, Albert	01031837 CAV
Capt	TORRES, Enrique L., Jr.	0888748 INF	1 Lt	SMITH, Frank B.	01324796 INF
Capt	UNCIANO, Isaac F.	01102098 INF	1 Lt	STAHL, Robert E.	01690905 SIG C
Capt	YOUNG, Frank H.	0888453 INF	1 Lt	TAMAYO, Claudio M.	01305526 INF
1 Lt	AGPAOA, Pacifico D.	01323652 INF	1 Lt	TEO, Konglam	0888892 AUS
1 Lt	ANCHETA, Carlos F.	01316979 INF	1 Lt	TISON, Sidney S., Jr.	0534061 CAC
1 Lt	BAKER, Henry L.	0534045 CAC	1 Lt	TUDTUD, Joe L.	01315532 INF
1 Lt	BONE, Alvin V.	01641447 SIG C	1 Lt	VALERA, Jose V.	0888893 AUS
1 Lt	CAMPBELL, Merle E.	01051505 CAC	1 Lt	VIVIT, David T.	01306774 INF
1 Lt	COHEN, Wilfred A.	0164472 SIG C	1 Lt	WOOD, Rhys C.	01081044 CAC
1 Lt	CRESPO, Toribio	0888253 CAC	1 Lt	YUHICO, Delfin C.	0888175 TC

WOJG	CAMPEAU, Lucien V.	W2114938 (Atchd)
2D LT	CASWELL	
2D LT	GEORGE	

*Pages missing.
**INF Infantry, CAC Coast Artillery Corp, SIG C Signal Corp, AUS Army of the United States, TC Transportation Corp, FA Field Artillery, CAV Cavalry

Source: U.S. National Archives and Records Administration.

Appendix B - 27

978th Signal Service Company
Company Strength As of 7 August 1945*

RANK / NAME	MISSION DATE	RANK / NAME	MISSION DATE
Major Croell, Harry T.		T/5 Marinas, Toby L.	
Capt Ferguson, Charles B.		T/5 Mateo, Carl M.	Sept 1944
1st Lt Hale, Edward H.		T/5 Nichols, Victor R.	
1st Lt McFarland, Clinton B.		T/5 Nuevo, Leodegario O.	June 1944
1st Lt Strickland, Lee A.		T/5 O'Gwin, Charles W.	
2nd Lt Becker, George T.		T/5 Olson, Clarence Y.	
2nd Lt Bone, Alvin V.		T/5 Padgett, Henry L.	
2nd Lt Cohen, Wilfred A.		T/5 Padilla, Bartolome M.	
2nd Lt McGivney, James B.		T/5 Palmeto, Daniel S.	June 1944
2nd Lt Oates, Kenneth H.		T/5 Pascual, Angelo E.	July 1944
2nd Lt Spire, Meyer J.		T/5 Paquarella, Joseph R.	
2nd Lt Wootin, Harold B.		T/5 Pena, Doroteo M.	
CWO Wise, Braynard L.	Nov 1943	T/5 Peredo, Joe E.	
WOJG Claunch, Alphie L.		T/5 Pino, Joseph E.	
WOJG Daley, Patrick		T/5 Pressnell, Richard M.	
WOJG Maxam, Harry A.		T/5 Robert, Louis	
M/Sgt Alberto, Alfred A.	Nov 1943	T/5 Ruiz, Julius B.	
M/Sgt Clark, Walter G.		T/5 Saladino, Selmo C.	Aug 1944
1/Sgt Villarta, Frederico B.		T/5 Sammons, William H.	Aug 1944
T/Sgt Almero, Emiliano A.	Sept 1944	T/5 Santos, Rudolph B.	June 1944
T/Sgt Arbis, Mariano V.	Dec 1944	T/5 Schmitt, Gerald J.	
T/Sgt Bugh, Marion W.		T/5 Sivila, Andres C.	Sept 1944
S/Sgt Cacabelos, Rufino F.		T/5 Esperando, William P.	
S/Sgt Camp, Lowell A.		T/5 Strow, Wallace F.	
S/Sgt Carter, Loy W.		T/5 Svobodny, Raymond D.	
S/Sgt Corpus, Aquilino M.		T/5 Tadeo, Carmine	
S/Sgt Ergas, Eugene N.		T/5 Tellier, Paul A.	
S/Sgt Faragan, Manuel F.	Dec 1944	T/5 Thompson, Clifford P.	
S/Sgt Fortun, Salvador B.	June 1944	T/5 Toribio, Alipio C.	July 1944
S/Sgt Gustilo, Conrado A.		T/5 Ulivas, Albert U.	
S/Sgt Ibea, Artemio I.	Sept 1944	T/5 Villalon, Hermenegildo	July 1944
S/Sgt Lynch, Griffin H.		T/5 Viray, Don J.	
S/Sgt McCleary, Dean W.		T/5 Ward Earl A.	
S/Sgt Nagle, David A.		T/5 Wheat, Jesse P.	
S/Sgt Nery, Gerardo B.	June 1944	PFC Aglubat, Al C.	Sept 1944
S/Sgt Palac, Juan M.	Sept 1944	PFC Ahola, John A.	
S/Sgt Rodriguez, Fred N.		PFC Alcantara, Marcelino C.	
S/Sgt Sarmiento, Angel T.		PFC Alcaras, Juan D.	
S/Sgt Tanato, Leandro P.		PFC Alvarado, John R.	
S/Sgt Thomas, James P.		PFC Ambrosio, Nicanor G.	Sept 1944
T/3 Africa, Robino		PFC Andres, Manuel V.	
T/3 DeTorres, Ronald A.	Sept 1944	PFC Aquino, Alejandro C.	Aug 1944
T/3 Ginsberg, Seymour		PFC Asuncion, Telesforo T.	Aug 1944
T/3 Giron, Leovigildo M.	Aug 1944	PFC Bargo, Querubin B.	Dec 1943
T/3 Gregory, George F.		PFC Bell, Harry W.	
T/3 Lopez, Marcos T.		PFC Bent, George F.	
T/3 Robinson, Irving		PFC Cahill, James J.	
T/3 Stahl, Robert E.	Dec 1943	PFC Carrasca, Romain S.	
T/3 Umipeg, Marcelo B.	Sept 1944	PFC Corpuz, Robert B.	Sept 1944
T/3 Visaya, Vincent L.		PFC De La Pena, Geronimo M.	Sept 1944
Sgt Arcenas, Semeniano D.	June 1944	PFC Eals, Orien N.	
Sgt Agcaoili, Reymundo E.	June 1944	PFC Enderiz, Victor P.	
Sgt Allen, David A.		PFC Felizar, Pete G.	
Sgt Ancheta, Alfonso		PFC Gangl, Leo J.	
Sgt Arce, Mariano T.	Sept 1944	PFC Gonzales, Juan L.	
Sgt Balcena, Fred	Sept 1944	PFC Guieb, Lorenzo M.	
Sgt Balleras, Julio V.	July 1944	PFC Haymore, Lester D.	
Sgt Barr, Charles N.		PFC Jose, Teodoro M.	
Sgt Begonia, Daniel B.	July 1944	PFC Lattanzio Salvatore	
Sgt Biado, Anacleto		PFC Laurina, Filomena A.	
Sgt Borillo, Pio B.	July 1944	PFC Liberato, Don J.	Aug 1944

RANK / NAME	MISSION DATE	RANK / NAME	MISSION DATE
Sgt Brown, James R.		PFC Maravel, Antonio L.	
Sgt Bueno, George B.	Sept 1944	PFC Marcelino, Juan C.	Sept 1944
Sgt Candelario, Jose E.		PFC Morales, Catalino U.	
Sgt Castro, Ray F.		PFC Morris, Joseph A.	
Sgt Collier, Durwood M.		PFC Moscoso, Anselmo S.	Aug 1944
Sgt Cortez, Ramon F.	June 1944	PFC Muncie, Victor P.	
Sgt Cutaran, Jacinto	Aug 1944	PFC Ovalles, Marcelo C.	
Sgt Goloyugo, Vincent C.	Aug 1944	PFC Page, Chester D.	
Sgt Gouldner, Frederick J.		PFC Pedro, Florendo R.	
Sgt Griggers, Willie D.		PFC Peralta, Estanislao C.	Aug 1944
Sgt Guyot, Gaudencio V.	June 1944	PFC Petro, Frank J.	
Sgt Herrera, Donato E.	June 1944	PFC Piana, Victor L.	
Sgt Ignacio, Fred C.	Sept 1944	PFC Proffetta, Stephen J.	
Sgt Ignacio, Maguel E.		PFC Ramos, Aquilino R.	Sept 1944
Sgt Legaspi, Alex B.		PFC Rimando, Juan F.	Sept 1944
Sgt Mauricio, Abamile L.		PFC Rodriguez, Feliciano S.	Aug 1944
Sgt Montero, Jack	June 1944	PFC Rosales, Doroteo A.	
Sgt Nagtalon, Alex A.		PFC Santos, Gregorio	
Sgt Ortiz, Paterno A.		PFC Sapon, Leoncio S.	
Sgt Padilla, Balbino R.	Aug 1944	PFC Savellano, Andres S.	Dec 1943
Sgt Parcasio, Joe B.		PFC Silverio, Ben V.	
Sgt Placido, Carlos S.	June 1944	PFC Stovall, Carrol M.	
Sgt Reynoso, Jaime R.	June 1944	PFC Tobofunda, Pedro P.	
Sgt Runtal, Henry E.	Sept 1944	PFC Thompson, James C.	
Sgt Salgado, Mauro B.		PFC Thompson, Robert C.	
Sgt Schuler, Marvin O.		PFC Trubiroha, Cyril M.	
Sgt Sibonga, Emil J.		PFC Ventura, Dionisio N.	
Sgt Sturgeon, Charles G.		PFC Wurst, Nelson A.	
Sgt Thys, Alfred P.		Pvt Abang, Bernardino B.	
Sgt Vite, Doroteo V.	Sept 1944	Pvt Aquino, Bernardino P.	
Sgt Wright, Carl	Sept 1944	Pvt Archuelata, Tobias	
Sgt Zola, Edward J.		Pvt Astronomo, Telesforo R.	
T/4 Agabayani, Vincent A.		Pvt Balinton, Vincente	Sept 1944
T/4 Boncavil, Domiciano O.	Sept 1944	Pvt Barret, James B.	
T/4 Calmorin, Angelo A.	Sept 1944	Pvt Beckman, Neil B.	
T/4 Carrillo, Patrocinio S.	July 1944	Pvt Bowyer, Frank E.	
T/4 Dagandan, Fortunato C.	Aug 1944	Pvt Bulatao, George M.	June 1944
T/4 Davin, Numeriano D.		Pvt Bumanglag, Arsenio B.	
T/4 Fernandez, Cipriano C.	Sept 1944	Pvt Bunte, Earl F.	
T/4 Gonzales, Jose T.	July 1944	Pvt Calip, Ronnie E.	
T/4 Logan, Doming L.	July 1944	Pvt Campos, Norberto C.	
T/4 Lopez, Cirilo G.	Sept 1944	Pvt Carpenter, Charles E.	
T/4 Luz, Pete L.	June 1944	Pvt Castillo, Rudolph	
T/4 Marquina, Leo	Aug 1944	Pvt Cayabyab, Adolfo V.	
T/4 Medina, Mariano A.	Sept 1944	Pvt Charvat, Ralph R.	
T/4 Pacificar, Nicomedes		Pvt Cortez, Calixto T.	Sept 1944
T/4 Pajara, George T.		Pvt Dennis, John D.	
T/4 Palafox, Rosendo M.	Sept 1944	Pvt Dulay, George R.	
T/4 Parong, Astor C.	Sept 1944	Pvt Espiritu, Artemio C.	
T/4 Pomarez, Leon V.		Pvt Foster, Dean H.	
T/4 Runes, Lucas D.	Sept 1944	Pvt Fulton, Harry L.	
T/4 Rasmussen, Paul O.		Pvt Gachallan, Hipolito M.	
T/4 Sabado, Daniel B.	Dec 1943	Pvt Girard, Stanley J.	
T/4 Thames, David M.		Pvt Glazar, Joseph P.	
T/4 Vergara, Thomas O.	Aug 1944	Pvt Gonzalo, Arthur V.	
T/4 Watson, Carl T.		Pvt Gurrea, Precilliano B.	
T/4 Victorio, Raymond D.		Pvt Guzman, Simeon R.	
T/4 Wright, Clarence L.		Pvt Goroza, Edward E.	Aug 1944
T/4 Young, Clarence R.		Pvt Hawrysio, Eugene J.	
Cpl Agustin, Pedro P.	Oct 1944	Pvt Hendelman, Lambert J.	
Cpl Alcala, Atanacio L.		Pvt Higgins, Donald V.	
Cpl Balino, Porfirio	Sept 1944	Pvt Hodges, Quincy M.	
Cpl Barcenas, Urbano B.	July 1944	Pvt Hosberg, Glen B.	
Cpl Buccat, Alejandro A.		Pvt Hufana, Frank A.	
Cpl Bugarin, Enrico M.	July 1944	Pvt Madariaga, Fred M.	
Cpl Denicola, Jasper		Pvt Madura, Edward J.	
Cpl Dodson, Hadley W.		Pvt Magale, Isidro P.	June 1944
Cpl Duran, Agripino J.	June 1944	Pvt Manalo, Armand	

RANK / NAME	MISSION DATE	RANK / NAME	MISSION DATE
Cpl Guico, Pete S.		Pvt Mapalo, Martin L.	Sept 1944
Cpl Ilusorio, Johnnie A.		Pvt Marcelo, Benjamin R.	Sept 1944
Cpl Navarro, Ambrosio A.	June 1944	Pvt Martin, Isidro C.	
Cpl Peralta, Materno P.		Pvt Martin, Leslie H.	
Cpl Rallojay, Teodoro J.	June 1944	Pvt Martin, Sammy C.	
Cpl Randall, Roberto A.		Pvt Marzan, Ricarte S.	
Cpl Reyes, Felix T.	July 1944	Pvt Mejia, Fulgencio L.	
Cpl Sensano, Daniel G.		Pvt Milana, Pilip G.	
Cpl Sheppard, Edward T.		Pvt Neighbors, John G.	
Cpl Torio, Isaias T.		Pvt Nichols, Howard M.	
Cpl Villote, Paul P.		Pvt Obrero, Dick V.	
T/5 Advincula, Julio C.	June 1944	Pvt Ogoy, Dale Y.	Sept 1944
T/5 Albalos, Segundo Q.	Sept 1944	Pvt Pabros, Ambrosio A.	Nov 1944
T/5 Alegre, Emilio A.		Pvt Patacsil, Regino T.	Sept 1944
T/5 Angcos, Robert O.	June 1944	Pvt Patubo, Gaudioso L.	Sept 1944
T/5 Arce, George U.		Pvt Penalvar, Pedro R.	
T/5 Beedy, Nowell J.		Pvt Powell, William L.	
T/5 Beeler, Richard N.		Pvt Riotock, Manuel E.	
T/5 Burgos, Edward M.	June 1944	Pvt Salvador, Frank A.	
T/5 Burns, Edward P.		Pvt Saria, Juan A.	
T/5 Candania, Amadeo C.		Pvt Scott, Clell D.	
T/5 Dacquel, Richie D.	June 1944	Pvt Schneider, Robert J.	
T/5 Estavillo, Sergio B.		Pvt Seares, Pascual P.	
T/5 Ferrell, John M.		Pvt Smith, James J.	
T/5 Filek, Joseph F.		Pvt Sullivan, Robert I.	
T/5 Green, James L.		Pvt Taban, Lorenzo F.	
T/5 Halla, Albert F.		Pvt Thrall, Gerald V.	
T/5 Herreria, George R.	Dec 1943	Pvt Udani, Carlos D.	Nov 1944
T/5 Holgado, Eddie C.	June 1944	Pvt Udaundo, Esteban C.	
T/5 Isaac, Cledonaldo	Sept 1944	Pvt Valmonte, Frank G.	
T/5 Kobliska, La Verne J.		Pvt Vicente, Victor L.	
T/5 Landrum, Jesse E.		Pvt Viduya, Jose F.	
T/5 Long, Edward J.		Pvt Willhoite, Charles B.	
T/5 Long, Howard D.		Pvt Woods, Alwin C.	
T/5 Malitovsky, Sydney M.			

*978th Signal men sent on missions. Others may have been assigned to KAZ station or Philippine Island Message Center (PIMC)

Source: The Filipino American Experience Research Project, October 1998

Appendix B - 28

1ST RECONNAISSANCE BATTALION (SPECIAL) AND 978TH SIGNAL SERVICE COMPANY AFTERACTION REPORTS*

RANK	FIRST	I	LAST	RANK	FIRST	I	LAST
Sgt	Seminiano	D	Acenas	T5	Isidro	P	Magale
Cpl	Baltazar		Agbunag	T5	Martin	I	Mapalo
Sgt	Raymundo	E	Agcaoili	T5	Benjamin	R	Marcelo
Sgt	Alfredo	R	Agron	Pvt	Carl	M	Mateo
Cpl	Pedro	P	Agustin	Lt	Paul	P	Mauricio
Sgt	Alfred	A	Alberto	Sgt	Mariano	A	Medina
Sgt	Thomas	T	Amante	Sgt	Cipriano	I	Miguel
Capt	Irineo	A	Ames	T3	Jack		Montero
Pvt	Martiniano	S	Atad	T4	Ambrosio		Navarro
Sgt	Zoil	S	Azcueta	Sgt	Gerardo	B	Nery
Capt	Andrew	P	Bahr	T5	Dale	Y	Ogoy
Sgt	Fred		Balcena	Pvt	Patricio		Ollada
Sgt	Julio	V	Balleras	Sgt	Paterno	A	Ortiz
Pvt	Jaime	A	Bernal	T4	Daniel	S	Palmeto
Cpl	Edilberto	S	Bibat	Pvt	Jerry	D	Pascua
Sgt	Domiciano	O	Boncavil	T4	Angelo	E	Pascual
T4	Enrico	M	Bugarin	Cpl	Amrafel	D	Pascual
T5	George	M	Bulatao	Sgt	Julian	A	Pilar
Lt	Bartolome	C	Cabangbang	Cpl	Lorenzo	U	Pimentel
Sgt	Eutiquio	B	Cabais	Sgt	Carlos	S	Placido
Sgt	Isidoro	D	Dacquel	Cpl	Mack	H	Pulido
Pvt	Andres	D	Darang	Cpl	Teodoro	J	Rallojay
Sgt	Ronald	A	De Torres	Sgt	Jerry		Regalado
Sgt	Esteban	M	Diputado	Sgt	Juan	A	Rodriguez
Sgt	Buenaventura	L	Dolendo	Sgt	Henry	E	Runtal
Sgt	Salvador	B	Fortun	Sgt	Daniel	B	Sabado
Sgt	Urbano	M	Francisco	Sgt	William	H	Sammons
Maj	Ricardo	C	Galang	Sgt	Fortunato		Santos
Sgt	Pedro	L	Garces	Lt Col	Charles	M	Smith
Sgt	Leovigildo	M	Giron	Lt	Claudio	M	Tamayo
Lt	Gaudencio	V	Guyot	Lt	Konglam		Teo
Sgt	Donato	E	Herrera	T4	Alipio	C	Toribio
T5	George		Herreria	Lt	Jose	V	Valera
Lt	Vicente	W	Labrador	Sgt	Hermenegildo	I	Villalon
Sgt	Domingo	L	Logan	WOJG	Brainard		Wise
				Lt	Delfin	C	Yuhico

*Only reports found.

Source:
Individual After Action Reports – 1st Reconn and 978th Signal Service Co October 5, 1998, Edited by Alex Fabros, Jr.
In Honor of Our Fathers, History of the 1st Reconnaissance Battalion U.S. Army, August 27, 1996, Edited by Alex S.
Fabros, Jr., Daniel P. Gonzales, Danilo T. Begonia
Filipino American National Historical Society Archives, Seattle WA
U.S. National Administration and Records Administration

Appendix B - 29

GENERAL HEADQUARTERS
UNITED STATES ARMY FORCES, PACIFIC

APO 500
29 September 1945

GENERAL ORDERS)
NOS. 201, 205, 206, 207, 208, 209, 210, 211, 212, 213, 214, 219

GENERAL ORDERS)
NOS. 239

8 October 1945

BRONZE STAR MEDAL
1st Reconnaissance Battalion (Special) and 978th Signal Service Company
Mission Men*

By direction of the President, under the provisions of Executive Order 9419, 4 February 1944 (Sec. 2, bulletin 3, WD, 1944), a Bronze Star Medal is awarded by the Commander-In-Chief, United States Army Forces, Pacific, to the following-named officers and enlisted men for heroic achievement in connection with military operations against the enemy in the Philippine Islands, during the period indicated, with citation for each as shown herein below:

1st Lt	Claudio C. Tamayo	Sgt	Mariano V. Arbis
1st Lt	Robert F. Stahl	Sgt	Maximo G. Cacas
1st Lt	Saturnino R. Silva	Sgt	Vincent Goloyugo
2nd Lt	Gaudencio V. Guyot	Sgt	William H. Sammons
WOJG	Braynard L. Wise	TEC 4	Alipio Toribio
M/Sgt	Alfred A. Alberto	TEC 4	Angelo E. Pascual
M/Sgt	Conrado A. Gustilo	TEC 4	Ambrosio Navarro
M/Sgt	Gerardo B. Nery	TEC 4	Carl M. Mateo
M/Sgt	Isay M. Real	TEC 4	Daniel S. Palmeto
T/Sgt	Angel T. Sarmiento	TEC 4	Eddie C. Holgado
T/Sgt	Doroteo V. Vite	TEC 4	Enrico M. Bugarin
T/Sgt	Emiliano A. Almero (KIA)	TEC 4	Felix T. Reyes
T/Sgt	Eustacio C. Corpuz	TEC 4	George R. Herreria
T/Sgt	Leovigildo M. Giron	TEC 4	Leodegario O. Nuevo
T/Sgt	Marcelo B. Umipeg	TEC 4	Porfirio Balino
T/Sgt	Ronald A. De Torres	TEC 4	Richie D. Dacquel
T/Sgt	Salvador B. Fortun	TEC 4	Robert O. Angcos
S/Sgt	Angelo A. Calmorin	TEC 4	Teodoro J. Rallojay
S/Sgt	Artemio I. Ibea (KIA)	TEC 4	Urbano B. Barcenas
S/Sgt	Astor C. Parong	Cpl	Agrifino Duran
S/Sgt	Balbino R. Padilla	Cpl	Pedro P. Agustin
S/Sgt	Carl Wright	TEC 5	Al C. Aglubat
S/Sgt	Cirilo G. Lopez	TEC 5	Alejandro C. Aquino
S/Sgt	Cipriano C. Fernandez	TEC 5	Ambrosio A. Pabros
S/Sgt	Daniel B. Begonia	TEC 5	Amedeo C. Cendania (KIA)
S/Sgt	Daniel B. Sabado	TEC 5	Andres S. Savellano (KIA)
S/Sgt	David C. Cardenas	TEC 5	Andres Sivila
S/Sgt	Domiciano O. Boncavil	TEC 5	Anselmo S. Moscoso
S/Sgt	Domingo L. Logan	TEC 5	Benjamin R. Marcelo
S/Sgt	Donato E. Herrera	TEC 5	Calixto T. Cortez
S/Sgt	Fortunato C. Dagandan	TEC 5	Carlos D. Udani
S/Sgt	Fred Balcena Jr.	TEC 5	Cledonaldo Isaac
S/Sgt	Henry Runtal	TEC 5	Dale Y. Ogoy
S/Sgt	Jacinto Cutaran	TEC 5	Edward M. Burgos
S/Sgt	Jaime R. Reynoso Jr.	TEC 5	Estanislao C. Peralta
S/Sgt	Jose T. Gonzales	TEC 5	Feliciano S. Rodriguez
S/Sgt	Julio V. Balleras	TEC 5	Gaudioso L. Patubo

S/Sgt	Leo Marquina	TEC 5	George M. Bulatao
S/Sgt	Lucas D. Runes	TEC 5	Juan C. Marcelino
S/Sgt	Marcelo G. Gonzales	TEC 5	Julio Advincula
S/Sgt	Mariano A. Medina	TEC 5	Martin L. Mapalo
S/Sgt	Paterno A. Ortiz	TEC 5	Querobin B. Bargo
S/Sgt	Patrocinio A. Carillo	TEC 5	Ramon D. Vitorio
S/Sgt	Pete L. Luz	TEC 5	Regino P. Patacsil
S/Sgt	Pio B. Borillo	TEC 5	Robert B. Corpuz
S/Sgt	Ramon F. Cortez	TEC 5	Rudolfo Santos
S/Sgt	Raymundo E. Agcaoili	TEC 5	Sammy C. Martin
S/Sgt	Rosendo M. Palafox	TEC 5	Segundo O. Albalos
S/Sgt	Semeniano D. Acenas	TEC 5	Selmo C. Saladino
S/Sgt	Thomas C. Vergara	TEC 5	Telesforo T. Asuncion
TEC 3	Jack Montero	TEC 5	Vincente Balinton
Sgt	Angelo P. Candasan	Pfc	Aquilino B. Ramos (KIA)
Sgt	Carlos S. Placido	Pfc	Don Liberato
Sgt	Fred C. Ignacio	Pfc	Geronimo De La Pena
Sgt	George B. Bueno (KIA)	Pfc	Nicanor G. Ambrosio
Sgt	Hermenegildo Villalon	Pvt	Edward E. Goroza
Sgt	Manuel F. Faragan	Pvt	Isidro P. Magale
Sgt	Mariano T. Arce	Pvt	Juan M. Palac

Citation for each of the above-named officers and enlisted men is as follows:

Volunteering for a secret and dangerous military intelligence mission, he was landed by submarine on the islands indicated where he assisted in successfully extending lines of communication, securing vital weather data and obtaining military information which proved of the greatest assistance to impending military operations. By his loyalty, daring, and skillful performance of duty under most hazardous conditions, he rendered a valuable service to operational planning and materially accelerated the campaign for the recapture of the Philippine Islands.

AG-PA 200.6

By command of General MacArthur:

R. K. SUTHERLAND
Lieutenant General, United States Army,
Chief of Staff

OFFICIAL:
B. M. FITCH
Brigadier General, U.S. Army
Adjutant General

*The General Orders stated Bronze Star medals awarded to the men who went on missions. All the 978th Signal men listed in the U.S. Submarine Missions to the Philippines, 1942-1945 are included.
** Received the Bronze Star posthumously.

Source: U.S. National Archives and the 978th Service personnel in the U.S. Submarine Missions to the Philippines, 1942-1945; MacArthur Archives, Norfolk VA

Appendix B - 30

2nd Salo Salo sponsored by the 1st Reconnaissance Battalion. Philippines.
December 1, 1945

SCHEDULE OF EVENTS

1100 – 1200	Mass (Chapel)
1300 – 1400	Reception (Rec Hall)
1400 – 1500	Announcements
1500 – 1600	Impromptu Program
1600 – 1800	Supper
1800 – 1900	Take Ten
1900 – 2100	Dance
2100 – 2200	Refreshment
2200 – 2400	Dance

"GOODNIGHT AND PLEASANT DREAMS"
*****

Our thanks for the cooperation of the Athletic and Recreation Office.
2d Lt JOHN J. REGAN, A R Officer, Tec 5 JIMMY BARROGA, and Tec 5 Alex Fabros

....

GENERAL HEADQUARTERS
UNITED STATES ARMY FORCES, PACIFIC
OFFICE OF THE COMMANDER IN CHIEF

A.P.O 500
15 August 1945

SUBJECT: Commendation
TO   :   Commanding Officer, 1st Reconnaissance Battalion (Sp), A.P.O. 75.

1.   On this day the 1st Reconnaissance Battalion (Sp) is to be de-activated inasmuch as its mission has been accomplished.

2.   I wish to express to all members of the Battalion my grateful appreciation for the splendid service they have rendered, both individually and collectively, toward the liberation of the Filipino people, a cause to which they have so long and so faithfully dedicated themselves. Those of them who were dispatched by me prior to the start of the Philippine campaign on secret missions within the Philippines proceeded in the path of their duty with magnificent courage and marked ability and resourcefulness.

Their service in large measure inspired in the Filipino people the will to continued resistance during their darkest hours and contributed in no small degree to the success of our Philippine operations.

3. Those of them who did not have the opportunity for this type of service, through their wholehearted support of the main purposes for which the Battalion was formed, provided an esprit de corps next to none in any unit in the American Army.

4. I shall always take deep satisfaction in the accomplishments of the 1st Reconnaissance Battalion and desire to command all of its officers and men for the high type of service which was responsible, therefore. They carry with them to their new assignments my very best wishes.

/s/Douglas MacArthur
/t/DOUGLAS MacARTHUR

DEDICATION

May the glorious memory, the true devotion to duty and the unselfish sacrifice of the following officers and enlisted men listed below and also of those men whose names we can not recall who gave all that others may live ... these unsung heroes who were once our leaders and companions at arms... be to us... a shining example of courage and resourcefulness in the face of the enemy..... that their deeds will ever remain fresh in our hearts no matter where fate and destiny may lead us henceforth...

May our prayers today be one of Divine mercy upon their souls, and to their kins and loved ones we offer our deepest sympathy and eternal gratitude for their mutual service to God and country.

				Chaplain Monleon	
Lt Col	Phillips, L. H.	Sgt	Cardenas, D.	W. O.	Wise
Lt	Padilla, L. P.	Sgt	Cacas, M	1st/Sgt	Peralta, G
1st/Sgt	Corpus, A.	Cpl	Bargo	T/Sgt	Baniares, A.
T/Sgt	Almero	Pfc	Ambrosio	S/Sgt	Ibea, A.
S/Sgt	Gregorio, Q.	Capt	Songco, R. P.	Sgt	Loma
Cpl	Anderson, M.	T/4	Fugose, J.		

OFFICERS OF THE 1ST RCN BN SP
Brig. Gen. Courtney Whitney

Lt. Col	Brown, L. III	Capt	Young, F. H.	1st Lt	Hondolero, F. S.
Lt Col	Smith, C. N.	Capt	Huerto, P. B.	1st Lt	Hope, C. F. Jr
Maj	O'Connor	Capt	Jamison, D. V.	1st Lt	Labrador, V. W.
Maj	Skundale, F. H.	Capt	Songco, R. P.	1st Lt	Lorenzo, S. B.
Maj	Galang, R. C.	Capt	Morgan, L.	1st Lt	Malan, J. B. Sr
Maj	Evans, J. L.	Capt	Stoddard, B.	1st Lt	Marfori, E. J.
Capt	Abanto, R	Capt	Toralba	1st Lt	Mauricio, P. F.
Capt	Arellano, A.	Capt	Walter, C. E. Jr	1st Lt	Mendoza, J. T.
Capt	Ames, I. A.	Capt	Ubarro	1st Lt	Quinto, E. F.
Capt	Arguelles, C. D.	1st Lt	Ancheta, C. F.	1st Lt	Padilla, L.
Capt	Behar, F. H.	1st Lt	Agpaoa, P. D.	1st Lt	Sarmiento, I. M.
Capt	Bove, J. E.	1st Lt	Allen, R. C. Jr	1st Lt	Schmaltz, A.
Capt	Cabangbang, B.	1st Lt	Abana, R.	1st Lt	Silva, S. R.
Capt	Dionolo, T.	1st Lt	Baker, H. L.	1st Lt	Smith, F. B.
Capt	Frias, G. D.	1st Lt	Campbell, M. E.	1st Lt	Teo, K.
Capt	Ignacio, R.	1st Lt	Crespo, T.	1st Lt	Tamayo, C. M.
Capt	Montesclaro, M. J.	1st Lt	Davis, W. J.	1st Lt	Tison, C. M.
Capt	Jones, E.	1st Lt	Elefano, A.	1st Lt	Tudtud, J. L.
Capt	Pickering	1st Lt	Ensor, R. A.	1st Lt	Valera, J. V.
Capt	Torres, F. L.	1st Lt	Hawley, M. C.	1st Lt	Vivit, D. T.
Capt	Unciano, I. F.	1st Lt	Hernandez, A.	1st Lt	Wood, R. C.
				1st Lt	Yuhico, D. C.

ENLISTED MENS' ROSTER OF THE
1ST RCN BN SP
MASTER SERGEANTS

Master Sergeants

Aguilar, P. B.
Bacus, F. B.
Berg, G. G.
Besid, R. J.
Bucol, S. G.
Bucol, S. G.
Buenaventura, N.
Cabais, E.
Dacquel, I. D.

Domalaog, B. N.
Garcia, V.
Goloyugo, V. C.
Guzman, L. O.
Harder, B.
Manzano, A. C.
Miguel, C. L.
Pilar, J. A.
Ramos, J. J.

Ramos, S. R.
Real, I. M.
Regala, W. M.
Reyes, E.
Robles, C. C.
Sanchez, G. A.
Santos, F.
Tabios, V. R.

First Sergeants

Corpus, A.
Reyes, G. V.

Peralta, C.

Luntao, C. D.

Tech Sergeants

Abellera, A. D.
Agron, A. R.
Alfabeto, O. A.
Amante, T. T.
Anacleto, V. H.
Arias, N. T.
Azcueta, Z. A.
Baniares, A.
Bello, C. S.
Bernal, J.
Bernardo, M. P.
Bolivar, C.
Brillante, L. T.
Bulatao, B.
Caoili, H. R.
Castro, C.
Constantino, J. R.

Corpuz, E. C.
Cortez, L.
Daelto, M. R.
De Los Santos, M.
Diputado, E. M.
Dolendo, P.
Dongallo, F. L.
Fortes, M. D.
Ginobiagon, F.
Gubatan, S. D.
Hilario, C. G.
Javonillo, J. O.
Jorge, P. D.
Jubane, V. C.
Lim, A. E.
Llapitan, C.
Marciano, J. S.

Padilla, L. P.
Palac, J. M.
Palmeyra, L. H.
Pascua, I. V.
Pinuela, V. A.
Quitoriano, E. Q.
Ramirez, J. A.
Regalado, J. V.
Rosete, P. M.
Ruiz, D. S.
SIson, C.
Solidarios, S. V.
Taluban, T. M.
Turqueza, A. A.
Velasco, R. M.
Yotoko, S. L.

Staff Sergeants

Adia, P. M.
Alfafara, G. C.
Alvarez, V. F.
Annaguey, E. D.
Baladad, J.
Baldos, F.
Balunes, P.
Balunes, R. F.
Basug, G. J.
Bautista, F. L.
Bibat, E. S.
Bolante, J. A.
Calix, P. M.
Capistrano, A. N.
De La Pena, A.
De Ocampo, M. L.
Diala, V. C.

Dumpit, F. N.
Ebarle, P.
Fermin, R. R.
Fontanez, M. P.
Francisca, F. M.
Gallemore, J.
Garcia, M.
Gaston, M. M.
Hipol, F. C.
Jaramilla, S. Q.
Kintanar, A. S.
Lagrimas, W. B.
Lopez, F. R
Magbual, F. D.
Malic, D.
Mapile, C. B.
Merca, J. A.

Mones, V. M.
Nietes, I.
Paler, L. B.
Paraiso, M. Q.
Quesada, A. M.
Quindiagan, L. C.
Quisada, A. M.
Rillera, P. A.
Rodriguez, J. A.
Rosales, P. A.
Santorio, C. D.
Sison, T. C.
Soliven, L. G.
Supnet, F.
Talaro, A. T.
Trinion, E. T.
Villanueva, A. F.
Yuponce, L.

Sergeants

Abad, O. V.
Agbunag, B.
Agustin, P. P.
Almojera, L. N.
Alvez, B. C.
Antikol, C.F.
Antonio, A.L.
Arciaga, E.
Arellano, M. Q.
Arquero, L.C.
Artiego, M.W.
Ayson, B.L.
Azcueta, B.S.
Baggao, M.B.
Baladad, J.C.
Bareng, G.F.
Bautista, S.B.
Bernal, J. A.
Buenavista, S.
Cariaga, P.F.
Carlitos, A.T.
Carpio, J.P.
Casia, M.C.
Clemente, L.C.
Concepcion, B.F.
Costales, G.D.
Curaming, E.B.
De La Pena, G.M.
Delfun, U.P.

Despy, A. H.
Dumag, G. L.
Dupitas, A. M.
Francisco, U. M.
Gallemore, J. R.
Generales, N. C.
Gobaleza, D. F.
Ibarra, E.S.
Ignacio, F.C.
Javier, B.E
Javier, F. O.
Lazo, E.C.
Lopez, A.
Machon, S.M.
Madarang, J.S.
Malangit, M.V.
Martin, P.
Mendoza, P.T.
Noel, R.B.
Oandasan, A.
Ovalles, M.
Pascua, J.D.
Pascual, A. D.
Peneyra, C.C.
Pilar, J.A.
Pimentel, L. U.
Pulido, M.H.
Quiaoit, E.R.
Quiming, P. C.

Quinito, B.A.
Rabang, T.V.
Racho, P.S.
Ramos, J. G.
Respecio, T. P.
Reyes, F.
Rimando, A. C.
Rivera, E.A.
Robles, R. D.
Sabado, H.D.
Sagun, F.C.
Salamanca, A.S.
Salvacion, A.L.
Sammons, W.H.
Sandoval, N.E.
Santos, I.L.
Santos, S.A.
Savellano, P.E.
Serrano, A.B.
Sison, A.R.
Tolentino, Z. Jr.
Trocio, P.M.
Tucay, H.T.
Tullo, A.P.
Valdez, C.R.
Valido, P.W.
Vicente, J.O.
Villalon, H.
Wendam, S.D.

Corporals

Adadiano, R. H.
Agbunag, B. C.
Aguila, I.Z.
Aguilar, H.
Alayu, C. D.
Alerta, P. M.
Aliacbar, S. M.
Altorino, E. B.
Arcia, N. N.
Arciaga, E. D.
Argueza, D. T.
Atad, M. S.
Bairulla, S. H.

Balderama, E. G.
Barba, C. A.
Basilio, P. L.
Bautista, M. L.
Bautista, P. A.
Bentillo, V. I.
Bermudez, F.
Catalina, E.S.
Clemente, L.C.
Dumbrigui, S. C.
Erana, M.
Fernandez, L.C.
Gaduang, I.D.

Liadjahasan, A.
Liadjahasan, M.
Lobendino, N.L.
Medrano, S.N.
Mosquera, B.F.
Nerez, M.M.
Niebres, F.
Oania, H.C.
Ollada, P.
Quiaoit, E.K.
Quismundo, R.Q.
Tolentino, J.T.

Technician 4

Agcaoili, J.S.
Calaustro, P. A.

Lucas, F. M.
Padilla, L.

Punsalan, R. M.

Citation for a secret and dangerous military intelligence mission, he was landed by submarine in the Philippine Islands, where he assisted in successfully extending lines of communications, securing vital weather data and obtaining military information which proved of the greatest assistance to impending military operations. By his loyalty, daring, and skillful performance of duty under the most hazardous conditions, he rendered a valuable service to operational planning and materially accelerated to campaign for the recapture of the Philippine Islands.

AG PA 200.6
>By command of General MacArthur

R. K. SUTHERLAND,
Lt. Gen. U. S. ARMY
Chief of Staff

COMMITTEES FOR 1ST RCN BN SP
SALO – SALO

INVITATION

Capt	Unciano, S.	T/Sgt	Velasco, R.	T/Sgt	Padilla, L.
M/Sgt	Ramos, J.	T/Sgt	Monico, L.	S/Sgt	Lagrimas, W.
T/Sgt	Arias, N.	T/Sgt	Jorge, P.	Sgt	Alfafara, G.
				M/Sgt	Santos, F.

PROGRAM

Lt	Mendoza, J.	M/Sgt	Harder, B.	S/Sgt	Lopez, F.
M/Sgt	Bacus, F. B.	T/Sgt	Javonillo, J.	Sgt	Trocio, F.
M/Sgt	Sanchez, G.	S/Sgt	De La Pena, A.	Sgt	Ignacio, F.
				M/Sgt	Goloyugo, V.

TRANSPORTATION

Lt	Frias, A.	T/Sgt	Marqueza, A	T/Sgt	Quitoriano, E.
M/Sgt	Pilar, J.	T/Sgt	Lim, A	S/Sgt	Supnet F.
M/Sgt	Bucol, S.	T/Sgt "	Gubatan, S.	Sgt	Pascual, A.
				Cpl	Rosario, G.

ENTERTAINMENT

Capt	Morgan, L.	S/Sgt	Soliven, L.	Sgt	Serrano, A.
1st/Sgt	Reyes, G.	S/Sgt	Bibat, E.	Cpl	Velayo, B.
T/Sgt	Agron, A.	Sgt	Despy, A.	Cpl	Aguilar, H.
				Cpl	De Leon, P.

RECEPTION

Capt	Montesclaros, M.	T/Sgt	Regalado, J.	S/Sgt	Kinvanar, A.
M/Sgt	Dumalaog, B.	T/Sgt	Ramirez, J.	Sgt	De La Pena, G.
M/Sgt	Garele, V.	T/Sgt	Vite, D.	Cpl	Ramos, P.
				Pvt	Cacho,

GUARD

Lt	Sarmiento, I.	T/Sgt	Amante, T.	T/Sgt	Ruiz, D
M/Sgt	Robles, C.	T/Sgt	Azcueta, Z.	S/Sgt	Quindiagan, L.
1st/Sgt	Luntao, C.	T/Sgt	Palac, J.	S/Sgt	Garcia, M.
				Cpl	Solar, M.

DECORATION

Lt	Smith, F.	T/Sgt	Dongallo, F.	S/Sgt	Balunes, R.
T/Sgt	Diputado	S/Sgt	Gaston, M	Sgt	Artiago, M.
T/Sgt	Alfabeto, Q.	S/Sgt	Fermin, R.	Sgt	Casia, M.
				Cpl	Arciaga, E.

SPECIAL DETAIL

Lt	Vivit, D.	S/Sgt	Adia, P.	Sgt	Alfonso, A.
M/Sgt	Santos, F.	S/Sgt	Paler, L.	Sgt	Carpio, J.
T/Sgt	De Los Santos, M.	S/Sgt	Calix, P.	Sgt	Sison, A.
				Cpl	Bermudez, F.

COMMISSARY

Capt	Frias, G.	T/Sgt	Pinuela, V.	S/Sgt	Mapili,
M/Sgt	Ramos, S.	T/Sgt	Yotoko, S.	Cpl	Medrano, D.
T/Sgt	Constantino, J.	S/Sgt	Villanueva, A.	Pfc	Caballes, P.
				Pfc	Reyno, J.

FINANCE

Lt	Labrador, V.	T/Sgt	Fortez, M.	Sgt	Pascua, J.
M/Sgt	Dacquel, I.	T/Sgt	Cortez, I.	Sgt	Gobaleza, D.
T/Sgt	Caoili, H.	S/Sgt	Dumpit, F.	Sgt	Rimando, A.
				Cpl	Costales, F.

MENU

PANSIT CELESTIAL
Canungra Special Improvisation
LECHON
Roasted to a Queens' taste
ROLLS
Fraser Commando Style
TEAS'
ICE CREAM
BEAUDESERT
COCA COLA

Prepared by. . . . . Yours Truly

Appendix B - 31

978TH SIGNAL SERVICE COMPANY
EIGHTH ARMY
APO 181

HISTORICAL REPORT FEBRUARY 1946

SUBJECT: Monthly Historical Report
TO: Commanding General, USASCOM-C, APO 44   4 March 1946
      Attn: Signal Section

1.   In compliance with par (c), USASCOM-C Regulation 67-30, dated 31 December 1945, the following Historical Report for the month of February 1946 is submitted.
2.   During the month of February the company continued to supply Personnel to assist in the operation of communication facilities for GHQ, AFPAC.

Due to the arrival of 200 replacements on February 1st from the 123d Infancy, 33d Division transferred to this organization per par 13, SO 12, Hq 123 Inf dated 28 January 1946, the strength of the company of 1 February 1946 was 2 officers, 1 warrant officer, and 230 enlisted men. On 12 February 1946, 4 more officers, 2nd Lt. Notterman, Joseph, 2nd Lt Depey, Roby H., 2nd Lt Bellitz, Bernard, 2nd Lt, O'Brien, Raimund F. Jr., were assigned per AFPAC Letter Order AG 210.31 AGPD, dated 10 Feb 1946, "SUBJECT: Order AGPD 41-19," During February the company lost 7 men through readjustment, so that on 2 February 1946 the strength was 6 officers, 1 warrant officer, and 223 enlisted men, although the T/O & E authorized a strength of 11 officers, 5 warrant officers, and 279 enlisted men.

Further precautions were initiated at the company motor pool and fire drills were held for the men quartered in the San Shin Building.

In line with the consolidation of this organization with the newly authorized 71st Signal Service Battalion living quarters were rearranged in the San Shin Building. The 978th now occupied the sixth floor.

Excess motor vehicles and equipment were turned in, and requisitions placed to bring the company supply to authorized T/O & E levels.

An increase in the number of venereal cases caused Lt Midgett to reappoint Asst. Unit Venereal Disease Officer to Capt. White. Det, 2, 4025th Sign Sv, Op. Medical Officer. Measures were then instituted to bring the venereal disease rate down to an authorized rate of 50 cases per 1000 men per annum.

A snack bar selling beer, cokes, and hamburgers was set up in the basement of the San Shin Building for the convenience of the personnel housed there.

During February, the 978th organized a Basketball Team under the direction of 1st Sergeant Kafer, Shen the month ended. The team was a part of the Detachment 2, 4025th Signal Service Group Basketball League and had won one game and lost another.

                                            RAYMOND W. MIDGETT
                                            2nd Lt. Sig C
                                            Historical Officer

Approved: HARLEY N. CLAUSSEN
              1ST Lt. Sig C, Commanding

Source: Filipino American Experience Research Project. November 6, 1998.

Appendix B - 32

978th SIGNAL SERVICE COMPANY
EIGHTH ARMY
APO 181

4 April 1946

HISTORICAL REPORT MARCH 1946

During the month of February, the company continued to supply personnel to assist in the operation of communication facilities for GHQ, AFPAC.

The strength of the company remained fairly constant during March. On 1 March the morning report showed the company strength to be 6 officers, 1 warrant officer, and 223 enlisted men. On 31 March there were 5 officers, 1 warrant officer, and 219 enlisted men: a net loss of only 1 officer and 4 enlisted men.

Four men were sworn into the Regular Army, an increase over the recruiting figure for February.

The sixth floor of the San Shin Building in Tokyo, the company billet, has been redecorated to provide some of the best living quarters in Tokyo.

The fire precautions were revised at the supply and motor pool warehouse.

Classes in venereal disease were instituted to instruct those men unfortunate enough to contract VD. This is a measure designed to educate the enlisted men and curb the rising VD rate.

The basketball team finished a fairly successful season, by tieing for third place in the GHQ Signal Service League.

When the month ended deactivation seemed imminent. The 978th headquarters and personnel is to assume the role of the Headquarters Company of the soon to be activated 71st Signal Service Battalion.

RAYMOND W. MIDGETT Jr.
2d Lt, 978th Sig Sv Co
Historical Officer

Source: Filipino American Experience Research Project. November 1998.

Appendix B – 33

Battle Honors, Decorations, and Service Records

1st Filipino Infantry Regiment – Army of the United States (AUS)
- Earned battle honors for New Guinea, the Southern Philippines, *Samar Island and Ormoc, Leyte Island*
- Philippine Presidential Unit citation

2nd Filipino Battalion (Separate) – Army of the United States (AUS)
- Earned battle honors for New Guinea.
- Philippine Presidential Unit citation

1st Reconnaissance Battalion (Special) – Army of the United State (AUS)
(Formerly designated as 5217th Reconnaissance Battalion (Provisional).
Included were the 978th Signal Service personnel.)
- Earned battle honors for New Guinea and the *secret missions to the Philippines*
- Philippine Presidential Unit citation

- Combat Infantry Badge (CIB) – Around 200
- Bronze Star – 281 *(Some with a V or an oak leaf cluster)*
- Distinguished Service Cross – 4
- Silver Star – 6
- Legion of Merit – 13
- Purple Heart *(Some with an oak leaf cluster)*

The men also received other medals.
- Army Good Conduct
- Asiatic-Pacific Campaign
- Philippine Liberation with Three Battle Stars
- World War II Victory

Note: The italics are the author's to add missing battle participations.

Source:

Filipino American Experience Research Project, October 1998
U.S. National Archives and Records Administration (NARA)
U.S. Army heritage & Education Center, U.S. Army Military History Institute, April 10, 2007

Appendix B – 34

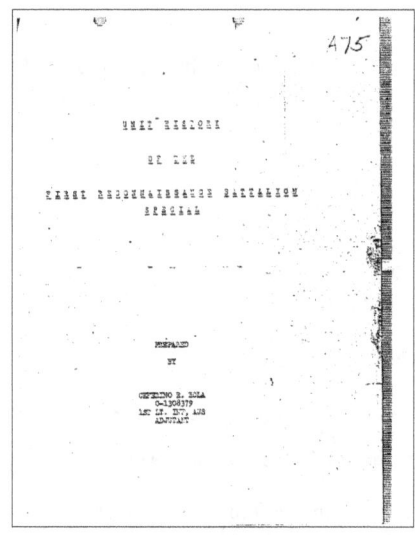

UNIT HISTORY
OF THE FIRST RECONNAISSANCE BATTALION
SPECIAL

PREPARED
BY

CEFERINO R. ROLA
0-1308379
1ST LT. INF, AUS
ADJUTANT

Foreword

This history is prepared in accordance with Army Regulations 345-105. It is as complete as all available records permit.

The reports of the Party Leaders the accomplishments of their parties while on secret mission in the Philippines paving the way for the return of General MacArthur with his invading forces include only the highlights. It is unfortunate that the rather early disbandment of the First Reconnaissance Battalion did not give a chance for the other Party Leaders to submit their reports on time for inclusion in this history. However, it is hoped that the citations for the various awards of decorations furnish adequate information regarding the heroic achievements of the individuals and the substantial contribution rendered by those parties in the liberation of the Philippines.

Aside from the fact that those who went on mission did not bring any cameras with them they were invariably under constant danger which made it impossible for them to take any photographs of the results of their work behind enemy lines. The only photographs appearing in this book, therefore, are those taken in station camps.

In the preparation of this work, considerable assistance was given by M/SGT Juan F. Dahilig, Sergeant Major of the Battalion, and S/SGT Apolinar G. Salvador of the Scout Company and Column.

Ceferino R. Rola*
1st Lieutenant, AUS

TABLE OF CONTENTS

*Ceferino R. Rola was instructed to write the history of the 1st Reconnaissance Battalion (Special).
**More reports on secret missions were gathered after this document was issued.

Source:
Filipino American Experience Research Project, October 5, 1998
U.S. National Archives and Records Administration (NARA)

Appendix B - 35

TIMELINE
ALLIED INTELLIGENCE BUREAU (AIB)
1ST RECONNAISSANCE BATTALION AND 978TH SIGNAL SERVICE COMPANY

1943

Jan to Aug -Submarine missions under Allied Intelligence Bureau (AIB) to the Philippines
May - Selected men from the 1st and 2nd Filipino Infantry Regiments began training for missions, Brisbane
Jul - 978th Signal Service Company activated, Brisbane.
Aug - 5217th Recon Battalion (Provisional) and 5218th Recon Battalion (Provisional) activated, Australia
Oct - 978th assigned to USAFFE. Attached to 5217th Recon Battalion
Nov 43 to Dec 43 – Philippine Regional Section (PRS) submarine missions to the Philippines

1944

Feb 44 to Dec – Mission men provided radio support and supplies to the guerrillas in major islands. Gathered intelligence to send back to Australia
Oct – KAZ and PIMC signalmen in Leyte
Nov - 5217th and 5218th Reconnaissance Battalions inactivated. 1st Recon Battalion (Special) activated

1945

Jan to Sep - Mission men provided radio support and supplies to the guerrillas. Sent intelligence to Australia
July - 1st Reconnaissance Battalion Headquarters departed for the Philippines.
Aug - 978th Signal men Headquarters, San Miguel, Tarlac
Aug - 1st Recon Battalion Headquarters, Camp Murphy, Manila
Aug - 1st Recon Battalion (Special) inactivated, Manila

1946

Jan to April - 978th Signal men in Tokyo, Japan
May - 978th Signal Service Company inactivated, Japan

Legend
FIR – First Filipino Infantry Regiment   KAZ – General MacArthur's radio station
2nd FIR – 2nd Filipino Infantry Regiment   PIMC – Philippine Intelligence Message Center
2nd FIB – 2nd Filipino Infantry Battalion

Source: Documents on the 1st Reconnaissance Battalion (Special), 97th Signal Service Company, 1st and 2nd Filipino Infantry Regiments, 2nd Filipino Infantry Battalion, Philippine Civil Affairs Unit (PCAU), Counter Intelligence Corp (CIC), Hawaii soldiers, U.S. Army Service Summary, and Abdon Llorente's 1st Filipino Infantry Regiment timeline.

Appendix B - 36

Recommendation for Legion of Merit Award to Jose Valera and his mission men (August 1944). Legion of Merit (LOM) is to the left of Valera's name and Bronze Star (BS) for his men.

Appendix B - 37

Recommendation for Legion of Merit Award to Eustaquio Cabais and his mission men (August 1944). Legion of Merit (LOM) is to the left of Cabais's name and Bronze Star (BS) for his men.

CHAPTER SOURCES

Research material primarily came from the U.S. National Archives and Records Administration, MacArthur Memo- rial Archives and Library, the U.S. Army Military History Institute, and the Filipino American Experience in World War II Research Project October 1998.

Chapter 1: U.S. Nationals

Anti-Miscegenation Act. 1930.

Begonia, Dan. "Asian American Studies 456, Filipinos in America." Filipino American Experience Research Project. Vol. 9. Unpublished. San Francisco State University. 2001.

Bernardo, Joseph A. "From 'Little Brown Brothers' to 'Forgotten Asian Americans' Race, Space, and Empire in Filipino." PhD. Diss., University of Washington, 2014.

Dumlao, Felipe, Leonard Aliwang, Mariano Angeles, Martin Mamuyac, and Saturnino Silva. Record. n.d., Filipino American National Historical Society (FANHS), Seattle, Washington.

Filipinos in California. Interview by Dionisio. President Manuel Quezon's WWII Exile Government.

Maram, Linda Espana. Creating Masculinity in Los Angeles's Little Manila. Columbia University Press, 2006.

"On this Day – January 19, 1930 White Mobs Attack Filipino Farmworkers in Watsonville, California." A History of Racial Injustice. Accessed October 24, 2021. https://calendar.eji.org/racial-injustice/jan/19.

Pedraza, Silvia and Ruben Rumbaut. Origins and Destinies: Immigration, Race, and Ethnicity in America. 1st ed. Cengage Learning, 1995.

Postcards from Salinas Photobook, curated by Alex S. Fabros. Stockton, CA: FANHS National Museum, 2013.

Shor, Franc. "See You in Manila." The American Legion Magazine (March 1943).

The Philippine National Defense Act. 1935.

Vallejo, Marie Silva. Collection. Author of The Battle of Ising. Daughter of Saturnino Ramos Silva.

Chapter 2: A Resistance

Acebes, Rodolfo Meim. Mindoro: the Stepping Stone. Manila: AV Manila Creative, 2016.

Agoncillo, Teodoro. The Fateful Years, Japan's Adventure in the Philippines, 1941-1945. Quezon City: R.P. Garcia Publication, 1965.

Aquino, Belinda. "Pearl Harbor Revisited." Philippine Daily Inquirer. December 14, 2012.

Buencamino, Victor Dr. "Manila Under Japanese Occupation (V)." Bulletin American Historical Collection 8, no. 3 (1980).

Campbell, Douglas A. Eight Survived. Lyon Press, 2011.

Chronology of the Allied Intelligence Bureau, GHQ, SWPA. June 1942-January 1946, Bobb Glenn Collection. AIB Chief Supply Officer.

Chua, Gregorio. Interview by Ken Wong. San Francisco Examiner, April 16, 1980.

Connaughton, Richard. MacArthur and Defeat in the Philippines. Woodstock, NY: The Overlook Press, 2001.

Daly, John F. III. An Officer in MacArthur's Court. Robertson, 2014.

Fabros, Alex S., Jr. "A History of the 978th Signal Service Company." Filipino American Experience Research Project. Unpublished. San Francisco State University. November 6, 1998.

Fairfield, William A. "Mactan Memoirs." The Philadelphia Inquirer (May 1943). Accessed October 24, 2021. http://ww2f.com/threads/the-fighting-mactan.21187/.

Full Speed Ahead: Negros Navigation at 75. Manila: Negros Navigation, 2008.

Geoffrey, Perrett. Old Soldiers Never Die: the Life of Douglas MacArthur. Random House, 1996.

Gibson, Emmett F. Interview by David Camelon. Chicago Herald-American, July 22, 1942.

Hara, Tameichi, Fred Saito and Roger Pineau. Japanese Destroyer Captain. Naval Institute Press, 2013.

Harding, Stephen. "Prelude to Pearl." History Net. Accessed October 26, 2021. https://www.historynet.com/prelude-to-pearl.htm.

Hart, Thomas C. 8 December 1941 to 15 February 1942. Commander in Chief, Asiatic Fleet. December 31, 2012, NARA Declassified, National Archives and Records Administration (NARA), College Park, Maryland.

Hemingway, Al. "Early Warning of Attack on Pearl Harbor." Charlotte Sun, December 7, 2012.

Hoyt, Edwin P. War in the Pacific. Vol 1. New York: Avon Books, 1990.

Ind, Allison W. "History of 5217th Reconnaissance Battalion." Bahala Na Anniversary Magazine (1944).

Jurado, Enrique L. "1934 Lucky Bag." U.S. Naval Academy Virtual Memorial Hall. Accessed October 26, 2021. https://usnamemorialhall.org/index.php/Enrique L. Jurado, Lt Col, PA.

Lapham, Robert and Bernard Norling. Lapham's Raiders: Guerrillas in the Philippines 1942-1945. Lexington, Kentucky: The University of Kentucky Press, 1996.

Maeda, Masami. 14th Army Chief of Staff. Interrogations of Japanese Officials (English Translations) Carlisle. GHQs.FarEast Command, May 10, 1947, G-2 Historical Section. U.S. Army Military History Institute (USAMHI), Carlisle, Pennsylvania.

Manchester, William. *American Caesar: Douglas MacArthur 1880-1964*. Boston: Little, Brown, 1978.

Messages from General MacArthur during the night to President Roosevelt and the War Department, February 11-12, 1942, Franklin D. Roosevelt Library.

Meyer, Milton Dr. "A World War II Vignette: The Darwin-Philippine Connection." *Bulletin American Historical Collection Foundation* 26, no. 2 (1998).

Milner, Samuel. *Victory in Papua*. Washington, D.C.: Office of the Chief of Military History, 1957.

Morningstar, James Kelley. *War and Resistance in the Philippines, 1942-1944*. Naval Institute Press, 2021.

Noyer, William L. *Mactan Ship of Destiny*. Fresno, California: Rainbow Press, 1979.

Ongpauco, Fidel. *They Refused to Die: True Stories about World War II Heroes in the Philippines 1941-1945*. Levesque Publications, 1982.

Pye, John. *The Tiwi Islands*. Darwin: Colemans Printing, 1977.

Rottman, Gordon L. *US Special Warfare Units in the Pacific Theatre 1941-1945*. Osprey Publishing, 1985.

Ruiz, Julius. "A Personal Remembrance." *Filipino American Experience Research Project*.

Stevens, Peter F. *The Twilight Riders*. Lyons Press, 2011.

Styles, Ralph E. USN. Interview by Earl Fowler, USN and Lucian L. Vestal, USMC. Oral History Program. Naval Historical Foundation, September-October 2001.

Taaffe, Stephen R. *MacArthur's Jungle War*. University Press of Kansas, 1998.

Valdez, Basilio J. Digitized Collection. n.d., Presidential Museum and Library, Malacañang Palace, Manila.

Villamin, Vicente. "Saving Prisoners in Japanese Hands." *The Philippine Mail* (September 1944).

Villanueva, James A. "Awaiting the Allies' Return: The Guerrilla Resistance Against the Japanese in the Philippines during WWII." PhD. Diss., 2019.

Whitney, Courtney. *MacArthur's Rendezvous with History*. LIFE, 1955.

Williams, Irving. Letter to Courtney Whitney. American Red Cross (1994).

Chapter 3 to 4: A Filipino Battalion, 1st Filipino Infantry Regiment

1st Filipino Infantry Camp Roberts Trainee 2, no 3, August 6, 1942 Angeles, Ranesto Bello. Document. Mariano Angeles.

"Army's New Citizens To Go on Air." *Shot 'N Shell*. June 24, 1942.

Astor, Gerald. *Crisis in the Pacific*. Dutton Adult, 1996.

Baldoz, Rick. "6. Another Mirage of Democracy': War, Nationality, and Asymmetrical Allegiance" *In The Third Asiatic Invasion: Empire and Migration in Filipino America, 1898-1946*, New York, USA: New York University Press (2011). pp. 194-236. Accessed October 24, 2021. https://doi.org/10.18574/9780814789889-008.

Bernardo, Joseph A. "From 'Little Brown Brothers' to 'Forgotten Asian Americans' Race, Space, and Empire in Filipino." PhD. Diss., University of Washington, 2014.

Bersamin, Manuel. Document on Paulino Bersamin.

"Big Reunion for Former Members of the Regiments at Place of Activation." *Poster* (1992).

Bob Luna. Document. Toribio Luna.

Brooks, Lester. *Behind Japan's Surrender*. McGraw-Hill, 1968.

Buaken, Iris Brown. "My Brave New World." *Philippine Mall*.

Buaken, Manuel. *I Have Lived with the American People*. Caldwell, Idaho: Caxton Printers, 1948.

Buaken, Manuel. "Our Fighting Love of Freedom." *Filipino American Experience Research Project*. February 20, 1943.

Budd, George. n.d., Ike Skelton Combined Arms Research Library Digital Library.

Budd, Talman C. "PCAU Experiences, Report No. 3, January 17, 1945." *Cases and Materials on Military Government*. Accessed October 24, 2021. https://cgsc.contentdm.oclc.org/digital/collection/p4013coll8/id/2221/rec/4.

Cacabelos, Rufino F. "Rufino Autobiography, Army Life: July 3, 1942 – January 3, 1946." *The Bayonet*, June 17, 1943.

Campos, Dixon. "California's Own." San Francisco, California (1990).

Campos, Dixon. Interview by Christie Cheatham, Spring 1983.

Castillo, Stephanie J. and Noel M. Izon, dirs. *Untold Triumph: America's Filipino Soldiers*. 2005; ICT Productions.

Chan, Sucheng. *Asian Americans: An Interpretive History*. Twayne, 1991.

Chua, Gregorio. "Former Regiments Barracks at Ft Ord Still There." Edited by Alex De Leon Fabros. *Bolo News Newsletter*. Salinas, CA: Filipino Infantry Association.

"Commanders May Select Officer Candidates." *Shot 'N Shell*. July 29, 1942.

Commanding General Eight U.S. Army on the Leyte-Samar Operation (Including Clearance of the Visayan Passages). Report. December 26, 1944-May 8, 1945, U.S. Army Military History Institute (USAMHI), Carlisle, Pennsylvania, pp. 13, 14, 23, 24.

Crossman, Edgar Gibson. "My Experience in WWII. Unpublished Memoir." *Wikipedia* (November 1966). Accessed October 24, 2021. https://en.wikipedia.org/w/index.php?title=File%3AEdgar_Gibson_Crossman_-_My_Experiences_in_WWII.PDF&page=1.

Dalisay, Jose Y. *Wash: Only a Bookkeeper, a Biography of Washington Z. Sycip*. Asia Rice Foundation and Metrobank Foundation, 2010.

Daly, John F., III. *An Officer in MacArthur's Court*. Robertson, 2014.

Daniel, Rhetta M. Document. Vaughan Moore.

"Dawn of the Philippine Freedom. A Scrapbook Commemorating the 50th Anniversary of the Liberation of the Philippines." Compiled by Cresencia Cespedes Gray. Self-Published. January 1, 1995.

De La Cruz, Jesselynn, ed. *Civilians in WWII*. James B. Reuter, S.J. Foundation, 1994.

"Discharges Eased for 38-Year-Olds." *Shot 'N Shell*. February 17, 1943.

Dollard, Charles and Donald Young. "In the Armed Forces." *Survey Graphic* 36 (1947).

Donesa, John. "The Mindanao Death March: Establishing a Historical Fact Through Online Research." *International Science Review* 1, no. 1 (2020).

Dowlen, Dorothy Dore. *Enduring What Cannot Be Endured*. McFarland, 2001.

Dumlao, Felipe, Leonard Aliwang, Mariano Angeles, Martin Mamuyac, and Saturnino Silva. Record. n.d., Filipino American National Historical Society (FANHS), Seattle, Washington.

Earle, Dixon. *Bahala Na*. Berkeley, California: Howell-North Books, 1961. Eichelberger, Robert L. *Our Jungle Road to Tokyo*. Independently Published, 2017.

Estrella, Cicero A. "Filipinos had to Fight for the Right to Serve their Adopted Home." SFGATE (2005).

Evangelista, Susan. "Aurelio Bulosan's Wartime Diary." *Philippine Studies* 37, no. 4 (1989): pp. 467-78.

Fabros, Alex De Leon. "Filipinos Receive Colors in Regimental Review: 2nd Filipino Infantry Regiment." *Cooke Clarion*, June 4, 1943.

Fabros, Alex De Leon. "We won't Forget Bataan Day Vow of Filipino Regiment." (1943).

Fabros, Alex S., Jr. "A Document Collection of the History of the 1st Filipino Infantry Regiment in World War II." Master's thesis, 1997.

Fabros, Alex S., Jr. "The Boogie Woogie Boys." *Filipinas Magazine* (September 1993). Accessed October 24, 2021. http://www.positivelyfilipino.com/magazine/the-boogie-woogie-boys.

Fabros, Michelle S. *A Christmas Present from our Fathers* (2011).

"Field Mass Celebrated for U.S. Filipino War Heroes." *Shot 'N Shell*. September 30, 1942.

"Filipinos Filmed in Technicolor." *Shot 'N Shell*. July 8, 1942.

"Filipino Soldiers at Ft. Ord Treated." *Panorama,* December 1942.

"First Filipino Officer in Our Army is Here." *The Bayonet*, November 12, 1942.

FlorCruz, Paul and Maria O'Toole. Document. Pedro Flor Cruz.

Francisco, Emiliano A. Interview by FANHS. Seattle, Washington.

Galang, Ricardo. *Hammer and Anvil*. Phoenix, 1989.

Ganley, Eugene F. and Robert H. Offley. Letters (March 6, 1959) and (March 9, 1959). Pennsylvania: USAMHI Carlisle.

Geiger, Jeffrey E. *Camp Cooke and Vandenburg Air Force Base 1941-1966*. Jefferson, North Carolina: McFarland, 2014.

Golden Anniversary Celebration of First/Second Filipino Combat Infantry Regiments, U.S. Army with a dinner dance. August 1, 1992, Salinas Community Center.

Gonzales, Carlos I. Bulacan Military Area (BMA). n.d. RG-401, Box 422., National Archives and Records Administration (NARA), College Park, Maryland.

Gregory, Jim. "California in World War II, The First Filipino Infantry Regiment and San Luis Obispo County."*Military Museum*. Accessed October 24, 2021. http://www.militarymuseum.org/1FIR-SLO.html.

Guittard, George. Interview by Marie S. Vallejo. San Jose, California, September 2014.

Hall, Lou. "World War II Diary of Robert T. Webber." Accessed October 24, 2021. http://home.pcisys.net/-pwebber/31_id/rtw_comment.htm.

Halsema, James J. "The Rest of the War." *Bulletin American Historical Collection* 20, no. 2 (1992).

Hazard, Henry. "Reminiscences of the Naturalization of Members of Our Armed Forces Overseas." *Monthly Review* 6, no. 6 (1948).

Hazard, Henry. "Valor Knows Neither Nationality, Race, Nor Creed." *Monthly Review* 2, no. 9 (1945): p. 110.

Historical Report. Operations on Northern Samar, Burias, and Ticao Islands in the Philippines Liberation Campaign, February 16, 1945-May 7, 1945, National Archives and Records Administration (NARA), College Park, Maryland.

Historical Report. Operations on Northern Samar, Burias, and Ticao Islands in the Philippines Liberation Campaign, May 8-20 and September 1945, National Archives and Records Administration (NARA), College Park, Maryland.

Ilustre, Ernest D. "Filipino Fighting Men." *The Washington Post,* July 28, 1942.

Javier, Jessie. "Society Notes: Soldiers from the Fourth Army, Camp Cooke and Camp Beale. Members of 2nd Filipino Infantry Band Played." November 24, 1943.

Johnson, Danny. "Historic California Posts, Camps Stations and Airfields." *California State Military*, issue 29. Beale Air Force Base (February 2012).

Krohn, Bubi. Interview by Rico Jose. Manila, Philippines.

Labrador, Albert. Mandirigma. Manila: JM Publishing House, 2021.

Lamata, Al. Phone Interview by Marie S. Vallejo. *2nd Filipino Infantry Regiment,* July 6, 2019.

Liban, Harold. Document. John A. Maraggay.

Liban, Harold. Document. Miguel L. Lauagan, PCAU.

Lim, Roberto H. *Pushing the Envelope: a Biography*. Pasay City, Philippines: Mapua House Publications, 2010.

Linderman, Gerald F. *The World Within War*. New York, NY: The Free Press, 1997.

Lindholm, Paul R. *Shadows From the Rising Sun*. Quezon City: New Day, 2009.

Los Banos, Domingo. Phone interview by Marie S. Vallejo. *2nd Filipino Infantry Regiment,* March 12, 2007.

Lutz, Stephen D. "The 93rd Infantry Division: The African-American Soldiers in the Pacific." *Warfare History Network*. Accessed October 24, 2021. https://warfarehistorynetwork.com/2019/01/19/the-93rd-infantry- division-the-african-american-soldiers-in-the-pacific/.

Maram, Linda Espana. *Creating Masculinity in Los Angeles's Little Manila*. Columbia University Press, 2006.

McPherson, Milton M. Infantry OCS Class 16-52. *The Ninety-Day Wonders, OCS and the Modern American Army*. Columbus, Georgia: United States Army Officer Candidate School Alumni Association, 2001.

Mullinix, Earl. Interview by Alex Fabros Jr. Salinas, California, 1972.

Offley, Robert H. Interview by Alex Fabros Jr. Salinas, California, 1971 and 1973.

Owens, William. "Eye-Deep in Hell: A Memoir of the Liberation of the Philippines, 1944-45." (1989). *Bulletin American Historical Collection* 23 (1995).

"PCAU 7." *Bulletin American Historical Collection* 22 (1994).

"Pacific States Groups Act to Form a Force, Provided War Department Sanctions One." *The New York Times,* January 13, 1942.

Presentation of US Army's 1st Filipino Infantry Regiment Regimental Colors by Gen. Morris at King City. Photo. July 13, 1943, The Wing Lake Museum of the Asian Pacific American Experience.

Rhoades, Weldon. *Flying MacArthur to Victory*. Texas A&M University Press, 2000.

Rosal, Raoul. Document. "PCAU #20", Cpl Toribio J. Rosal.

Ruiz, Julius B. "Report to Quezon: Pinoys in Wartime." Recalled by Ruiz 1942, August 18, 1946.

Ruiz, Julius B. "'Salinas Nite' to Honor Men from Bataan." *Filipino American Experience Research Project.* November 6, 1942.

Sanchez, Helen Ragsac. "Fil-Am Pinays and the 'Filipino Sports Parade.'" *Positively Filipino*. Accessed October 25, 2014. http://www.positivelyfilipino.com/magazine/fil-am-pinays-and-the-filipino-sports-parade.

Santos, Agustin, Dixon Campos, Philip Ordono, and Clemente J. Felipe. Interview by Cecilia Gaerlan and Fred Basconcillo. *First Filipino Infantry Men*. San Francisco, 1994.

Scharrenberg, Paul. "The Philippine Problem: Attitude of American Labor Toward Filipino Immigration and Philippine Independence." *Pacific Affairs* 2, no. 2 (February 1929).

Shor, Franc. "See You in Manila." *The American Legion Magazine* (March 1943).

Suatengco, Remi. Document. Eduardo T. Suatengco.

Taaffe, Stephen R. *MacArthur's Jungle War*. University Press of Kansas, 1998.

Taggaoa, Fernando A. "No Cause for Regret." *Filipino American Experience Research Project*. October 1942.

Takaki, Ronald. *Double Victory*. Time Warner Publishing's Online Publishing Program, 2000.

Talaugon, Nora and Federico Talaugon. Collection. 1944, John Spoor Broome Library, California State University, Channel Islands.

"There'll be Martial Music for the Avengers of the Philippines." *Shot 'N Shell*. September 2, 1942.

Turner, Thomas M.D. "US Army Medical Department. The Philippines and Okinawa. Chapter XVI." *A MEDD Center of History & Heritage*. Accessed October 24, 2021. http://history.amedd.army.mil/booksdocs/wwii/civilaf-fairs/chapter16.htm. Moved to: https://achh.army.mil/.

U.S. Army photograph 924-570-46/AM-51-811. Training Photo. n.d., Wing Lake Museum of the Asian Pacific American Experience.

Valdez, Pelagio. "Camp Roberts Trainer." *Bulletin* 2, no. 3.

Velasco, Andres L. Autobiography by Harold Liban. n.d.

Velasquez, Jaime. "Quezon's Aide at Benning." *The Bayonet,* February 18, 1943.

Villanueva, James A. "Awaiting the Allies' Return: The Guerrilla Resistance Against the Japanese in the Philippines during WWII." PhD. Diss., 2019.

Visaya, Ted. Document. Pedro Visaya.

Vite, Doroteo V. "A Filipino in Uncle Sam's Arm." *Filipino American Experience Research Project*.

Vitorio, Ramon. Letter to Mr. Rodriguez in San Francisco. October 16, 1942. Waite, Elmonte. "Jose Filipino Runs His Own Show." *Dawn of the Philippine Freedom* (1995).

Whitney, Courtney. *Report on Philippine Civil Affairs: 32nd Infantry Division 'Red Arrow*. Vol. 2. Carlisle, Pennsylvania: USAMHI, 1942.

Willoughby, Charles. *MacArthur, 1941-1945*. McGraw-Hill, 1954.

Wingo, James G. "The First Filipino Regiment." *Filipino American Experience Research Project*. Unpublished. San Francisco State University. October 1942.

Wingo, James G. "Two Filipinos Here." Arizona R epublic, January 10, 1943.

Appendix: A-1 to A-9, A-13, A-19, A-20 to A-24

Chapter 5: A Filipino Battalion, 2nd Filipino Infantry Regiment

Appendix: A-10 to A-12

Chapter 6: Top Secret

Alcala, Aurora. Document. Atanacio Alcala.

Arguelles, Carlos. *Panorama*. November 13, 1942.

Bahala Na First Anniversary 1, no. 39. Special Edition (1944).

Bahala Na Newsletter. Newsletter First Generated (1944).

Bowler, Robert. "History of the Mindanao Guerrillas." W est P oint. Accessed October 24, 2021. https://www.west-point. org/family/japanese-pow/Guerrillas/History%20of%20the%20Mindanao%20Guerrillas.pdf.

Cacabelos, Rufino F. "Rufino Autobiography, Army Life: July 3, 1942 – January 3, 1946." T he Bayonet, June 17, 1943.

Childress, Clyde. "A Critical Review of They Fought Alone." Bulletin of American H istorical Collection 31, no. 1 (2003).

Chronology of Events in Mindanao 1943-1944, Bobb Glenn Collection. AIB Chief Supply Officer.

Dacquel, Isidoro. Reports. n.d., MacArthur Archives (RG16 Box 46 Fol 21003, RG16 Box 15 Fol 2, RG16 Box 46 Fol 30004). MacArthur Memorial Archives and Library, Norfolk, Virginia.

Eisner, Peter. *MacArthur's Spies*. Penguin Books, 2018.

Fabros, Alex S., Jr. "A History of the 1st Reconnaissance Battalion United States Army." *Filipino American Experience Research Project*. Unpublished. San Francisco State University. October 5, 1998.

"Fraser Commando School." Accessed October 24, 2021. https://www.ozatwar.com/sigint/frasercommandoschool.htm.

Gangl, Theresa. Document. Gangl Leo.

Ilustre, Ernesto D. "Filipino Commandos in America." *Panorama,* December 7, 1975.

Loomis, Anna. The Guerrilla Resistance Movement in Northern Luzon. Paper. January 1945, RG-16, GHQ SWPA, G-2 Information Bulletin,

MacArthur, Douglas. *Reminiscences*. New York: McGraw-Hill Book, 1964.

Mashbir, Sysdey F. I *Was an American Spy*. Vantage Press, 1953.

Memo Lee Telesco to Ceferino Rola. Records, n.d., MacArthur Memorial Archives and Library, Norfolk, Virginia.

Mitsos, Tom. "Guerrilla Radio by American Guerrillas of Mindanao (AGOM)."

Morningstar, James Kelley. *War and Resistance in the Philippines, 1942-1944*. Naval Institute Press, 2021.

Murray, Mary. *Hunted: a Coastwatcher's Story*. Rigby, 1973.

Orders of 5217th Reconnaissance Battalion Men Attach to 11th Airborne. n.d. MacArthur Memorial Archives and Library, Norfolk, Virginia.

Orders of men to AIB, RG16 Box 15 Fol 4012. n.d., MacArthur Memorial Archives and Library, Norfolk, Virginia.

Parsons, Peter. "Special Mission Submarines in the Philippines." Bulletin American H istorical Collection 29, no. 4 (2001).

Rhoades, Weldon. *Flying MacArthur to Victory*. Texas A&M University Press, 2000.

Rolley, Ailsa. "Australia Remembers 1945-1995." Beaudesert Shire Council and Beaudesert Historical Society (1996).

Rottman, Gordon L. *US Special Warfare Units in the Pacific Theatre 1941-1945*. Osprey Publishing, 1985.

Ruiz, Julius. "A Personal Remembrance." *Filipino American Experience Research Project*.

Thompson, George Raynor and Dixie R. Harris. *The Signal Corps: the Outcome (Mid – 1943 T hrough 1945)*. Create Space Independent, 2015. p. 273.

USAFFE Check List. January 24, 1944, RG-16 B15 F2011, MacArthur Memorial Archives and Library, Norfolk, Virginia.

Villanueva, James A. "Awaiting the Allies' Return: The Guerrilla Resistance Against the Japanese in the Philippines during WWII." PhD. Diss., 2019.

Villarta, Fred. "Memorial of Julius Ruiz."

Vivero, Joe. Interview by Marie Vallejo. Carlos Arguelles.

Walters, Earl. "Rescue From Shangri La." Interview by Patrick O'Donnell. November 25, 1998. Accessed October 24, 2021. http://www.thedropzone.org/pacific/walters.htm.

Whitney, Courtney. *MacArthur's Rendezvous with History*. LIFE, 1955.

Williams, Irving. Letter to Courtney Whitney. *American Red Cross* (1994).

Willoughby, Charles. *The Guerrilla Resistance Movement in the Philippines*. Vantage Press, 1972.

Worcester, Dean C. Letter to Charles Parsons. October 6, 1943.

Appendix: B-1 to B-20, B-22 to B-28, B-30 to B-37

Chapter 7: The Radio

Fabros, Alex S., Jr. "A History of the 978th Signal Service Company." *Filipino American Experience Research Project*. Unpublished. San Francisco State University. November 6, 1998.

Miller, Donald L. *D-Days in the Pacific*. 1st ed. Simon & Schuster, 2005.

Morningstar, James Kelley. *War and Resistance in the Philippines, 1942-1944*. Naval Institute Press, 2021.

Appendix: B-21

Chapter 8: 1st Reconnaissance Battalion Paratroopers

Agullana, Melissa. Document. Domingo 'Don' S. Ruiz.

Hughes, Les. "The Philippine Airborne." *The 503d P .R.C.T. Heritage Battalion Onlin*e. Accessed October 24, 2021. https://corregidor.org/heritage_battalion/hughes/hughes.html.

"Modern Legend of Shangri-La." *Far East Asian Service Command: Jungle Journal* 1 (1945).

Ruiz, Don. Collection. n.d, Commendation of Shangri-La Rescue.

Solidarios, Serni. Document. Sergio Solidarios.

Velasco, Andres L. Autobiography by Harold Liban. n.d. Walter, Cecil Earl. Letter to Les Hughes. May 27, 1994

Walters, Earl. "Rescue From Shangri La." Interview by Patrick O'Donnell. November 25, 1998. Accessed October 24, 2021. http://www.thedropzone.org/pacific/walters.htm.

Zuckoff, Michael. Lost in Shangri-La. Harper Collins, 2011.

Chapter 9: Negros Island

Breuer, William B. *MacArthur's Undercover War*. Edison, NJ: Castle Books, 2005.

Campbell, Douglas A. *Eight S urvived*. Lyon Press, 2011.

Chronology of the Allied Intelligence Bureau, GHQ, SWPA, June 1942-January 1946, Bobb Glenn Collection. AIB Chief Supply Officer.

Dawn of Freedom. Films. 1944.

Dissette, Hans C. and Edward Adamson. *Guerrilla Submarines*. Bantam Books, 1980.

Doromal, Jose D. *The War in Panay*. Diamond Historical Publications, 1952.

Duque, Suzanne. *Soldiers as Guerrillas*. Raleigh, North Carolina: Lulu, 2015.

Elphick, James. "Life Aboard WWII Submarines was Brutal." *We are the Almighty* (July, 2021). Accessed October 24, 2021. https://www.wearethemighty.com/popular/life-aboard-wwii-submarines-was-brutal/.

Fabros, Alex S., Jr., ed. "A History of the 978th Signal Service Company." *Filipino American Experience Research Project*. November 6, 1998.

Glenn, Bobb. "Account of Submarine Mission Landing (January 1943)." Told by Jesus Villamor.

Holian, Thomas. "Saviors and Suppliers, World War II Submarine Special Operations in the Philippines." Undersea *Warfare Magazine*, no. 23 (Summer, 2004).

Ind, Allison W. Check Sheet from Philippine Regional Section (PRS) to Chief of Staff through AIB. Memo. December 6, 1943, MacArthur Memorial Archives and Library, Norfolk, Virginia.

Ind, Allison W. *Secret War Against Japan: the Allied Intelligence Bureau in World War II*. United States: Curtis, 1970.

Jorge, Patricio. Interview by Marie S. Vallejo. 1st Reconnaissance Battalion (Special). December 2011.

Keats, John. *They Fought Alone*. New York: Pocket Books, 1965.

Lapham, Robert and Bernard Norling. *Lapham's Raiders: Guerrillas in the Philippines 1942-1945*. Lexington, Kentucky: The University of Kentucky Press, 1996.

Lindholm, Paul R. *Shadows From the Rising Sun*. Quezon City: New Day, 2009.

Lozano, Mark. "The Untold Story of Delfin Yuhico." Accessed July 10, 2022. https://cantilanhistory.weebly.com/collection.html.

Manikan, Gamaliel L. *Guerrilla Warfare on Panay Island in the Philippine*s. Sixth Military District Foundation, 1977.

Maritime Museum. "Don Isidro a WWII 'blockade-runner' sunk off Darwin February 1942." Created by Peter Rout (blog). Australian Maritime Museums Council (August 9, 2017). Accessed October 18, 2022.http://maritimemuseumsaustralia.com/profiles/blogs/don-isidro-a-wwii-blockade-runner-sunk-off-darwin-february-1942.

Meider, Henry. Interview by GHQ, 1945.

Meyer, Milton Dr. "A World War II Vignette: The Darwin-Philippine Connection." *Bulletin American Historical Collection Foundation* 26, no. 2 (1998).

Morningstar, James Kelley. *War and Resistance in the Philippines, 1942-1944*. Naval Institute Press, 2021.

Mossman, Helen Madamba. *A Letter to My Father: Growing Up Filipina and American*. University of Oklahoma Press, 2008

Noyer, William L. *Mactan Ship of Destiny*. Fresno, California: Rainbow Press, 1979.

Parsons, Peter. "Special Mission Submarines in the Philippines." *Bulletin American Historical Collection* 29, no. 4 (2001).

Schmidt, Larry S. "American Involvement in the Filipino Resistance Movement on Mindanao During the Japanese Occupation, 1942-1945." Master's Thesis, Command and General Staff College, 1982.

Smith, Steven Trent. *The Rescue*. Wiley, 2003.

Villamor, Jesus. "They Never Surrendered, Hero as Critic." *Bulletin American Historical Collection* 12, no. 2 (1985).

Worcester, Dean. Report on Negros 1943-1944. Negros.

Yuhico, Delfin C. Interview by Marie S. Vallejo. 1st Reconnaissance Battalion (Special), August 2011.

Yuhico, Delfin C. Report. n.d., Our Mission in the Philippines. 1st Reconnaissance Battalion History.

Chapter 10: Mindanao Island February 1943 – March 1944

Bowler, Robert. "History of the Mindanao Guerrillas." *West Point*. Accessed October 24, 2021. https://www.west- point. org/family/japanese-pow/Guerrillas/History%20of%20the%20Mindanao%20Guerrillas.pdf.

Celdran, Jesus. Interview by Marie S. Vallejo, 2009.

Fabros, Alex S., Jr. "A History of the 1st Reconnaissance Battalion United States Army." *Filipino American Experience Research Project*. Unpublished. San Francisco State University. October 5, 1998.

Fabros, Alex S., Jr. "A History of the 978th Signal Service Company." *Filipino American Experience Research Project*. Unpublished. San Francisco State University. November 6, 1998.

Fajardo, Edgar. Interview by Marie S. Vallejo. Edgar Fajardo, November 5, 2008.

Fertig to Gen. MacArthur. Radio Message. June 3, 1944, MacArthur Memorial Archives and Library, Norfolk, Virginia.

Fertig, Wendell. Diary. August-September 1944, MacArthur Memorial Archives and Library, Norfolk, Virginia.

Filipinas Foundation. *Zamboanga Hermosa: Memories of the Old Town*. 1984.

Gangl, Theresa. Document. Gangl Leo.

Grashio, Samuel C. and Bernard Norling. *Return to Freedom*. MCN Press, 1982.

Hamner, Jordan A. "Hamner's War." *Bulletin American Historical Collection* I-V (2004-2005).

Holmes, Kent. *Wendell Fertig and His Guerrilla Forces in the Philippines: Fighting the Japanese Occupation, 1942-1945*. Jefferson, N.C. McFarland. 2015.

Intelligence Activities in the Philippines during the Japanese Occupation. GHQ, United States Army Forces, Pacific, n.d., Military Intelligence Section, U.S. Army Military History Institute (USAMHI), Carlisle, Pennsylvania.

Jose, Rico T. "Preparing for Emergency: The Civilian Emergency Administration 1941." *Diliman Review* 38, no. 2 (1991): pp. 37-46.

Keats, John. *They Fought Alone*. New York: Pocket Books, 1965.

Kuder, Edward M. "The Philippines Never Surrendered." *Saturday Evening Post*, February 10, 1943.

Levine, Alan. *Captivity, Flight , and Survival in World War II*. Praeger, 2000.

McCarroll. "The Memories Live on Forever." The Odessa American, January 2, 1983.

Military Intelligence Section. U.S. Army Military History Institute (USAMHI), Carlisle, Pennsylvania.

Mills, Scott A. *Stranded in the Philippines: Professor Bell's Private War Against the Japanese*. Annapolis, Maryland: The Naval Institute Press, 2009.

Morningstar, James Kelley. *War and Resistance in the Philippines, 1942-1944*. Naval Institute Press, 2021.

Morton, Louis. *The Fall of the Philippines, the War in the Pacific (United States Army in World War II)*. Washington, D.C.: The U.S. Army Center of Military History, 1989.

Parsons, Peter. "Chick Parsons in Occupied Manila." *Bulletin American Historical Collection* 36, no. 1 (2008).

Preston, William B. "Mission Naval." *History Magazine*. Accessed October 24, 2021. https://news.usni.org/tag/uss-william-b-preston.

Rivera, Juan A. "Enemy Invasion of Mindanao, 1941-1945: a Philippine Account."

Schmidt, Larry S. "American Involvement in the Filipino Resistance Movement on Mindanao During the Japanes Occupation, 1942-1945." Master's Thesis, Command and General Staff College, 1982.

Smith, Steven Trent. *The Rescue*. Wiley, 2003.

USAFFE to the Adjutant General and General Douglas MacArthur. February 25, 1942, National Archives and Records Administration (NARA), College Park, Maryland.

United States Army in World War II: The Technical Services, Signal Corps, n.d., U.S. Army Military History Institute (USAMHI), Carlisle, Pennsylvania.

Villanueva, James A. "Awaiting the Allies' Return: The Guerrilla Resistance Against the Japanese in the Philippines during WWII." PhD. Diss., 2019.

Young, Frank H. "The Sulu Sharpshooter." *World War II Heroes Project*. Bamban Historical Society, Bamban Museum of History (2020).

Chapter 11: Mindanao Island March 1944 – June. 1945

Bowler, Robert. "History of the Mindanao Guerrillas." *West Point*. Accessed October 24, 2021. https://www.west- point. org/family/japanese-pow/Guerrillas/History%20of%20the%20Mindanao%20Guerrillas.pdf.

Childress, Clyde. "A Critical Review of They Fought Alone." *Bulletin of American Historical Collection* 31, no. 1 (2003).

"Darwin and the Philippines Submarine Run." *Journal of Northern Territory History* (1943)., Bobb Glenn Collection. AIB Chief Supply Officer.

Dissette, Hans C. and Edward Adamson. *Guerrilla Submarines*. Bantam Books, 1980.

Fertig, Wendell. Diary. August-September 1944, MacArthur Memorial Archives and Library, Norfolk, Virginia.

Fertig, Wendell. "Letter to Harold Rosenquist." *MacArthur Archives* (1944).

G-2 Staff Study of Philippine Islands Situation. Appendix XV. February 24-25, 1944, Military Intelligence Section, U.S. Army Military History Institute (USAMHI), Carlisle, Pennsylvania.

Gangl, Theresa. Document. Gangl Leo.

Grashio, Samuel C. and Bernard Norling. *Return to Freedom*. MCN Press, 1982.

Gustilo, Araceli. Interview by Marie S. Vallejo, June 30, 2013.

Intelligence Summary, 10th MD, USFIP, No. 9, October 1944. The Ising Operation, May 1, 1945, MacArthur Memorial Archives and Library, Norfolk, Virginia.

Keats, John. *They Fought Alone*. New York: Pocket Books, 1965.

McCracken, Alan. *Very Soon Now, Joe*. New York: Hobson Book Press, 1947.

Mitsos, Tom. "Guerrilla Radio by American Guerrillas of Mindanao (AGOM)."

Morningstar, James Kelley. *War and Resistance in the Philippines, 1942-1944*. Naval Institute Press, 2021.

Pompea, Edward T. "Galapagos-Long and Lonesome." *Cross Country: News and Views Around the World*. Air Force Association. September 1945.

Rosenquist, Harold A. An Estimate of the Enemy Situation. April 30, 1945, Military Intelligence Section, U.S. Army Military History Institute (USAMHI), Carlisle, Pennsylvania.

Rosenquist, Harold A. Letter to Wendell Fertig, October 4, 1944.

Rosenquist, Harold A. The 10th MD. Report. March-August 1944, MacArthur Memorial Archives and Library, Norfolk, Virginia.

Schmidt, Larry S. "American Involvement in the Filipino Resistance Movement on Mindanao During the Japanese Occupation, 1942-1945." Master's Thesis, Command and General Staff College, 1982.

Stuckenschneider, Placid. "The Last Campaign Mindanao – March '99 World War II Feature." *History Net*. Accessed October 24, 2021. https://www. historynet.com/the-last-campaign-mindanao-march-99-world-war-ii-feature.htm.

Velasquez, M. A. "History of the Sulu Area Command 1945: From The Library of Romulo Espaldon." *Internet Archives*. Accessed October 24, 2021. https://archive.org/details/sulu-area-command.

Whitney, Courtney. *MacArthur's Rendezvous with History*. LIFE, 1955.

Chapter 12: Panay Island

Agoncillo, Teodoro. *The Fateful Years, Japan's Adventure in the Philippines, 1941-1945*. Quezon City: R.P. Garcia Publication, 1965.

Ames, Irineo A. Interview by Ed Crespo with Marie S. Vallejo, September 10, 2018.

Anderson, Bernard. York Group. July 21, 1944, National Archives and Records Administration (NARA), College Park, Maryland.

Crespo, Toribio. Interview by Ed Crespo, September 18, 2018.

Dacquel, Isidoro. Reports. n.d., MacArthur Archives (RG16 Box 46 Fol 21003, RG16 Box 15 Fol 2, RG16 Box 46 Fol 30004). MacArthur Memorial Archives and Library, Norfolk, Virginia.

Doromal, Jose D. *The War in Panay*. Diamond Historical Publications, 1952.

Fabros, Alex S., Jr. "A History of the 1st Reconnaissance Battalion United States Army." *Filipino American Experience Research Project*. Unpublished. San Francisco State University. October 5, 1998.

Manikan, Gamaliel L. *Guerrilla Warfare on Panay Island in the Philippines*. Sixth Military District Foundation, 1977.

Morningstar, James Kelley. *War and Resistance in the Philippines, 1942-1944*. Naval Institute Press, 2021.

Parsons, Charles. Papers. n.d., RG-58, Spyron 1943-1945, MacArthur Memorial Archives and Library, Norfolk, Virginia.

Parsons, Peter C. "The Panay Guerrillas / USS Narwhal Debacle at Lipata Point." Accessed October 24, 2021. http://myphilippinelife.com/the-panay-guerrillasuss-narwhal-debacle-at-lipata-point/.

Villanueva, James A. "Awaiting the Allies' Return: The Guerrilla Resistance Against the Japanese in the Philippines during WWII." PhD. Diss., 2019.

War Reports of the USS Narwhal and Messages between Col. Macario Peralta and SWPA GHQ, n.d., MacArthur Memorial Archives and Library, Norfolk, Virginia.

Chapter 13: Mindoro Island

Acebes, Rodolfo Meim. *Mindoro: the Stepping Stone*. Manila: AV Manila Creative, 2016.

Crossman, Edgar Gibson. "My Experience in WWII. Unpublished Memoir." *Wikipedia* (November 1966). Accessed October 24, 2021. https://en.wikipedia.org/w/index.php?title=File%3AEdgar_Gibson_Crossman_-_My_Experiences_in_WWII.PDF&page=1.

Fabros, Alex S., Jr. "A History of the 1st Reconnaissance Battalion United States Army." *Filipino American Experience Research Project*. Unpublished. San Francisco State University. October 5, 1998.

Fabros, Alex S., Jr. "A History of the 978th Signal Service Company." *Filipino American Experience Research Project*.

Unpublished. San Francisco State University. November 6, 1998.

Galang, Ricardo. *Secret Mission to the Philippines*. University, 1948.

Hemingway, Al. "Early Warning of Attack on Pearl Harbor." *Charlotte Sun,* December 7, 2012.

Ind, Allison W. *Secret War Against Japan: The Allied Intelligence Bureau in World War II*. United States: Curtis, 1970.

Linderman, Gerald F. *The World Within War*. New York, NY: The Free Press, 1997.

MacArthur, Douglas. *Reminiscences*. New York: McGraw-Hill Book, 1964.

Marquardt, Frederic. "'I Shall Return:' A Footnote to History." *Bulletin American Historical Collection,* no. 1.

Maynard, Mary. *My Faraway Home*. Isis Large Print, 2002.

Morison, Samuel E. *History of United States Naval Operations in World War II*. Vol. 13. The Liberation of the Philippines, 1944-45.

Morningstar, James Kelley. *War and Resistance in the Philippines, 1942-1944*. Naval Institute Press, 2021. Ongpauco, Fidel. *They Refused to Die: True Stories about World War II Heroes in the Philippines 1941-1945*. Levesque Publications, 1982.

Ortiz, Pacifico S. J. "Letter to Aurora Quezon." December 17, 19-.

Parsons, Peter. "Me and the AGOM." *Bulletin American Historical Collection*.

Patton, Howard. Interview by William J. Alexander. University of Texas Oral History Collection. No. 1304, January 5, 1999.

Rowe, George. Radio Messages to General MacArthur. January 1945, U.S. Army Military History Institute (USAMHI), Carlisle, Pennsylvania.

"SS167 Narwhal". Accessed October 25, 2021. https://fleetsubmarine.com/ss-167.html.

Villanueva, James A. "Awaiting the Allies' Return: The Guerrilla Resistance Against the Japanese in the Philippines during WWII." PhD. Diss., 2019.

Chapter 14: Samar Island

Boggs, Charles W. Jr. *Maj. Marine Aviation in the Philippines*. Washington, D.C.: U.S. Government Printing Office, 1951.

Campeau, Lucien. "My Airforce Weather Mission from April 1944 through April 1945 with the American Guerrillas of Mindanao." *Guerrilla Radio by American Guerrillas of Mindanao* (1992).

Fabros, Alex S., Jr. "A History of the 1st Reconnaissance Battalion United States Army." *Filipino American Experience Research Project*. Unpublished. San Francisco State University. October 5, 1998.

Fabros, Alex S., Jr., ed. "A History of the 978th Signal Service Company." *Filipino American Experience Research Project*. November 6. 1998.

Gangl, Theresa. Document. Gangl Leo.

Herras, Domingo. Documents. Gaudencio Vera, Tayabas Guerilla.

Keats, John. *They Fought Alone*. New York: Pocket Books, 1965.

Mitsos, Tom. "Guerrilla Radio by American Guerrillas of Mindanao (AGOM)."

Morningstar, James Kelley. *War and Resistance in the Philippines, 1942-1944*. Naval Institute Press, 2021.

Ozawa, Jisaburo. Interrogations of Japanese Officials. OPNAV-P-03-100. Naval Analysis Division. Interrogation Nav No. 3 USSBS No. 32. *The Battle of the Philippine Sea* (June 19-20, 1944). Interrogation October 16, 1945.

"Philippines Leyte-Samar. Guerrilla and Civilians on Samar Wage Heroic Fight Against Japanese." Free Philippines. November 23, 1944.

Rottman, Gordon L. *US Special Warfare Units in the Pacific Theatre 1941-1945*. Osprey Publishing, 1985.

Smith, Charles. Interview by Murray Maj, February 11, 1947.

Stahl, Bob. *You're No Good to Me Dead: Behind the Japanese Lines in the Philippines*. Annapolis, Maryland: Naval Institute Press, 1997

Villanueva, James A. "Awaiting the Allies' Return: The Guerrilla Resistance Against the Japanese in the Philippines during WWII." PhD. Diss., 2019.

Watari Group Intelligence Report B. No. 146. July 1944, National Archives and Records Administration (NARA), College Park, Maryland.

White, Michael. *Australian Submarines*. Vol 2. Australian Teachers of Media, 2015.

Chapter 15: Palawan Island

Campbell, Douglas A. *Eight Survived*. Lyon Press, 2011.

Elphick, James. "Life Aboard WWII Submarines was Brutal." *We are the Almighty* (July, 2021). Accessed October 24, 2021. https://www.wearethemighty.com/popular/life-aboard-wwii-submarines-was-brutal/.

Hughes, Rebekah. *Surviving the Flier*. Phoenix Flair Press, 2010.

Moore, Stephen L. "The Heroes of Palawan – How Survivors of a Japanese Massacre Live to Tell the Tale of Atrocities in the Philippines." (December 2016). Accessed October 24, 2021. https://militaryhistorynow.com/2016/12/02/the-heroes- of-palawan-how-survivors-of-a-japanese-massacre-lived-to-tell-the-tale-of-atrocities-in-the-philippines/.

Morningstar, James Kelley. *War and Resistance in the Philippines, 1942-1944*. Naval Institute Press, 2021.

Philippine Sea Frontier Command Letter: Massacre of American Prisoners of War at Puerto Princesa, Palawan, January 23, 1945, Ike Skelton Combined Arms Research Library Digital Library.

Rallojay, Teodoro J. "Diary: May 23 – November 2, 1944."

Sides, Hampton. Ghost Soldiers. New York, NY: Doubleday, 2001.

Smith, Rufus W. "World War II POW." Interview by Rufus W. Smith. *Humanities Texas* (March 2015). Accessed October 24, 2021. https://www.humanitiestexas.org/news/articles/interview-rufus-w-smith-world-war-ii-pow.

Chapter 16: Leyte Island

MacArthur, Douglas. *Reminiscences*. New York: McGraw-Hill Book, 1964.

Morningstar, James Kelley. *War and Resistance in the Philippines, 1942-1944*. Naval Institute Press, 2021.

Villanueva, James A. "Awaiting the Allies' Return: The Guerrilla Resistance Against the Japanese in the Philippines during WWII." PhD. Diss., 2019.

Chapter 17: Luzon Island, Northern Area

Acebes, Rodolfo Meim. *Mindoro: the Stepping Stone*. Manila: AV Manila Creative, 2016.

Agoncillo, Teodoro. *The Fateful Years, Japan's Adventure in the Philippines, 1941-1945*. Quezon City: R.P. Garcia Publication, 1965.

"American Forces Secure Balete Pass." *Free Philippines*. May 15, 1945.

Breuer, William B. *MacArthur's Undercover War*. Edison, NJ: Castle Books, 2005.

Cal, Ben. "The Heroism and Sacrifice of Gen. Fortunato Abat." *Philippine News Agency* (March 2018). Accessed November 14, 2021. https://www.pna. gov.ph/articles/1028265.

Cal, Ben. *Victory at Bessang Pass*. Manila, Philippines. 2012.

Carlisle, John M. *Red Arrow Men: the 32nd Division on the Villa Verde*. Detroit: Arnold-Powers, 1945.

Cuesta, Karla Dela. Document. Ignacio Dela Cuesta.

Duque, Suzanne. *Soldiers as Guerrillas*. Raleigh, North Carolina: Lulu, 2015.

Fabros, Alex S., Jr. "A History of the 1st Reconnaissance Battalion United States Army." *Filipino American Experience Research Project*. Unpublished. San Francisco State University. October 5, 1998.

Fitzpatrick, Georgina, Timothy L.H. McCormack and Narrelle Morris. *Australia's War Crimes Trials*. Martinus Nijhoff, 2016.

Garcia, Arturo P. "The Real Heroes of Bessang Pass." *Bulatlat* (June 2005). Accessed October 24, 2021. https://www.bulatlat.com/2005/06/25/the-real-heroes-of-bessang-pass/.

"Guerrillas Help Capture Ipo Dam. (Miller/Stoddard)." *Free Philippines*. May 19, 1945.

Halsema, James J. "The End of the War 1945." *Bulletin American Historical Collection* 20, nos. 3-4 (1992).

Ind, Allison W. "Six Came Back". Original typewritten document from Delfin Yuhico.

Jerry, Grant E. "All Those Who Remained: the American-Led Guerrillas in the Philippines, 1942-1945." United States Army, School of Advanced Military Studies. Fort Leavenworth, Kansas: United States Army Command and General Staff College, 2014.

Laubenthal, Sanders A. "A History of John Hay Air Base." *Bulletin American Historical Collection* 22, no. 2 (1994).

Maram, Linda Espana. *Creating Masculinity in Los Angeles's Little Manila*. Columbia University Press, 2006.

"On this Day – January 19, 1930 White Mobs Attack Filipino Farmworkers in Watsonville, California." *A History of Racial Injustice*. Accessed October 24, 2021. https://calendar.eji.org/racial-injustice/jan/19.

Pimentel, Lorenzo. Interview by Cynthia Mejia. Transcribed by Carolina Apostol. Washington State Oral History Program, November 14, 1971.

Santes, Jose M. *Bataan, Corregidor, Bessang Pass, Mindanao Battlefields*. Davao City: Joana A. Santes, 1992.

Sixth United States Army. "Report of the Luzon Campaign, 9 January 1945 – 30 June 1945. Vol. IV." *Ike Skelton Combined Arms Research Library Digital Library*. Accessed October 29, 2021. https://cgsc.contentdm.oclc.org/digital/collection/p4013coll8/id/2289.

Smith, Robert Ross. "Chapter XXVIII. Action at the Northern Apex." *U.S. Army in World War II: Triumph in the Philippines*. Accessed October 29, 2021. https://www.ibiblio.org/hyperwar/USA/USA-P-Triumph/USA-P-Triumph-28.html.

Villanueva, James A. "Awaiting the Allies' Return: The Guerrilla Resistance Against the Japanese in the Philippines during WWII." PhD. Diss., 2019.

Volckmann, Russell W. *We Remained: Three Years Behind the Enemy Lines in the Philippines*. New York: W.W. Norton, 1954.
Whitlock, Anthony. "Secret Record Gave Clue, Japanese Tactics on Luzon." *The Sydney Morning Herald (NSW 1842 – 1954)*, October 24, 1945.

Chapter 18: Luzon Island, Central Area

Breuer, William B. *MacArthur's Undercover War*. Edison, NJ: Castle Books, 2005.

Earle, Dixon. *Bahala Na*. Berkeley, California: Howell-North Books, 1961.

Elphick, James. "Life Aboard WWII Submarines was Brutal." *We are the Almighty* (July, 2021). Accessed October 24, 2021. https://www.wearethemighty.com/popular/life-aboard- wwii-submarines-was-brutal/.

Fabros, Alex S., Jr. "A History of the 1st Reconnaissance Battalion United States Army." *Filipino American Experience Research Project*. Unpublished. San Francisco State University. October 5, 1998.

Lapham, Robert and Bernard Norling. *Lapham's Raiders: Guerrillas in the Philippines 1942-1945*. Lexington, Kentucky: The University of Kentucky Press, 1996.

Miculka, Cameron. "Living Legacy: Al Lim, 016, Reflects on Award Serving in WWII." W est H awaii T oday (2017).

Morningstar, James Kelley. *War and Resistance in the Philippines, 1942-1944*. Naval Institute Press, 2021.

Ogawa, Tetsuro. *Terraced Hell, Tokyo*. Charles E. Tuttle Company, 1972.

Raquepo, Veronica, Aegina Festin and KarenDonila. "Anderson Guerrillas." University of the Philippines, 2012.

Schmidt, Larry S. "American Involvement in the Filipino Resistance Movement on Mindanao During the Japanese Occupation, 1942-1945." Master's Thesis, Command and General Staff College, 1982.

Stahl, Bob. *You're No Good to Me Dead: Behind the Japanese Lines in the Philippines*. Annapolis, Maryland: Naval Institute Press, 1997.

Villanueva, James A. "Awaiting the Allies' Return: The Guerrilla Resistance Against the Japanese in the Philippines during WWII." PhD. Diss., 2019.

Chapter 19: Cebu Island

Morningstar, James Kelley. *War and Resistance in the Philippines, 1942-1944*. Naval Institute Press, 2021.

Chapter 20: Philippine Civil Affairs Unit (PCAU) and Counter Intelligence Corps (CIC)

Agoncillo, Teodoro. *The Fateful Years, Japan's Adventure in the Philippines, 1941-1945*. Quezon City: R.P. Garcia Publication, 1965.

Alexander, Larry. *Shadows in the Jungle: the Alamo Scouts Behind Japanese Lines in World War II*. London, England: Penguin, 2010.

Bamont, Nick. Document. John Bamont.

Boggs, Charles W. Jr. Maj. *Marine Aviation in the Philippines*. Washington, D.C.: U.S. Government Printing Office, 1951.

Counterintelligence Corps, History and Mission in World War II. n.d., U.S. Army Military History Institute (USAMHI), Carlisle, Pennsylvania.

Daly, John F., III. *An Officer in MacArthur's Court*. Robertson, 2014.

Evangelista, Susan. "Aurelio Bulosan's Wartime Diary." *Philippine Studies* 37, no. 4 (1989): pp. 467-78.

Francisco, Emiliano A. Interview by FANHS. Seattle, Washington.

Galang, Ricardo. *Hammer and Anvil*. Phoenix, 1989.

Gangl, Theresa. Document. Gangl Leo.

Japanese Plans for Defense of Manila, January 1-21, 1945, February 2, 1945, Allied Translator and Interpreter Section (ATIS), National Archives and Records Administration (NARA), College Park, Maryland.Labrador, Juan. *A Diary of the Japanese Occupation December 7, 1941-May 7, 1945*. Manila, Philippines: Santo Tomas Press, 1989.

Laubenthal, Sanders A. "A History of John Hay Air Base." *Bulletin American Historical Collection* 22, no. 2 (1994).

Lergards, Benito, Jr. "Double Take." *The Philippine Press,* February 6, 2010.

"Leyte-Samar." *Free Philippines*. November 19, 1944.

"Leyte-Samar." *Free Philippines*. November 23, 1944.

Mashbir, Sysdey F. *I Was an American Spy*. Vantage Press, 1953.

Mcmicking, Joseph. "Leyte Landing and the Battle of Leyte Gulf." Speech at the Insular Life Assurance Co. and Tacloban Lions Club, Leyte, Philippines, October 19, 1956.

Miller, Donald L. *D-Days in the Pacific*. 1st ed. Simon & Schuster, 2005.

Morningstar, James Kelley. *War and Resistance in the Philippines, 1942-1944*. Naval Institute Press, 2021.

Mydans, Carl. "Return to Sto. Tomas. US Wins Heart of the Philippines." *Life* 18, no. 8 (1945).

Parsons, Peter. "Chick Parsons in Occupied Manila." *Bulletin American Historical Collection* 36, no. 1 (2008).

"Psychological Warfare Branch., GHQ., Office of War Information Unit. Jap Orders to Massacre All Filipino Civilians Captured." *Free Philippines*.

Quezon, Aurora A. Free Philippines Vol 1-3, n.d., Digitized Collection. Presidential Museum and Library, Malacañang Palace, Manila.

Rhoades, Weldon. *Flying MacArthur to Victory*. Texas A&M University Press, 2000.

Richards, Peter. "After 3 February 1945." *Bulletin American Historical Collection* 26, no. 2.

Rowe, George. Radio Messages to General MacArthur. January 1945, U.S. Army Military History Institute (USAMHI), Carlisle, Pennsylvania.

Smollar, David. "Hard, Bitter, Unpleasantly Necessary Duty, A Little-Known World War II Story of the Philippines."*Philippine E-Journals* 61, no. 1 (2015).

"Somewhere in the Philippines." *Free Philippines.* February 12, 1944.

"The Army Weekly 1942-1945: Part VI, March 2, 1945-December 28, 1945." *Yank.* New York: Arno Press (1967).

Turner, Thomas M.D. "US Army Medical Department. The Philippines and Okinawa. Chapter XVI." *A MEDD Center of History & Heritage.* Accessed October 24, 2021. http://history.amedd.army.mil/booksdocs/wwii/civilaffairs/ chapter16. htm. Moved to: https://achh.army.mil/.

U.S. Army Center of Military History. "Landing Units and Beaches at Leyte." Accessed October 25, 2021. https://history. army.mil/books/wwii/Beachhd_Btlefrnt/ChapterXX.html.

Valdez, Basilio J. Digitized Collection. n.d., Presidential Museum and Library, Malacañang Palace, Manila.

Whitney, Courtney. *MacArthur's Rendezvous with History.* LIFE, 1955.

Whitney, Courtney. *Report on Philippine Civil Affairs: 32nd Infantry Division 'Red Arrow* . Vol. 2. Carlisle, Pennsylvania: USAMHI, 1942.

Appendix: A-16 - A-17

Chapter 21: Luzon Island, Cabanatuan Prisoners of War Camp

Alexander, Larry. *Shadows in the Jungle: the Alamo Scouts Behind Japanese Lines in World War II.* London, England: Penguin, 2010.

Black, Robert W. *Rangers in World War II.* Presidio Press, 1992.

Breuer, William B. *The Great Raid on Cabanatuan.* John Wiley, 1994.

Dwyer, John B. "The Untold Story Behind the Great Raid." *American Thinker* (2005).

Henderson, Bruce. *Rescue at Los Banos.* New York, NY: Harper-Collins, 2015.

Hughes, Les. "The Alamo Scouts 1986." Accessed October 29, 2021. http://www.insigne.org/alamo-scouts.htm.

Krueger, Walter. *From Down Under to N ippon: the Story of the Sixth Army in World War II.* Combat Forces Press, 1953.

Laubenthal, Sanders A. "A History of John Hay Air Base." *Bulletin American Historical Collectio*n 22, no. 2 (1994).

Legarda, Benito Jr. *Occupation: 1942-1945.* Vibal Foundation, 2016.

Sasser, Charles W. *Raider.* New York: St. Martin's Press, 2002.

Sides, Hampton. *Ghost Soldiers.* New York, NY: Doubleday, 2001.

Sitter, Larry. "Bill Nellist Remembered." Accessed October 29, 2021. http://www.alamoscouts.com/news/ol_newsletter_ july2006.pdf.

Strausbaugh, Leo V. "Ranger History. 6th Battalion. Strausbaugh Remembers." *Descendants of WWII Rangers, Inc.* (2020) Accessed October 25, 2021. https://wwiirangers.org/our-history/ranger-history/6th-btn/.

U.S. Press Corps. Ironwoods house of Dionesio Ompod y Serana. Photo. Loreto, Dinagat Island, Pentagon Archives.

Zedric, Lance. *Silent Warriors of World War II: the Alamo Scouts Behind Enemy Lines.* Pathfinder, 2020.

Appendix: A-18

Chapter 22: Luzon Island, Los Banos Prisoners of War Camp

Parsons, Charles. Papers. n.d., RG-58, Spyron 1943-1945, MacArthur Memorial Archives and Library, Norfolk, Virginia.

Giangreco, D. M. *Hell to Pay: Operation Downfall and the Invasion of Japan 1945-1947.* U.S. Naval Institute, 2010.

Krivido, Michael E. "Major Jay D. Vanderpool: Advisor to the Philippine Guerrillas." *Veritas* 9, no. 1 (2013).

McGowan, Sam. "Liberating Los Banos Internment Camp." Accessed October 29, 2021. https://www.historynet.com/ world-war-ii-liberating-los-banos-internment-camp.htm.

Morningstar, James Kelley. *War and Resistance in the Philippines, 1942-1944.* Naval Institute Press, 2021.

Santos, Terry. Interview by Eddie Graham. 11th Airborne, Provisional Reconnaissance Platoon. National Museum of the Pacific War. Center for Pacific War Studies. Fredericksburg, TX.

Villanueva, James A. "Awaiting the Allies' Return: The Guerrilla Resistance Against the Japanese in the Philippines during WWII." PhD. Diss., 2019.

Chapter 23: Aftermath

Agullana, Melissa. Document. Don S. Ruiz.

Baldoz, Rick. "6. 'Another Mirage of Democracy': War, Nationality, and Asymmetrical Allegiance" In *The Third Asiatic Invasion: Empire and Migration in Filipino America, 1898-1946,* New York, USA: New York University Press (2011). pp. 194-236. Accessed October 24, 2021. https://doi.org/10.18574/9780814789889-008.

Bernardo, Joseph A. "From 'Little Brown Brothers' to 'Forgotten Asian Americans' Race, Space, and Empire in Filipino." PhD. Diss., University of Washington, 2014.

Campbell, Douglas A. *Eight Survived*. Lyon Press, 2011.

Endy, Clarence E., Jr. "U.S. Army Recognition Program of Philippine Guerrillas." HQS, Philippines Command United States Army circa 1949, National Archives and Records Administration (NARA), College Park, Maryland.

Fabros, Alex S., Jr., ed. "A History of the 978th Signal Service Company." *Filipino American Experience Research Project*. November 6, 1998.

Galang, Ricardo. *Hammer and Anvil*. Phoenix, 1989.

Gangl, Theresa. Document. Gangl Leo.

Linderman, Gerald F. *The World Within War*. New York, NY: The Free Press, 1997.

Maram, Linda Espana. *Creating Masculinity in Los Angeles's Little Manila*. Columbia University Press, 2006.

Morningstar, James Kelley. *War and Resistance in the Philippines, 1942-1944*. Naval Institute Press, 2021.

"PA to Let Out 100,000 this Month." *Manila Post*. Dec 1-23, 1945.

Showalter, Dennis E. *If the Allies Had Fallen*. Skyhorse, 2012.

Stahl, Bob. *You're No Good to Me Dead: Behind the Japanese Lines in the Philippines*. Annapolis, Maryland: Naval Institute Press, 1997.

Valdez, Pelagio. "1st and 2nd Filipino Infantry Regiments, U.S. Army, 1942-1946" (Facebook Group). Accessed May 12, 2022. https://www.facebook.com/1st-and-2nd-Filipino-Infantry-Regiments-185129034874357/.

Appendix: A-14, A-15, A-23, B-29

SOURCES
ARCHIVAL SOURCES, GOVERNMENT, AND MILITARY RECORDS

MacArthur Memorial Archives and Library:

RG-3 Records of Headquarters, Southwest Pacific Area (SWPA), 1942-1945; RG-4 Records of Headquarters, U.S. Army Forces Pacific (USAFPAC), 1942-1947; RG-16 Papers of Major General Courtney Whitney, USA, Philippine Section, SWPA; RG-23b Selected Papers of Major General Charles A. Willoughby, USA, 1943-1954; RG-30 Papers of Lieutenant General Richard K. Sutherland, USA, Chief of Staff, SWPA, 1941-1945; RG-53 Papers of Lt. Col. Wendell W. Fertig, USA, Commanding Officer, 10th Military District, SWPA; RG-58 Papers of Commander Charles "Chick" Parsons, USNR, SPYRON, 1943-1945; RG-106 Papers of Lt. Col. Charles M. Smith, USA; 5th Military District, American Guerrillas of Mindanao; RG-109 Papers of Colonel Clyde C. Childress, USA; 10th Military District, American Guerrillas of Mindanao; RG-146 Papers of Brigadier General Lee A. Telesco, USA; Philippine Section, G-2, SWPA / USAFPAC

1st and 2nd Filipino Infantry Regiment Disposition Reports, n.d., National Archives and Records Administration (NARA), College Park, Maryland.

1st Reconnaissance Battalion (Special) and 978th Signal Service Co. Individual After Action Reports. n.d., National Archives and Records Administration (NARA), College Park, Maryland.

Allied Intelligence Bureau, G-2, GHQ, SWPA. Philippine Sea. Appendix 1. n.d., NARA Declassified. NND39471, Nation- al Archives and Records Administration (NARA), College Park, Maryland.

Allied Translator and Interpreter Section (ATIS), n.d., U.S. Army Military History Institute (USAMHI), Carlisle, Pennsylvania.

American Guerrillas of Mindanao (AGOM) Scrapbook, n.d., U.S. Army Military History Institute (USAMHI), Carlisle, Pennsylvania.

Anderson, Bernard. York Group. July 21, 1944, National Archives and Records Administration (NARA), College Park, Maryland.

Anti-Miscegenation Act. 1930.

Army Intelligence Files of Philippine Guerrilla Papers. Messages. n.d., Files of Individual Guerrilla Districts and Submarine Evacuation Reports (Military Civilian), RG-319, Records of the Army Staff, National Archives and Records Administration (NARA), College Park, Maryland.

Budd, George. n.d., Ike Skelton Combined Arms Research Library Digital Library.

Castillo, Stephanie. Collection. Untold Triumph, the story of the 1st and 2nd Filipino Infantry Regiment, U.S. Army.

Chronology of Events in Mindanao 1943-1944, Bobb Glenn Collection. AIB Chief Supply Officer.

Chronology of the Allied Intelligence Bureau, GHQ, SWPA. June 1942-January 1946, Bobb Glenn Collection. AIB Chief Supply Officer.

Civil Affairs Handbook: the Philippines. n.d., Section 14: Public Safety, Ike Skelton Combined Arms Research Library Digital Library.

Clement, Patricia McDermott. May 8, 1981, Philippine Archives Collection. National Archives and Records Administration (NARA), College Park, Maryland.

Commanding General Eight U.S. Army on the Leyte-Samar Operation (Including Clearance of the Visayan Passages). Report. December 26, 1944-May 8, 1945, U.S. Army Military History Institute (USAMHI), Carlisle, Pennsylvania, pp. 13, 14, 23, 24.

Counterintelligence Corps, History and Mission in World War II. n.d., U.S. Army Military History Institute (USAMHI) Carlisle, Pennsylvania.

Cordova, Fred and Dorothy Cordova. Collection. n.d., Filipino American National Historical Society (FANHS), Seattle, Washington.

Current Translations No. 147. January 1-21, 1945 and February 2, 1945, Allied Translator and Interpreter Section (ATIS) Southwest Pacific Area. U.S. Army Military History Institute (USAMHI), Carlisle, Pennsylvania.

Dacquel, Isidoro. Reports. n.d., MacArthur Archives (RG16 Box 46 Fol 21003, RG16 Box 15 Fol 2, RG16 Box 46 Fol 30004). MacArthur Memorial Archives and Library, Norfolk, Virginia.

"Darwin and the Philippines Submarine Run." Journal of Northern Territory History (1943). Bobb Glenn Collection. AIB Chief Supply Officer.

Destruction of Manila and Japanese Atrocities. Report. February 1945, Military Intelligence Sections, U.S. Army Military History Institute (USAMHI), Carlisle, Pennsylvania.

Dumlao, Felipe, Leonard Aliwang, Mariano Angeles, Martin Mamuyac, and Saturnino Silva. Record. n.d., Filipino American National Historical Society (FANHS), Seattle, Washington.

Endy, Clarence E., Jr. "U.S. Army Recognition Program of Philippine Guerrillas." HQS, Philippines Command United States Army circa 1949, National Archives and Records Administration (NARA), College Park, Maryland.

Fabros, Alex S., Jr. Collection. 1st and 2nd Filipino Infantry Regiment, 1st Reconnaissance Battalion, and 978th Signal Service.

Fertig to Gen. MacArthur. Radio Message. June 3, 1944, MacArthur Memorial Archives and Library, Norfolk, Virginia.

Fertig, Wendell. Diary. August-September 1944, MacArthur Memorial Archives and Library, Norfolk, Virginia.

G-2 Staff Study of Philippine Islands Situation. Appendix XV. February 24-25, 1944, Military Intelligence Section, U.S. Army Military History Institute (USAMHI), Carlisle, Pennsylvania.

General Headquarters, Far East Command. n.d., Military Intelligence Section. MacArthur Memorial Archives and Library, Norfolk, Virginia.

General MacArthur. The Campaigns of MacArthur in the Pacific. Vol I. General Staff. Report. 1994, Center of Military History, Washington D.C.

Golden Anniversary Celebration of First/Second Filipino Combat Infantry Regiments, U.S. Army with a dinner dance. August 1, 1992, Salinas Community Center.

Gonzales, Carlos I. Bulacan Military Area (BMA). n.d. RG-401, Box 422., National Archives and Records Administration (NARA), College Park, Maryland.

Hart, Thomas C. 8 December 1941 to 15 February 1942. Commander in Chief, Asiatic Fleet. December 31, 2012, NARA Declassified, National Archives and Records Administration (NARA), College Park, Maryland.

Historical Report. Operations on Northern Samar, Burias, and Ticao Islands in the Philippines Liberation Campaign, February 16, 1945-May 7, 1945, National Archives and Records Administration (NARA), College Park, Maryland.

Historical Report. Operations on Northern Samar, Burias, and Ticao Islands in the Philippines Liberation Campaign, May 8-20 and September 1945, National Archives and Records Administration (NARA), College Park, Maryland.

Ind, Allison W. Check Sheet from Philippine Regional Section (PRS) to Chief of Staff through AIB. Memo. December 6, 1943, MacArthur Memorial Archives and Library, Norfolk, Virginia.

Individual After-Action Reports: 1st Reconnaissance Battalion (Special) and 978th Signal Service Co., n.d., Filipino American National Historical Society (FANHS), Seattle, Washington.

Individual After Action Reports. 1st Reconnaissance Battalion (Special) and 978th Signal Service Co., n.d., National Archives and Records Administration (NARA), College Park, Maryland.

Intelligence Activities in the Philippines during the Japanese Occupation. GHQ, United States Army Forces, Pacific, n.d., Military Intelligence Section, U.S. Army Military History Institute (USAMHI), Carlisle, Pennsylvania.

Intelligence Summary, 10th MD, USFIP, No. 9, October 1944, The Ising Operation, May 1, 1945, MacArthur Memorial Archives and Library, Norfolk, Virginia.

Japanese Monographs, n.d., RG-550, National Archives and Records Administration (NARA), College Park, Maryland.

Japanese Plans for Defense of Manila, January 1-21, 1945, February 2, 1945, Allied Translator and Interpreter Section (ATIS), National Archives and Records Administration (NARA), College Park, Maryland.

Lee Telesco to Ceferino Rola. Memo. n.d., RG-16, Box 15 Fol 3011, MacArthur Memorial Archives and Library, Norfolk Virginia.

Loomis, Anna. The Guerrilla Resistance Movement in Northern Luzon. Paper. January 1945, RG-16, GHQ SWPA, G-2 Information Bulletin.

Maeda, Masami. 14th Army Chief of Staff. Interrogations of Japanese Officials (English Translations) Carlisle. GHQs. FarEast Command, May 10, 1947, G-2 Historical Section. U.S. Army Military History Institute (USAMHI), Carlisle, Pennsylvania.

Memoir, Brisbane. WWII. Collection. September 1942-June 1946, Bobb Glenn Collection. AIB Chief Supply Officer.

Messages from General MacArthur during the night to President Roosevelt and the War Department, February 11-12, 1942, Franklin D. Roosevelt Library.

Mindanao: Historical Report of the 24th Infantry Division, V-5 Operations. April 17, 1945-June 30, 1945, U.S. Army Military History Institute (USAMHI), Carlisle, Pennsylvania.

Official Japan Order to Kill All POWs: Doc 2701, Exhibit "O." War Crimes, Japan, n.d., RG-24, Box 2015, National Archives and Records Administration (NARA), College Park, Maryland.

Orders of 5217th Reconnaissance Battalion Men Attach to 11th Airborne. n.d. MacArthur Memorial Archives and Library, Norfolk, Virginia.

Orders of men to AIB, RG16 Box 15 Fol 4012. n.d., MacArthur Memorial Archives and Library, Norfolk, Virginia.

PCAU Experiences, Report No. 3. Case and Materials on Military Government, January 17, 1945, Ike Skelton Combined Arms Research Library Digital Library.

Pardo, Clay. 1st Filipino Infantry Regiment. n.d., Army Services Experiences Questionnaire, U.S. Army Military History Institute (USAMHI), Carlisle, Pennsylvania.

Parsons, Charles. Papers. n.d., RG-58, Spyron 1943-1945, MacArthur Memorial Archives and Library, Norfolk, Virginia.

Pelagio Valdez. Collection. 1st and 2nd Filipino Infantry Regiments, 1st Reconnaissance Battalion, and 978th Signal Service. 1942-1946.

Philippine Sea Frontier Command Letter: Massacre of American Prisoners of War at Puerto Princesa, Palawan, January 23, 1945, Ike Skelton Combined Arms Research Library Digital Library.

Planet Party: First AIB Submarine Mission to the Philippines, n.d., Bobb Glenn Collection. AIB Chief Supply Officer.

Presentation of US Army's 1st Filipino Infantry Regiment Regimental Colors by Gen. Morris at King City. Photo. July 13, 1943, The Wing Lake Museum of the Asian Pacific American Experience.

Quezon, Aurora A. Free Philippines Vol 1-3, n.d., Digitized Collection. Presidential Museum and Library, Malacañang Palace, Manila.

Records of the Adjutant General's Office. Collection. n.d. Personnel Files on Guerrilla Leaders, Daily Summary Reports of Individual Guerrilla Districts, and Other Files, RG-407, Philippines Archive Collection, National Archives and Records Administration (NARA), College Park, Maryland.

Rod Hall. Collection, n.d., Filipinas Heritage Library, Manila, Philippines.

Rola, Ceferino. Unit History of the First Reconnaissance Battalion Special. January 21, 1995, NARA Declassified NND735017, National Archives and Records Administration (NARA), College Park, Maryland.

Rosenquist, Harold A. An Estimate of the Enemy Situation. April 30, 1945, Military Intelligence Section, U.S. Army Military History Institute (USAMHI), Carlisle, Pennsylvania.

Rosenquist, Harold A. The 10th MD. Report. March-August 1944, MacArthur Memorial Archives and Library, Norfolk, Virginia.

Rowe, George. Radio Messages to General MacArthur. January 1945, U.S. Army Military History Institute (USAMHI), Carlisle, Pennsylvania.

Ruiz, Don. Collection. n.d., 1st Reconnaissance Battalion. Member of the Shangri-La Rescue.

Ruiz, Don. Collection. n.d, Commendation of Shangri-La Rescue.

Ruiz, Julius. Papers. n.d., Filipino American National Historical Society (FANHS), Seattle, Washington

SS167/A16-3 US Narwhal Report of Pearl Harbor Attack December 12, 1941, to the Commander in Chief, U.S. Pacific Fleet. Enclosure (E) CINCPAC Action Report Serial 0479, February 15, 1942, National Archives and Records Administration (NARA), College Park, Maryland.

Southwest Pacific Area and United States Army Forces, Pacific (World War II), 1941-1947. Record. n.d., Records of General Headquarters, RG-496, National Archives and Records Administration (NARA), College Park, Maryland.

Southwest Pacific Area (SWPA GHQ). Radio Messages. n.d., Files of the Allied Intelligence Bureau, Lists of Personnel (201) Files, RG-338, Records of General Headquarters, National Archives and Records Administration (NARA), College Park, Maryland.

Submarine Intelligence and Supplies Missions to the Philippines 1944-1945, n.d., National Archives and Records Administration (NARA), College Park, Maryland.

Surrender in the Far East, USAFFFE Annex VI, n.d., RG-496, National Archives and Records Administration (NARA), College Park, Maryland.

Talaugon, Nora and Federico Talaugon. Collection. 1944, John Spoor Broome Library, California State University, Channel Islands.

The Philippine National Defense Act. 1935.

United States Army in World War II: The Technical Services, Signal Corps, n.d., U.S. Army Military History Institute (USAMHI), Carlisle, Pennsylvania.

U.S. Army photograph 924-570-46/AM-51-811. Training Photo. n.d., Wing Lake Museum of the Asian Pacific American Experience.

U.S. Army Recognition Program of Philippine Guerrillas. Headquarters Philippines Command United States Army. n.d., NARA Declassified NND39471, National Archives and Records Administration (NARA), College Park, Maryland.

U.S. Press Corps. Ironwoods house of Dionesio Ompod y Serana. Photo. Loreto, Dinagat Island, Pentagon Archives.

USAFFE Check List. January 24, 1944, RG-16 B15 F2011, MacArthur Memorial Archives and Library, Norfolk, Virginia.

USAFFE to the Adjutant General and General Douglas MacArthur. February 25, 1942, National Archives and Records Administration (NARA), College Park, Maryland.

Valdez, Basilio J. Digitized Collection. n.d., Presidential Museum and Library, Malacañang Palace, Manila.

Vallejo, Marie Silva. Collection. Author of The Battle of Ising. Daughter of Saturnino Ramos Silva.

Velasco, Andres L. Autobiography by Harold Liban. n.d.

War Reports of the USS Narwhal and Messages between Col. Macario Peralta and SWPA GHQ, n.d., MacArthur Memorial Archives and Library, Norfolk, Virginia.

Watari Group Intelligence Report B. No. 146. July 1944, RG-406, National Archives and Records Administration (NARA), College Park, Maryland.

Wendell W. Fertig Papers, n.d., U.S. Army Military History Institute (USAMHI), Carlisle, Pennsylvania.

Whitney, Courtney. Report on Philippine Civil Affairs, Vol II. 32nd Division, 'Red Arrow,' Appendices. August 25, 1945, U.S. Army Military History Institute (USAMHI), Carlisle, Pennsylvania.

WWII. Collection. n.d., Hoover Institute, Stanford University, California.

WWII. Collection. n.d., Ortigas Library, Manila, Philippines.

Worcester, Dean. Report on Negros 1943-1944. Negros.

Yuhico, Delfin C. Report. n.d., Our Mission in the Philippines. 1st Reconnaissance Battalion History.

WEBSITES

"26th Cavalry (PS)". Accessed April 9, 2022. https://www.philippinescouts.org/the-scouts/ps-wall-of-heroes?rq=%20silva.

Arzaga, Joseph. "Filipino Scout Met Japanese Invasion." *Joseph's Blog* (blog). November 10, 2009.

Baldoz, Rick. "6. Another Mirage of Democracy': War, Nationality, and Asymmetrical Allegiance" *In The Third Asiatic Invasion: Empire and Migration in Filipino America, 1898-1946*, New York, USA: New York University Press (2011). pp. 194-236. Accessed October 24, 2021. https://doi.org/10.18574/9780814789889-008.

Bowler, Robert. "History of the Mindanao Guerrillas." *West Point*. Accessed October 24, 2021. https://www.west-point.org/family/japanese-pow/Guerrillas/History%20of%20the%20Mindanao%20Guerrillas.pdf.

Budd, Talman C. "PCAU Experiences, Report No. 3, January 17, 1945." *Cases and Materials on Military Government*. Accessed October 24, 2021. https://cgsc.contentdm.oclc.org/digital/collection/p4013coll8/id/2221/rec/4.

Cal, Ben. "The Heroism and Sacrifice of Gen. Fortunato Abat." *Philippine News Agency* (March 2018). Accessed November 14, 2021. https://www.pna.gov.ph/articles/1028265.

Creel, George. "The Heroes: Truman Hemingway." *Collier's. Old Magazine* Articles. Accessed October 24, 2021. http://www.oldmagazinearticles.com/Truman-Heminway-ww2-hero-pdf.

Crossman, Edgar Gibson. "My Experience in WWII. Unpublished Memoir." *Wikipedia* (November 1966). Accessed October 24, 2021. https://en.wikipedia.org/w/index.php?title=File%3AEdgar_Gibson_Crossman_My_Experiences_in_WWII.PDF&page=1.

Elphick, James. "Life Aboard WWII Submarines was Brutal." *We are the Almighty* (July, 2021). Accessed October 24, 2021. https://www.wearethemighty.com/popular/life-aboard-wwii-submarines-was-brutal/.

Fabros, Alex S., Jr. "The Boogie Woogie Boys." Filipinas Magazine (September 1993). Accessed October 24, 2021. http://www.positivelyfilipino.com/magazine/the-boogie-woogie-boys.

Fairfield, William A. "Mactan Memoirs." The Philadelphia Inquirer (May 1943). Accessed October 24, 2021. http://ww2f.com/threads/the-fighting-mactan.21187/.

"Fighting World War II in the Pacific." Cebu: Hostile Beach (1945). Accessed March 1, 2022. http://www.182ndinfantry.org/cebu-hostile-beach-1945/.

"Fighting World War II in the Pacific." Leyte: Mop-up Turns to Slaughter (1945). Accessed March 1, 2022. http://www.182ndinfantry.org/leyte-mop-up-turns-to-slaughter-1945/.

Fisher, William P. "Talk to G-4 Logistics, WDGDS." (March 20, 1942). Accessed Accessed October 24, 2021. http://docs.fdrlibrary.marist.edu/psf/box6/folo71.html.

"Fraser Commando School." Accessed October 24, 2021. https://www.ozatwar.com/sigint/frasercommandoschool.htm.

Garcia, Arturo P. "The Real Heroes of Bessang Pass." Bulatlat (June 2005). Accessed October 24, 2021. https://www.bulatlat.com/2005/06/25/the-real-heroes-of-bessang-pass/.

Gregory, Jim. "California in World War II: The First Filipino Infantry Regiment and San Luis Obispo County." *Military Museum*. Accessed October 24, 2021. http://www.militarymuseum.org/1FIR-SLO.html.

Hall, Lou. "World War II Diary of Robert T. Webber." Accessed October 24, 2021. http://home.pcisys.net/-pwebber/31_id/rtw_comment.htm.

Harding, Stephen. "Prelude to Pearl." History Net. Accessed October 26, 2021. https://www.historynet.com/prelude-to-pearl.htm.

Hughes, Les. "The Alamo Scouts 1986." Accessed October 29, 2021. http://www.insigne.org/alamo-scouts.htm.

Hughes, Les. "The Philippine Airborne." The 503d P.R.C.T. Heritage Battalion Online. Accessed October 24, 2021. https://corregidor.org/heritage_battalion/hughes/hughes.html.

Jurado, Enrique L. "1934 Lucky Bag." U.S. Naval Academy Virtual Memorial Hall. Accessed October 26, 2021. https://usnamemorialhall.org/index.php/Enrique L. Jurado, Lt Col, PA.

Lozano, Mark. "The Untold Story of Delfin Yuhico." Accessed July 10, 2022. https://cantilanhistory.weebly.com/collection.html.

Lutz, Stephen D. "The 93rd Infantry Division: The African-American Soldiers in the Pacific." Warfare History Network. Accessed October 24, 2021. https://warfarehistorynetwork.com/2019/01/19/the-93rd-infantry-division-the-african-american-soldiers-in-the-pacific/.

Maritime Museum. "Don Isidro a WWII 'blockade-runner' sunk off Darwin February 1942." Created by Peter Rout (blog). *Australian Maritime Museums Council* (August 9, 2017). Accessed October 18, 2022. http://maritimemuseumsaustralia.com/profiles/blogs/don-isidro-a-wwii-blockade-runner-sunk-off-darwin-february-1942.

McGowan, Sam. "Liberating Los Banos Internment Camp." Accessed October 29, 2021. https://www.historynet.com/world-war-ii-liberating-los-banos-internment-camp.htm.

Metraux, Daniel A. "Teaching Pearl Harbor: A New Japanese Perspective." Education About Asia 17, no. 3 (Winder 2012): US, Asia, and the World: 1914-2012 (Winter 2012). Accessed Dec 23, 2022. https://www.asianstudies.org/publications/eaa/archives/teaching-pearl-harbor-a-new-japanese-perspective/.

Michigan. Department of the Army, Field Operations of Military Government Units. "Operations of Civil Affairs Units in the Philippine Islands." (January 1945). Accessed October 24, 2021. https://play.google.com/store/books/

author?id=United+States.+Army.+Civil+Affairs+Division.

Moore, Stephen L. "The Heroes of Palawan – How Survivors of a Japanese Massacre Live to Tell the Tale of Atrocities in the Philippines." (December 2016). Accessed October 24, 2021. https://militaryhistorynow.com/2016/12/02/the-heroes-of-palawan-how-survivors-of-a-japanese-massacre-lived-to-tell-the-tale-of-atrocities-in-the-philippines/.

Naval History and Heritage Command. "Submarine Activities Connected with Guerrilla Organizations." (November 2017). Accessed October 24, 2021. https://www.history.navy.mil/research/library/online-reading-room/title-list-alphabetically/s/submarine-activities-connected-with-guerrilla-organizations.html.

"On this Day – January 19, 1930 White Mobs Attack Filipino Farmworkers in Watsonville, California." *A History of Racial Injustice*. Accessed October 24, 2021. https://calendar.eji.org/racial-injustice/jan/19.

Parsons, Peter C. "The Battle of Manila – Myth and Fact." *Battle of Manila Online*. Accessed October 24, 2021. https://corregidor.org/mnl/Parsons/htm/parsons_01.htm.

Parsons, Peter C. "The Panay Guerrillas / USS Narwhal Debacle at Lipata Point." Accessed October 24, 2021. http://myphilippinelife.com/the-panay-guerrillasuss-narwhal-debacle-at-lipata-point/.

Preston, William B. "Mission Naval." *History Magazine*. Accessed October 24, 2021. https://news.usni.org/tag/uss-william-b-preston.

Quesada, Frank B. "Ordeal in War's Hell Part I." Accessed October 25, 2021. https://filipinos-ww2usveterans-4equity.tripod.com/id42.html.

Sanchez, Helen Ragsac. "Fil-Am Pinays and the 'Filipino Sports Parade.'" Positively Filipino. Accessed October 25, 2014. http://www.positivelyfilipino.com/magazine/fil-am-pinays-and-the-filipino-sports-parade.

Sitter, Larry. "Bill Nellist Remembered." Accessed October 29, 2021. http://www.alamoscouts.com/news/ol_newsletter_july2006.pdf.

Sixth United States Army. "Report of the Luzon Campaign, 9 January 1945 – 30 June 1945. Vol. IV." *Ike Skelton Combined Arms Research Library Digital Library*. Accessed October 29, 2021. https://cgsc.contentdm.oclc.org/digital/collection/p4013coll8/id/2289.

Smith, Robert Ross. "Chapter XXII. The Reduction of the *Shimbu Group*. Phase II: The Seizure of Wawa and Ipo Dams." *U.S. Army in World War II: Triumph in the Philippines*. Accessed October 25, 2021. https://www.ibiblio.org/hyperwar/USA/USA-P-Triumph/USA-P-Triumph-22.html.

Smith, Robert Ross. "Chapter XXVIII. Action at the Northern Apex." *U.S. Army in World War II: Triumph in the Philippines*. Accessed October 29, 2021. https://www.ibiblio.org/hyperwar/USA/USA-P-Triumph/USA-P-Triumph-28.html.

Smith, Rufus W. "World War II POW." Interview by Rufus W. Smith. Humanities Texas (March 2015). Accessed October 24, 2021. https://www.humanitiestexas.org/news/articles/interview-rufus-w-smith-world-war-ii-pow.

"SS167 Narwhal". Accessed October 25, 2021. https://fleetsubmarine.com/ss-167.html.

Strausbaugh, Leo V. "Ranger History. 6th Battalion. Strausbaugh Remembers." *Descendants of WWII Rangers, Inc.* (2020). Accessed October 25, 2021. https://wwiirangers.org/our-history/ranger-history/6th-btn/.

Stuckenschneider, Placid. "The Last Campaign Mindanao – March '99 World War II Feature." *History Net*. Accessed October 24, 2021. https://www.historynet.com/the-last-campaign-mindanao-march-99-world-war-ii-feature.htm.

Turner, Thomas M.D. "US Army Medical Department. The Philippines and Okinawa. Chapter XVI." *A MEDD Center of History & Heritage*. Accessed October 24, 2021. http://history.amedd.army.mil/booksdocs/wwii/civilaffairs/chapter16.htm. Moved to: https://achh.army.mil/.

U.S. Army Center of Military History. "Distinctive Unit Insignia & Coat of Arms for the 1st Filipino Regiment." Accessed October 24, 2021. https://history.army.mil/html/topics/apam/filipino_regt/DUI_and_COA.html.

U.S. Army Center of Military History. "Landing Units and Beaches at Leyte." Accessed October 25, 2021. https://history.army.mil/books/wwii/Beachhd_Btlefrnt/ChapterXX.html.

Valdez, Pelagio. "1st and 2nd Filipino Infantry Regiments, U.S. Army, 1942-1946" (Facebook Group). Accessed May 12, 2022. https://www.facebook.com/1st-and-2nd-Filipino-Infantry-Regiments-185129034874357/.

Velasquez, M. A. "History of the Sulu Area Command 1945: From The Library of Romulo Espaldon." *Internet Archives*. Accessed October 24, 2021. https://archive.org/details/sulu-area-command.

Walters, Earl. "Rescue From Shangri La." Interview by Patrick O'Donnell. November 25, 1998. Accessed October 24, 2021. http://www.thedropzone.org/pacific/walters.htm.

Zedric, Lance Q. "The Miracle Mission – 75th Anniversary of DOVE Team's rescue of the Sycip family from Fuga Island July 28-30, 1945." (2003). Accessed July 16, 2022. https://www.facebook.com/dynamicheartyoga/posts/celebrating-the-75th-anniversary-of-the-miracle-mission-of-my-familys-rescue-fro/593789861334681/.

DISSERTIONS, ACADEMIC PAPERS, AND MONOGRAPHS

Amor, Simeon, Jr. "A Brief History of the 1st Filipino Infantry." Hawaii Filipino Veterans Club. September 1981.

Begonia, Dan. "Asian American Studies 456, Filipinos in America." Filipino American Experience Research Project. Vol. 9. Unpublished. San Francisco State University. 2001.

Bernardo, Joseph A. "From 'Little Brown Brothers' to 'Forgotten Asian Americans' Race, Space, and Empire in Filipino." PhD. Diss., University of Washington, 2014.

Buaken, Manuel. "Our Fighting Love of Freedom." *Filipino American Experience Research Project*. February 20, 1943.

Buaken, Manuel. "The First Filipino on the Mainland, Educational Problems of the Filipino." *Filipino American Experience Research Project*. February 20, 1943.

Caceres, Michael V. "Rising Sun in the Southern Land: Destruction and Resistance in Sulu and Tawi-Tawi Archipelago (1941-1945)." Ateneo de Zamboanga University.

"Dawn of the Philippine Freedom. A Scrapbook Commemorating the 50th Anniversary of the Liberation of the Philippines." Compiled by Cresencia Cespedes Gray. Self-Published. January 1, 1995.

Fabros, Alex S., Jr. "Bahala Na Newsletter, 1st Reconnaissance Battalion." *Filipino American Experience Research Project*. Unpublished. San Francisco State University. October 5, 1998.

Fabros, Alex S., Jr. "A Document Collection of the History of the 1st Filipino Infantry Regiment in World War II." Master's thesis, 1997.

Fabros, Alex S., Jr. "A History of the 1st & 2nd Filipino Infantry Regiments, Philippine Civil Affairs Unit." Vol. VII. Unpublished. (1992).

Fabros, Alex S., Jr. "A History of the 1st Reconnaissance Battalion United States Army." *Filipino American Experience Research Project*. Unpublished. San Francisco State University. October 5, 1998.

Fabros, Alex S., Jr. "A History of the 978th Signal Service Company." *Filipino American Experience Research Project*. Unpublished. San Francisco State University. November 6, 1998.

Fabros, Alex S., Jr., Daniel P. Gonzales and Daniel T. Begonia. "The Filipino American Experience in World War II." In *Honor of Our Fathers*. Vols. 1-14. Unpublished. San Francisco State University. November 6, 1998.

Fabros, Alex S., Jr., Daniel P. Gonzales, Danilo T. Begonio. "Regimental Records of the 1st & 2nd Filipino Infantry Regiments and the 1st Reconnaissance Battalion." *Filipino American Experience Research Project*. Unpublished. San Francisco State University. November 6, 1998.

Fabros, Alex S., Jr., ed. "A History of the 978th Signal Service Company." *Filipino American Experience Research Project*. November 6, 1998.

Fabros, Alex S., Jr. "Individual Action Reports of the 1st Reconnaissance Battalion and 978th Signal Service Company." Filipino American Experience Research Project . Unpublished. San Francisco State University. October 5, 1998.

Fabros, Alex S., Jr. "News Articles of the Regiments, Filipino American Experience Research Project." *World War II - Research*. Vol. 9. Unpublished. San Francisco State University. November 6, 1998.

Jerry, Grant E. "All Those Who Remained: the American-Led Guerrillas in the Philippines, 1942-1945." United States Army, School of Advanced Military Studies. Fort Leavenworth, Kansas: United States Army Command and General Staff College, 2014.

Magdua, Jeannie M. "Filipina War Brides." Master's Thesis, University of Hawaii, 2020.

Morgan, Matthew T. "Predictors of Success at the U.S. Army Officer Candidate School." Master's Thesis, Arizona State University, 1997.

Raquepo, Veronica, Aegina Festin and KarenDonila. "Anderson Guerrillas." University of the Philippines, 2012.

Rola, Ceferino R. "1st Lt. AUS, "A History of the 1st Reconnaissance Battalion." United States Army. 1946.

Ruiz, Julius B. "A Personal Remembrance." *Filipino American Experience Research Project*.

Ruiz, Julius B. "'Salinas Nite' to Honor Men from Bataan." Filipino American Experience Research Project . November 6, 1942.

Schmidt, Larry S. "American Involvement in the Filipino Resistance Movement on Mindanao During the Japanese Occupation, 1942-1945." Master's Thesis, Command and General Staff College, 1982.

Sinclair, Peter T., II. "Men of Destiny: The American and Filipino Guerillas During the Japanese Occupation of the Philippines." Master's Thesis, Command and General Staff College, 2011.

Taggaoa, Fernando A. "No Cause for Regret." *Filipino American Experience Research Project*. October 1942.

Villanueva, James A. "Awaiting the Allies' Return: The Guerrilla Resistance Against the Japanese in the Philippines during WWII." PhD. Diss., 2019.

Vite, Doroteo V. "A Filipino in Uncle Sam's Arm." *Filipino American Experience Research Project* .

Wingo, James G. "Two Filipino Medics Here." *Filipino American Experience Research Project*. Arizona Republic. January 10, 1943.

Wingo, James G. "The First Filipino Regiment." *Filipino American Experience Research Project*. Unpublished. San Francisco State University. October 1942.

JOURNALS, ARTICLES AND LETTERS

1st Filipino Infantry Camp Roberts Trainee 2, no 3, August 6, 1942.

Adler, Walter H. "From the Allied Intelligence Bureau to Mindanao: The 'Free Philippines' Guerrilla Stamps." *Philippine-Philatelic Journal* 49, no. 3 (1991).

Agullana, Melissa. Document. Domingo 'Don' S. Ruiz.

Alcala, Aurora. Document. Atanacio Alcala.

Angeles, Ranesto Bello. Document. Mariano Angeles.

Ara, Satoshi. "Food Supply Problem in Leyte, Philippines during the Japanese Occupation 1942-1944." *Journal of Southeast Asian Studies* 39, no. 1 (2008).

Bahala Na First Anniversary 1, no. 39. Special Edition (1944).

Bahala Na Newsletter. Newsletter First Generated (1944).

Bamont, Nick. Document. John Bamont.

Bersamin, Manuel. Document on Paulino Bersamin.

"Big Reunion for Former Members of the Regiments at Place of Activation." *Poster* (1992).

Bob Luna. Document. Toribio Luna.

Buencamino, Victor Dr. "Manila Under Japanese Occupation (V)." *Bulletin American Historical Collection* 8, no. 3 (1980).

Campeau, Lucien. "My Airforce Weather Mission from April 1944 through April 1945 with the American Guerrillas of Mindanao." *Guerrilla Radio by American Guerrillas of Mindanao* (1992).

Campos, Dixon. "California's Own." San Francisco, California (1990).

Capistrano, Robert. "Bahala Na: the 5217th 1st Reconnaissance Battalion and the Intelligence Penetration of the Philippines." *The Trading Post: American Society of Military Insignia Collectors* (1994).

Childress, Clyde. "A Critical Review of They Fought Alone." *Bulletin of American Historical Collection* 31, no. 1 (2003).

Chua, Gregorio. "Former Regiments Barracks at Ft Ord Still There." Edited by Alex De Leon Fabros. *Bolo News Newsletter.* CA: Filipino Infantry Association.

Cressman, Robert. "Historic Fleets – The Saga of the Williebee." *Naval History Magazine* (2012).

Cuesta, Karla Dela. Document. Ignacio Dela Cuesta.

Daniel, Rhetta M. Document. Vaughan Moore.

Dollard, Charles and Donald Young. "In the Armed Forces." *Survey Graphic* 36 (1947).

Donesa, John. "The Mindanao Death March: Establishing a Historical Fact Through Online Research." *International Science Review* 1, no. 1 (2020).

Dwyer, John B. "The Untold Story Behind the Great Raid." *American Thinker* (2005).

Endy, Clarence E., Jr. "The Gentleman from the Philippines, The Foreign Cadet Program." *Bulletin American Historical Collection* 11, no. 3 (1983).

Estrella, Cicero A. "Filipinos had to Fight for the Right to Serve their Adopted Home." SFGATE (2005).

Evangelista, Susan. "Aurelio Bulosan's Wartime Diary." *Philippine Studies* 37, no. 4 (1989): pp. 467-78.

Fabros, Alex De Leon. "We won't Forget Bataan Day Vow of Filipino Regiment." (1943).

Fabros, Alex S., Jr. "No Filipino War Heroes Allowed." *Filipina Magazine* (1994).

Fabros, Michelle S. *A Christmas Present from our Fathers* (2011).

Fertig, Wendell. "Letter to Harold Rosenquist." *MacArthur Archives* (1944).

FlorCruz, Paul and Maria O'Toole. Document. Pedro Flor Cruz.

Gangl, Theresa. Document. Gangl Leo.

Ganley, Eugene F. and Robert H. Offley. Letters (March 6, 1959) and (March 9, 1959). Pennsylvania: USAMHI Carlisle.

Glenn, Bobb. "Account of Submarine Mission Landing (January 1943)." Told by Jesus Villamor.

"Guerrilla Days in North Luzon." *Historical Records Section. Luna,* La Union: USAFIP (July 1948).

Halsema, James J. "The End of the War 1945." *Bulletin American Historical Collection* 20, nos. 3-4 (1992).

Halsema, James J. "The Rest of the War." *Bulletin American Historical Collection* 20, no. 2 (1992).

Hamner, Jordan A. "Hamner's War." *Bulletin American Historical Collection* I-V (2004-2005).

Hazard, Henry. "Reminiscences of the Naturalization of Members of Our Armed Forces Overseas." *Monthly Review* 6, no.6 (1948).

Hazard, Henry. "Valor Knows Neither Nationality, Race, Nor Creed." *Monthly Review* 2, no. 9 (1945): p. 110.

Holian, Thomas. "Saviors and Suppliers, World War II Submarine Special Operations in the Philippines." *Undersea Warfare Magazine*, no. 23 (Summer, 2004).

Ind, Allison W. "History of 5217th Reconnaissance Battalion." *Bahala Na Anniversary Magazine* (1944).

Ind, Allison W. "Six Came Back". Original typewritten document from Delfin Yuhico.

Johnson, Danny. "Historic California Posts, Camps Stations and Airfields." *California State Military*, issue 29. Beale Air Force Base (February 2012).

Jose, Rico T. "Preparing for Emergency: The Civilian Emergency Administration 1941." *Diliman Review* 38, no. 2 (1991): pp. 37-46.

Krivido, Michael E. "Major Jay D. Vanderpool: Advisor to the Philippine Guerrillas." *Veritas* 9, no. 1 (2013).

Kuder, Edward M. "The Philippines Never Surrendered." *Saturday Evening Post,* February 10, 1943.

Laubenthal, Sanders A. "A History of John Hay Air Base." *Bulletin American Historical Collection* 22, no. 2 (1994).

Liban, Harold. Document. John A. Maraggay.

Liban, Harold. Document. Miguel L. Lauagan, PCAU.

Marquardt, Frederic. "'I Shall Return:' A Footnote to History." *Bulletin American Historical Collection,* no. 1.

Mcmicking, Joseph. "Leyte Landing and the Battle of Leyte Gulf." Speech at the Insular Life Assurance Co. and Tacloban Lions Club, Leyte, Philippines, October 19, 1956.

Meyer, Milton Dr. "A World War II Vignette: The Darwin-Philippine Connection." *Bulletin American Historical Collection Foundation* 26, no. 2 (1998).

Miculka, Cameron. "Living Legacy: Al Lim, 016, Reflects on Award Serving in WWII." *West Hawaii Today* (2017).

Mitsos, Tom. "Guerrilla Radio by American Guerrillas of Mindanao (AGOM)."

"Modern Legend of Shangri-La." *Far East Asian Service Command: Jungle Journal* 1 (1945).

Mydans, Carl. "Return to Sto. Tomas. US Wins Heart of the Philippines." *Life* 18, no. 8 (1945).

Ortiz, Pacifico S. J. "Letter to Aurora Quezon." December 17, 19⁻.

Owens, William. "Eye-Deep in Hell: A Memoir of the Liberation of the Phils, 1944-45." (1989). *Bulletin American Historical Collection* 23 (1995).

Parsons, Peter. "Chick Parsons in Occupied Manila." *Bulletin American Historical Collection* 36, no. 1 (2008).

Parsons, Peter. "Commander Chick Parsons and the Japanese." *Bulletin American Historical Collection* 35, no. 35 (2007).

Parsons, Peter. "In Search of a Biography." *American History Collection* 19, no. 1 (2001).

Parsons, Peter. "Me and the AGOM." *Bulletin American Historical Collection.*

Parsons, Peter. "Special Mission Submarines in the Philippines." *Bulletin American Historical Collection* 29, no. 4 (2001).

Payumo, Marcelino. "Farm Seven." *50th Anniversary Liberation of Davao* (1945).

"PCAU 7." *Bulletin American Historical Collection* 22 (1994).

Pompea, Edward T. "Galapagos-Long and Lonesome." *Cross Country: News and Views Around the World.* Air Force Association. September 1945.

Rallojay, Teodoro J. "Diary: May 23-November 2, 1944."

Revilla, Linda A. "'Pineapples', 'Hawayanos', and 'Loyal Americans': Local Boys in the First Filipino Infantry Regiment, US Army." *Social Process in Hawaii* 37 (1996).

Reyes, Vince. "War Brides." *Filipinas Magazine* (1995).

Richards, Peter. "After 3 February 1945." *Bulletin American Historical Collection* 26, no. 2.

Rivera, Juan A. "Enemy Invasion of Mindanao, 1941-1945: a Philippine Account."

Rolley, Ailsa. "Australia Remembers 1945-1995." Beaudesert Shire Council and Beaudesert Historical Society (1996).

Rosal, Raoul. Document. "PCAU #20", Cpl Toribio J. Rosal.

Rosenquist, Harold A. Letter to Wendell Fertig, October 4, 1944.

Ruiz, Julius B. "Report to Quezon: Pinoys in Wartime." Recalled by Ruiz 1942, August 18, 1946.

Scharrenberg, Paul. "The Philippine Problem: Attitude of American Labor Toward Filipino Immigration and Philippine Independence." *Pacific Affairs* 2, no. 2 (February 1929).

Shor, Franc. "See You in Manila." *The American Legion Magazine* (March 1943).

Smollar, David. "Hard, Bitter, Unpleasantly Necessary Duty, A Little-Known World War II Story of the Philippines." *Philippine E-Journals* 61, no. 1 (2015).

Solidarios, Serni. Document. Sergio Solidarios.

Suatengco, Remi. Document. Eduardo T. Suatengco.

Valdez, Pelagio. "Camp Roberts Trainer." *Bulletin* 2, no. 3.

Villamin, Vicente. "Saving Prisoners in Japanese Hands." *The Philippine Mail* (September 1944).

Villamor, Jesus. "They Never Surrendered, Hero as Critic." *Bulletin American Historical Collection* 12, no. 2 (1985).

Villarta, Fred. "Memorial of Julius Ruiz."

Visaya, Ted. Document. Pedro Visaya.

Vitorio, Ramon. Letter to Mr. Rodriguez in San Francisco. October 16, 1942.

Volckmann, Russell W. "Role of Airpower in Counterinsurgency and Unconventional Warfare: Allied Resistance to the Japanese on Luzon, World War II." Symposium (1963).

Waite, Elmonte. "Jose Filipino Runs His Own Show." *Dawn of the Philippine Freedom* (1995).

Walter, Cecil Earl. Letter to Les Hughes. May 27, 1994.

Williams, Irving. Letter to Courtney Whitney. *American Red Cross* (1994).

Worcester, Dean C. Letter to Charles Parsons. October 6, 1943.

Young, Frank H. "The Sulu Sharpshooter." *World War II Heroes Project.* Bamban Historical Society, Bamban Museum of History (2020).

BOOKS

Acebes, Rodolfo Meim. *Mindoro: the Stepping Stone.* Manila: AV Manila Creative, 2016.

Agoncillo, Teodoro. *The Fateful Years, Japan's Adventure in the Philippines, 1941-1945.* Quezon City: R.P. Garcia Publication, 1965.

Alexander, Larry. *Shadows in the Jungle: the Alamo Scouts Behind Japanese Lines in World War II.* London, England: Penguin, 2010.

Astor, Gerald. *Crisis in the Pacific.* Dutton Adult, 1996.

Baclagon, Uladarico S. *The Philippine Resistance Movement Against Japan, 10.12.1941-14.06.1945.* Quezon City: Munoz Press, 1965.

Baldoz, Rick. "6. Another Mirage of Democracy': War, Nationality, and Asymmetrical Allegiance." *In The Third Asiatic Invasion: Empire and Migration in Filipino America, 1898-1946.* New York, USA: New York University Press, 2011. pp. 194-236.

Black, Robert W. *Rangers in World War II.* Presidio Press, 1992.

Boggs, Charles W. Jr. *Maj. Marine Aviation in the Philippines.* Washington, D.C.: U.S. Government Printing Office, 1951.

Bohannan, Charles T. *Air-Sea Support, The Philippine Guerrilla, World War II.* Manila, Philippines: Ateneo University, 1966.

Breuer, William B. *MacArthur's Undercover War.* Edison, NJ: Castle Books, 2005.

Breuer, William B. *The Great Raid on Cabanatuan.* John Wiley, 1994.

Brooks, Lester. *Behind Japan's Surrender.* McGraw-Hill, 1968.

Buaken, Manuel. *I Have Lived with the American People.* Caldwell, Idaho: Caxton Printers, 1948.

Cal, Ben. *Victory at Bessang Pass.* Manila, Philippines. 2012.

Campbell, Douglas A. *Eight Survived.* Lyon Press, 2011.

Campbell, Douglas E. *Save Our Souls: Rescues Made by U.S. Submarines During World War II.* Lulu.com, 2016.

Carlisle, John M. *Red Arrow Men: the 32nd Division on the Villa Verde.* Detroit: Arnold-Powers, 1945.

Chan, Sucheng. *Asian Americans: An Interpretive History.* Twayne, 1991.

Connaughton, Richard. *MacArthur and Defeat in the Philippines.* Woodstock, NY: The Overlook Press, 2001.

Connaughton, Richard, John Pimlott and Duncan Anderson. *The Battle for Manila.* Novato, California: Presidio Press, 1995.

Cook, Haruko Taya and Theodore F. Cook, eds. *Japan at War: An Oral History.* New York: The New Press, 1992.

Cortesi, Lawrence. *Valor at Samar.* New York, N.Y.: Kensington, 1980.

Cronin, Francis D. *Under the Southern Cross: the Saga of the American Division.* Combat Forces Press, 1981.

Dalisay, Jose Y. *Wash: Only a Bookkeeper, a Biography of Washington Z. Sycip.* Asia Rice Foundation and Metrobank Foundation, 2010.

Daly, John F., III. *An Officer in MacArthur's Court.* Robertson, 2014.

Daroy, Ester Vallado. *Flight of the Hermit Crab.* Quezon City, Philippines: Central Book Supply, 2007.

De La Cruz, Jesselynn, ed. *Civilians in WWII.* James B. Reuter, S.J. Foundation, 1994.

Dissette, Hans C. and Edward Adamson. *Guerrilla Submarines.* Bantam Books, 1980.

Doromal, Jose D. *The War in Panay.* Diamond Historical Publications, 1952.

Dowlen, Dorothy Dore. *Enduring What Cannot Be Endured.* McFarland, 2001.

Duffy, James P. *War at the End of the World.* Dutton Caliber, 2016.

Duque, Suzanne. *Soldiers as Guerrillas.* Raleigh, North Carolina: Lulu, 2015.

Earle, Dixon. *Bahala Na.* Berkeley, California: Howell-North Books, 1961.

Egeberg, Roger Olaf. *The General, MacArthur and the Man He Called Doc.* Washington D.C.: Oak Mountain Press, 1993.

Eichelberger, Robert L. *Our Jungle Road to Tokyo.* Independently Published, 2017.

Eisner, Peter. *MacArthur's Spies.* Penguin Books, 2018.

Fertig, Wendell. *Fighting the Japanese Occupation 1942-1945.* McFarland, 2015.

Filipinas Foundation. *Zamboanga Hermosa: Memories of the Old Town.* 1984.

Fitzpatrick, Georgina, Timothy L.H. McCormack and Narrelle Morris. *Australia's War Crimes Trials.* Martinus Nijhoff, 2016.

Flanagan, E. M. *Los Banos Raid.* Presidio Press, 1986.

Full Speed Ahead: Negros Navigation at 75. Manila: Negros Navigation, 2008.

Galang, Ricardo. *Hammer and Anvil.* Phoenix, 1989.

Galang, Ricardo. *Secret Mission to the Philippines.* University, 1948.

Geiger, Jeffrey E. *Camp Cooke and Vandenburg Air Force Base 1941-1966.* Jefferson, North Carolina: McFarland, 2014.

Geoffrey, Perrett. *Old Soldiers Never Die: the Life of Douglas MacArthur.* Random House, 1996.

Giangreco, D. M. *Hell to Pay: Operation Downfall and the Invasion of Japan 1945-1947.* U.S. Naval Institute, 2010.

Gilbert, Cant. *The Great Pacific Victory.* New York: John Day, 1945.

Grashio, Samuel C. and Bernard Norling. *Return to Freedom.* MCN Press, 1982.

Gray, Edwyn. *Submarine Warriors.* Novato, CA: Presidio Press, 1988.

Hanrahan, Gene Z. *Japanese Operations Against Guerrilla Forces.* Chevy Chase, Maryland: The Johns Hopkins University,

Operations Research Office, 1954.

Hara, Tameichi, Fred Saito and Roger Pineau. *Japanese Destroyer Captain*. Naval Institute Press, 2013.

Harding, Stephen. *Voyage to Oblivion*. Amberley, 2010.

Harkins, Philip. *Blackburn's Headhunters*. Vail-Ballou Press, 1955.

Heisinger, Duane. *Father Found*. Xulon Press, 2003.

Henderson, Bruce. *Rescue at Los Banos*. New York, NY: Harper-Collins, 2015.

Herras, Domingo. Documents. Gaudencio Vera, Tayabas Guerilla.

Holmes, Kent. *Wendell Fertig and His Guerrilla Forces in the Philippines: Fighting the Japanese Occupation, 1942-1945*. Jefferson, N.C. McFarland. 2015.

Holmes, Virginia H. *Guerrilla Daughter*. Kent State University Press, 2008.

Hoyt, Edwin P. *War in the Pacific*. Vol 1. New York: Avon Books, 1990.

Hughes, Rebekah. *Surviving the Flier*. Phoenix Flair Press, 2010.

Ileto, Reynaldo Clemena. *Knowledge and Pacification*. Quezon City, Philippines: Ateneo de Manila University Press, 2017.

Ind, Allison W. *Secret War Against Japan: the Allied Intelligence Bureau in World War II*. United States: Curtis, 1970.

Ingham, Travis. *Rendezvous By Submarine: the Story of Charles Parsons and the Guerrilla-Soldiers in the Philippines*. Garden City, New York: Doubleday, Doran, 1945.

Javellana, Dominador, III. *Bolomen: Stories of the Civilian Resistance in Wartime Ilocos*. Manila, Philippines: Interformat, 2003.

Keats, John. *They Fought Alone*. New York: Pocket Books, 1965.

Krueger, Walter. *From Down Under to Nippon: the Story of the Sixth Army in World War II*. Combat Forces Press, 1953.

Labrador, Albert. Mandirigma. Manila: JM Publishing House, 2021.

Labrador, Juan. *A Diary of the Japanese Occupation December 7, 1941-May 7, 1945*. Manila, Philippines: Santo Tomas Press, 1989.

Lapham, Robert and Bernard Norling. *Lapham's Raiders: Guerrillas in the Philippines 1942-1945*. Lexington, Kentucky: The University of Kentucky Press, 1996.

Legarda, Benito Jr. *Occupation: 1942-1945*. Vibal Foundation, 2016.

Lelyveld, Joseph. *His Final Battle, the Last Months of Franklin Roosevelt*. Vintage, 2017.

Levine, Alan. *Captivity, Flight, and Survival in World War II*. Praeger, 2000.

Lim, Roberto H. *Pushing the Envelope: a Biography*. Pasay City, Philippines: Mapua House Publications, 2010.

Linderman, Gerald F. *The World Within War*. New York, NY: The Free Press, 1997.

Lindholm, Paul R. *Shadows From the Rising Sun*. Quezon City: New Day, 2009.

Lukacs, John D. *Escape From Davao: the Forgotten Story of the Most Daring Prison Escape of the Pacific War*. New York: NAL Caliber, 2011.

Luvaas, Jay. Dear Miss Em: *General Eichelberger's War in the Pacific, 1942-1945 (Contributions in Military History)*. Praeger, 1972.

MacArthur, Douglas. *Reminiscences*. New York: McGraw-Hill Book, 1964.

Manchester, William. *American Caesar: Douglas MacArthur 1880-1964*. Boston: Little, Brown, 1978.

Manikan, Gamaliel L. *Guerrilla Warfare on Panay Island in the Philippines*. Sixth Military District Foundation, 1977.

Maram, Linda Espana. *Creating Masculinity in Los Angeles's Little Manila*. Columbia University Press, 2006.

Marloff, Maurice, ed. *World War II: Adapted from American Military History, Office of the Chief of Military History, United States Army*. New York: Galahad Books, 1982.

Mashbir, Sysdey F. *I Was an American Spy*. Vantage Press, 1953.

Maynard, Mary. *My Faraway Home*. Isis Large Print, 2002.

McCracken, Alan. *Very Soon Now, Joe*. New York: Hobson Book Press, 1947.

McPherson, Milton M. *Infantry OCS Class 16-52. The Ninety-Day Wonders, OCS and the Modern American Army*. Columbus, Georgia: United States Army Officer Candidate School Alumni Association, 2001.

Meixsel, Richard B. *Frustrated Ambition: General Vicente Lim and the Philippine Military Experience 1910-1944*. University of Oklahoma Press, 2018.

Meixsel, Richard B. *Philippine-American Military History 1902-1942: an Annotated Bibliography*. McFarland, 2003.

Mellnik, Stephen. *Philippine Diary, 1939-1945*. New York: Van Nostrand Reinhold, 1969.

Miller, Donald L. *D-Days in the Pacific*. 1st ed. Simon & Schuster, 2005.

Mills, Scott A. *Stranded in the Philippines: Professor Bell's Private War Against the Japanese*. Annapolis, Maryland: The Naval Institute Press, 2009.

Milner, Samuel. *Victory in Papua*. Washington, D.C.: Office of the Chief of Military History, 1957.

Morison, Samuel E. *History of United States Naval Operations in World War II*. Vol. 13. The Liberation of the Philippines, 1944-45.

Morningstar, James Kelley. *War and Resistance in the Philippines, 1942-1944*. Naval Institute Press, 2021.

Morton, Louis. *The Fall of the Philippines, the War in the Pacific (United States Army in World War II)*. Washington, D.C.: The U.S. Army Center of Military History, 1989.

Mossman, Helen Madamba. *A Letter to My Father: Growing Up Filipina and American*. University of Oklahoma Press, 2008.

Murray, Mary. *Hunted: a Coastwatcher's Story*. Rigby, 1973.

Norling, Bernard. *The Intrepid Guerrillas of Northern Luzon*. Lexington, Kentucky: The University Press of Kentucky, 1999.

Noyer, William L. *Mactan Ship of Destiny*. Fresno, California: Rainbow Press, 1979.

Ogawa, Tetsuro. *Terraced Hell, Tokyo*. Charles E. Tuttle Company, 1972.

Ongpauco, Fidel. *They Refused to Die: True Stories about World War II Heroes in the Philippines 1941-1945*. Levesque Publications, 1982.

Panlilio, Yay. *The Crucible: an Autobiography by Colonel Yay, Filipina American Guerrilla*, ed. by Denise Cruz. New Brunswick, New Jersey: Rutgers University Press, 2010.

Pedraza, Silvia and Ruben Rumbaut. *Origins and Destinies: Immigration, Race, and Ethnicity in America*. 1st ed. Cengage Learning, 1995.

Peralta, Laverne Y. *Who Is Who: Philippine Guerrilla Movement 1942-1945*. Quezon City, 1972.

Postcards from Salinas Photobook, curated by Alex S. Fabros. Stockton, CA: FANHS National Museum, 2013.

Poweleit, Alvin C. *USAFFE, the Loyal Americans and Faithful Filipinos*. Poweleit, 1975.

Prados, John. *Combined Fleet Decoded*. 1st ed. Naval Institute Press, 2001.

Pye, John. *The Tiwi Islands*. Darwin: Colemans Printing, 1977.

Ramsey, Edwin Price and Stephen J. Rivele. *Lieutenant Ramseys War: From Horse Soldier to Guerrilla Commander*. Washington, D.C.: Brassey's Books, 1990.

Rhoades, Weldon. *Flying MacArthur to Victory*. Texas A&M University Press, 2000.

Roscoe, Theodore. *United States Submarine Operations in World War II*. Annapolis, Maryland: United States Naval Institute, 1949.

Rottman, Gordon L. *US Special Warfare Units in the Pacific Theatre 1941-1945*. Osprey Publishing, 1985.

Rottman, Gordon L. *World War II Pacific Island Guide: A Geo-Military Study*. Westport, Connecticut: Greenwood Press, 2002.

Rutherford, Ward. *Fall of the Philippines*. New York, NY: Ballantine Books, 1971.

Santes, Jose M. *Bataan, Corregidor, Bessang Pass, Mindanao Battlefields*. Davao City: Joana A. Santes, 1992.

Sasser, Charles W. *Raider*. New York: St. Martin's Press, 2002.

Segura, Manuel F. *The Koga Papers: Stories of World War II*. Cebu City, Philippines: MF Segura Publications, 1992.

Showalter, Dennis E. *If the Allies Had Fallen*. Skyhorse, 2012.

Sides, Hampton. *Ghost Soldiers*. New York, NY: Doubleday, 2001.

Smith, Robert Ross. *Triumph in the Philippines (The United States Army in World War II: the War in the Pacific)*. Washington, D.C.: Center of Military History, 1933.

Smith, Steven Trent. *The Rescue*. Wiley, 2003.

Snyder, Louis L. *Historical Guide to World War II*. Westport, Connecticut: Greenwood Press, 1982.

Stahl, Bob. *You're No Good to Me Dead: Behind the Japanese Lines in the Philippines*. Annapolis, Maryland: Naval Institute Press, 1997.

Stevens, Peter F. *The Twilight Riders*. Lyons Press, 2011.

Sturma, Michael. *Death at a Distance*. Naval Institute Press, 2013.

Sturma, Michael. *Fremantle's Submarines: How the Allied Submarines and Western Australia Helped to Win the War in the Pacific*. Annapolis, Maryland: Naval Institute Press, 2015.

Taaffe, Stephen R. *MacArthur's Jungle War*. University Press of Kansas, 1998.

Takaki, Ronald. *Double Victory*. Time Warner Publishing's Online Publishing Program, 2000.

Thompson, George Raynor and Dixie R. Harris. *The Signal Corps: the Outcome (Mid-1943 Through 1945)*. CreateSpace Independent, 2015. p. 273.

Valeriano, Napoleon D. and Charles T. Bohannan. *Counter-Guerrilla Operations: the Philippine Experience (Psi Classics of the Counterinsurgency Era)*. Annotated ed. Praeger, 2006.

Vallejo, Marie S. *The Battle of Ising*. 2nd ed. New Day, 2015.

Volckmann, Russell W. *We Remained: Three Years Behind the Enemy Lines in the Philippines*. New York: W.W. Norton, 1954.

White, Michael. *Australian Submarines*. Vol 2. Australian Teachers of Media, 2015.

Whitney, Courtney. *MacArthur's Rendezvous with History*. LIFE, 1955.

Whitney, Courtney. *Report on Philippine Civil Affairs: 32nd Infantry Division 'Red Arrow'*. Vol. 2. Carlisle, Pennsylvania: USAMHI, 1942.

Willoughby, Charles. *MacArthur, 1941-1945*. McGraw-Hill, 1954.

Wise, William. *Secret Mission to the Philippines: The Story of 'Spyron' and the American-Filipino Guerrillas of World War II*. Lincoln, Nebraska: iUniverse.com, 2001.

Zedric, Lance. *Silent Warriors of World War II: the Alamo Scouts Behind Enemy Lines*. Pathfinder, 2020.

Zuckoff, Michael. *Lost in Shangri-La*. Harper Collins, 2011.

INTERVIEWS AND INTERROGATION

Agbanawag, Joseph. Interview by Alex Fabros. California, 1993.

Aguinid, Alexander Ramos. Interview by James Vaughn. FANHS Sacramento Chapter, California, April 9, 2003.

Ames, Irineo A. Interview by Ed Crespo with Marie S. Vallejo, September 10, 2018.

Campos, Dixon. Interview by Christie Cheatham, Spring 1983.

Celdran, Jesus. Interview by Marie S. Vallejo, 2009.

Chua, Gregorio. Interview by Ken Wong. San Francisco Examiner, April 16, 1980.

Crespo, Toribio. Interview by Ed Crespo, September 18, 2018.

Dumlao, Felipe G. Interview by Cynthia Mejia. Washington State Oral/Aural History Program, November 21, 1975.

Fajardo, Edgar. Interview by Marie S. Vallejo. Mindanao, November 5, 2008.

Filipinos in California. Interview by Dionisio. President Manuel Quezon's WWII Exile Government.

Francisco, Emiliano A. Interview by Caroline D. Koslowsky. Washington State Oral/Aural History Program, June 30, 1975.

Francisco, Emiliano A. Interview by FANHS. Seattle, Washington.

Gibson, Emmett F. Interview by David Camelon. Chicago Herald-American, July 22, 1942.

Guittard, George. Interview by Marie S. Vallejo. San Jose, California, September 2014.

Gustilo, Araceli. Interview by Marie S. Vallejo, June 30, 2013.

Jorge, Patricio. Interview by Marie S. Vallejo. 1st Reconnaissance Battalion (Special), December 2011.

Krohn, Bubi. Interview by Rico Jose. Manila, Philippines.

Lamata, Al. Phone Interview by Marie S. Vallejo. 2nd Filipino Infantry Regiment, July 6, 2019.

Los Banos, Domingo. Phone interview by Marie S. Vallejo. *2nd Filipino Infantry Regiment,* March 12, 2007.

Meider, Henry. Interview by GHQ, 1945.

Mullinix, Earl. Interview by Alex Fabros Jr. Salinas, California, 1972.

Offley, Robert H. Interview by Alex Fabros Jr. Salinas, California, 1971 and 1973.

Ozawa, Jisaburo. Interrogations of Japanese Officials. OPNAV-P-03-100. Naval Analysis Division. Interrogation Nav No.3 USSBS No. 32. *The Battle of the Philippine Sea* (June 19-20, 1944). Interrogation, October 16, 1945.

Patton, Howard. Interview by William J. Alexander. University of Texas Oral History Collection. No. 1304, January 5, 1999.

Pimentel, Lorenzo. Interview by Cynthia Mejia. Transcribed by Carolina Apostol. Washington State Oral History Program, November 14, 1971.

Santos, Agustin, Dixon Campos, Philip Ordono, and Clemente J. Felipe. Interview by Cecilia Gaerlan and Fred Basconcillo. *First Filipino Infantry Men.* San Francisco, 1994.

Santos, Terry. Interview by Eddie Graham. 11th Airborne, Provisional Reconnaissance Platoon. National Museum of the Pacific War. Center for Pacific War Studies. Fredericksburg, TX.

Smith, Charles. Interview by Murray Maj, February 11, 1947.

Styles, Ralph E. USN. Interview by Earl Fowler, USN and Lucian L. Vestal, USMC. Oral History Program. Naval Historical Foundation, September-October 2001.

Vivero, Joe. Interview by Marie Vallejo. Carlos Arguelles.

Yuhico, Delfin C. Interview by Marie S. Vallejo. 1st Reconnaissance Battalion (Special), August 2011.

Veterans, children and descendants' communication with Marie S. Vallejo, the Author

Carla Ramsay, daughter of Carlos Arguelles; Marie Galang, daughter of Ricardo Galang; Tess Velasquez, daughter of Jaime Velasquez; Serni Solidarios, son of Sergio Solidarios; Edward Brotonel, son of Sgt. Herman Brotonel; Richard Arca, son of Ricardo Arca; Theresa Gangl, daughter of Leo Gangl; Melissa Agullana, grand- daughter of Domingo 'Don' S. Ruiz; Pelagio Valdez, son of Pablo S. Valdez; Ranesto Bello Angeles, son of Mari- ano Angeles; Aurora Alcala, daughter of Atanacio Alcala; Alex S. Fabros, son of Alex de Leon Fabros; Diana and Ramon Estella, children of Patricio Jorge; Ciedelle Yuhico, daughter of Delfin Yuhico; Maria O'Toole, daughter of Gregorio Chua; Victoria Rallojay, daughter of Teodoro J. Rallojay; Talman Budd, son of George Budd; Peter Parsons, son of Charles Parsons; Paul FlorCruz, son of Pedro Flor Cruz; Rhetta M. Daniel, daughter of Vaughan Moore; Manuel Bersamin, nephew of Paulino Bersamin; Remi Suatengco, son of Eduardo Suateng- co; Raoul Rosal, son of Toribio J. Rosal; Bob Luna, son of Toribio Luna; Reena Ilumin, daughter of Mamerto Ilumin; Harold Liban, nephew of Jovito V. Liban, Arcadio B. Cabang, and Miguel L. Lauagan; Richard Arca, son of Ricardo T. Arca; Clifton Trinidad, son of Mike Trinidad; Aurora Alcala, daughter of Atanasio L. Alcala; Melissa Agullana, daughter of Don Ruiz; Katherine J. Balanza, daughter of Eugene Balanza, Sr.: Nick Bamont, son of John Bamont; Gary Jacobs, son of Leon Jacobs; Bob Capistrano, son of Alberto Capistrano; Ed Crespo, son of Toribio Crespo; Thelma Morgan, daughter of Marcelino S. Laureta; Dio Montero, grandson of Alfonso B. Montero; Deborah Fajardo-Zouai, daughter of Tirso Fajardo; Mark P. Cazem, son of Ponce Cazem; Ted Visaya, son of Pedro Visaya; Lawrence Acosta son of Telesforo Acosta; Tina and Rey Ileto, children of Rafael Il- eto; Corinne Lamata daughter of Al Lamata; Nieves Lim daughter of Roberto Lim; Domingo Los Banos; Alex Aguinid; Joe Vivero about

Carlos Arguelles 2020; Jesus Celdran 2009; Araceli Gustilo wife of Conrado Gustilo June 2013; Ed Crespo, son of Irineo A. Ames, AIB September 10, 2018; Edgar Fajardo, son of Angel Fajardo; Domingo Herras nephew of Gaudencio Vera, commander of Tayabas Guerrillas 2020; Nick Bamont, son of John Bamont; Gary Jacobs, son of Leon Jacobs; Edmund Goroza, son of Edward Goroza; Karla deLa Cuesta granddaughter of Ignacio Dela Cuesta; Jessie Valdez grandson of Teodorico Galicia.

FILMS

Castillo, Stephanie J., dir. *Remember the Boys, A Story of Hawaii Filipino Boys in WWII*. April 6, 2010. DVD.
Castillo, Stephanie J. and Noel M. Izon, dirs. *Untold Triumph: America's Filipino Soldiers. 2005; ICT Productions. Dawn of Freedom*. Films. 1944.
Filipino Troops in U.S. Army. 1943; Movietone News, U.S. Archives and Records Administration.
On To Bataan. 1st and 2nd Filipino Infantry Regiment. RG-11; Film 288: U.S. Archives and Records Administration.
Shelton, T., A. S. Fabros, and R. Pike, dirs. *Unsung heroes: Hawaii's Fighting Filipinos; An Untold Triumph: America's Filipino Soldiers; On To Bataan*. 2011; 1st Filipino Infantry. Wong Audiovisual Center, University of Hawaii at Manoa.
U.S. Army Filipino Regiment – Receives Bolo Knives in Ceremony. 1943; U.S. Archives and Records Administration.

NEWSPAPERS

"A Mass was Held to Celebrate the 8th Anniversary of the Philippine Commonwealth." *Shot 'N Shell*.
"American Forces Secure Balete Pass." *Free Philippines*. May 15, 1945.
Aquino, Belinda. "Pearl Harbor Revisited." *Philippine Daily Inquirer*. December 14, 2012.
Arguelles, Carlos. *Panorama*. November 13, 1942.
"Army's New Citizens To Go on Air." *Shot 'N Shell*. June 24, 1942.
Buaken, Iris Brown. "My Brave New World." *Philippine Mail*.
Cacabelos, Rufino F. "Rufino Autobiography, Army Life: July 3, 1942 – January 3, 1946." *The Bayonet*, June 17, 1943.
"Commanders May Select Officer Candidates." *Shot 'N Shell*. July 29, 1942.
"Discharges Eased for 38-Year-Olds." *Shot 'N Shell*. February 17, 1943.
Fabros, Alex De Leon. "Filipinos Receive Colors in Regimental Review: 2nd Filipino Infantry Regiment." *Cooke Clarion*, June 4, 1943.
Farolan, Ramon. "Message from Nakar." *Philippine Daily Inquirer*, October 1, 2012.
"Filipino Infantrymen Get Sleeve Insignia." *Shot 'N Shell. August* 26, 1942.
"Filipino Soldiers at Ft. Ord Treated." *Panorama*, December 1942.
"Filipinos Filmed in Technicolor." *Shot 'N Shell*. July 8, 1942.
"Field Mass Celebrated for U.S. Filipino War Heroes." *Shot 'N Shell*. September 30, 1942.
"First Filipino Infantry Regiment Crests were made for Officers that Contributed Private Silver." *Shot 'N Shell*.
"First Filipino Officer in Our Army is Here." *The Bayonet*, November 12, 1942.
"Guerrillas Help Capture Ipo Dam. (Miller/Stoddard)." *Free Philippines*. May 19, 1945.
Hemingway, Al. "Early Warning of Attack on Pearl Harbor." *Charlotte Sun*, December 7, 2012.
Ilustre, Ernesto D. "Filipino Commandos in America." *Panorama*, December 7, 1975.
Ilustre, Ernest D. "Filipino Fighting Men." *The Washington Post*, July 28, 1942.
Ilustre, Ernest D. "Filipino Trains to Drive Japs out of Islands."
"It's Mayon in Violent Temper." *Shot 'N Shell*. August 26, 1942.
"It's Barbecue Time Among Soldiers of the 1st Filipino Infantry." *Shot 'N Shell*.
Javier, Jessie. "Society Notes: Soldiers from the Fourth Army, Camp Cooke and Camp Beale. Members of 2nd Filipino Infantry Band Played." *Philippine Mail*, November 24, 1943.
Lergards, Benito, Jr. "Double Take." *The Philippine Press*, February 6, 2010.
"Leyte-Samar." *Free Philippines*. November 19, 1944.
"Leyte-Samar." *Free Philippines*. November 23, 1944.
McCarroll. "The Memories Live on Forever." *The Odessa American*, January 2, 1983.
"Moros Aiding U.S. Drive in Jolo Sectors." *Free Philippines*. July 1945.
"PA to Let Out 100,000 this Month." *Manila Post* . Dec 1-23, 1945.
"Pacific States Groups Act to Form a Force, Provided War Department Sanctions One." *The New York Times*, January 13, 1942.
"Part 1 and 3. Free Philippines. From Collection of Mrs. A. A. Quezon." *Free Philippines*.
"Philippines Leyte-Samar. Guerrilla and Civilians on Samar Wage Heroic Fight Against Japanese." *Free Philippines*. November 23, 1944.
"Psychological Warfare Branch., GHQ., Office of War Information Unit. Jap Orders to Massacre All Filipino Civilians

Captured." *Free Philippines.*

"Radiomen Worked 5 Months on Mindanao." *Free Philippines.* March 30, 1945.

Santos, Angelito L., Joan Orendain, Helen N. Mendoza, Bernard L. Karganilla. "Under Japanese Rule: Memoir & Reflections." *The Daily Manila Shimbun,* 2012.

"Somewhere in the Philippines." *Free Philippines.* February 12, 1944.

"There'll be Martial Music for the Avengers of the Philippines." *Shot 'N Shell.* September 2, 1942.

"Underground Radio in Luzon Aided Liberation Strategy." *Manila Post.* October 25, 1945.

Velasquez, Jaime. "Quezon's Aide at Benning." *The Bayonet,* February 18, 1943.

Villamin, Vicente. "Saving Prisoners in Japanese Hands." *Philippine Mail.* September 28, 1944.

Waite, Elmonte. "Jose Filipino Runs His Own Show." *Dawn of the Philippine Freedom,* January 1, 1995.

Whitlock, Anthony. "Secret Record Gave Clue, Japanese Tactics on Luzon." *The Sydney Morning Herald (NS W 1842 – 1954),* October 24, 1945.

Wingo, James G. "Two Filipinos Here." *Arizona Republic,* January 10, 1943.

IMAGES

Pfc Larry Benitez, Oakland, California and Pfc E Pepito, Wilmington, California. Company E, 1st Filipino Regiment. Demonstrate the art of bolo fighting in the camp area. Oro Bay, New Guinea. May 18, 1944. Photo by T4 Ernani D'Emidio.

1st Filipino Infantry Regiment Company E. 1942

1st Filipino Infantry Regiment Headquarters Company. 1942

1st Filipino Infantry Regiment Service Company. 1942

1st Filipino Infantry Regiment. Herman Brotonel is on the right. Source: Edward Brotonel

1st Filipino Infantry Regiment. San Luis Obispo 1942. Source: Don Ruiz Collection

1st Filipino Infantry Regiment. Salinas 1942. Source: Don Ruiz Collection

1st Filipino Infantry Regiment. Fort Ord 1942. Source: Don Ruiz Collection

1st Filipino Infantry Regiment. San Luis Obispo. 1942. Isidoro Dacquel on the left. Source: Cary Dacquel

1st Filipino Infantry Regiment. San Luis Obispo 1942. Source: Don Ruiz Collection

1st Filipino Infantry Regiment. San Luis Obispo 1942. Source: Don Ruiz Collection

1st Filipino Infantry Regiment, Anti-Tank Crack Gunners. San Luis Obispo 1942. Source: Don Ruiz Collection

1st Filipino Infantry Regiment 1943

1st Filipino Infantry Regiment Company C. Camp Beale 1943

1st Filipino Regiment Company M. Camp Beale 1943

1st Filipino Infantry Regiment Company K. Camp Beale 1943

1st Filipino Regiment – Cannon Company. Inscription on cannon, "Inang Bayan" means Mother Land or Home Land. 1942. Photo by Carl Gaston. Signal Corps.

1st Filipino Regiment. Training in San Luis Obispo. 1942. Photo by Carl Gaston. Signal Corps.

Training of 1st Filipino Infantry Battalion at Camp San Luis Obispo, CA....Three soldiers of the 1st Filipino Battalion are shown charging up the beach in simulated bayonet attack. The Japs have already felt the steel of Filipino bayonets. April 1942. Signal Corps photo by BPR (Carl Gaston).

Training of 1st Filipino Infantry Battalion at Camp San Luis Obispo, CA.... Sgt Anthony Colosino of Buffalo, N.Y. instructs Pvts Roberto Sarmiento and Patenciano Piroco of the 1st Filipino Battalion in the use of the hand grenade. April 1942. Signal Corps photo by BPR (Carl Gaston)

Training of 1st Filipino Infantry Battalion at Camp San Luis Obispo, CA.... Leaping from their assault boat, the soldiers rush for protective cover on the beach where they are training for invasion tactics. Signal Corps photo by BPR (Carl Gaston).

2nd Filipino Infantry Regiment. Camp Cooke 1943-1944

2nd Filipino Infantry Regiment. Secundio Bucol is in the middle. Camp Cooke 1943-44.
Source: Secundio Bucol Collection

2nd Filipino Infantry Regiment, Camp Cooke 1943-44. Source: Larry Acosta

Filipino Infantry Regiments. Pedro Santos is seated on the left. Source: Manny Santos

2nd Filipino Infantry Regiment pup tent inspection. Camp Cooke 1943-44

1st Reconnaissance Paratroopers. New Guinea 1944. Source: Joel Vivero Rico

978th Signal Service Co. Orderly – Hollandia 1944-1945.
Source: Unit History of the 1st Reconnaissance Battalion (Special) by Ceferino Rola. 1946

Battalion dispensary & Staff – Hollandia 1944-1945.
Source: Unit History of the 1st Reconnaissance Battalion (Special) Rola. 1946

CHAPLAIN EDMUND DIETZEL & CHAPEL -- Hollandia,

Chaplain Edmund Dietzel & chapel – Hollandia 1944-1945.
Source: Unit History of the 1st Reconnaissance Battalion (Special) by Ceferino Rola. 1946

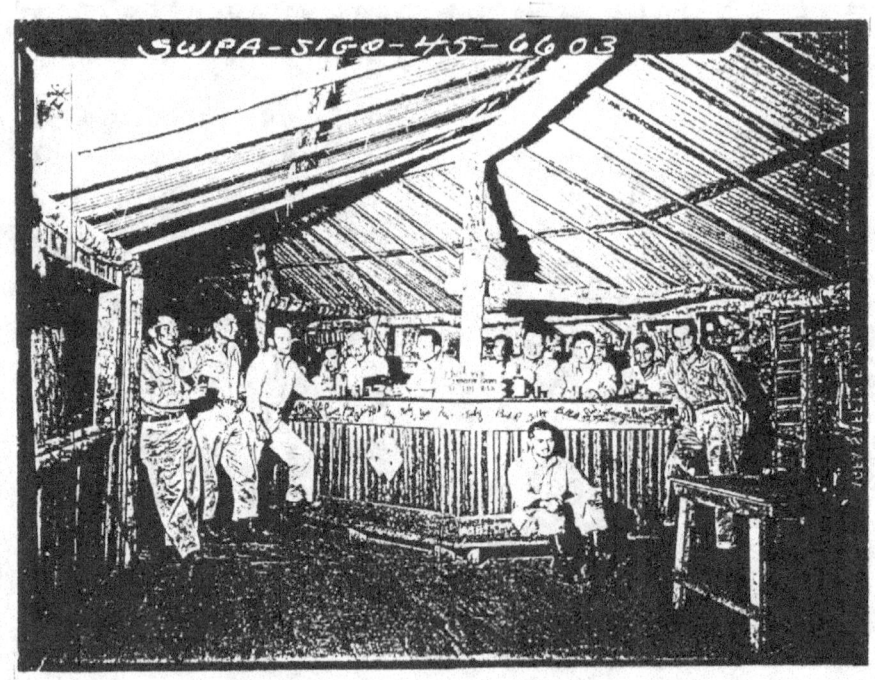

Interior view of Officers' Club – Hollandia, 1944-1945.
Source: Unit History of the 1st Reconnaissance Battalion (Special) by Ceferino Rola. 1946

Officers' Club by creek – Hollandia 1944-1945.
Source: Unit History of the 1st Reconnaissance Battalion (Special) by Ceferino Rola. 1946

Enlisted men recreation hall – Hollandia 1944-1945.
Source: Unit History of the 1st Reconnaissance Battalion (Special) by Ceferino Rola. 1946

OFFICERS' CLUB BY ROAD -- Hollandia, 1944-1945.

Officers' Club by road – Hollandia 1944-1945.
Source: Unit History of the 1st Reconnaissance Battalion (Special) by Ceferino Rola. 1946

S-4 Warehouse and staff – Hollandia 1944-1945.
Source: Unit History of the 1st Reconnaissance Battalion (Special) by Ceferino Rola. 1946

Guerrillas in the Philippines carrying the American and Philippine flags. WWII 1943-1945.
Source: US National Archives and Records Administration

Guerrillas in the Philippines WWII 1943. Source: US National Archives and Records Administration